日本淡水化石珪藻図説
―関連現生種を含む―

田中 宏之 著

Atlas of Freshwater Fossil Diatoms in Japan
— Including related recent taxa —

by Hiroyuki TANAKA

内田老鶴圃
Uchida Rokakuho Publishing Co., Ltd.

Author:
Hiroyuki TANAKA, Ph.D.
Maebashi Diatom Institute
57-3 Kawamagari, Maebashi City
Gunma 371-0823
Japan
E-mail: guntana@green.ocn.ne.jp
Fax: +81-(0)27-251-9096

Published by:
Uchida Rokakuho Publishing Co., Ltd.
3-34-3 Otsuka, Bunkyo-Ku
Tokyo 112-0012
Japan
Published on March 10, 2014

本書の全部あるいは一部を断わりなく転載または
複写(コピー)することは，著作権および出版権の
侵害となる場合がありますのでご注意下さい．

序：Preface

　誰でもプレパラートを作り，地層から最初に化石珪藻を見出したとき，その美しさに感動し，次にこの珪藻の名前はなんだろうと好奇心を抱くと思う．私もその一人であったので，そのような折に参考になる本と思い，日本淡水化石珪藻図説を出版することにした．

　日本の淡水生化石珪藻は，一般的に湖沼性堆積物から多く産出し，中心類が多産し羽状類は少ない．このため本書の掲載珪藻は珪藻類全体の割合からすると中心類が比較的多くなっており，化石として産出するが，現生種としても生育している羽状類珪藻の多くは割愛してある．本書を使用しての珪藻の同定は，すでに現生種を収録した2種類の珪藻図鑑（渡辺ら 2005，小林ら 2006）が刊行されているので，それらを併用すると効果的と思われる．しかしながら本書にはそれらの図鑑に収録されていない現生種，化石種と混同しやすい現生種は収録したので，現生種の調査・研究にも役立つと考えている．

　本書の作成にあたり，電子顕微鏡の使用，文献入手等のご援助をいただいた日本歯科大学教授南雲　保博士，同大学で電子顕微鏡観察の支援をしていただいた三橋扶佐子助教，松岡孝典博士，その他ご協力いただいた多くの方々，本書の出版を快く引き受けていただいた内田老鶴圃社長内田　学氏，編集にご尽力いただいた笠井千代樹編集長に深くお礼申し上げます．

　　2014年3月

　　　　　　　　　　　　　　　　　　　　　　　　　　　　　　　　田中　宏之

　The beautiful shapes and varied structures of fossil diatoms from sediment never fail to make a strong impression on all who see them for the first time. It is then one may begin to take an interest in their names and what it is that distinguishes one from another. I have put this book together in order to share the fascination that I felt when I first saw the beauty of diatoms and made their study my life work.

　This book mostly contains information on fossil freshwater diatoms found mainly in lacustrine deposits thoughout Japan, but includes information on some recent diatoms as well. And, since centric diatoms tend to dominate these deposits, the focus is on relatively centric diatoms to the exclusion of many pennate diatoms. For more information on pennate and recent diatoms in Japan, please refer to two previous volumes: Watanabe *et al*. (2005) and Kobayasi *et al*. (2006).

　I am grateful to Dr. Tamotsu Nagumo for his generous support as well as to assistant lecturer Ms. Fusako Mitsuhashi and Dr. Takanori Matsuoka for their kind assistance in SEM observations at The Nippon Dental University as well as the other collaborators including Manabu Uchida, president of Uchida Rokakuho Publishing, and Chiyoki Kasai, editor, for their help in editing and publishing this Atlas.

　　March, 2014

　　　　　　　　　　　　　　　　　　　　　　　　　　　　　　　　Hiroyuki Tanaka

凡例：Explanatory Notes

1. 本書は，日本の淡水成堆積物から見出された淡水生化石珪藻の図説である．一部に汽水種，および化石として産するが，化石からは適当な写真が得られなかった分類群を中心に，現生の淡水珪藻も収録してある（計236分類群）．
2. 掲載は分類群の写真を中心とし，短文の解説（和文），図の説明（英文）を記した．
3. 分類群解説の文献欄には，原記載，または標記の所属に組み合わせを行った文献，あるいは殻の形態，種の特性等を理解する上で参考になると思われる入手しやすい文献を記した．
4. 末尾の引用文献欄には，各分類群解説における"ノート"，"産出"の項目で引用した文献を記してある．
5. 計測値は，特に記していない場合は，図に使用した産地の分類群のものである．
6. 産出報告における他の論文の引用は，図あるいは解説文で当該分類群であることが確認できたもののみを使用した．現生試料からの報告は，特に参考になると思われるものにとどめた．
7. 試料は前期中新世から完新世までの地層から採取したが，一部は現在の湖沼等から採取したものを含んでおり，産出欄等への記入は，現生の浮遊・付着性試料は現生（Recent），底泥の場合は完新世（Holocene）と記した．
8. 命名者名は，文章中に使用する場合を除いて省略しないで記した．省略等の場合は，後藤（2003）で報告・提案されている略号等を使用した（たとえば原記載文の著者名がSkvortzowであってもSkvortsovとした）．
9. 用語は，Ross *et al.*(1979)，小林弘珪藻図鑑（小林ら 2006）で示されたものを基本的に使用した．おもな用語は図示した（pp. 35-36）．

目次：Contents

序：Preface ·· **iii**
凡例：Explanatory Notes ·· **v**
Ⅰ．新分類群・新組み合わせ：16 New Species and Three Normenclatural Changes ········· **1**
Ⅱ．試料：Samples ·· **19**
Ⅲ．試料処理・プレパラート作成：Method of Cleaning Samples and Preparations ········ **23**
Ⅳ．収録分類群の配列：Taxa Arrangement ·· **25**
Ⅴ．収録分類群一覧：List of Taxa ·· **27**
Ⅵ．記述用語：Terms ·· **35**
Ⅶ．分類群解説・欧文図版説明・図版：Explanation of Diatoms and Plates ·············· **37**
　1．中心類：Centric diatoms ··· **38**
　2．無縦溝羽状類：Araphid, pennate diatoms ····································· **282**
　3．単縦溝羽状類：Monoraphid, pennate diatoms ································ **336**
　4．双縦溝羽状類：Biraphid, pennate diatoms ···································· **350**
Ⅷ．引用文献：List of References ·· **581**
Ⅸ．学名索引：Index of Scientific Names ··· **593**
Ⅹ．地層・産地別一覧：List of Formations or Localities and Taxa ····················· **599**

Ⅰ．新分類群・新組み合わせ：
16 New Species and Three Normenclatural Changes

1. New Species

Centric diatoms
Aulacoseira fukushimae H. Tanaka sp. nov. (Plates 4–5)

Description: Valves cylindrical, 6.5–13 μm in diameter, 4–9 μm height. Valve faces slightly convex or concave without areolae. Mantle with round areolae, irregularly spaced, ca. 12 in 10 μm. Areolae rows on mantle ca. 12–16 in 10 μm. Spines are different between sibling valves, with one pointed and the other short and spatulate, both occuring atop every interstria. Rimoportulae located on mantle, ca.4 per valve, with straight canals parallel to apical axis usually ending at valve face/mantle boundary, but sometimes just before the boundary. Short collar without ringleist.

Holotype: MPC-25049. Micropaleontology Collection, National Museum of Nature and Science, Japan.

Isotype: TA-3541. Maebashi Diatom Institute, Gunma Prefecture, Japan.

Type locality: Shichiku, Iwaki City, Fukushima Prefecture, Japan. 38°8.35′N, 140°54.22′E (Fig. 1).

Type material: SCH-102, collected by author on 25 February 2007. Mudstone of the Shichiku Formation (Early Miocene).

Etymology: The species is named in honor of Dr. Hiroshi Fukushima, honorary President of the Japanese Diatom Society.

Remarks: The new species is similar to *Aulacoseira pfaffiana* (Reinsch) Krammer and *Orthseira americana* (Kützing) Thwaites. It is distingushed from both by its lack of oronamentation on its valve face. It is further distinguished from the former by the two different shapes of its spines, the location of its rimoportulae and its tubes.

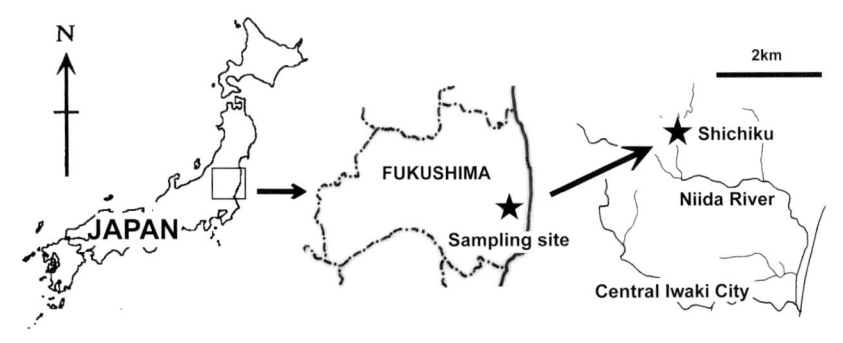

Fig. 1. Location of sampling site (★), Shichiku, Iwaki City, Fukushima, Japan.

Aulacoseira miosiris H. Tanaka sp. nov. (Plates 13-14)

Description: Valves cylindrical, 8-25 μm in diameter, 4-11 μm height. Valve faces flat and without ornamentation except for occasional areolae on marginal area. Straight areolae rows on mantle, 8-10 in 10 μm, areolae in row also 8-10 in 10 μm. All areolae, round, star-like velum externally and curved rica internally. Spines of separating valves are pointed while linking valves are zippered, both types located every interstria at mantle/valve face boundary. Initial valves have partitions, or pillars, from ringleist to valve face that gradually shorten and almost disappear on succeeding valves. Rimoportulae located on ringleist, 7-13 per valve, with very short straight tubes lifting labium up from ringleist.

Holotype: MPC-25050. Micropaleontology Collection, National Museum of Nature and Science, Japan.

Isotype: TA-3542. Maebashi Diatom Institute, Gunma Prefecture, Japan.

Type locality: The outcrop of north shore in Tazawa-ko (Lake Tazawa) Senboku City, Akita, Japan. 39°49′N, 140°41′E (Fig. 2).

Type material: TAZ-F01, collected by author on 5 August 1994. Siltstone of the Miyata Formation (Mio-Pliocene).

Etymology: The species is named for similar feature to genus *Miosira*.

Remarks: The original description of genus *Miosira* in Krammer (1997) states that it has partitions and pillars between valve face and ringleist and no rimoportulae. Houk & Klee (2007), however, showed rimoportulae in *Miosira jouseana*, which was transfered from *Melosira jouseana* to *Miosira* by the former report. This particular species, having partitions, or pillars, in its early valves closely resembles those of *Miosira*, but as these characteristics gradually disappear in its later valves and always have rimoportulae, the grater number of valves with these characteristics leads the author to belive it suitably belongs to the genus *Aulacoseira*.

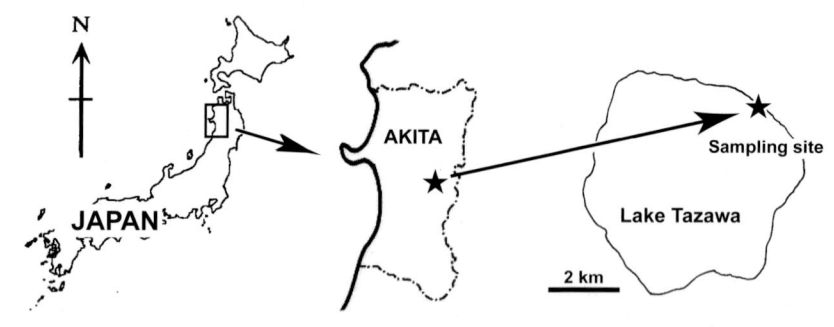

Fig. 2. Location of sampling site (★), north shore in Tazawa-ko, Senboku City, Akita, Japan.

Aulacoseira polispina H. Tanaka sp. nov. (Plates 17–19)

Description: Valves cylindrical, 13–31 μm in diameter, 17–28 μm height in separating valves and 10–19 μm in linking valves. Valve faces flat, no ornamentation on separating valves though linking valves have small areolae located on marginal area of valve face and sometimes small areolae scattered around valve center. Areolae on mantle, forming straight rows, ca. 6 in 10 μm, usually square or polygonal (sometimes oval), 5–6 in 10 μm in rows. Inner shape of areolae of separating valves are usually round or oval, whereas linking valves are square or polygonal or oval. Rimoportulae located on mantle, estimated about 10 with curved canals. On separating valves, spines cover two interstria at their bases with every to every third spine long, thick and inserted into grooves that extend to the collar of the sibling valve. On linking valves, spines are short spatulated and located at the end of every interstriae on the valve face/mantle boundary. All valves have relatively short collar and ringleist.

Holotype: MPC-25051. Micropaleontology Collection, National Museum of Nature and Science, Japan.

Isotype: TA-3543. Maebashi Diatom Institute, Gunma Prefecture, Japan.

Type locality: Imai, Ajimu Town, Oita Prefecture, Japan. 33°22.4′N, 131°20.7′E (Fig. 3).

Type material: OIT-212, collected by author on 28 April 2001. Siltstone of the Tsubusagawa Formation (Late Pliocene).

Etymology: The species is named for having many long spines on its valve face/mantle boundary.

Remarks: The new species is distingushed from *Aulacoseira granulata* by its many long spines. This new taxon was reported by Tanaka & Matsuoka (1985) as *Melosira* sp. 2. This author has confirmed that *Melosira* sp. 2 is the same as *A. polispina* using a sample taken from the same Koga Formation.

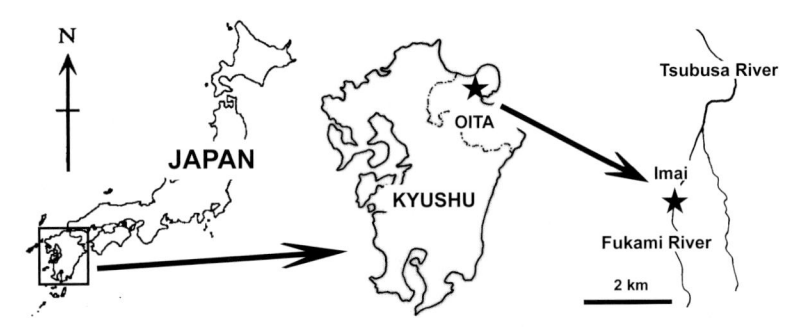

Fig. 3. Location of sampling site (★), Imai, Ajimu Town, Oita, Japan.

Aulacoseira tsugaruensis H. Tanaka sp. nov. (Plates 25-26)

Description: Valves cylindrical, 5.5-24 μm in diameter, 10.5-18.5 μm height. Valve faces flat, no areolae or processes. Areolae on mantle usually form straight rows though those of especially small valves are slightly angled, 10-21 in 10 μm, round areolae 14-20 in 10 μm in rows. Probably one rimoportula located on mantle, in a slanted canal toward ringleist. Opening of rimoportula located toward proximal end of areolae row, one, or two, areolae between valve rim. All spines are large and spatulate, located at valve face/mantle boundary and cover about three interstriae at their bases. Valves have a relatively short ringleist.

Holotype: MPC-25052. Micropaleontology Collection, National Museum of Nature and Science, Japan.

Isotype: TA-3544. Maebashi Diatom Institute, Gunma Prefecture, Japan.

Type locality: Fukaura, Fukaura Town, Aomori Prefecture, Japan (Fig. 4).

Type material: ROK-001, Mudstone of the Rokkakuzawa Formation (Early Miocene).

Etymology: The species name refers to the local name of type locality, Tsugaru, the western half of present day Aomori Prefecture, Japan.

Remarks: The new species is distingushed from other similar *Aulacoseira* taxa having large spatulated spines by its one rimoportula always located on mantle in a slanted canal to ringleist. The relative fruquency of the *A. tsugaruensis* in type material was 92%.

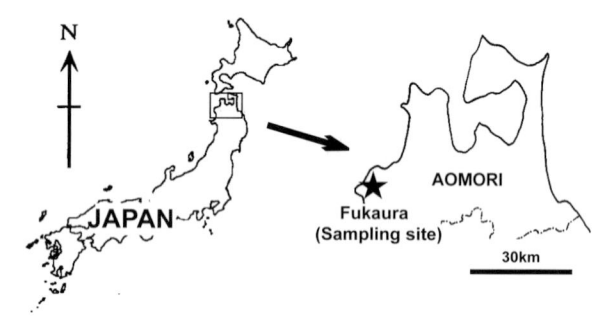

Fig. 4. Location of sampling site (★), Fukaura, Fukaura Town, Aomori, Japan.

Cyclotella nogamiensis H. Tanaka sp. nov. (Plates 57–58)

Description: Valves circular with large convex or concave central area of valve face, diameter 11–36 μm. Striae on marginal area, 14–17 in 10 μm, continue to mantle. Dark lines (thick costae with mantle fultoportulae in SEM observation) are 4–6 in 10 μm or every 3–4(5) interstria. SEM observation shows central area with foramina of areolae, openings of valve face fultoportulae, many bumps and small granules. Bumps are also on interstriae of marginal area. Though difficult to discern in LM views, areolae rows on central area are arranged in an approximately radial pattern. The striae are made up of two rows of areolae on marginal area of valve face and becoming three (four) rows on mantle. Short spines or granules located on the valve face/mantle boundary with granules on the mantle as well. SEM internal view shows 6–18 valve face fultoportulae with three satellite pores located near center, one fultoportula per areolae row if present. Mantle fultoportulae are located on every thick costa with two satellite pores laterally. One, or two rimoportuae, is located on marginal area of valve face. The cingulum includes a ligula-like segment.

Holotype: MPC-25053. Micropaleontology Collection, National Museum of Nature and Science, Japan.

Isotype: TA-3545. Maebashi Diatom Institute, Gunma Prefecture, Japan.

Type locality: Okunameshi, Nogami, Kokonoe Town, Oita Prefecture, Japan. 33°12.9′N, 131°13.5′E (Fig. 5).

Type material: OIT-229, collected by author on 29, April 2001. Diatomite of Nogami Formation (Middle Pleistocene).

Etymology: The species epithet refers to Nogami, the type locality.

Remarks: The new species is similar to *Cyclotella pantanelliana* Castracane and good LM photographs of the latter can be found in Houk *et al.* (2010). *C. nogamiensis* differs from *C. pantanelliana* in the size of its areolae on central area of valve face in LM, location pattern of valve face fultoportulae and interstria that do not branch on mantle.

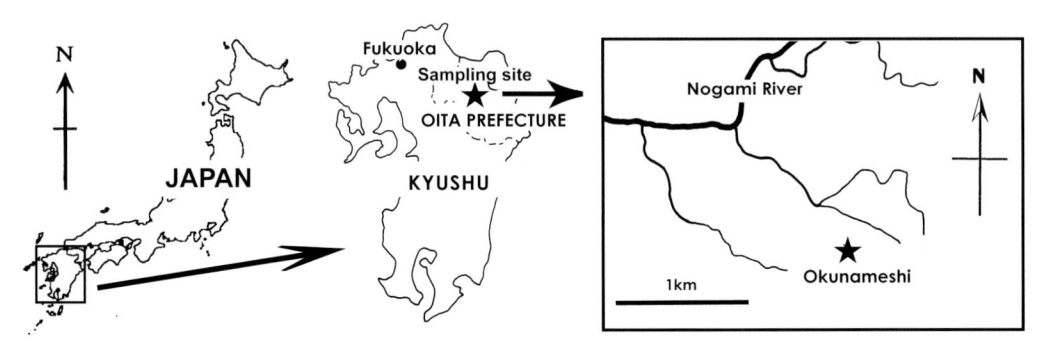

Fig. 5. Location of sampling site (★), Okunameshi, Kokonoe Town, Oita, Japan.

Cyclotella oitaensis H. Tanaka sp. nov. (Plates 61-62)

Description: Valves circular with blisters on central area of valve face, diameter 8.5-35.5 μm. Large central area with stellate pattern of punctae (valve face fultoportulae with two satellite pores by SEM observation). Striae on marginal area of valve face, 13-14 in 10 μm, continue to mantle and some interstriae branched on valve face or mantle. The striae are made up of two rows of areolae, sometimes with one smaller row of areolae in between. Usually every second (third) interstria does not branch and has an opening of mantle fultoportula. These interstriae form thick costae on the internal valve margin. All other interstriae branch one or two times and none have any openings of fultoportulae. SEM internal view shows mantle fultoportulae with two satellite pores laterally. A single rimoportula is located off center of a thick costa together with a mantle fultoportula, a little more toward the valve face side than the fultoportula. The cingulum includes a ligula-like segment.

Holotype: MPC-25054. Micropaleontology Collection, National Museum of Nature and Science, Japan.

Isotype: TA-3546. Maebashi Diatom Institute, Gunma Prefecture, Japan.

Type locality: Omoto, Kitsuki City, Oita Prefecture, Japan. 33°23.9′N, 131°38.2′E (Fig. 6).

Type material: OIT-020, collected by Mr. E. Kitabayashi on 12, June 2010. Siltstone of Omoto Formation (Early Pleistocene).

Etymology: The species name refers to Oita Prefecture, the type locality.

Remarks: The new species is similar to *Cyclotella astraea* (Ehrenb.) Kütz. or *Cyclotella schambica* Alesch. & Pirum. It differs in its large central area and openings of mantle fultoportulae on interstriae that do not branch.

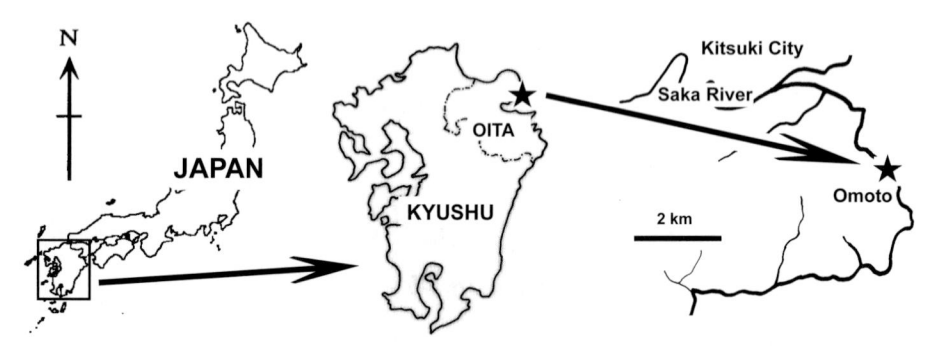

Fig. 6. Location of sampling site (★), Omoto, Kitsuki City, Oita, Japan.

Discostella kitsukiensis H. Tanaka sp. nov. (Plates 75–76)

Description: Valves circular, diameter 9–20 μm, valve face divided into an extremely bumpy central area and striated marginal area. Striae, 8–13 in 10 μm at valve face/mantle boundary. Alternating striae and interstriae continue to mantle with a single row of areolae surrounding the valve on the lowest part of mantle. Striae composed of small areolae ca. 7 areolae rows at the widest place. The interstriae are raised. Internally, smooth central area of valve face with no fultoportulae. Thick costae (interstriae) on marginal area. Mantle fultoportulae with two laterally placed satellite pores located on the end of every stria intersecting the areolae row surrounding the valve. A single rimoportula replaces a fultoportula at the same position. Every stria forms an alveolus.

Holotype: MPC-25055. Micropaleontology Collection, National Museum of Nature and Science, Japan.

Isotype: TA-3547. Maebashi Diatom Institute, Gunma Prefecture, Japan.

Type locality: Kanuki, Kitsuki City, Oita Prefecture, Japan. 33°22.5′N, 131°38.1′E (Fig. 7).

Type material: OIT-511, collected by H. Tanaka on 2, April 2011. Siltstone of Kanuki Formation (Early-Middle Pleistocene).

Etymology: The species named refers to Kitsuki City, the type locality.

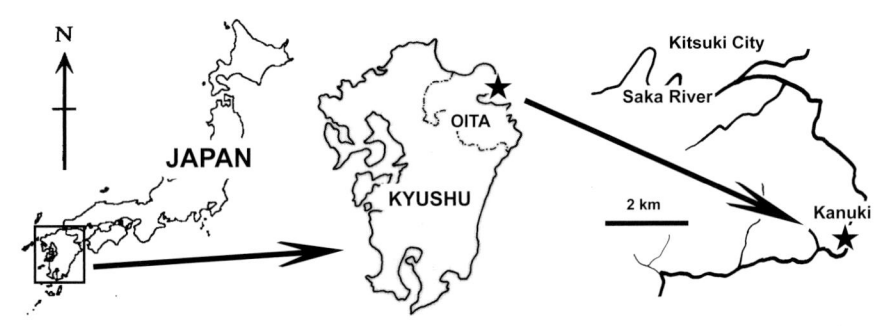

Fig. 7. Location of sampling site (★), Kanuki, Kitsuki City, Oita, Japan.

Stephanodiscus iwatensis H. Tanaka sp. nov. (Plates 95–96)

Description: Valves small, and either elevated or depressed in the center, diameter 11–23 μm. The areolae of the valve face tend to be disorganized in the very center, but then radiate in fascicles to the valve face/mantle boundary. Fascicles consist of usually two areolae rows at the valve face/mantle boundary. Interfascicles are 9–11 in 10 μm at the valve face/mantle boundary and at the end of every interfascicle occurs a spine or its remnant. Single valve face fultoportula located almost at the very valve center and in a heterotopic position with a rimoportula, always accompanied by three satellite pores internally. The mantle is shallow with an increasing number of small areolae. Mantle fultoportulae with three satellite pores located every three to five interfascicles. A single rimoportula appears at the valve face/mantle boundary, with a thick tube in the ring of spines externally.

Holotype: MPC-25056. Micropaleontology Collection, National Museum of Nature and Science, Japan.

Isotype: TA-3548. Maebashi Diatom Institute, Gunma Prefecture, Japan.

Type locality: Masuzawa, Shizukuishi Town, Iwate, Japan, 39°39′N, 140°56′E (Fig. 8).

Type material: MAS-117, siltstone of Masuzawa Formation (Mio-Pliocene).

Etymology: The species name refers to the type locality, Iwate Prefecture.

Remarks: In LM, the new species appears similar to *Stephanodiscus minutulus* (Kütz.) Round which also has one valve face fultoportula, but *S. minutulus* always has two satellite pores while *S. iwatensis* always has three. Also diameter of *S. minutulus* is 2–12 μm (Håkansson 2002) while that of *S. iwatensis* is 11–23 μm.

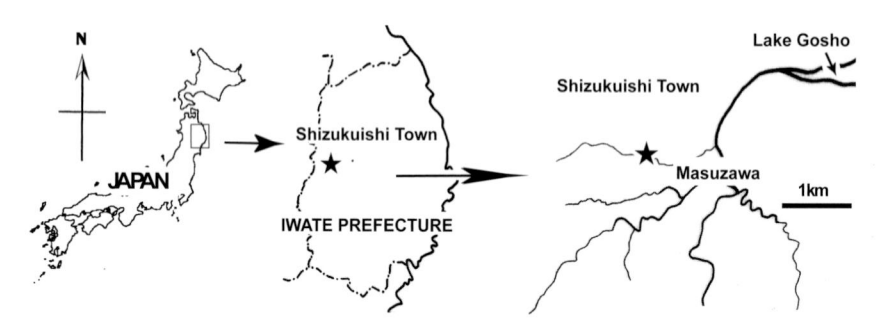

Fig. 8. Location of sampling site (★), Masuzawa, Shizukuishi Town, Iwate, Japan.

Tertiarius agunensis H. Tanaka sp. nov. (Plates 118-119)

Description: Valves circular, diameter 9-24 μm, with a flat valve face divided into a central area with random areolae and fultoportulae, and marginal area with striae. Striae number 10-18 in 10 μm and continue to valve edge. One stria consists of two rows of small areolae sometimes with smaller areolae between them. Many interstriae branch on marginal valve face or mantle. 4-9 valve face fultoportulae with two satellite pores arranged in a circle and areolae have domed cribra. Single rimoportula is situated in alveolus. Alveoli have both centrifugal and centripetal roofing over. Cingulum consists of four bands including a ligula-like segment.

Holotype: MPC-25057. Micropaleontology Collection, National Museum of Nature and Science, Tokyo, Japan. Single specimen slide (Figs 2, 3).

Isotype: TA-3549. Maebashi Diatom Institute, Gunma Prefecture, Japan.

Type locality: Fudenzaki, Aguni Island, Okinawa Prefecture, Japan. 26°34.5′N, 127°13′E (Fig. 9).

Type material: AGU-001, Coll. by Dr. Kazuhiko Uemura. Muddy tuff of the Fudenzaki Formation (Pliocene).

Etymology: The species name refers to the type locality, Aguni Island.

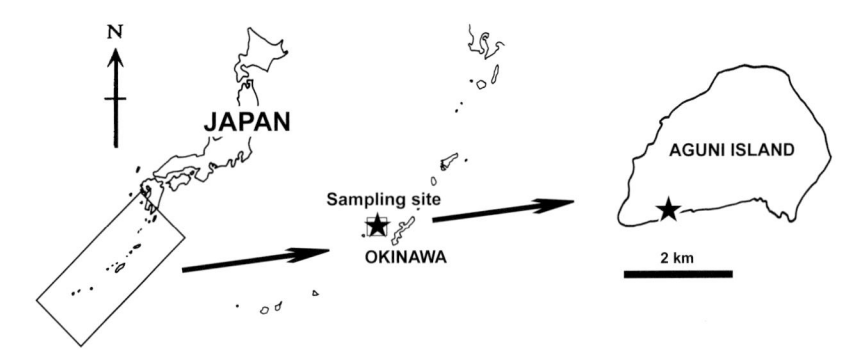

Fig. 9. Location of sampling site (★), Fudenzaki, Aguni Island, Okinawa, Japan.

Remarks: The new species described here is characterized by the difference in ornamentation between the central area of the valve face with its random areolae and fultoportulae and the marginal area of the valve face with its striae, typical features of the Cyclotelloid species, in particular the *Cyclotella comta* group (Khursevich & Kociolek 2002). The *C. comta* group, however, always has valve face fultoportulae with three satellite pores and the fultoportulae of this new species always has two. More notably the location of the rimoportula of the *C. comta* group is always on the valve face while that of the new species is located in an alveolus. As the authors believe the location of the rimoportula to be an important factor in classification, this new taxon is proposed to belong to the genus *Tertiarius*.

Among the nine established taxa of *Tertiarius*, the only species having random areolae on central area of valve face similar to *T. agunensis* is *Tertiarius transilvanicus* var. *disseminatepunctatus* (Pant.) Håk. & Khur. while the punctae of the others are radial or approximately radial. *T. transilvanicus* var. *disseminatepunctatus* and *T. agunensis* differ, however, in that the former has a hyaline area between the central area and the marginal area of valve face while the latter does not. Also the interstriae of *T. transilvanicus* var. *disseminatepunctatus* do not branch and number only 6–8 in 10 μm (measured from Figures of Håkansson & Khursevich (1997) by this authors) while those of *T. agunensis* branch and greatly outnumber those of *T. transilvanicus* var. *disseminatepunctatus* at 10–18 in 10 μm.

Araphid, pennate diatoms
Tabellaria japonica H. Tanaka sp. nov. (Plates 139–140)

Description: Valves linear lanceolate-fusiform with round or slightly capitate apices, 39–90 μm in length to 4–5.5 μm in width. Both apices have apical pore fields. Axial area is narrowly linear. Transapical striae are parallel, 10–18 in 10 μm, spacing irregular. Single rimoportula located near valve center on inward terminal of a stria. External opening of rimoportula is a slit. Septum with band, theca with at least two bands.

Holotype: MPC-25058. Micropaleontology Collection, National Museum of Nature and Science, Japan.

Isotype: TA-3550. Maebashi Diatom Institute, Gunma Prefecture, Japan.

Type locality: Shichiku, Iwaki City, Fukushima Prefecture, Japan. 37°8.35′N, 140°54.22′E (Fig. 1).

Type material: SCH-102, collected by author on 25 February 2007. Mudstone of the Shichiku Formation (Early Miocene).

Etymology: The species is named for Japan, the type locality.

Remarks: Because of its shape and its septum, this species belongs to the genus *Tabellaria*. The new species differs from all existing *Tabellaria* taxa in the almost parallel outline of its valve center and round or slightly capitate ends. The new species has only been found in the Shichiku Formation, an Early Miocene freshwater sediment.

The type material of the new species and the sampling site is the same as *Aulacoseira fukushimae* H.Tanaka sp. nov.

Biraphid, pennate diatoms
Cymbella ocellata H. Tanaka sp. nov. (Plates 159–160)

Description: Valves semilanceolate having a wide axial and central area and tapering to small apiculate valve apices. Length 26–32 μm, breadth 11–14 μm. Valve faces almost flat with a large transapical expansion in the center. Axial areas are very broad near the central area and narrow near the apices. Valve apices with helictoglossae internally and apical pore fields consisting of parallel small areolae rows. Raphe nearly midline, lateral, becoming filiform near the proximal and distal ends. Terminal fissures dorsally bent, while central fissures slightly ventrally bent with no central pores. Stigmata not present and alveoli without silica struts. Striae almost parallel increasing in density and becoming slightly radial toward the apices, 8–10 in 10 μm in center, 10–14 in 10 μm near apices.

Holotype: MPC-25059, Micropaleontology Collection, The National Museum of Nature and Science, Tokyo, Japan. Single specimen slide.

Isotype: TA-3551, Maebashi Diatom Institute, Gunma Prefecture, Japan.

Type locality: Kaminomaru, Koriyama, Kagoshima City, Kagoshima Prefecture, Japan. 31°43'55"N, 130°27'15"E (Fig. 10).

Type material: KAG-506, diatomite of the Koriyama Formation, Late Pliocene, collected by author on 27 November 2007.

Etymology: The species name refers to the valve shape, eye (ocellus).

Remarks: The new species reported here has all the characteristics of the genus *Cymbella*, the unique characters of its valve shape, broad central and axial area, no stigmata, areolae without silica strips and central fissures bent slightly toward the ventral side. Other taxa with similar valve shape include *Encyonema formosum* (Hust.) D.G. Mann, *Encyonema rugosum* (Hust.) D.G. Mann and *Cymbella leptoceroides* Hust. The terminal fissures of the first two species, however, are bent toward the ventral side (character of genus *Encyonema*), different from the dorsal side bent (character of genus *Cymbella*) of the new species.

LM photographs of *Cymbella leptoceroides*, reported from Hawaii by Hustedt (1942), are provided by Simonsen (1987) and Krammer (2002). According to them, *C. leptoceroides* has narrowly rounded valve edges and narrower central and axis areas than the new species. The species *Cymbella gemeinhardtii* Metzeltin & Krammer has both wide central and axial areas, but it has broadly rostrate valve apices, slightly radiate striae as well as stigmata, all different from the new species.

The author compared this taxon to other taxa in the genus *Cymbella*, but were unable to find any taxon that had the following combination of characteristics observaed in this specimens: a semilanceolate valve having a wide center and tapering to small apiculate valve apices, central and axial area very broad, valve edges with apical pore fields, raphe lateral and becoming filiform near the center, terminal fissures dorsally bent while central fissures ventrally bent, stigmata not present and striae almost parallel and increasing in density and becoming slightly radial near the apices. I therefore propose *Cymbella ocellata* as a new species.

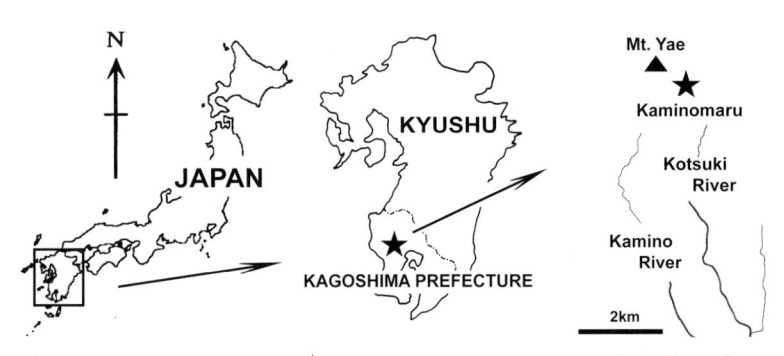

Fig. 10. Location of sampling site (★), Kaminomaru, Kagoshima City, Kagoshima, Japan.

Cymbella okunoi H. Tanaka sp. nov. (Plates 161-162)

Description: Valves triangular, dorsal margin with a strongly expanded central portion, ventral margin nearly straight with a slightly convex center. Valve apices are narrow. Length 43-64 µm, width 15-17 µm. Axial area narrow and central area slightly expanded. Raphe bent to the dorsal side in the apices. Striae run slightly radial in the center becoming more radial toward the apices, 7-8 in 10 µm in the middle of ventral side, puncta 16-20 in 10 µm. Two or three stigmata located on the top of striae on the ventral side of the central area.

SEM observations show the raphe proximal endings strongly bent to the ventral side ending at small central pores. Both apices have apical pore fields and helictoglossae internally. Two or three stigmata, round pores externaly, on the ventral side of the valve. Areolae foramina are mostly slits, but in rare cases, X or Y shaped. Internally, the stigmata are narrow furrows on top of areolae rows. Though only some mantle areolae separated by silica struts are observable, probably all areolae are separated by silica struts. The internal areolae rows of the apical pore fields are parallel.

Holotype: MPC-25060. Micropaleontology Collection, National Museum of Nature and Science, Japan.

Isotype: TA-3552. Maebashi Diatom Institute, Gunma Prefecture, Japan.

Type locality: Okunameshi, Nogami, Kokonoe Town, Oita Prefecture, Japan. 33°12.9′N, 131°13.5′E (Fig. 5).

Type material: OIT-229, collected by author on 29 April 2001. Nogami Formation (Middle Pleistocene).

Etymology: The species is named in honor of the late Dr. H. Okuno who first reported this taxon from the Nogami Formation.

Remarks: Distinguished from *Cymbella cistula* var. *insignis* Meister by its strongly arched valve shape and wide apices, and *Cymbella kemiana* Krammer by its narrow apices and strongly expanded central dorsal side.

The type material of the new species and the sampling site is the same as *Cyclotella nogamiensis* H. Tanaka sp. nov.

Cymbella tsumurae H. Tanaka sp. nov. (Plates 167-169)

Description: Valves large, semicircularly dorsiventral, ventral margin with slightly expanded central portion, dorsal margin arched. Valve apices are wide with angular edges. Length 123-188 μm, width 26-30 μm. Axial area relatively wide, about 1/5 of the total valve breadth and central area slightly expanded. Raphe almost midline, slightly bent toward the ventral side in the central area, but bent to the dorsal side in the apices. Striae run slightly radial in the center and then parallel toward the apices, in the middle 8-10 in 10 μm, ends ca. 12 in 10 μm, puncta 10-12 in 10 μm. Many stigmata located on the top of striae on the ventral side of the central area, but sometimes difficult to discern in LM.

SEM observations show the raphe endings bent to the ventral side of the central area are central pores. Both apices have apical pore fields and helictoglossae internally. Up to 14 stigmata, narrow slits, on the ventral side of the central area. Areolae foramina are mostly slits, but in rare cases, X or Y shaped. Distal end of areolae rows are irregular. Internally, the stigmata are narrow furrows on top of areolae rows. Silica struts separate the areolae. The areolae rows of the apical pore fields are radially arranged from the helictoglossae.

Holotype: MPC-25061. Micropaleontology Collection, National Museum of Nature and Science, Japan.

Isotype: TA-3553. Maebashi Diatom Institute, Gunma Prefecture, Japan.

Type locality: Yokogo, Numata City, Gunma Prefecture, Japan. 36°37′N, 139°03′E (Fig. 11).

Type material: YOK-101, collected by author on 06 May 1979. Diatomite of the Yokogo (Early Pleistocene).

Etymology: The species name refers to the late Dr. K. Tsumura who provided the information of the Yokogo diatomite and the sampling site.

Remarks: Distinguished from *Cymbella praerupta* Hust. by the number of stigmata and direction of striae near apices and *Cymbella australica* (A. Schmidt) Cleve by shape of central area, number of stigmata and direction of striae.

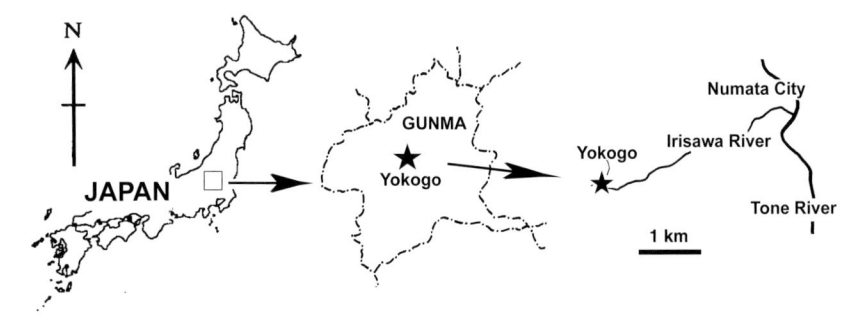

Fig. 11. Location of sampling site (★), Yokogo, Numata City, Gunma, Japan.

Didymosphenia nipponica H. Tanaka sp. nov. (Plates 178-179)

Description: Valves asymmetrical and slightly curved in valve view. Valve face is flat and head-pole is broadly cuneate while the foot-pole is rounded. Valves 93-191 μm long, 26-38 μm wide, striae uniseriate, ca. 8 in 10 μm, puncta 7-12 in 10 μm. At the valve center toward the head-pole the striae are at first radial becoming parallel and then finally slightly convergent, while toward the foot-pole, they start radial, become parallel and finally radial again. The apices endings of raphe are toward the opposit side of stigmata. Usually two to five stigmata (rarely one or six). All areolae surrounded by vertical walls with occluded volae. Foot-pole with apical pore field.

Holotype: MPC-25062. Micropaleontology Collection, National Museum of Nature and Science, Tokyo, Japan.

Type locality: Matamizu, Kitsuki City, Oita Prefecture, Japan. 33°29′N, 131°35′E (Fig. 12).

Type material: OIT-104, collected by author on 26 March 1999. Diatomite of the Matamizu Formation (Early Pleistocene).

Etymology: The species name refers to Japan, Nippon, as the country is known to the Japanese.

Remarks: The combination of features of *Didymosphenia nipponica* which distinguish it from other species of the same genus are the cuneate head-pole of valve face, including the broadly rounded valve rim, the radial to convergent striae direction, areolae with vertical walls at valve surface and the lack of ridges or spines on the valve face/mantle boundary.

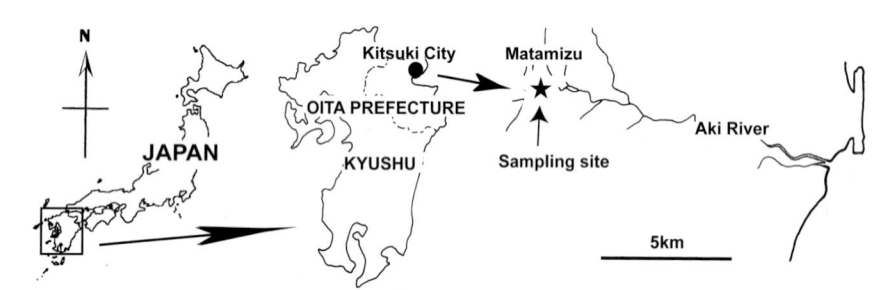

Fig. 12. Location of sampling site (★), Matamizu, Kitsuki City, Oita, Japan.

Oricymba cunealjaponica H. Tanaka sp. nov. (Plates 192-193)

Description: Valves slightly dorsiventral to linear lanceolate with slightly expanded valve center and cuneate apices. Length: 38-67 μm, width 10-13 μm. Plicate raphe, external raphe fissures reflected to ventral side near center and near apices, bent to dorsal side in apices. Inner raphe fissure straight. Axial area almost linear lanceolate. Central area expanded toward ventral side where one stigma between the central nodule and ventral striae is located. Striae slightly radiate, becoming moderately radiate near the apices, 7-8 in 10 μm in center and puncta in striae 20-24 in 10 μm.

SEM observations show external valve faces flat, central raphe endings slightly distended while endings of apices reflected to the ventral side of valve face and then finally to the dorsal side on the mantle. Apical pore fields are present on both apices. Ridges and grooves located on both ventral and dorsal sides of valve face/mantle boundary except near apices. One stigma visible in external view, though two stigma openings can be seen in internal view. Areolae openings are wide slits externally and volae internally. Internal raphe of apices end in helictoglossae.

Holotype: MPC-25063. Micropaleontology Collection, National Museum of Nature and Science, Japan.

Isotype: TA-3554. Maebashi Diatom Institute, Gunma Prefecture, Japan.

Type locality: Tsumori, Mashiki Town, Kumamoto Prefecture, Japan. 130°51.5′E, 32°47.6′N (Fig. 13).

Type material: KUM-102-2, collected by author on 29 April 2001. Tsumori Formation (Middle Pleistocene).

Etymology: The species name is a combination of the cuneate shape of its valve and the location where it was found, Japan.

Remarks: Distinguished from the other *Oricymba* species by the cuneate shape of its valve.

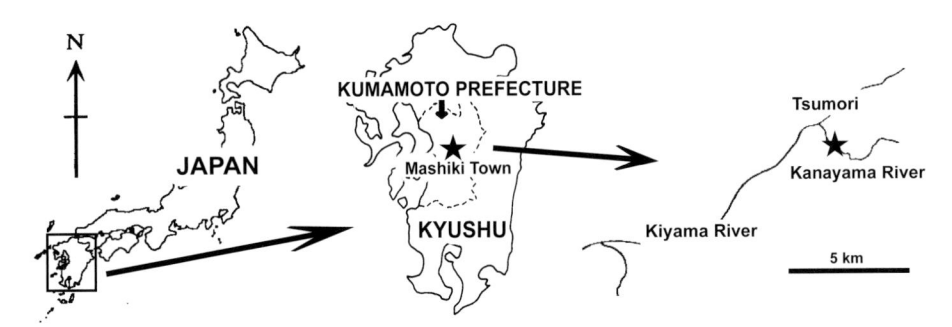

Fig. 13. Location of sampling site (★), Tsumori, Mashiki Town, Kumamoto, Japan.

Epithemia numatensis H. Tanaka sp. nov. (Plates 248-249)

Description: Valves linear-arculate, having a weakly convex dorsal margin and also a weakly concave ventral margin, the two margins lying almost parallel to each other. The apices are extremely capitate and have very narrow necks. Length 49-88 μm, breadth 10-14 μm. 3-4 transapical costae in 10 μm, (2)3-7(8) striae in between.13-14 striae in 10 μm and ca. 12 puncta in rows in 10 μm. Raphe is biarculate and lies along the ventral margin with the center and the apices of the valve slightly toward the ventral side at the midline position. The central raphe endings are expanded externally into small central pores. The internal raphe is interrrupted by the central nodule. There are internal thickend transapical costae and sometimes weakly thickend costae. Areolae are closed domed caps externally.

Holotype: MPC-25064. Micropaleontology Collection, National Museum of Nature and Science, Japan.

Isotype: TA-3555. Maebashi Diatom Institute, Gunma Prefecture, Japan.

Type locality: Yokogo, Numata City, Gunma Prefecture, Japan. 36°37′N, 139°03′E (Fig. 11).

Type material: YOK-101, collected by author on 06 May 1979. Diatomite of the Yokogo (Early Pleistocene).

Etymology: The species name refers to Numata City, the type locality.

Remarks: The most pronounced characteristic of this new species is its extremely capitate apices with very narrow necks. A similar characteristic has been displayed, although rarely, in *Epthemia adnata*, but it is always presence in *E. numatensis*.

   The type material of the new species and the sampling site is the same as *Cymbella tsumurae* H.Tanaka sp. nov.

2. Normenclatural Changes

Achnanthes okunoi (Hustedt) H. Tanaka comb. nov. (Plates 152-153)

Basionym: *Navicula okunoi* Hustedt, in Dr. Rabenhorst's Kryptogamen-Flora, Band Ⅶ, 3. Teil. Die Kieselalgen, p. 740. Fig. 1718. 1961-1966.

Remarks: Hustedt (1961-1966) described the taxon as being a new species in genus *Navicula*. The author observed the taxon from the sample of the type locality, Setana, Hokkaido, Japan and found one valve having raphe and others with not of the same frustule, only the fiber like remains of raphe which do not peneprate silica wall of valve. The author considers the taxon suitable to genus *Achnanthes*.

Fragilariforma fossilis **(Pantocsek) H. Tanaka comb. nov. et stat. nov.**
(Plate 128)

Basionym: *Diatoma anceps* var. *fossils* Pantocsek, Beiträge zur Kenntniss der fossilen Bacillarien Ungarns. Ⅲ. p. 46. pl. 8, f. 141. 1905.

Synonym: *Fragilaria bicapitata* A.Mayer sensu Okuno, Journ. Jap. Bot. **34**(9): 16. f. 1, a. 1959.

Neotype: MPC-25065. Micropaleontology Collection, National Museum of Nature and Science, Tokyo, Japan.

Material: SET-306, collected by author on 25 April 2012, Futoro Formation (Diatomite of Setana), Early Miocene.

Remarks: Pantocsek (1905) reported *Diatoma anceps* var. *fossils* Pant. from diatomite as Sentenai (exact name is Setana) in insula Jesso, Japan. Okuno (1959) reported it without cross costa and he identified it with *Fragiraria bicapitata* A.Mayer (now *Fragilariforma bicapitata* (A.Mayer) D.M.Williams & Round). According to Williams & Round (1988), *Fragilariforma* species have one rimopotula and apical pore fields. According to the author's SEM observation using the material from same locality of Pantocsek (1905), *D. anceps* var. *fossils* have neither rimoportulae nor apical pore fields. Though the morphology of *D. anceps* var. *fossils* does not fit perfectly to genus *Fragilariforma* mentioned above, its most similar genus is *Fragilariforma*.

As the slides and materials were lost during World War Ⅱ, I designated the neotype specimen slide using material from the type locality, Setana, the same as Sentenai in Pantocsek (1905).

Fragilariforma kamczatica **(Lupikina) H. Tanaka comb. nov. et stat. nov.**
(Plate 129)

Basionym: *Fragilaria nitzschioides* var. *kamczatica* Lupikina, Novitates Systematicae Plantorum non Vascularium **2**: 20, Table 4, Fig. 4. Illustoration 5. 1965.

Remarks: Kupikina (1965) reported this taxon as a variety of *Fragilaria nitzschoides*. As the genus *Fragilaria* was later divided into some other genera (Williams & Round 1987, 1988), the taxon is suitable to genus *Fragilariforma* as a noninal variety.

II. 試料：Samples

　試料は次の中新世・鮮新世・更新世試料46地点，完新世・現生試料35地点で採取した（表と図14，図15に付してある採取地点の番号は一致する）．

1. 中新世・鮮新世・更新世試料採取地

番号	地　層	地質年代	地　域
1	小松沢層	鮮新世	北見市留辺蘂（北海道）
2	網走湖底ボーリングコア	更新世	網走市（北海道）
3	注1 太櫓層（瀬棚の珪藻土）	前期中新世	瀬棚町（北海道）
4	六角沢層	前期中新世	深浦町（青森県）
5	宮田層	注2 中新-鮮新世	仙北市（秋田県）
6	舛沢層	中新-鮮新世	雫石町（岩手県）
7	鬼首層	後期更新世	大崎市鳴子（宮城県）
8	白沢層	後期中新世	仙台市（宮城県）
9	向山層	鮮新世	仙台市（宮城県）
10	円田層	前期更新世	蔵王町（宮城県）
11	紫竹層	前期中新世	いわき市（福島県）
12	塩原湖成層	前期更新世	那須塩原市塩原（栃木県）
13	小野上層	前期更新世	渋川市小野上（群馬県）
14	沼田湖成層	中期更新世	沼田市（群馬県）
15	横子珪藻土	前期更新世	沼田市横子（群馬県）
16	中之条湖成層	中期更新世	中之条町（群馬県）
17	嬬恋湖成層	中期更新世	嬬恋村（群馬県）
18	大戸湖沼性堆積物	中期更新世	東吾妻町（群馬県）
19	倉渕湖沼性堆積物	中期更新世	高崎市倉渕（群馬県）
20	兜岩層	前期鮮新世	南牧村星尾（群馬県）
21	稲城層	前期更新世	府中市（東京都）
22	渋沢湖沼性堆積物	更新世	上田市真田（長野県）
23	香坂礫岩層	鮮新世	佐久市香坂東地（長野県）
24	瓜生坂層	前期更新世	小諸市（長野県）
25	美ヶ原層	前期更新世	松本市美ヶ原（長野県）
26	注3 大鷲湖沼性堆積物	鮮新世	郡上市高鷲（岐阜県）
27	蜂屋層	前期中新世	八百津町蜂屋（岐阜県）
28	平牧層	前期中新世	可児市（岐阜県）
29	伊賀層	鮮新世	伊賀市（三重県）
30	大阪層群	鮮新世	河合町（奈良県）
31	春来層	鮮新世	新温泉町（兵庫県）
32	三徳層	後期中新世	三朝町（鳥取県）
33	人形峠層	後期中新世～鮮新世	人形峠（岡山-鳥取県境）
34	長者原層	中期中新世	壱岐市芦辺（長崎県）

注1：太櫓層は瀬棚として多くの文献に記され，海外にも知られているので（瀬棚の珪藻土）を加えた．
注2：中新-鮮新世は中新世と鮮新世の境界の年代を示す．
注3：大鷲湖沼性堆積物は，東方の阿多岐に分布する湖沼性堆積物から鮮新世を指示する植物化石が見出され，Matsuo(1968)はこの地域に広く分布する湖沼性堆積物を鮮新統とした．しかし，高橋・下野(1980)は大鷲より西の鮎鈷に分布する湖沼性堆積物は更新統である可能性を示唆している．大鷲の湖沼性堆積物からは時代を指示する資料は得られていないので，年代はMatsuo(1968)のとおり鮮新世とした．

番号	名称		地域
35	津房川層	後期鮮新世	宇佐市安心院（大分県）
36	俣水層	前期更新世	杵築市俣水（大分県）
37	野原層	前期更新世	杵築市山鹿（大分県）
38	野上層	中期更新世	九重町野上（大分県）
39	阿蘇野層	中期更新世	由布市庄内（大分県）
40	尾本層	前期更新世	杵築市尾本（大分県）
41	加貫層	中期～前期更新世	杵築市加貫（大分県）
42	津森層	中期更新世	益城町（熊本県）
43	人吉層	後期鮮新世	人吉市（熊本県）
44	永野層	後期鮮新世	さつま町（鹿児島県）
45	郡山層	後期鮮新世	鹿児島市郡山（鹿児島県）
46	筆ん崎層	鮮新世	粟国島（沖縄県）

2. 完新世・現生試料採取地

番号	名称	地域
1	パンケ沼	幌延町（北海道）
2	屈斜路湖	弟子屈町（北海道）
3	阿寒湖	釧路市（北海道）
4	八郎潟調整池	潟上市（秋田県）
5	鎌沼	福島市（福島県）
6	毘沙門沼	北塩原村（福島県）
7	元温泉小屋温泉	檜枝岐村（福島県）
8	尾瀬沼・尾瀬地域	片品村（群馬県）
9	中禅寺湖	日光市（栃木県）
10	大沼（赤城山）	前橋市（群馬県）
11	池の岳の池塘	魚沼市（新潟県）
12	利根川源流	水上町（群馬県）
13	奥平鉱泉	新治村（群馬県）
14	桐生川	桐生市（群馬県）
15	波志江沼	伊勢崎市（群馬県）
16	多々良沼	館林市（群馬県）
17	城沼	館林市（群馬県）
18	磯部鉱泉	安中市（群馬県）
19	赤久縄鉱泉	神流町（群馬県）
20	星尾鉱泉	南牧村（群馬県）
21	神流川	上野村（群馬県）
22	北浦	鹿嶋市沖（茨城県）
23	霞ヶ浦	稲敷市沖（茨城県）
24	手賀沼	我孫子市（千葉県）
25	諏訪湖	諏訪市（長野県）
26	山中湖	山中湖村（山梨県）
27	琵琶湖	長浜市沖（滋賀県）
28	沢の池	京都市（京都府）
29	藺牟田池	薩摩川内市（鹿児島県）
30	吹上浜（泥炭層）	日置市（鹿児島県）
31	池田湖	指宿市（鹿児島県）
32	泥染公園の泥田	奄美大島（鹿児島県）
33	上嘉鉄湧泉池	喜界島（鹿児島県）
34	大正池	粟国島（沖縄県）
35	福上湖	東村（沖縄県）

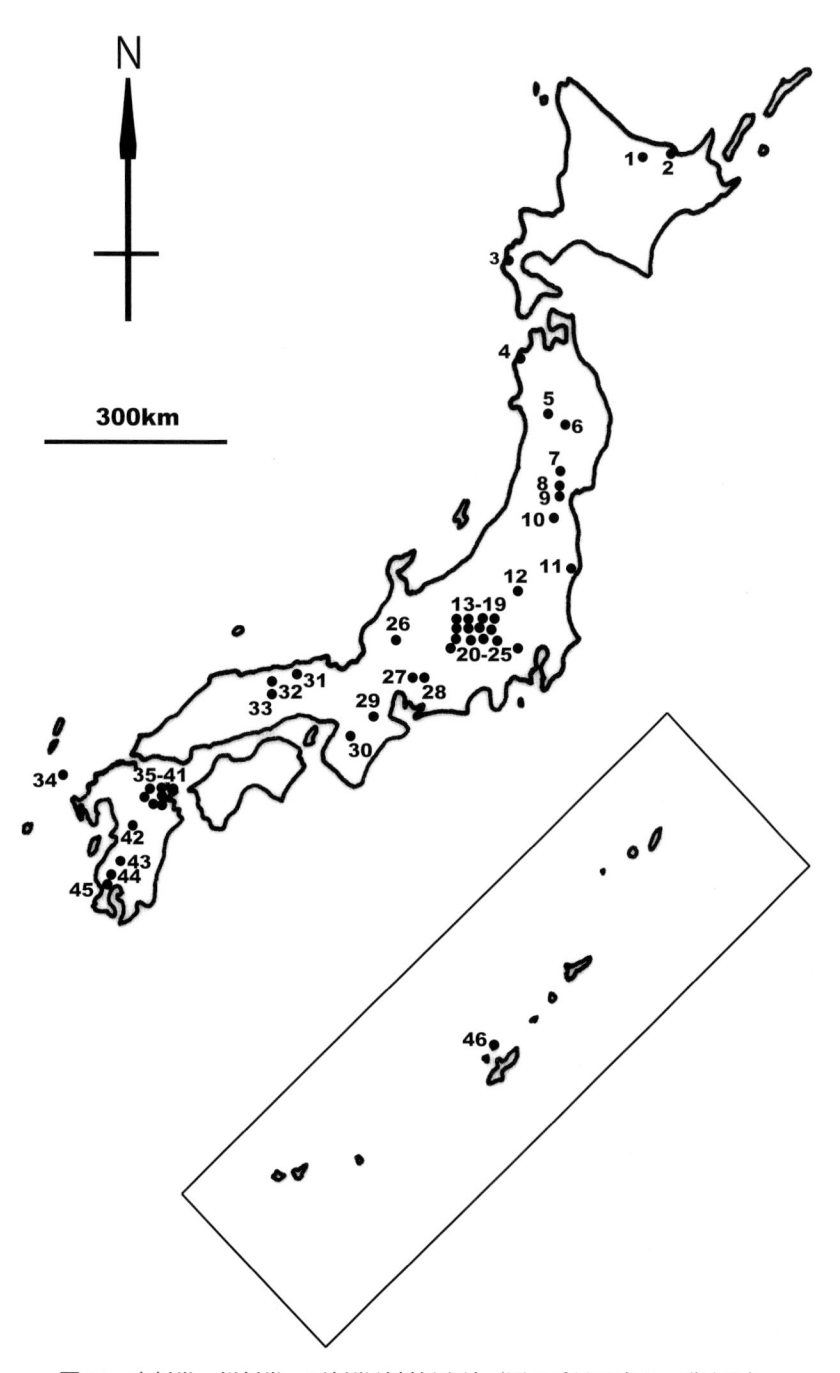

図14　中新世・鮮新世・更新世試料採取地（図の番号は表と一致する）.

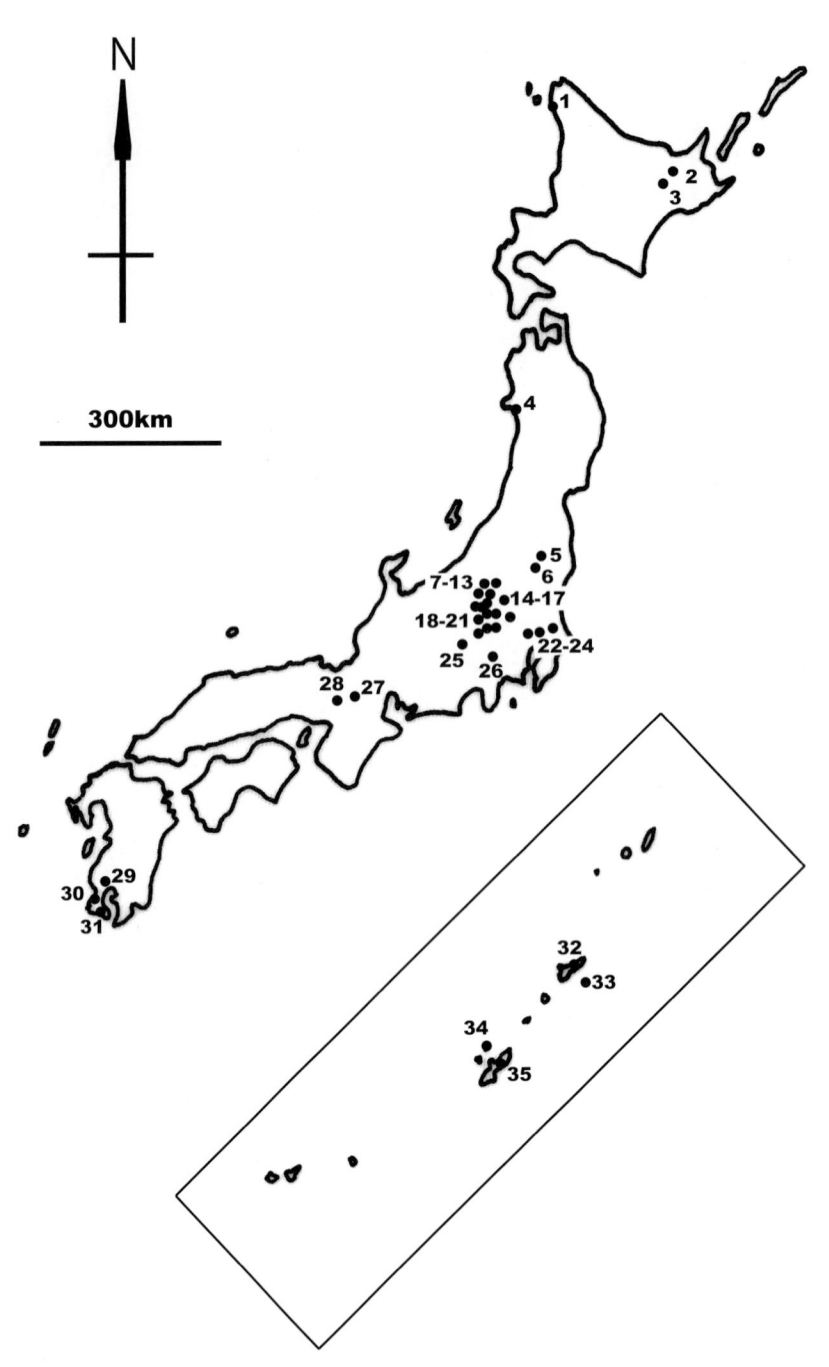

図15 完新世・現生試料採取地（図の番号は表と一致する）．

III. 試料処理・プレパラート作成：
Method of Cleaning Samples and Preparations

1. 試料の処理

化石試料
　試料は，1グラム程度に砕いた（あるいは厚刃のナイフで削り取った）後，乾燥し，30％過酸化水素水で煮沸した．第四紀程度の軟質試料は，この後，精製水で洗浄し，光顕・電顕用の試料とした．固結が進んでいる試料は，上記の後，塩酸・硝酸の混合液で煮沸し，精製水で洗浄し，光顕・電顕用の試料とした．

現生試料
　硫酸を使用して有機物の除去を行い，硝酸カリウムで漂白の後，精製水で洗浄し，光顕・電顕用の試料とした．

2. プレパラート作成

　混種（散布）スライドは，カバーガラス上に水と混合させた状態の試料を滴下し，乾燥の後，プルーラックス，スティラックスあるいはエンテランニューでスライドガラスと固着させ，検鏡用のスライドとした．
　単種スライドは，新種記載用のタイプスライドのために製作した．スライドガラスに丸く穴を開けたアルミホイルを貼付し，その中に目的とする珪藻を置き，カバーガラスをかけて封入した．封入剤はスティラックスを使用した．ただし目的とする珪藻が小型でアルミホイルの穴に置く作業が難しい場合は，薄めの混合液を用いて混種スライドを製作し（封入剤はスティラックスを使用する）鏡下で目的とするのにふさわしい珪藻を探して，その珪藻がわかるようにカバーガラス上から丸印をつけて，特定の珪藻を指定した．
　走査電子顕微鏡（SEM）で観察した標本を光学顕微鏡（LM）で観察する場合は，上記単種スライド作製と同様に，SEM観察で使用した載台上の標本を穴あきアルミホイルが貼付してあるスライドガラスの穴へ移動し，封入した．

Ⅳ. 収録分類群の配列：Taxa Arrangement

　本書における分類群の配列は，縦溝の有無により 4 大別した．それらは必要に応じて以下のとおり細区分してある．細区分に際しては Round *et al.*(1990)の分類体系の「目（order）」を使用したが，近年多くの属が新設されているので，適宜追加記入してある．本書に収録した分類群の配列を次に示す．目，および目に区分された属，種はアルファベット順に記した．

1. 中心類（Centric diatoms）

　条線は中心から放射状に配列する．殻面は円形またはそれに類似している．縦溝はない．唇状突起・有基突起（ない場合もある）がある．

Aulacoseirales: *Aulacoseira, Brevisira*
Biddulphiales: *Hydrosera, Stoermeria*
Coscinodisales: *Actinocyclus*
Orthoseirales: *Orthoseira*
Melosirales: *Melosira*
Paraliales: *Ellerbeckia*
Thalassiosirales: *Cyclostephanos, Cyclotella, Cyclotubicoalitus, Dimidialimbus, Discostella, Mesodictyon, Mesodictyopsis, Pliocaenicus, Spicaticribra, Stephanodiscus, Tertiariopsis, Tertiarius, Thalassiocyclus, Thalassiosira*
Triceratiales: *Pleurosira*

2. 無縦溝羽状類（Araphid, pennate diatoms）

　条線は中肋（南雲・真山 2000）から平行，放射状あるいは反放射状に配列する．殻面は一般に棒状〜皮針形．被殻の両方に縦溝がない．

Fragilariales: *Diatoma, Fragilaria, Fragilariforma, Hannaea, Meridion, Pseudostaurosira, Staurosira, Staurosirella, Ulnaria*
Tabellariales: *Tabellaria, Tetracyclus*

3. 単縦溝羽状類（Monoraphid, pennate diatoms）

　条線は中肋または縦溝中肋（南雲・真山 2000）から平行，放射状あるいは反放射状に配列する．殻面は一般に棒状〜皮針形．被殻の片方に縦溝がない．

Achnanthales: *Achnanthes, Cocconeis*

4. 双縦溝羽状類（Biraphid, pennate diatoms）

　条線は縦溝中肋から平行，放射状あるいは反放射状に配列する．殻面は一般に棒状〜皮針形．被殻の両方に縦溝がある．

Cymbellales: *Cymbella, Cymbopleura, Didymosphenia, Encyonema, Gomphoneis, Gomphonema, Gomphopleura, Gomphosphenia, Oricymba*

Eunotiales: *Actinella, Eunotia*
Naviculales: *Aneumastus, Caloneis, Cavinula, Craticula, Diadesmis, Diploneis, Frustulia, Gyrosigma, Navicula, Neidium, Pinnularia, Plagiotropis, Sellaphora, Stauroneis*
Rhopalodiales: *Epithemia, Rhopalodia*
Surirellales: *Campylodiscus, Cymatopleura, Surirella*
Thalassiophysales: *Amphora, Denticula, Halamphora, Nitzschia*

V. 収録分類群一覧：List of Taxa

（太字は新種，または新組み合わせを行った分類群）

1. 中心類（Centric diatoms）

Aulacoseirales

Aulacoseira ambigua (Grunow) Simonsen ·· 38
Aulacoseira cataractarum (Hustedt) Simonsen ····································· 40
Aulacoseira crassipunctata Krammer ··· 42
Aulacoseira fukushimae H. Tanaka sp. nov. ································· 1, 44, 46
Aulacoseira granulata (Ehrenberg) Simonsen var. *granulata* ················ 48
Aulacoseira granulata var. *angustissima* (O. Müller) Simonsen ············ 50
Aulacoseira hachiyaensis H. Tanaka ·· 52
Aulacoseira houki H. Tanaka ·· 54
Aulacoseira italica (Ehrenberg) Simonsen ·· 56
Aulacoseira iwakiensis H. Tanaka ··· 58
Aulacoseira longispina (Hustedt) Simonsen ·· 60
Aulacoseira miosiris H. Tanaka sp. nov. ····································· 2, 62, 64
Aulacoseira nipponica (Skvortsov) Tuji ·· 66
Aulacoseira nivalis (W. Smith) English & Potapova ······························ 68
Aulacoseira polispina H. Tanaka sp. nov. ································ 3, 70, 72, 74
Aulacoseira pusilla (F. Meister) Tuji & Houki ······································· 76
Aulacoseira satsumaensis H. Tanaka ··· 78, 80
Aulacoseira subarctica (O. Müller) Haworth ··· 82
Aulacoseira tenella (Nygaard) Simonsen ··· 84
Aulacoseira tsugaruensis H. Tanaka sp. nov. ······························ 4, 86, 88
Aulacoseira valida (Grunow) Krammer ··· 90
Brevisira arentii (Kolbe) Krammer ·· 92
Miosira tscheremissinovae (Khursevich) Khursevich ····························· 94

Biddulphiales

Hydrosera whampoensis (A.F. Schwarz) Deby ······································ 96
Stoermeria trifoliata (Cleve) Kociolek, Escobar & Richardson ················ 98

Coscinodisales

Actinocyclus normanii f. *subsalsa* (Juhlin-Dannfelt) Hustedt ················ 100
Actinocyclus octonarius Ehrenberg s.l. ·· 102, 104

Orthoseirales

Orthoseira asiatica (Skvortsov) H. Kobayasi ······································ 106

Melosirales

Melosira undulata (Ehrenberg) Kützing var. *undulata* ························ 108

Melosira undulata var. *producta* A. Schmidt ·· 110
Melosira varians C. Agardh ··· 112

Paraliales
Ellerbeckia arenaria (Moore) R.M. Crawford f. *arenaria* ···················· 114
Ellerbeckia arenaria f. *teres* (Brun) R.M. Crawford ·························· 116, 118

Thalassiosirales
Cyclostephanos costatilimbus (H. Kobayasi & Kobayashi)
    Stoermer, Håkansson & Theriot ··· 120
Cyclostephanos dubius (Fricke) Round ····································· 122
Cyclostephanos kyushuensis H. Tanaka ································· 124, 126
Cyclostephanos numataensis H. Tanaka & Nagumo ······················ 128
Cyclotella atomus Hustedt ··· 130
Cyclotella cyclopuncta Håkansson & Carter ································ 132
Cyclotella iris Brun & Héribaud s.l. ··· 134
Cyclotella iwatensis H. Tanaka ·· 136
Cyclotella kitabayashii H. Tanaka ·· 138, 140
Cyclotella kohsakaensis H. Tanaka & H. Kobayasi ························ 142
Cyclotella meneghiniana Kützing ·· 144
Cyclotella mesoleia (Grunow) Houk, Klee & Tanaka ··················· 146, 148
Cyclotella nogamiensis H. Tanaka sp. nov. ·························· 5, 150, 152
Cyclotella notata Loseva ·· 154
Cyclotella ocellata Pantocsek ·· 156
Cyclotella oitaensis H. Tanaka sp. nov. ·························· 6, 158, 160
Cyclotella ozensis (H. Tanaka & Nagumo) H. Tanaka ················· 162, 164
Cyclotella praetermissa Lund ··· 166
Cyclotella radiosa (Grunow) Lemmermann ································· 168
Cyclotella rhomboideo-elliptica Skuja var. *rhomboideo-elliptica* ······· 170, 172
Cyclotella rhomboideo-elliptica var. *rounda* Qi & Yang ·················· 174
Cyclotella satsumaensis H. Tanaka & Houk ································· 176
Cyclotella schumannii (Grunow) Håkansson ································ 178
Cyclotubicoalitus undatus Stoermer, Kociolek & Cody ···················· 180
Dimidialimbus bungoensis H. Tanaka ··· 182
Discostella asterocostata (Xie, Lin & Cai) Houk & Klee ··················· 184
Discostella kitsukiensis H. Tanaka sp. nov. ······················ 7, 186, 188
Discostella pliostelligera (H. Tanaka & Nagumo) Houk & Klee ······· 190, 192
Discostella pseudostelligera (Hustedt) Houk & Klee ······················ 194
Discostella stelligera (Cleve & Grunow) Houk & Klee ····················· 196
Discostella woltereckii (Hustedt) Houk & Klee ······························ 198
Mesodictyon yanagisawae H. Tanaka (in press) ··························· 200
Mesodictyopsis akitaensis H. Tanaka & Nagumo ·························· 202
Mesodictyopsis miyatanus H. Tanaka ·· 204

Pliocaenicus costatus (Loginova, Lupikina & Khursevich)
　Flower, Ozornina & A.I. Kuzmina ·· 206, 208
Pliocaenicus nipponicus H. Tanaka & Nagumo ································ 210
Pliocaenicus omarensis (Kuptsova) Stachura-Suchoples & Khursevich ·············· 212
Pliocaenicus radiatus H. Tanaka (in press) ····································· 214
Pliocaenicus tanimurae H. Tanaka & Saito-Kato ······························ 216
Spicaticribra kingstonii Johansen, Kociolek & Lowe ··························· 218
Stephanodiscus akanensis Tuji, Kawashima, Julius & Stoermer ············ 220, 222
Stephanodiscus hashiensis H. Tanaka (in press) ································ 224
Stephanodiscus iwatensis H. Tanaka sp. nov. ·················· 8, 226, 228
Stephanodiscus kobayasii H. Tanaka (in press) ···························· 230, 232
Stephanodiscus komoroensis H. Tanaka ·································· 234, 236
Stephanodiscus kusuensis Julius, Tanaka & Curtin ························ 238, 240
Stephanodiscus kyushuensis H. Tanaka (in press) ························· 242, 244
Stephanodiscus minutulus (Kützing) Cleve & Möller ·························· 246
Stephanodiscus miyagiensis H. Tanaka & Nagumo ······················· 248, 250
Stephanodiscus nagumoi H. Tanaka (in press) ····························· 252, 254
Stephanodiscus rotula (Kützing) Hendey ································· 256, 258
Stephanodiscus suzukii Tuji & Kociolek ··································· 260, 262
Stephanodiscus tenuis Hustedt ·· 264
Stephanodiscus uemurae H. Tanaka ·· 266
Tertiariopsis costatus H. Tanaka ·· 268
Tertiariopsis nipponicus H. Tanaka ··· 270
Tertiarius agunensis H. Tanaka sp. nov. ···················· 9, 272, 274
Thalassiocyclus pankensis H. Tanaka & Nagumo ······························ 276
Thalassiosira lacustris (Grunow) Hasle ·· 278

Triceratiales
Pleurosira laevis (Ehrenberg) Compère ··· 280

2. 無縦溝羽状類（Araphid, pennate diatoms）

Fragilariales
Diatoma anceps (Ehrenberg) Kirchner ·· 282
Diatoma ehrenbergii Kützing ·· 284
Diatoma hyemalis (Roth) Heiberg ··· 286
Diatoma mesodon (Ehrenberg) Kützing ··· 286
Diatoma vulgaris Bory ··· 284
Fragilaria neoproducta Lange-Bertalot ·· 288
Fragilaria vaucheriae (Kützing) Petersen ······································· 290
Fragilariforma fossilis (Pantocsek) H. Tanaka comb. nov. et stat. nov. ······ 17, 292
Fragilariforma kamczatica (Lupikina) H. Tanaka comb. nov. et stat. nov.
　·· 17, 294
Hannaea arcus (Ehrenberg) R.M. Patrick var. *arcus* ·························· 296

Hannaea arcus var. *hattoriana* (F. Meister) Ohtsuka ·· 296
Hannaea arcus var. *recta* (Cleve) M. Idei ·· 296
Meridion circulare var. *constrictum* (Ralfs) Van Heurck ·· 298
Pseudostaurosira brevistriata var. *nipponica* (Skvortsov) H. Kobayasi ························ 300
Staurosira construens Ehrenberg var. *construens* ·· 302
Staurosira construens var. *binodis* (Ehrenberg) Hamilton ·· 304
Staurosira construens var. *triundulata* (H. Reichelt) H. Kobayasi ······························ 306
Staurosirella lapponica (Grunow) D.M. Williams & Round ··· 308
Ulnaria capitata (Ehrenberg) Compère ·· 310

Tabellariales
Tabellaria fenestrata (Lyngbye) Kützing ·· 312
Tabellaria japonica H. Tanaka sp. nov. ·· 10, 314, 316
Tetracyclus castellum (Ehrenberg) Grunow ·· 318
Tetracyclus cruciformis Andrews ·· 320
Tetracyclus ellipticus (Ehrenberg) Grunow var. *ellipticus* ·· 322
Tetracyclus ellipticus var. *constricta* Hustedt ·· 324
Tetracyclus ellipticus var. *lancea* f. *subrostrata* Hustedt ··· 326
Tetracyclus ellipticus var. *latissima* f. *minor* Hustedt ··· 328
Tetracyclus emarginatus (Ehrenberg) W. Smith ··· 330
Tetracyclus glans (Ehrenberg) Mills ·· 332
Tetracyclus lacustris Ralfs ·· 334

3. 単縦溝羽状類 (Monoraphid, pennate diatoms)

Achnanthales
Achnanthes exigua var. *angustirostrata* (Krasske) Lange-Bertalot ······························ 336
Achnanthes obliqua (Gregory) Hustedt ·· 338
Achnanthes okunoi (Hustedt) H. Tanaka comb. nov. ·························· 16, 340, 342
Cocconeis jimboites VanLandingham ··· 344
Cocconeis placentula Ehrenberg var. *placentula* ·· 346
Cocconeis placentula var. *lineata* (Ehrenberg) Van Heurck ·· 348

4. 双縦溝羽状類 (Biraphid, pennate diatoms)

Cymbellales
Cymbella cymbiformis C. Agardh ·· 350
Cymbella neoleptoceros Krammer ··· 352
Cymbella ocellata H. Tanaka sp. nov. ·· 10, 354, 356
Cymbella okunoi H. Tanaka sp. nov. ·· 12, 358, 360
Cymbella orientalis Lee ··· 362
Cymbella peraspera Krammer ··· 364
Cymbella proxima Reimer ·· 366
Cymbella stuxbergii var. *robusta* Okuno ··· 368

Cymbella tsumurae **H. Tanaka sp. nov.** ················· 13, 370, 372, 374
Cymbopleura apiculata Krammer ·· 376
Cymbopleura inaequalis (Ehrenberg) Krammer ······························ 378, 380
Cymbopleura naviculiformis (Auerswald) Krammer ····························· 382
Cymbopleura subaequalis (Grunow) Krammer ····································· 384
Didymosphenia curvata (Skvortsov & Meyer) Metzeltin & Lange-Bertalot ············ 386
Didymosphenia fossils Horikawa & Okuno ·· 388
Didymosphenia geminata (Lyngbye) M. Schmidt ····································· 390
***Didymosphenia nipponica* H. Tanaka sp. nov.** ··················· 14, 392, 394
Encyonema geisslerae Krammer ·· 396
Encyonema vulgare Krammer ··· 398
Gomphoneis okunoi Tuji ··· 400
Gomphoneis tumida (Skvortsov) Kociolek & Stoermer ···························· 402
Gomphonema augur var. *gautieri* Van Heurck ·· 404
Gomphonema biceps F. Meister ·· 406
Gomphonema coronatum Ehrenberg ·· 408
Gomphonema nipponicum Skvortsov ·· 410
Gomphonema truncatum Ehrenberg ··· 412
Gomphonema vastum Hustedt ·· 414
Gomphopleura frickei Reichelt ·· 416
Gomphosphenia grovei var. *lingulata* (Hustedt) Lange-Bertalot ··············· 418
***Oricymba cunealjaponica* H. Tanaka sp. nov.** ················ 15, 420, 422

Eunotiales

Actinella brasiliensis Grunow ··· 424
Eunotia arcus Ehrenberg ··· 426
Eunotia biareofera f. *linearis* H. Kobayasi ·· 428
Eunotia bidens Ehrenberg ·· 430
Eunotia clevei Grunow ·· 432
Eunotia diadema Ehrenberg ·· 434
Eunotia duplicoraphis H. Kobayasi, Ando & Nagumo ····························· 436
Eunotia epithemioides Hustedt ··· 438
Eunotia formica Ehrenberg ··· 440
Eunotia incisa W. Gregory ·· 442
Eunotia monodon var. *tropica* (Hustedt) Hustedt ···································· 444
Eunotia nipponica Skvortsov ··· 446
Eunotia serra Ehrenberg ·· 448

Naviculales

Aneumastus tusculus (Ehrenberg) D.G. Mann & Stickle ·························· 450
Caloneis schumanniana (Grunow) Cleve ·· 452
Cavinula pseudoscutiformis (Hustedt) D.G. Mann & Stickle ···················· 454
Craticula ambigua (Ehrenberg) D.G. Mann ·· 456
Craticula cuspidata (Kützing) D.G. Mann ··· 458

Diadesmis confervacea Kützing ⋯⋯⋯⋯⋯⋯⋯⋯⋯⋯⋯⋯⋯⋯⋯⋯⋯⋯⋯⋯⋯⋯⋯⋯⋯⋯⋯⋯⋯ 460
Diploneis elliptica (Kützing) Cleve ⋯⋯⋯⋯⋯⋯⋯⋯⋯⋯⋯⋯⋯⋯⋯⋯⋯⋯⋯⋯⋯⋯⋯⋯⋯⋯ 462
Diploneis finnica (Ehrenberg) Cleve ⋯⋯⋯⋯⋯⋯⋯⋯⋯⋯⋯⋯⋯⋯⋯⋯⋯⋯⋯⋯⋯⋯⋯⋯⋯ 464
Diploneis ovalis (Hilse) Cleve ⋯⋯⋯⋯⋯⋯⋯⋯⋯⋯⋯⋯⋯⋯⋯⋯⋯⋯⋯⋯⋯⋯⋯⋯⋯⋯⋯⋯⋯ 466
Diploneis smithii var. *rhombica* Mereschkowsky ⋯⋯⋯⋯⋯⋯⋯⋯⋯⋯⋯⋯⋯⋯⋯⋯ 468
Diploneis subovalis Cleve ⋯⋯⋯⋯⋯⋯⋯⋯⋯⋯⋯⋯⋯⋯⋯⋯⋯⋯⋯⋯⋯⋯⋯⋯⋯⋯⋯⋯⋯⋯⋯ 470
Frustulia rhomboides var. *amphipleuroides* (Grunow) De Toni ⋯⋯⋯⋯⋯⋯ 472
Frustulia rhomboides var. *saxonica* (Rabenhorst) De Toni ⋯⋯⋯⋯⋯⋯⋯⋯ 474
Gyrosigma spencerii (Quekett) Griffith & Henfrey ⋯⋯⋯⋯⋯⋯⋯⋯⋯⋯⋯⋯⋯⋯ 476
Navicula americana Ehrenberg ⋯⋯⋯⋯⋯⋯⋯⋯⋯⋯⋯⋯⋯⋯⋯⋯⋯⋯⋯⋯⋯⋯⋯⋯⋯⋯⋯ 478
Navicula anthracis Cleve & Brun ⋯⋯⋯⋯⋯⋯⋯⋯⋯⋯⋯⋯⋯⋯⋯⋯⋯⋯⋯⋯⋯⋯⋯⋯⋯ 480
Navicula cari Ehrenberg ⋯⋯⋯⋯⋯⋯⋯⋯⋯⋯⋯⋯⋯⋯⋯⋯⋯⋯⋯⋯⋯⋯⋯⋯⋯⋯⋯⋯⋯⋯⋯ 482
Navicula hasta Pantocsek ⋯⋯⋯⋯⋯⋯⋯⋯⋯⋯⋯⋯⋯⋯⋯⋯⋯⋯⋯⋯⋯⋯⋯⋯⋯⋯⋯⋯⋯⋯ 484
Navicula radiosa Kützing ⋯⋯⋯⋯⋯⋯⋯⋯⋯⋯⋯⋯⋯⋯⋯⋯⋯⋯⋯⋯⋯⋯⋯⋯⋯⋯⋯⋯⋯⋯ 486
Navicula reinhardtii (Grunow) Grunow ⋯⋯⋯⋯⋯⋯⋯⋯⋯⋯⋯⋯⋯⋯⋯⋯⋯⋯⋯⋯⋯ 488
Navicula tanakae Fukushima, Ts. Kobayashi & Yoshitake ⋯⋯⋯⋯⋯⋯⋯⋯⋯ 490
Neidium ampliatum (Ehrenberg) Krammer ⋯⋯⋯⋯⋯⋯⋯⋯⋯⋯⋯⋯⋯⋯⋯⋯⋯⋯ 492
Neidium gracile Hustedt ⋯⋯⋯⋯⋯⋯⋯⋯⋯⋯⋯⋯⋯⋯⋯⋯⋯⋯⋯⋯⋯⋯⋯⋯⋯⋯⋯⋯⋯⋯ 494
Pinnularia episcopalis Cleve ⋯⋯⋯⋯⋯⋯⋯⋯⋯⋯⋯⋯⋯⋯⋯⋯⋯⋯⋯⋯⋯⋯⋯⋯⋯⋯⋯ 496
Pinnularia esoxiformis Fusey ⋯⋯⋯⋯⋯⋯⋯⋯⋯⋯⋯⋯⋯⋯⋯⋯⋯⋯⋯⋯⋯⋯⋯⋯⋯⋯⋯ 498
Pinnularia higoensis Okuno ⋯⋯⋯⋯⋯⋯⋯⋯⋯⋯⋯⋯⋯⋯⋯⋯⋯⋯⋯⋯⋯⋯⋯⋯⋯⋯⋯⋯ 500
Pinnularia lignitica Cleve ⋯⋯⋯⋯⋯⋯⋯⋯⋯⋯⋯⋯⋯⋯⋯⋯⋯⋯⋯⋯⋯⋯⋯⋯⋯⋯⋯⋯⋯ 502
Pinnularia macilenta (Ehrenberg) Ehrenberg ⋯⋯⋯⋯⋯⋯⋯⋯⋯⋯⋯⋯⋯⋯⋯⋯⋯ 504
Pinnularia rivularis Hustedt ⋯⋯⋯⋯⋯⋯⋯⋯⋯⋯⋯⋯⋯⋯⋯⋯⋯⋯⋯⋯⋯⋯⋯⋯⋯⋯⋯ 506
Pinnularia senjoensis H. Kobayasi ⋯⋯⋯⋯⋯⋯⋯⋯⋯⋯⋯⋯⋯⋯⋯⋯⋯⋯⋯⋯⋯⋯⋯ 508
Pinnularia subgibba var. *lanceolata* Gaiser & Johansen ⋯⋯⋯⋯⋯⋯⋯⋯⋯ 510
Plagiotropis lepidoptera var. *proboscidea* (Cleve) Reimer ⋯⋯⋯⋯⋯⋯⋯⋯ 512
Sellaphora bacillum (Ehrenberg) D.G. Mann ⋯⋯⋯⋯⋯⋯⋯⋯⋯⋯⋯⋯⋯⋯⋯⋯⋯ 514
Sellaphora laevissima (Kützing) D.G. Mann ⋯⋯⋯⋯⋯⋯⋯⋯⋯⋯⋯⋯⋯⋯⋯⋯⋯ 516
Stauroneis acuta W. Smith var. *acuta* ⋯⋯⋯⋯⋯⋯⋯⋯⋯⋯⋯⋯⋯⋯⋯⋯⋯⋯⋯⋯⋯ 518
Stauroneis acuta var. *terryana* Tempère ⋯⋯⋯⋯⋯⋯⋯⋯⋯⋯⋯⋯⋯⋯⋯⋯⋯⋯⋯ 520
Stauroneis phoenicenteron (Nitzsch) Ehrenberg var. *phoenicenteron* ⋯⋯⋯⋯⋯⋯⋯ 522
Stauroneis phoenicenteron var. *hattorii* Tsumura ⋯⋯⋯⋯⋯⋯⋯⋯⋯⋯⋯⋯⋯ 524

Rhopalodiales

Epithemia adnata (Kützing) Brébisson ⋯⋯⋯⋯⋯⋯⋯⋯⋯⋯⋯⋯⋯⋯⋯⋯⋯⋯⋯⋯⋯ 526
Epithemia cistula (Ehrenberg) Ralfs ⋯⋯⋯⋯⋯⋯⋯⋯⋯⋯⋯⋯⋯⋯⋯⋯⋯⋯⋯⋯⋯⋯ 528
Epithemia hyndmanii W. Smith ⋯⋯⋯⋯⋯⋯⋯⋯⋯⋯⋯⋯⋯⋯⋯⋯⋯⋯⋯⋯⋯⋯⋯⋯⋯ 530
Epithemia numatensis H. Tanaka sp. nov. ⋯⋯⋯⋯⋯⋯⋯⋯⋯⋯⋯ 16, 532, 534
Epithemia reticulata Kützing ⋯⋯⋯⋯⋯⋯⋯⋯⋯⋯⋯⋯⋯⋯⋯⋯⋯⋯⋯⋯⋯⋯⋯⋯⋯⋯⋯ 536
Epithemia sorex Kützing ⋯⋯⋯⋯⋯⋯⋯⋯⋯⋯⋯⋯⋯⋯⋯⋯⋯⋯⋯⋯⋯⋯⋯⋯⋯⋯⋯⋯⋯⋯ 538
Epithemia turgida var. *porcellus* Héribaud ⋯⋯⋯⋯⋯⋯⋯⋯⋯⋯⋯⋯⋯⋯⋯⋯⋯⋯ 540
Rhopalodia gibba (Ehrenberg) O. Müller ⋯⋯⋯⋯⋯⋯⋯⋯⋯⋯⋯⋯⋯⋯⋯⋯⋯⋯⋯ 542
Rhopalodia gibberula (Ehrenberg) O. Müller ⋯⋯⋯⋯⋯⋯⋯⋯⋯⋯⋯⋯⋯⋯⋯⋯⋯ 544

Rhopalodia novae-zelandiae Hustedt ·· 546
Rhopalodia rupestris (W. Smith) Krammer ·································· 548

Surirellales
Campylodiscus echeneis Ehrenberg ·· 550
Campylodiscus levanderi Hustedt ·· 552
Cymatopleura elliptica (Brébisson) W. Smith ······························ 554
Cymatopleura solea (Brébisson) W. Smith ·································· 556
Surirella bifrons Ehrenberg ··· 558
Surirella splendida (Ehrenberg) Kützing ···································· 560
Surirella robusta var. *splendida* f. *constricta* Hustedt ·················· 562
Surirella tenera W. Gregory ·· 564

Thalassiophysales
Amphora copulata (Kützing) Schoeman & Archibald ·················· 566
Amphora veneta Kützing ··· 568
Denticula elegans f. *valida* Pedicino ·· 570
Denticula tenuis Kützing ··· 572
Halamphora normannii (Rabenhorst) Levkov ····························· 574
Nitzschia heidenii (F. Meister) Hustedt ······································ 576
Nitzschia tabellaria (Grunow) Grunow ······································· 578

VI. 記述用語：Terms

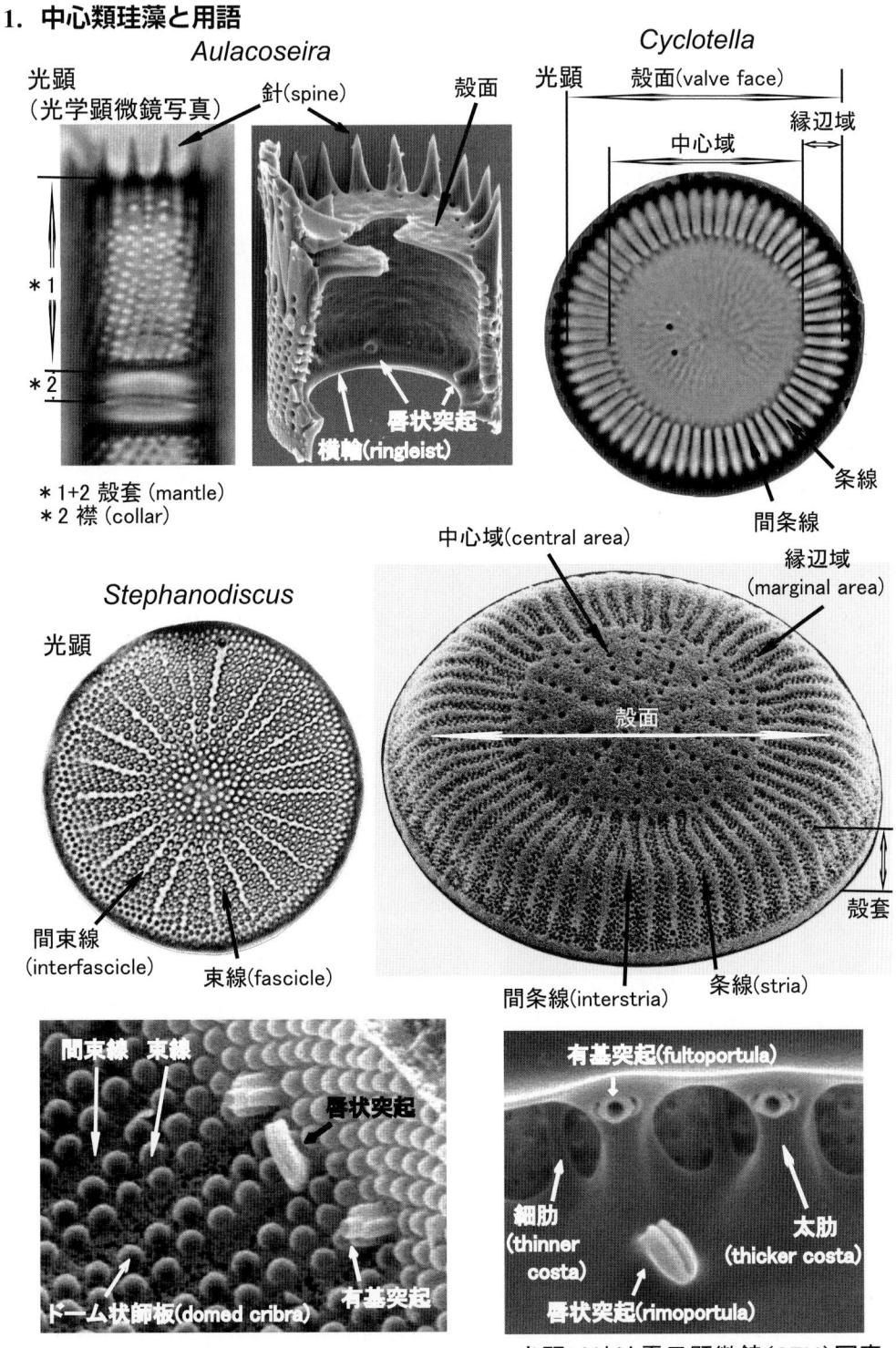

1. 中心類珪藻と用語

Aulacoseira

光顕（光学顕微鏡写真）　針(spine)　殻面

*1+2 殻套 (mantle)
*2 襟 (collar)

唇状突起　横輪(ringleist)

Cyclotella

光顕　殻面(valve face)　縁辺域　中心域

条線　間条線

Stephanodiscus

光顕

間束線(interfascicle)　束線(fascicle)

中心域(central area)　縁辺域(marginal area)　殻面　殻套　間条線(interstria)　条線(stria)

間束線　束線　唇状突起
ドーム状師板(domed cribra)　有基突起

有基突起(fultoportula)　細肋(thinner costa)　太肋(thicker costa)　唇状突起(rimoportula)

光顕以外は電子顕微鏡(SEM)写真

2. 羽状類珪藻と用語

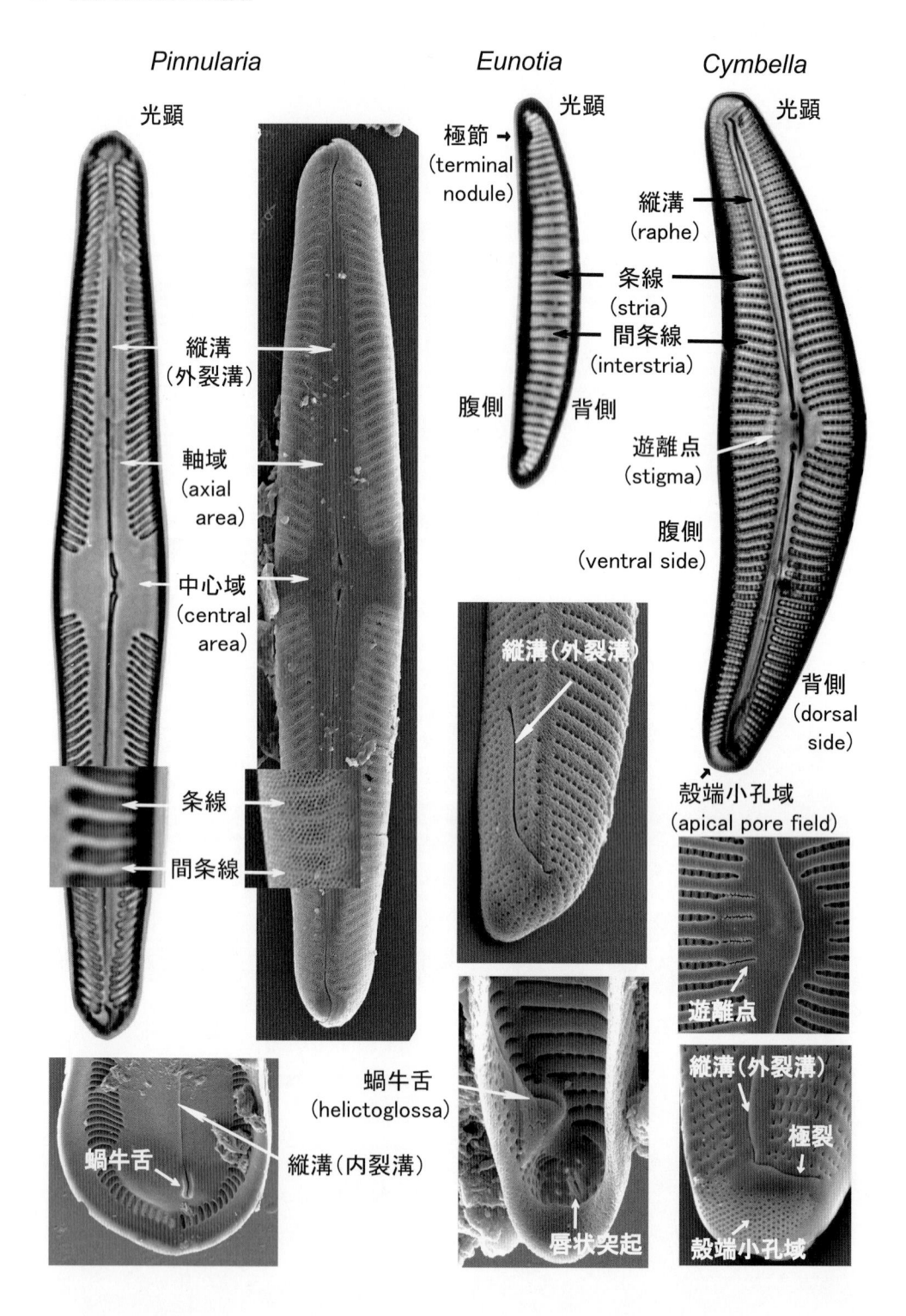

Pinnularia

- 光顕
- 縦溝（外裂溝）
- 軸域（axial area）
- 中心域（central area）
- 条線
- 間条線
- 蝸牛舌（helictoglossa）
- 蝸牛舌
- 縦溝（内裂溝）

Eunotia

- 光顕
- 極節→（terminal nodule）
- 条線（stria）
- 間条線（interstria）
- 腹側　背側
- 縦溝（外裂溝）
- 唇状突起

Cymbella

- 光顕
- 縦溝（raphe）
- 条線（stria）
- 間条線（interstria）
- 遊離点（stigma）
- 腹側（ventral side）
- 背側（dorsal side）
- 殻端小孔域（apical pore field）
- 遊離点
- 縦溝（外裂溝）
- 極裂
- 殻端小孔域

VII. 分類群解説・欧文図版解説・図版：
Explanation of Diatoms and Plates

Plate 1.　Centric diatoms: Aulacoseirales
Aulacoseira ambigua (Grunow) Simonsen 1979

基礎異名：*Melosira crenulata* var. *ambigua* Grunow 1882
文　　献：Simonsen, R. 1979. The diatom system: ideas on phylogeny. Bacillaria **2**: 9-71.
　　　　小林　弘・野沢美智子 1981. 淡水産中心類ケイソウ *Aulacoseira ambigua* (Grun.) Sim. の微細構造について．藻類 **29**: 121-128.
形　　態：殻は円筒形，ふつう殻面には点紋が所在しない．殻套の点紋列は斜めに走るが，殻面近くになるとしばしば直線状になり，わずかに点紋が大きく粗くなる．図に使用した長者原層と大鷲湖沼性堆積物産の個体は，殻径 5-15.5 μm，殻套長 7.5-16 μm，点紋列は 10 μm に 14-18 本であった．横輪が中空になっているのが大きな特徴である．分離針は短い先頭形，結合針は柄の短い先広形である．唇状突起は横輪にあるが，まれに横輪近くの殻套にあることがある（田中ら 2008）．
ノ ー ト：殻面には点紋が存在しないのが一般的であるが，縁辺部に所在する場合がある．および殻面全域に分布する場合もある（岩橋 1936）．
産　　出：山戸田層（出井・鈴木 1999）：中期中新世・石川県，長者原層：中期中新世・長崎県，人形峠層（田中ら 2008）：後期中新世～鮮新世・岡山-鳥取県境，大鷲湖沼性堆積物（田中ら 2011）：鮮新世・岐阜県．
図　　版：長者原層標本（Figs 1-3, 7-9），大鷲湖沼性堆積物標本（Figs 4-6, 10）．

Figs 1-10.　*Aulacoseira ambigua* (Grunow) Simonsen
LM. Figs 1-6. SEM. Figs 7-10. Materials, Figs 1-3, 7-9, from Chojabaru Formation (Middle Miocene), Iki Island, Nagasaki, Japan and Figs 4-6, 10, from lacustrine deposit of Owashi (Pliocene), Gujo City, Gifu, Japan. Scale bars: Fig. 7＝5 μm, Figs 8-10＝2 μm.

Figs 1-6.　Different size valves.
Fig. 7.　External view of linking valves with triangular bicuspid end linking spines, opening of rimoportula (arrowhead).
Fig. 8.　Internal view of broken valve, rimoportula (arrowhead).
Fig. 9.　External view of separating valve face and short pointed spines.
Fig. 10.　Oblique view of valve bottom showing rimoportula (arrowhead) and external opening of rimoportula (arrow).

Plate 1

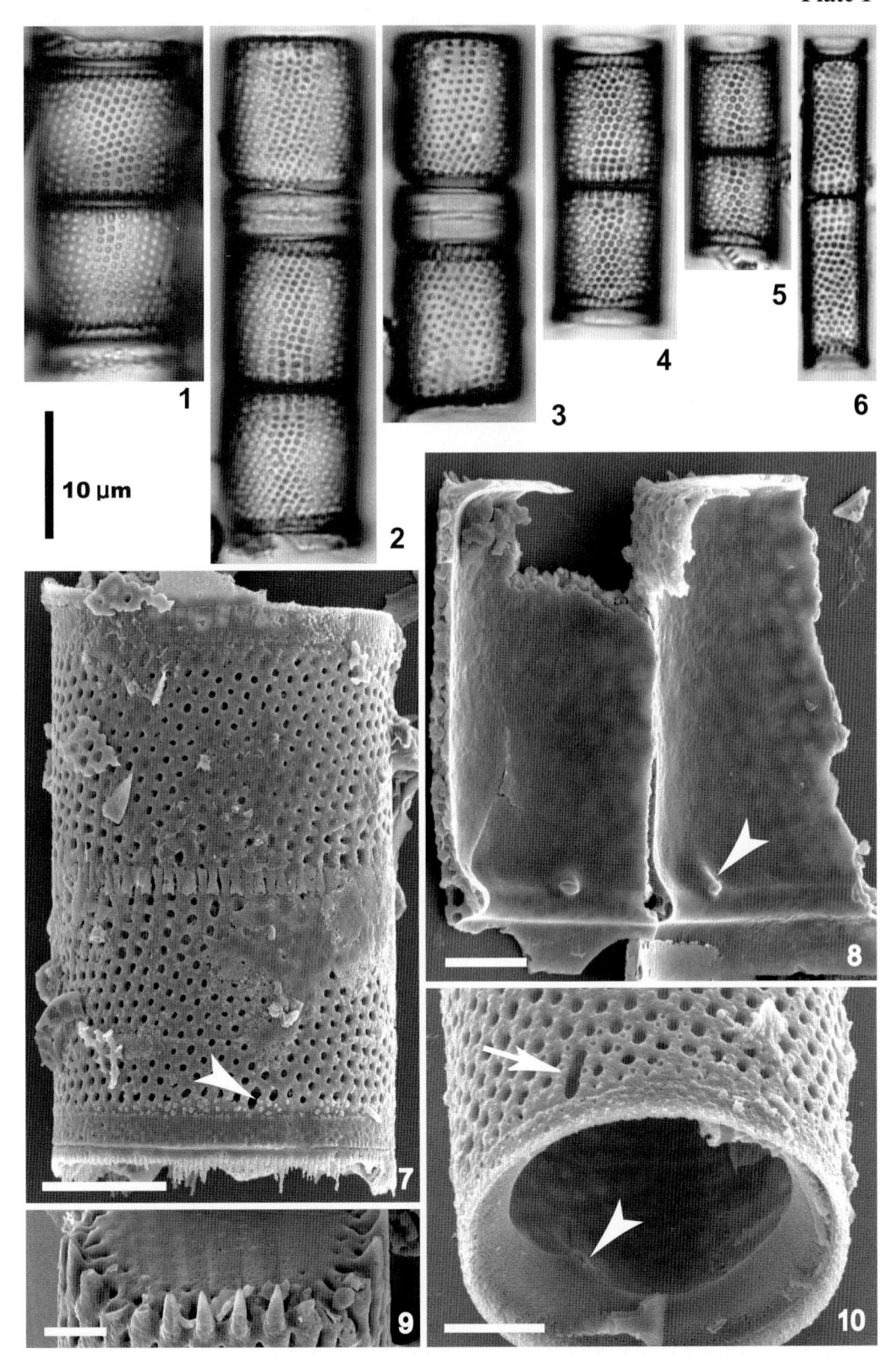

10 μm

Plate 2.   Centric diatoms: Aulacoseirales
Aulacoseira cataractarum (Hustedt) Simonsen 1979

基礎異名：*Melosira cataractarum* Hustedt 1938
文　　献：Simonsen, R. 1979. The diatom system: ideas on phylogeny. Bacillaria **2**: 9-71.
　 Simonsen, R. 1987.   Atlas and catalogue of the diatom types of Friedrich Hustedt. Vol. **1**: Catalogue. pp. 1-525. Vol. **2**: Atlas, Taf. 1-395. Vol. **3**: Atlas, Taf. 396-772. J. Cramer Berlin/Stuttgart.
形　　態：殻はきわめて小形の円筒形，殻面の条線は縁辺部に所在するが長さが異なり，殻面中心部はやや楕円形である．殻套の条線は長軸に平行である．針は殻面／殻套境界にある，殻径 4-5.5 μm，殻套条線は 10 μm に約 20 本であった．横輪は認められず，唇状突起は観察できなかった．
ノ　ー　ト：殻面縁辺部に条線があるが，中心部が完全な円形でないためしばしば羽状類的な形態と思えることがある．胞紋や針などの形態と合わせて所属する属の検討が必要な種と考えられる．
産　　出：元温泉小屋温泉：福島県（現生）．
図　　版：元温泉小屋温泉標本．

Figs 1-8.   *Aulacoseira cataractarum* (Hustedt) Simonsen
LM. Figs 1-2. SEM. Figs 3-8. Material from Motoonsengoya mineral spring (Recent), Oze, Fukushima, Japan. Scale bars: Figs 3-8＝1 μm.

Fig. 1.　Whole valve view.
Fig. 2.　Girdle view of two frustules.
Fig. 3.　External view of valve face.
Fig. 4.　Internal view of valve face.
Fig. 5.　Oblique view of Fig. 3.
Fig. 6.　Oblique view of Fig. 4.
Fig. 7.　Two frustules.
Fig. 8.　Oblique view of two sibling valves.

Plate 2

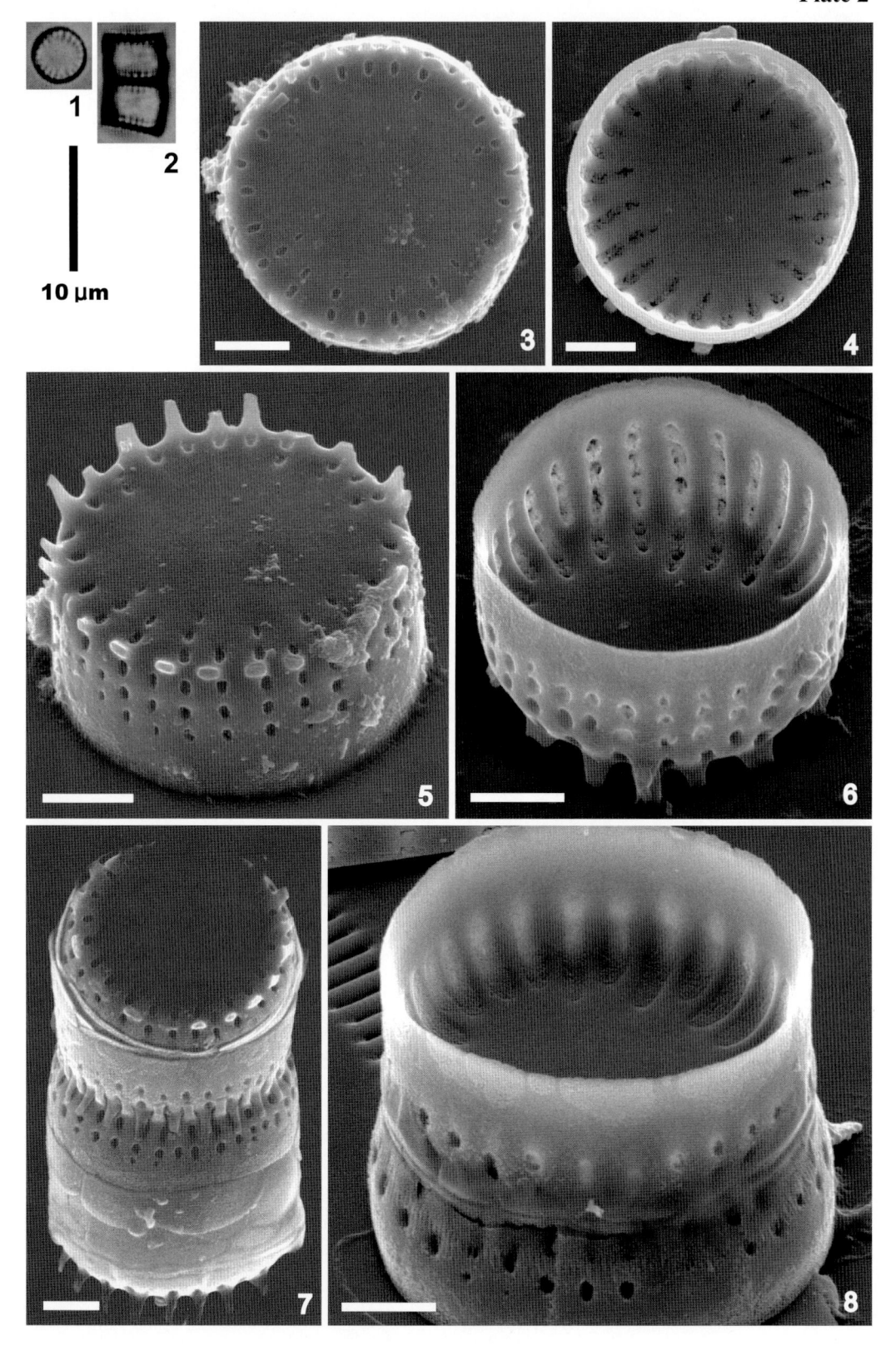

10 μm

Plate 3.　Centric diatoms: Aulacoseirales
Aulacoseira crassipunctata Krammer 1991

文　　献：Krammer, K. 1991. Morphology and taxonomy in some taxa in the genus *Aulacoseira* Thwaites (Bacillariophyceae). Ⅱ. Taxa in the *A. granulata-, italica-* and *lirata-* groups. Nova Hedwigia **53**: 477-796.

　　Tanaka, H. & Nagumo, T. 2010. Fine structure of the *Aulacoseira crassipunctata* Krammer in Japan. Diatom **26**: 40-43.

形　　態：殻は円筒形，殻面は円形，一般的には殻面に点紋が存在しないが，ときどき殻面縁辺に観察されることもある．殻径 4.5-29 μm，殻套高 6.5-26 μm，長い襟がある．条線はまっすぐで 10 μm に 6-12 本であり，条線を構成する点紋は 10 μm に 4-9 個である．点紋は楕円形で大きさは様々である．唇状突起は存在しない．

ノ　ー　ト：Likhosway & Crawford(2001)は日本産（Lake Bishamon, Japan）の試料を用いて本種に唇状突起が存在しないことを報告した．Tanaka & Nagumo(2010)は本邦の3箇所から見出した本種について内部を観察し唇状突起を欠くことを確認している．および Houk & Klee(2007)，Tanaka & Nagumo(2010) が記しているように，本種と類似している *Melosira pensacolae* A. Schmidt との識別が明瞭ではない．

産　　出：沼田湖成層：中期更新世・群馬県，沢の池（Yoshikawa 2007）：完新世・京都府，屈斜路湖（Tanaka & Nagumo 2010）：北海道（現生），毘沙門沼（Tanaka & Nagumo 2010）：福島県（現生）．

図　　版：毘沙門沼（Figs 1-2, 5-7），屈斜路湖（Figs 3, 8），沢の池標本（Fig. 4）．

Figs 1-8.　*Aulacoseira crassipunctata* Krammer
LM. Figs 1-5. SEM. Figs 6-8. Materials, Figs 1-2, 5-7, from Bishamon Pond (Recent), Fukushima, Japan, Figs 3, 8, from Lake Kussharo (Recent), Hokkaido, Japan and Fig. 4, from Sawano-ike Pond (Recent), Kyoto, Japan. Scale bars: Figs 6-8＝5 μm.

Figs 1-5.　Different size valves.
Fig. 6.　External oblique view of linking valve showing valve face without areolae.
Fig. 7.　External girdle view showing irregular size areolae in rows and linking spines.
Fig. 8.　Internal view of two broken valves.

Plate 3

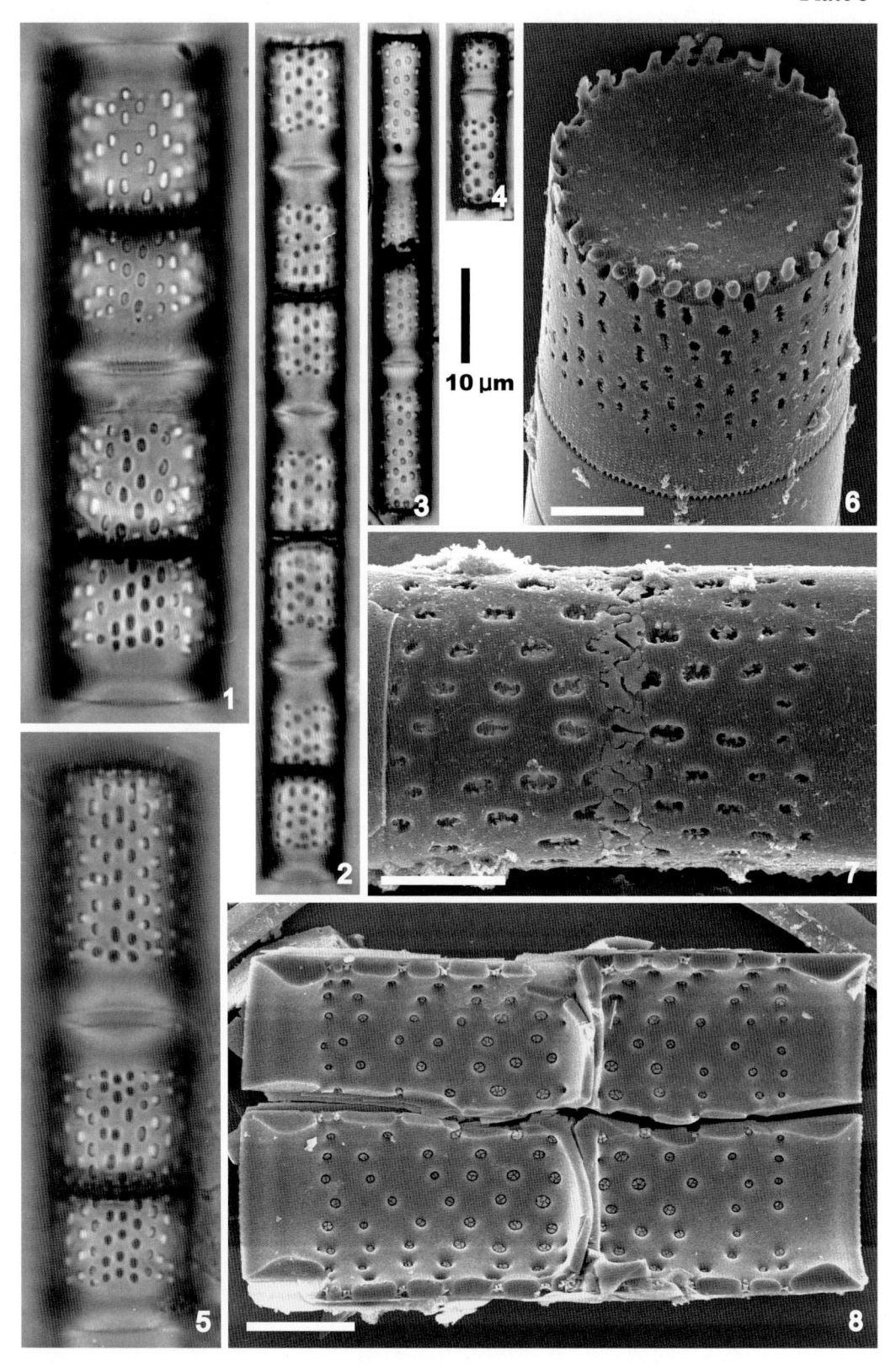

10 µm

43

Plates 4-5. Centric diatoms: Aulacoseirales
Aulacoseira fukushimae H. Tanaka sp. nov.

新　　種：記載文（英文）は 1 頁参照．
形　　態：殻は円筒形，殻面はわずか凹または凸状の円形で，点紋は存在しない．殻径 6.5-13 μm，殻套の高さ 4-9 μm，襟は短い．針は結合している殻において，片方は常に柄の短いスパチュラ形で他方は先頭形である．条線は直線状で 10 μm に約 12-16 本．条線を構成する胞紋の間隔は不規則に配列しており，開口は円形で，10 μm に約 12 個．唇状突起は約 4 個／殻で殻套に所在し，管はふつう殻面／殻套境界へ伸びるが，手前で終了してしまうこともある．明瞭な横輪は存在しない．
ノ ー ト：本分類群は *Aulacoseira pfaffiana*（Reinsch）Krammer や *Orthoseira americana*（Kütz.）Thwaites と類似するが，殻面の装飾，針の形，唇状突起の所在とその管などを比較すれば識別できる．種小名は日本珪藻学会名誉会長の福嶋　博博士を記念して付けられた．
産　　出：紫竹層：前期中新世・福島県．
図　　版 4-5：紫竹層標本．

Plate 4, Figs 1-8.   *Aulacoseira fukushimae* H. Tanaka sp. nov.
LM. Figs 1-5. SEM. Figs 6-8. Material from type material (SCH-102), Shichiku Formation (Early Miocene), Iwaki City, Fukushima, Japan. Scale bars: Figs 6-7＝2 μm, Fig. 8＝1 μm.

Figs 1-5.   Girdle view of four different size valves.
Figs 1-2.   Holotype, at different focal planes.
Fig. 6.   External view, valve chain, one side with convex valve face.
Fig. 7.   Oblique view of valve face without ornamentation at end of Fig. 6.
Fig. 8.   Broken valve showing collar without ringleist.

Plate 4

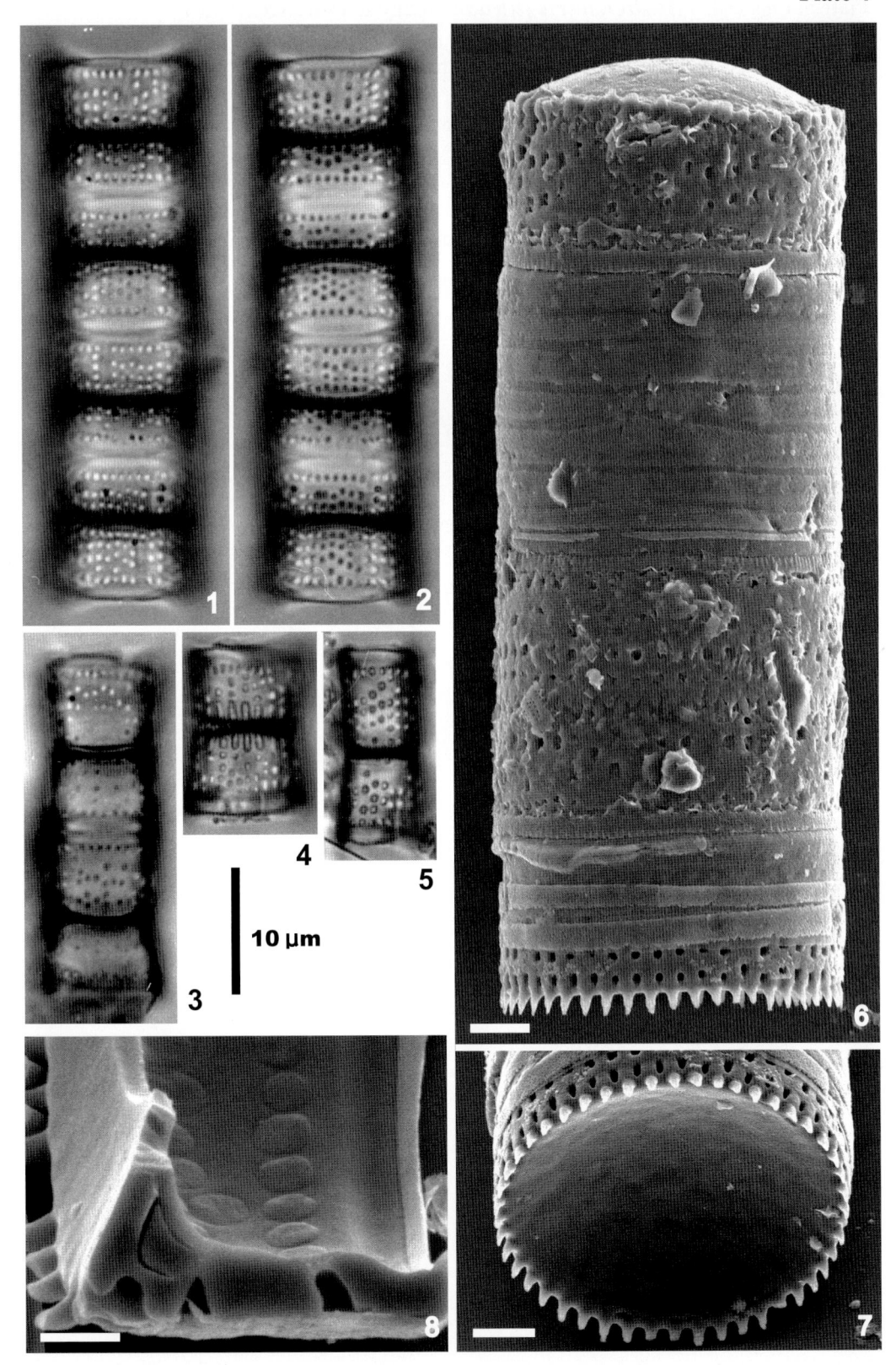

45

Plate 5, Figs 1–4. *Aulacoseira fukushimae* **H. Tanaka** sp. nov.
SEM. Figs 1–4. Material from type material (SCH-102), Shichiku Formation (Early Miocene), Iwaki City, Fukushima, Japan. Scale bars: Figs 1–3=2 μm, Figs 2, 4=0.5 μm.

Fig. 1. Internal view of broken chain of five valves showing rimoportulae and their canals.
Fig. 2. Detailed view of rimoportula (arrowhead) and its canal (arrow).
Fig. 3. External slightly oblique view of two sibling valves.
Fig. 4. Enlarged view of part of Fig. 3 showing linking spines, upper valve pointed, lower valve spatulate.

Plate 5

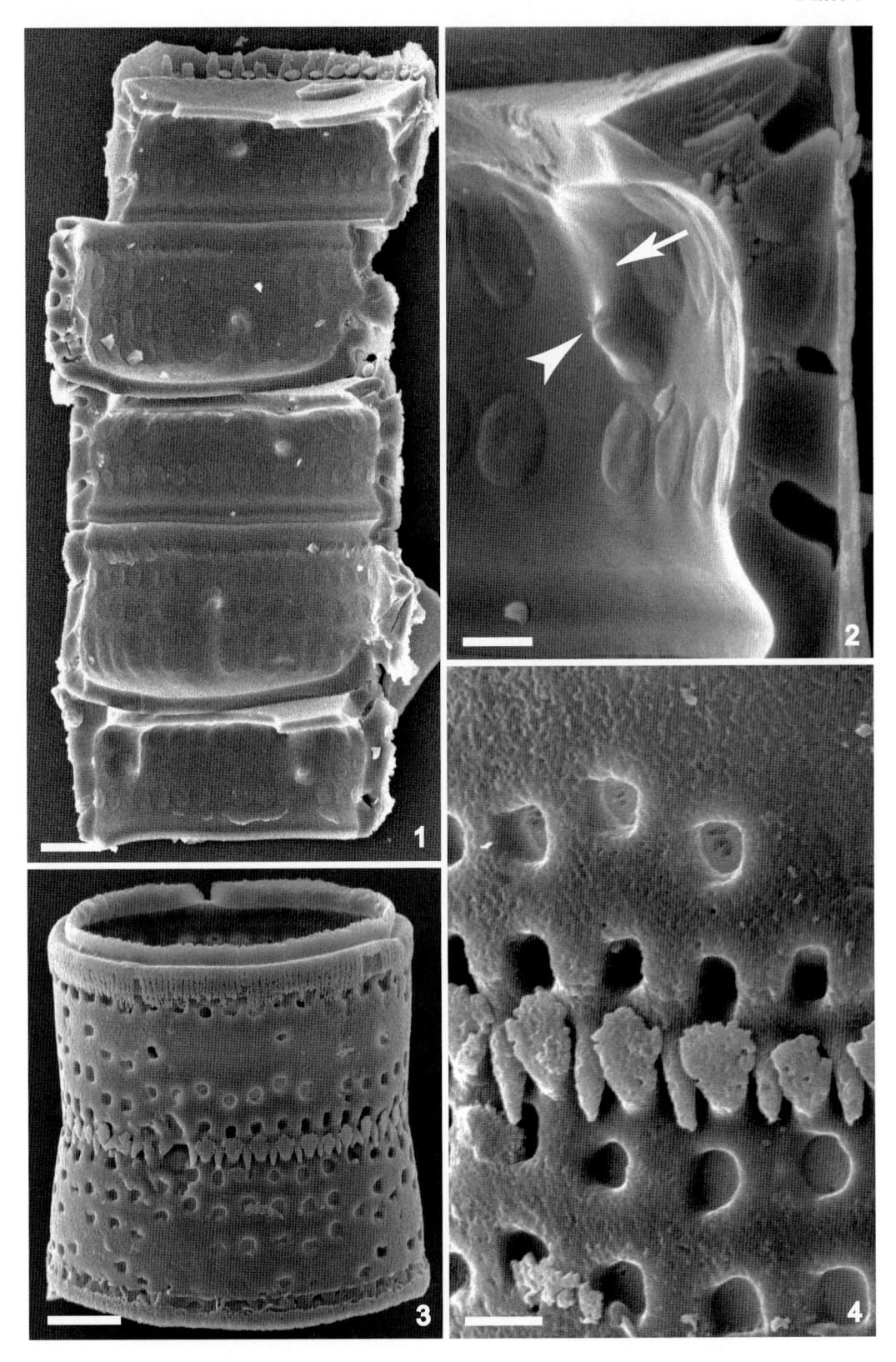

Plate 6.　Centric diatoms: Aulacoseirales
Aulacoseira granulata (Ehrenberg) Simonsen var. *granulata* 1979

基礎異名：*Gallionella granulata* Ehrenberg 1843
文　　献：Simonsen, R. 1979. The diatom system: ideas on phylogeny. Bacillaria **2**: 9-71.
　　田中宏之・南雲　保 2000. 化石珪藻 *Stephanodiscus komoroensis* Tanaka のタイプ試料（前期更新世）の珪藻群集と古環境. 地学研究 **49**: 67-75.
形　　態：殻は円筒形，殻面に点紋は存在しない．殻径 4.5-12 μm，殻套長 10-21 μm，点紋列は 10 μm に 10 本，点紋列を構成する点紋は列に沿って 10 μm に 6-12 個であった．殻套は点紋が粗く分布し，分離殻では直線状，結合殻では斜行する．殻面／殻套境界にある針は，分離殻では 1～数本が太く長い尖頭形であり，間条線 2 本から 1 本の針が生じているが，結合針は柄の短い先広形で各間条線に針が生じている．唇状突起は殻面／殻套境界付近から横輪までの間の殻套にあり，柄は長くカーブしている．
ノ　ー　ト：現生の試料では各地から報告があるが，時代をさかのぼるにつれて少なくなる傾向にある．第三紀からの産出として報告されているものの多くは，再検討が必要である．田中・南雲（2000）には形態が比較的詳述されている．
産　　出：円田層（円田珪藻土）：前期更新世・宮城県，瓜生坂層（田中・南雲 2000）：前期更新世・長野県．
図　　版：円田層標本．

Figs 1-8.　*Aulacoseira granulata* (Ehrenberg) Simonsen var. *granulata*
LM. Figs 1-4. SEM. Figs 5-8. Material from Enda diatomite (Early Pleistocene), Zao Town, Miyagi, Japan. Scale bars: Figs 5, 7＝2 μm, Figs 6, 8＝1 μm.

Figs 1-4.　Girdle view of different size valves.
Fig. 5.　External view of whole valve, openings of rimoportulae (arrowheads).
Fig. 6.　External view of linking spines and areolae.
Fig. 7.　Oblique view of Fig. 5 showing valve face without areolae, pointed spines and opening of rimoportula (arrowhead).
Fig. 8.　Internal view showing a rimoportula with curved stalked canal on mantle (arrowhead).

Plate 6

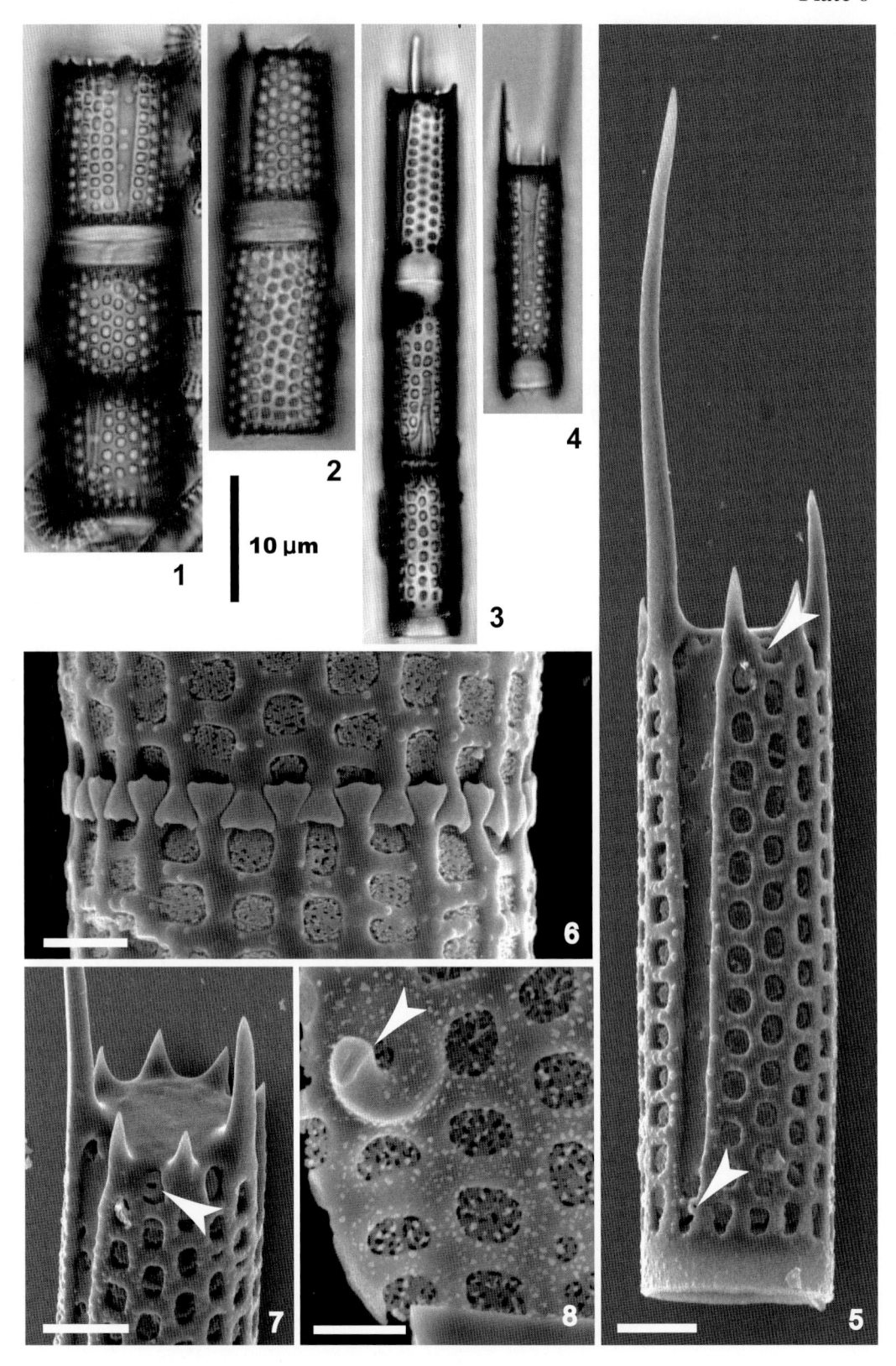

10 μm

Plate 7. Centric diatoms: Aulacoseirales
Aulacoseira granulata var. *angustissima* (O. Müller) Simonsen 1979

基礎異名：*Melosira granulata* var. *angustissima* O. Müller 1899

文　　献：Simonsen, R. 1979. The diatom system: ideas on phylogeny. Bacillaria **2**: 9-71.

形　　態：殻は円筒形，殻面に点紋は存在しない．図に使用した野上層の個体は，殻径約3 μm，殻套長 14-17 μm，条線は 10 μm に約 16 本，点紋の密度は 10 μm に約 16 個であった．分離殻は 1-2 本の長い針をもつが，結合殻では柄の短い先広形である．唇状突起の柄は長くカーブしている．

ノート：本分類群は基本種（var. *granulata*）に含められることもあるが，殻のサイズの違いが水域の環境等の推定に役立つことがあるかもしれないと考え，渡辺ら（2005）に倣い別分類群として示した．野上層の個体は結合殻しか見出せなかったため，参考として波志江沼（現生種，群馬県伊勢崎市）産の写真を Fig. 1 に掲載した．波志江沼の本分類群の形態は田中・南雲（2007）に記されている．

産　　出：太櫓層（瀬棚の珪藻土）（奥野 1958）：前期中新世・北海道，塩原湖成層（Akutsu 1964）：前期更新世・栃木県，野上層：中期更新世・大分県，波志江沼（田中・南雲 2007）：群馬県（現生）．

図　　版：波志江沼（Fig. 1），野上層（Figs 2-6）標本．

Figs 1-6. *Aulacoseira granulata* var. *angustissima* (O. Müller) Simonsen
LM. Figs 1-3. SEM. Figs 4-6. Materials, Fig. 1, from Hashie Pond (Recent), Isesaki City, Gunma, Japan and Figs 2-6, from Nogami Formation (Middle Pleistocene), Kokonoe Town, Oita, Japan. Scale bars: Fig. 4=5 μm, Figs 5-6=1 μm.

Fig. 1.　A chain showing separating valve (arrowhead) and linking valve (arrow), photomontage.
Figs 2-3.　Linking valves.
Fig. 4.　External view of two linking valves.
Fig. 5.　A broken valve showing rimoportula on mantle (arrowhead).
Fig. 6.　Enlarged view of Fig. 4 showing linking spines.

Plate 7

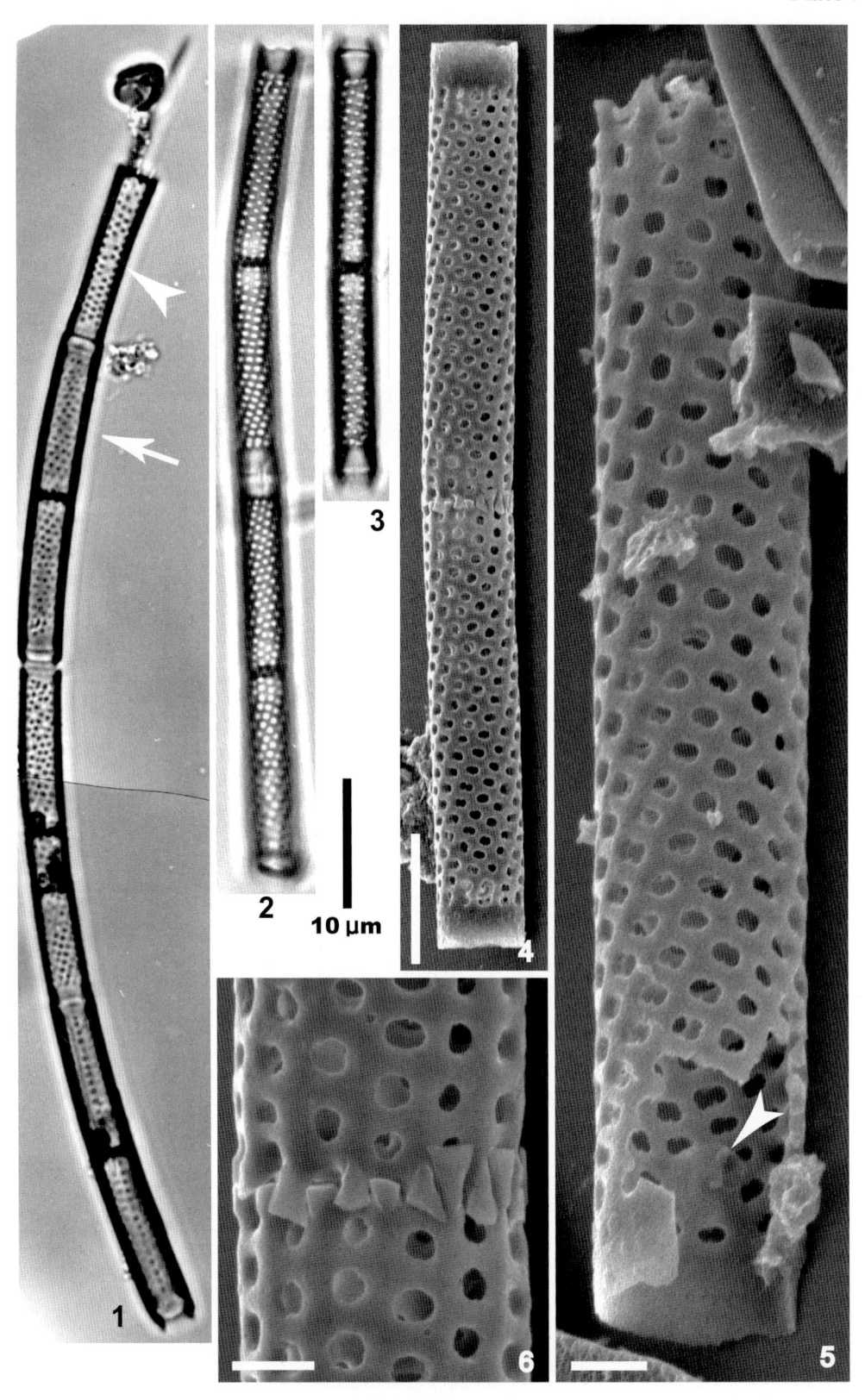

10 µm

Plate 8. Centric diatoms: Aulacoseirales
Aulacoseira hachiyaensis H. Tanaka in Tanaka *et al.* 2008

文　　献：Tanaka, H., Nagumo, T. & Akiba, F. 2008. *Aulacoseira hachiyaensis*, sp. nov., a new Early Miocene freshwater fossil diatom from the Hachiya Formation, Japan. *In*: Likhoshway, Y.(ed.) Proceedings of the 19th International Diatom Symposium, Listvyanka, Russia, August 28-September 3, 2006. pp. 115-123. Biopress, Bristol.

形　　態：殻面は楕円形で，点紋が存在する場合と存在しない場合とがある．殻径は，長径 11-19 μm，短径 8.5-12 μm，殻套の高さは 6-11 μm である．条線はまっすぐで 10 μm に 8-10 本であり，条線を構成する点紋は 10 μm に 10-12 個である．針は短いが各間条線の殻面／殻套境界にある．殻套部の内側には縦に細長い肥厚部が 1-2 本存在する．唇状突起は通常横輪上だが，まれに殻套にある場合がある．

ノート：殻面が楕円形であることに加えて，内側に肥厚部が存在することが本種の際立った特徴である．

産　　出：蜂屋層（Tanaka *et al.* 2008）：前期中新世・岐阜県．

図　　版：蜂屋層標本．

Figs 1-10. *Aulacoseira hachiyaensis* H. Tanaka
LM. Figs 1-7. SEM. Figs 8-10. Type material, HAC-002, Hachiya Formation (Early Miocene), Yaotsu Town, Gifu, Japan. Figs 4-5. Copy of "*Aulacoseira hachiyaensis*" from Tanaka *et al.* (2008), original report. Scale bars: Figs 8-10＝5 μm.

Figs 1-3.　Different size valves.
Figs 6-7.　Same valve at different focal planes (Fig. 6 ringleist, Fig. 7 valve face).
Fig. 8.　Oblique external view of valve face.
Fig. 9.　External view of valve bottom showing ringleist.
Fig. 10.　Internal view of broken valve showing ridge (arrow) and canal of rimoportula (arrowhead).

Plate 8

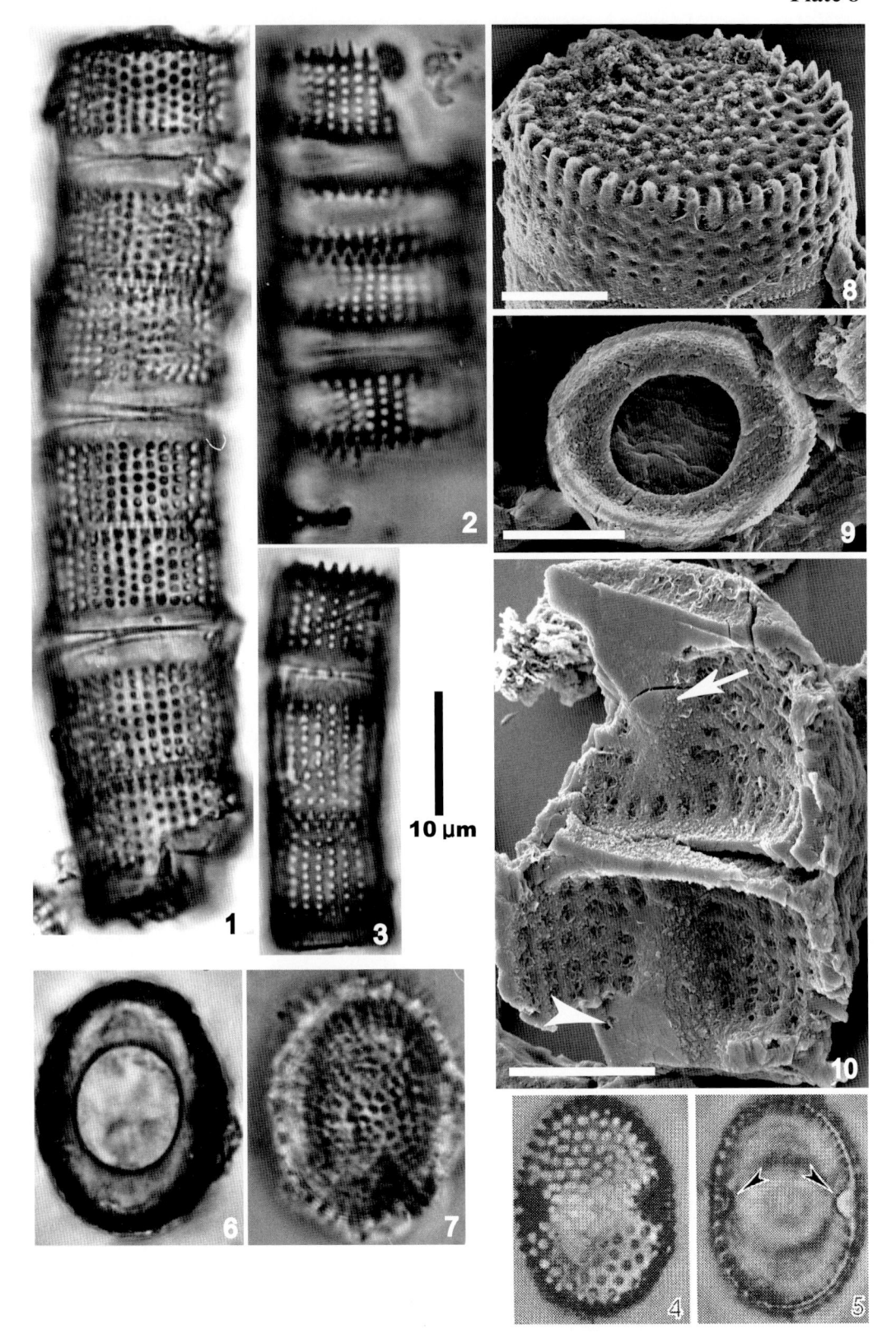

10 μm

Plate 9.　Centric diatoms: Aulacoseirales
Aulacoseira houki **H. Tanaka** in Tanaka & Nagumo 2011

文　　献：Tanaka, H. & Nagumo, T. 2011.　*Aulacoseira houki*, a new Early Miocene freshwater diatom from Hiramaki Formation, Gifu Prefecture, Japan. Diatom Research **26**: 161-165.

形　　態：殻は円筒形，殻面にはふつう全域にわたって点紋が分布するが，殻面縁辺部のみの場合もある．殻径 7.5-20 μm，殻套高は 2.5-9 μm で比較的短く，条線はまっすぐで 10 μm に約 16 本，条線を構成する点紋も 10 μm に約 16 本である．針は短く，各間条線の殻面／殻套境界にある．唇状突起は横輪の直上と殻面に存在する．

ノート：唇状突起が横輪の直上と殻面の両方に存在することが際立った本種の特徴である．種小名は最近 *Aulacoseira* 属の図鑑を出版した，チェコ国の珪藻研究者 Dr. Václav Houk 氏を記念して付けられた．

産　　出：平牧層（Tanaka & Nagumo 2011）：前期中新世・岐阜県．

図　　版：平牧層標本．

Figs 1-11.　*Aulacoseira houki* H. Tanaka

LM. Figs 1-5. SEM. Figs 6-11. From the type material, HIR-102, Hiramaki Formation (Early Miocene), Kani City, Gifu, Japan. Scale bars: Figs 6-7＝5 μm, Figs 8-9＝2 μm, Fig. 10＝1 μm, Fig. 11＝0.5 μm.

Figs 1, 4-5.　Girdle views.
Figs 2-3.　Valve views.
Fig. 6.　External view of whole valve.
Fig. 7.　Oblique view of Fig. 6.
Fig. 8.　Internal view of broken valve showing two valve face rimoportulae (arrowheads).
Fig. 9.　Oblique view of small valve.
Fig. 10.　Broken valve showing rimoportulae on valve face (arrowhead) and near ringleist (arrow).
Fig. 11.　Internal detailed view of valve face rimoportula (arrowhead).

Plate 9

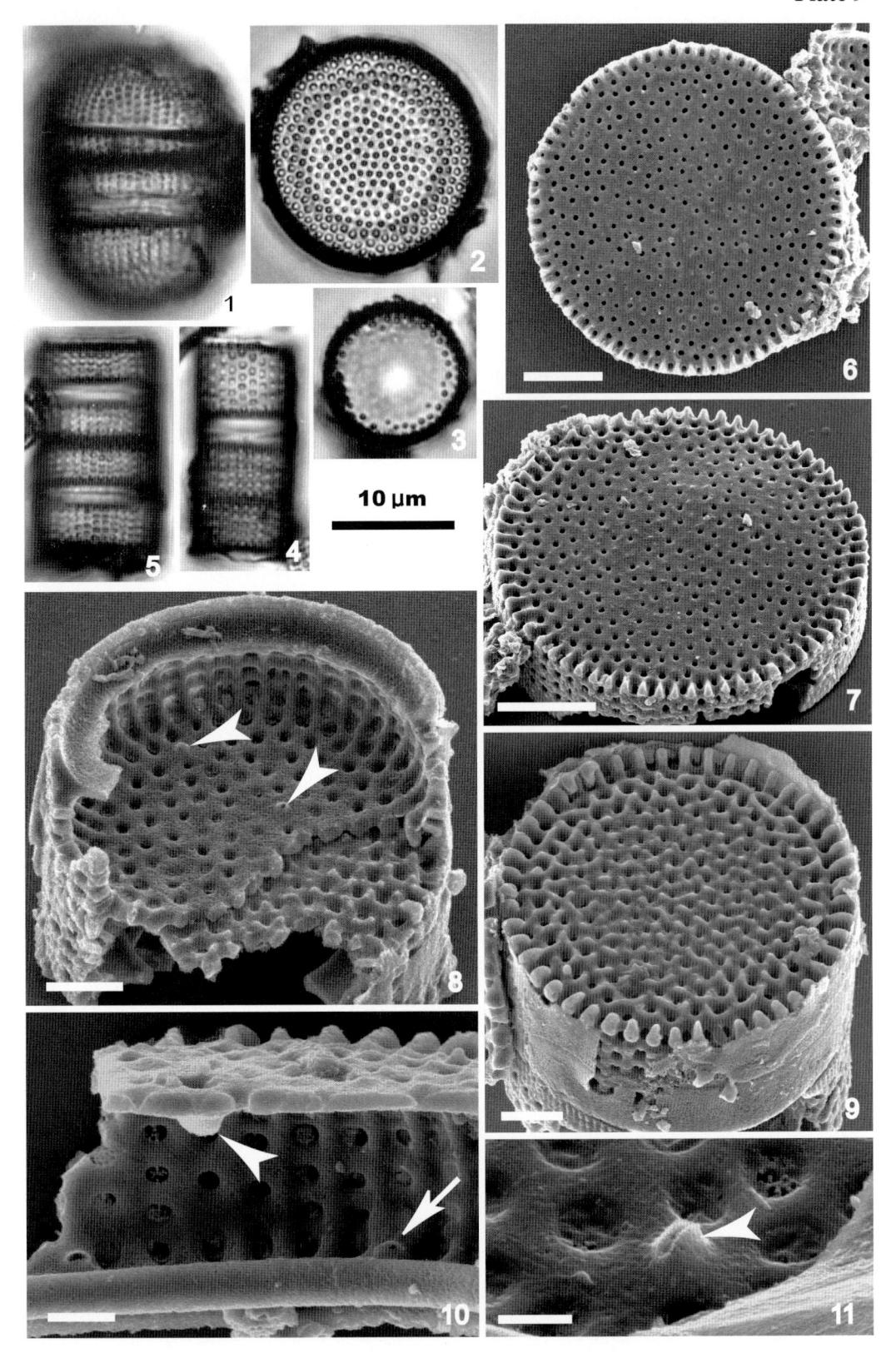

10 µm

Plate 10. Centric diatoms: Aulacoseirales
Aulacoseira italica (Ehrenberg) Simonsen 1979

基礎異名：*Gallionella italica* Ehrenberg 1838

文　　献：Simonsen, R. 1979. The diatom system: ideas on phylogeny. Bacillaria **2**: 9-71. Crawford, R.M., Likhoshway, Y.V. & Jahn, R. 2003. Morphology and identity of *Aulacoseira italica* and typification of *Aulacoseira* (Bacillariophyta). Diatom Research **18**: 1-19.

形　　態：殻は円筒形で殻径 6-13 μm，殻套長 8-13 μm。条線は直線～右らせん状に配列し，10 μm に約 24 本，条線を構成する点紋はふつう楕円形で，10 μm に 16-24 個である。結合殻は間条線のおよそ 3 本から 1 本の針が生じている。針は柄が長くスパチュラ状である。唇状突起は内側殻套に所在するが，管は横輪（殻套の端，あるいは殻面）と平行（胞紋列と直交）である。

ノート：イタリアの Santa Fiora に分布する珪藻土から Ehrenberg(1838) によって原記載された種である。Crawford *et al.*(2003) は，その原試料が紛失しているので，epitype 試料を用いて SEM 観察を行った。その SEM 写真によると，唇状突起は殻套に所在し管は横輪と平行である点で，紫竹層の本種とよく一致している。

産　　出：太櫓層（瀬棚の珪藻土）：前期中新世・北海道，紫竹層：前期中新世・福島県，三徳層（Tanaka & Nagumo 2006）：後期中新世・鳥取県。

図　　版：紫竹層標本。

Figs 1-6. *Aulacoseira italica* (Ehrenberg) Simonsen

LM. Figs 1-2. SEM. Figs 3-6. Material from Shichiku Formation (Early Miocene), Iwaki City, Fukushima, Japan. Scale bars: Figs 3-4, 6=5 μm, Fig. 5=1 μm.

Fig. 1.　Girdle view, slightly curved mantle striae.
Fig. 2.　Girdle view, staight mantle striae.
Fig. 3.　External view of sibling valves, lower half broken.
Fig. 4.　Oblique view of Fig. 3, rimoportula (arrowhead).
Fig. 5.　Enlarged view of Fig. 4, rimoportula (arrowhead).
Fig. 6.　Broken valve internal view, showing rimoportulae with canals, parallel ringleist lying on mantle (arrowheads).

Plate 10

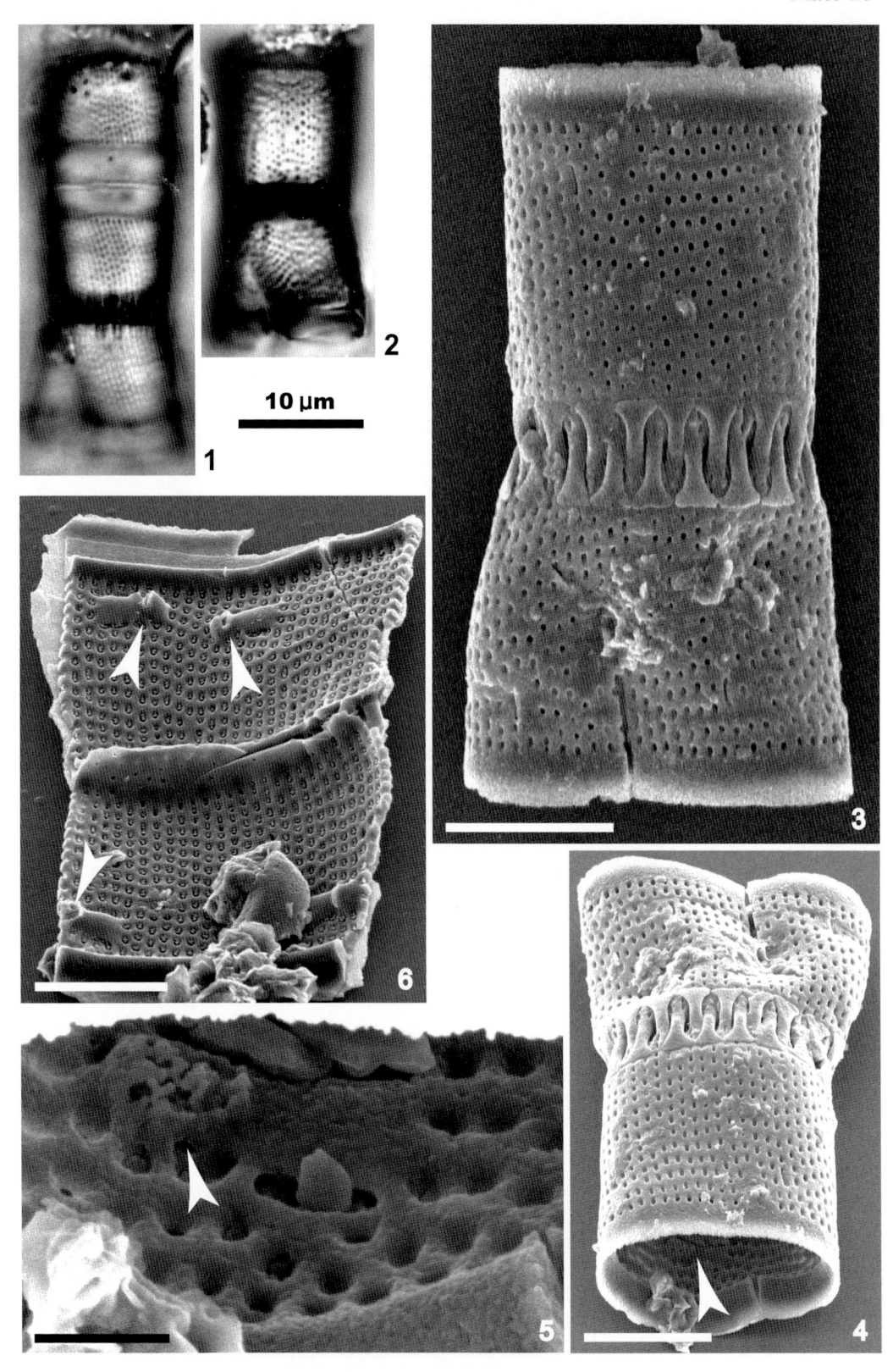

10 µm

57

Plate 11.   Centric diatoms: Aulacoseirales
Aulacoseira iwakiensis H. Tanaka in Tanaka & Nagumo 2011

文　　献：Tanaka, H. & Nagumo, T. 2011. *Aulacoseira iwakiensis* sp. nov., a new elliptical *Aulacoseira* species, from an Early Miocene sediment, Japan. Diatom **27**: 1-8.

形　　態：殻は殻面が楕円形の筒形で，帯面観において分離殻ではいろいろな程度にカーブしているが，結合殻はほぼ直線である．殻面は長径 8-15 μm，短径 6.5-13 μm，殻套長 3-6 μm．分離殻は殻面に多数の点紋が観察できるが，結合殻では少ない．殻套の点紋列は 10 μm に約 14 本，点紋列を構成する点紋は 10 μm に約 16 個である．針は短く各間条線にあり，分離殻では先頭形，結合殻ではへら状である．唇状突起は 5, 6 個で横輪にあり，外側への開口は胞紋の開口と共同している．

ノート：楕円形殻面の *Aulacoseira* 属は *A. elliptica, A. hachiyaensis, A. ovata, A. iwakiensis* の 4 種が報告されているが，すべて前期中新世からの産出である．および *Melosira distans* var. *ovata* は *Aulacoseira* 属へ組み合わせすべきと考えられるが産出場所についての吟味が必要である．

産　　出：紫竹層（Tanaka & Nagumo 2011）：前期中新世・福島県．

図　　版：紫竹層標本．

Figs 1-11.   *Aulacoseira iwakiensis* H. Tanaka

LM. Figs 1-6. SEM. Figs 7-11. From the type material, SCH-101, Shichiku Formation (Early Miocene), Iwaki City, Fukushima, Japan. Scale bars: Figs 7-11＝2 μm.

Figs 1, 6.   Girdle views.
Figs 2-3, 4-5.   Valve views. Two valves at different focal planes.
Fig. 7.   External oblique view of four separating valves.
Fig. 8.   External oblique view of two linking valves showing linking spines and ringleist.
Fig. 9.   Two separating valves showing ringleist.
Fig. 10.   Broken linking valve showing linking spines (arrow), canal of rimoportula (arrowhead) and valve face with few areolae.
Fig. 11.   Broken separating valve showing spines (arrow), rimoportulae (arrowhead) and valve face with many areolae.

Plate 11

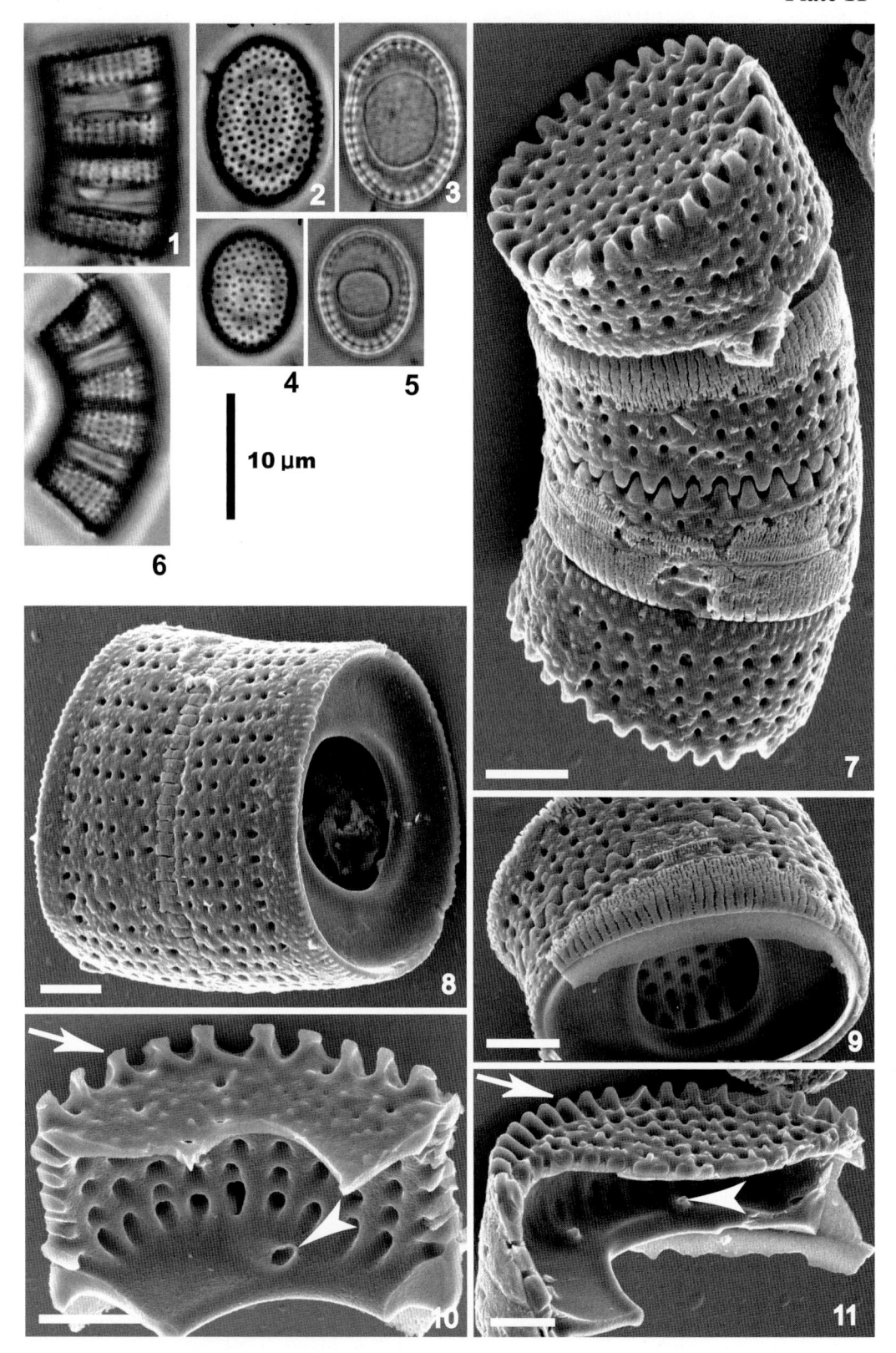

Plate 12.　Centric diatoms: Aulacoseirales
Aulacoseira longispina (Hustedt) Simonsen 1979

基礎異名：*Melosira longispina* Hustedt 1942

文　　献：Hustedt, F. 1942.　Diatomeen.　*In*: G. Huber-Pestalozzi, Das phytoplankton des Süsswassers. Systematic und Biologie. Die Binnengewässer. Band 16, Teil 2, Hälfte 2. 367-549. Taf. 118-178. E. Schweizerbant'sche Verlagsbuchhandlung, Stuttgart.

　　Simonsen, R. 1979.　The diatom system: ideas on phylogeny. Bacillaria **2**: 9-71.

形　　態：殻は円筒形，条線はゆるいらせん状であるが殻面近くになると直線状になる．間条線2本から1本の針が生じている．針は長くLMで明瞭に観察できる．殻面には胞紋は見当たらない．唇状突起は横輪上または横輪の直上に数個存在する．殻径6.5-17 μm，殻套長10-15 μm，条線は10 μmに約12本，条線を構成する点紋は10 μmに約15個である．

ノート：Hustedt(1942)により中禅寺湖の試料から*Melosira longispina*として記載され，Simonsen(1979)により*Aulacoseira*属へ組み合わせになった．筆者は中禅寺湖から本種の初生殻と思われるものを見出した．殻套長が従来報告されている数字よりかなり小さいので除外してあるが，その計測値を上記に挿入すると殻径6.5-18.5 μm，殻套長3-15 μmとなる．
　　Tuji & Houki(2004)によって*Aulacoseira subarctica* var. *longispina* (Hust.) Tuji & Houkiとして組み合わせされているが，本書ではその後に発行された小林ら(2006)に倣って*A. longispina*とした．

産　　出：中禅寺湖（Hustedt 1942）：栃木県（現生）．

図　　版：中禅寺湖標本．

Figs 1-7.　*Aulacoseira longispina* (Hustedt) Simonsen

LM. Figs 1-3. SEM. Figs 4-7. Material from the type locality, Lake Chuzenji (Recent), Nikko City, Tochigi, Japan. Scale bars: Figs 4-6＝5 μm, Fig. 7＝1 μm.

Figs 1-3.　Different size valves.
Fig. 4.　Broken valve, rimoportulae (arrowheads).
Fig. 5.　External oblique view of whole valve.
Fig. 6.　Internal view of broken sibling valves, rimoportulae (arrowheads).
Fig. 7.　Detailed view of opening of rimoportula (arrowhead).

Plate 12

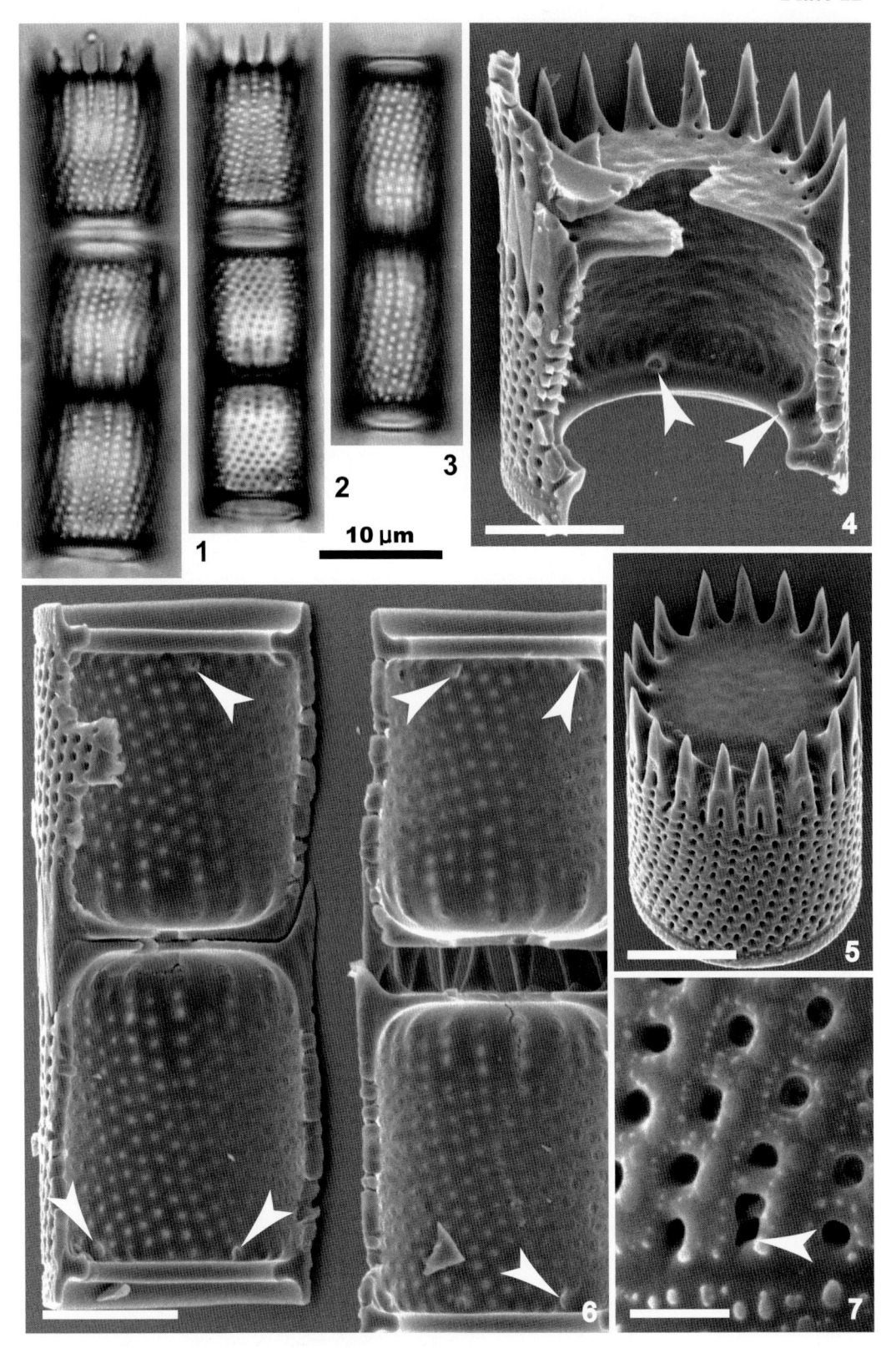

10 µm

Plates 13-14. Centric diatoms: Aulacoseirales
Aulacoseira miosiris H. Tanaka sp. nov.

新　　種：記載文（英文）は 2 頁参照．
形　　態：殻は円筒形．殻面には縁辺部に胞紋が所在する場合と，所在しない場合がある．殻径が大きい殻では内側の間条線部分が肥厚して内部へ張り出し，仕切り～柱あるいは肋を形成し，多くが殻面から横輪まで達するが，殻径が小さくなるにつれて徐々に肥厚が少なくなり，殻面と殻套の接合部，および横輪と殻套の接合部のみになり，さらに SEM 観察でわかる程度に小さくなる．肥厚部は光顕では黒い肥厚線として観察することができる．殻径 8-25 μm，殻套長 4-11 μm，分離針は短い先頭形，結合針は短い先広形である．胞紋の外側は星形をした師板で仕切られており，内側は外側へ膨らんだ師皮で覆われている．胞紋列の数は 10 μm に 8-10 本，胞紋列を構成する胞紋は 10 μm に 8-10 個である．唇状突起は横輪の面上に突出して 7-13 個所在する．
ノ　ー　ト：初生殻および初生殻に近い殻は肥厚線は殻面から横輪に達し *Miosira* 属の特徴をもつが，細胞分裂が進むに従い肥厚が少なくなり殻面と殻套の接合部，および横輪と殻套の接合部のみになってしまう，さらに肥厚が少なくなり光顕では識別できなくなり *Aulacoseira* 属の形態を示す．Houk & Klee(2007)は *Miosira* 属の 2 分類群に唇状突起が存在することを写真で示しているが，*Miosira* 属の記載文（Krammer 1997）では本属には唇状突起が存在しないと記してある．宮田層からの本種は *Miosira* 属の特徴を示す殻が *Aulacoseira* 属の特徴を示す殻より少ないこと，唇状突起が明瞭であることから *Aulacoseira* 属に所属するのが適当と思われる．
産　　出：宮田層：中新-鮮新世・秋田県，耶馬溪層：鮮新世・大分県．
図　　版 13-14：宮田層標本．

Plate 13, Figs 1-9.　*Aulacoseira miosiris* H. Tanaka sp. nov.
LM. Figs 1-9. Figs 1-6, 9: girdle views; Figs 7-8: valve views. Type material, TAZ-F01, Miyata Formation (Mio-Pliocene), Tazawako, Senboku City, Akita, Japan.

Figs 1-2.　Holotype. Same valve at different focal planes.
Figs, 3-6, 9.　Valve chains, initial valves with partitions (pillars or costae) between valve face and ringleist, gradually becoming short partitions (or bulges) only at junction of valve face and mantle or ringleist and mantle on succeeding valves and then finally no partition-like structures on valves that follow.
Fig. 4.　Valves chain including initial valve, note: partition (arrow) between valve face and ringleist.
Fig. 7.　Ringleist, canal of rimoportula (arrowhead).
Fig. 8.　Valve face, areolae located on marginal area.

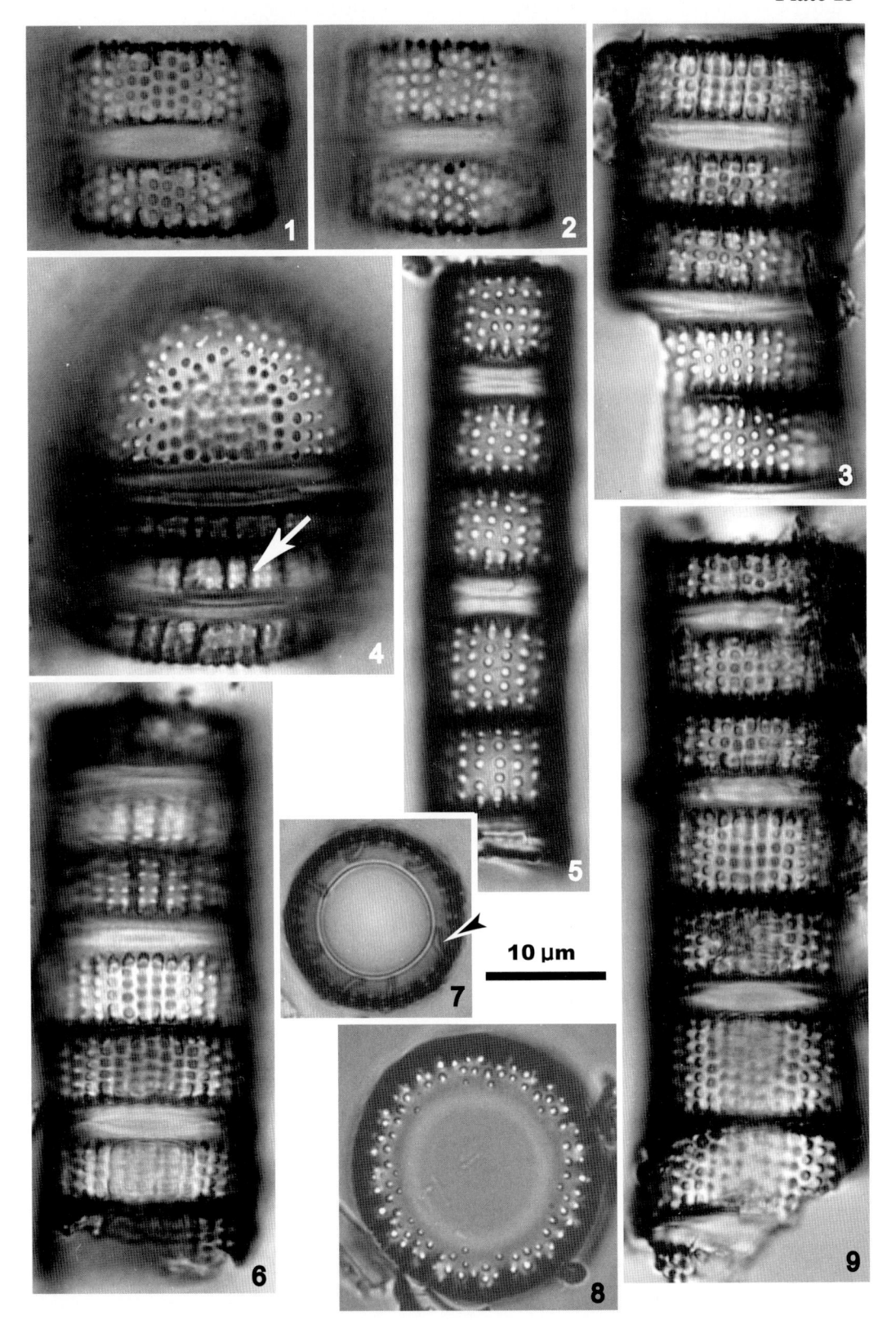

Plate 13

10 µm

Plate 14, Figs 1–10. *Aulacoseira miosiris* **H. Tanaka** sp. nov.

SEM. Figs 1–10. Type material, TAZ-F01, Miyata Formation, Tazawako, Senboku City, Akita, Japan. Scale bars: Figs 1, 4, 8=5 μm, Figs 6, 9–10=2 μm, Figs 2, 7=1 μm, Figs 3, 5=0.5 μm.

Fig. 1.   External view of a chain.
Fig. 2.   Enlarged view of part of Fig. 1, showing linking spines, circular areolae foramina and veluma.
Fig. 3.   Broken valve mantle, showing velum (arrow) and rica.
Fig. 4.   Oblique view of valve face of Fig. 1.
Fig. 5.   Detailed view of one rimoportula of Fig. 7.
Fig. 6.   Internal view, showing rimoportulae on ringleist.
Fig. 7.   Internal oblique view, showing rimoportulae on ringleist and costae forming partitions.
Fig. 8.   Internal oblique view, showing costae forming partitions or pillars between valve face and ringleist.
Fig. 9.   Internal view, no partitions.
Fig. 10. Every second to third costae forming short partitions located at valve face/mantle boundary, rimoportula (arrowhead).

Plate 14

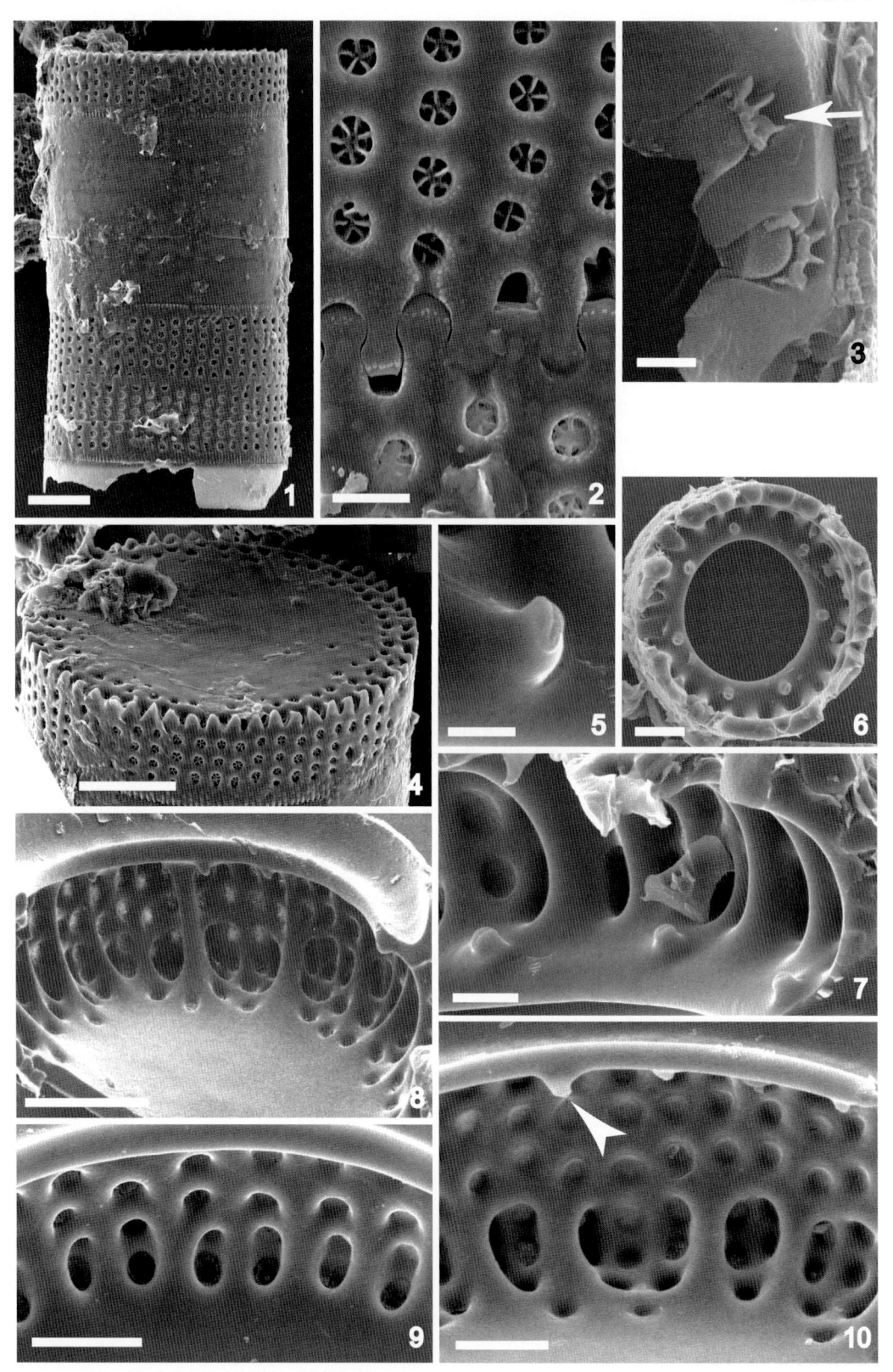

Plate 15.　Centric diatoms: Aulacoseirales
Aulacoseira nipponica (Skvortsov) Tuji 2002

基礎異名：*Melosira solida* var. *nipponica* Skvortsov 1936
文　　献：辻　彰洋・伯耆晶子 2001．琵琶湖の中心目珪藻．琵琶湖研究モノグラフ **7**: 1-90．滋賀県琵琶湖研究所．
　　　Tuji, A. 2002．Observations on *Aulacoseira nipponica* from Lake Biwa, Japan, and *Aulacoseira solida* from North America (Bacillariophyceae). Phycological Research **50**: 313-316.
形　　態：殻はふつう厚く肥厚している．殻面は円形で，点紋を欠く．殻径 5.5-16 μm，殻套長 8-10 μm，条線は貫殻軸に平行な直線で 10 μm に約 12 本．条線を構成する点紋は 10 μm に約 12 個．針は比較的長く 2 本の間条線に 1 本の割合で存在する．唇状突起は横輪上にある．
ノ　ー　ト：Skvortzow(1936)は琵琶湖の試料を用いて，殻套の点紋の分布の違いから *Melosira solida* var. *solida* と *Melosira solida* var. *nipponica* を識別したが，Tuji(2002)は，両者は同一分類群でタイプの *Melosira solida*（現在の *Aulacoseira solida*（Eulenst.）Krammer）とは異なるので，琵琶湖の分類群を *Aulacoseira nipponica*（Skvortsov）Tuji とした．Tanimura *et al.*(2006)によれば本種に殻の薄いものも見出されている．
産　　出：琵琶湖（Skvortzow 1936，辻・伯耆 2001, Tuji 2002）：滋賀県（現生）．
図　　版：琵琶湖標本．

Figs 1-9.　*Aulacoseira nipponica* (Skvortsov) Tuji
This author should use the authors of scientific names according to Brummitt & Powell (1992) and Gotoh (2003), e.g. Skvortzow → Skvortsov.
LM. Figs 1-5. SEM. Figs 6-9. Material from the type locality, Lake Biwa (Recent), Shiga, Japan. Scale bars: Figs 6-7＝5 μm, Figs 8-9＝2 μm.

Figs 1-2, 3-4.　Two valves at different focal planes.
Fig. 5.　Underside valve view showing ringleist and rimoportulae tubes.
Fig. 6.　External oblique view of whole valve.
Fig. 7.　Internal oblique view of broken valve, rimoportula (arrowhead).
Fig. 8.　Internal view of broken valve showing canal of rimoportula (arrowhead).
Fig. 9.　External oblique view of two sibling valves.

Plate 15

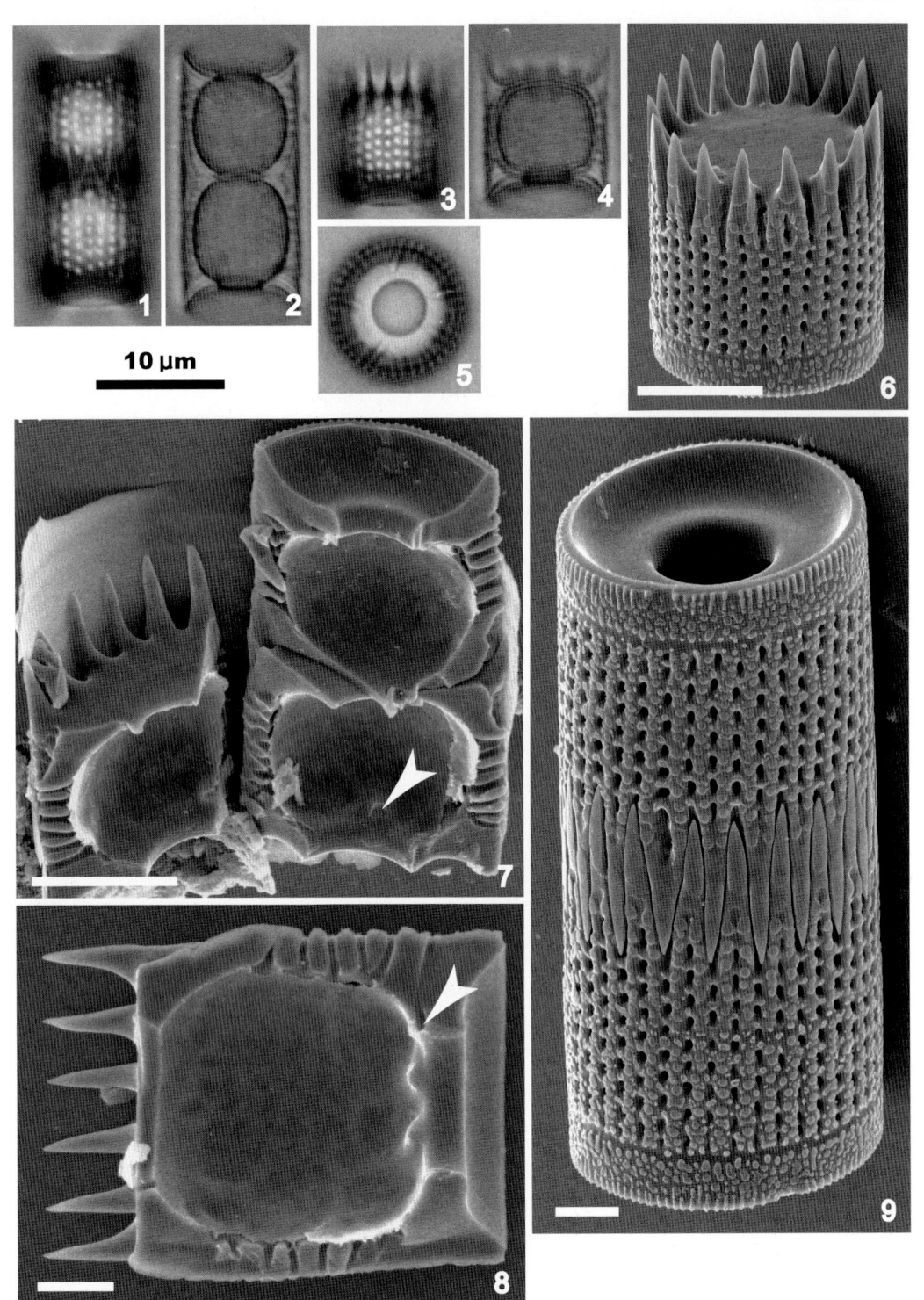

10 μm

Plate 16. Centric diatoms: Aulacoseirales
Aulacoseira nivalis (W. Smith) English & Potapova 2009

基礎異名：*Melosira nivalis* W. Smith 1855

文　　献：English, J. & Potapova, M. 2009. *Aulacoseira pardata* sp. nov., *A. nivalis* comb. nov., *A. nivaloides* comb. et stat. nov., and their occurrences in western North America. Proceeding of the Academy of Natural Sciences of Philadelphia **158**: 37-48.

　　　　　Haworth, E.Y. 1988. Distribution of diatom taxa of the old genus *Melosira* (now mainly *Aulacoseira*) in Cumbrian waters. *In*: Round, F.E.(ed.) Algae and the aquatc environment. pp. 138-169. Biopress, Bristol.

　　　　　田中宏之・南雲　保 2007. 福島県鎌沼から見出した *Aulacoseira distans* (Ehrenb.) Simonsen var. *nivalis* (W. Sm.) E.Y. Haw. の形態について．Diatom **23**: 113-116.

形　　態：殻は比較的背の低い円筒形，殻面は円形で，分離殻・結合殻とも殻面に点紋が存在する．殻径8.5-17.5 μm，殻套の高さ4-7.5 μm，襟は短い．条線は直線ないしわずかに左らせん状に配列し10 μmに12-16本．条線を構成する点紋は10 μmに14-16個．唇状突起は1個で横輪に所在する．横輪は非常に浅い．

ノ ー ト：本分類群は *Aulacoseira pfaffiana*(Reinsch)Krammer と同一分類群である可能性があるが，この関係については田中・南雲(2007)に記述がある．*Aulacoseira distans* とは横輪が浅いこと，唇状突起が1個であることにより区別できる．

産　　出：鎌沼（田中・南雲 2007）：福島県（現生）．

図　　版：鎌沼標本．

Figs 1-13. *Aulacoseira nivalis* (W. Smith) English & Potapova
LM. Figs 1-8. SEM. Figs 9-13. Material from Kama-numa Pond (Recent), Fukushima, Japan. Scale bars: Figs 9-11＝2 μm, Figs 12-13＝1 μm.

Figs 1-4.　Valve view, three different size valves.
Figs 1-2.　Same valve at different focal planes.
Figs 5-8.　Girdle view.
Fig. 9.　External view of whole valve.
Fig. 10.　Oblique view of Fig. 9.
Fig. 11.　Internal oblique view of whole valve, rimoportula (arrowhead).
Fig. 12.　Enlarged view of rimoportula (arrowhead).
Fig. 13.　Enlarged girdle view of sibling valves showing linking spines and a rimoportula opening (arrow).

Plate 16

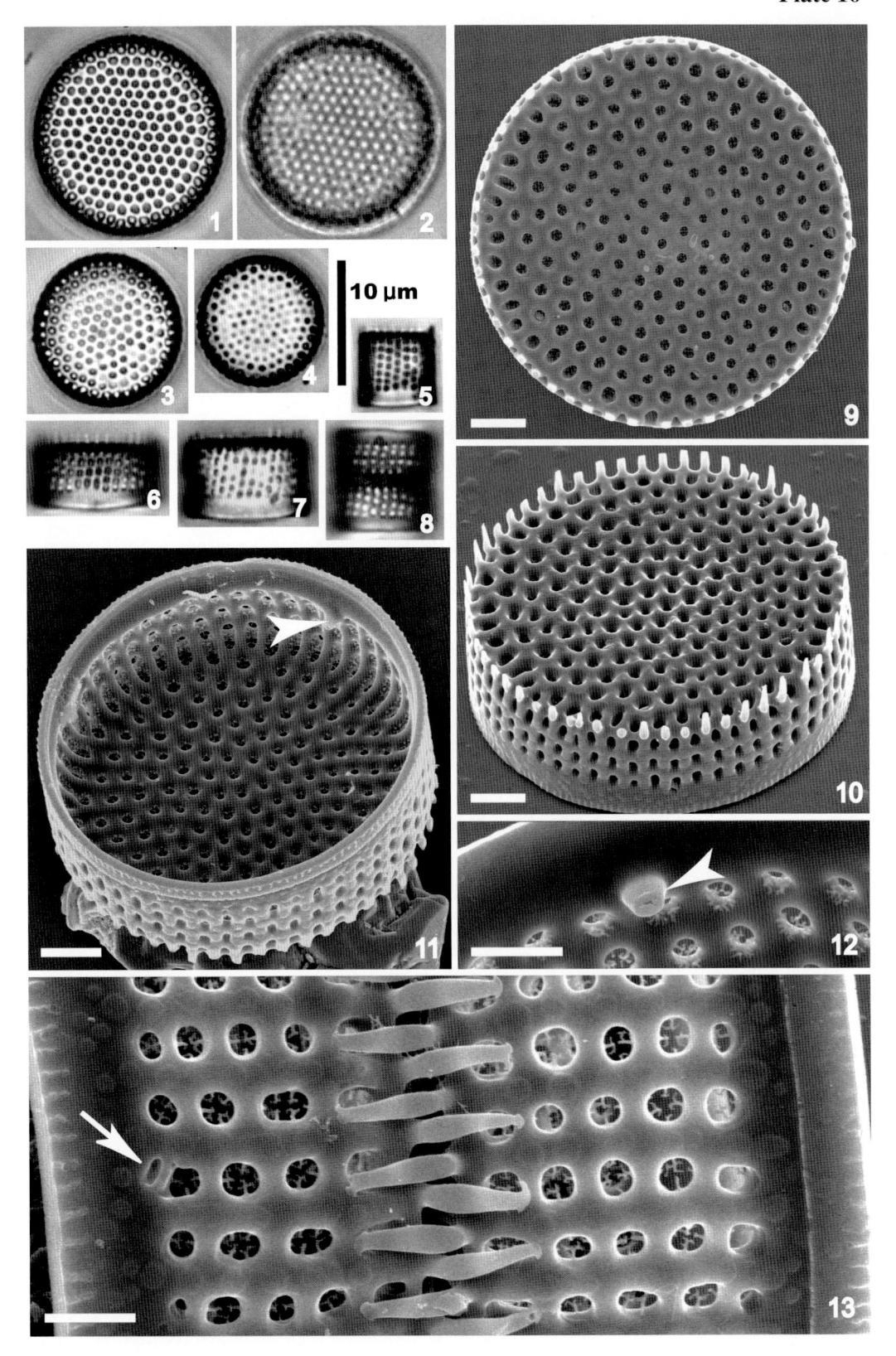

Plates 17-19. Centric diatoms: Aulacoseirales
Aulacoseira polispina H. Tanaka sp. nov.

新　　種：記載文（英文）は 3 頁参照.
形　　態：殻は円筒状，殻面は平らで円形，殻径 13-31 μm. 結合殻は小胞紋が縁辺部に分布すると共に，中心部には散在する．殻套長は分離殻で 17-28 μm，結合殻で 10-19 μm，条線は貫殻軸に平行な直線で 10 μm に約 6 本．分離殻の針は 2 本の間条線に 1 本の割合で存在し，毎 1-3 本ごとに太く長くなり接合する殻の襟まで達する．結合殻では柄の短いスパチュラ形で各間条線にある．条線を構成する点紋は 10 μm に 5-6 個．点紋は外側では多角形～楕円形であるが内側では分離殻はふつう丸い．唇状突起は殻套にあり，管はカーブし，1 殻に約 10 個である.
ノ ー ト：田中・松岡（1985）は古琵琶湖層群甲賀層から，長い針が多い分類群を見出し *Melosira* sp. 2 として報告している．筆者は同地域の試料から田中・松岡（1985）の *M.* sp. 2 と同じ分類群を見出し，*A. polispina* であることを確認した.
産　　出：甲賀層（田中・松岡 1985）：鮮新世・滋賀県，津房川層：後期鮮新世・大分県.
図　　版 17-19：津房川層標本.

Plate 17, Figs 1-5. *Aulacoseira polispina* H. Tanaka sp. nov.
LM. Figs 1-5, from the type material, OIT-212, Tsubusagawa Formation (Late Pliocene), Imai, Usa City, Oita, Japan.

Fig. 1.　Holotype specimen.
Fig. 2.　Separating valve with long spines and grooves (arrowheads) and linking valves (arrows).
Figs 1-3, 5.　Girdle views.
Fig. 4.　Valve view showing ringleist.
Fig. 5.　Separating valve with long spines.

Plate 17

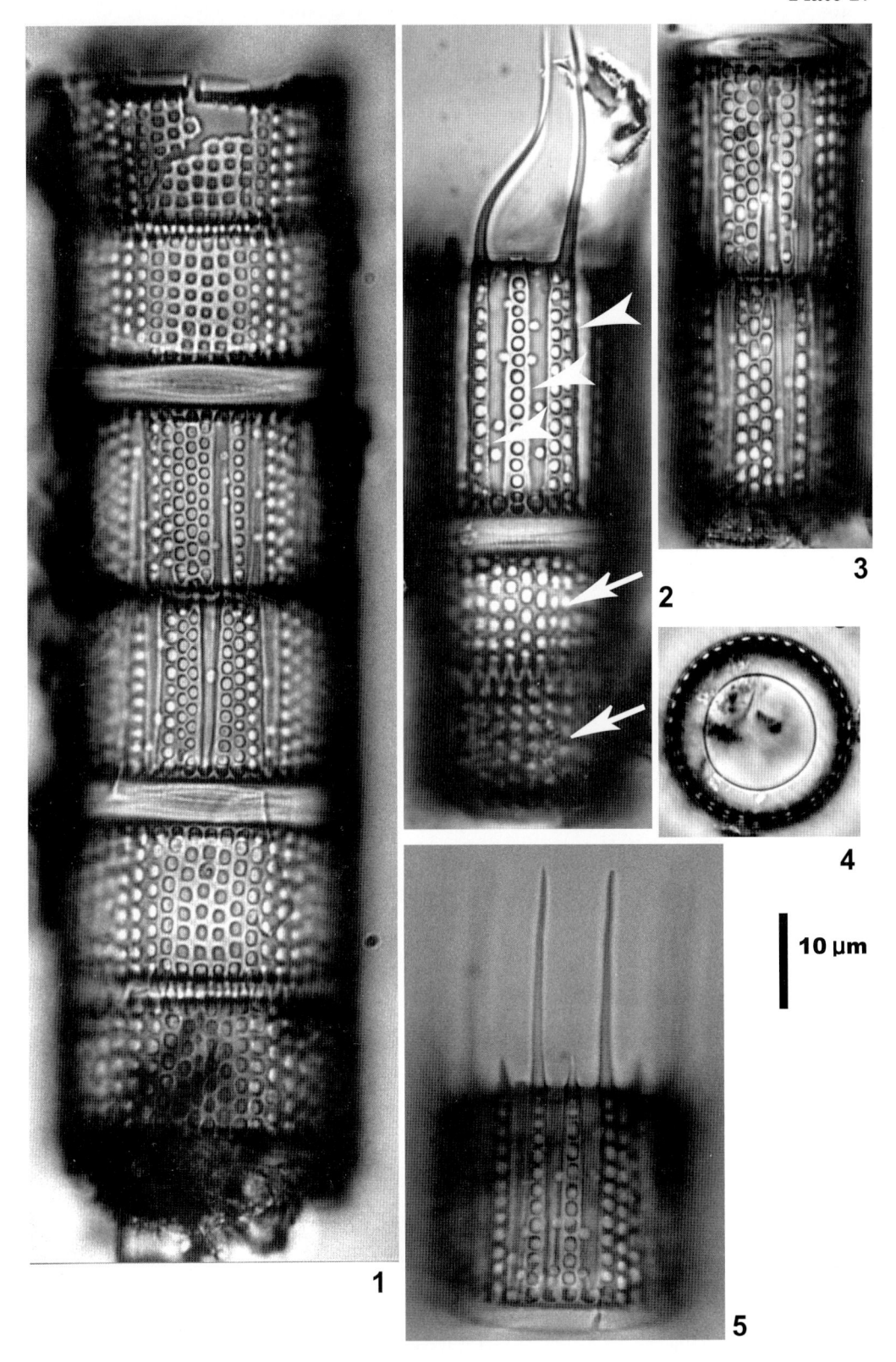

10 µm

Plate 18, Figs 1–6. *Aulacoseira polispina* **H. Tanaka** sp. nov.

SEM external views. Figs 1–6, from the type material, OIT-212, Tsubusagawa Formation (Late Pliocene), Imai, Usa City, Oita, Japan. Scale bars: Figs 1, 3=10 μm, Figs 2, 6=5 μm, Fig. 4=2 μm, Fig. 5=1 μm.

Fig. 1. Chained valve, both ends are separating valves.

Fig. 2. Enlarged view of linking valve of Fig. 1, linking spines (arrow) and opening of rimoportula (arrowhead).

Fig. 3. Oblique view of separating valve with no ornamentation on valve face, groove for spine (arrow).

Fig. 4. Detailed view of areolae, spine in the groove (arrow).

Fig. 5. Detailed view of opening of rimoportula and curved canal (arrowhead).

Fig. 6. Oblique view of linking valve face showing small areolae on marginal area and some on central area (arrows).

Plate 18

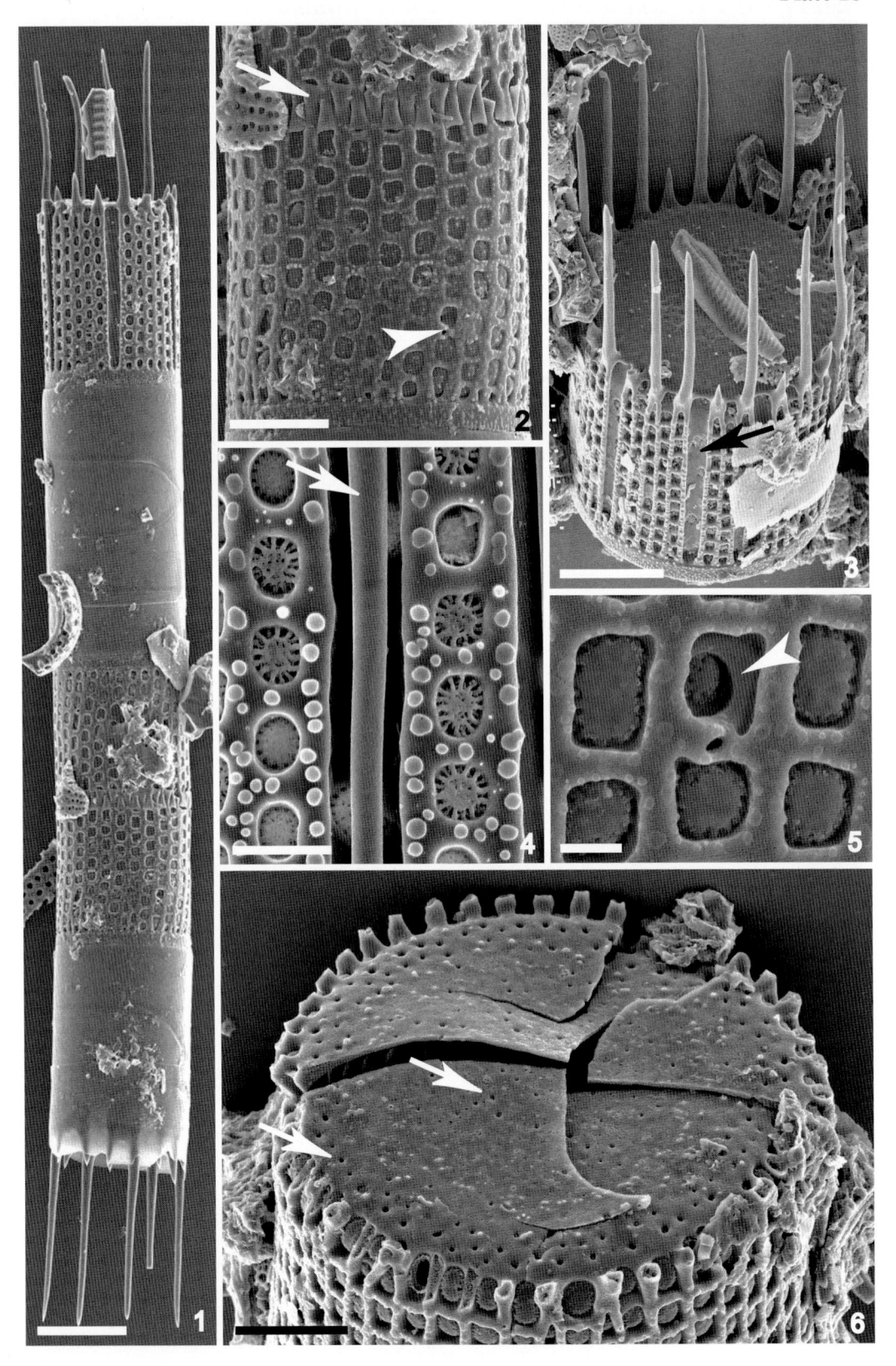

Plate 19, Figs 1-6. *Aulacoseira polispina* **H. Tanaka** sp. nov.

SEM internal views. Figs 1-6, from the type material, OIT-212, Tsubusagawa Formation (Late Pliocene), Imai, Usa City, Oita, Japan. Scale bars: Figs 1-2=10 μm, Figs 3-4=5 μm, Fig. 5=2 μm, Fig. 6=0.5 μm.

Fig. 1.　Broken linking valves showing rimoportulae on mantle (arrowheads).
Fig. 2.　Oblique view of Fig. 1.
Fig. 3.　Broken separating valve showing round shaped areolae, rimoportulae (arrowheads) and some areolae in grooves.
Fig. 4.　Oblique enlarged view of Fig. 3, broken valve face, and two rimoportulae on mantle.
Fig. 5.　Enlarged view of linking valve showing quadrilateral to polygonal areolae and two rimoportulae.
Fig. 6.　Detailed view of round areolae of separating valve.

Plate 19

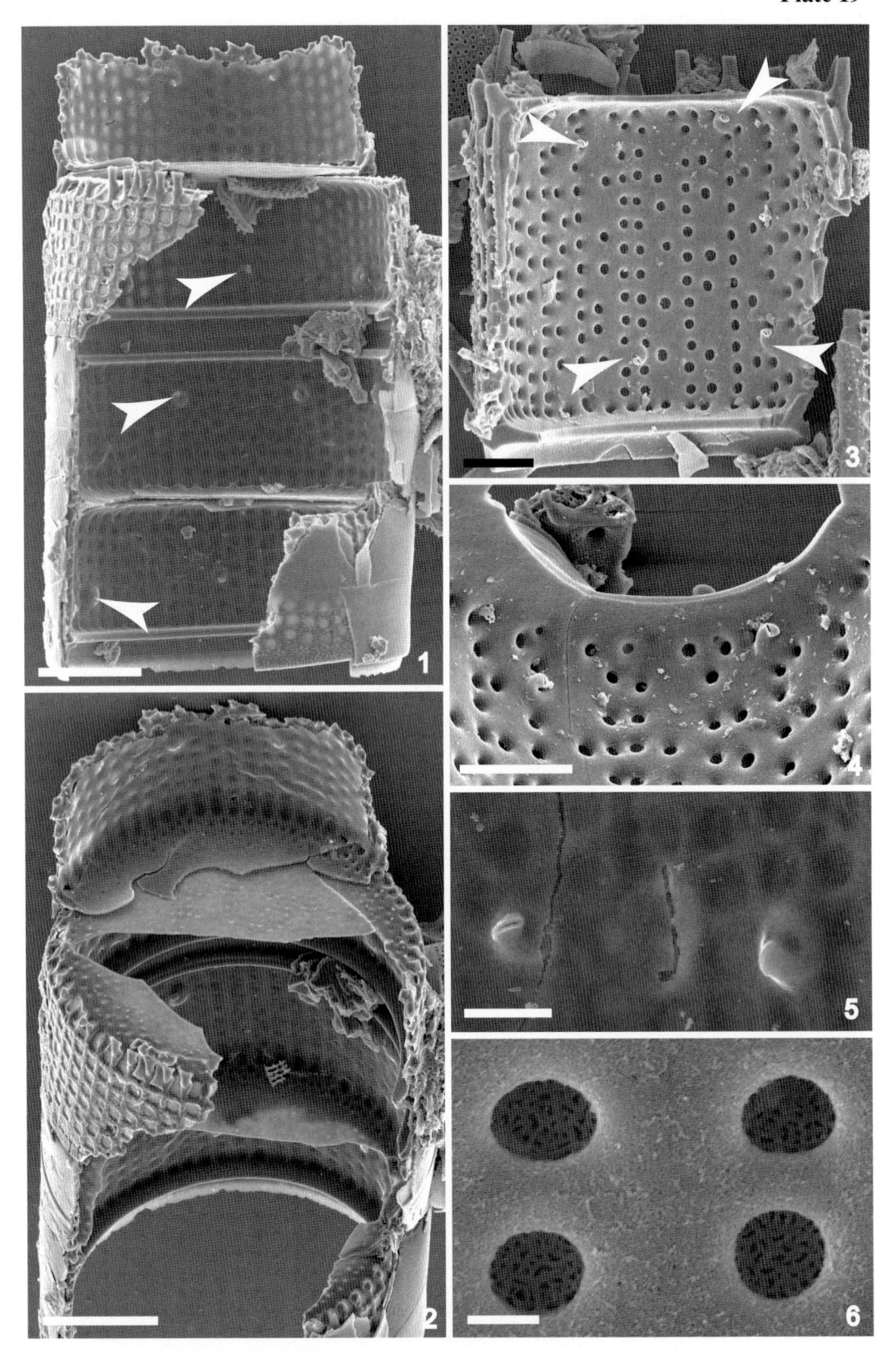

Plate 20. Centric diatoms: Aulacoseirales
Aulacoseira pusilla (F. Meister) Tuji & Houki 2004

基礎異名：*Melosira pusilla* F. Meister 1913

文　献：Tuji, A. & Houki, A. 2004. Taxonomy, ultrastructure, and biogeography of the *Aulacoseira subarctica* species complex. Bulletin of the National Museum of Nature and Science, Series B **30**: 35-54.

　　Tuji, A. & Williams, D.M. 2007. Type examination of Japanese diatoms described by Friedrich Meister (1913) from Lake Suwa. Bulletin of the National Museum of Nature and Science, Series B **33**: 69-79.

形　態：殻は小形で円筒形，殻面には点紋が全面～縁辺に所在する．殻径 4.5-8 μm，殻套高 3-5.5 μm である．殻套の点紋は左らせんを描き（まれにほぼ直線的なものも見られた）10 μm に約 24 本．条線を構成する点紋は 10 μm に約 30 個である．針は短いが太く 2-3 本の間条線に 1 本の割合で存在する．唇状突起は横輪上にある．

ノート：本種は Meister(1913)により諏訪湖の試料を用いて，*Melosira pusilla* として記載されたが，Tuji & Houki(2004)により *Aulacoseira* 属へ組み合わせになった．本書の記事，図はタイプ地（諏訪湖）から採取した試料に基づいている．

産　出：人形峠層（田中ら 2008）：後期中新世～鮮新世・岡山-鳥取県境，諏訪湖（Meister 1913）：長野県（現生）．

図　版：諏訪湖標本．

Figs 1-11. *Aulacoseira pusilla* (F. Meister) Tuji & Houki

LM. Figs 1-5. SEM. Figs 6-11. Material from the type locality, Lake Suwa (Recent), Nagano, Japan. Scale bars: Figs 6-9＝2 μm, Figs 10-11＝0.5 μm.

Figs 1-2, 4-5.　Two valve chains at different focal planes.
Fig. 3.　Girdle view of a frustule.
Fig. 6.　Girdle view of a frustule.
Fig. 7.　Oblique view showing valve face of left side of valve of Fig. 6.
Fig. 8.　Oblique view showing valve face of right side of valve of Fig. 6.
Fig. 9.　Slightly oblique view of valve, rimoportula (arrowhead).
Fig. 10.　Enlarged view of Fig. 9 showing rimoportula on the ringleist (arrowhead).
Fig. 11.　Opening of rimoportula associated with an areola (arrowhead).

Plate 20

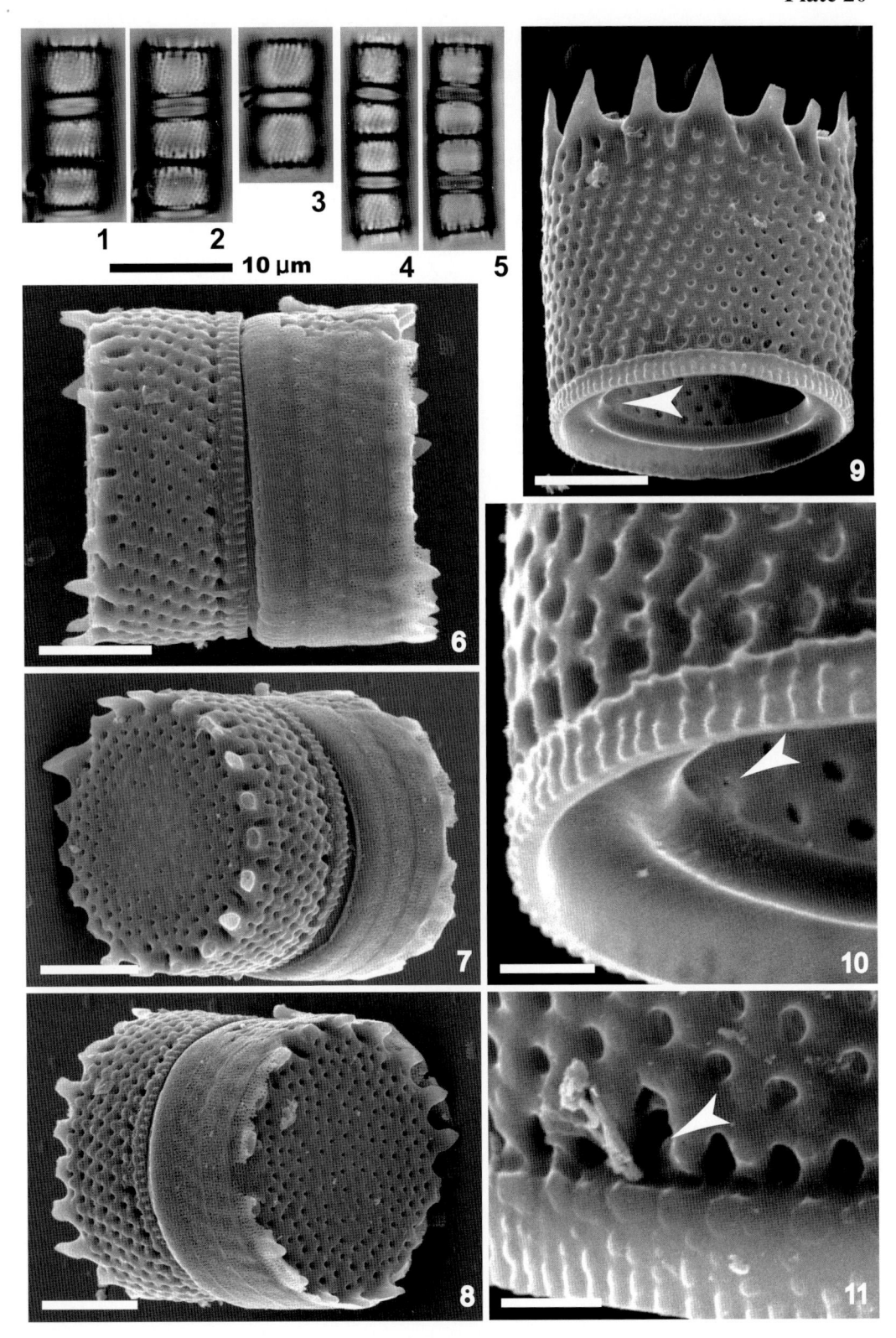

Plates 21-22. Centric diatoms: Aulacoseirales
Aulacoseira satsumaensis H. Tanaka in Tanaka & Nagumo 2010

文　献：Tanaka, H. & Nagumo, T. 2010. *Aulacoseira satsumaensis*, a new Pliocene diatom species with two morphotypes from Kagoshima, Japan. Diatom Research **25**: 163-174.

形　態：殻の厚さの違いにより厚殻タイプと薄殻タイプに区分できる．厚殻の殻径 4-9 μm，殻套長 3.5-20 μm，条線中の点紋 12-20 μm．薄殻は殻径 2.5-5 μm，殻套長 9-26.5 μm，条線中の点紋 16-24 μm である．条線は両タイプとも 10 μm に 14-20 本である．針は 2 本の間条線ごとに 1 個ある．SEM 観察によると両タイプは唇状突起の位置が，厚殻タイプは横輪上にあるのに対して薄殻タイプは内側殻套にあることでも異なっている．また一般的に薄殻タイプは横輪を欠く．

ノート：*Aulacoseira* 属では，同一分類群で厚殻タイプと薄殻タイプに区分できる分類群がいくつか報告されているが，本種は殻の厚さだけでなく唇状突起の所在場所，および薄殻タイプは横輪が（普通）欠如することが重要な特徴である．

産　出：郡山層（Tanaka & Nagumo 2010）：鮮新世・鹿児島県．

図　版 21-22：郡山層標本．

Plate 21, Figs 1-10.　*Aulacoseira satsumaensis* H. Tanaka

LM. Figs 1-6. SEM. Figs 7-10. From the type material, KAG-512, Koriyama Formation (Late Pliocene), Kagoshima City, Kagoshima, Kyushu, Japan. Scale bars: Figs 8-9＝5 μm, Figs 7, 10＝2 μm.

Figs 1-3, 5-6.　Girdle view of thicker valves.
Figs 5-6.　Same valve chain at different focal planes.
Fig. 4.　Ringleist with canals of rimoportulae.
Fig. 7.　External girdle view of three valves.
Fig. 8.　External girdle view of sibiling valves.
Fig. 9.　Initial cell.
Fig. 10.　Internal broken valve view showing rimoportulae on ringleist (arrowheads).

Plate 21

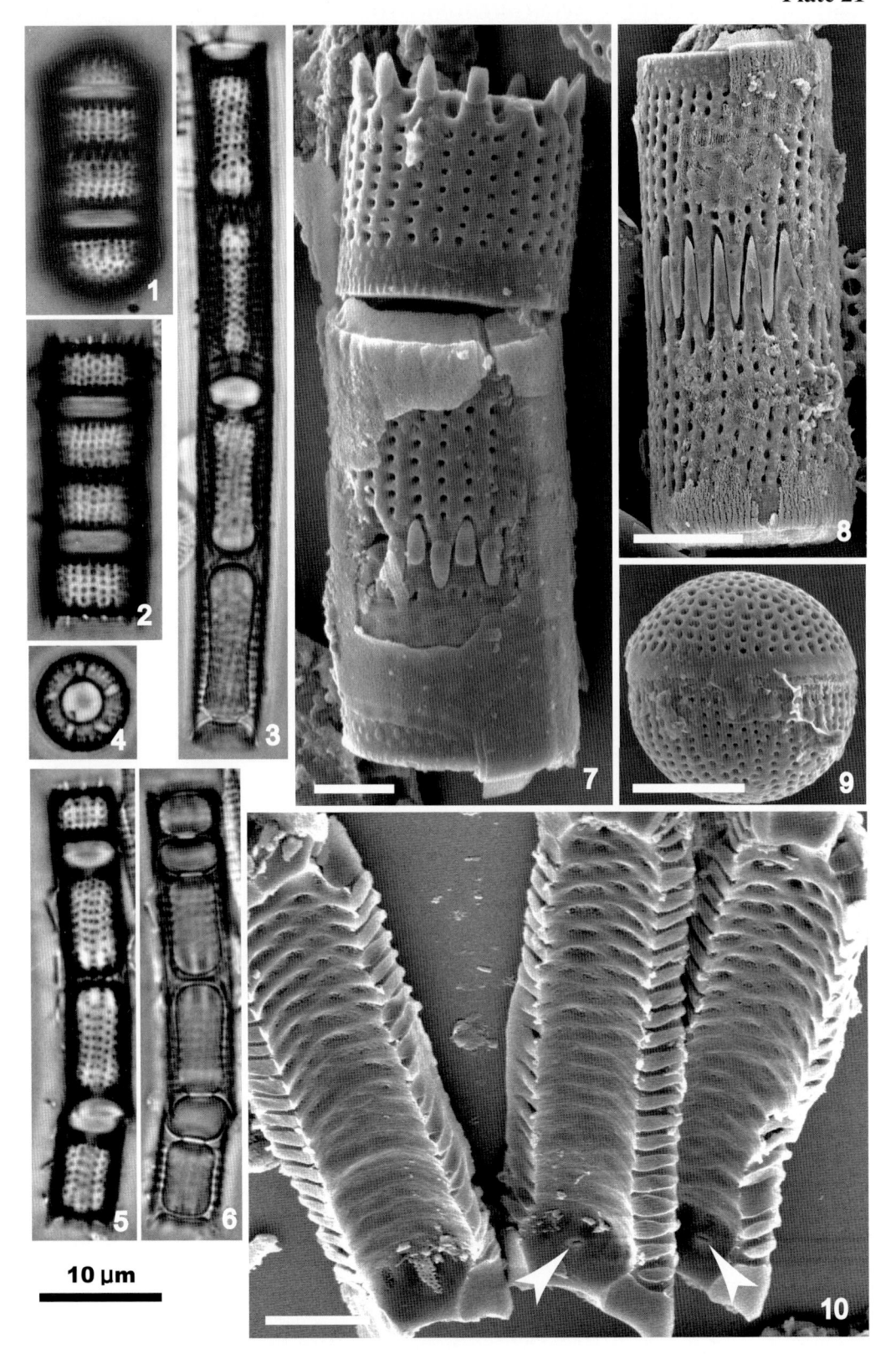

Plate 22, Figs 1-7. *Aulacoseira satsumaensis* **H. Tanaka**

LM. Figs 1-3. SEM. Figs 4-7, thinner valves. From the type material, KAG-512, Koriyama Formation (Late Pliocene), Kagoshima City, Kagoshima, Kyushu, Japan. Scale bars: Fig. 4=2 μm, Fig. 5=1 μm, Figs 6-7=0.5 μm.

Figs 1-3. Girdle view, each upper valve is thicker valve with ringleist and others are thinner valves without ringleist or very shallow ringleist.

Fig. 4. Two thinner sibiling valves.

Fig. 5. Valve face and spines.

Fig. 6. Detailed view of internal mantle showing two rimoportulae on areolae rows (arrowheads).

Fig. 7. Detailed view of inner collar and mantle showing rimoportula (arrowhead) and no ringleist.

Plate 22

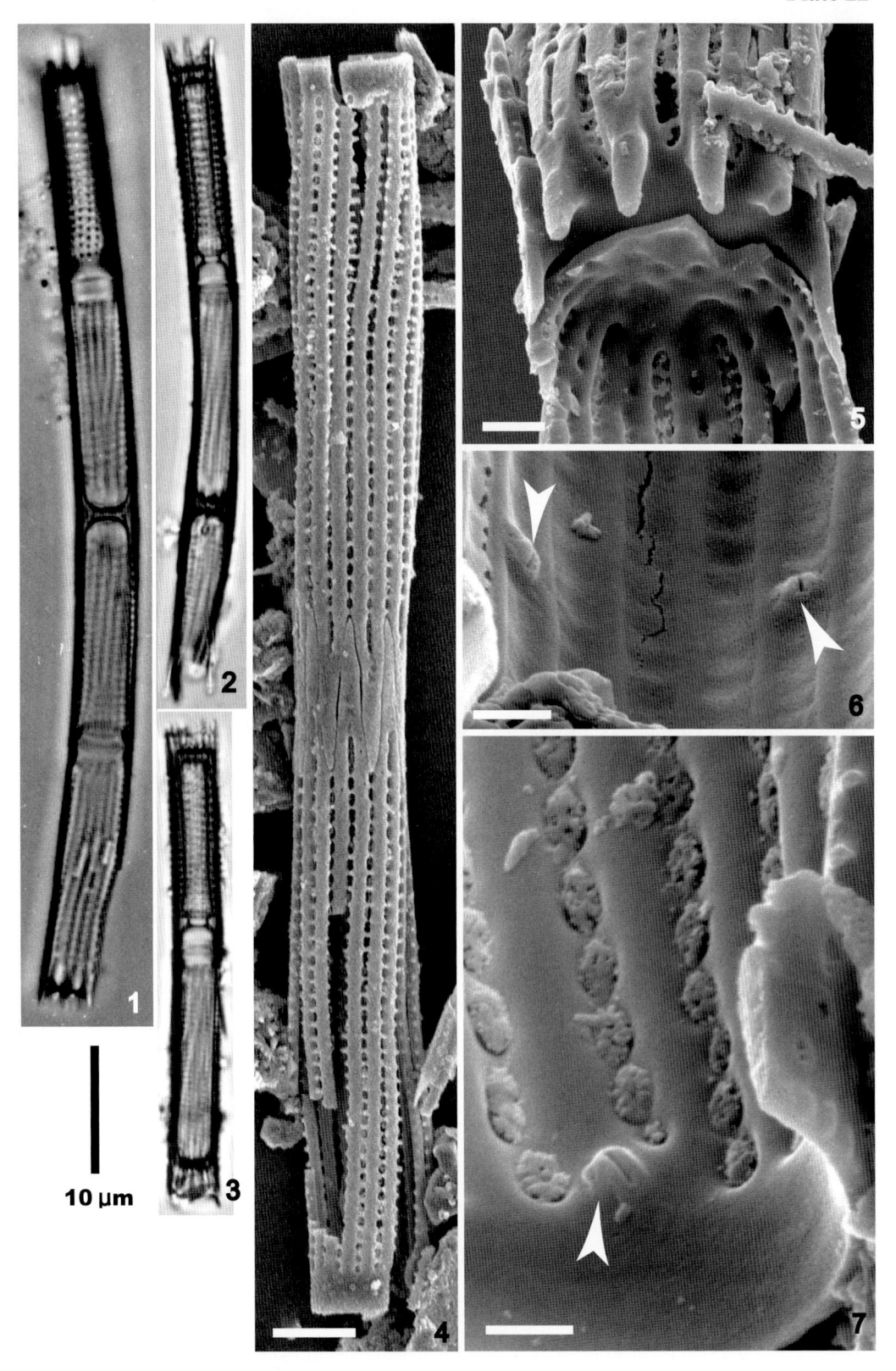

10 µm

Plate 23.　Centric diatoms: Aulacoseirales
Aulacoseira subarctica (O. Müller) Haworth 1988

基礎異名：*Melosira italica* subsp. *subarctica* O. Müller 1906
文　　献：Haworth, E.Y. 1988. Distribution of diatom taxa of the old genus *Melosira* (now mainly *Aulacoseira*) in Cumbrian waters. *In*: Round, F.E.(ed.) Algae and the aquatic environment. pp. 138-167. Biopress, Bristol.
　　河島綾子，小林　弘 1993. 阿寒湖の珪藻（1. 中心類）．自然環境科学研究 **6**: 41-58.
形　　態：殻は円筒形，殻面には点紋が散在ないし殻縁にわずか存在する．殻径 4.5-10 μm，殻套長 5-11 μm，条線は斜めに走るが，殻面近くになると直線的になる．条線は 10 μm に 12-16 本で，構成する点紋の密度は 10 μm に約 18 個であった．針は長く先頭形である．唇状突起は殻套および横輪と殻套の境付近に所在する．
ノ ー ト：中之条湖成層では多くの地点から産出するが，特に中之条町八幡地区の試料からは 93％の頻度で出現した（田中・小林 1992）．
産　　出：中之条湖成層(田中・小林 1992)：中期更新世・群馬県，沼田湖成層：中期更新世・群馬県．
図　　版：中之条湖成層標本．

Figs 1-8.　*Aulacoseira subarctica* (O. Müller) Haworth
LM. Figs 1-4. SEM. Figs 5-8. Material from Nakanojo Lacustrine Deposit (Middle Pleistocene), Nakanojo Town, Gunma, Japan. Scale bars: Figs 5, 8＝2 μm, Figs 6-7＝1 μm.

Figs 1-4.　Girdle views, different size chains, rimoportulae (arrowheads).
Fig. 5.　Broken valve, showing valve face, spines and inner valve.
Fig. 6.　Enlarged view of part of Fig. 5, showing two rimoportulae on mantle (arrowheads).
Fig. 7.　Internal view of mantle showing two rimoportulae (arrowheads).
Fig. 8.　External view of sibling valves.

Plate 23

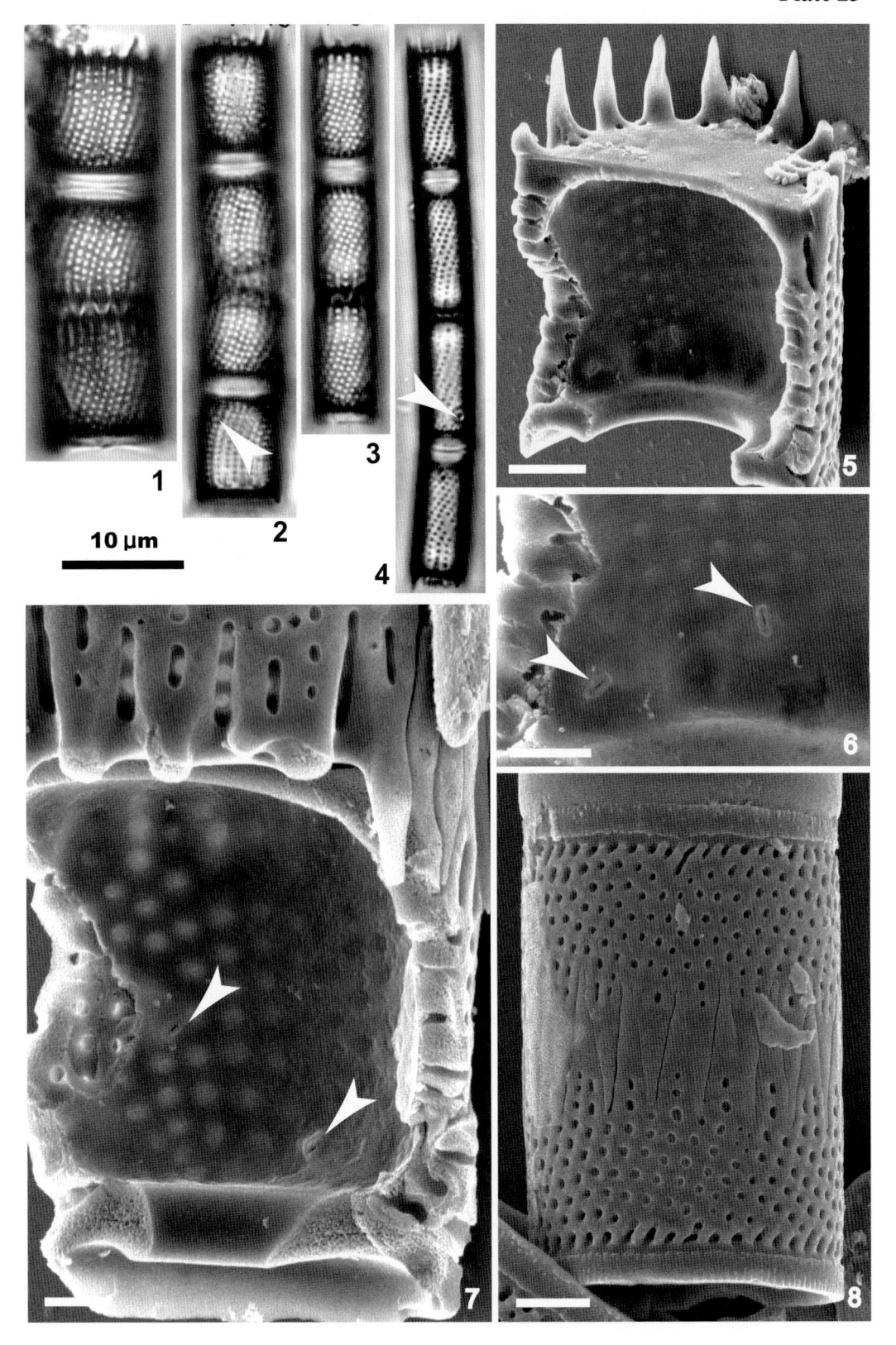

10 μm

Plate 24. Centric diatoms: Aulacoseirales
Aulacoseira tenella (Nygaard) Simonsen 1979

基礎異名：*Melosira tenella* Nygaard 1956

文　献：Nygaard, G. 1956. Ancient and recent flora of diatoms and Chrysophyceae in Lake Gribso. Folia Limnol. Scand. **8**: 32-262, pls 1-12.

　　　Siver, P.A. & Kling, H. 1997. Morphological observations of *Aulacoseira* using scanning electron microscopy. Canadian Journal of Botany **75**: 1807-1835.

　　　Simonsen, R. 1979. The diatom system: ideas on phylogeny. Bacillaria **2**: 9-71.

形　態：殻は小形で浅い円筒形，殻面全面に点紋が所在する．殻径約 7.5 µm，殻套長約 2 µm．条線は直線的であるがしばしば斜めに走る，10 µm に約 20 本，条線を構成する点紋は各列に 2-3 個である．横輪は認められず，点紋列が襟に接する部分に唇状突起が 2-3 個所在する．

ノート：Nygaard(1956)によりデンマークにおけるボーリング試料から *Melosira tenalla* Nygaard として記載された種である．Siver & Kling(1997)によればアメリカ合衆国コネチカット州では pH 5-7 で貧栄養～初期の中栄養水域で出現し，殻径 5-7 µm，殻套長平均 2 µm と報告されている．SEM 写真による針の形態もほぼ一致しているので本種に同定した．

産　出：福上湖：沖縄県（現生）．

図　版：福上湖標本．

Figs 1-14. *Aulacoseira tenella* (Nygaard) Simonsen

LM. Figs 1-7. SEM. Figs 8-14. Material from Lake Fukugami (Recent), Higashi-son Village, Okinawa, Japan. Scale bars: Figs 8, 12＝2 µm, Figs 9, 11＝1 µm, Figs 10, 13-14＝0.5 µm.

Figs 1-3.　Girdle views.
Figs 4-6.　Valve views.
Figs 5-6.　Same valve at different focal planes.
Fig. 7.　Oblique valve view.
Fig. 8.　External view of whole valve.
Fig. 9.　Oblique view of Fig. 8.
Fig. 10.　Enlarged view of Fig. 9 showing external opening of rimoportula (arrow).
Fig. 11.　Internal oblique view, rimoportulae (arrowheads).
Fig. 12.　Girdle view of valves.
Fig. 13.　Internal view showing a rimoportula between mantle and collar (arrowhead).
Fig. 14.　Marginal view of rimoportula (arrowhead) and its outer opening (arrow).

Plate 24

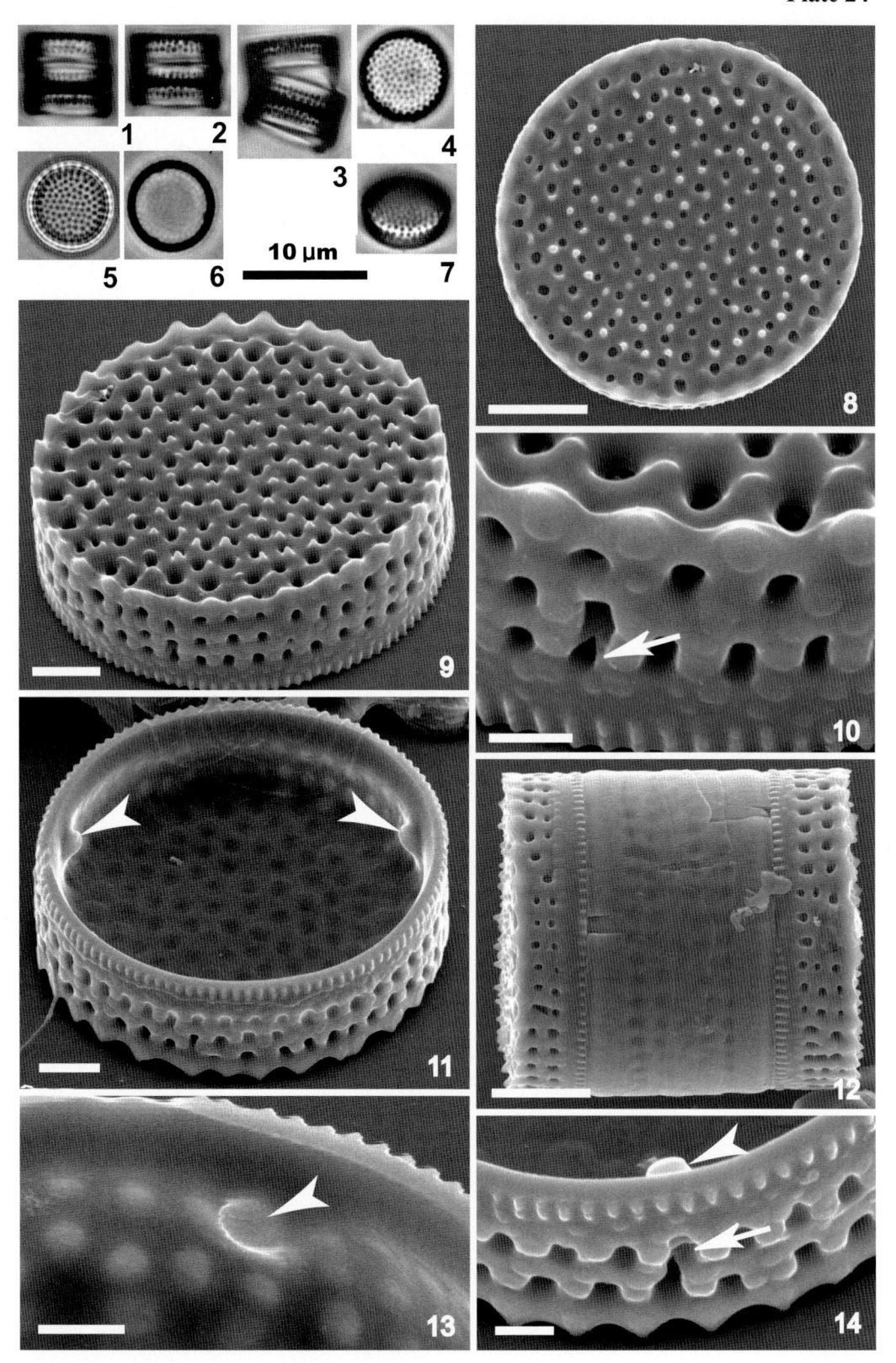

Plates 25-26.   Centric diatoms: Aulacoseirales
Aulacoseira tsugaruensis H. Tanaka sp. nov.

新　　種：記載文（英文）は4頁参照.
形　　態：殻は円筒形，殻面は平らで円形，胞紋，突起等は所在しない．殻径5.5-24 μm，殻套長は10.5-18.5 μm．条線はふつう貫殻軸に平行で10 μmに10-21本であるが，殻形が小さくなると斜めになる．針は殻面／殻套境界にあり，3本の間条線に1本の割合で，大形のスパチュラ形である．条線を構成する点紋は比較的小形で，開口は円形であり10 μmに14-20個．唇状突起（おそらく1個／殻）は殻套にあり，管は横輪へ向かって斜めに下がる．唇状突起の外部への開口は細長いスリット状で，条線の末端にあるが，襟との間に1-2個の胞紋が存在することもある．
ノ　ー　ト：大きいスパチュラ形の結合針をもつ種類は他にも存在するが，本種は唇状突起は殻套に所在するが，その管は斜めに横輪に向かい条線の末端あるいは末端近くに外部への開口があることで他種と異なる.
産　　出：六角沢層：前期中新世・青森県.
図　　版 25-26：六角沢層標本.

Plate 25, Figs 1-7.   *Aulacoseira tsugaruensis* H. Tanaka sp. nov.
LM. Figs 1-6. SEM Fig. 7. From the type material, ROK-001. Rokkakuzawa Formation (Early Miocene), Fukaura Town, Aomori, Japan. Scale bar: Fig. 7＝5 μm.

Fig. 1.　　Initial cell.
Figs 2, 4-5.　　Valves with straight striae.
Figs 3, 6.　　Valves with slightly curved striae.
Fig. 4.　　Holotype specimen. MPC-25052.
Fig. 7.　　Sibling valves showing areolae rows, large spatulated spines and a slanted canal with rimoportula (arrowhead).

Plate 25

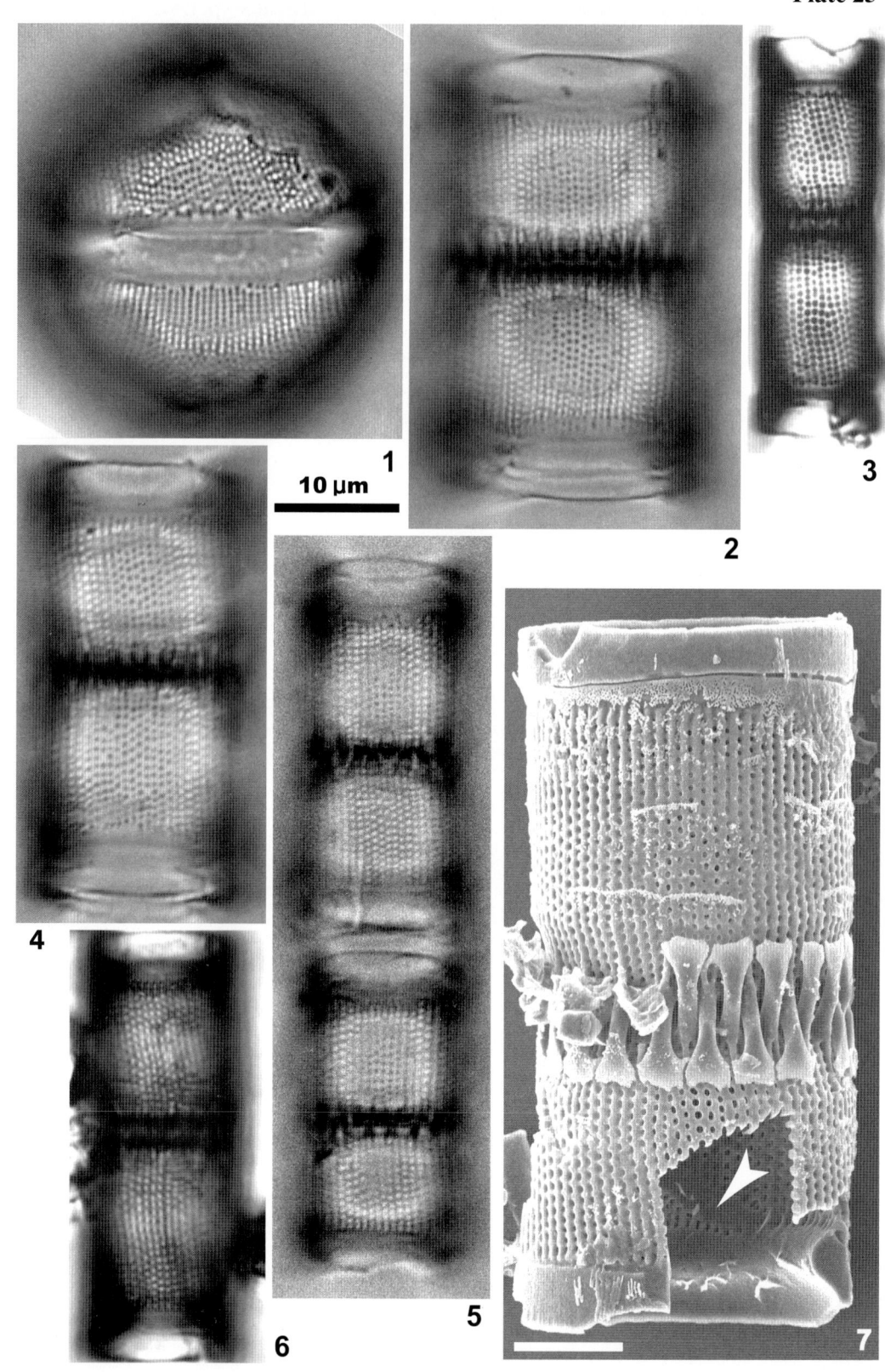

Plate 26, Figs 1–6. *Aulacoseira tsugaruensis* **H. Tanaka** sp. nov.

SEM. Figs 1–6. From the type material, ROK-001. Rokkakuzawa Formation (Early Miocene), Fukaura Town, Aomori, Japan. Scale bars: Fig. 3=5 μm, Figs 1–2, 4, 6=2 μm, Fig. 5=1 μm.

Fig. 1. Internal oblique view of Plate 26, Fig. 7, showing rimoportula (arrowhead) in a slanted canal.

Fig. 2. Broken valve showing rimoportula (arrowhead) and slanted canal.

Fig. 3. Linking valve showing large spatulate spines, areolae rows and opening of rimoportula (arrowhead).

Fig. 4. Oblique view showing valve face without ornamentation (arrow).

Fig. 5. Part of enlarged view of Fig. 3 showing opening of rimoportula (arrowhead).

Fig. 6. Internal valve face showing smooth surface.

Plate 26

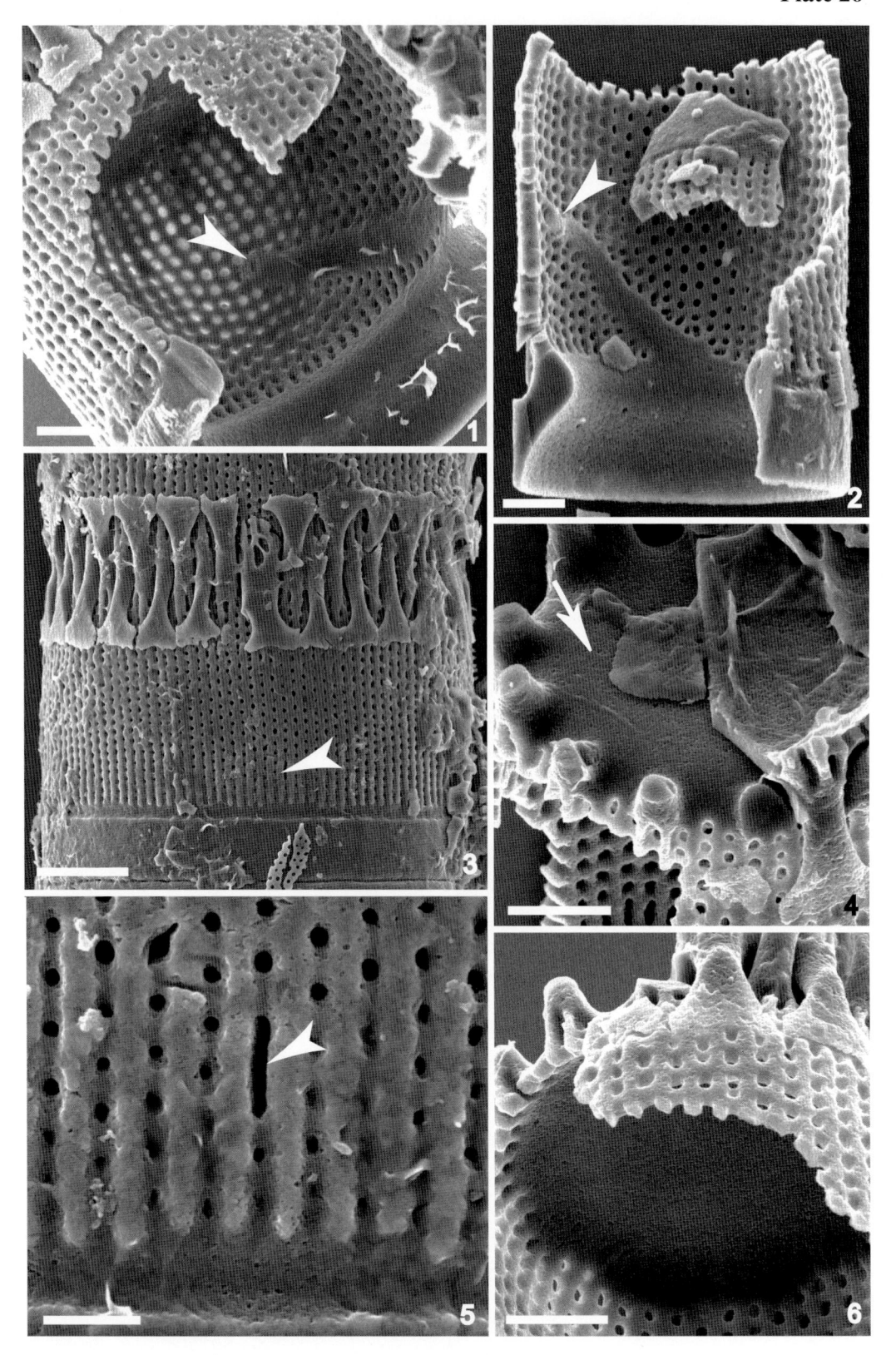

Plate 27. Centric diatoms: Aulacoseirales
Aulacoseira valida (Grunow) Krammer 1991

基礎異名：*Melosira crenulata* var. *valida* Grunow in Van Heurck 1882
文　　献：佐竹俊子・小林　弘 1991. 淡水産中心類珪藻 *Aulacoseira valida*（Grunow in Van Heurck）Krammer の微細構造. 自然環境科学研究 **4**: 45-57.
　　　 Krammer, K. 1991. Morphology and taxonomy in some taxa of the genus *Aulacoseira* Thwaites (Bacillariophyceae). 1. *Aulacoseira distans* and similar taxa. Nova Heedwigia **52**: 89-112.
形　　態：殻は円筒形で殻壁は厚い．殻径約 10-32 μm，殻套長 7-21 μm，点紋列は斜行し 10 μm に約 13 本，点紋は殻面近くが大きく，襟に向かって小さくなる．点紋は点紋列に沿って 10 μm に 8-14 個である．針は約 2 本の間条線に 1 個の割合であり，大きく頑丈で先端が歯状である．唇状突起は横輪にあり，外部への開口は点紋列の襟に接する部分に点紋の数個と置き換わって存在する．
ノ ー ト：本種の形態は上記文献に詳述されている．
産　　出：下末吉層（佐竹・小林 1991）：更新世・神奈川県，尾瀬沼：完新世・群馬県．
図　　版：尾瀬沼標本.

Figs 1-6. *Aulacoseira valida* (Grunow) Krammer
LM. Figs 1-2. SEM. Figs 3-6. Material from bottom sediment of Lake Oze (Holocene), Gunma, Japan. Scale bars: Fig. 3＝5 μm, Figs 4-6＝2 μm.

Figs 1-2.　Girdle view of sibling valves at different focal planes.
Fig. 3.　Girdle view, opening of rimoportula (arrowhead).
Fig. 4.　Internal view of broken valve showing rimoportula (arrowhead) on ringleist.
Figs 5-6.　Oblique views at different angles of same valve.

Plate 27

10 µm

Plate 28.　Centric diatoms: Aulacoseirales
Brevisira arentii (Kolbe) Krammer 2001

基礎異名：*Cyclotella arentii* Kolbe 1948
文　　献：Krammer, K. 2001. Taxonomie und morphologie von *Brevisira arentii* (Kolbe) Krammer gen. nov., comb. nov. *In*: Jahn, R., Kociolek, J.P., Witkowski, A. & Compère, P. (eds) Lange-Bertalot-Festschrift. pp. 9-20. A.G.G. Gantner K.G., Ruggell.
　　南雲　保・小林　弘 1977. 光顕及び電顕的研究に基く *Melosira arentii* (Kolbe) comb. nov. について. 藻類 **25**: 182-183.
形　　態：殻は殻套が比較的短い円筒形で，殻面はほぼ平ら，中心部には不規則に点紋があり縁辺部には放射状に点紋が配列する．SEM で被殻の外側を観察すると，しばしば上殻と下殻でかなり形態が異なるように見えるが（Figs 5-7），内側を観察すると同じ構造であることがわかる（Fig. 4）．殻径 11-16 μm，殻套の条線は直線で 10 μm に約 22 本.
ノート：半球形の殻は初生殻の可能性がある．使用した試料からはしばしば出現した．今のところ化石の報告は見当たらない．
産　　出：蘭牟田池（南雲・小林 1977）：鹿児島県（現生）.
図　　版：蘭牟田池標本.

Figs 1-8.　*Brevisira arentii* (Kolbe) Krammer

LM. Figs 1-2. SEM. Figs 3-8. Materials from Imuta Pond (Recent), Satsumasendai City, Kagoshima, Kyushu, Japan. Scale bars: Figs 2-6, 8＝2 μm, Fig. 7＝1 μm.

Figs 1-2.　Two different size valves.
Fig. 3.　Internal view of whole valve.
Fig. 4.　Oblique view of Fig. 3.
Fig. 5.　Girdle view of a frustule. Left valve: possibility of initial valve.
Fig. 6.　Oblique view of Fig. 5 showing one side of valve face.
Fig. 7.　Oblique view of Fig. 5 showing opposite side of Fig. 6.
Fig. 8.　Enlarged view of mantle of Fig. 5, different angle.

Plate 28

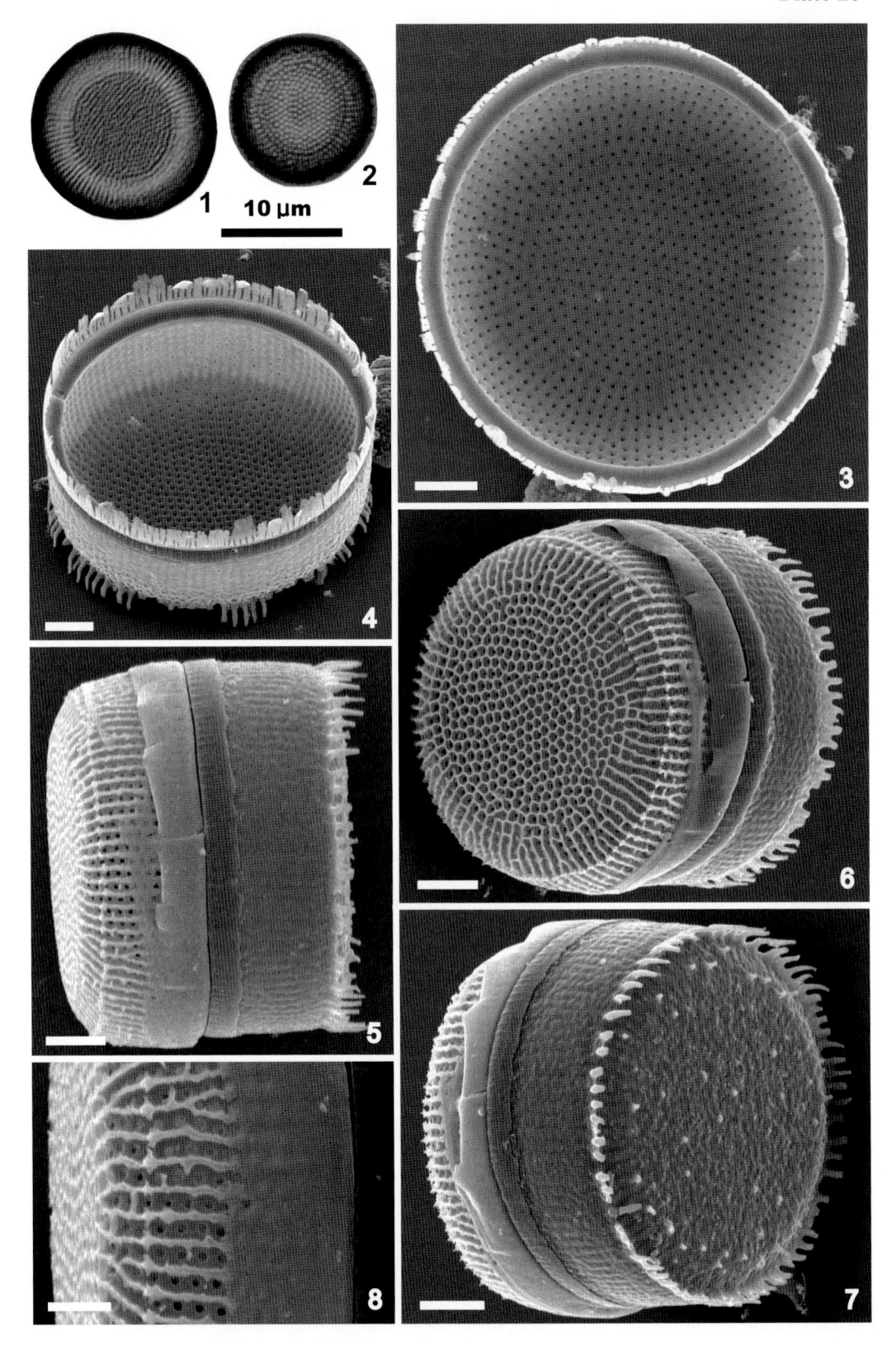

Plate 29.　Centric diatoms: Aulacoseirales
Miosira tscheremissinovae (Khursevich) Khursevich 2008

基礎異名：*Alveolophora tscheremissinovae* Khursevich 1884
文　　献：Khursevich, G.K. 1994. Morphology and taxonomy of some centric diatom species from the Miocene sediments of the Dzhilinda and Tunkin Hollows. *In*: Kociolek, J.P. (ed.) Proceedings of the 11th International Diatom Symposium. pp. 269-280. California Academy of Sciences, San Francisco.
　　　　　Kozyrenko, T.F., Strelnikova, N.I., Khursevich, G.K., Tsoy, I.B., Jakovschikova, T.K., Mukhina, V.V., Olshtynskaya, A.P. & Semina, G.I. 2008. The diatoms Russia and adjacent countries. Fossil and recent. Vol. 2. Issue 5. 171 pp. St. Petersburg University Press, St. Petersburg.
形　　態：殻は小形で浅い円筒形．殻面にはふつう縁辺域に点紋が所在するが，中心近くまで分布することもあり，殻中心から放射状に配列しているように観察できることもある．殻径10-18 μm, 殻套長2.5-3 μm. 殻套条線は10 μmに約12本．横輪は認められない．唇状突起は殻面の中心域と縁辺域の境付近に4-7個所在し，管は内側表面を殻套へ向かい，殻縁近くの殻套に外側への開口がある．唇状突起と殻面の管はピントにより光顕でも観察することができる．
ノ ー ト：Khursevich(1994)により*Alveolophora*属の1種として記載されたが，後に*Miosira*属へ組み合わせになった（Kozyrenko *et al.* 2008）．しかし著者による詳細なSEM観察によって殻套内側の柱状構造物は唇状突起の管であることがはっきりしたので，既存の属の中では*Aulacoseira*属が最も近いと思われるが，ここでは最も最近の組み合わせであるKozyrenko *et al.*(2008)に揃える．
産　　出：太櫓層（瀬棚の珪藻土）：前期中新世・北海道，六角沢層：前期中新世・青森県，柳田層：前期中新世・石川県，白沢層：後期中新世・宮城県，宮田層：中新-鮮新世・秋田県．
図　　版：宮田層標本．

Figs 1-9.　*Miosira tscheremissinovae* (Khursevich) Khursevich
LM. Figs 1-4. SEM. Figs 5-9. Material from Miyata Formation (Mio-Pliocene), Senboku City, Akita, Japan. Scale bars: Figs 5-7＝2 μm, Figs 8-9＝1 μm.

Figs 1-2.　Valve views of two different size valves, Fig. 1 showing rimoportula (arrowhead) and its canal (arrow).
Figs 3-4.　Girdle views of two different size valves.
Fig. 5.　External oblique view of whole valve.
Fig. 6.　Internal oblique view of whole valve showing rimoportulae (small arrowheads), and large arrowhead, rimoportula, and its corresponding opening (arrow).
Fig. 7.　Enlarged view of part of Fig. 6 showing rimoportula (arrowhead) and its external opening (arrow).
Fig. 8.　Internal valve face view, rimoportulae (arrowheads) and large arrowhead, rimoportula, and its external opening (arrow).
Fig. 9.　Girdle view showing spines.

Plate 29

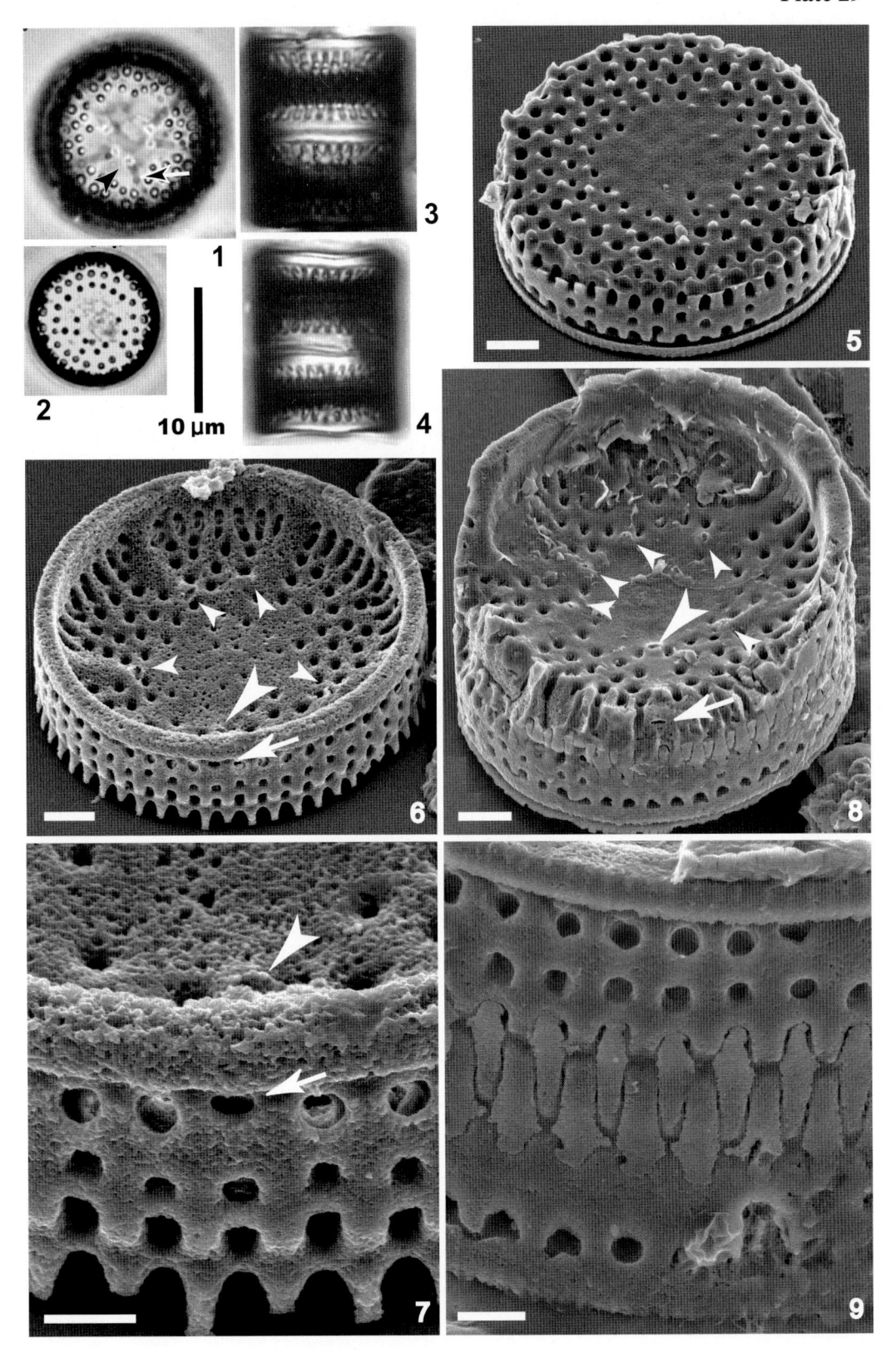

Plate 30. Centric diatoms: Biddulphiales
Hydrosera whampoensis (A.F. Schwarz) Deby 1891

基礎異名：*Triceratium whampoensis* A.F. Schwarz 1874

文　　献：Qi, Yu-zao, Reimer, C.W. & Mahoney, R.K. 1984. Taxonomic studies of the genus *Hydrosera*. 1. Comparative morphology of *H. triquetra* Wallich and *H. whampoensis* (Schwarz) Deby, with ecological remarks. In: Mann, D.G.(ed.) Proceedings of the 7th International Diatom Symposium, Philadelphia (22-27 August, 1982). pp. 213-224. Otto Koeltz, Koenigstein.

形　　態：殻は大形で光顕では三角形を2個組み合わせたような，大きな三角形の辺から別の三角形の頂部が見えているような殻面をしており，突き出ている三角部の先端には偽眼域が存在する．内側での観察では，突き出ている三角部と基本となる三角形とは偽隔壁で区切られている．基部の三角形には1(2)個の唇状突起がある．殻径約90 μm．

ノ　ー　ト：本種は従来 *Hydrosera triquetra* Wallich と同定されてきた（原口ら1998）．*H. triquetra* としてではあるが秋田県の八郎潟調整池から報告があり，本邦での北限とされる（加藤ら 1977）．

産　　出：吹上浜（泥炭層）：完新世（縄文時代）・鹿児島県

図　　版：吹上浜（泥炭層）標本．

Figs 1-6.　*Hydrosera whampoensis* (A.F. Schwarz) Deby
LM. Fig. 1. SEM. Figs 2-6. Material from the peat deposit of Fukiage-Hama (Holocene), Kagoshima, Kyushu, Japan. Scale bars: Figs 3-4=20 μm, Figs 2, 5=10 μm, Fig. 6=2 μm.

Fig. 1.　Whole valve view.
Fig. 2.　Oblique view of part of Fig. 4, rimoportulae (arrowheads).
Fig. 3.　External view of whole valve.
Fig. 4.　Internal view of whole valve, rimoportulae (arrowheads).
Fig. 5.　Enlarged oblique view of part of Fig. 3 showing a pseudocellus.
Fig. 6.　Enlarged view showing one rimoportula of Fig. 4.

Plate 30

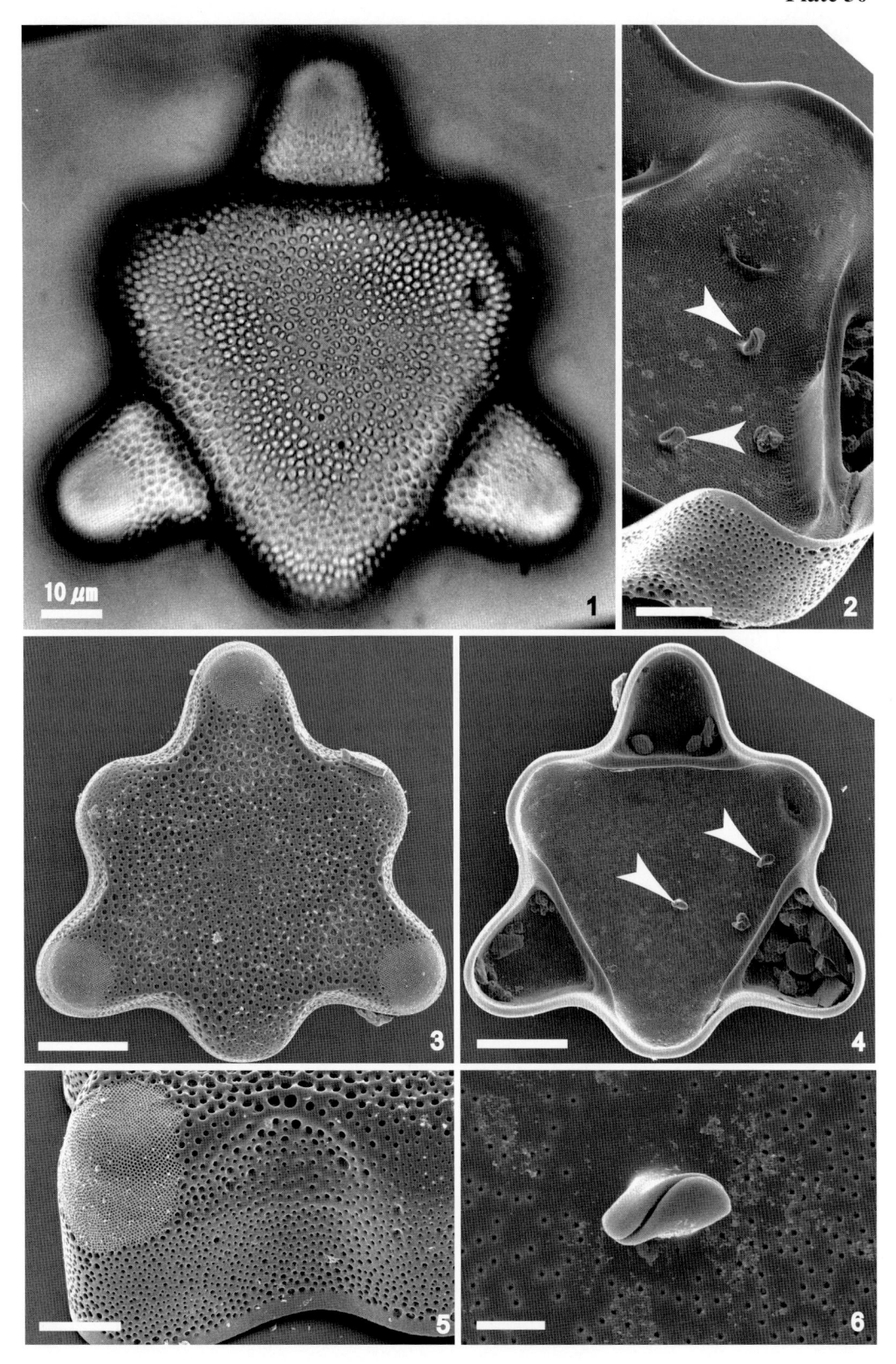

Plate 31. Centric diatoms: Biddulphiales
Stoermeria trifoliata (Cleve) Kociolek, Escobar & Richardson 1996

基礎異名：*Triceratium trifoliatum* Cleve 1881

文　献：Kociolek, J.P., Escobar, L. & Richardson, S. 1996. Taxonomy and ultrastructure of *Stoermeria*, a new genus of diatoms (Bacillariophyta). Phycologia **35**: 70-78.

形　態：殻面はほぼ平ら，中心には三角形の頂部が欠けた形の基部があり，その長辺から3方向に伸長している．伸長した部分はそれぞれ先端がさらに3方向に分岐する．全体の殻形はほぼ三角形である．表面は小刺と点紋で覆われている．伸長した部分には伸長方向へ配列した点紋が分布する．基部には点紋が非常に少ない．SEM観察によると胞紋は外側表面に師板がある．殻径42-100 μm．

ノート：宮田層産の分類群はBrun(1891)によって仙台の亜炭から記載された*Terpsinoe inflata*に類似するが，3方向へ伸長した部分がさらに3方向に分岐した中心の突出が太いこと，SEM観察により唇状突起が見当たらないことから，*Stoermeria*に所属する方が適切と思われる．*Stoermeria*属はアメリカ合衆国カリフォルニア州の鮮新世の地層から見出された種をタイプとして標記の文献で設立された．

産　出：宮田層：中新-鮮新世・秋田県．

図　版：宮田層標本．

Figs 1-5. *Stoermeria trifoliata* (Cleve) Kociolek, Escobar & Richardson
LM. Fig. 1. SEM. Figs 2-5. Material from Miyata Formation (Mio-Pliocene), Senboku City, Akita, Japan. Scale bars: Fig. 2＝10 μm, Figs 3-5＝5 μm.

Fig. 1.　Whole valve view.
Fig. 2.　Internal oblique view of whole valve.
Fig. 3.　External oblique view of whole valve.
Fig. 4.　Enlarged view of part of Fig. 3.
Fig. 5.　Enlarged view of part of Fig. 2 showing inner an apex.

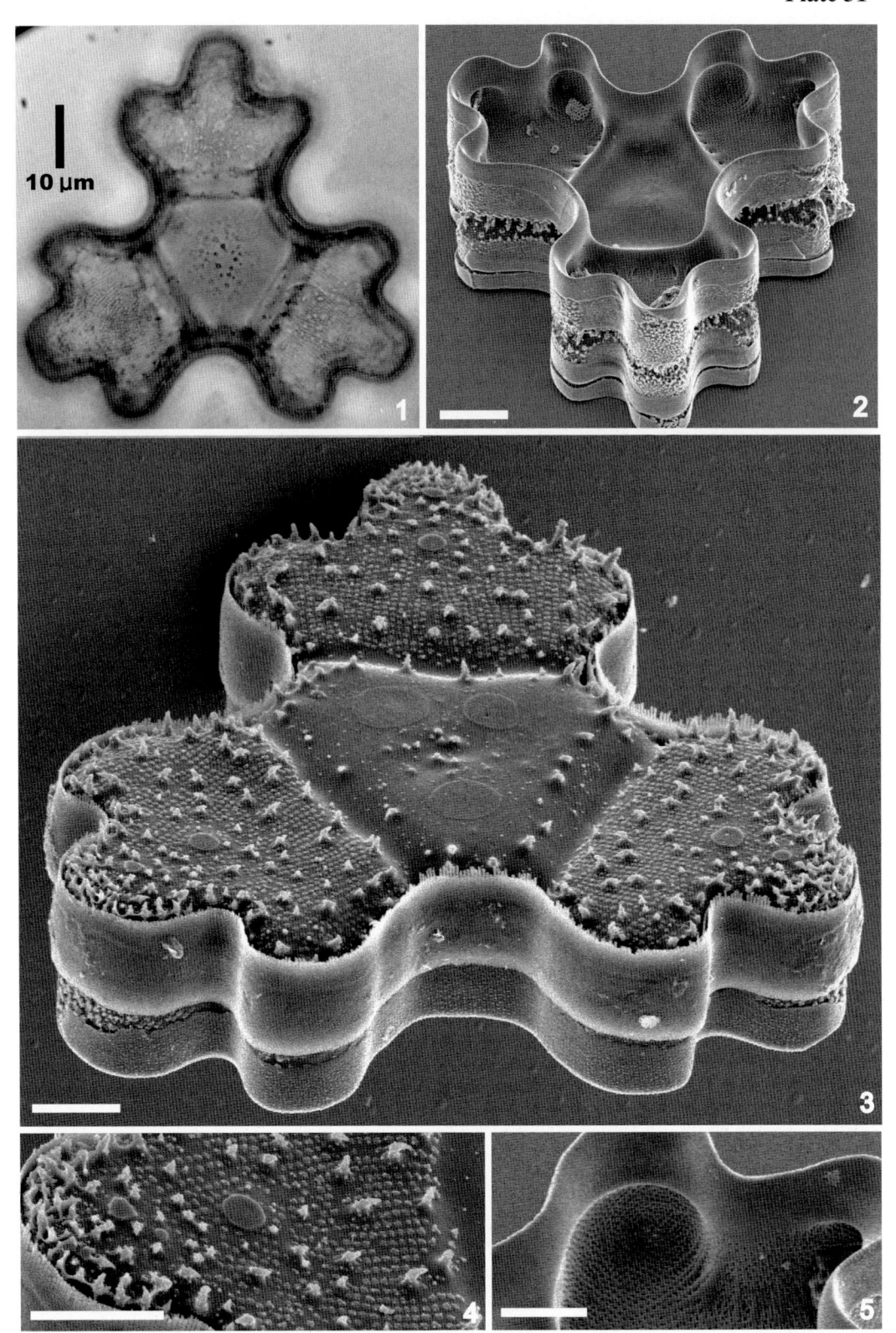

Plate 31

Plate 32.　Centric diatoms: Coscinodisales
Actinocyclus normanii f. *subsalsa* (Juhlin-Dannfelt) Hustedt 1957

基礎異名：*Coscinodiscus subsalsus* Juhlin-Dannfelt 1882

文　　献：Hustedt, F. 1957. Die Diatomeenflora des Fluss-systems der Weser im Gebiet der Hansestadt Bremen. Abhandlungen herausgegeben vom Naturwissenschaftlichen Verein zu Bremen **34**: 181-440, Taf. 1.

　　Naya, T., Tanimura, Y., Nakazato, R. & Amano, K. 2007. Modern distribution of diatoms in the surface sediments of Lake Kiraura, central Japan. Diatom **23**: 55-70.

形　　態：殻は円盤状で小形．殻径 15-28 μm，殻面には中心から殻縁に向かい束線状に点紋列が分布する，その数 10 μm に約 14 本．殻面中央では点紋が欠ける場合がある．唇状突起は 4-6 個／殻で，唇は殻面に平行であり，柄は長く伸張し LM でも観察することができる．偽節は殻外側では明瞭であるが内側では不明瞭である．胞紋は外側では平らな師板，内側ではドーム状の師板で覆われる．

ノート：本属に所属するか否かは偽節の有無，胞紋の閉塞および唇状突起の形態等によるが，本種は LM では偽節の確認が難しい，また内側は SEM によっても観察しづらい．南雲・安藤 (1984) によると，Hustedt (1928, 1957) は殻の大きさで *A. normanii* f. *normanii* と f. *subsalsa* を区分し，後者は前者より小形で殻径 25-40 μm である．また，f. *subsalsa* の基礎異名である *Coscinodiscus subsalsus* Juhl-Dannf. の殻径は 35-45 μm である (Juhlin-Dannfelt 1882)．この区分は大きさが異なるだけで他は同じであるとして品種を区分しない見解もある（たとえば Krammer & Lange-Bertalot 1991）．北浦・霞ヶ浦産の本種はこれらよりも小形であり，著者が調査を行った北浦および霞ヶ浦の個体はしばしば中心に点紋を欠き，小さいが無紋の中心域を形成している．無紋の中心域を考慮すると *A. normanii* s.l. とは同定しづらいが，中心に点紋のある個体とない個体の形態が連続していること，点紋のある個体の方が多いことを考慮し，最近の研究である Naya et al. (2007) に従って，北浦産の本分類群を *A. normanii* f. *subsalsa* に同定した．しかし，Hustedt による f. *normanii* および f. *subsalsa* の殻径よりかなり小さいことと，胞紋の内・外に師板があり *Actinocyclus octonarius* s.l. に似た形態の殻が観察できることから，再検討が必要であろう．および *Lobodiscus* Lupikina & Khursevich に属する可能性もある．

産　　出：北浦（Naya *et al.* 2007）・霞ヶ浦：茨城県（現生）．

図　　版：北浦（Figs 1-2, 4, 6-10），霞ヶ浦（Figs 3, 5）標本．

Figs 1-10.　*Actinocyclus normanii* f. *subsalsa* (Juhlin-Dannfelt) Hustedt
LM. Figs 1-5. SEM. Figs 6-10. Materials, Figs 1-2, 4, 6-10, from Lake Kita (Kita-ura) (Recent), Ibaraki, Japan and Figs 3, 5, from Lake Kasumiga-ura (Recent), Ibaraki, Japan. Scale bars: Fig. 9＝5 μm, Figs 6-7＝2 μm, Figs 8, 10＝1 μm.

Figs 1-5.　Five different size valves.
Fig. 1.　Rimoportula (arrowhead).
Fig. 6.　External view of whole valve, pseudonodulus (arrow).
Fig. 7.　Oblique view of a frustule, pseudonodulus (arrow), opening of rimoportula (arrowhead).
Fig. 8.　Enlarged view of part of Fig. 7, pseudonodulus (arrow), opening of rimoportula (arrowhead).
Fig. 9.　Internal oblique view of whole valve showing broken internal domed cribra, rimoportulae (arrowheads), note: usually broken internal veluma.
Fig. 10.　Detailed view of internal valve showing domed cribra.

Plate 32

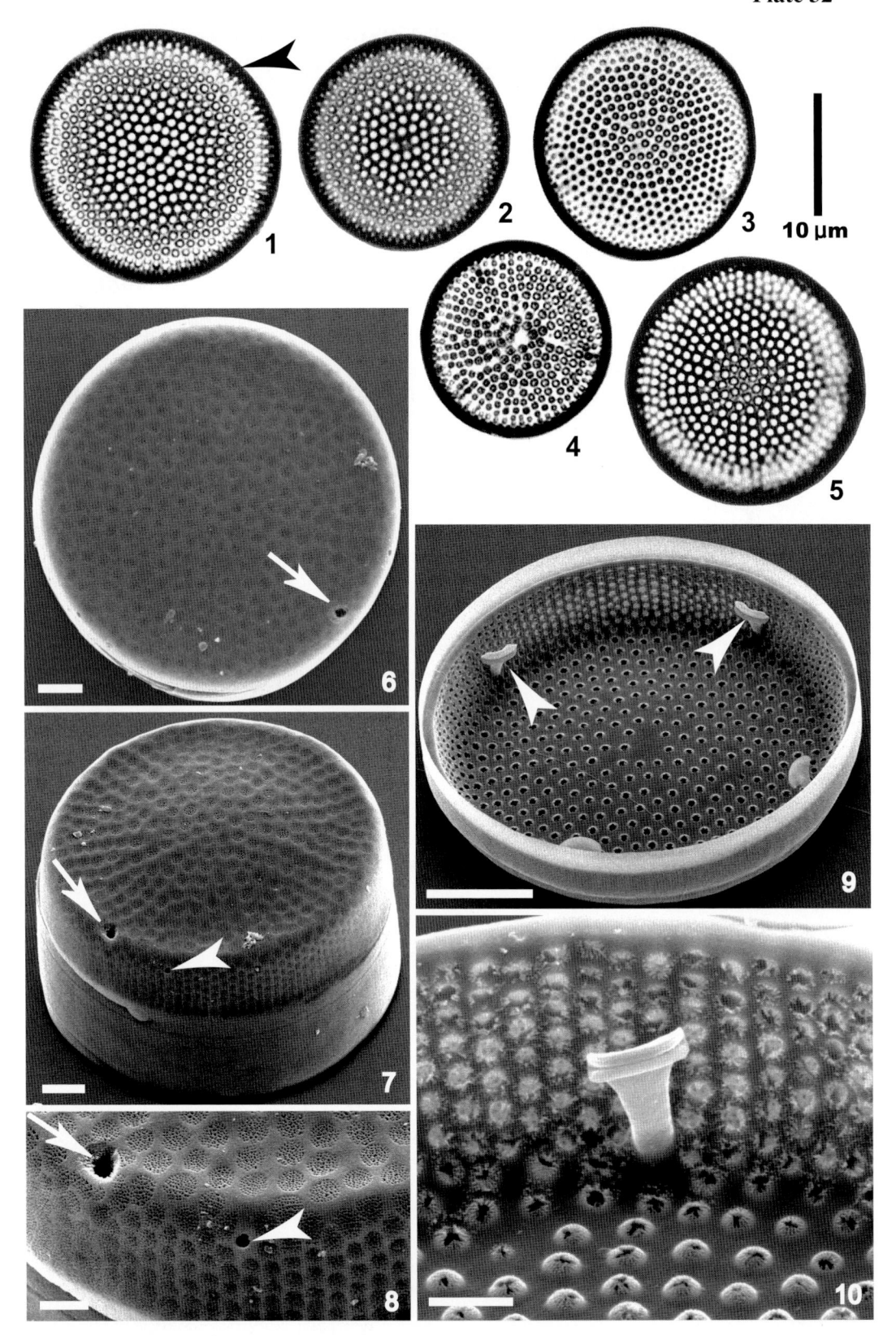

10 µm

Plates 33-34. Centric diatoms: Coscinodisales
Actinocyclus octonarius Ehrenberg 1838 s.l.

文　　献：Ehrenberg, C.G. 1838. Die Infusiosthierchen als volkommene Organismen. Leipzig. Hendey, N. I. 1964.　An introductory account of the smaller algae of British coastal waters. Part Ⅴ：Bacillariophyceae（Diatoms）317 pp. 45 pls. Her Majesty's Stationery Office, London.

形　　態：殻は円盤状，殻径 25-55 μm，殻面には中心から放射状に点紋列がある，点紋列は数を増しながら殻端に達する．そのうち殻面中心から走る 4-9 本は殻端に唇状突起が観察される．点紋列数 10 μm に 10-14 本，点紋列を構成する点紋は 10 μm に約 10 個であった．点紋の外側は平らな師板で覆われており，内側は破損している殻が多いがドーム状師板で覆われている．唇状突起の唇は殻面にほぼ平行であり，柄は伸張し LM でも観察することができる．偽節は小さく観察しづらい．

ノート：類似種に *Actinocyclus ehrenbergii* Ralfs があるが，本種の異名である（Hendey 1964）．胞紋の外・内側とも師板で覆われることや唇状突起の形態から *Lobodiscus* Lupikina & Khursevich と類似性がある．

　本種は最近の研究である Kozyrenko *et al.*(2008)によっても形態が広く示されているが，将来は細分化が必要と思われる．

産　　出：阿寒湖（河島・小林 1993）：北海道（現生），太田川および極楽寺山の溜水（岩橋 1935，*A. ehrenbergii* として）：広島県（現生），吹上浜（泥炭層）：完新世・鹿児島県．

図　版 33-34：吹上浜（泥炭層）標本．

Plate 33, Figs 1-4. *Actinocyclus octonarius* Ehrenberg s.l.
LM. Figs 1-4. Material from peat of Fukiagehama (Holocene), Kagoshima, Kyushu, Japan.

Figs 1-2, 3-4.　Two different size valves, at different focal planes.

Plate 33

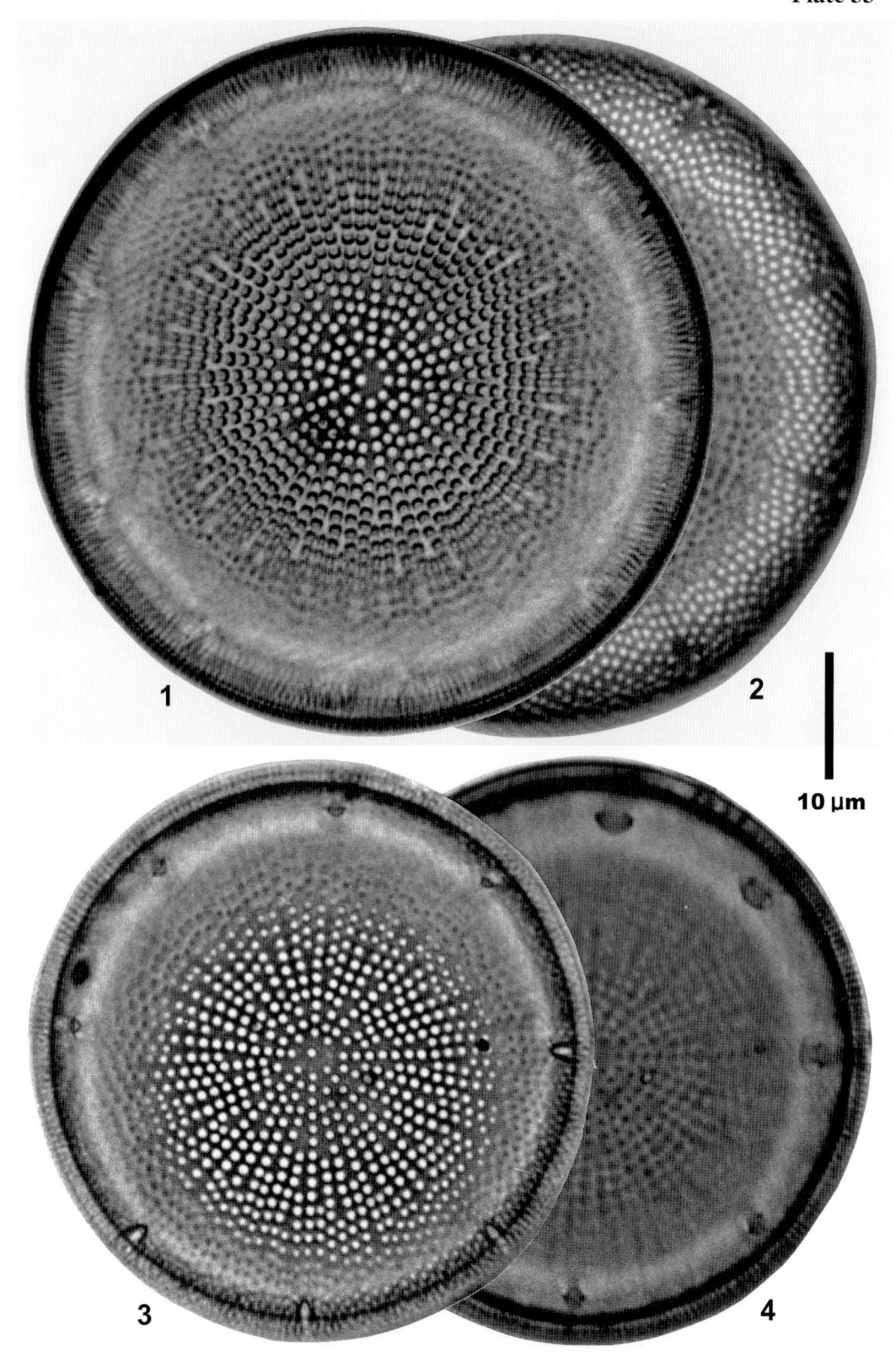

10 µm

Plate 34, Figs 1-6. *Actinocyclus octonarius* **Ehrenberg** s.l.
SEM. Figs 1-6. Material from peat of Fukiagehama (Holocene), Kagoshima, Kyushu, Japan.
Scale bars: Fig. 1=10 μm, Figs 2-3, 5=5 μm, Figs 4, 6=1 μm.

Fig. 1. External oblique view of whole valve.
Fig. 2. Enlarged view of part of Fig. 1 showing opening of pseudonodulus (arrow) and openings of rimoportulae (arrowheads).
Fig. 3. Internal view of whole valve: rimoportulae (arrowheads).
Fig. 4. Enlarged view of part of Fig. 1 showing external veluma of areolae.
Fig. 5. Internal oblique view of whole valve.
Fig. 6. Detailed view of valve margin showing inner veluma (broken) and rimoportula.

Plate 34

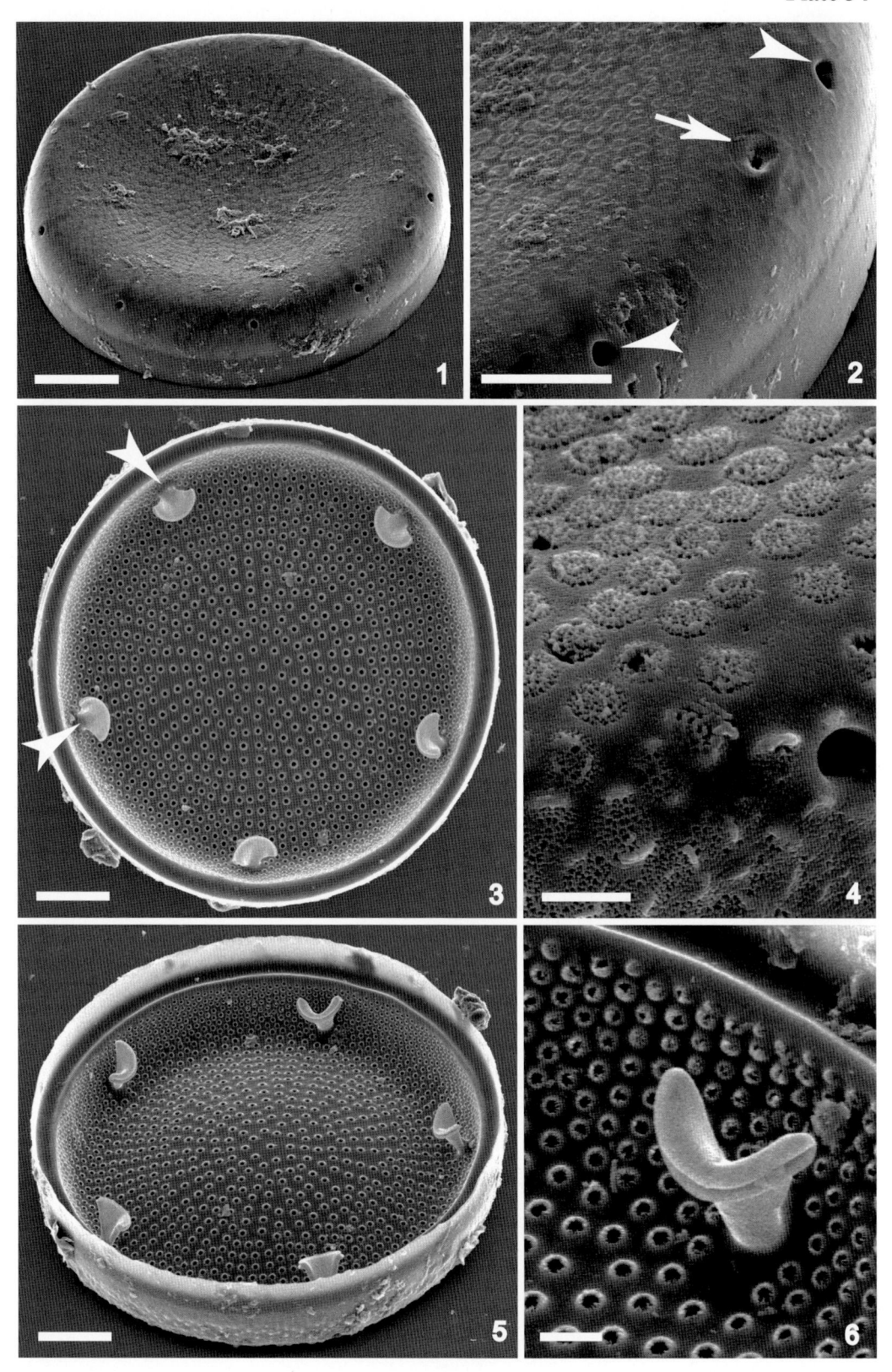

Plate 35.　Centric diatoms: Orthoseirales
Orthoseira asiatica (Skvortsov) H. Kobayasi in Mayama *et al.* 2002

基礎異名：*Melosira roseana* var. *asiatica* Skvortsov 1938

文　　献：Skvortsov, B.W. 1938. Subaerial diatom from Pin-Chiang-Sheng Province, Manchoukuo. Philippine Journal of Sciences **65**: 263-281, 4 pls.
　　Mayama, S., Idei, M., Osada, K. & Nagumo, T. 2002. Normenclatural changes for 20 diatom taxa occurring in Japan. Diatom **18**: 89-91.

形　　態：被殻は円筒形で，殻径 7.5-19 μm．中心部には 2-3 個の孔状構造がある．殻面から殻縁にかけて外側が板状に肥厚し，放射状の仕切りを形成している．仕切りは，殻縁近くの殻套で二叉して殻縁に達する．殻面／殻套境界では仕切りから刺が伸長している．仕切りは 10 μm に 5-7 本，光学顕微鏡での観察では仕切りは間条線が肥厚しているように思えるが，内側では明瞭ではない．仕切りの間には殻中心付近で 1(2)列，殻面／殻套境界では約 3 列の胞紋列がある．

ノ ー ト：群馬県の野殿層から，中島・南雲(1999)により *Orthoseira* sp. として報告・図示されている分類群は本種と思われる．

産　　出：神流川（上流）：群馬県（現生），橋立鍾乳洞（小林ら 2006）：埼玉県（現生）．

図　　版：神流川標本．

Figs 1-8.　*Orthoseira asiatica* (Skvortsov) H. Kobayasi

LM. Figs 1-2. SEM. Figs 3-8. Material from Kanna River (Recent), Gunma, Japan. Scale bars: Fig. 8＝5 μm, Figs 3, 5-7＝2 μm, Fig. 4＝1 μm.

Fig. 1.　Whole valve view.
Fig. 2.　Girdle view of frustule.
Fig. 3.　External view of whole valve.
Fig. 4.　Enlarged view of Fig. 3 showing central area.
Fig. 5.　Oblique view of Fig. 3.
Fig. 6.　Internal view of whole valve.
Fig. 7.　Oblique view of Fig. 6.
Fig. 8.　Enlarged view of central area of Fig. 6.

Plate 35

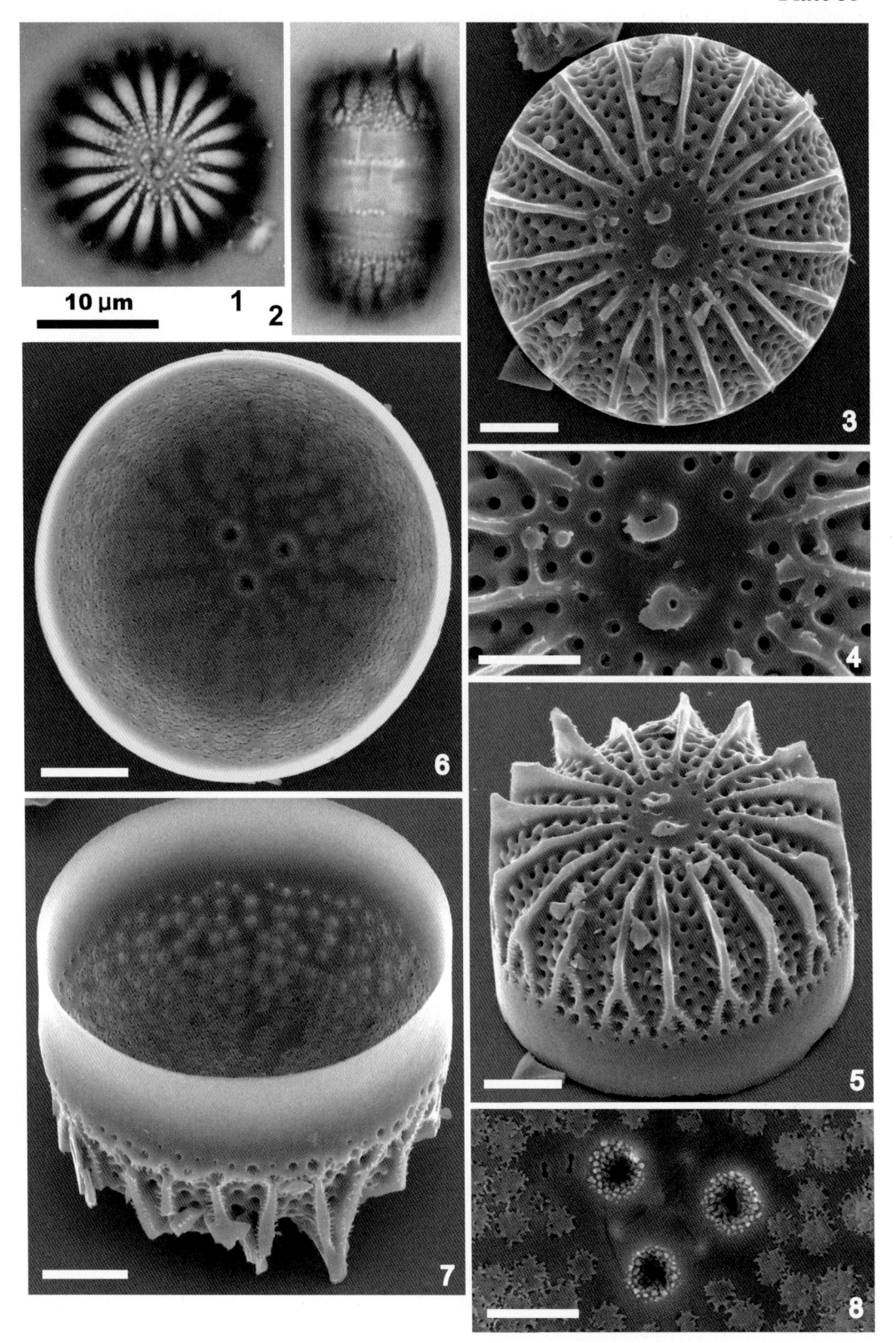

Plate 36.   Centric diatoms: Melosirales
Melosira undulata (Ehrenberg) Kützing var. *undulata* 1844

基礎異名：*Gallionella undulata* Ehrenberg 1840
文　　献：Kützing, F. T. 1844. Die kieselschaligen Bacillarien oder Diatomeen. 152 pp. Tafs 1-30. Nordhausen.
形　　態：殻は大形の円筒形，図に使用した人形峠層，大鷲湖沼性堆積物からの個体は，殻径 20-78 μm，殻套長 15-20 μm，殻面の中心には小円形に点紋が不規則に配列し，その小円から点紋列が放射状に直線～らせん状で殻縁へ配列する．殻套はほぼ直線的に点紋が配列する．殻面・殻套とも表面に顆粒が観察される．結合針は先広で（Fig. 5 矢印），しっかり組み合わさっているが，分離針は先頭形である（Fig. 6 矢印）．
ノ ー ト：殻面に唇状突起が存在するもの（田中ら 2011）と，ないもの（田中ら 2008）が観察される．これらを同じ種としてよいかどうか検討したが，今までの研究者が同じ分類群としているのでこれに倣った．
産　　出：中新世から現生まで多くの産出報告がある．太櫓層（瀬棚の珪藻土）：前期中新世・北海道，長者原層（石田ら 1970）：中期中新世・長崎県，人形峠層（田中ら 2008）：後期中新世～鮮新世・岡山-鳥取県境，伊賀層（Tanaka *et al.* 1984, 田中・松岡 1985）：鮮新世・三重県，蒲生層（Negoro 1981）：前期更新世・滋賀県，大鷲湖沼性堆積物（田中ら 2011）：鮮新世・岐阜県．
図　　版：人形峠層（Fig. 1），大鷲湖沼性堆積物（Figs 2-6）標本．

Figs 1-6.   *Melosira undulata* (Ehrenberg) Kützing var. *undulata*
LM. Figs 1-2. SEM. Figs 3-6. Materials, Fig. 1, from Ningyo-Toge Formation (Late Miocene-Pliocene), boundary of Okayama and Tottori Prefectures, Japan and Figs 2-6, from lacustrine deposit of Owashi (Pliocene), Gujo City, Gifu, Japan. Scale bars: Figs 3-4=10 μm, Figs 5-6=2 μm.

Figs 1-2.   Two different size valves.
Fig. 3.   External oblique view of valve.
Fig. 4.   Internal oblique view of valve, rimoportula (arrowhead).
Fig. 5.   Detailed view of mantle showing linking spines (arrows).
Fig. 6.   Detailed view of mantle of separating valve, spine (arrow).

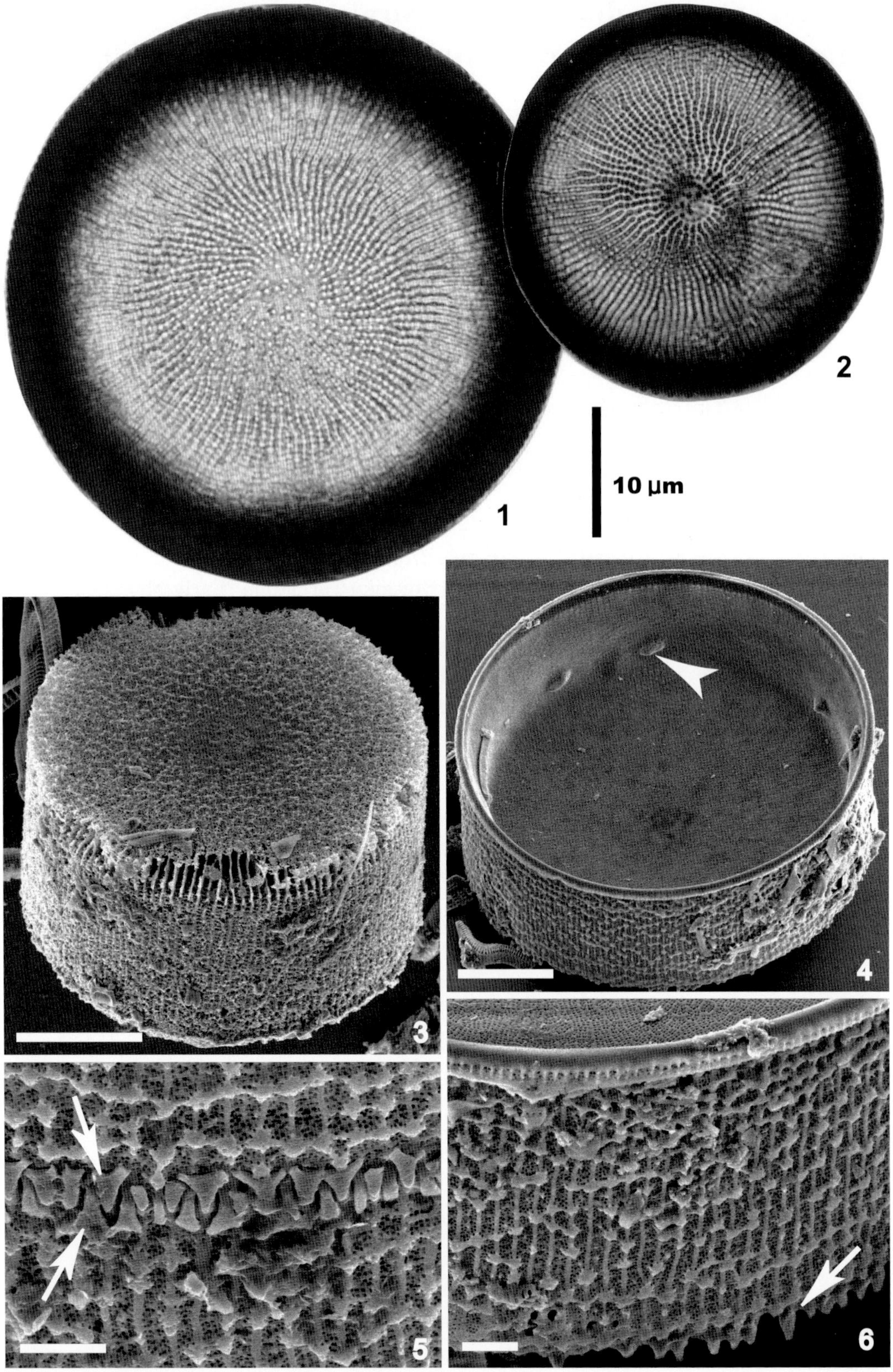

Plate 36

10 μm

Plate 37.　Centric diatoms: Melosirales
Melosira undulata var. *producta* A. Schmidt in A. Schmidt *et al.* 1892

文　　献：Schmidt, A. 1892. *In*: A. Schmidt's Atlas der Diatomaceen-Kunde(1874-1959), pl. 180, Fig. 18. O.R. Reisland, Leipzig.

形　　態：殻は比較的長い円筒形，殻径 14-39 μm，殻套長 22-33 μm，殻面の中心には小円形に点紋が，少ないあるいは観察しづらい部分があるが，その小円から点紋列が放射状に直線（らせん）状で殻縁へ走る．殻套の点紋列は，ゆるく右へカーブすることが多いが，直線的のこともある．10 μm に約 10 本．殻面・殻套とも外側表面に小顆粒が観察される．結合針は先広で組み合わさっている．分離針は先頭形である．殻面に近い殻套内側に唇状突起が円状に配列しているが，殻面中心付近にも所在することが多い．

ノ　ー　ト：基本種（var. *undulata*）と同じく殻面に唇状突起が存在するものとないものが観察できた．殻套長は，殻の直径とほぼ同じか殻径より長い（小形の殻になるほどこの傾向は顕著である）．

産　　出：紫竹層：前期中新世・福島県．

図　　版：紫竹層標本．

Figs 1-7.　*Melosira undulata* var. *producta* A. Schmidt

LM. Figs 1-3. SEM. Figs 4-7. Material from Shichiku Formation (Early Miocene), Iwaki City, Fukushima, Japan. Scale bars: Figs 4, 7＝5 μm, Figs 5-6＝2 μm.

Fig 1.　Valve view showing radially arranged puncta rows.
Figs 2-3.　Girdle view showing different size valves, rimoportula (arrowhead).
Fig. 4.　External girdle view of mantle.
Fig. 5.　Enlarged oblique view of valve face of Fig. 4, openings of rimoportulae (arrowheads).
Fig. 6.　Detailed view of external surface of mantle and linking spines (arrows).
Fig. 7.　Oblique view of valve face of opposite side of Fig. 4.

Plate 37

10 μm

Plate 38.   Centric diatoms: Melosirales
Melosira varians C. Agardh 1827

文　　献：Crawford, R.M. 1978. The taxonomy and classification of the diatom genus *Melosira* C.A. Ag. *Melosira lirata* (Dillw) C.A. Ag. and *M. varians* C.A. Ag. Phycologia **17**: 237-250.

形　　態：殻は円筒形で殻径 33-40 μm, 殻面外側には顆粒および不定形の小針が多数分布しているが, 特に中心付近には密に所在する. 殻套はほとんど顆粒である. 唇状突起は殻面および殻套に散在する. 胞紋は殻全体に密に分布している.

ノ ー ト：珪藻群集の産出表に記されていても, 図・写真が掲載されている報告は少ない. 産出リストへの記載のみであるが中新世の山戸田層 (Ichikawa *et al*. 1955), 投石堂凝灰角礫岩層 (赤木ら 1984) で記載があるので, 中新世には日本に分布していたと考えられる.

産　　出：瓜生坂層 (窪田ら 1976)：前期更新世・長野県, 吹上浜 (泥炭層)：完新世・鹿児島県.

図　　版：吹上浜 (泥炭層) 標本.

Figs 1-7.   *Melosira varians* C. Agardh

LM. Fig. 1. SEM. Figs 2-7. Material from the peat deposit of Fukiage-Hama (Holocene), Kagoshima, Kyushu, Japan. Scale bars: Figs 2-3=10 μm, Figs 4, 6=2 μm, Fig. 5=1 μm, Fig. 7=0.5 μm.

Fig. 1.   Whole valve view.
Fig. 2.   External oblique view of valve.
Fig. 3.   Internal oblique view of valve, rimoportula (arrowhead).
Fig. 4.   Enlarged view of part of mantle of Fig. 2.
Fig. 5.   Detailed view of part of valve face, large hole (arrow).
Fig. 6.   Enlarged view of part of Fig. 3, rimoportula (arrowhead).
Fig. 7.   Detailed view of a rimoportula (arrowhead).

Plate 38

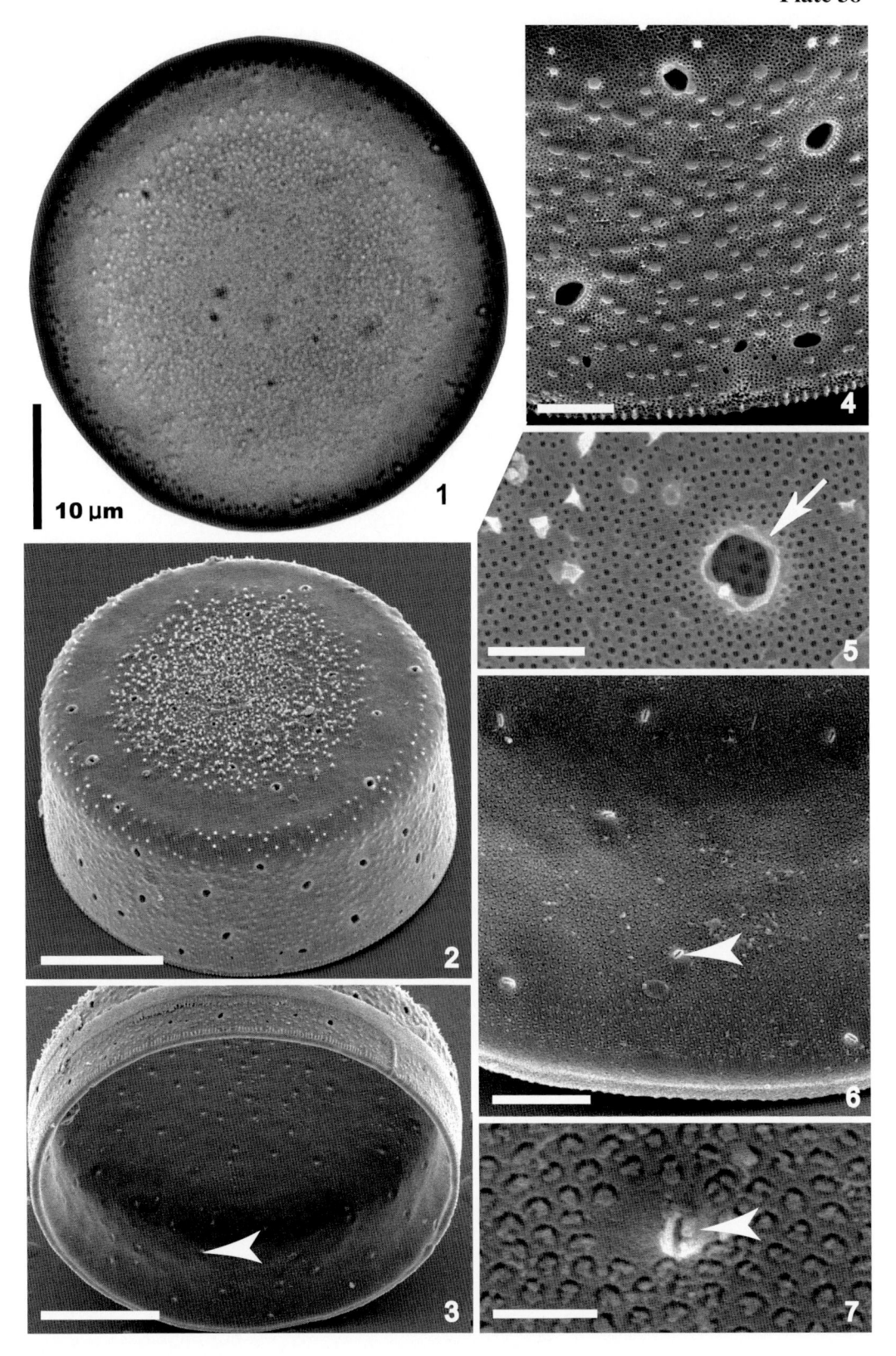

10 μm

Plate 39.　Centric diatoms: Paraliales
Ellerbeckia arenaria (Moore) R.M. Crawford f. *arenaria* 1988

基礎異名：*Melosira arenaria* Moore 1843

文　　献：Crawford, R.M. 1988. A reconsideration of *Melosira arenaria* and *M. teres*; resulting in a proposed new genus *Ellerbeckia*. *In*: Round, F.E. (ed.) Algae and the aquatic environment. pp. 413-433. Biopress, Bristol.

形　　態：殻は大形で殻面は円形，殻径 40-70 μm，殻面には殻面／殻套境界から中心に向かう放射状のすじ模様が観察される．殻套長 10-15 μm．SEM 観察によると殻面はわずか凹または凸状で，放射状模様を構成する刻み目と不規則な形の殻面／殻套境界にある針で二つの殻が接着している．殻套には胞紋が貫殻軸方向へ並んで密に分布する．また，小孔が殻ごとに多くは二重にリング状に分布し，内側の管状突起へ連続している．殻套内側には貫殻軸に平行な多数のすじが分布するが，管状突起はこの2本のすじの間に立ち上がり，上部には2個の微小孔がある．胞紋の内側開口も2本のすじの間に存在する．すじはときおり方向が乱れるのが観察された．

ノ　ー　ト：本分類群は日本で化石での産出記録はあるが，図が添えられていても，帯面観では f. *teres* との区別ができない．著者が調査した化石はすべて f. *teres* だったので f. *arenaria* は化石としては産出していない可能性が強い．現生種では山中湖のみから見出すことができた．

産　　出：山中湖：山梨県（現生）．

図　　版：山中湖標本．

Figs 1-7.　*Ellerbeckia arenaria* (Moore) R.M. Crawford f. *arenaria*
LM: Figs 1-2. SEM: Figs 3-7. Materials from Lake Yamanaka (Recent), Yamanakako Village, Yamanashi, Japan. Scale bars: Figs 1-2, 5-6＝10 μm, Figs 4, 7＝2 μm, Fig. 3＝0.5 μm.

Figs 1-2.　Girdle views.
Fig. 3.　Detailed internal oblique view of tube-process (arrow).
Fig. 4.　External mantle view showing spines and opening of tube-process (arrow).
Fig. 5.　External view of valve chain.
Fig. 6.　Broken valve showing internal view of valve face and part of external valve face of sibling valve.
Fig. 7.　Oblique enlarged view of Fig. 6 showing inner mantle, tube-processes (arrows).

Plate 39

Plates 40–41. Centric diatoms: Paraliales
Ellerbeckia arenaria f. *teres* (Brun) R.M. Crawford 1988

基礎異名：*Melosira teres* Brun 1892

文　　献：Crawford, R.M. 1988. A reconsideration of *Melosira arenaria* and *M. teres*; resulting in a proposed new genus *Ellerbeckia*. *In*: Round, F.E.(ed.) Algae and the aquatic environment. pp. 413-433. Biopress, Bristol.

形　　態：殻は大形で円筒形，殻面はゆるく凹または凸状で，縁には殻が組み合う放射状の刻み目がある．殻径38-110 μm，殻套長20-52 μm．殻套には胞紋が貫殻軸方向へ直線状に密に分布する．また，小孔がふつう1殻につき二重にリング状に分布し，内側の管状突起へ連続している．殻套内側には貫通軸に平行で直線状に多数のすじがあるが，しばしば部分的に方向が乱れることがあり，甚だしい場合は盛り上がりをすることもある．このすじ4本から1個の管状突起が立ち上がり，すじは管状突起を支持するように管状突起の外側へ殻套から連続する，管状突起の上部には微小孔がある．

ノ ー ト：本種はCrawford(1988)が*Ellerbeckia*属の設立と同時に，同じ群体に*E. arenaria*と*E. teres*の両者の特徴を有する分類群があることから，*E. teres*を*E. arenaria*の品種にすることを提案したものである．本邦の化石からは*E. teres*としての報告は見当たらず，*E. arenaria* f. *teres*として，田中・小林(1995, 1996)，南雲・田中(2001)，Tanaka *et al.*(2004)，田中ら(2008)などから報告されている．現生では河島・小林(1993)に詳しく記述がある．

産　　出：太櫓層（瀬棚の珪藻土）：前期中新世・北海道，蜂屋層：前期中新世・岐阜県，六角沢層：前期中新世・青森県，宮田層：中新-鮮新世・秋田県，人形峠層：後期中新世〜鮮新世・岡山-鳥取県境，香坂礫岩層：鮮新世・長野県，津房川層：鮮新世・大分県，大鷲湖沼性堆積物：鮮新世・岐阜県，人吉層：後期鮮新世・熊本県，円田珪藻土：前期更新世・宮城県，渋沢湖沼性堆積物：更新世・長野県，小野上層：前期更新世・群馬県，野原層：前期更新世・大分県，鬼首層：後期更新世・宮城県．

図　　版40：人吉層（Fig. 1），小野上層（Figs 2-3），野原層（Fig. 4），香坂礫岩層（Fig. 5）標本．

Plate 40, Figs 1-5. *Ellerbeckia arenaria* f. *teres* (Brun) R.M. Crawford

Stereomicroscope. Fig. 1. LM. Figs 2-5. Materials, Fig. 1, from Hitoyoshi Formation (Late Pliocene), Hitoyoshi City, Kumamoto, Japan, Figs 2-3, from Onogami Formation (Early Pleistocene), Shibukawa City, Gunma, Japan, Fig. 4, from Nobaru Formation (Early Pleistocene), Kitsuki City, Oita, Japan and Fig. 5, from Kohsaka Conglomerate Member (Pliocene), Saku City, Nagano, Japan. Scale bar: Fig. 1=200 μm.

Fig. 1.　Stereomicroscopic assemblage view.
Figs 2-4.　Girdle views. Tube-processes visible as black points (arrows).
Fig. 5.　Oblique view of valve face with radial markings at valve rim (relief or intaglio structure).

Plate 40

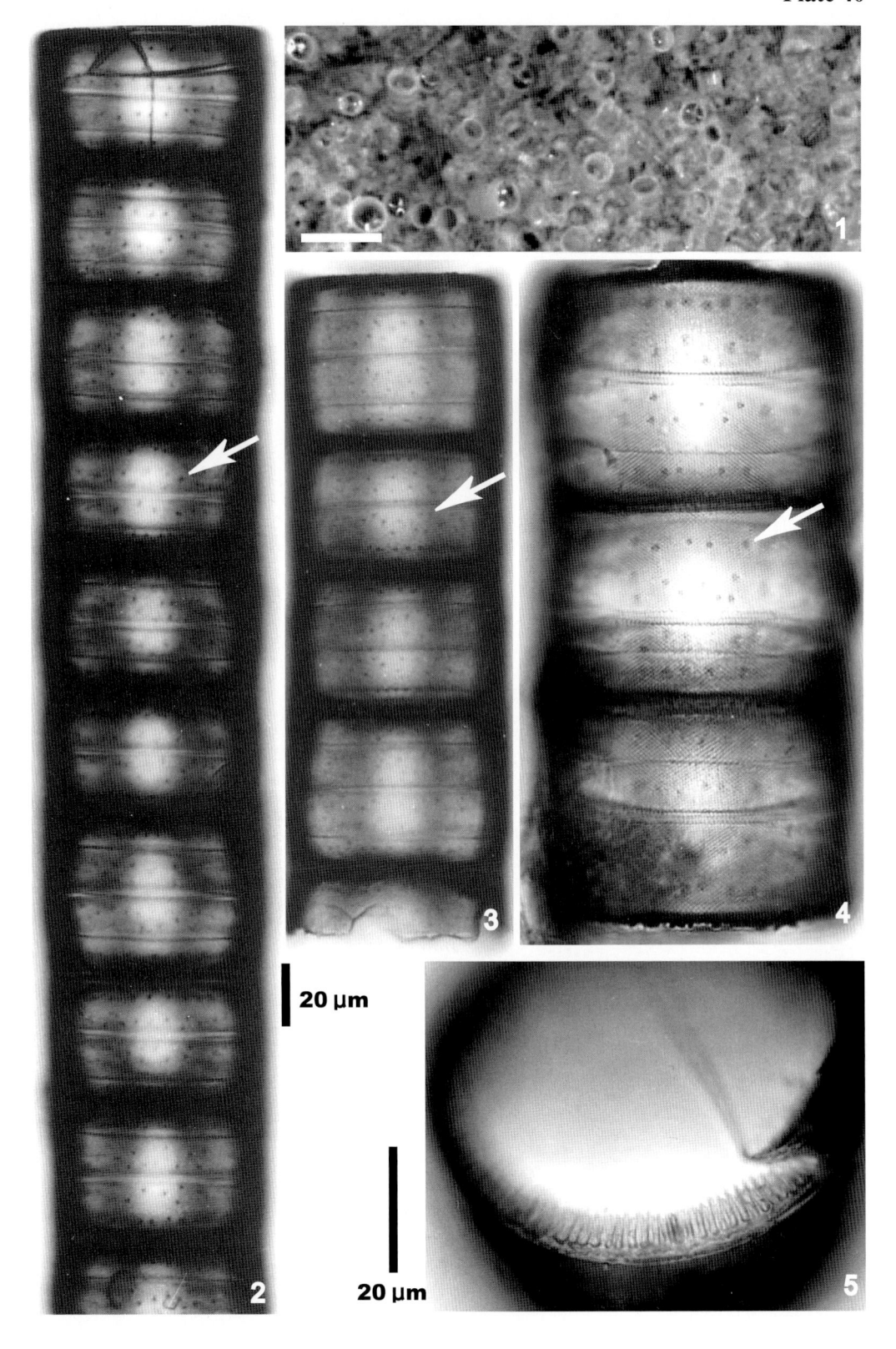

20 μm

20 μm

図　版 41：人形峠層（Figs 1, 4），野原層（Figs 2-3, 5），宮田層（Fig. 6）標本．

Plate 41, Figs 1-6. *Ellerbeckia arenaria* **f.** *teres* **(Brun) R.M. Crawford**

SEM. Figs 1-6. Materials, Figs 1, 4, from Ningyo-toge Formation (Late Miocene-Pliocene), boundary of Okayama and Tottori Prefectures, Japan, Figs 2-3, 5, from Nobaru Formation (Early Pleistocene), Kitsuki City, Oita, Japan and, Fig. 6, from Miyata Formation (Mio-Pliocene), Senboku City, Akita, Japan. Scale bars: Figs 1, 3＝10 μm, Figs 2, 4＝5 μm, Fig. 6＝2 μm, Fig. 5＝0.5 μm.

Fig. 1.　Oblique view of intaglio valve.
Fig. 2.　External oblique view of mantle.
Fig. 3.　Oblique view of intaglio valve face.
Fig. 4.　Internal oblique view of mantle showing tube-processes (arrows) and ribs.
Fig. 5.　Detailed view of a tube-process with internal apertures (arrows).
Fig. 6.　Internal view of mantle with irregular ribs (arrow).

Plate 41

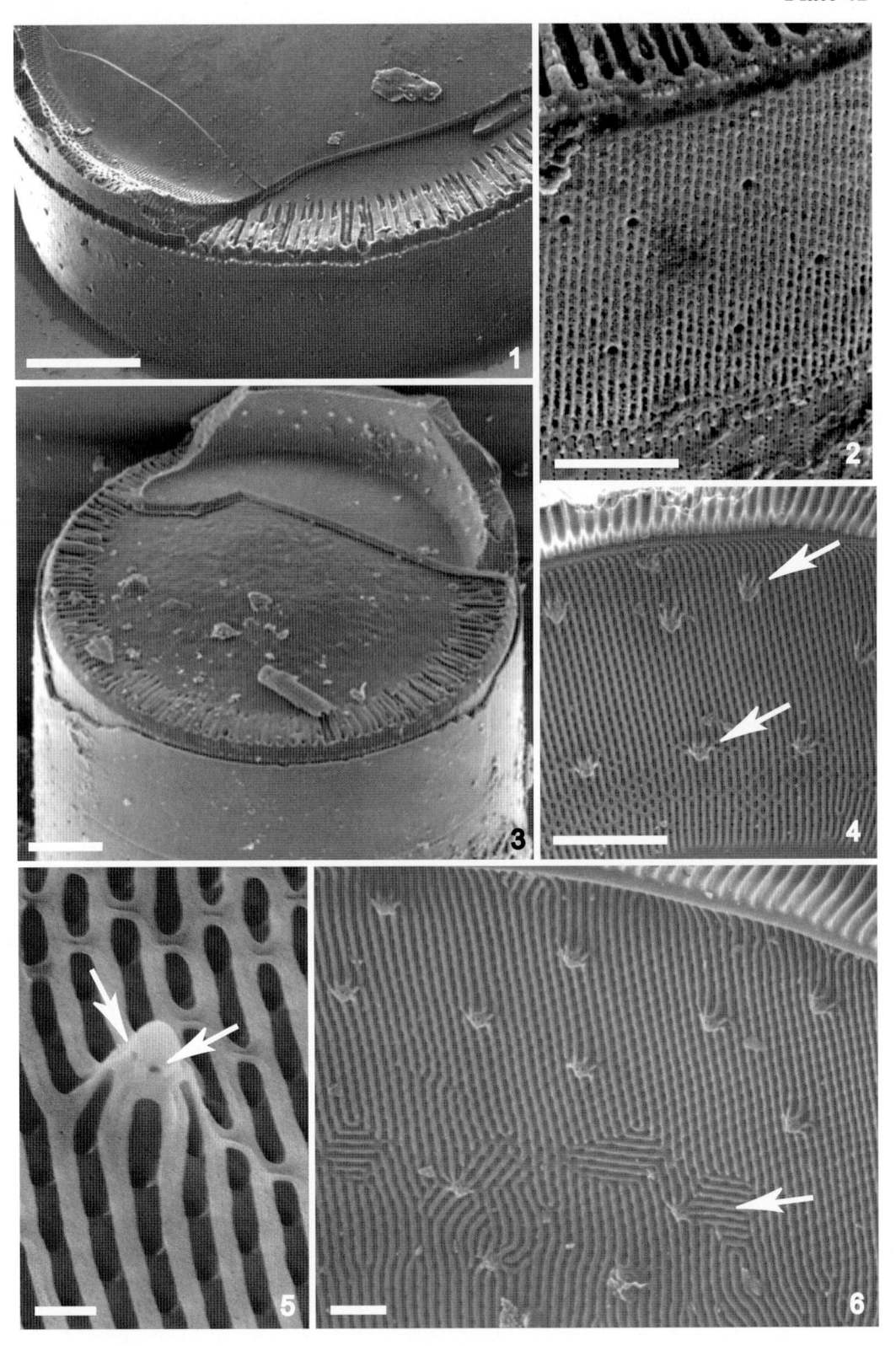

Plate 42. Centric diatoms: Thalassiosirales
Cyclostephanos costatilimbus (H. Kobayasi & Kobayashi) Stoermer, Håkansson & Theriot 1987

基礎異名：*Stephanodiscus costatilimbus* H. Kobayasi & Kobayashi 1986

文　献：Kobayasi, H. & Kobayashi, H. 1986. Fine structure and taxonomy of the small and tiny *Stephanodiscus* (Bacillariophyceae) species in Japan 4. *Stephanodiscus costatilimbus* sp. nov. The Japanese Journal of Phycology **34**: 8-12.

　　Stoermer, E.F., Håkansson, H. & Theriot, E.C. 1987. *Cyclostephanos* species newly reported from North America: *C. tholiformis* sp. nov. and *C. costatilimbus* comb. nov. British Phycological Journal **22**: 349-358.

形　態：殻面はほぼ平らで，殻中心から束線が放射状に分布している．殻径7-11 μm，束線数は殻面／殻套境界で 10 μm に 14-18 本である．SEM 観察によると中心付近には 1 個の殻面有基突起が所在し，束線・間束線はともに殻面から殻套へ延長し殻端で終了する．一般的に束線は殻面／殻套境界で 2 列の胞紋列であるが，殻套では 3 列になる．殻面／殻套境界のすべての間束線には針がある．3-8 本（非常にまれに 1）ごとの針の下には殻套有基突起の開口，およびやや殻面よりには 1 個の唇状突起の開口がある．内側では殻面の胞紋は平ら～ドーム状の師板で覆われる．殻縁には間束線内側が肥厚した肋があり，殻套有基突起と唇状突起は殻套上部で肋に所在する．殻面有基突起・殻套有基突起とも付随孔は 2 個である．

ノート：Kobayasi & Kobayashi(1986)では明確にタイプ試料として指定していないが，使用した試料は八郎潟調整池サンプル番号 N-1005 と記してある．筆者はこの N-1005 試料を用いて観察を行い *C. costatilimbus* と同定できるいくつかの殻を見出したが，いずれも殻套有基突起の付随孔数は 2 個であった．Kobayasi & Kobayashi(1986)の記載文では 3 個と記されているが，同論文の図では 2 個である．

産　出：八郎潟調整池（Kobayasi & Kobayashi 1986）：秋田県（現生）．

図　版：八郎潟調整池標本．

Figs 1-7. *Cyclostephanos costatilimbus* (H. Kobayasi & Kobayashi) Stoermer, Håkansson & Theriot

LM. Figs 1-2. SEM. Figs 3, 5-7. TEM. Fig. 4. Material, N-1005, from Hachirogata Regulation Pond (Recent), Katagami City, Akita, Japan, housed in Nagumo Laboratory. Scale bars: Fig. 3=2 μm, Figs 4-5=1 μm, Figs 6-7=0.5 μm.

Figs 1-2.　Two different size valves.
Fig. 3.　External oblique view of broken valve.
Fig. 4.　Part of valve, valve face fultoportula (arrow).
Fig. 5.　Enlarged view of valve margin, openings of mantle fultoportulae (arrows).
Fig. 6.　Detailed internal oblique view of valve face, valve face fultoportula with two satellite pores (arrow).
Fig. 7.　Detailed internal view of valve margin, rimoportula (arrowhead) and mantle fultoportula with two satellite pores (arrow).

Plate 42

10 µm

Plate 43.　Centric diatoms: Thalassiosirales
Cyclostephanos dubius (Fricke) Round in Theriot *et al.* 1987

基礎異名：*Cyclotella dubius* Fricke 1900
文　　献：Round, F.E. 1982.　*Cyclostephanos*—a new genus within the Sceletomemaceae. Archiv für Protistenkunde **125**: 323-329.
　　　Theriot, E., Håkansson, H., Kociolek, J.P., Round, F.E. & Stoermer, E.F. 1987.　Validation of the centric diatom genus name *Cyclostephanos*. British Phycological Journal **22**: 345-347.
形　　態：殻面は強く同心円状に凹凸し，殻径 6.5-27 µm，殻中心から束線が放射状に分布し，その数は殻面／殻套境界で 10 µm に 10-12 本である．SEM 観察によると中心付近には 1-4 個の殻面有基突起があり，殻面／殻套境界の外側間束線上には 2-3 本ごとに針がある．針の下には殻套有基突起，針はないが間束線上で殻套有基突起よりやや殻面よりには 1 個の唇状突起の開口がある．内側では殻面の胞紋はドーム状篩板で覆われる．殻縁には間束線内側が肥厚した肋があり，殻套有基突起と唇状突起が所在する．殻面有基突起・殻套有基突起とも付随孔は 2 個である．
ノ ー ト：*Cyclostephanos* 属は Round(1982) で提案になったが不備があり，Theriot *et al.* (1987) で正式な属名になった．今まで本邦から報告された本種の殻面有基突起数は 1 個であるが，ここで示す網走湖（ボーリングコア）の分類群は 1-4 個である．しかし Håkannson (2002) は殻面有基突起の数は 1-数個と記しており，6 個の写真も示しているので，網走湖底ボーリングコアの分類群は *C. dubius* に同定された．
産　　出：野殿層（中島・南雲 1999）：中期更新世・群馬県，網走湖底ボーリングコア：更新世・北海道．
図　　版：網走湖底ボーリングコア標本．

Figs 1-13.　*Cyclostephanos dubius* (Fricke) Round

LM. Figs 1-8. SEM. Figs 9-13. Material from boring core of Lake Abashiri, -10 meters (Pleistocene), Abashiri City, Hokkaido, Japan. Scale bars: Figs 9-10＝2 µm, Figs 11-13＝1 µm.

Figs 1-8.　Four different size valves.
Figs 1-2, 3-4, 5-6, 7-8.　Four valves at different focal planes.
Fig. 9.　External view of whole concave valve, opening of valve face fultoportula (arrow).
Fig. 10.　Oblique view of Fig. 9, opening of valve face fultoportula (arrow).
Fig. 11.　Enlarged view of mantle of Fig. 10, opening of rimoportula (arrowhead), opening of mantle fultoportula (arrow).
Fig. 12.　Internal oblique view of valve, valve face fultoportula with two satellite pores (arrow).
Fig. 13.　Enlarged view of Fig. 12, rimoportula (arrowhead), mantle fultoportulae with two satellite pores (arrows).

Plate 43

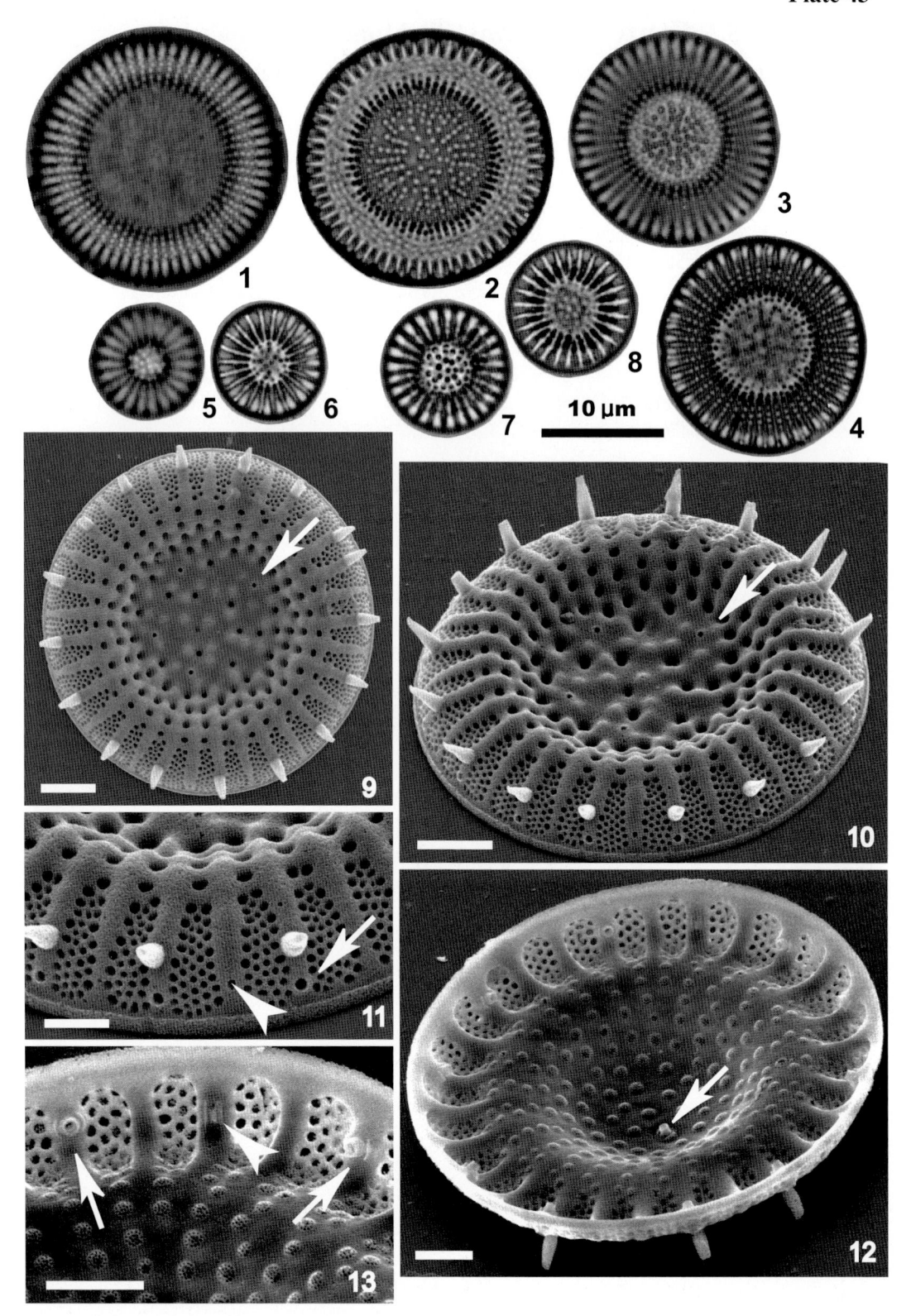

Plates 44-45.　Centric diatoms: Thalassiosirales
Cyclostephanos kyushuensis H. Tanaka 2003

文　　献：Tanaka, H. 2003. *Cyclostephanos kyushuensis* sp. nov., from Pliocene sediments in southwestern Japan. Diatom Research **18**: 357-364.

形　　態：殻面は強く同心円状に波打ち，殻径 5.5-15 μm，殻套の肋は 10 μm に 10-14 本である．殻面有基突起（付随孔2個）は1個，殻套有基突起（付随孔3個）は（1）2-3ごとの肋，唇状突起は1個で殻套有基突起と同じ高さにある．有基突起，唇状突起とも外管を欠く．殻面中心が凸面の殻と，凹面の殻では肋の形態が異なり，凸殻では唇状突起・殻套有基突起が所在する肋は殻端まで届くが，凹殻では突起で終了し肋と殻縁の間に胞紋がある．

産　　出：大鷲湖沼性堆積物（田中ら 2011）：鮮新世・岐阜県，野原層（原記載論文の Tanaka(2003)では津房川層（鮮新統）と記されているが，石塚ら(2005)によって当地の地層は津房川層とは異なることが判明し，野原層と命名された）：前期更新世・大分県，津森層（田中ら 2005）；中期更新世・熊本県．

図　　版 44-45：野原層標本．

Plate 44, Figs 1-11.　*Cyclostephanos kyushuensis* H. Tanaka
LM. Figs 1-6. SEM. Figs 7-11. Material from Nobaru Formation (Early Pleistocene), Kitsuki City, Oita, Japan. Scale bars: Figs 7-8, 10=5 μm, Fig. 9=1 μm, Fig. 11=0.5 μm.

Figs 1-2, 3-4, 5-6.　Three valves at different focal planes.
Figs 7.　External view of whole convex valve.
Fig. 8.　Internal view of convex valve, valve face fultoportula with two satellite pores (arrow) and rimoportula (arrowhead).
Fig. 9.　External oblique view of valve.
Fig. 10.　Internal oblique view of valve, valve face fultoportula (arrow), rimoportula (arrowhead).
Fig. 11.　Detailed view of valve margin showing rimoportula (arrowhead) and mantle fultoportula with three satellite pores (arrow).

Plate 44

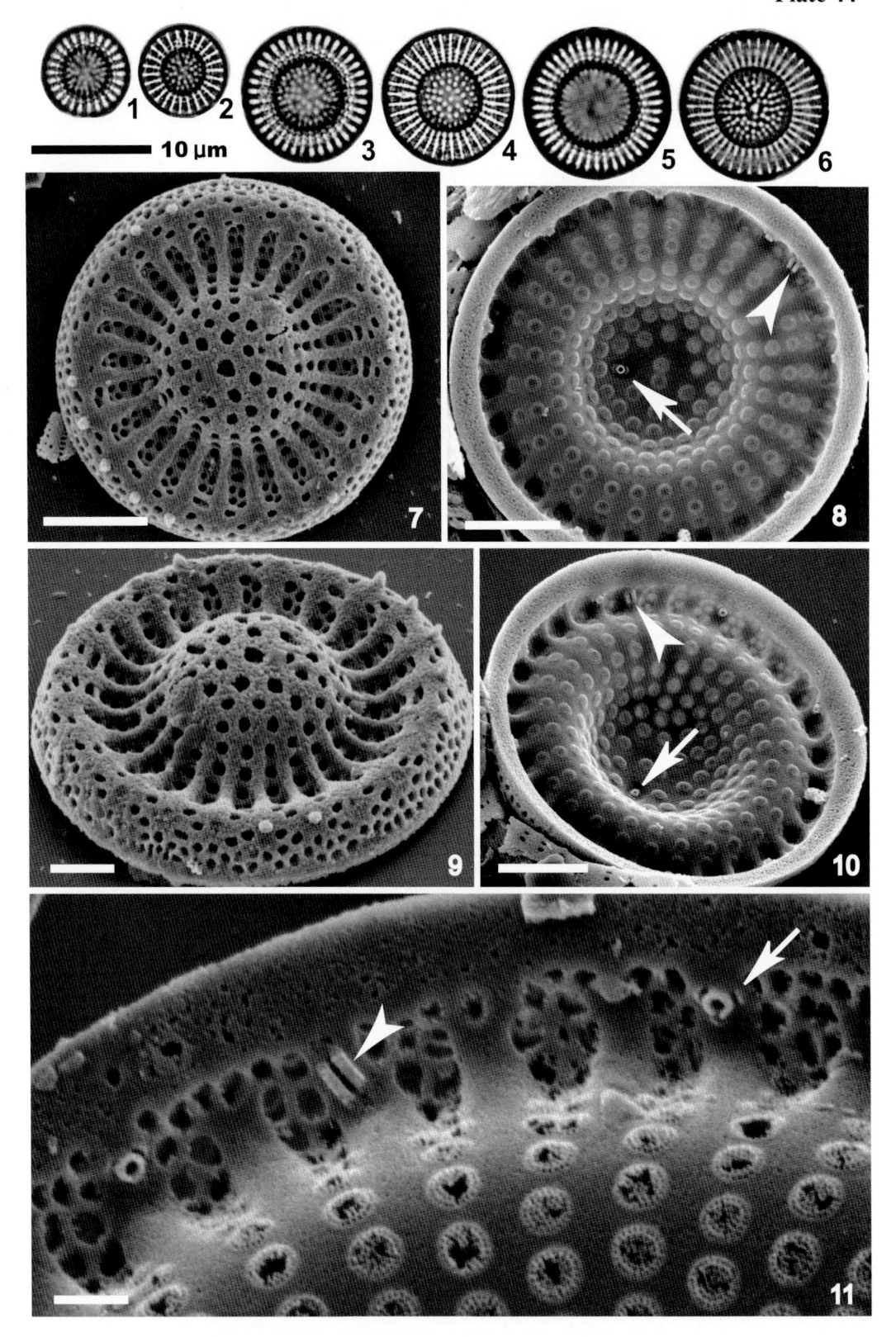

125

Plate 45, Figs 1-6. *Cyclostephanos kyushuensis* **H. Tanaka**
SEM. Figs 1-6. Material from Nobaru Formation, Kitsuki City, Oita, Japan. Scale bars: Figs 1-2=2 μm, Figs 3-4, 6=1 μm, Fig. 5=0.5 μm.

Fig. 1. External view of whole concave valve.
Fig. 2. External oblique view of concave valve.
Fig. 3. Enlarged view of Fig. 2, opening of rimoportula (arrowhead) and opening of mantle fultoportula (arrow).
Fig. 4. Internal oblique view of valve, rimoportula (arrowhead) and valve face fultoportula with two satellite pores (arrow).
Fig. 5. Enlarged view of Fig. 4, rimoportula (arrowhead).
Fig. 6. Detailed view of valve margin showing mantle fultoportulae with three satellite pores (arrow).

Plate 45

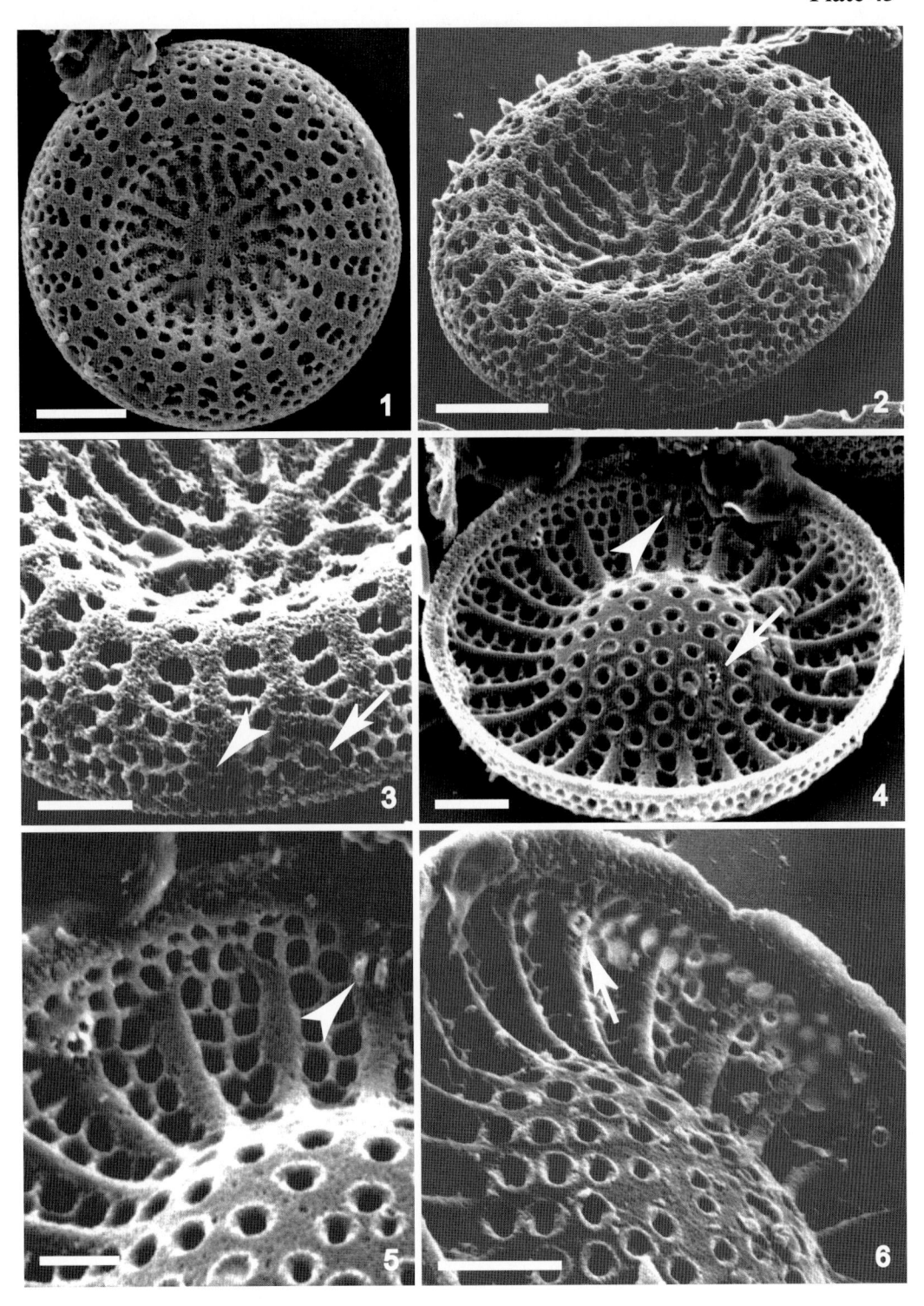

Plate 46.　Centric diatoms: Thalassiosirales
Cyclostephanos numataensis H. Tanaka & Nagumo 2000

文　　献：Tanaka, H. & Nagumo, T. 2000.　*Cyclostephanos numataensis* sp. nov., a new Pleistocene diatom from central Japan. Diatom **16**: 19-25.

形　　態：殻面は強く同心円状に二重に波打つ．殻径 5.5-24.5 μm，束線は 10 μm に 9-11 本，殻中心ではしばしば胞紋が少なくなる．殻套有基突起は 3 個の付随孔を伴い 10 μm に約 4 個．有基突起，唇状突起の外管と針を欠く．殻面有基突起はない．唇状突起は殻縁で殻套有基突起よりも外側にあるが，開口は有基突起の開口よりも殻中心寄りにある．殻套有基突起が所在する間束線は殻端まで連続する場合としない場合があるが，唇状突起を通る間束線は常に殻端まで連続する．

ノ ー ト：*Cyclostephanos* 属としては殻面内側の肋が存在しないが，唇状突起・有基突起の外管がないこと，唇状突起が所在する間束線は常に殻端まで連続することから本属へ所属するのが適当である．類似した分類群は芳野層からも見出されている（田中・北林 2011）．Tanaka & Nagumo(2000)では沼田湖成層は後期更新世とされているが，中期更新世が適切である．

産　　出：沼田湖成層（Tanaka & Nagumo 2000）：中期更新世・群馬県，芳野層（田中・北林 2011）：中期更新世・熊本県．

図　　版：沼田湖成層標本．

Figs 1-12.　*Cyclostephanos numataensis* H. Tanaka & Nagumo

LM. Figs 1-6. SEM. Figs 7-12. From the type material, NUM-106, Numata Lacustrine Deposit (Middle Pleistocene), Numata City, Gunma, Japan. Scale bars: Figs 7, 11＝5 μm, Figs 8-10＝2 μm, Fig. 12＝1 μm.

Figs 1-6.　Three different size valves.
Figs 1-2, 3-4, 5-6.　Three valves at different focal planes.
Fig. 7.　External view of whole concave valve.
Fig. 8.　Oblique view of Fig. 7.
Fig. 9.　External view of whole convex valve.
Fig. 10.　Detailed view of valve margin, opening of mantle fultoportula (arrow).
Fig. 11.　Internal oblique view of valve, rimoportula (arrowhead).
Fig. 12.　Detailed view of internal valve margin showing rimoportula (arrowhead) and mantle fultoportulae with three satellite pores (arrows).

Plate 46

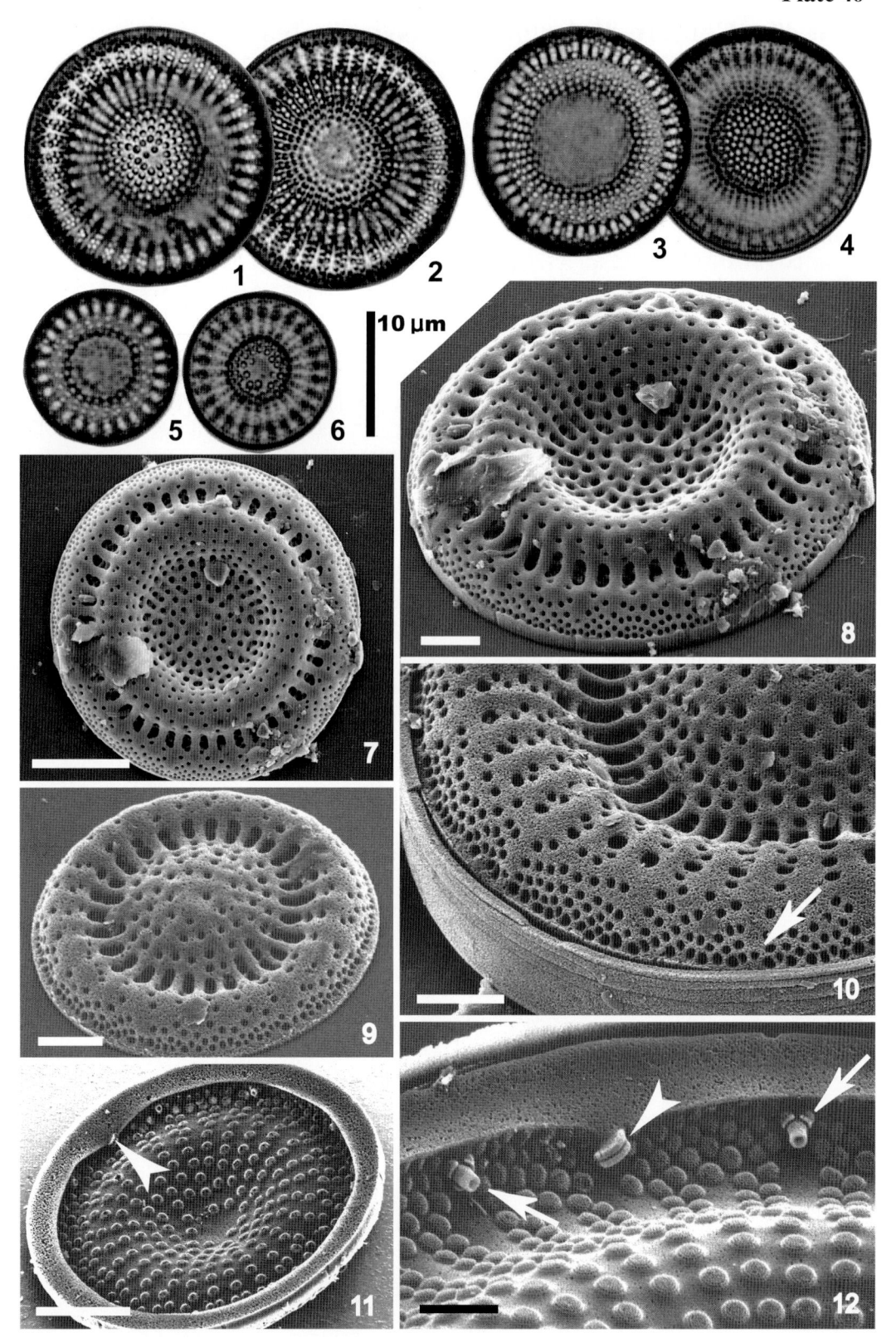

Plate 47.　Centric diatoms: Thalassiosirales
Cyclotella atomus Hustedt 1937

文　　献：Hustedt, F. 1937. Systematische und ökologische Untersuchungen über die Diatomeen-flora von Java, Bali und Sumatra nach dem Material der Deutschen Limnologischen Sunda-Expedition. Archiv für Hydrobiologie, Supplement Band **15**: 131-177.

形　　態：殻は小形円盤状で殻径 4.0-8.5 μm，条線は殻面縁辺部で放射状に配列する，10 μm に 12-20 本．殻面中心近くには黒点が 1 個観察されるがない場合もある．SEM での観察では殻面は平ら〜わずか横に波打つ．LM 観察で殻面に観察された黒点は，内側では 2-3 個の付随孔を伴った殻面有基突起であり，外面ではその開口である．殻套有基突起は 3-4 本ごとの肋にあり，常に 2 個の付随孔を伴う．唇状突起は 1 個で肋にあるが，殻套有基突起よりもわずか殻面側にある．

ノ ー ト：Hustedt(1937)によりインドネシアから記載された種である．日本産の試料を用いた形態研究は南雲・小林(1985)，辻・伯耆(2001)，Tanaka(2007)等で詳しい．

産　　出：現在までの報告はすべて現生の個体である．八郎潟・涸沼・中川・荒川（南雲・小林 1985）：秋田県・茨城県・東京都（現生），琵琶湖（辻・伯耆 2001）：滋賀県（現生），手賀沼（小川 1990）：千葉県（現生），城沼・多々良沼・波志江沼・印旛沼・北浦（Tanaka 2007）：群馬県・茨城県・千葉県（現生）．

図　　版：手賀沼標本．

Figs 1-11.　*Cyclotella atomus* Hustedt
LM. Figs 1-6. SEM. Figs 7-11. Material from Tega Pond (Recent), Chiba, Japan. Scale bars: Figs 7-10＝1 μm, Fig. 11＝0.5 μm.

Figs 1-2, 3-4, 5-6.　Three valves at different focal planes.
Fig. 7.　External view of whole valve.
Fig. 8.　Oblique view of Fig. 7, opening of valve face fultoportula (arrow).
Fig. 9.　Internal view of whole valve, valve face fultoportula with three satellite pores (arrow) and rimoportula (arrowhead).
Fig. 10.　Oblique view of Fig. 9, mantle fultoportula with two satellite pores oriented radially (arrow) and rimoportula (arrowhead).
Fig. 11.　Enlarged view of part of Fig. 8, mantle fultoportula opening (arrow) and rimoportula opening (arrowhead).

Plate 47

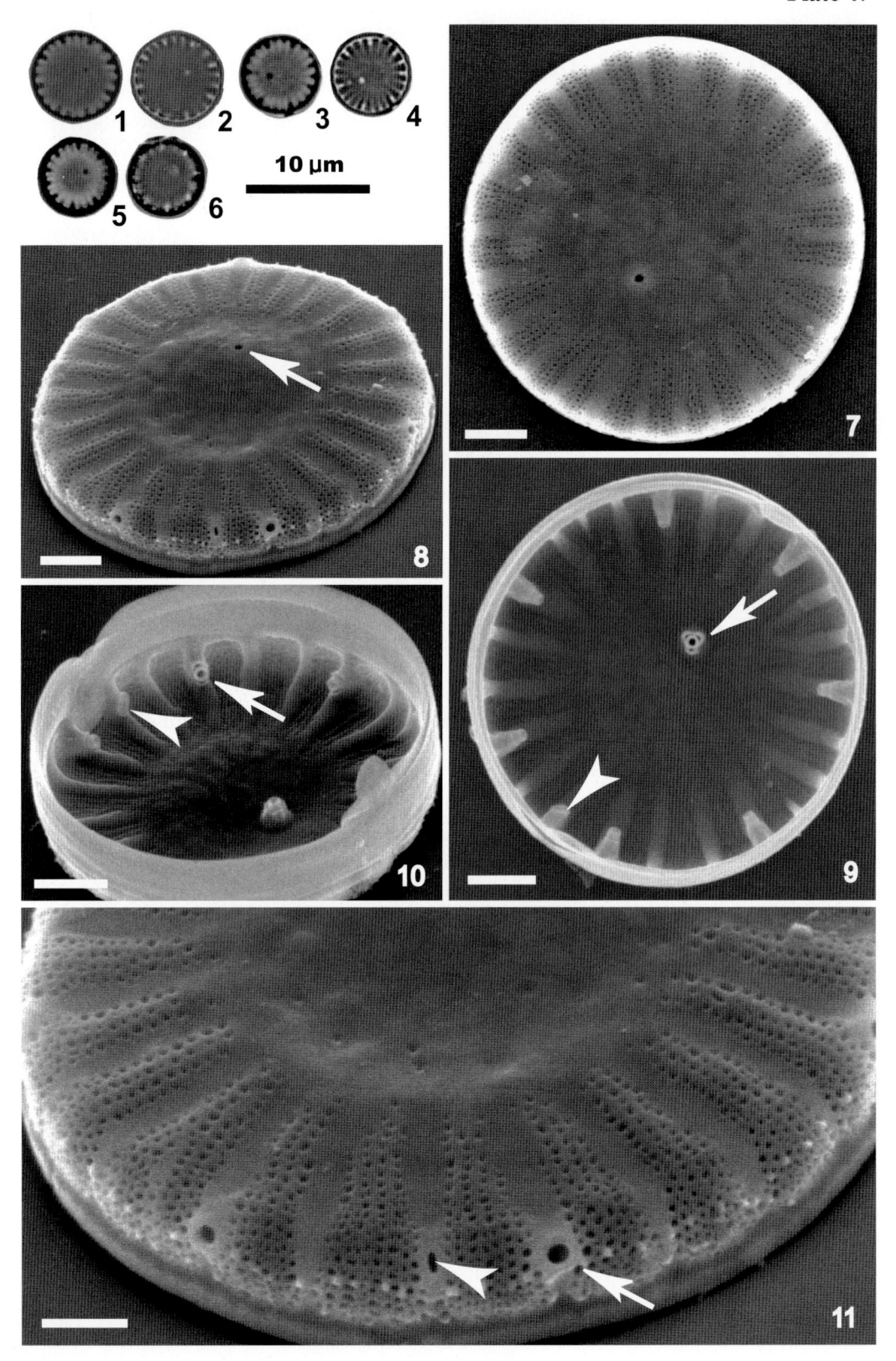

Plate 48.　Centric diatoms: Thalassiosirales
Cyclotella cyclopuncta Håkansson & Carter 1990

文　　献：Håkansson, H. & Carter, J.R. 1990. An interpretation of Hustedt's terms "Schattenlinien", "Perlenreihe" and "Höcker" using specimens of the *Cyclotella radiosa*-complex, *C. distinguenda* Hust., and *C. cyclopuncta* nov. sp. Jour. Iowa Academy of Sciences **97**: 153-156.

形　　態：殻は小形円盤状で殻径 8-22 μm，殻面は平らで中心域には 1-3 個の黒点が観察できる．条線は縁辺域に分布し 10 μm に 16-18 本．間条線の(1)2-3(4)ごとに間条線の殻端部が白く輝いて見える．SEM での観察では殻面中心域の黒点は殻面有基突起である．殻縁には肋があるが，殻端の白い輝きは窪肋であり，これには殻套有基突起がある．殻面・殻套有基突起とも付随孔は 2 個である．唇状突起は 1 個で殻面縁辺域にある．

産　　出：渋沢湖沼性堆積物（南雲・田中 2001）：更新世・長野県．

図　　版：渋沢湖沼性堆積物標本．

Figs 1-12.　*Cyclotella cyclopuncta* Håkansson & Carter
LM. Figs 1-6. SEM. Figs 7-12. Material from Shibusawa lacustrine deposit (Pleistocene), Ueda City, Nagano, Japan. Scale bars: Figs 7-8, 10＝2 μm, Figs 9, 12＝1 μm, Fig. 11＝0.5 μm.

Figs 1-6.　Four different size valves.
Figs 3-4, 5-6.　Two valves at different focal planes.
Fig. 7.　External view of whole valve.
Fig. 8.　Oblique view of valve.
Fig. 9.　Enlarged view of part of Fig. 8 showing outer openings of mantle fultoportulae (arrows).
Fig. 10.　Internal oblique view of whole valve, valve face fultoportula (arrow).
Fig. 11.　Detailed view of valve face fultoportula with two satellite pores.
Fig. 12.　Detailed view of internal valve margin, rimoportula (arrowhead) and mantle fultoportulae with two lateral satellite pores (arrows).

Plate 48

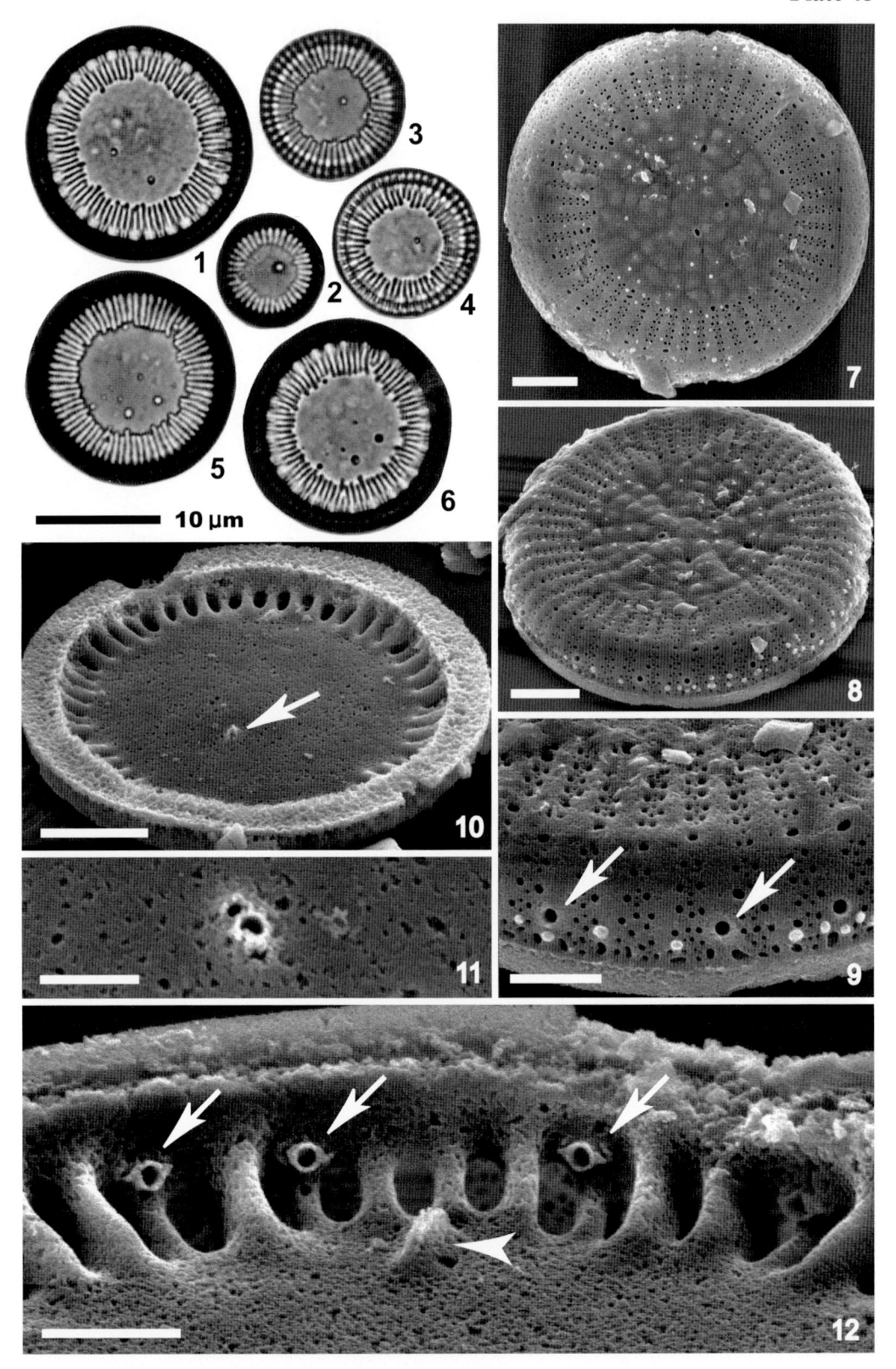

133

Plate 49.　Centric diatoms: Thalassiosirales
Cyclotella iris Brun & Héribaud 1893 s.l.

文　　献：Houk, V., Klee, R. & Tanaka, H. 2010. Atlas of freshwater centric diatoms with a brief key and descriptions. Part Ⅲ. Stephanodiscaceae A: *Cyclotella, Tertiarius, Discostella*. Fottea **10** Supplement, 498 pp, 215 pls.

形　　態：殻は円形からやや楕円形，長径 8-36 μm，短径 7-36 μm．殻面は長径方向へ横に波打ち，中心域は楕円形～長皮針形である．条線は 10 μm に 12-16 本．殻面および間条線には多数のいぼ状構造があり LM でも観察できる．間条線の 2-6 本ごとに間条線の殻端部が白く輝いて見える．SEM 観察によると間条線の内側殻縁は肋を形成しているが，殻端の白い輝きは窪肋であり，これには 3 個の付随孔を伴う殻套有基突起が所在する．唇状突起は 1 個で窪肋にあり，外側への開口は殻套有基突起の開口よりも殻面よりにある．

ノート：人形峠層産の本分類群は，殻面観は円形～やや楕円形，中心域は楕円形～長皮針形で，特に中心域は従来報告のある *C. iris* よりやや広い．本種には変種が多くありそれらとも類似しているので sensu lato とした．

産　　出：人形峠層（田中 2012）：後期中新世～鮮新世・岡山-鳥取県境．

図　　版：人形峠層標本．

Figs 1-8.　*Cyclotella iris* Brun & Héribaud s.l.

LM. Figs 1-4. SEM. Figs 5-8. Materials from Ningyo-toge Formation (Late Miocene-Pliocene): Figs 1-3, 6-8. Onbara, Saibara Village, Okayama, Japan and Figs 4-5 from Tatsumi-toge, Saji Village, Tottori, Japan. Scale bars: Figs 5-6, 8＝5 μm, Fig. 7＝1 μm.

Figs 1-4.　Four different size valves.
Fig. 5.　External view of whole valve, opening of rimoportula (arrowhead) and openings of mantle fultoportulae (arrows).
Fig. 6.　External oblique view of valve, opening of rimoportula (arrowhead).
Fig. 7.　Detailed view of internal mantle showing two fultoportulae with three satellite pores on recessed costae.
Fig. 8.　Internal oblique view of valve, rimoportula (arrowhead).

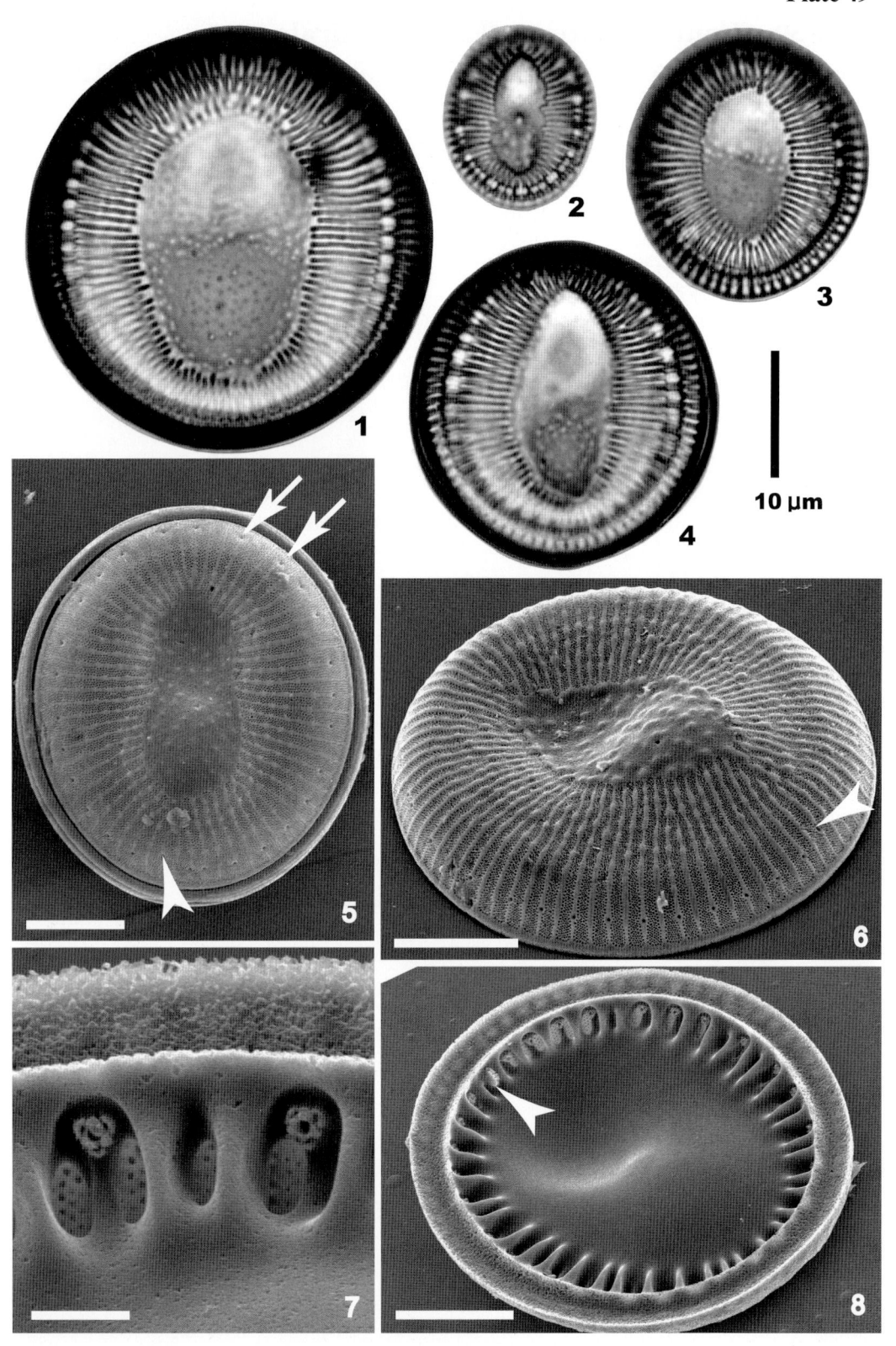

Plate 49

10 µm

Plate 50.   Centric diatoms: Thalassiosirales
Cyclotella iwatensis H. Tanaka in Tanaka & Nagumo 2012

文　　献：Tanaka, H. & Nagumo, T. 2012.　*Cyclotella iwatensis* sp. nov. from Mio-Pliocene freshwater sediment, Iwate Prefecture, Japan. Diatom Research **27**: 121-126.

形　　態：殻径 19-38 μm，殻面は平らで，中心域には放射状のすじ模様があり，条線が分布する縁辺域との境はジグザグしている．条線は 10 μm に 13-16 本で，間条線はしばしば枝分かれをし，いぼ状の凹凸がある．殻套有基突起は 2-5 本ごとの肋に所在し 3 個の付随孔を伴う．唇状突起も殻套の肋にあり，有基突起と同じく外管を欠く．殻帯は Ligula-like segment を含む 4 帯片から構成される．

ノ ー ト：本種は *Cyclotella iris* Brun & Hérib. に類似するが，殻面中心域の筋模様およびいぼ状構造の有無で区別できる．

産　　出：舛沢層（Tanaka & Nagumo 2012）：中新-鮮新世・岩手県．

図　　版：舛沢層標本．

Figs 1-7.　*Cyclotella iwatensis* H. Tanaka

LM. Figs 1-3. SEM. Figs 4-7. From the type material, MAS-107, Masuzawa Formation (Mio-Pliocene), Shizukuishi Town, Iwate, Japan. Scale bars: Figs 4, 7＝5 μm, Fig. 5＝2 μm, Fig. 6＝1 μm.

Figs 1-3.　Three different size valves.
Fig. 4.　External oblique view of valve, opening of rimoportula (arrowhead).
Fig. 5.　Enlarged view of mantle showing opening of rimoportula (arrowhead), openings of mantle fultoportulae and detail of striae and interstriae.
Fig. 6.　Enlarged internal view of mantle showing rimoportula (arrowhead), mantle fultoportulae with three satellite pores and openings of alveoli.
Fig. 7.　Internal view of whole valve.

Plate 50

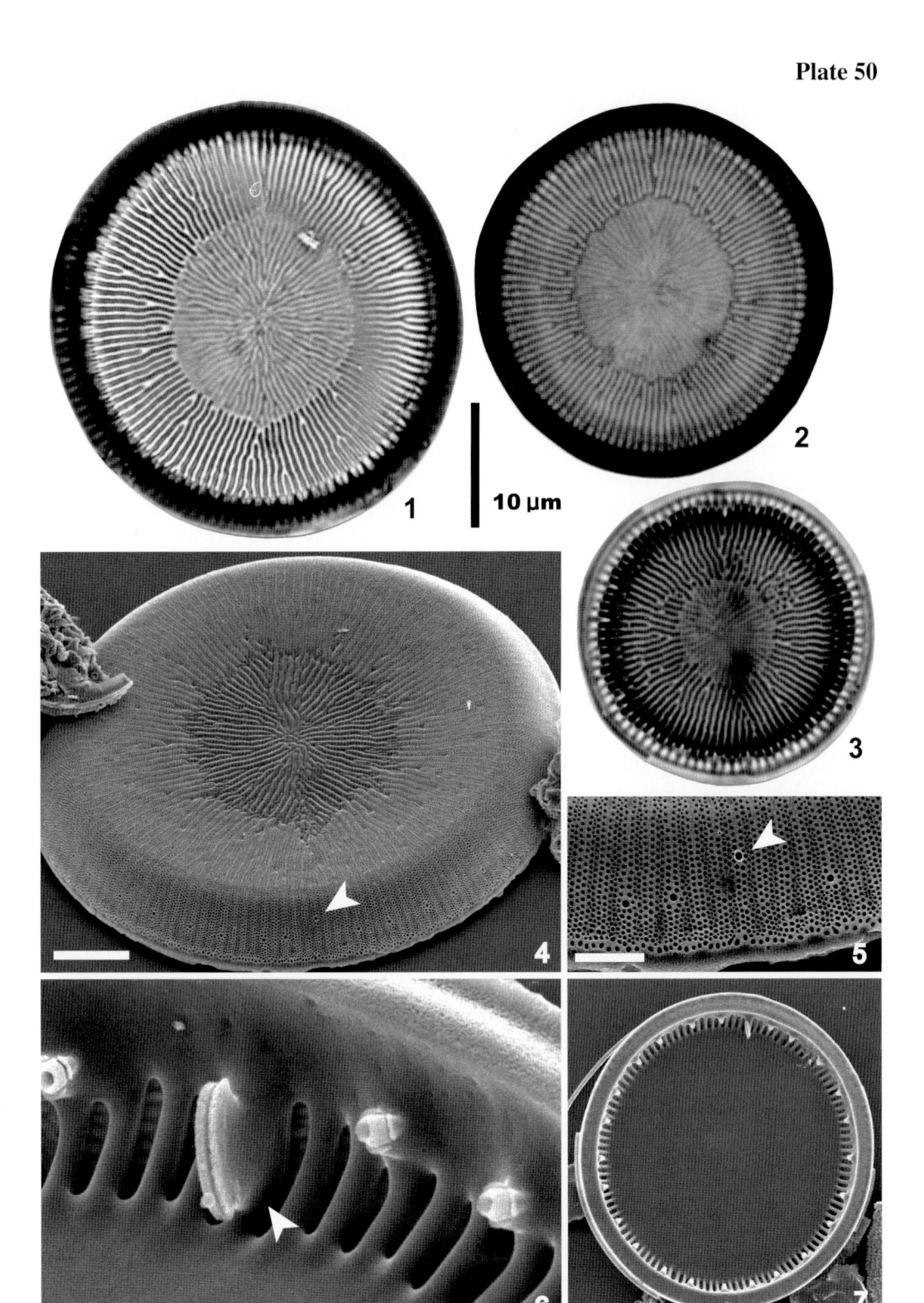

10 µm

Plates 51-52.　Centric diatoms: Thalassiosirales
Cyclotella kitabayashii H. Tanaka in Tanaka & Kashima 2010

文　　献：Tanaka, H. & Kashima, K. 2010.　*Cyclotella kitabayashii* sp. nov., a new fossil diatom species from Pliocene sediment in southwestern Japan. Diatom **26**: 10-16.

形　　態：殻は大形で，殻面はふつう円形であるが小形のものは楕円形になる，殻径 8-49 μm．殻面は強く横に波打つ．中心域には多数の胞紋があり，縁辺域には条線が分布する．間条線は 10 μm に 7-10 本で，殻面／殻套境界または殻套有基突起まで連続する．殻面および殻套には付随孔が 3 個の有基突起があり，殻套有基突起は毎〜2(3)本ごとの間条線（内側では窪肋）に所在する．唇状突起は 1 個で殻面の突出している側の殻套近くにある．内側殻縁の長胞は遠心性および求心性の覆いがある．

ノ　ー　ト：本種は *Pliocaenicus* 属の特徴ももつが，*Pliocaenicus* には内側殻縁部に求心性の覆いがあるものは所属していないので，*Cyclotella* に所属することになった．種小名は試料採集の協力者である大分県地質研究者の北林栄一氏を記念して付けられた．

産　　出：津房川層（Tanaka & Kashima 2010）：後期鮮新世・大分県．

図　　版 51-52：津房川層標本．

Plate 51, Figs 1-8.　*Cyclotella kitabayashii* H. Tanaka

LM. Figs 1-4. SEM. Figs 5-8. From the type material, OIT-207, Tsubusagawa Formation (Late Pliocene), Usa City, Oita, Japan. Scale bars: Figs 6-7 = 10 μm, Fig. 8 = 2 μm, Fig. 5 = 1 μm.

Figs 1-4.　Four different size valves.
Fig. 5.　Broken valve margin showing alveolus.
Fig. 6.　Whole valve view.
Fig. 7.　Oblique view of Fig. 6.
Fig. 8.　Detailed view of valve margin showing opening of rimoportula (arrowhead) and openings of mantle fultoportulae (arrows).

Plate 51

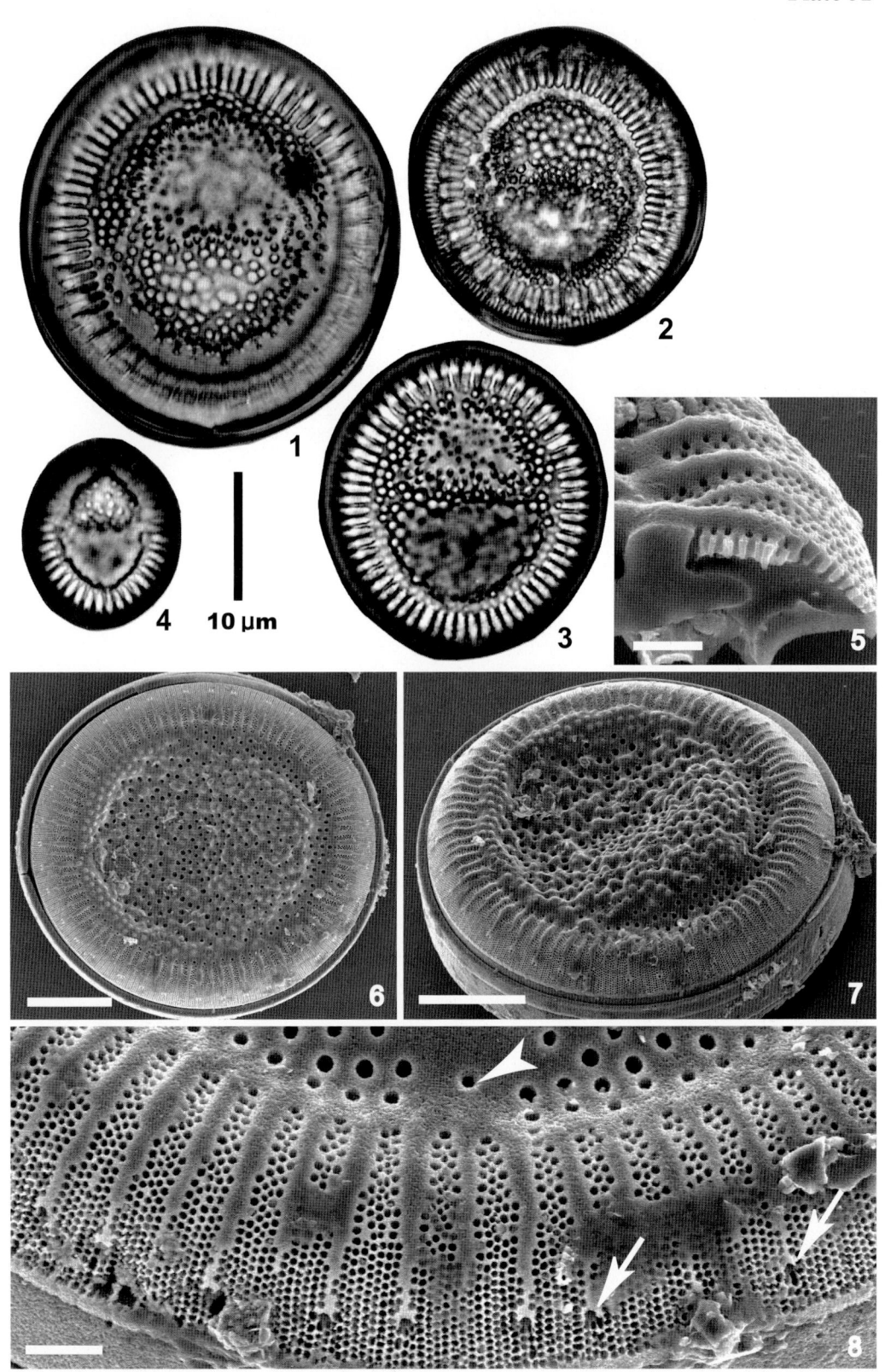

Plate 52, Figs 1-5. *Cyclotella kitabayashii* **H. Tanaka**

SEM. Figs 1-5. From the type material, OIT-207, Tsubusagawa Formation (Late Pliocene), Usa City, Oita, Japan. Scale bars: Figs 2-3=5 μm, Figs 1, 5=2 μm, Fig. 4=0.5 μm.

Fig. 1. Enlarged view of cingulum showing ligula-like segment (arrow).
Fig. 2. External view of slightly elliptical valve.
Fig. 3. Internal view of whole valve, rimoportula (arrowhead).
Fig. 4. Internal view of valve margin showing mantle fultoportula with three satellite pores.
Fig. 5. Enlarged internal view of valve, valve face fultoportulae with three satellite pores (arrows) and rimoportula (arrowhead).

Plate 52

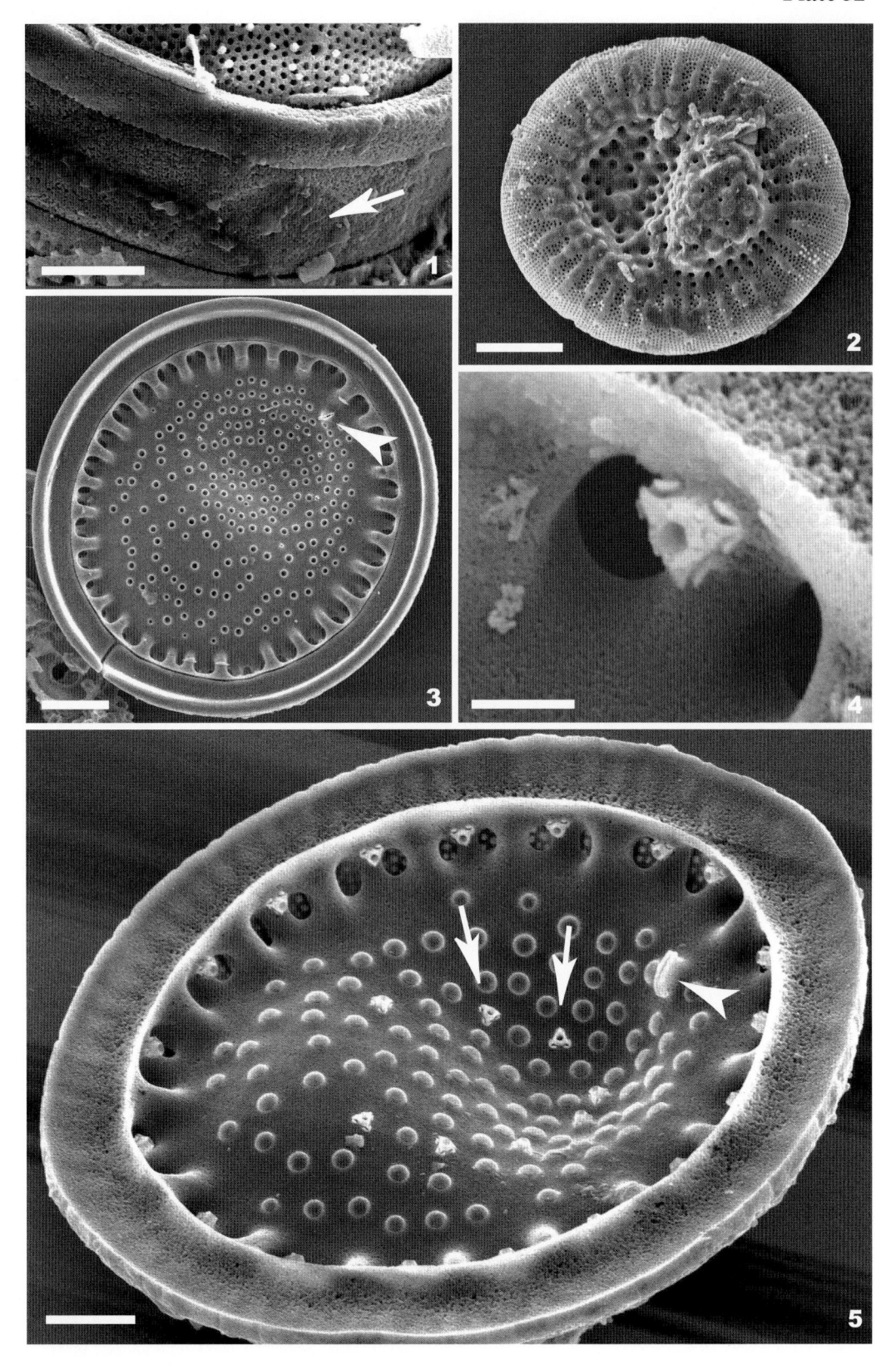

141

Plate 53. Centric diatoms: Thalassiosirales
Cyclotella kohsakaensis H. Tanaka & H. Kobayasi 1996

文　　献：Tanaka, H. & Kobayasi, H. 1996. A new species of *Cyclotella*, *C. kohsakaensis* sp. nov., from a Pliocene deposit, Central Japan. Diatom **12**: 1-6.

形　　態：殻は円盤形で殻面中心域は強く横に波打つ，殻径 11-28 μm．殻面中心域には胞紋・突起は存在しない．縁辺域には条線があり 10 μm に 10-14 本である．条線を構成する胞紋はすべて同じ大きさで 3 方向斜行型の配列をしている．殻套有基突起は毎 3-5 本ごとの窪肋にあり付随孔は 3 個である．唇状突起は 1 個で窪肋にある．

ノ　ー　ト：種小名は原産地の長野県佐久市香坂に由来する．

産　　出：三徳層（Tanaka & Nagumo 2006）：後期中新世・鳥取県，香坂礫岩層（Tanaka & Kobayasi 1996）：鮮新世・長野県．

図　　版：香坂礫岩層標本．

Figs 1-10. *Cyclotella kohsakaensis* H. Tanaka & H. Kobayasi

LM. Figs 1-5. SEM. Figs 6-10. From the type material, KOS-203, Kohsaka Conglomerate Member (Pliocene), Kohsakahigashichi, Saku City, Nagano, Japan. Scale bars: Figs 6-7 = 5 μm, Figs 8, 10 = 1 μm, Fig. 9 = 0.5 μm.

Figs 1-5.　Five different size valves.

Fig. 6.　Internal oblique view of entire valve with cingulum, rimoportula (arrowhead) and ligula-like segment (arrow).

Fig. 7.　External oblique view of entire valve showing transversally undulate central area as well as striae and interstriae in marginal area.

Fig. 8.　Detailed view of internal valve margin, rimoportula on recessed costa (arrowhead).

Fig. 9.　Detailed view of internal valve margin showing mantle fultoportula with three satellite pores on recessed costa.

Fig. 10.　Enlarged view of Fig. 7 showing marginal area of valve face and mantle, openings of mantle fultoportulae on interstriae (arrows).

Plate 53

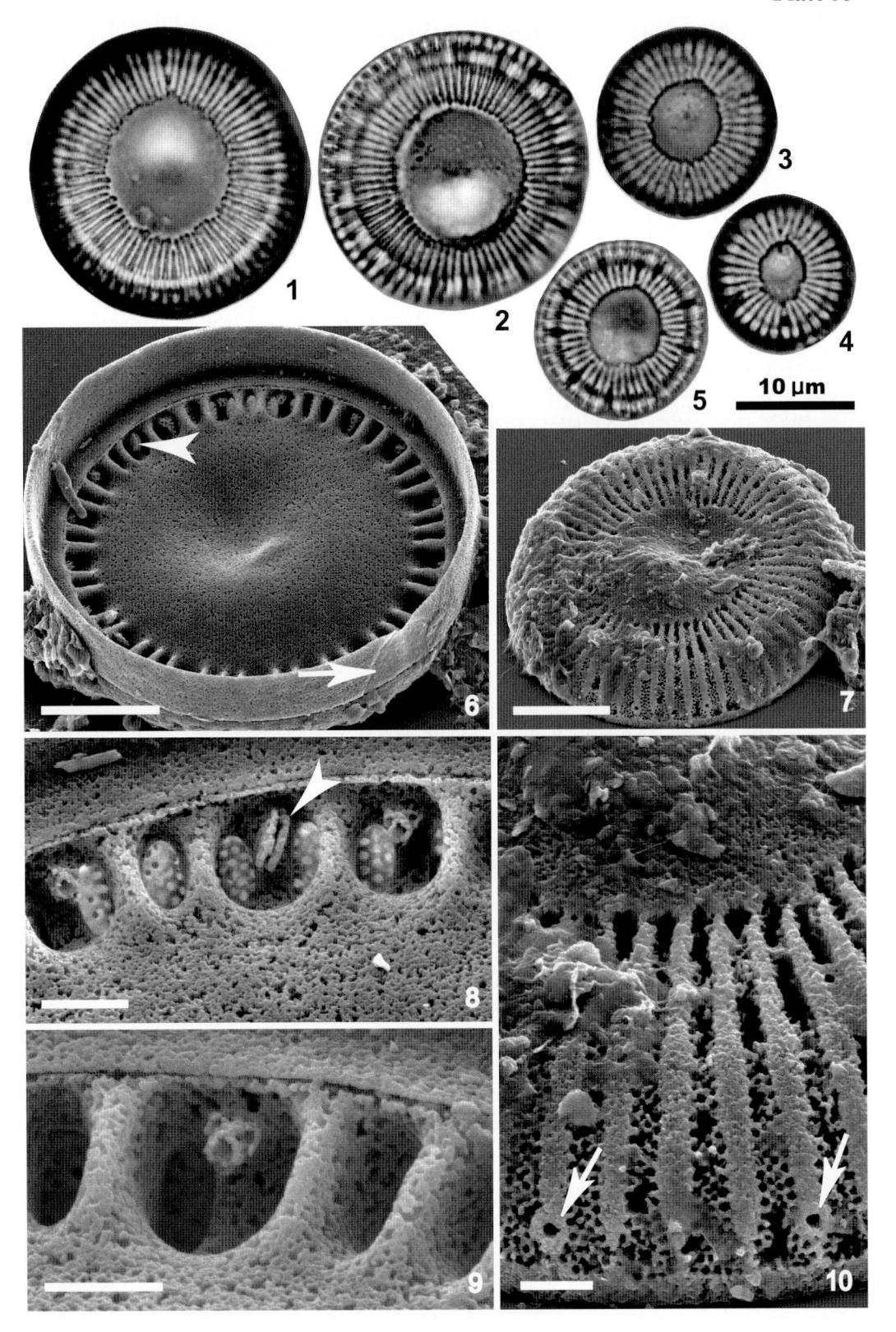

10 µm

Plate 54.　Centric diatoms: Thalassiosirales
Cyclotella meneghiniana Kützing 1844

文　　献：Kützing, F.T. 1844. Die kieselschaligen Bacillarien oder Diatomeen. Nordhausen, W. Kohne. 152 pp, pls 1-30.

形　　態：殻は円盤状，殻径 14.5-18 μm，殻面中心域は横に波打つ．殻面凸部に 1-数個の黒点がある（SEM 観察によれば殻面有基突起）．および殻中心から放射される細かいすじ模様が観察される．縁辺域には太い条線が放射状に配列している，10 μm に 8-10 本．SEM 観察によれば，縁辺域の条線は盛り上がっており，間条線の殻面／殻套境界には短い針と殻套有基突起の開口がある．唇状突起の開口は殻套有基突起の開口よりわずか殻縁側にある．内側では殻面・殻套有基突起とも付随孔は 3 個である．殻套有基突起，唇状突起はともに肋に所在する．

ノ ー ト：本書で使用した大鷲湖沼性堆積物（岐阜県郡上市）産個体の計測値は上記のとおりであるが，多数報告のある現生個体の計測値を考慮するとやや小形で，殻径 10-30 μm，条線密度は 10 μm に 7-10 本程度が一般的と思われる．

産　　出：大鷲湖沼性堆積物（田中ら 2011）：鮮新世・岐阜県．

図　　版：大鷲湖沼性堆積物標本．

Figs 1-9.　*Cyclotella meneghiniana* Kützing

LM. Figs 1-3. SEM. Figs 4-9. Material from lacustrine deposit of Owashi (Pliocene), Gujo City, Gifu, Japan. Scale bars: Figs 4-6＝5 μm, Figs 7-8＝2 μm, Fig. 9＝1 μm.

Figs 1-3.　Three different size valves.
Fig. 4.　External view of whole valve.
Fig. 5.　Oblique view of Fig. 4, opening of rimoportula (arrowhead).
Fig. 6.　Internal oblique view of valve, rimoportula (arrowhead).
Fig. 7.　Detailed view of internal valve margin showing rimoportula (arrowhead) and mantle fultoportulae (broken) with three satellite pores (arrows).
Fig. 8.　View of marginal valve face and mantle of Fig. 5, opening of rimoportula (arrowhead).
Fig. 9.　Detailed view of two valve face fultoportulae with three satellite pores.

Plate 54

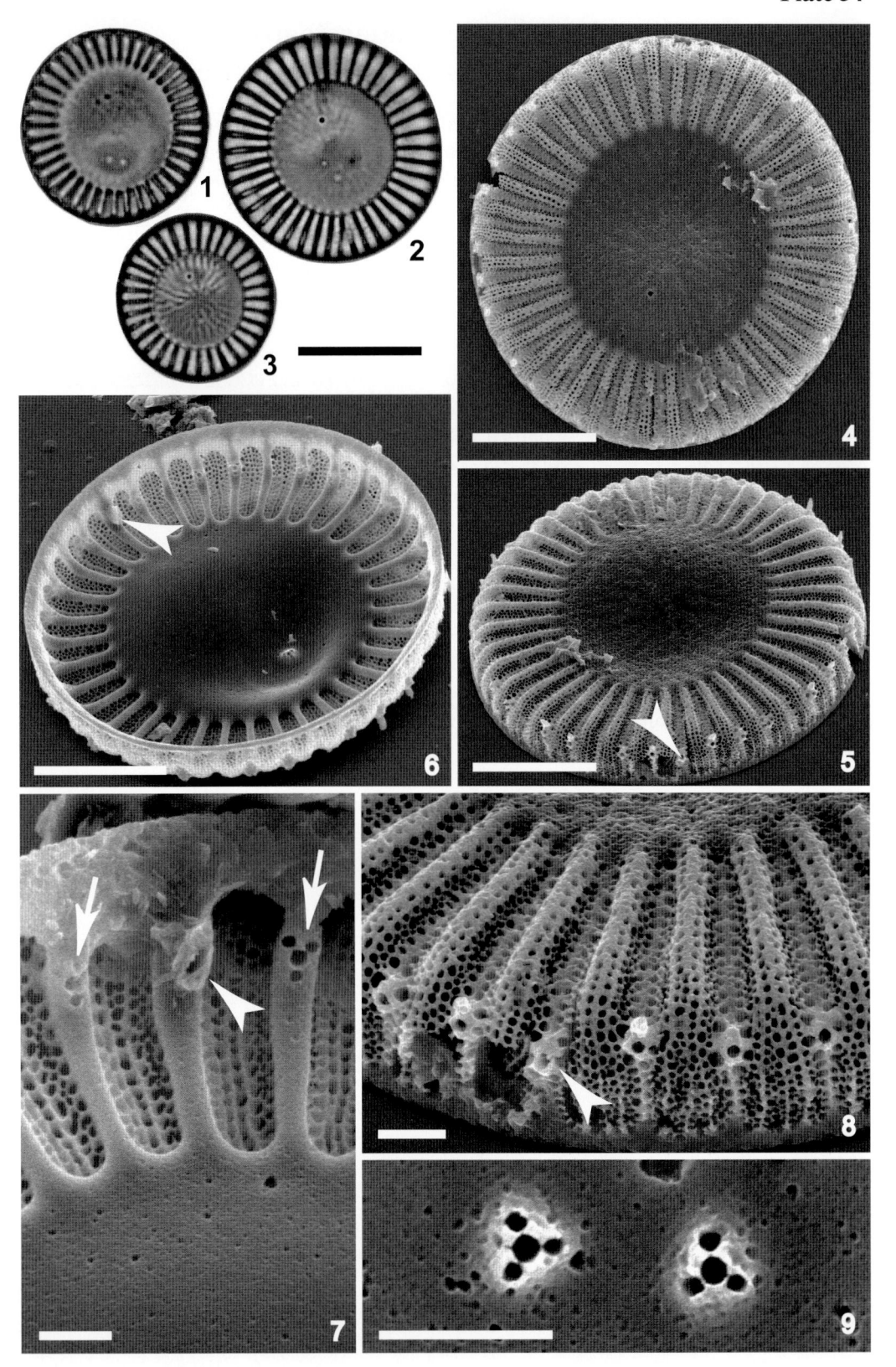

Plates 55-56.   Centric diatoms: Thalassiosirales
Cyclotella mesoleia (Grunow) Houk, Klee & Tanaka 2010

基礎異名：*Cyclotella striata* var. *mesoleia* Grunow 1882

文　　献：Houk, V., Klee, R. & Tanaka, H. 2010. Atlas of fershwater centric diatoms with a brief key and descriptions. Part Ⅲ. Stephanodiscaceae A: *Cyclotella, Tertiarius, Discostella*. Fottea **10** Supplement, 498 pp, 215 pls.

形　　態：殻は円盤形で殻面中心域は強く横に波打つ，殻径 19-45 μm．縁辺域の条線は，10 μm に約 9 本．殻面中心域の凸部域には 3 個の付随孔をもつ 3-7 個の殻面有基突起がある．殻套有基突起は窪肋にあり，上・下に付随孔をもつ．唇状突起は 1 個で窪肋にあり殻套有基突起と同じ高さにある．窪肋は肋の 1-2 本ごとにあり，殻套有基突起か唇状突起が存在する．

ノート：本種は *Cyclotella striata* の変種であったが，*C. striata* は殻套有基突起が 3 個の付随孔を伴うのに対し，2 個であることから変種から独立し種に格付けされたものである (Houk et al. 2010)．*Cyclotella baltica* と類似しているが，本書では Houk et al.(2010) のとおり，複長胞構造が光顕でも観察できること，殻径が 25 μm より大きいことから本種に同定した．

産　　出：大阪層群：鮮新世・奈良県，網走湖底ボーリングコア：更新世・北海道．

図　　版 55-56：大阪層群標本．

Plate 55, Figs 1-5.   *Cyclotella mesoleia* (Grunow) Houk, Klee & Tanaka
LM. Figs 1-5. Material from Lower Osaka Group (Pliocene), Maminooka park, Nara, Japan.

Figs 1-5.   Three different size valves.

Figs 1-2, 3-4.   Two valves shown at different focal planes.

Plate 55

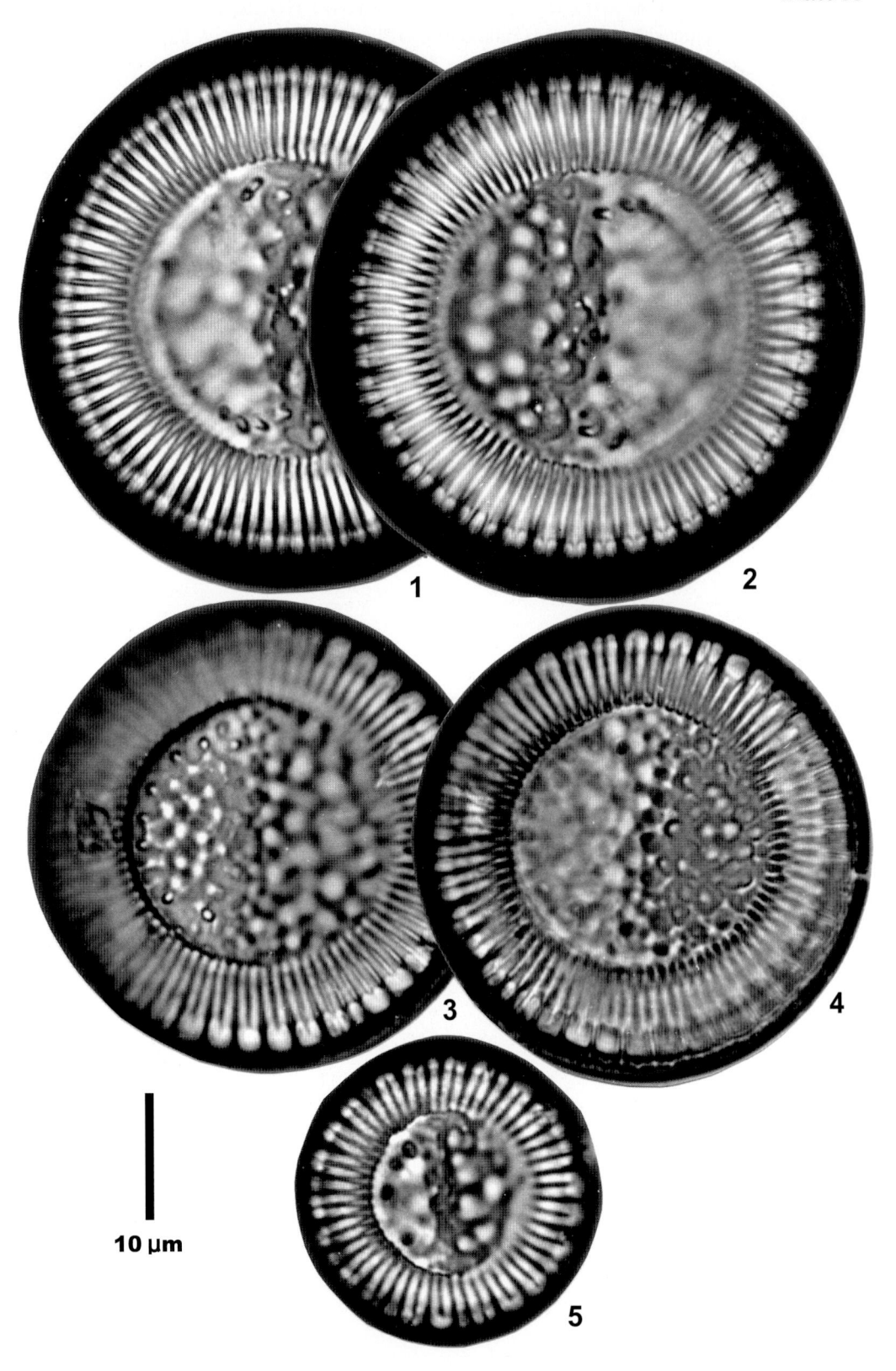

Plate 56, Figs 1–7. *Cyclotella mesoleia* **(Grunow) Houk, Klee & Tanaka**
SEM. Figs 1–7. Material from Lower Osaka Group (Pliocene), Maminooka Park, Nara, Japan. Scale bars: Figs 1–2, 4=5 μm, Fig. 3=2 μm, Figs 6–7=1 μm, Fig. 5=0.5 μm.

Fig. 1. External view of whole valve.
Fig. 2. Oblique view of Fig. 1.
Fig. 3. Enlarged view of part of Fig. 1 showing outer openings of valva face fultoportulae (arrows) as well as striae and interstriae of mantle.
Fig. 4. Internal oblique view, valve face fultoportulae (arrows).
Fig. 5. Detailed view of valve face fultoportula with three satellite pores.
Fig. 6. Detailed view of two mantle fultoportulae with two radial satellite pores.
Fig. 7. Enlarged view of part of Fig. 4 showing rimoportula (arrowhead) and mantle fultoportulae (arrows).

Plate 56

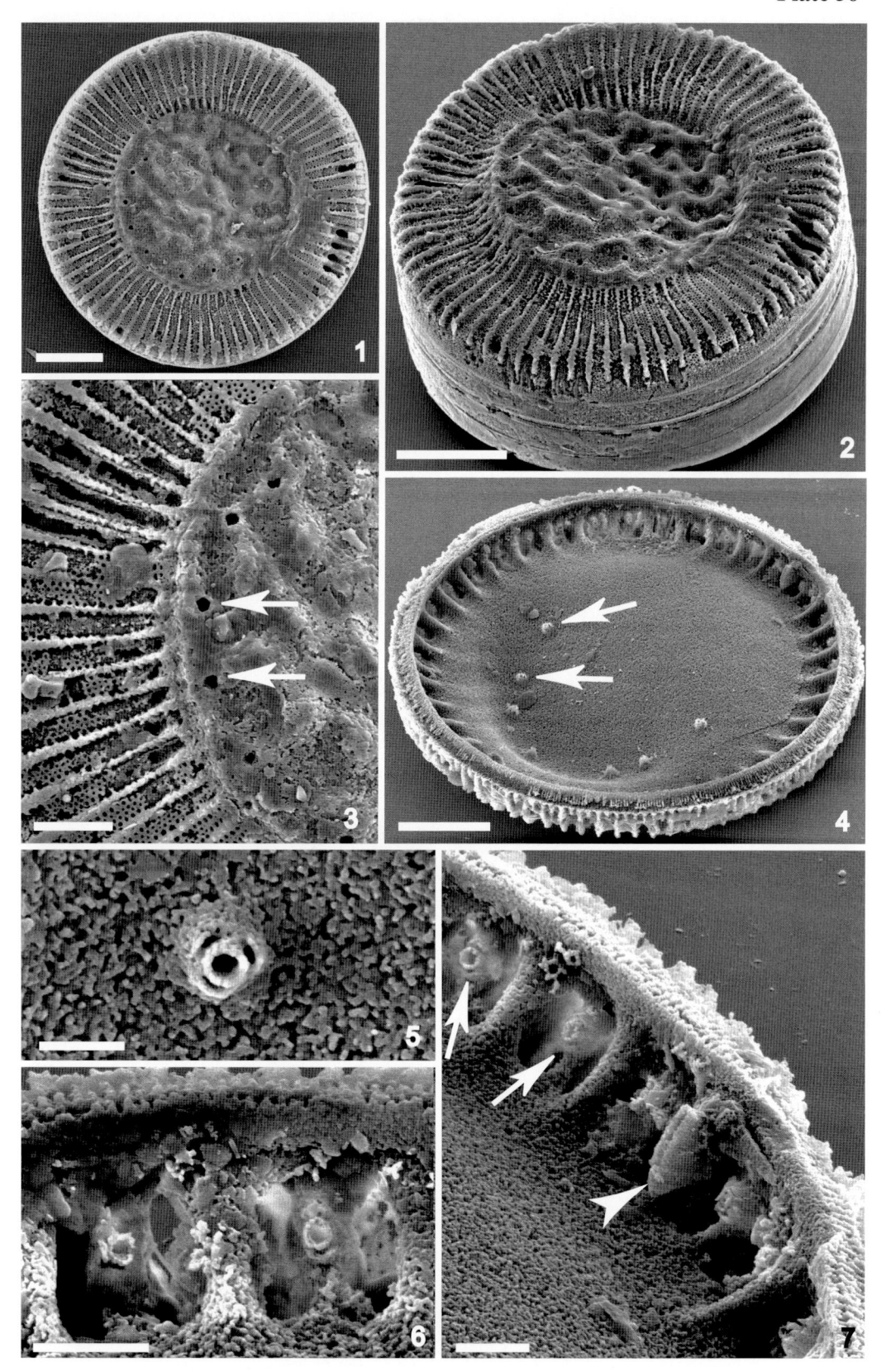

Plates 57-58.　Centric diatoms: Thalassiosirales
Cyclotella nogamiensis H. Tanaka sp. nov.

新　　種：記載文（英文）は5頁参照.
形　　態：殻は円形で殻径11-36 μm. 殻面中心域は広く，多数のこぶ状の突出物および小粒物があり，胞紋がほぼ放射状に配列している．縁辺部の条線は10 μmに14-17本で殻套へ連続する．間条線の枝分かれは少なく，殻套での枝分かれは認められなかった．条線は2本の胞紋列から構成されるが，殻套では3(4)列になる．縁辺部の間条線3-4(5)本ごとに黒線が存在するが，SEM観察によれば殻套有基突起が付随した内側の太い肋である．殻面縁辺部外側の間条線にはこぶ状突出物が存在し，殻面／殻套境界には小針または小粒物，殻套には多数の小粒物が観察できる．殻面中心域の有基突起は6-18個で3個の付随孔を伴い，普通は胞紋列中にあり，1本の胞紋列に1個存在する．殻套有基突起は2個の付随孔をもつ．唇状突起は1-2個で殻面縁辺域にある．殻帯にはligula-like segmentが含まれる．
ノ　ー　ト：*Cyclotella pantanelliana* Castracane に類似するが，光顕観察における殻面中心域の胞紋の大きさ，殻面有基突起の分布，および間条線が殻套で枝分かれをするかしないかで区別できる．
産　　出：野上層：中期更新世・大分県.
図　　版57-58：野上層標本.

Plate 57, Figs 1-7.　*Cyclotella nogamiensis* H. Tanaka sp. nov.
LM. Figs 1-7, from type material (OIT-229), Nogami Formation (Middle pleistocene), Kokonoe Town, Oita, Japan.

Figs 1-2.　Holotype, same valve at different focal planes.
Figs 3-4, 5-6.　Two valves, different focal planes.
Fig. 7.　Oblique view of mantle.

Plate 57

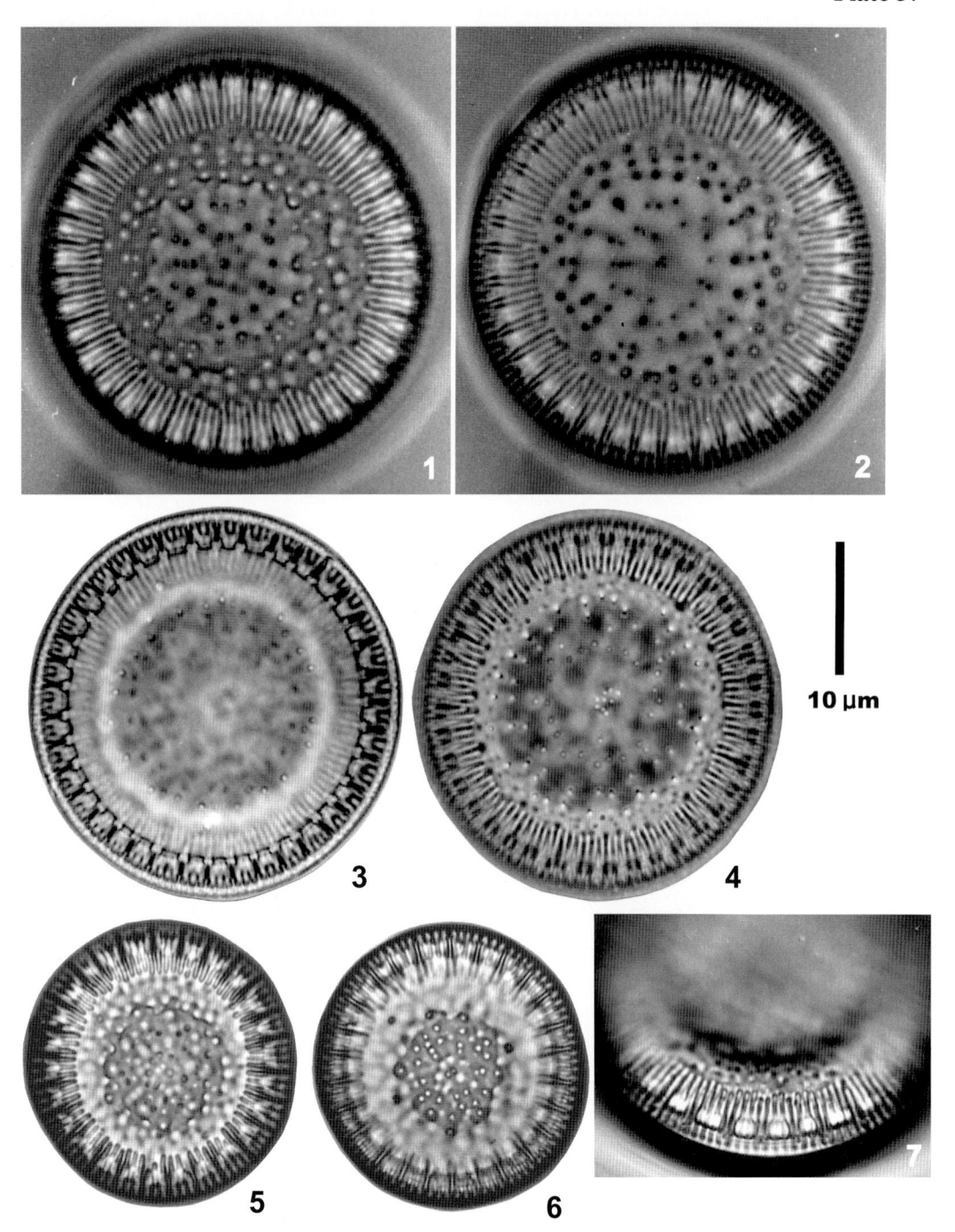

Plate 58, Figs 1–7. *Cyclotella nogamiensis* **H. Tanaka** sp. nov.

SEM. Figs 1–3, 7. External views. Figs 4–6. Internal views. Material from type material (OIT-229), Nogami Formation (Middle Pleistocene), Kokonoe Town, Oita, Japan. Scale bars: Figs 1–2, 4 = 5 μm, Figs 5, 7 = 2 μm, Figs 3, 6 = 1 μm.

Fig. 1. Whole large convex valve view, many bumps and granules on the surface.

Fig. 2. Oblique view of small concave valve, many bumps and granules on the surface.

Fig. 3. Part of enlarged oblique view of Fig. 1 showing bumpy interstriae, granules or their remains on mantle and openings of mantle fultoportulae (arrows).

Fig. 4. Internal oblique view of whole small valve, two rimoportulae (arrowheads).

Fig. 5. Internal view showing smooth surface of large valve face with areolae rows arranged radially, valve face fultoportulae with three satellite pores inserted in areolae rows (arrows).

Fig. 6. Enlarged view of part of Fig. 4 showing mantle fultoportulae with two satellite pores on thick costae and a rimoportula.

Fig. 7. Band with ligula-like segment (arrow).

Plate 58

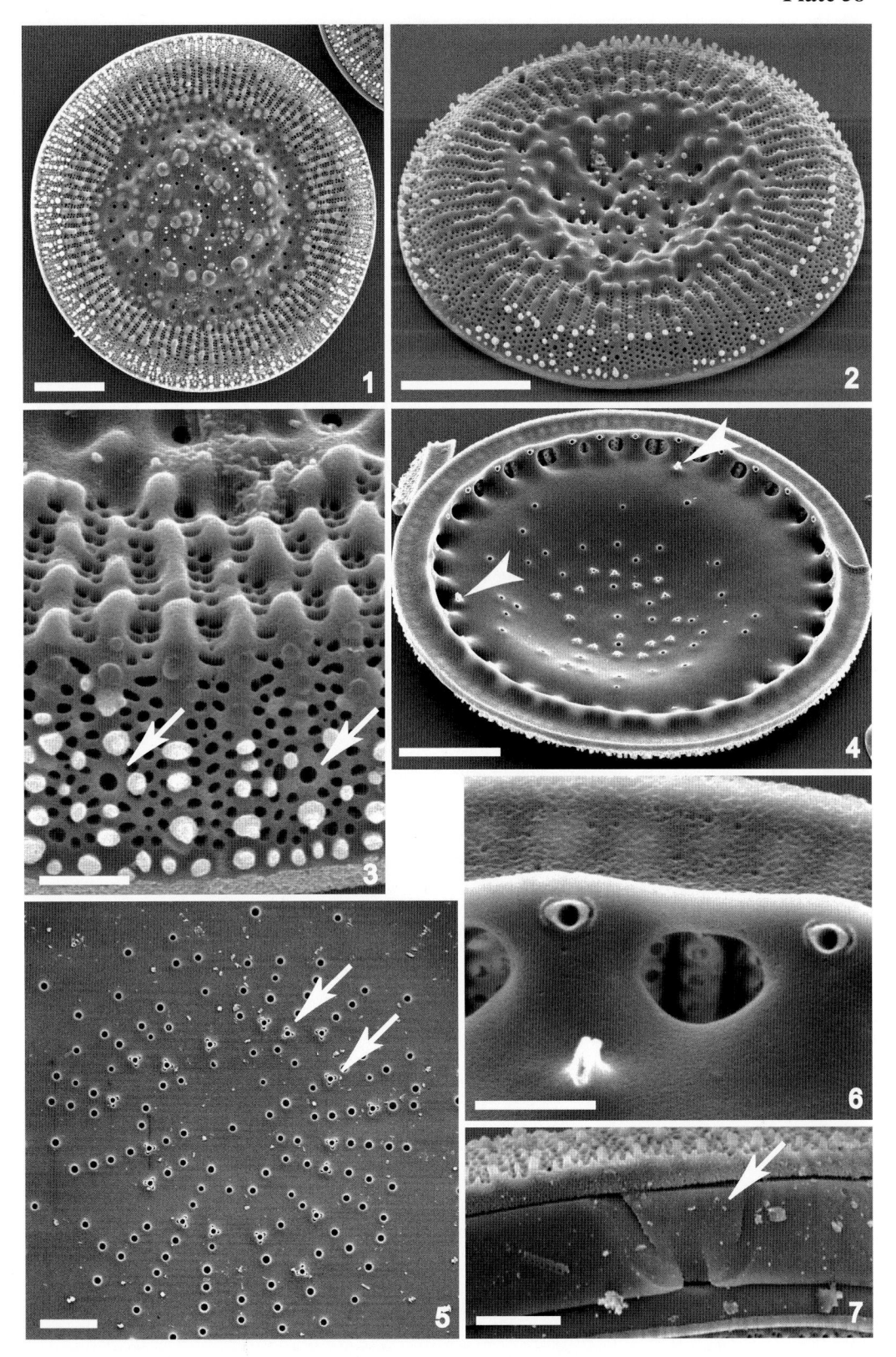

Plate 59. Centric diatoms: Thalassiosirales
Cyclotella notata Loseva 1980

文　　献：Loseva, E.I. 1980. Novye dannye o strukture pantsirya dvukh predstavitelei roda *Cyclotella* Kütz. iz verkhnepliotsenovikh otlozhenii basseina r. Kamy. Botanitjeskij Zjurnal SSSR **65**: 1618-1622.

形　　態：殻は円盤状で殻径 5.0-13.5 μm．殻面中心域には，光顕で 3-7 個の丸い白く輝く部分がある．縁辺部の条線は 10 μm に 16-18 本．間条線は 3-5 本ごとに黒線になる．SEM 観察によると，光顕で観察された輝く部分は内部まで貫通していない殻面の凹みである．凹みの間にはふつう，小乳頭状の突出物がある．殻中心から凹みに向かうようにそれぞれ数個の殻面有基突起が存在し，その数は 1 殻に 4-12 個である．殻套有基突起開口が所在しない間条線は，しばしば殻面あるいは殻套で枝分かれをしている．光顕で黒線に観察された間条線は，内側では太い肋であることを示しており，すべての太肋に殻套有基突起が所在する．殻面・殻套有基突起とも付随孔は 2 個である．唇状突起は 1 個で太肋に殻套有基突起と共にあり，開口は間条線上で有基突起開口より殻中心側にある．

ノ　ー　ト：日本からの産出は大分県からのみである．

産　　出：尾本層：前期更新世・大分県，龍原層：鮮新世・大分県．

図　　版：尾本層標本．

Figs 1-9. *Cyclotella notata* **Loseva**
LM. Figs 1-3. SEM. Figs 4-9. Material from Omoto Formation (Early Pleistocene), Kitsuki City, Oita, Japan. Scale bars: Figs 4-5, 7＝2 μm, Figs 6, 8-9＝0.5 μm.

Figs 1-3.　Three different size valves.
Fig. 4.　External view of whole valve.
Fig. 5.　Oblique view of Fig. 4.
Fig. 6.　Enlarged view of central area of Fig. 4 showing openings of valve face fultoportulae (arrows).
Fig. 7.　Internal oblique view, valve face fultoportulae (arrows).
Fig. 8.　Detailed view of valve margin showing mantle fultoportulae with two satellite pores and rimoportula (arrowhead) together with one mantle fultoportula (arrow) located on same thicker costa.
Fig. 9.　Detailed view of valve margin showing opening of rimoportula (arrowhead) together with an opening of mantle fultoportula (arrow) on same thicker interstriae.

Plate 59

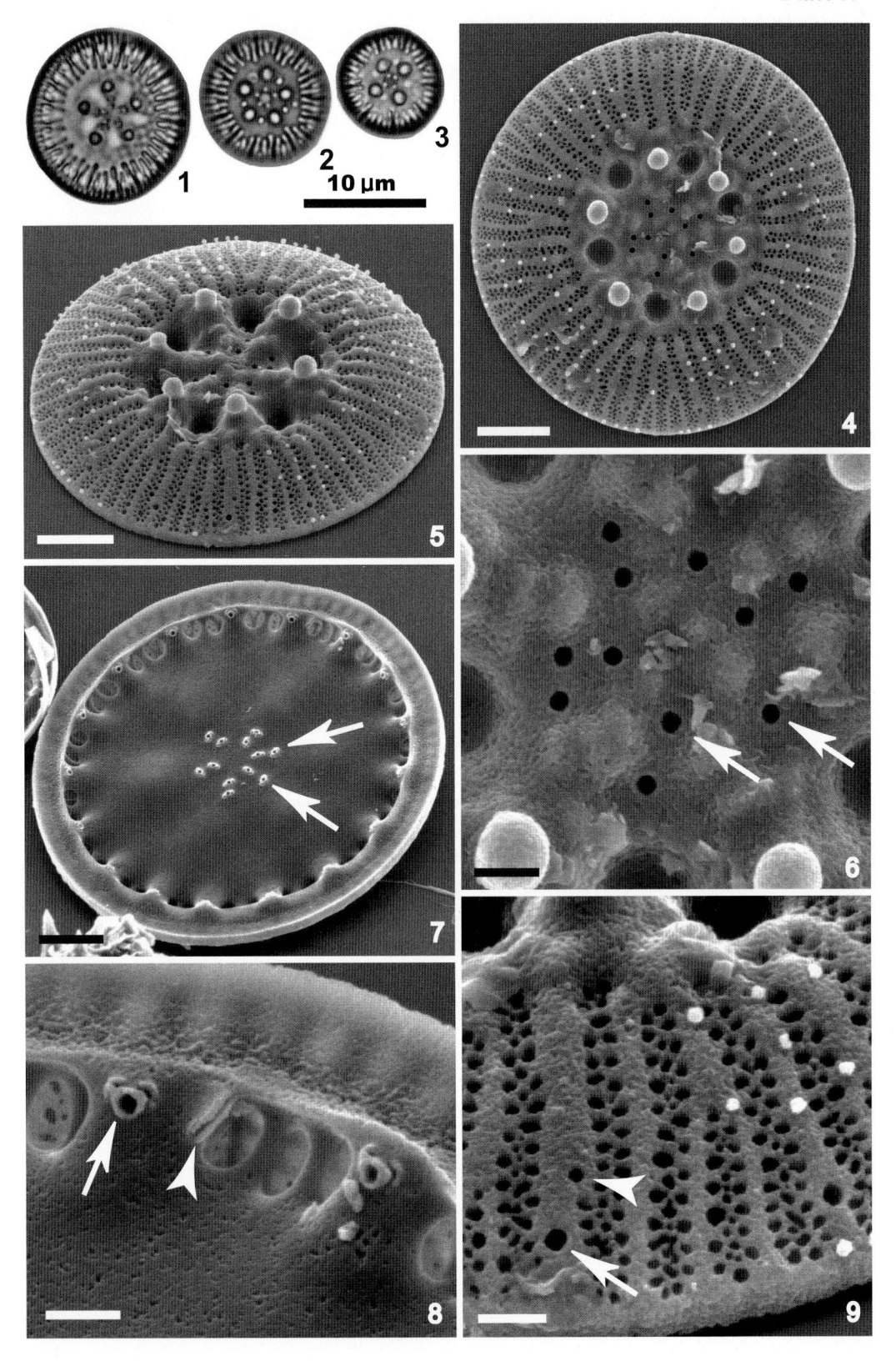

Plate 60. Centric diatoms: Thalassiosirales
Cyclotella ocellata Pantocsek 1901

文　　献：Pantocsek, J. 1901. Die kieselalgen oder Bacillarien des Balaton. 143 pp, 17 pls. Budapest.

形　　態：殻は円盤状で殻径 5.5-18.5 μm．殻面はほぼ平ら，中心域には光顕で 3(4, 5)個の丸い白く輝く部分が観察できる．縁辺部の条線は 10 μm に 15-24 本．SEM 観察によると，光顕で観察された輝く部分は殻面の凹みであり，内部まで貫通していない．殻面中心域には1-2個の殻面有基突起がある．唇状突起は 1 個で外部への開口は間条線の枝分かれをする部分にあり，殻面縁辺域にある．殻套有基突起は毎 2-4(5)本ごとの肋にあり，殻面・殻套有基突起とも付随孔は 2 個である．

ノート：日本からの産出は比較的少ないが，世界的には各地から出現し形態変異も大きい種である（Edlund *et al.* 2003）．

産　　出：塩原湖成層（Akutsu 1964）：前期更新世・栃木県，鬼首層：後期更新世・宮城県，野尻湖層（野尻湖珪藻グループ 1980）：更新世・長野県，牛ケ谷湖成堆積物：更新世・岐阜県．

図　　版：鬼首層標本.

Figs 1-12. *Cyclotella ocellata* Pantocsek
LM. Figs 1-6. SEM. Figs 7-12. Material from Onikobe Formation (Late Pleistocene), Osaki City, Miyagi, Japan. Scale bars: Figs 7-8, 10=2 μm, Fig. 9=1 μm, Figs 11-12=0.5 μm.

Figs 1-6.　Four different size valves.

Figs 2-3, 5-6.　Two valves at different focal planes.

Fig. 7.　External view of whole valve, opening of rimoportula (arrowhead).

Fig. 8.　Oblique view of Fig. 7, opening of mantle fultoportula (arrow) and opening of rimoportula (arrowhead).

Fig. 9.　Detailed view of valve margin, opening of mantle fultoportula (arrow) and opening of rimoportula (arrowhead).

Fig. 10.　Internal oblique view, valve face fultoportula (arrow) and rimoportula (arrowhead).

Fig. 11.　Detailed view of two valve face fultoportulae with two satellite pores.

Fig. 12.　Detailed view of internal valve margin showing mantle fultoportulae with two lateral satellite pores (arrows) and rimoportula (arrowhead).

Plate 60

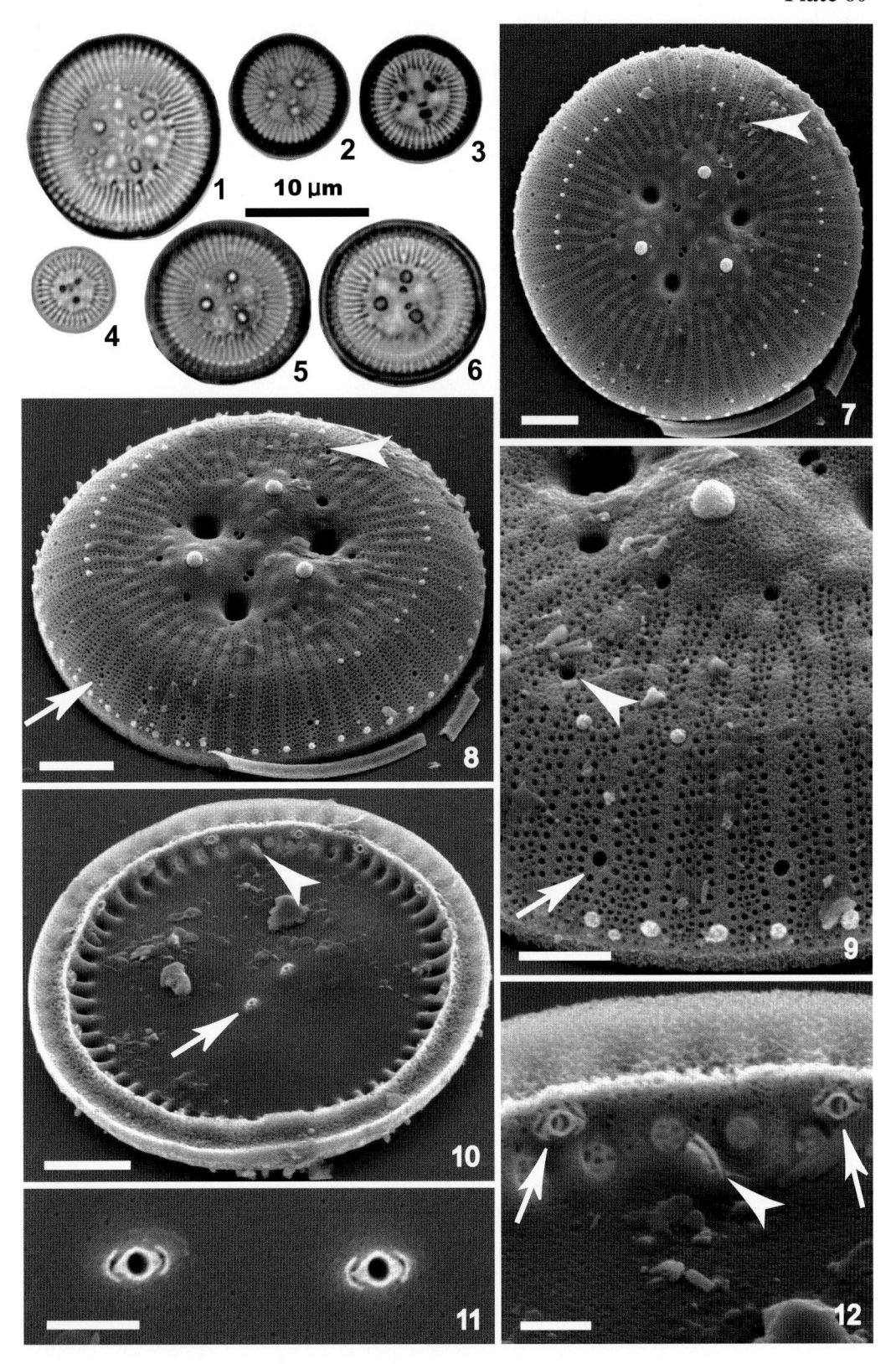

Plates 61-62.　Centric diatoms: Thalassiosirales
Cyclotella oitaensis H. Tanaka sp. nov.

新　　種：記載文（英文）は6頁参照.
形　　態：殻は円盤状で殻径 8.5-35.5 μm. 殻面中心域は直径の半分程度を占め，細かく凹凸している．中心域では点紋がふつう 4-7 本程度細長く集まって，殻中心から星形に配列している（この点紋はSEM観察によると，2個の付随孔を伴った殻面有基突起である）．縁辺部の条線は 10 μm に約 14 本で殻套へ連続する．条線は2本の胞紋列から構成されるが，ときどき間に細かい胞紋の列が存在する．間条線は 1(2) 本おきに中心域との境界から殻縁まで1本のまま殻套へ続き，殻套では殻套有基突起の開口があり，内側では太い肋を形成している．他の間条線は 1-2 回枝分かれをしながら殻縁に達する．殻套有基突起の付随孔は2個である．唇状突起は殻套有基突起と共に太い肋にあり，唇の向きは斜め横である．殻帯には ligula-like segment を含む.
ノ ー ト：*Cyclotella astraea* (Ehrenb.) Kütz. または *Cyclotella stellaris* Alesch. & Pirum. 等に類似するが殻面中心域の広さ，殻套有基突起が枝分かれをしない間条線にあること，長胞や肋の形態が異なり別種である．
産　　出：尾本層：前期更新世・大分県，龍原層：鮮新世・大分県．
図　　版 61-62：尾本層標本．

Plate 61, Figs 1-8.　*Cyclotella oitaensis* H. Tanaka sp. nov.
LM. Figs 1-5. SEM. Figs 6-8. From the type material (OIT-020), Omoto Formation (Early Pleistocene), Kitsuki City, Oita, Japan. Scale bars: Fig. 6=5 μm, Figs 7-8=1 μm.

Figs 1-2.　Holotype, same valve at different focal planes.
Figs 3-5.　Three different size valves.
Fig. 6.　External view of whole valve.
Fig. 7.　Enlarged view, part of central area of Fig. 6.
Fig. 8.　Cingulum, trace of ligula-like segment (arrow).

Plate 61

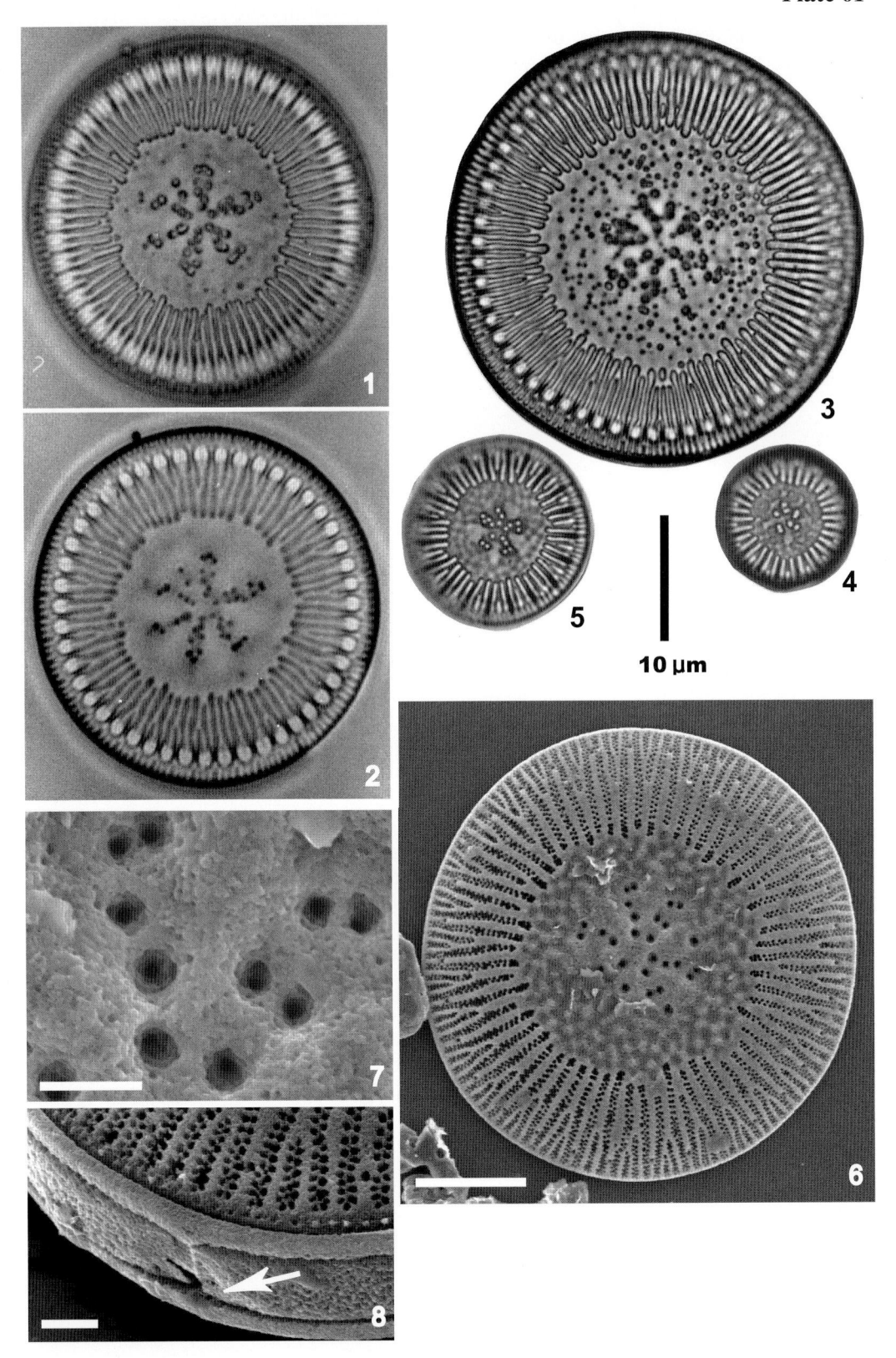

Plate 62, Figs 1–6. *Cyclotella oitaensis* **H. Tanaka** sp. nov.

SEM. Figs 1–6. From the type material (OIT-020), Omoto Formation (Early Pleistocene), Kitsuki City, Oita, Japan. Scale bars: Figs 1, 3–4=5 μm, Figs 2, 5–6=1 μm.

Fig. 1. Oblique view of Plate 61, Fig. 6.

Fig. 2. Enlarged view of part of Fig. 1 showing openings of mantle fultoportulae on non-branching interstriae (arrows) and opening of rimoportula (arrowhead).

Fig. 3. Internal view of whole valve.

Fig. 4. Oblique view of Fig. 3.

Fig. 5. Enlarged view of valve center of Fig. 3 showing valve face fultoportulae with two satellite pores arranged in a stellate pattern.

Fig. 6. Detailed view of valve margin showing mantle fultoportulae with two lateral satellite pores on thick costae (arrows) and a rimoportula off center of a thick costa (arrowhead).

Plate 62

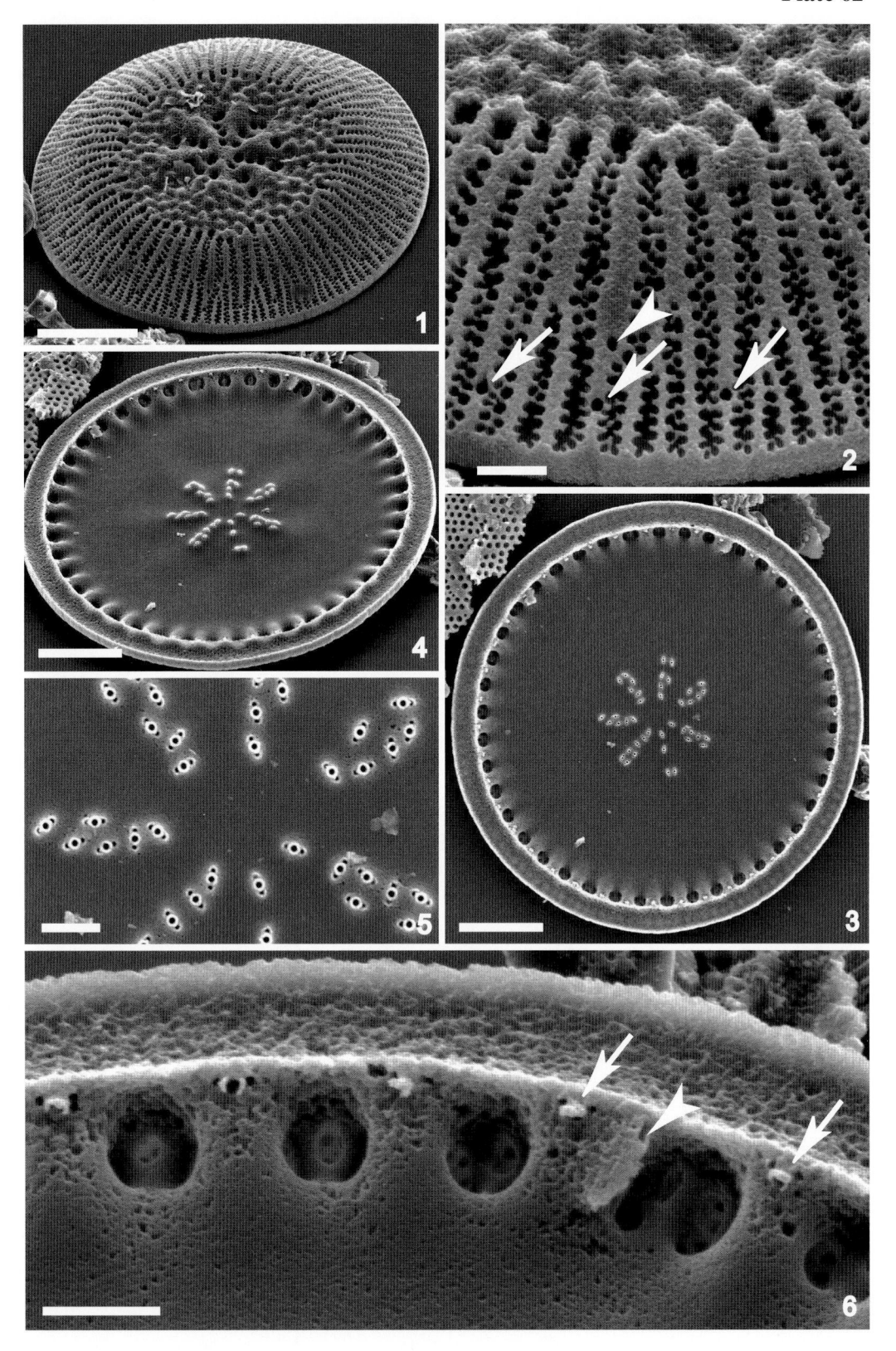

Plates 63-64. Centric diatoms: Thalassiosirales
Cyclotella ozensis (H. Tanaka & Nagumo) H. Tanaka
in Houk *et al.* 2010

基礎異名：*Puncticulata ozensis* H. Tanaka & Nagumo 2005
文　　献：Tanaka, H. & Nagumo, T. 2005. *Puncticulata ozensis* sp. nov., a new freshwater diatom in Lake Oze, Japan. Diatom **21**: 47-55.
　　Houk, V., Klee, R. & Tanaka, H. 2010. Atlas of freshwater centric diatoms with a brief key and descriptions. Part Ⅲ. Stephanodiscaceae A: *Cyclotella, Tertiarius, Discostella*. Fottea **10** Supplement, 498 pp, 215 pls.
形　　態：殻は円盤状で，殻径10-35 μm，殻面中心域の点紋はほぼ放射状に配列する．条線は殻縁で10 μmに9-14本．殻面有基突起は4-19個，付随孔は3個である．殻套有基突起開口は2(3)本ごとの間条線にあるが，間条線の内側は肥厚して肋を形成している．肋は1本おきに太くなり付随孔が2個の殻套有基突起が所在する．唇状突起は(1)2-3個で，条線中心域側のやや短い条線先端にある．
ノート：本種は *Puncticulata ozensis* H. Tanaka & Nagumo として（原産地は尾瀬沼）記載されたが（Tanaka & Nagumo 2005），Houk *et al.*(2010)により *Puncticulata* 属設立に関する命名規約上の問題点が指摘され，同時に *Cyclotella* 属へ組み合わせになったものである．類似する *Cyclotella affinis* (Grunow) Houk, Klee & Tanaka（以前の *Cyclotella bodanica* var. *affinis*）とは，殻面／殻套境界の殻面の間条線数と殻套の間条線数がほぼ同じか（*C. ozensis* は間条線が殻套で枝分かれをしないので同数），殻套の間条線数が増加しているか（*C. affinis* は有基突起が所在しない間条線が殻套で枝分かれをしている）でLMにおいても識別できる．
産　　出：尾瀬沼（Tanaka & Nagumo 2005）：群馬-福島県境（完新世），赤城山大沼（田中・南雲 2011）：群馬県（現生）．
図　　版 63-64：尾瀬沼標本．

Plate 63, Figs 1-9.　*Cyclotella ozensis* (H. Tanaka & Nagumo) H. Tanaka
LM. Figs 1-9. Material from the type locality, bottom deposit of Lake Oze (Oze-numa) (Holocene), boundary of Gunma and Fukushima, Japan.

Figs 1-3, 5-9.　Six different size valves.
Fig. 4.　Oblique view of valve margin.
Figs 5-6, 7-8.　Two valves at different focal planes.
Figs 7, 9.　Rimoportulae (arrowheads).

Plate 63

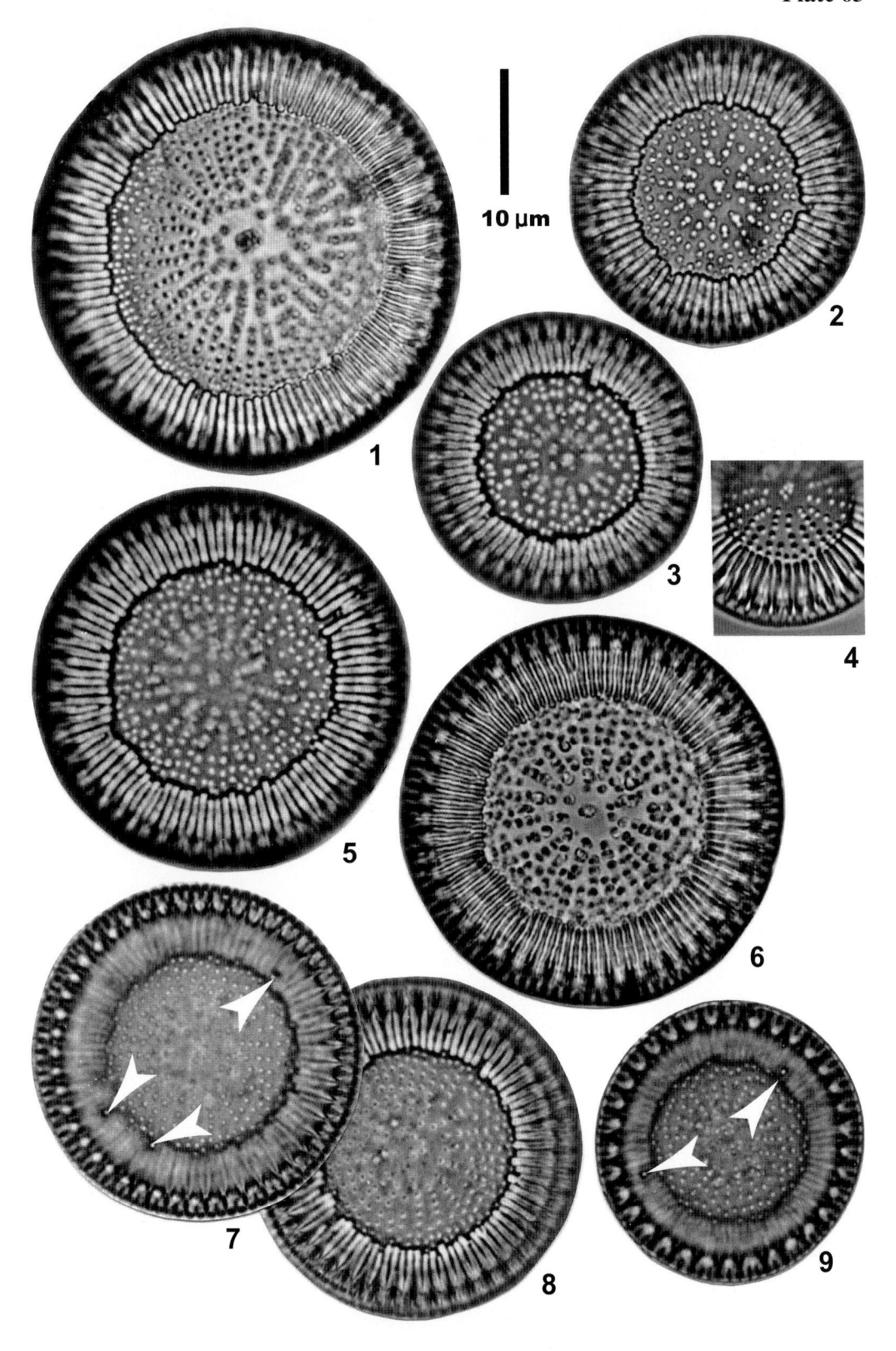

10 μm

Plate 64, Figs 1-7. *Cyclotella ozensis* **(H. Tanaka & Nagumo) H. Tanaka**
SEM. Figs 1-7. Material from the type locality, bottom deposit of Lake Oze (Oze-numa) (Holocene), boundary of Gunma and Fukushima, Japan. Scale bars: Figs 1-2, 4=5 μm, Fig. 5=2 μm, Figs 3, 7=1 μm, Figs. 6=0.5 μm.

Fig. 1. View of entire valve.
Fig. 2. Oblique view of whole valve.
Fig. 3. Detailed view of marginal area of valve face, opening of rimoportula (arrowhead).
Fig. 4. Internal oblique view of valve, rimoportulae (arrowheads).
Fig. 5. Detailed view of external marginal area of valve face and mantle, opening of mantle fultoportula (arrow).
Fig. 6. Detailed view of valve face fultoportulae with three satellite pores.
Fig. 7. Detailed view of internal marginal area, mantle fultoportula with two lateral satellite pores (arrow) and rimoportula (arrowhead).

Plate 64

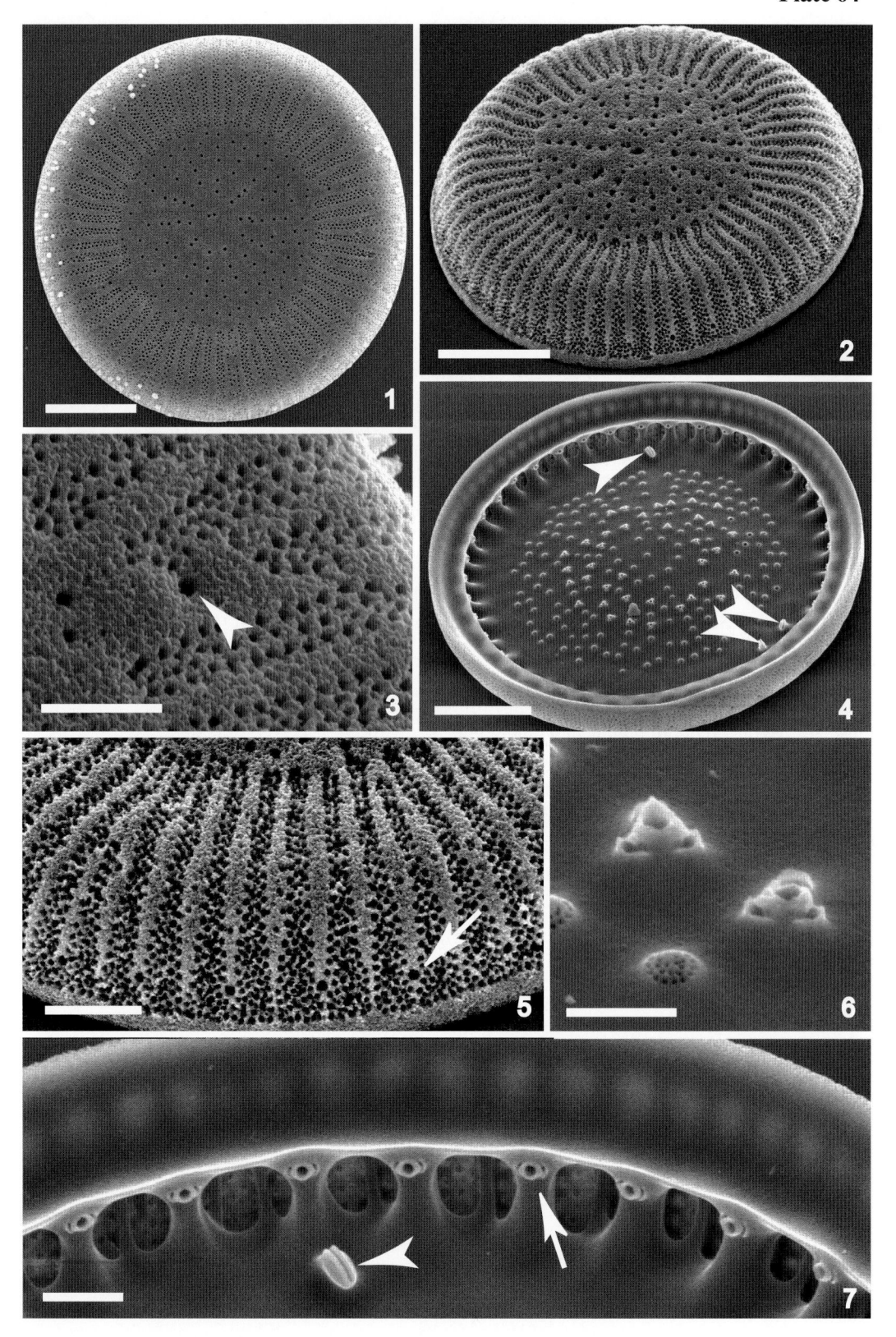

Plate 65.　Centric diatoms: Thalassiosirales
Cyclotella praetermissa Lund 1951

文　献：Lund, J.W.G. 1951. Contributions to our knowledge of British algae. Hydrobiologica **3**: 93-100.

形　態：殻は円盤形で，殻径 8.5-21.5 μm，中心域にはほぼ放射状に点紋が配列する．縁辺域の条線は 10 μm に 16-22 本で，間条線 3-6 本ごとに殻端に黒線が観察される．SEM 観察によると殻面中心域には 6-9 個の殻面有基突起がある．殻縁の肋には太肋（光顕観察による殻端の黒線）と細い肋があり，すべての太肋に殻套有基突起が所在する．殻面有基突起は付随孔が 3 個であるが，殻套有基突起は 2 個である．唇状突起は 1-2(3)個で殻面縁辺部の短い条線の殻中心方向の先端にある．殻帯は Ligula-like segment を含む．

ノート：図版に使用した大沼（赤城山）の分類群は，Houk et al.(2010)が示しているタイプ試料からの写真と比較すると殻面有基突起の数が少ないので，筆者らが大沼（赤城山）の珪藻類を報告した際（田中・南雲 2011）にはこの分類群を confer *praetermissa* としたが，原記載論文である Lund(1951)の試料採取地の一つであるイギリス湖水地方に所在する Bassenthwaite 湖の本種を検討したところ，殻面有基突起は一重の円状に配列しその数は約 14 個で大沼（赤城山）の個体より多いが他は一致したので本種に同定した．

産　出：大沼（赤城山）（田中・南雲 2011）：群馬県（現生）．

図　版：大沼（赤城山）標本．

Figs 1-12.　*Cyclotella praetermissa* Lund

LM. Figs 1-8. SEM. Figs 9-12. Materials from Lake Ono of Mt. Akagi (Recent), Gunma, Japan. Scale bars: Figs 9-10＝2 μm, Figs 11-12＝0.5 μm.

Figs 1-2, 3-4, 5-6, 7-8.　Four valves at different focal planes.
Fig. 9.　External oblique view of whole valve.
Fig. 10.　Internal view of valve showing valve face fultoportulae with three satellite pores (arrows) and a rimoportula (arrowhead).
Fig. 11.　Detailed external view of marginal area of valve face and mantle, opening of mantle fultoportula (arrow).
Fig. 12.　Detailed view of internal marginal area, rimoportula (arrowhead) and mantle fultoportula with two lateral satellite pores (arrow).

Plate 65

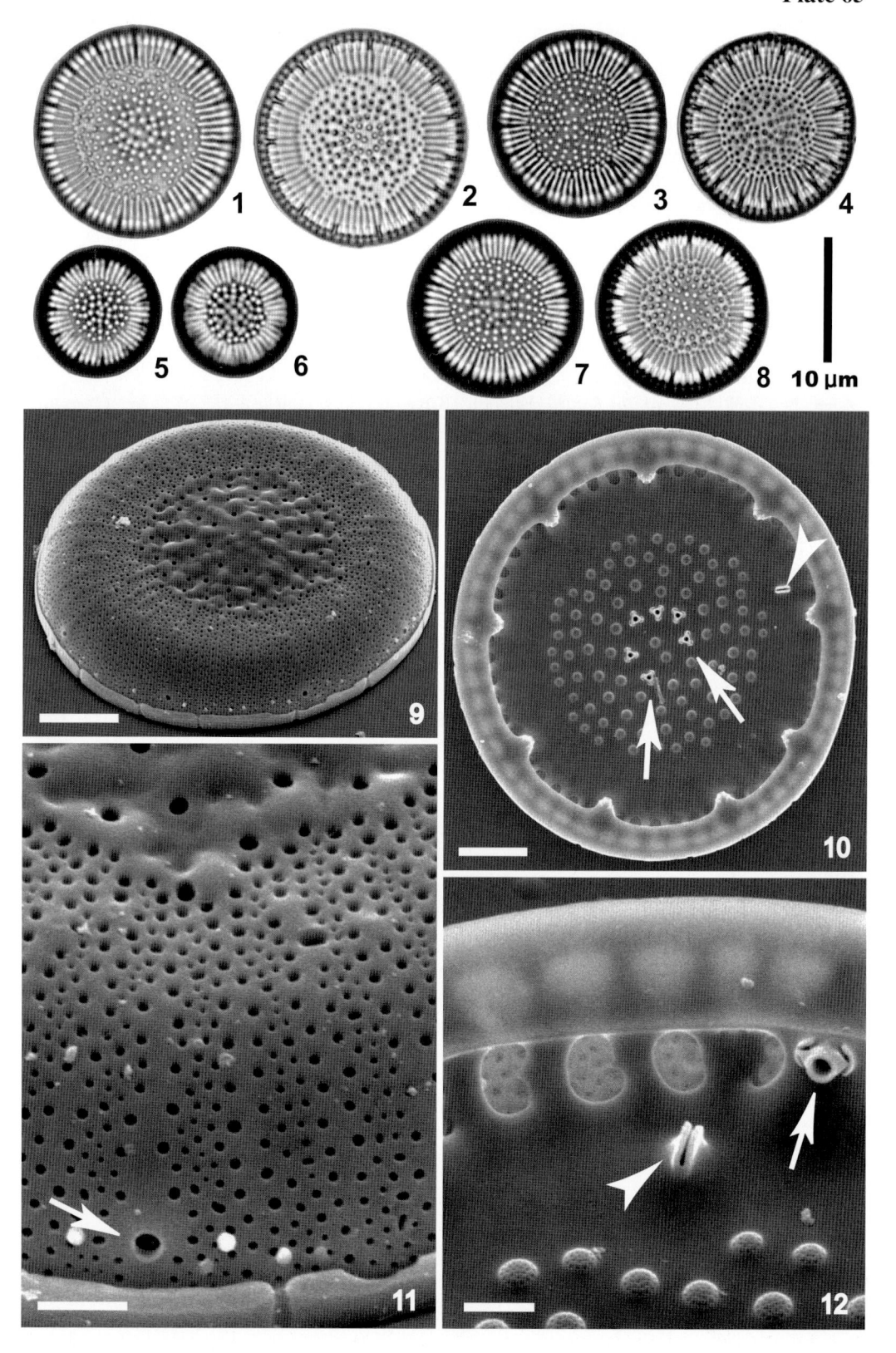

10 µm

Plate 66. Centric diatoms: Thalassiosirales
Cyclotella radiosa (Grunow) Lemmermann 1900

基礎異名：*Cyclotella comta* var. *radiosa* Grunow 1882
文　　献：Van Heuruk, H. 1880-1885. Synopsis des diatomées de Belgique. 235 pp. Atlas: Ducaju & Cie.; Index: J.F. Dieltjens; Text: Mtin. Brouwers & Co., Anvers.
形　　態：殻は円盤形で，殻径9.5-29 μm，殻面中心域には放射状に点紋が配列するが，中心付近で円状に分布する小点も観察される．縁辺域の条線は10 μmに20-22本で，間条線3-4本ごとに殻端に黒線が観察される．SEM観察によると，殻面中心域の放射状点紋列間は盛り上がっている．光顕で観察できた円状に分布する小孔は3個の付随孔を伴った殻面有基突起であり，7-14個存在する．殻套有基突起はすべての太肋（光顕観察による殻縁の黒線）にあり2個の付随孔を伴う．唇状突起は1個で殻面縁辺部の短い条線の先端にある．殻帯はLigula-like segmentを含む．
ノート：本種は*Cyclotella comta*(Ehrenb.)Kütz. としばしば混同されてきたが，*C. comta*の外側中心域は平らで殻面有基突起は散在するが，*C. radiosa*は凹凸のある殻面中心域と殻面有基突起は点紋列中に所在することから区別することができる．タイプ地の*C. comta*の観察と，他地域および多くの報告を検討したHouk *et al.*(2010)は，*C. comta*は化石のみであったと記している．
産　　出：小野上層（田中・小林 1995）：前期更新世・群馬県，中之条湖成層（田中・小林 1992）：中期更新世・群馬県．
図　　版：小野上層標本．

Figs 1-12. *Cyclotella radiosa* (Grunow) Lemmermann
LM. Figs 1-6. SEM. Figs 7-12. Material from Onogami Formation (Early Pleistocene), Shibukawa City, Gunma, Japan. Scale bars: Figs 7-8, 10＝2 μm, Fig. 9＝1 μm, Figs 11-12＝0.5 μm.

Figs 1-5.　Five different size valves.
Fig. 6.　Oblique view of valve margin.
Fig. 7.　External view of whole concave valve, opening of rimoportula (arrowhead).
Fig. 8.　External oblique view of convex valve.
Fig. 9.　Enlarged oblique view of Fig. 7, opening of valve face fultoportula (arrow) and opening of rimoportula (arrowhead).
Fig. 10.　Internal oblique view of whole valve showing valve face fultoportulae with three satellite pores (arrows) and a rimoportula (arrowhead).
Fig. 11.　Detailed view of rimoportula (arrowhead) and opening of alveoli.
Fig. 12.　Detailed view of two mantle fultoportulae with two lateral satellite pores.

Plate 66

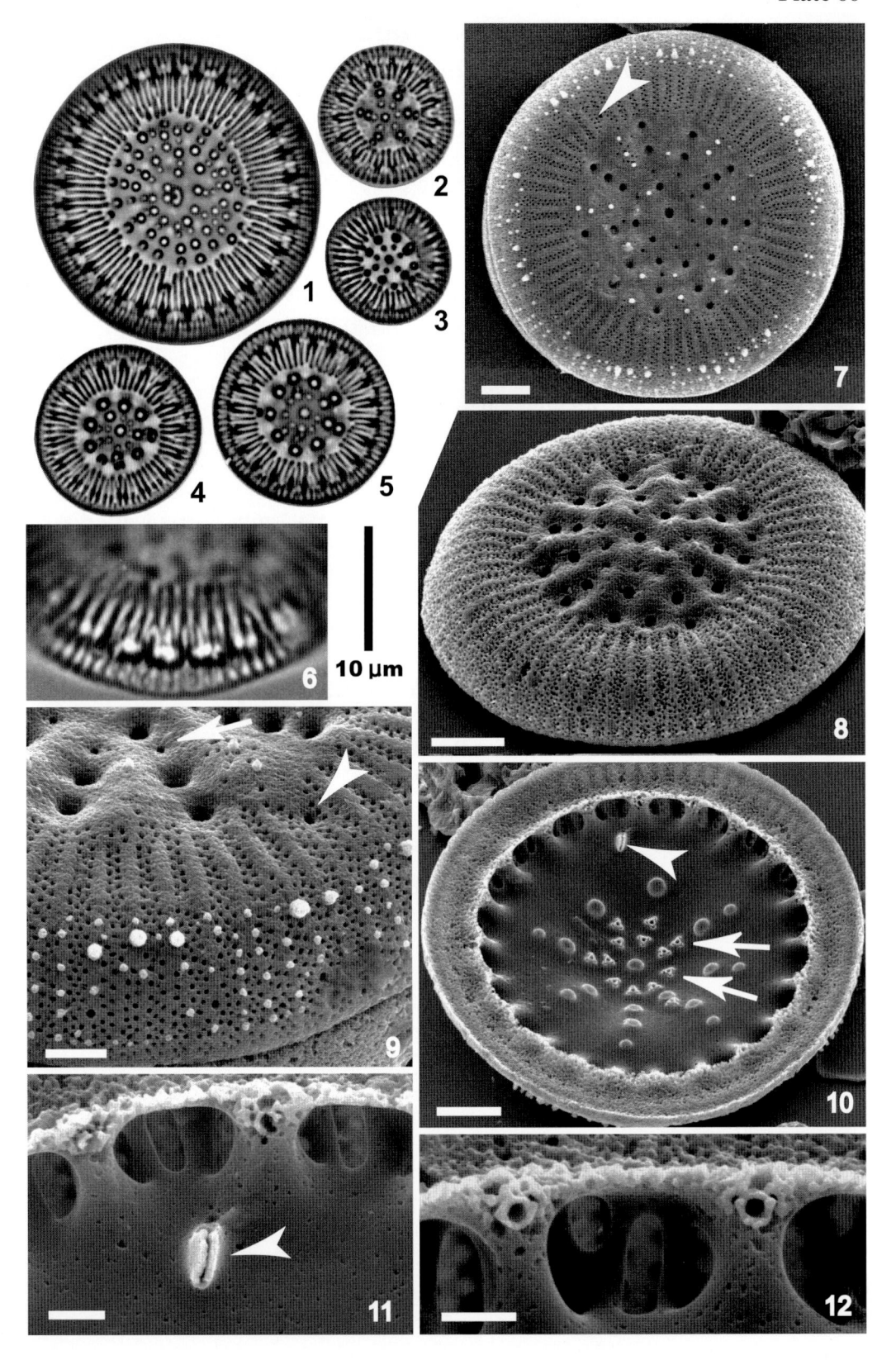

169

Plates 67-68. Centric diatoms: Thalassiosirales
Cyclotella rhomboideo-elliptica Skuja var. *rhomboideo-elliptica* 1937

文　献：Skuja, H. 1937. Algae. *In*: Heinrich H.(ed.) Symbolae Sinicae. Botanische Ergebnisse der Expedition der Akademie der Wissenschaften in Wien nach Südwest China 1914-1918. Sieben Teilen, Ⅰ. Wien.

形　態：殻は帯面観では平行四辺形，殻面観では菱形～楕円形で，長径10-54 μm，短径8-44 μm，条線は 10 μm に(8)10-12(14)本．殻面中心域はいぼ状の表面を呈し，短軸方向に強く横に波打つ．中心域の胞紋の数はゼロから100以上まで変化する．付随孔が3個の殻面有基突起は5-20個で，リング状に分布する．殻面縁辺域には条線と間条線が分布するが，間条線は条線の2(3)本おきに太くなり，内側では殻縁部で太い肋を形成している．殻套有基突起はすべての太い肋にあり，付随孔は2個である．唇状突起は光顕でも明瞭に観察でき，中心域の突出した部分と縁辺条線域の境に位置している．接殻帯片は厚く，殻の長胞域まで入り込んで組み合わさっている．

ノート：Tanaka & Kobayasi(1996)で中之条湖成層の本種が詳述されているが英文である．和文では田中・小林(1992)に詳細が記されている．

産　出：青麻山珪藻土：前期更新世・宮城県，中之条湖成層（田中・小林 1992）：中期更新世・群馬県，倉渕湖沼性堆積物（南雲ら 1998）：中期更新世・群馬県，芳野層（田中・北林 2011）：中期更新世・熊本県，関東平野ボーリングコア（納谷 2008）：更新世・関東平野．

図　版67：中之条湖成層標本．

Plate 67, Figs 1-7. *Cyclotella rhoemboido-elliptica* Skuja var. *rhomboideo-elliptica*

LM. Figs 1-6. SEM. Fig. 7. Material from Nakanojo Lacustrine Deposit (Middle Pleistocene), Nakanojo Town, Gunma, Japan. Scale bar: Fig. 7=2 μm.

Figs 1-6. Six different size valves.
Fig. 1. Rimoportula (arrowhead).
Fig. 7. External oblique view of whole valve, opening of rimoportula (arrowhead) and openings of mantle fultoportulae (arrows).

Plate 67

10 μm

図　版 68：中之条湖成層（Figs 1-5, 7），倉渕湖沼性堆積物（Fig. 6）標本.

Plate 68, Figs 1-7.　*Cyclotella rhomboideo-elliptica* Skuja var. *rhomboideo-elliptica*

SEM. Figs 1-7. Materials, Figs 1-5, 7, from Nakanojo Lacustrine Deposit (Middle Pleistocene), Nakanojo Town, Fig. 6, from lacustrine deposit of Kurabuchi (Middle Pleistocene), Takasaki City, Gunma, Japan. Scale bars: Fig. 7＝10 μm, Fig. 3＝5 μm, Figs 1, 4＝2 μm, Figs 2, 5-6＝1 μm.

Fig. 1.　Enlarged view of central area of valve face, opening of rimoportula (arrowhead).
Fig. 2.　Enlarged view of marginal area of valve face and mantle, opening of rimoportula (arrowhead) and opening of mantle fultoportulae (arrows).
Fig. 3.　Internal view of whole valve, rimoportula (arrowhead).
Fig. 4.　Enlarged view of Fig. 3 showing rimoportula (arrowhead) and valve face fultoportulae with three satellite pores.
Fig. 5.　Detailed view of valve margin with valvocopula (arrow) showing mantle fultoportulae with two lateral satellite pores on thick costae.
Fig. 6.　Detailed view of valve margin without valvocopula and thick costae with mantle fultoportulae.
Fig. 7.　View of the hook-like structure of the valvocopula.

Plate 68

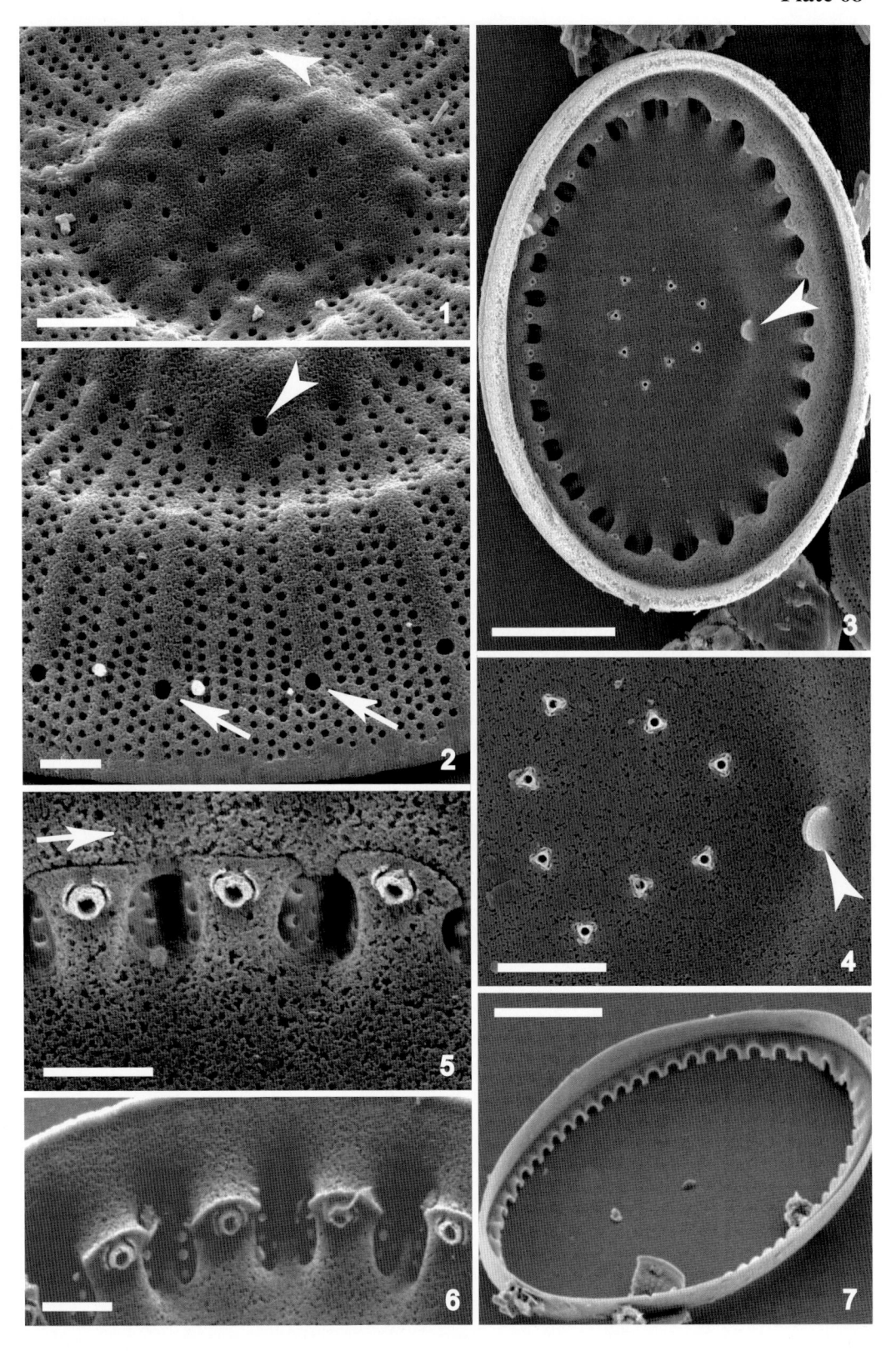

173

Plate 69. Centric diatoms: Thalassiosirales
Cyclotella rhomboideo-elliptica var. *rounda* Qi & Yang 1985

文　　献：Qi, Y. & Yang, J. 1985. New data on the early Pleistocene fossil diatoms from Miyi, Sichuan, China. Acta Micropalaeontologica Sinica **2**: 284-288, pls 1-2.

形　　態：殻は円盤形で殻面は横に波打つ，殻径 9-33 μm，条線は 10 μm に 11-13 本．間条線は 1-2(3) 本おきに太い間条線と細い間条線が交互に配列する．殻面有基突起は 4-20 個でリング状に分布する．殻套有基突起はすべての太い肋にあり，付随孔は 2 個である．唇状突起は光顕でも明瞭に観察でき，中心域の突出した部分と縁辺条線域の境に位置している．接殻帯片は厚く殻の長胞域まで入り込んで組み合わさっている．

産　　出：大戸湖沼性堆積物（田中 1991）：中期更新世・群馬県，芳野層（田中・北林 2011）：中期更新世・熊本県．

図　　版：大戸湖沼性堆積物標本．

Figs 1-12. *Cyclotella rhomboideo-elliptica* var. *rounda* Qi & Yang
LM. Figs 1-7. SEM. Figs 8-12. Material from lacustrine deposit of Odo (Middle Pleistocene), Higashiagatsuma Town, Gunma, Japan. Scale bars: Figs 8-9＝5 μm, Figs 10-12＝1 μm.

Figs 1-6.　Six different size valves.
Fig. 7.　Oblique view of valve margin.
Fig. 8.　Internal oblique view of valve with cingulum, rimoportula (arrowhead).
Fig. 9.　External oblique view, opening of rimoportula (arrowhead).
Fig. 10.　Enlarged view of part of Fig. 8 showing thinner costae and mantle fultoportulae with two lateral satellite pores on thicker costae.
Fig. 11.　Enlarged view of valve margin, opening of rimoportula (arrowhead).
Fig. 12.　Internal oblique view of valve margin, mantle fultoportulae (arrows).

Plate 69

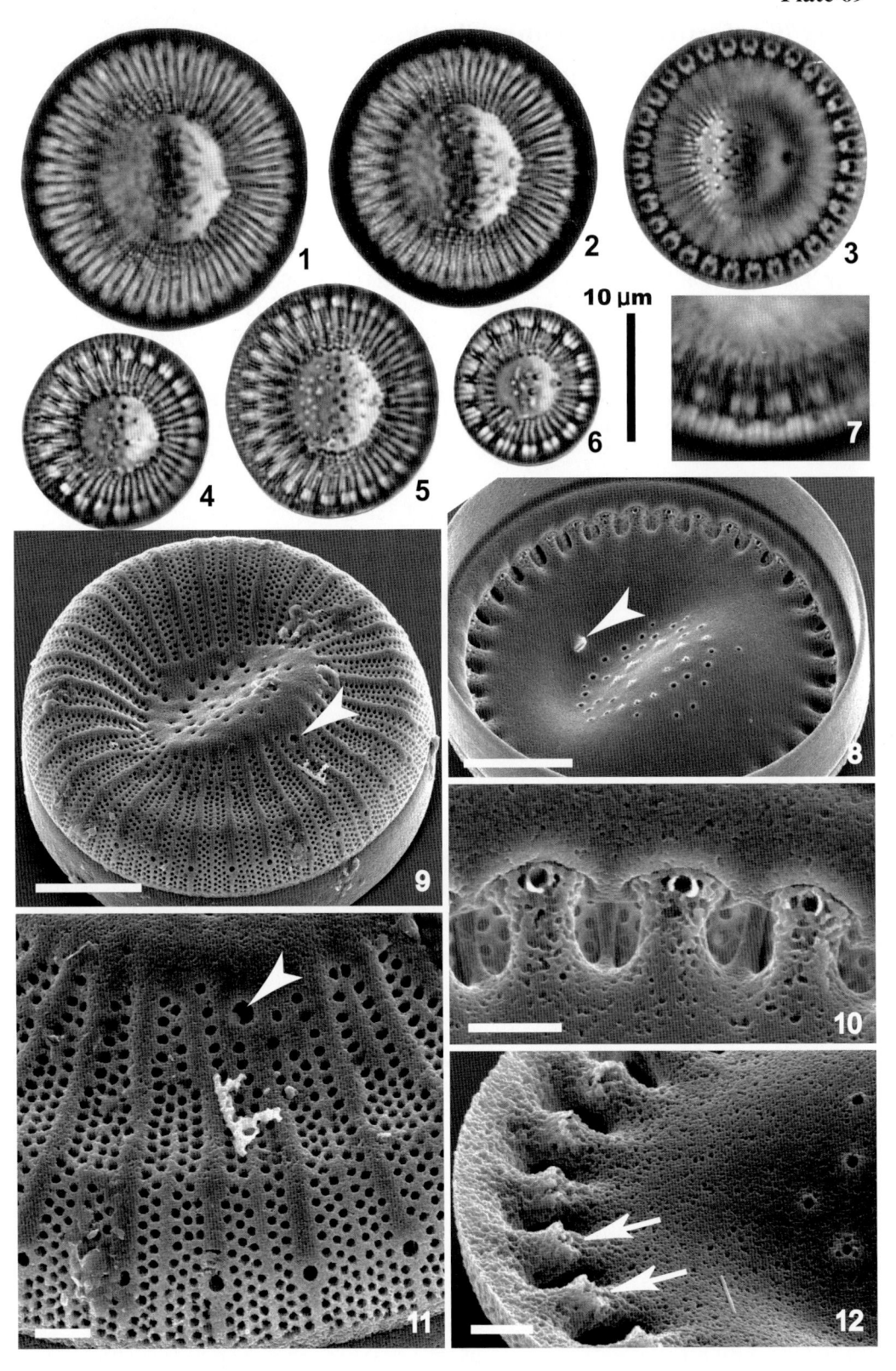

Plate 70. Centric diatoms: Thalassiosirales
Cyclotella satsumaensis H. Tanaka & Houk in Houk *et al.* 2010

文　献：Houk, V., Klee, R. & Tanaka, H. 2010. Atlas of fershwater centric diatoms with a brief key and descriptions. Part Ⅲ. Stephanodiscaceae A: *Cyclotella, Tertiarius, Discostella*. Fottea **10** Supplement, 498 pp, 215 pls.

形　態：殻は円盤状で小形，殻径 4.5-9.5 μm．殻面はほぼ平らで中心から 3(4) 方向に点紋状の放射状模様がある．縁辺域には 10 μm に 16-24 本の密度で条線が分布する．SEM 観察によれば殻面の放射状の模様は殻面を穿った穴であり，内部へは貫通していない．またふつう，放射状の模様 1 列に 1 個の殻面有基突起が存在する．内側殻縁の肋は (3) 4-8 (9) 本ごとに太くなり，これには殻套有基突起が所在する．唇状突起は 1 個で殻套有基突起がある肋に唇が横向きに所在する．殻面・殻套有基突起はともに付随孔は 2 個である．

ノート：光顕による殻面観の観察では，他の類似種との識別が難しい．特に *Cyclotella tripartita* Håkansson とはよく類似している．しかし，唇状突起が肋にあるのは類似種の中では *C. satsumaensis* のみである．

産　出：郡山層（Houk *et al.* 2010）：後期鮮新世・鹿児島県．

図　版：郡山層標本．

Figs 1-19. *Cyclotella satsumaensis* H. Tanaka & Houk

LM. Figs 1-12. SEM. Figs 13-19. From the type material, KAG-511, Koriyama Formation (Late Pliocene), Satsumasendai City, Kagoshima, Japan. Scale bars: Figs 13-17=1 μm, Figs 18-19=0.5 μm.

Figs 1-10.　Five different size valves.

Figs 1-2, 3-4, 5-6, 7-8, 9-10.　Five valves at different focal planes.

Figs 11-12.　Holotype specimen. Same valve at differernt focal planes. MPC-03019. Micropaleontology Collection, National Museum of Nature and Science, Tokyo, Japan.

Fig. 13.　External view of whole valve, opening of valve face fultoportula (arrow).

Fig. 14.　Internal view of whole valve with two valve face fultoportulae with two satellite pores each.

Fig. 15.　Oblique view of Fig. 13.

Fig. 16.　Oblique view of Fig. 14.

Fig. 17.　Detailed view of valve margin, opening of rimoportula (arrowhead), opening of mantle fultoportula (small arrow) and remains of ligula-like segment (large arrow).

Fig. 18.　Detailed view of internal valve showing rimoportula (arrowhead).

Fig. 19.　Detailed view of internal valve, rimoportula (arrowhead) and mantle fultoportulae with two satellite pores.

Plate 70

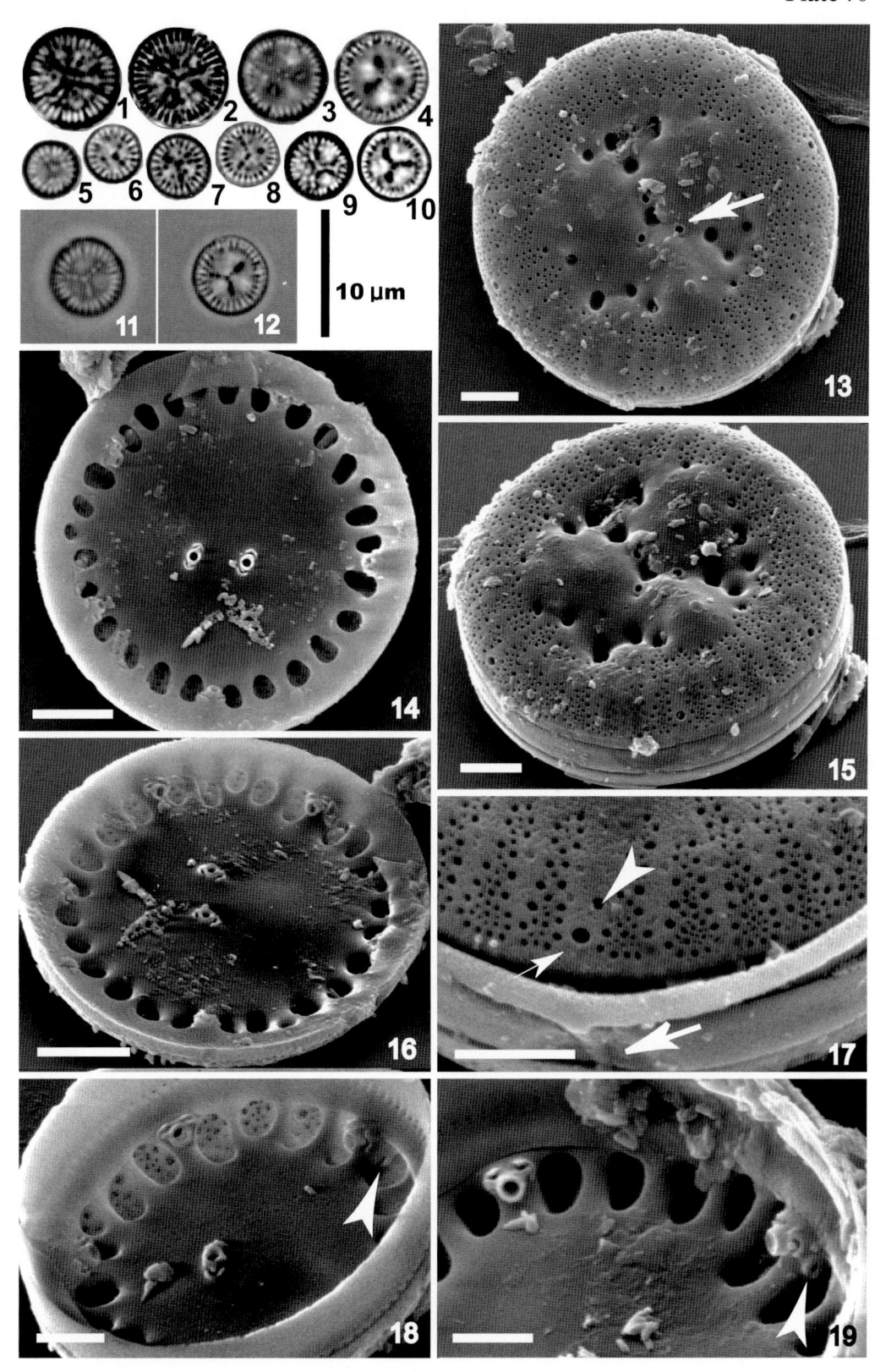

Plate 71.　Centric diatoms: Thalassiosirales
Cyclotella schumannii (Grunow) Håkansson 1990

基礎異名：*Cyclotella kuetzingiana* var. *schumannii* Grunow 1878

文　　献：Houk, V., Klee, R. & Tanaka, H. 2010. Atlas of fershwater centric diatoms with a brief key and descriptions. Part Ⅲ. Stephanodiscaceae A: *Cyclotella*, *Tertiarius*, *Discostella*. Fottea **10** Supplement, 498 pp, 215 pls.

形　　態：殻は円盤状で小形，殻径 7.5-15 μm．殻面中心域は丸みを帯びた楕円形で，細かい内部まで貫通していない凹凸があり横に波打つ．縁辺域の条線はやや長さが不規則で，殻面中心域との境がジグザグしている．条線密度は殻縁で 10 μm に 16-20 本である．SEM 観察によれば，殻面有基突起は 3-8 個で 2 個の付随孔を伴うが，まれに 1 個の場合がある．殻套有基突起は付随孔が 2 個で，毎 3-7 本ごとの肋にあり，この肋の太さは変わらないがわずか窪む場合が多い．唇状突起は 1 個で殻面縁辺部に所在し，外部への開口は間条線上にある．

ノ ー ト：本種のタイプスライドを検鏡した研究はいくつかあるが，Håkansson(2002)は併せてスウェーデンとフィンランド北方の湖から採取した材料を使用して SEM 観察を行い，本種の殻套有基突起は付随孔が 3 個であると報告した．他方 Houk *et al*.(2010)はタイプスライドのほか SEM 観察をタイプ試料，フィンランドの材料で行い，殻套有基突起は付随孔が 2 個であると報告した．本書では最近の研究である Houk *et al*.(2010) を引用し，阿蘇野層から見出した分類群の殻套有基突起は 2 個の付随孔をもつので本種に同定した．

産　　出：阿蘇野層：中期更新世・大分県．

図　　版：阿蘇野層標本．

Figs 1-11.　*Cyclotella schumannii* (Grunow) Håkansson

LM. Figs 1-4. SEM. Figs 5-11. Material from Asono Formation (Middle Pleistocene), Yufu City, Oita, Japan. Scale bars: Figs 5-6, 8-9＝2 μm, Figs 7, 11＝1 μm, Fig. 10＝0.5 μm.

Figs 1-4.　Four different size valves.

Fig. 5.　External view of whole valve.

Fig. 6.　Oblique view of Fig. 5.

Fig. 7.　Enlarged view of marginal valve face and mantle, opening of rimoportula (arrowhead).

Fig. 8.　Internal view of whole valve, valve face fultoportula (arrow) and rimoportula (arrowhead).

Fig. 9.　Oblique view of Fig. 8, rimoportula (arrowhead).

Fig. 10.　Detailed view of a mantle fultoportula with two lateral satellite pores located on a slightly recessed costa.

Fig. 11.　Detailed view of valve face fultoportulae with two satellite pores (arrow).

Plate 71

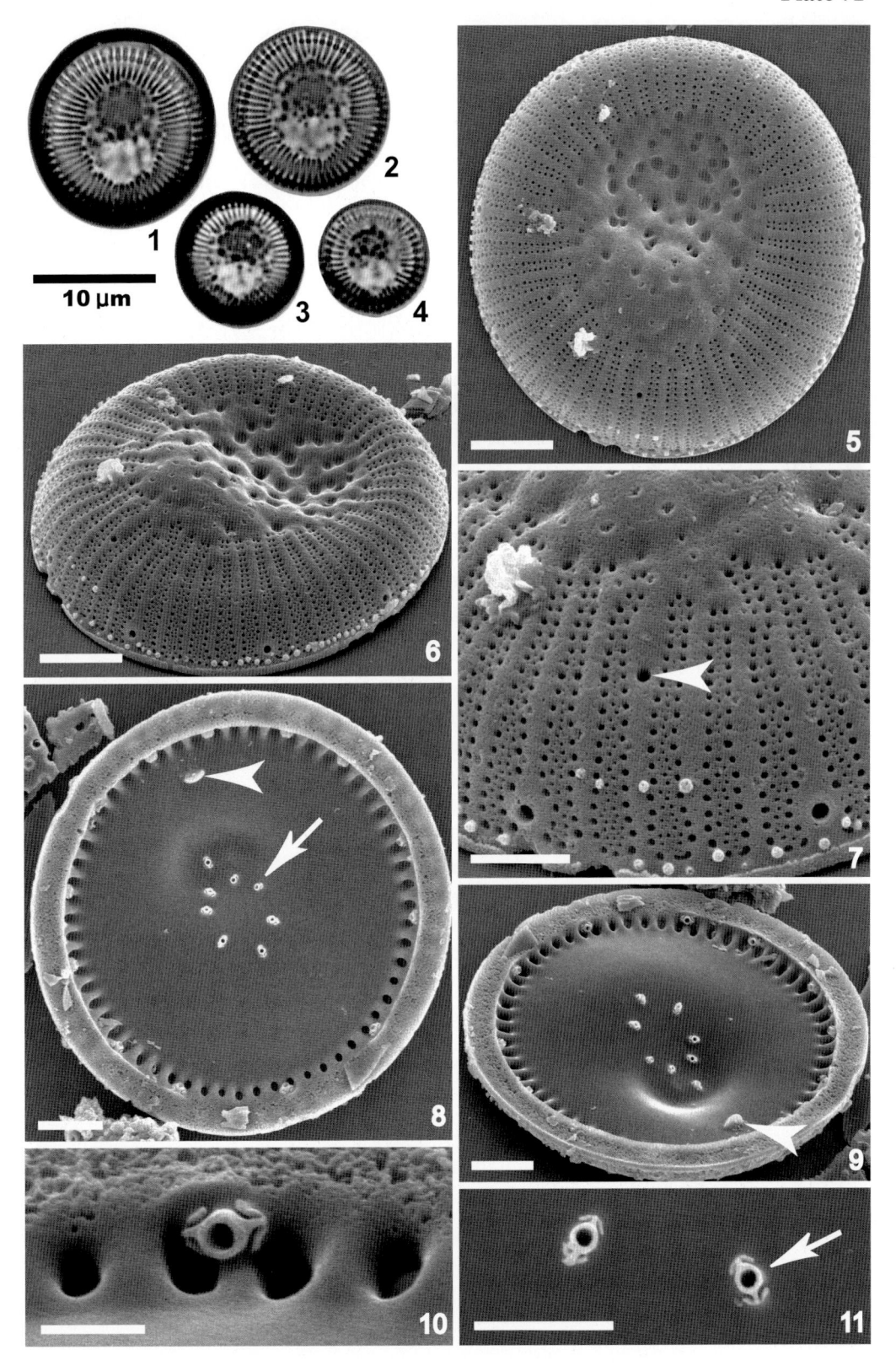

Plate 72. Centric diatoms: Thalassiosirales
Cyclotubicoalitus undatus Stoermer, Kociolek & Cody 1990

文　　献：Stoermer, E.F., Kociolek, J.P. & Cody, W. 1990. *Cyclotubicoalitus undatus*, genus et species nova. Diatom Research **5**: 171-177.
　　田中宏之・南雲　保 2000. 本邦新産属珪藻 *Cyclotubicoalitus undatus* Stoermer, Kociolek & Cody (Centrales, Bacillariophyceae). 藻類 **48**: 105-108.

形　　態：殻は円盤状，殻面は偏心的にうねる．殻径 12.5-15 μm．条線列は放射状に射出されるが，放射中心は殻中心とは一致しない．条線数は 10 μm に 19-26 本．条線を構成する胞紋の数は 10 μm に 18-25 個である．光顕観察で殻縁付近に黒く見えるのが殻套有基突起の開口の管とへら状の針が癒着したものである．殻面有基突起は存在しない．殻套有基突起は 2 個の付随孔をもつ．唇状突起は 1 個で殻套有基突起と同じ高さにある．

ノ　ー　ト：類似する *Stephanodiscus*, *Cyclostephanos* 属とは殻面の条線が単列であること，殻縁の針の形態から区別できる．*Cyclostephanos* 属とは *C. undatus* に肋がないことも区別点である．*Pliocaenicus* 属とは殻面が同心円状にうねることで区別できる．

産　　出：波志江沼・多々良沼・城沼（田中・南雲 2000）：群馬県（現生），高田城外堀（田中 2009）：新潟県（現生）．

図　　版：波志江沼（Figs 1-4, 9-14），城沼（Figs 5-8）標本．

Figs 1-14.　*Cyclotubicoalitus undatus* Stoermer, Kociolek & Cody
LM. Figs 1-8. SEM. Figs 9-14. Materials, Figs 1-4, 9-14, from Hashie-numa Pond (Recent), Isesaki City, Gunma and Figs 5-8, from Jo-numa Pond (Recent), Tatebayashi City, Gunma, Japan. Scale bars: Figs 9-10＝5 μm, Figs 11-14＝1 μm.

Figs 1-2, 3-4, 5-6, 7-8.　Four valves at different focal planes.
Fig. 9.　External oblique view of whole valve.
Fig. 10.　Internal oblique view of whole valve, rimoportula (arrowhead).
Fig. 11.　Detailed view of external valve margin, spatulated spine (arrow) and opening of mantle fultoportula (arrowhead).
Fig. 12.　Internal view of valve margin, rimoportula (arrowhead) and mantle fulutoportula with two lateral satellite pores (arrow).
Fig. 13.　Opening of mantle fultoportulae (arrowheads) and spatulated spines (arrows).
Fig. 14.　Internal view of valve margin.

Plate 72

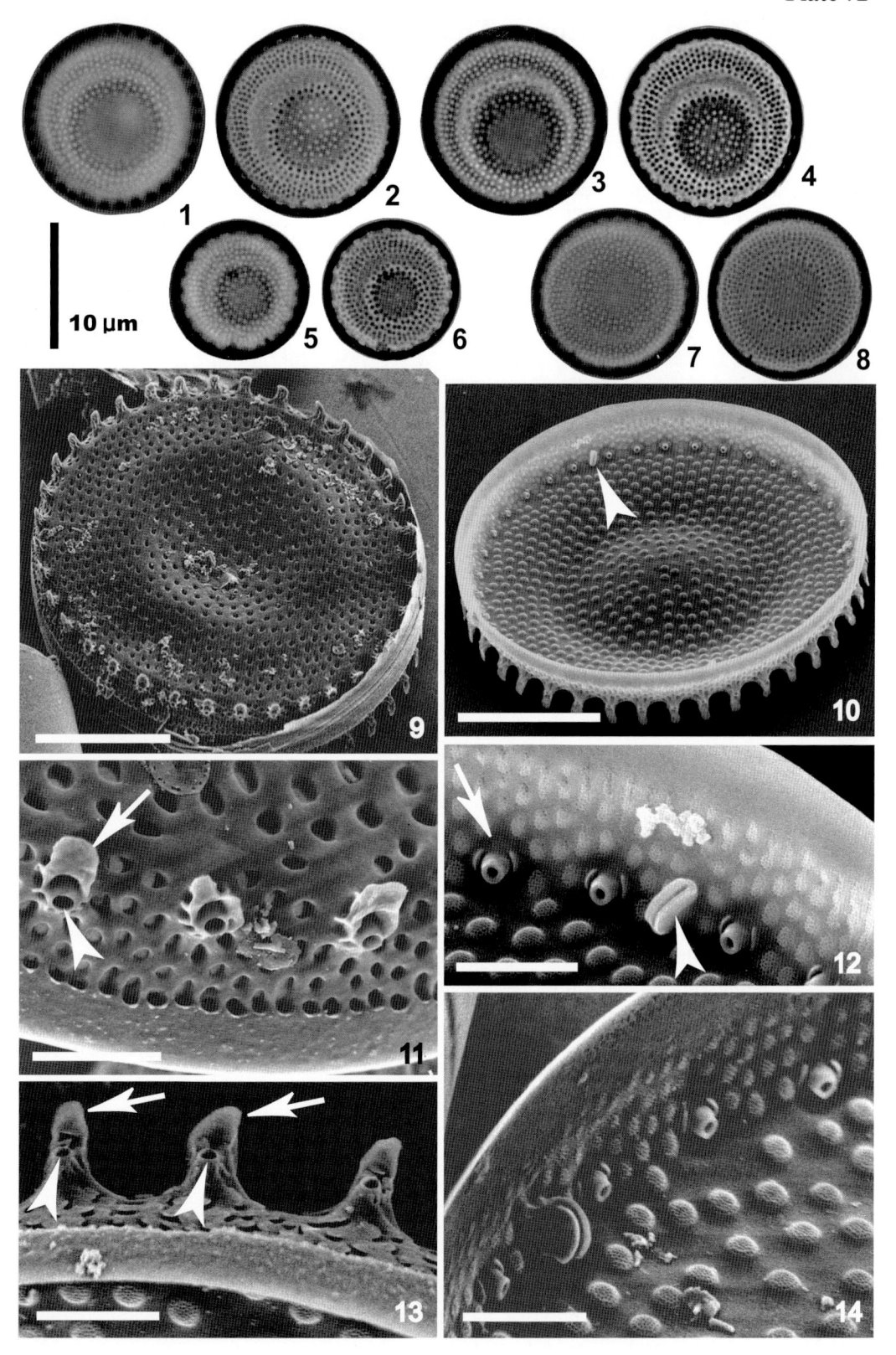

10 µm

Plate 73.　Centric diatoms: Thalassiosirales
Dimidialimbus bungoensis H. Tanaka in Tanaka & Nagumo 2013

文　　献：Tanaka, H. & Nagumo, T. 2013. *Dimidialimbus bungoensis* gen. nov. and sp. nov. (Stephanodiscaceae) a new diatom genus from Early Pleistocene sediment, Kyushu, Japan. Diatom **29**: 13-19.

形　　態：殻は小形で，大形の殻は円形に近くなるが一般的には楕円形である，長径 7-21 μm，短径 5-21 μm．殻面はわずか横に波打ち，殻套の存在する側が高くなり，ない側に向かって斜めに下がる．中心から放射状に配列する点紋列は殻縁で 10 μm に約 14 本．中心部に胞紋とは異なる小さい点が認められるが，SEM での観察によると殻面有基突起である，5-9 個．殻套部は唇状突起が存在する側だけにあり，唇状突起付近で最も高く，徐々に低くなり殻の半分で消滅する．殻套の反対側は殻の縁が広く厚く存在する．殻套有基突起は殻套にある肋の下方に所在する．殻面・殻套有基突起とも付随孔は2個である．唇状突起は1個で殻面／殻套境界に位置するが，唇状突起が所在する肋（外側では殻套の胞紋列を区分する縦長の無紋域）は幅が広く，しばしば隣の有基突起が所在する肋と融合している．

ノ ー ト：属名は殻套が殻の半分しかないことに由来する．現在のところ1属1種である．

産　　出：尾本層（Tanaka & Nagumo 2013）：前期更新世・大分県．

図　　版：尾本層標本．

Figs 1-13.　*Dimidialimbus bungoensis* H. Tanaka

LM. Figs 1-4. SEM. Figs 5-9. From the type material, OIT-330, Omoto Formation (Early Pleistocene), Kitsuki City, Oita, Japan. Scale bars: Figs 5-8＝2 μm, Fig. 9＝1 μm.

Figs 1-4.　Three different size valves.
Figs 2-3.　Same valve at different focal planes.
Fig. 5.　External view of whole valve, openings of mantle fultoportulae (arrows) and opening of rimoportula (arrowhead).
Fig. 6.　Oblique view of Fig. 5.
Fig. 7.　Oblique view showing slanted valve face and mantle only on the side with the opening of rimoportula (arrowhead).
Fig. 8.　Internal view of whole valve, valve face fultoportulae with two satellite pores in center, mantle fultoportulae (arrows) and rimoportula (arrowhead).
Fig. 9.　Oblique view showing diminishing mantle (arrowhead) and mantle fultoportulae with two lateral satellite pores (arrows).

Plate 73

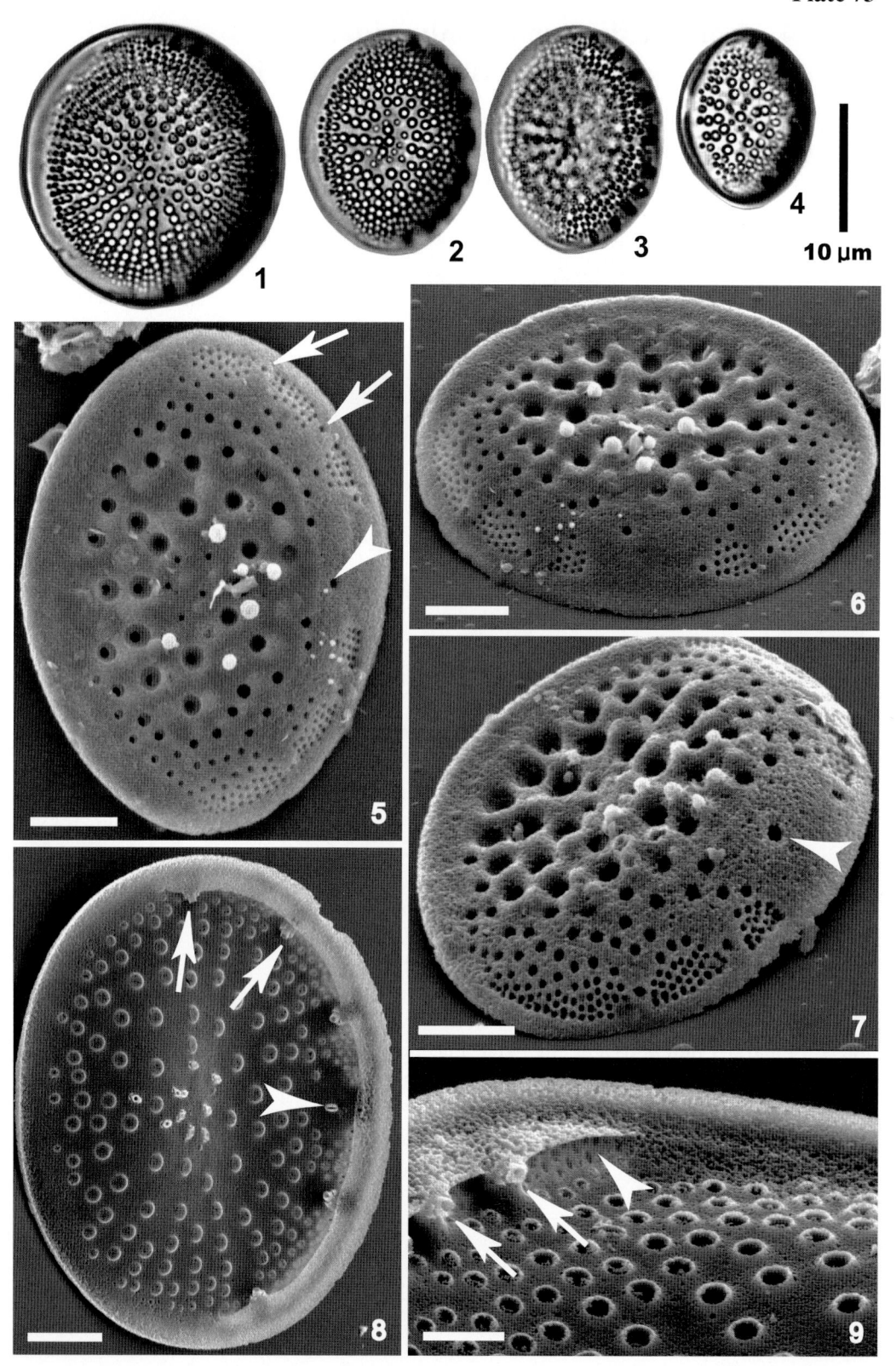

10 µm

Plate 74. Centric diatoms: Thalassiosirales
Discostella asterocostata (Xie, Lin & Cai) Houk & Klee 2004

基礎異名：*Cyclotella asterocostata* Xie, Lin & Cai 1985

文　　献：Houk, V. & Klee, R. 2004. The stelligeroid taxa of the genus *Cyclotella* (Kützing) Brébisson (Bacillariophyceae) and their transfer into the new genus *Discostella* gen. nov. Diatom Research **19**: 203-228.

形　　態：殻は円盤状で，殻径 11.5-24 μm，中心域には星形の模様があり，突出あるいは凹む．縁辺域の条線分布域との間には中心域を取り巻く無紋域の部分で区別される．SEM 観察によると，中心域が突出する殻では星形の模様は条線からできているが，凹殻では殻面の凹凸による模様である．内側殻套には毎 3-4 本ごとの肋の間に，2 個の付随孔を伴う殻套有基突起が殻面から連続する条線外側末端にある．殻套有基突起から殻端に向かって短い肋が生じているので，有基突起から外側は条線が 2 本になる．唇状突起は 1 個で条線上にあり，殻套有基突起よりも外側に所在する．

ノート：Xie *et al.*(1985) によって中国の 5 地点から見出され *Cyclotella asterocostata* として原記載されたが，Houk & Klee(2004) によって，*Discostella* 属設立と同時に本属へ組み合わせになった種である．
　光顕観察において類似する *Discostella stelligera* とは，比較的広い殻中心の無紋域，明瞭な殻套有基突起とそこから殻端までの短肋の存在などから区別することができる．

産　　出：大沼（渡島半島）：北海道（現生），波志江沼（田中・南雲 2007）・多々良沼・城沼：群馬県（現生），琵琶湖（辻・伯耆 2001）：滋賀県（現生）．このほか日本各地の湖沼から産する．

図　　版：波志江沼（Figs 1-3, 5-10），多々良沼（Fig. 4），城沼（Fig 11）標本．

Figs 1-11.　*Discostella asterocostata* (Xie, Lin & Cai) Houk & Klee

LM. Figs 1-5. SEM. Figs 6-11. Materials, Figs 1-3, 5-10, from Hashie-numa Pond (Recent), Isesaki City, Gunma, Fig. 4, from Tatara-numa Pond (Recent), Tatebayashi City, Gunma, Fig. 11, from Jo-numa Pond (Recent), Tatebayashi City, Gunma, Japan. Scale bars: Figs 6-8=5 μm, Fig. 11=2 μm, Figs 9-10=1 μm.

Figs 1-5.　Three different size valves.
Figs 1-3.　Same valve at different focal planes.
Fig. 6.　External oblique view of whole convex valve.
Fig. 7.　External oblique view of whole concave valve.
Fig. 8.　Internal oblique view of whole convex valve showing striae on central area.
Fig. 9.　Enlarged view of part of Fig. 7 showing outer tubes of mantle fultoportulae (arrows).
Fig. 10.　Detailed internal view of valve margin showing rimoportula (arrowhead) and fulutoportula with two lateral satellite pores (arrow).
Fig. 11.　Internal oblique view of whole concave valve showing smooth central area.

Plate 74

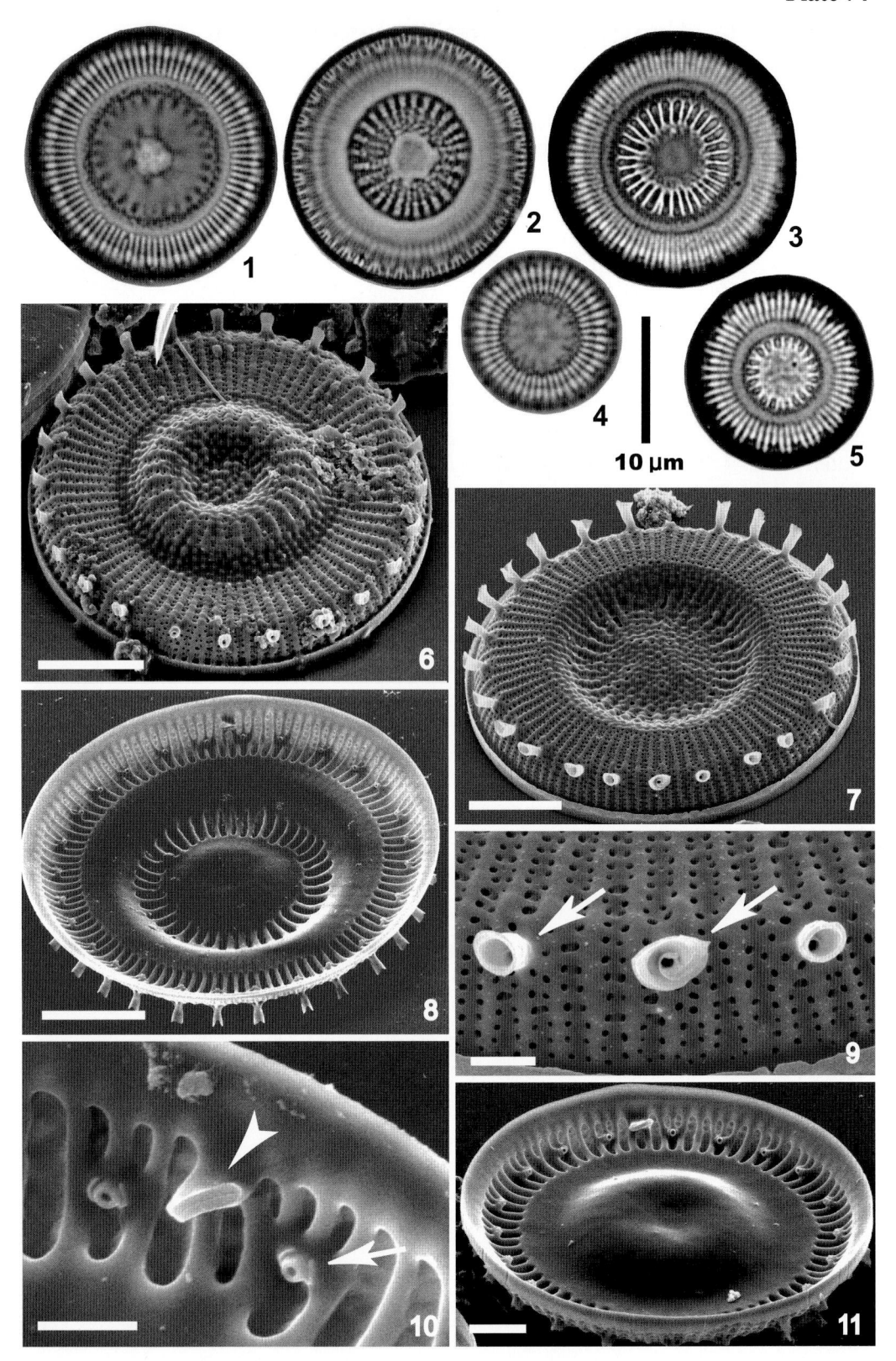

10 μm

Plates 75-76. Centric diatoms: Thalassiosirales
Discostella kitsukiensis H. Tanaka sp. nov.

新　　種：記載文（英文）は 7 頁参照．
形　　態：殻は円盤形で，殻径 9-20 μm，中心域は全体的に平らであるが，多数の垂直に立ち上がるでこぼこがある．縁辺域の条線は多数の小胞紋列からなり 10 μm に 8-13 本である．条線は殻套まで連続し，この末端には殻套有基突起の開口が所在するが，殻縁と殻套との境界に沿って円状に 1 列の胞紋列があり，殻套有基突起の開口はこの列の中にある．内側では，殻套有基突起は 2 個の付随孔を伴い，各条線の末端側にある．殻面有基突起は存在しない．唇状突起は 1 個で，条線の末端側で殻套有基突起と同じ位置に所在する．
ノ ー ト：種小名はタイプ地の杵築市に因んで付けられた．本種は大分県杵築市の海岸に露出する加貫層の露頭から採取された試料から見出された．
産　　出：加貫層：前期～中期更新世・大分県．
図　　版 75-76：加貫層標本．

Plate 75, Figs 1-9.　*Discostella kitsukiensis* H. Tanaka sp. nov.

LM. Figs 1-5. SEM. Figs 6-9. From the type material, OIT-511, Kanuki Formation (Early-Middle Pleistocene) from an outcrop of a sea cliff of Kanuki in Kitsuki City, Oita, Japan. Scale bars: Figs 6-7＝2 μm, Fig. 8＝1 μm, Fig. 9＝0.5 μm.

Figs 1-2.　Holotype specimen at different focal planes.
Figs 3-5.　Three different size valves.
Fig. 6.　View of entire valve.
Fig. 7.　Oblique view of Fig. 6.
Fig. 8.　Detailed view of marginal area of valve face showing striae and interstriae.
Fig. 9.　Detailed view of mantle showing one areolae row along the valve rim and openings of mantle fultoportulae (arrows).

Plate 75

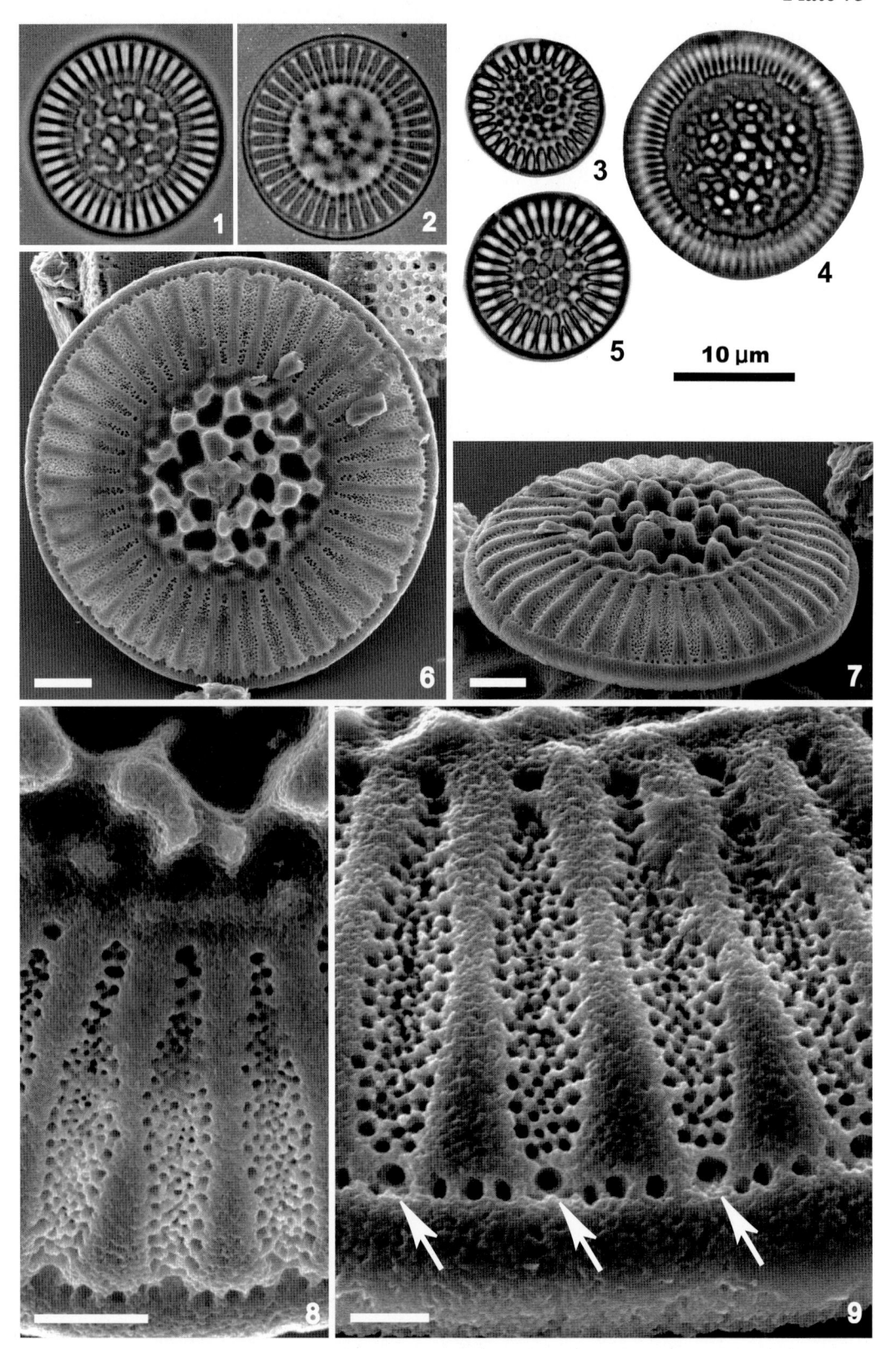

10 µm

Plate 76, Figs 1-5. *Discostella kitsukiensis* **H. Tanaka** sp. nov.

SEM Figs 1-6. Internal views. From the type material, OIT-511, Kanuki Formation (Early-Middle Pleistocene) from an outcrop of a sea cliff of Kanuki in Kitsuki City, Oita, Japan. Scale bars: Figs 1, 3=2 μm, Figs 2, 4-6=0.5 μm.

Fig. 1. Internal view of whole valve showing smooth central area of valve face.

Fig. 2. Detailed view of alveoli openings, mantle fultoportula (arrow).

Fig. 3. Oblique view of Fig. 1, rimoportula (arrowhead).

Fig. 4. Detailed view of valve margin, mantle fultoportula (arrow) and rimoportula (arrowhead).

Fig. 5. Enlarged view of part of Fig. 3, mantle fultoportulae with two lateral satellite pores (arrows) and rimoportula (arrowhead). Note: mantle fultoportulae occur in the ring of areolae on the lowest part of mantle.

Plate 76

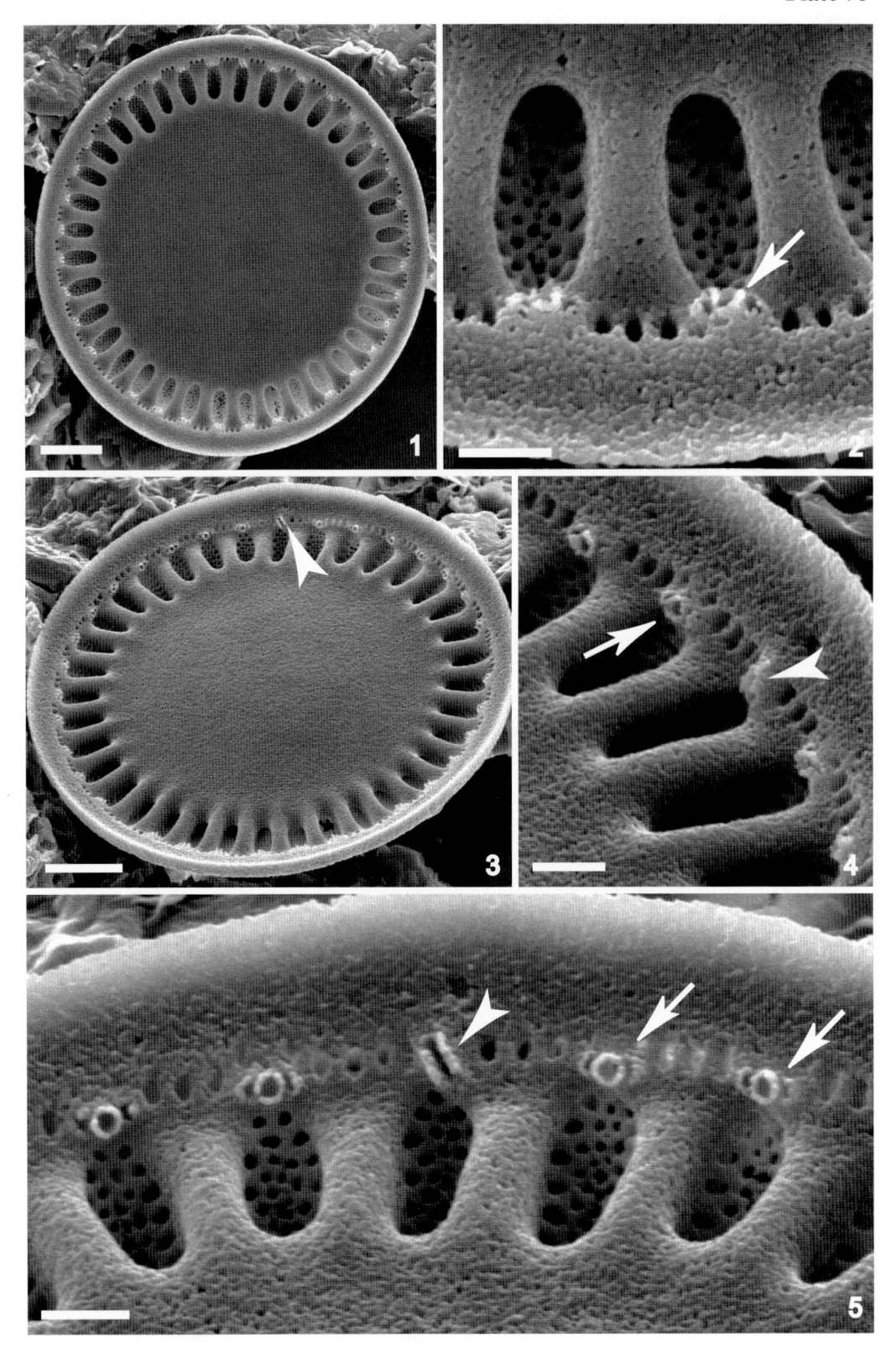

Plates 77-78. Centric diatoms: Thalassiosirales
Discostella pliostelligera (H. Tanaka & Nagumo) Houk & Klee 2004

基礎異名：*Cyclotella pliostelligera* H. Tanaka & Nagumo 2002

文　　献：Tanaka, H. & Nagumo, T. 2002.　*Cyclotella pliostelligera* sp. nov., a new fossil freshwater diatom from Japan. *In*: John, J.(ed.) Proceedings of the 15th International Diatom Symposium, p. 351-358, Perth 1998, Gantner Verlag, Liechtenstein.

　　　Houk, V. & Klee, R. 2004.　The stelligeroid taxa of the genus *Cyclotella* (Kützing) Brébisson (Bacillariophyceae) and their transfer into the new genus *Discostella* gen. nov. Diatom Research **19**: 203-228.

形　　態：殻は円盤形で，殻径 8.5-24.5 μm，中心域はゆるく凹または凸状で，多数の内部まで穿孔していない小穴が所在する．殻面有基突起は存在しない．縁辺域の条線は 10 μm に 10-16 本である．殻套有基突起は 2 個の付随孔を伴い，毎〜2 本ごとの条線の延長上で，殻面／殻套境界の殻套側にある．唇状突起は 1 個で殻套有基突起よりやや殻縁寄りにある．

ノ ー ト：秋田県田沢湖北岸に露出する宮田層から見出され *Cyclotella pliostelligera* として記載された (Tanaka & Nagumo 2002)．現在のところ本地点のみの産出である．後に Houk & Klee(2004) によって *Discostella* 属設立と同時に本属へ組み合わせになった．

産　　出：宮田層 (Tanaka & Nagumo 2002)：中新-鮮新世・秋田県．

図　　版 77-78：宮田層標本．

Plate 77, Figs 1-11.　*Discostella pliostelligera* (H. Tanaka & Nagumo) Houk & Klee

LM. Figs 1-7. SEM. Figs 8-11. From the type material, TAZ-101, Miyata Formation (Mio-Pliocene). Outcrop of northeast shore of Lake Tazawa (Tazawa-ko), Senboku City, Akita, Japan. Scale bars: Figs 8, 10＝5 μm, Figs 9, 11＝2 μm.

Figs 1-7.　Five different size valves.

Figs 1-2, 3-4.　Two valves at different focal planes.

Fig. 8.　View of entire valve.

Fig. 9.　Oblique view of convex valve.

Fig. 10.　Oblique view of concave valve of Fig. 8, external openings of mantle fultoportulae with short tubes (arrows).

Fig. 11.　Detailed view of external valve face.

Plate 77

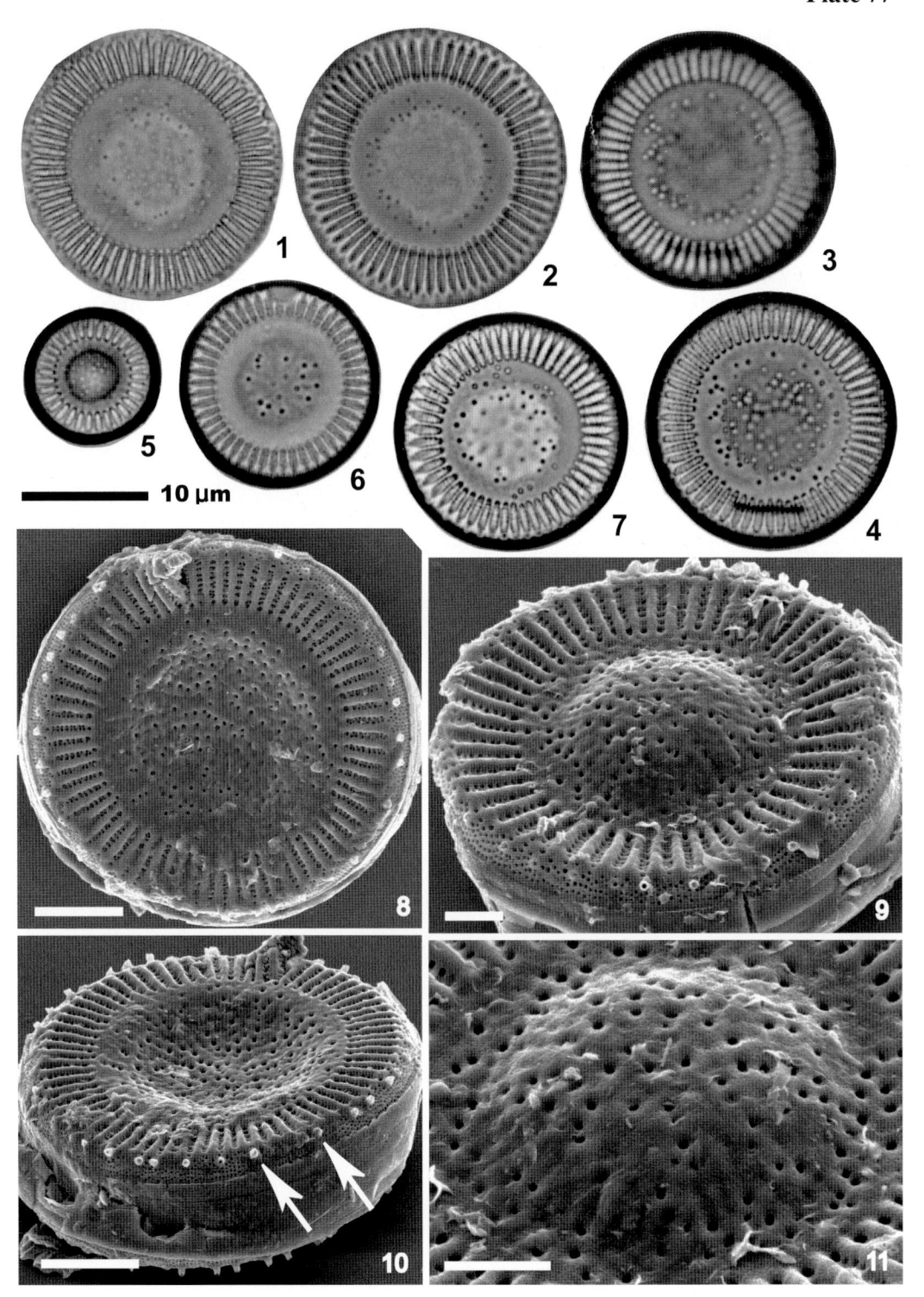

Plate 78, Figs 1-7. *Discostella pliostelligera* **(H. Tanaka & Nagumo) Houk & Klee**

SEM. Figs 1-3, 5, 7. Internal views. Figs 4, 6. External views. From the type material, TAZ-101, Miyata Formation (Mio-Pliocene), Senboku City, Akita, Japan. Scale bars: Figs 1-2 =5 μm, Fig. 3=2 μm, Figs 4, 6=1 μm, Figs 5, 7=0.5 μm.

Fig. 1. Internal view of whole concave valve, smooth central area of valve face, rimoportula (arrowhead).

Fig. 2. Oblique view of convex valve, rimoportula (arrowhead).

Fig. 3. Oblique view of concave valve, smooth central area of valve face.

Fig. 4. Detailed external view of marginal area of valve face with striae and interstriae.

Fig. 5. Detailed view of internal alveous of broken valve.

Fig. 6. Detailed external view of mantle showing small areolae in parallel rows and outer openings of mantle fultoportulae with tubes (arrows).

Fig. 7. Internal valve margin, rimoportula (arrowhead) and mantle fultoportulae with two lateral satellite pores (arrows).

Plate 78

Plate 79. Centric diatoms: Thalassiosirales
Discostella pseudostelligera (Hustedt) Houk & Klee 2004

基礎異名：*Cyclotella pseudostelligera* Hustedt 1939

文　　献：Houk, V., Klee, R. & Tanaka, H. 2010. Atlas of freshwater centric diatoms with a brief key and descriptions. Part Ⅲ. Stephanodiscaceae A: *Cyclotella, Tertiarius, Discostella*. Fottea **10** Supplement, 498 pp, 215 pls.

形　　態：殻は小形円盤状で，殻径 4.5-12 μm，条線は 10 μm に 18-24 本，殻面には有基突起は存在せず，殻面中心域は直径の 3 分の 1～2 分の 1 程度であるが，さらに狭い場合もある．殻套有基突起は 3-8 本ごとの条線に所在し 2 個の付随孔を伴い，外管は特徴的な装飾があり，本種の特徴の一つになっている（Haworth & Hurley 1986）．

ノ ー ト：変異が大きい種であり，Houk *et al.*(2010) も述べているように，類似する *D. wolterekii* 等と識別が難しい．Houk *et al.*(2010) は，両種は生態が異なっていると記しているが，これに加えて本種は殻套有基突起外管の装飾が特徴的なので（Haworth & Hurley 1986）併せて検討すると識別が容易である．

産　　出：波志江沼：群馬県（現生）．

図　　版：波志江沼標本．

Figs 1-11. *Discostella pseudostelligera* (Hustedt) Houk & Klee
LM. Figs 1-5. SEM. Figs 6-11. Material from Hashie Pond (Recent), Isesaki City, Gunma, Japan. Scale bars: Fig. 7=2 μm, Figs 6, 10=1 μm, Figs 8-9, 11=0.5 μm.

Figs 1-5.　Five different size valves.

Fig. 6.　External oblique view of convex valve, unique ornamentation of outer-tube of mantle fultoportula (arrow).

Fig. 7.　Oblique view of concave valve.

Fig. 8.　Detailed view of marginal area of valve face and mantle with two outer-tubes of mantle fultoportulae.

Fig. 9.　Enlarged internal view, mantle fultoportula with two lateral satellite pores (arrow) and rimoportula (arrowhead).

Fig. 10.　Internal oblique view of convex valve, mantle fultoportula (arrow) and rimoportula (arrowhead).

Fig. 11.　Detailed view of internal central area of convex valve.

Plate 79

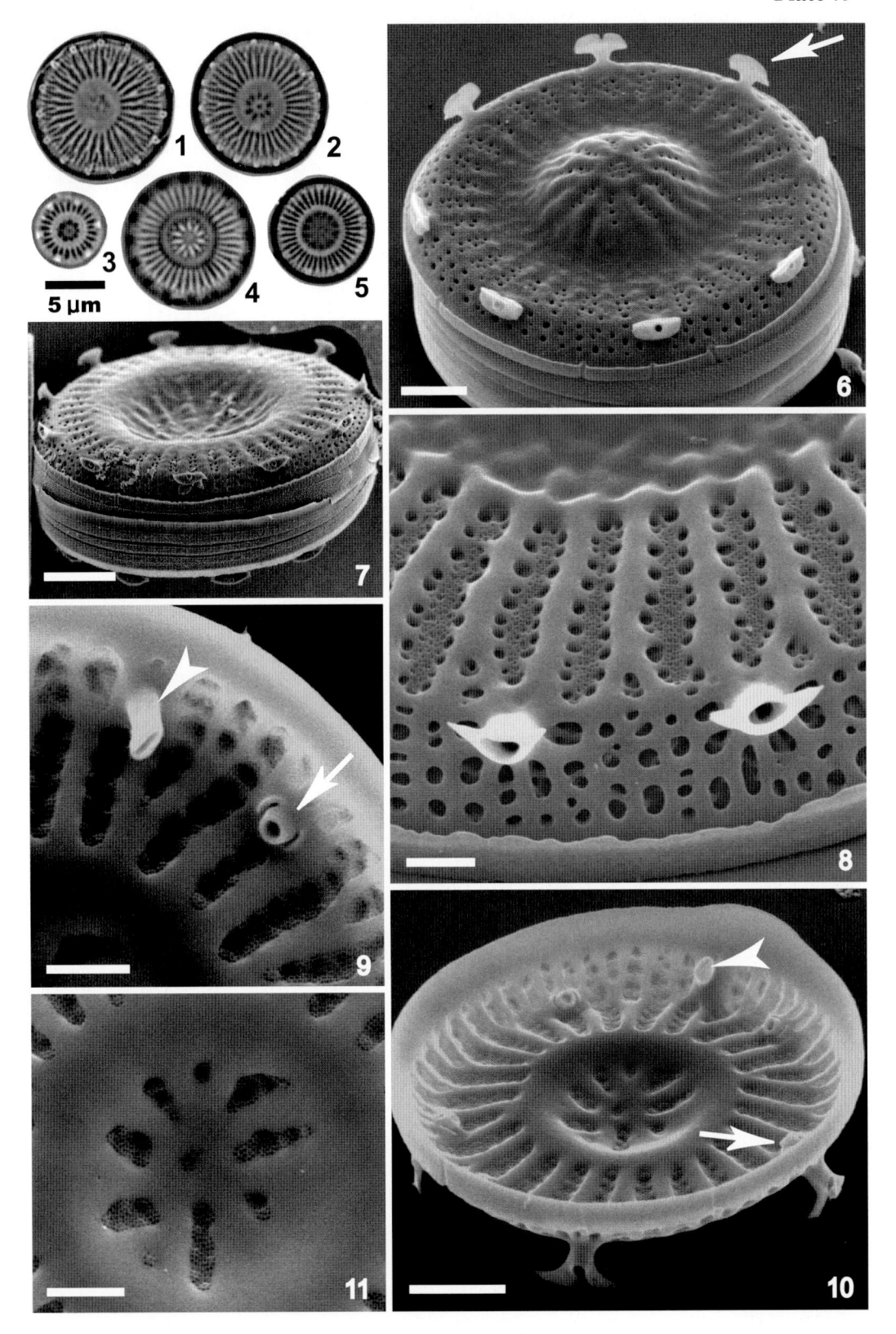

Plate 80.   Centric diatoms: Thalassiosirales
Discostella stelligera (Cleve & Grunow) Houk & Klee 2004

基礎異名：*Cyclotella meneghiniana* var.？*stelligera* Cleve & Grunow 1881
文　　献：Houk, V., Klee, R. & Tanaka, H. 2010. Atlas of freshwater centric diatoms with a brief key and descriptions. Part Ⅲ. Stephanodiscaceae A: *Cyclotella*, *Tertiarius*, *Discostella*. Fottea **10** Supplement, 498 pp, 215 pls.
形　　態：殻は円盤状で中心は突出または凹む．殻径 7.5-14 μm，条線は 10 μm に約 12 本．殻面には有基突起はなく，中心域は凸形殻では星形に条線が配列するが，凹殻では条線はなく外側表面に星形の凹凸が観察される．殻套有基突起は（毎）2-4 本ごとの殻縁条線の殻套延長上（肋の間）に所在し 2 個の付随孔を伴う．一般に殻套有基突起の開口には短管があるが，肥厚のみの場合もある．唇状突起は 1 個で殻套有基突起と同じ位置にある．
ノート：宮田層から見出された本種は，比較的小形殻も含むことから *D. stelligera* var. *tenuis*(Hustedt)Houk & Klee とも考えられるが，*D. stelligera* var. *stelligera* のタイプスライドおよびタイプ試料を調査した Houk *et al.*(2010)は本種の殻径を 5-40 μm と報告している．Houk *et al.*(2010)は *D. stelligera* var. *tenuis* の選定基準標本スライドも観察し，Hustedt(1937)では 5-7 μm と記されているが，殻径が 4.3 μm であったと記している．殻径から考えても宮田層産の分類群は *D. stelligera* var. *stelligera* に同定できると思われる．
産　　出：宮田層：中新-鮮新世・秋田県，中之条湖成層：中期更新世・群馬県，芳野層(田中・北林 2011)：中期更新世・熊本県．
図　　版：宮田層標本．

Figs 1-11.   *Discostella stelligera* (Cleve & Grunow) Houk & Klee
LM. Figs 1-4. SEM. Figs 5-11. Material from Miyata Formation (Mio-Pliocene), Senboku City, Akita, Japan. Scale bars: Figs 5, 7＝5 μm, Figs 6, 8, 11＝2 μm, Fig. 9＝1 μm, Fig. 10 ＝0.5 μm.

Figs 1-4.   Four different size valves.
Fig. 5.   External view of whole convex valve with bands.
Fig. 6.   Oblique view of Fig. 5.
Fig. 7.   Enlarged view of part of Fig. 6, opening of mantle fultoportula (arrow).
Fig. 8.   Oblique view of concave valve.
Fig. 9.   Internal oblique view of convex valve.
Fig. 10.   Internal oblique view of concave valve, rimoportula (arrowhead).
Fig. 11.   Detailed view of internal mantle showing rimoportula (arrowhead) and mantle fultoportulae with two lateral satellite pores (arrows).

Plate 80

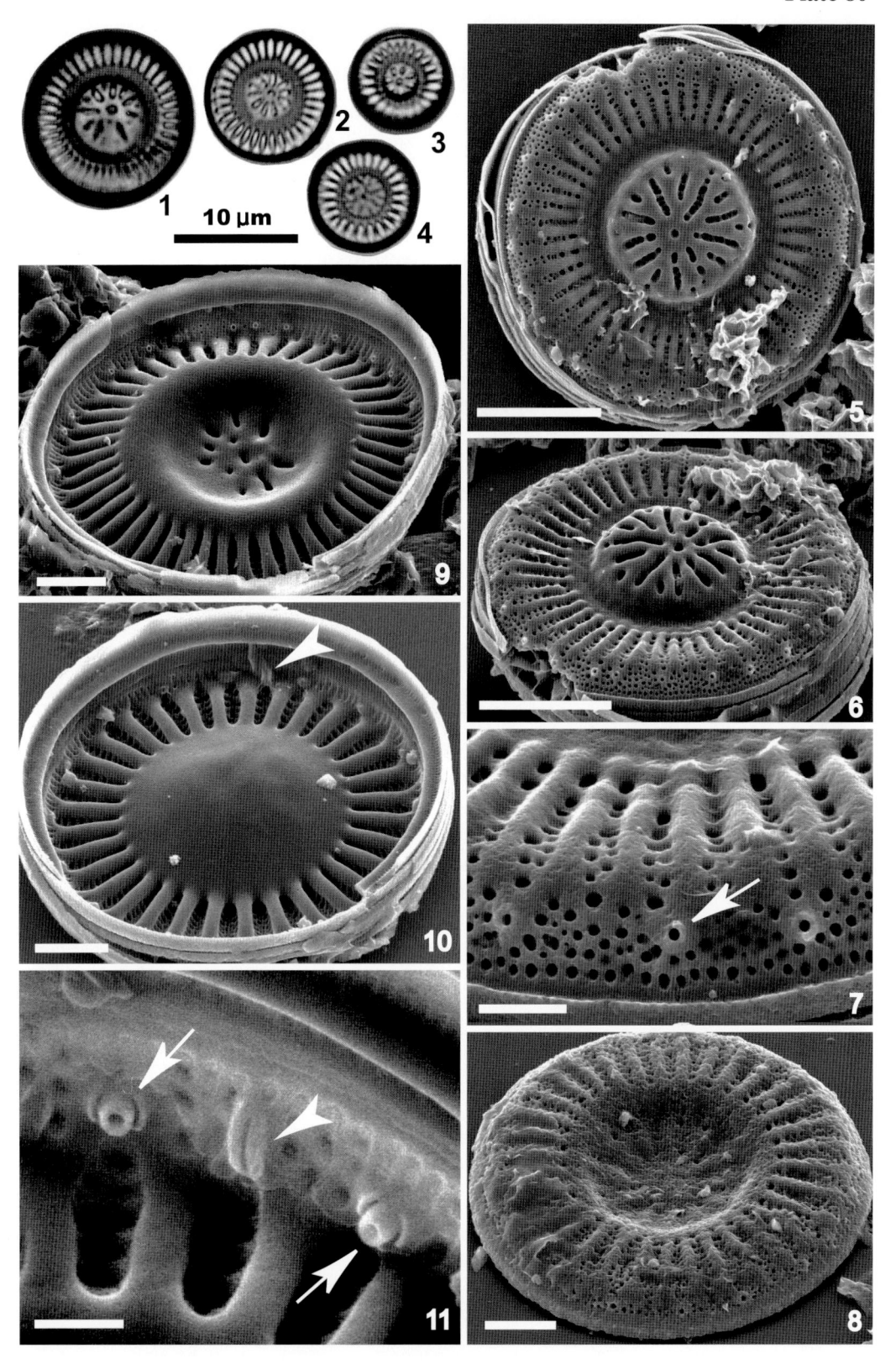

197

Plate 81.　Centric diatoms: Thalassiosirales
Discostella woltereckii (Hustedt) Houk & Klee 2004

基礎異名：*Cyclotella woltereckii* Hustedt 1942
文　　献：Houk, V., Klee, R. & Tanaka, H. 2010. Atlas of freshwater centric diatoms with a brief key and descriptions. Part Ⅲ. Stephanodiscaceae A: *Cyclotella*, *Tertiarius*, *Discostella*. Fottea **10** Supplement, 498 pp, 215 pls.
形　　態：殻径3.5-9 μm，条線は10 μmに20-22本．殻面に有基突起は存在しない．殻套有基突起は(5)6-8(10)本ごとの条線に所在する．殻面中心域は非常にわずかから直径の2分の1程度までを占め，非常に狭い場合は構造物（装飾）は見当たらないが，広くなると星形模様が観察できる．間条線はしばしば枝分かれをする．一般に殻套有基突起の開口には管があるが，肥厚のみの場合もある．有基突起は2個の付随孔を伴う．
ノ ー ト：変異が大きい種であり，Hustedt(1942)のタイプ試料を調査したKlee & Houk (1996)は，Hustedt(1942)の原記載の図と同様の個体を見出したが，殻の中心に星形模様のある殻に形態が連続することを見出し，これらをすべて *D. woltereckii* とした．しかし，類似する *D. pseudostelligera* 等と形態からは識別が難しくなるが，Houk *et al.*(2010) は，*D. pseudostelligera* とは生態が異なっていると記している．筆者の観察によっても被殻において両側の殻の形態がかなり異なるものも見出された（Figs 2-3, 4-5, 8-9）．
産　　出：北浦（Tanaka, H. 2007）：茨城県（現生）．
図　　版：北浦標本．

Figs 1-11.　*Discostella woltereckii* (Hustedt) Houk & Klee
LM. Figs 1-5. SEM. Figs 6-11. Material from Lake Kita (Kita-ura) (Recent), Ibaraki, Japan. Scale bars: Figs 8-9=2 μm, Figs 6-7, 10=1 μm, Fig. 11=0.5 μm.

Figs 1-5.　Three different valves.
Figs 2-3, 4-5.　Two frustules, at different focal planes.
Figs 6-7.　One frustule with same valve face ornamentation on opposite sides, outer-tube of mantle fultoportula (arrow).
Figs 8-9.　One frustule with different valve face ornamentation on opposite sides.
Fig. 10.　Internal oblique view of whole valve, mantle fultoportula with two lateral satellite pores (arrow) and rimoportula (arrowhead).
Fig. 11.　Detailed view of mantle fultoportula with two lateral satellite pores (arrow) and rimoportula (arrowhead).

Plate 81

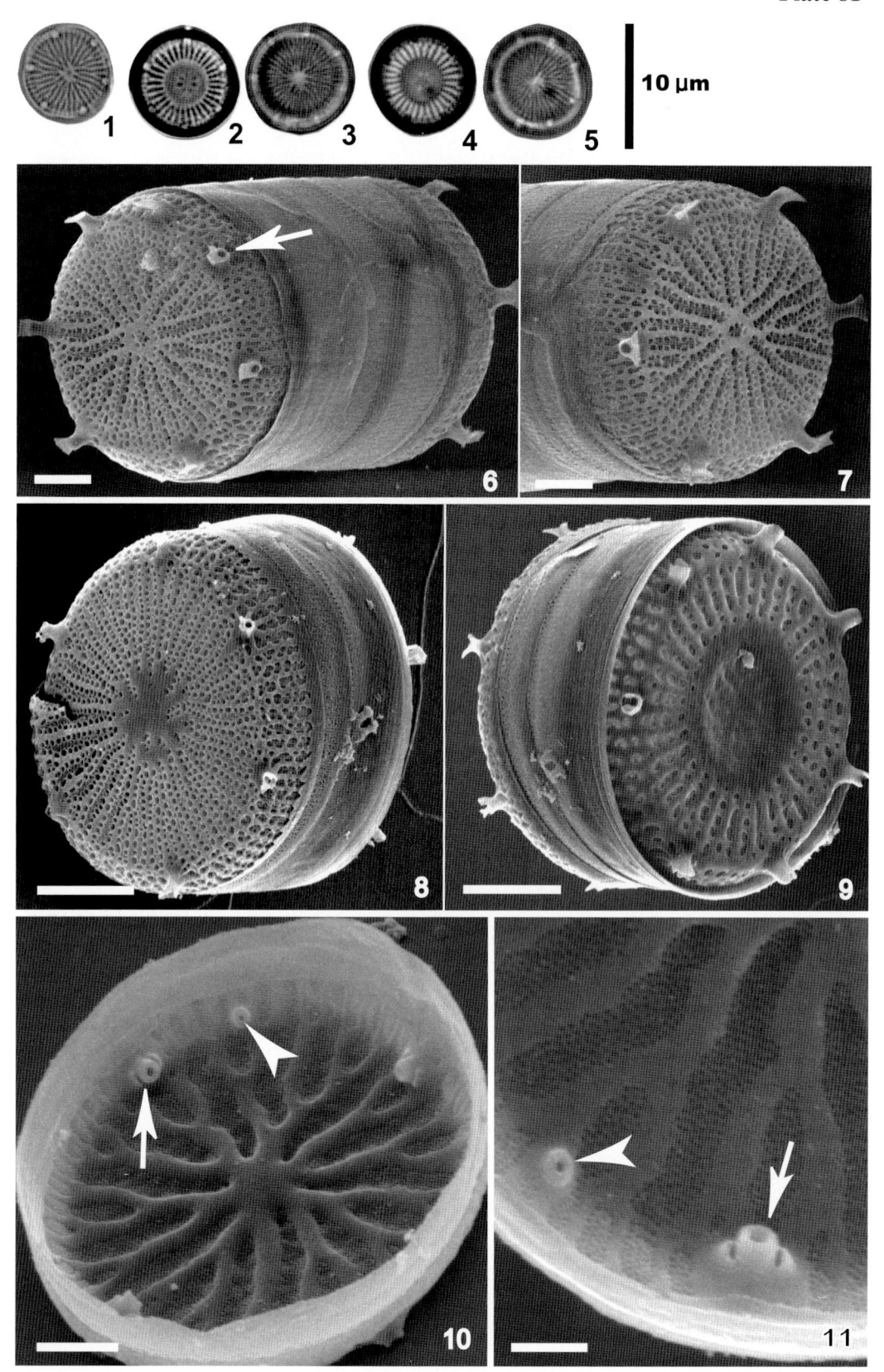

Plate 82.　Centric diatoms: Thalassiosirales
Mesodictyon yanagisawae H. Tanaka in Houk *et al.* 2014 (in press)

文　　献：Houk, V., Klee, R. & Tanaka, H. 2014. Atlas of freshwater centric diatoms with a brief key and descriptions. Part Ⅳ. Stephanodiscaceae B. *In*： Poulíčková, A.(ed.)： Fottea **14** Supplement.(in press)

形　　態：殻は円盤状で，殻径 19-30 μm．殻面は同心円状に波打ち，いぼ状の小隆起がある．殻中心から単列の点紋列が放射状に配列する，点紋列は殻縁で 10 μm に 7-10 本．点紋列を構成する点紋は 10 μm に約 10 個である．殻中心付近には表面のみで内側まで穿孔しない小孔が分布するが，中心部が凸あるいは凹状になっているので，焦点の位置によっては見逃す場合がある．有基突起は殻套のみに所在し付随孔は 2 個である．唇状突起は 1 個で殻套にある．殻面の胞紋を閉塞する師板は胞紋内部に存在するが，殻套の師板は胞紋の内側表面に所在する．

ノ ー ト：本邦での *Mesodictyon* 属の産出記録は柳沢ら（2003a, 2003b, 2004, 2010），平中ら(2004)，柳沢・工藤(2011)，柳沢・渡辺(2011)により *Mesodictyon* ? sp. あるいは *Mesodictyon* sp. として新潟県，秋田県の新第三系および第四系また茨城県の新第三系から報告されているが，これらは本種とは別種である．種小名は日本から *Mesodictyon* 属が産出することを最初に報告した，産業技術総合研究所 柳沢幸夫博士を記念して付けられた．

産　　出：白沢層（Houk *et al.* 2013）：後期中新世・宮城県．

図　　版：白沢層標本．

Figs 1-9.　*Mesodictyon yanagisawae* H. Tanaka
LM. Figs 1-2. SEM. Figs 3-9. From the type material, SND-310, Shirasawa Formation (Late Miocene), Sendai City, Miyagi, Japan. Scale bars: Figs 4-7＝5 μm, Fig. 9＝1 μm, Figs 3, 8＝0.5 μm.

Figs 1-2.　Two different size valves.
Fig. 3.　Detailed view of cross section of valve face showing cribrum inside an areola (arrowhead).
Fig. 4.　External view of whole valve, small holes on center of valve face (arrow).
Fig. 5.　Oblique view of Fig. 4.
Fig. 6.　Internal view of valve.
Fig. 7.　Oblique view of Fig. 6.
Fig. 8.　Detailed view of cribrum inside an areola (arrowhead).
Fig. 9.　Valve margin showing mantle fultoportula with two lateral satellite pores (arrow) and rimoportula (arrowhead).

Plate 82

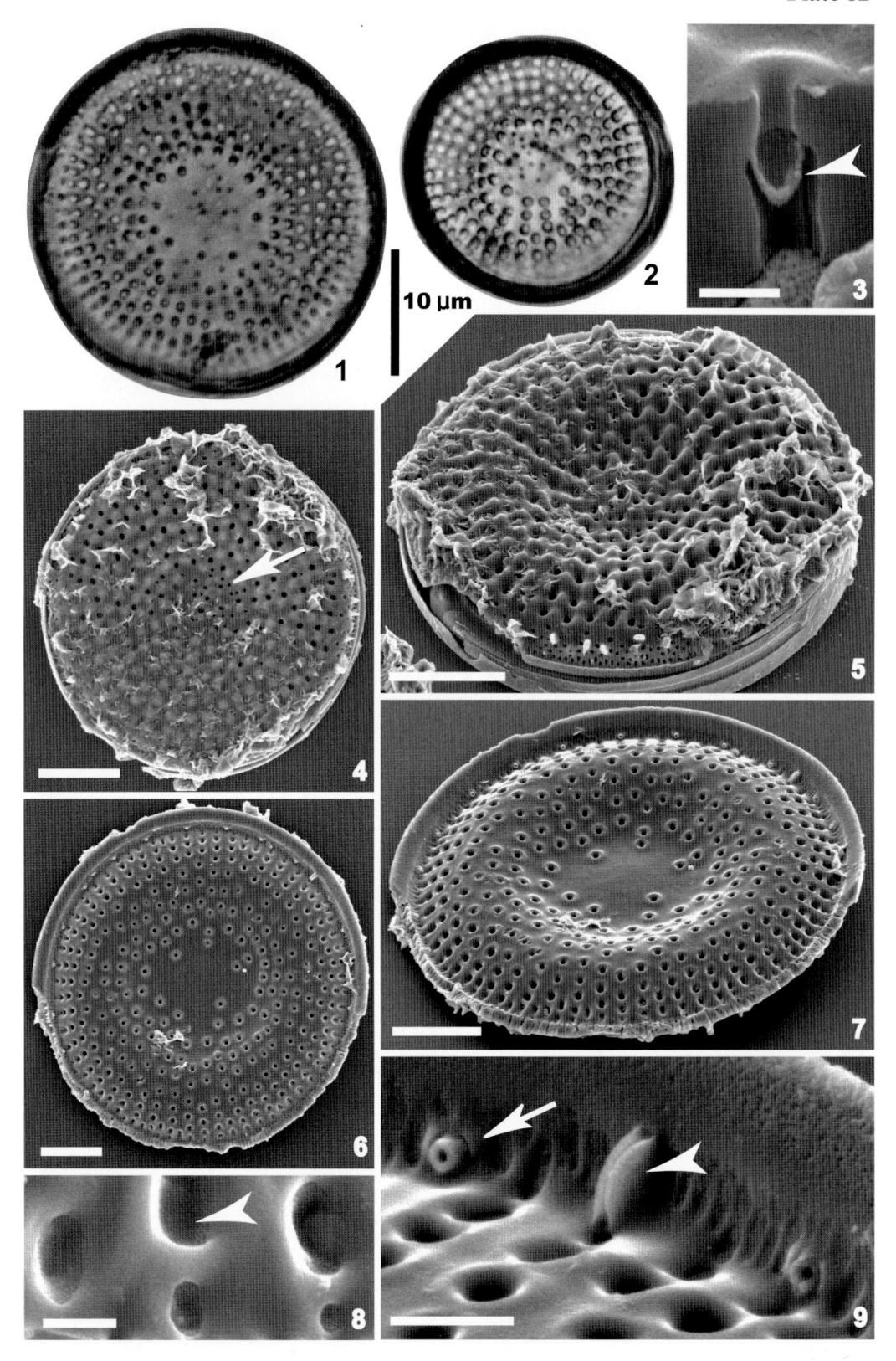

Plate 83.　Centric diatoms: Thalassiosirales
Mesodictyopsis akitaensis H. Tanaka & Nagumo 2009

文　献：Tanaka, H. & Nagumo, T. 2009.　Two new *Mesodictyopsis* species, *M. akitaensis* sp. nov. and *M. miyatanus* sp. nov., from a Late Miocene to Pliocene freshwater sediment, Japan. Acta Botanica Croatica **68**: 221-230.

形　態：殻は小形円盤状，殻径6.5-8.0 μm．殻面は平らで，中心から放射状に分布する点紋列は10 μmに約24本．SEMでの観察によると，胞紋の師板は胞紋の内部に所在する．殻面中心近くには，2個の付随孔を伴う1(2)個の殻面有基突起がある．殻套は浅く，殻套有基突起は3個の付随孔をもち殻套上部に所在し，外側では短い管があり，6-8個／殻である．唇状突起は1個で殻面縁辺部の点紋列の中にあり，外側では管を欠く．

ノート：胞紋の師板が胞紋の内部にある属は*Mesodictyon*属と*Mesodictyopsis*属があるが，殻面有基突起が存在すること，殻套有基突起が3個の付随孔をもつことから*Mesodictyopsis*属に所属するのが適当である．本種の種小名は秋田県に由来する．

　　Tanaka & Nagumo(2006)により三徳層（鳥取県）から報告された*Mesodictyon* sp. は，本属に所属するのが適切である．

産　出：宮田層（Tanaka & Nagumo 2009）：中新-鮮新世・秋田県．
図　版：宮田層標本．

Figs 1-11.　*Mesodictyopsis akitaensis* H. Tanaka & Nagumo
LM. Figs 1-5. SEM. Figs 6-11. From the type material, TAZ-F03, Miyata Formation (Mio-Pliocene), Hinokinaimata-zawa, Senboku City, Akita, Japan. Scale bars: Figs 6-10 =1 μm, Fig. 11=0.5 μm.

Figs 1-3.　Three different valves.
Figs 4-5. Same valve at different focal planes.
Fig. 6.　External view of whole valve, outer opening of valve face fultoportula (arrow).
Fig. 7.　External oblique view of whole valve.
Fig. 8.　Enlarged view of Fig. 6 showing outer opening of rimoportula (arrowhead).
Fig. 9.　Internal view of valve, valve face fultoportula (arrow).
Fig. 10.　Internal view, rimoportula (arrowhead).
Fig. 11.　Enlarged internal view of valve showing two mantle fultoportulae with three satellite pores and the valve face fultoportula (arrow).

Plate 83

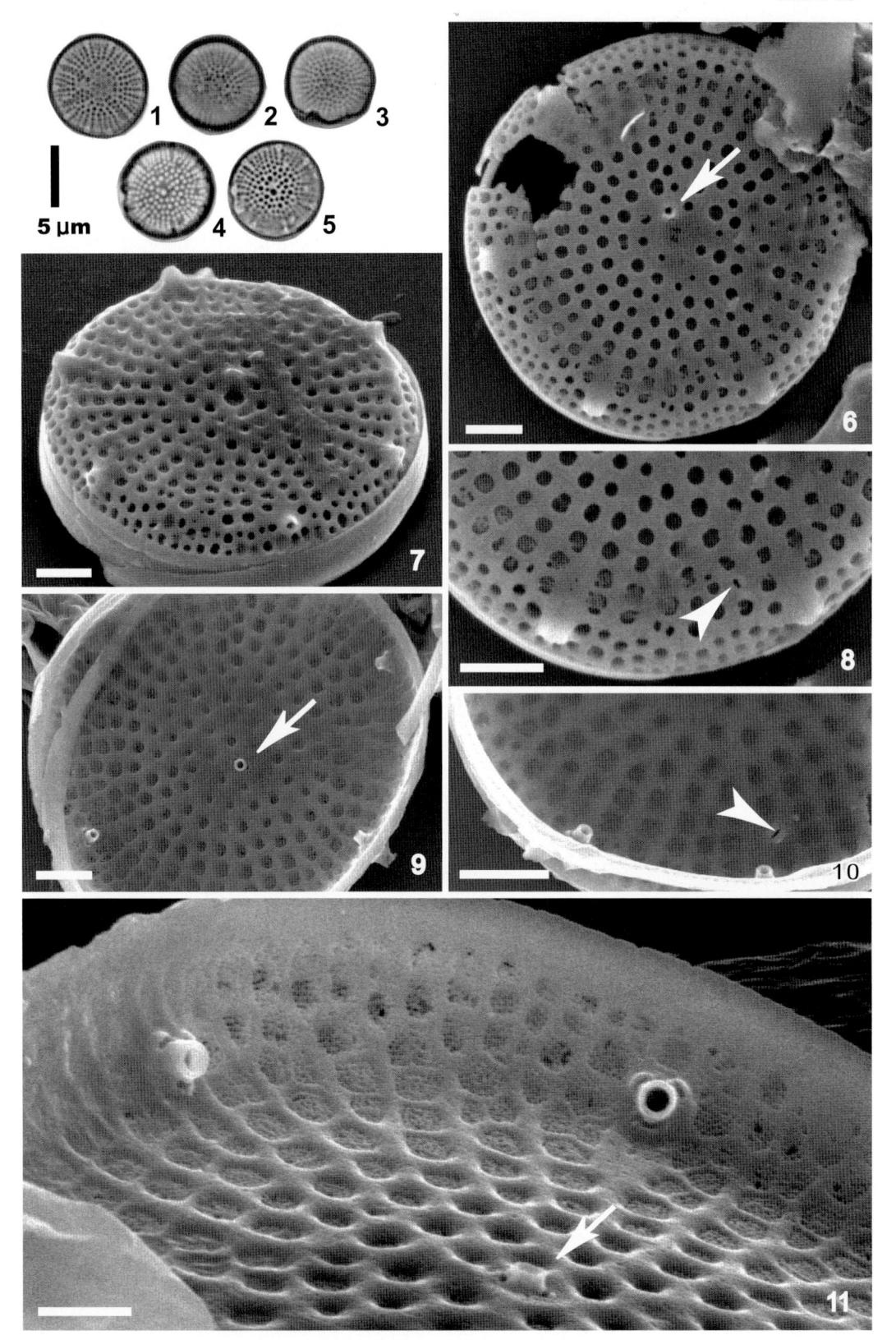

Plate 84.　Centric diatoms: Thalassiosirales
Mesodictyopsis miyatanus H. Tanaka in Tanaka & Nagumo 2009

文　　献：Tanaka, H. & Nagumo, T. 2009. Two new *Mesodictyopsis* species, *M. akitaensis* sp. nov. and *M. miyatanus* sp. nov., from a Late Miocene to Pliocene freshwater sediment, Japan. Acta Botanica Croatica **68**: 221-230.

形　　態：殻は小形円盤状で，殻径 4.0-8.0 μm．殻面は平らで中心から放射状に配列する点紋列は 10 μm に約 20 本．SEM での観察によると，胞紋の師板は胞紋の内部に所在する．殻面中心近くには，2個（まれに3個）の付随孔を伴う1個の殻面有基突起がある．殻套は浅く，殻套有基突起は3個の付随孔をもち殻套上部に所在し，外側では短い管があり，5-7個/殻である．唇状突起は1個で胞紋列の間で殻面/殻套境界にあり，外側では管を欠く．

ノ ー ト：種小名は本種を産出した宮田層に由来する．

産　　出：宮田層（Tanaka & Nagumo 2009）：中新-鮮新世・秋田県．

図　　版：宮田層標本．

Figs 1-13. *Mesodictyopsis miyatanus* H. Tanaka

LM. Figs 1-8. SEM. Figs 9-13. From the type material, TAZ-F03, Miyata Formation (Mio-Pliocene), Hinokinaimata-zawa, Senboku City, Akita, Japan. Scale bars: Figs 9-10, 13＝1 μm, Figs 11-12＝0.5 μm.

Figs 1-2, 3-4, 5-6, 7-8.　Four valves at different focal planes.
Fig. 9.　External oblique view of valve.
Fig. 10.　Whole valve view.
Fig. 11.　Internal detailed view of valve center, valve face fultoportula with two satellite pores (arrow) and cribra inside of areolae.
Fig. 12.　Internal detailed view of valve margin showing mantle fultoportula with three satellite pores and rimoportula (arrowhead).
Fig. 13.　Frustule, one side valve breaking off, opening of valve face fultoportula (arrow).

Plate 84

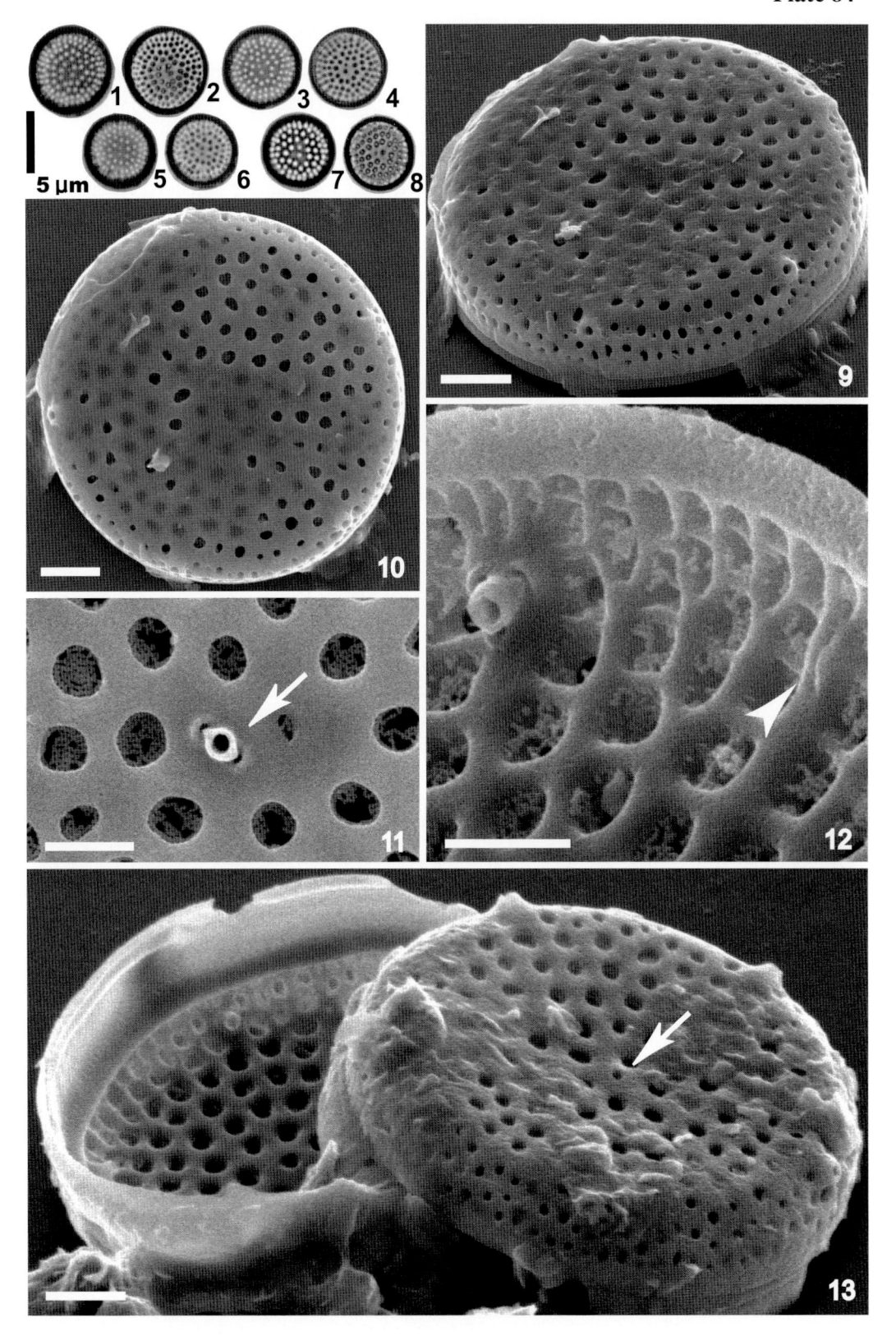

Plates 85-86. Centric diatoms: Thalassiosirales
Pliocaenicus costatus (Loginova, Lupikina & Khursevich) Flower, Ozornina & A.I. Kuzmina 1998

基礎異名：*Cyclostephanos costatus* Loginova, Lupikina & Khursevich 1989

文　　献：Cremer, H. & Vijver, B.V. 2007. *Pliocaenicus costatus*: an emended species description. *In*: Kusber, W.-H. & Jahn, R.(eds) Proceedings of the 1st Central European Diatom Meeting 2007. pp. 35-38. Botanic Garden and Botanical Museum Berlin-Dahlem, Freie Universität Berlin, Berlin-Dahlem.

形　　態：殻は円形～楕円形，殻径12.5-40 μm．中心域は強く横にうねる．中心域の胞紋は放射状～不規則に配列し，10 μm に約14本．殻縁にある黒線（SEM 観察によれば肋）は 10 μm に7-8本である．SEM 観察によると，胞紋は内側ではドーム状師板で覆われる．殻面有基突起の数は3-20個である．殻縁の肋にはすべて1個の殻套有基突起がある．殻面・殻套有基突起とも付随孔は2個である．唇状突起は1個で殻面突出部の縁辺にある．長胞には遠心性覆いがある．

ノ　ート：本属の設立当時（Round & Håkansson 1992）はすべて化石から知られていた種類のみであったが，*P. costatus* は極東アジアの淡水湖に生育していることが見出されている（たとえば Cremer & Vijver 2006）．Cremer & Vijver(2007)は現生の本種の形態に基づき *P. costatus* の記述を修正している．日本では美ヶ原層から産出している．

産　　出：美ヶ原層（Tanaka & Kobayasi 1999）：前期更新世・長野県．

図　　版 85-86：美ヶ原層標本．

Plate 85, Figs 1-10. *Pliocaenicus costatus* (Loginova, Lupikina & Khursevich) Flower, Ozornina & A.I. Kuzmina

LM. Figs 1-9. SEM. Fig. 10. Material from Utsukushigahara Formation (Early Pleistocene), Matsumoto City, Nagano, Japan. Scale bar: Fig. 10=5 μm.

Figs 1-9.　Nine different size valves.
Fig. 2.　Rimoportula (arrowhead).
Fig. 10.　External view of whole valve, opening of rimoportula (arrowhead).

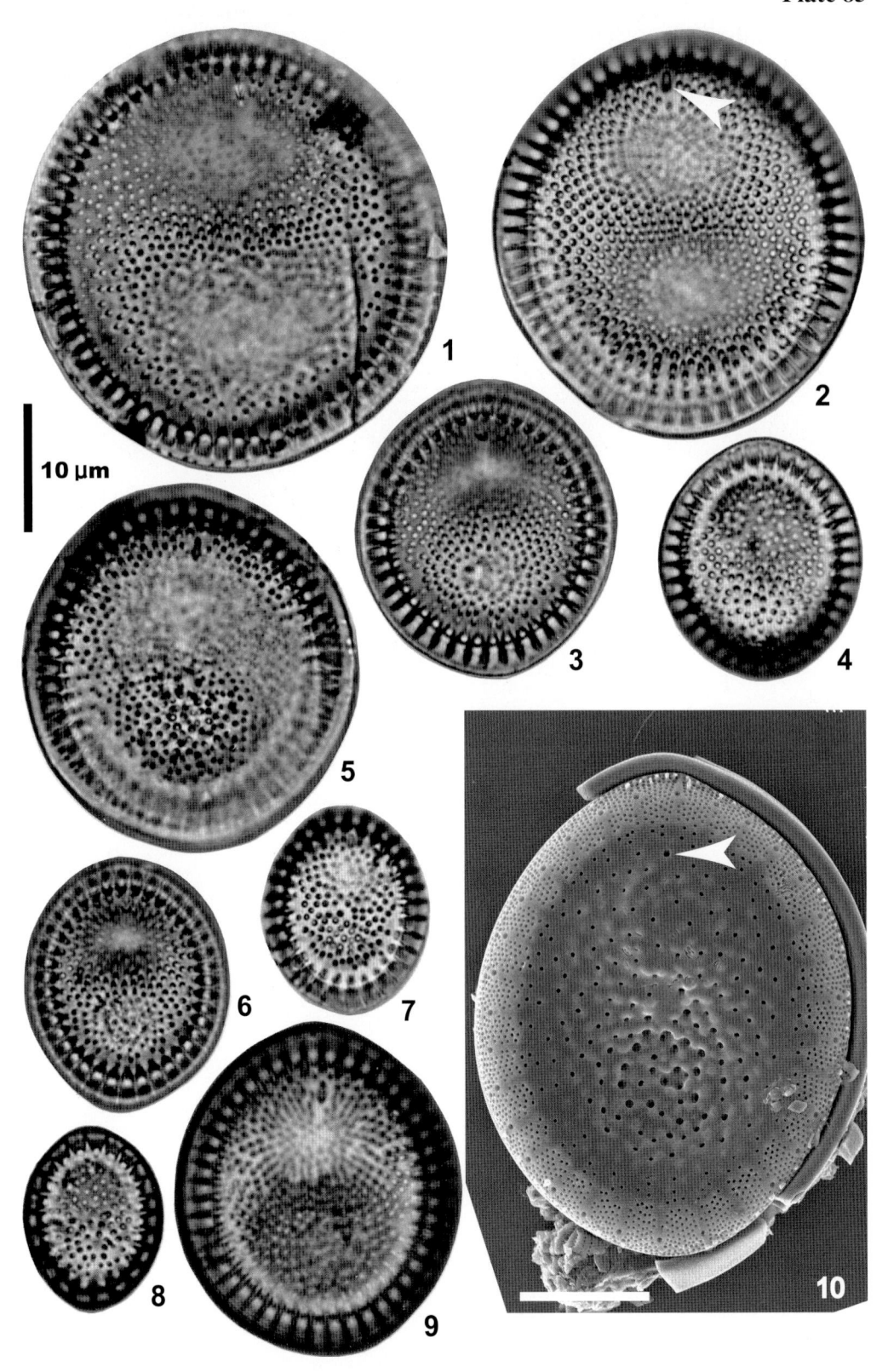

Plate 85

Plate 86, Figs 1–7. *Pliocaenicus costatus* (Loginova, Lupikina & Khursevich) Flower, Ozornina & A.I. Kuzmina

SEM. Figs 1–7. Materials from Utsukushigahara Formation (Early Pleistocene), Matsumoto City, Nagano, Japan. Scale bars: Figs 1–2, 4=5 µm, Fig. 3=2 µm, Figs 5–7=1 µm.

Fig. 1. External oblique view of whole valve, opening of rimoportula (arrowhead).
Fig. 2. Internal view of whole valve, rimoportula (arrowhead).
Fig. 3. View of cingulum, trace of ligula-like segment (arrow).
Fig. 4. Oblique view of Fig. 2, rimoportula (arrowhead).
Fig. 5. Detailed internal view of valve face fultoportulae with two satellite pores.
Fig. 6. Enlarged view of Fig. 1 showing outer opening of rimoportula (arrowhead) and openings of mantle fultoportulae (arrows).
Fig. 7. Detailed view of Fig. 4, rimoportula (arrowhead) and mantle fultoportula with two lateral satellite pores (arrow).

Plate 86

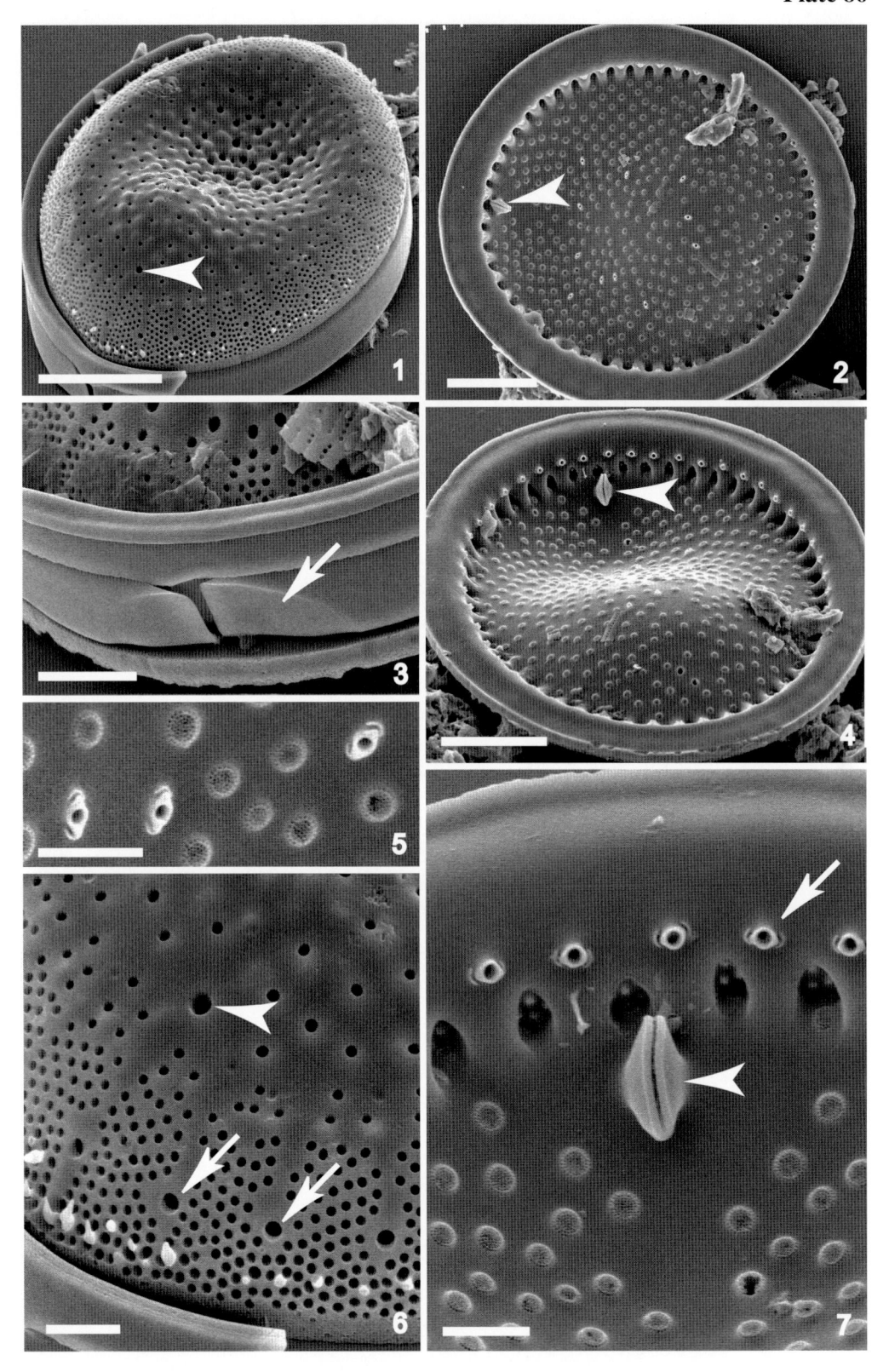

209

Plate 87.　Centric diatoms: Thalassiosirales
Pliocaenicus nipponicus H. Tanaka & Nagumo 2004

文　　献：Tanaka, H. & Nagumo, T. 2004. *Pliocaenicus nipponicus* sp. nov., a new freshwater fossil diatom from central Japan. Diatom **20**: 105-111.

形　　態：殻は円盤形，殻径 11-21 μm．殻面は強く横に波打つ．胞紋列は1列の胞紋から構成され，中心から放射状に配列し，殻縁で 10 μm に 12-14 本である．胞紋列を構成する胞紋は 10 μm に 12-16 個である．SEM 観察によれば，殻套では細かい胞紋列が存在するが，幅の広い間線によって 6-9 本ごとに区分されている．間線は 10 μm に 5-6 本である．殻面有基突起は 6-11 個で円状に分布し，殻套有基突起は各肋にある．殻面・殻套有基突起とも付随孔は 2 個である．唇状突起は 1 個で殻套近くの殻面縁辺にある．長胞には遠心性覆いがある．殻帯には Ligula-like segment が含まれる．

ノ　ー　ト：本種が見出された兜岩層の年代についてはいくつかの異なった年代を示す研究があるが，放射年代値に基づく最新の研究である佐藤(2007)に従い鮮新世とした．兜岩層は長野県と群馬県にわたって分布するが，試料を採取したのは群馬県側である．

産　　出：兜岩層（Tanaka & Nagumo 2004）：鮮新世・群馬県，宮田層（田中 2007）：中新-鮮新世・秋田県．

図　　版：兜岩層標本．

Figs 1-12.　*Pliocaenicus nipponicus* H. Tanaka & Nagumo
LM. Figs 1-6. SEM. Figs 7-12. From the type material, KAB-101, Kabutoiwa Formation (Pliocene), Hoshio, Nanmoku Village, Gunma, Japan. Scale bars: Figs 7-11＝2 μm, Fig. 12＝1 μm.

Figs 1-6.　Six different size valves.
Fig. 7.　External view of whole valve.
Fig. 8.　Oblique view of valve.
Fig. 9.　Detailed internal view of valve face, remains of valve face fultoportulae with two satellite pores (arrows).
Fig. 10.　View of valve margin and cingulum, remains of ligula-like segment (arrowhead) and openings of mantle fultoportulae (arrows).
Fig. 11.　Internal oblique view of valve.
Fig. 12.　Detailed internal view of valve margin, rimoportula (arrowhead) and mantle fultoportula with two lateral satellite pores (arrow).

Plate 87

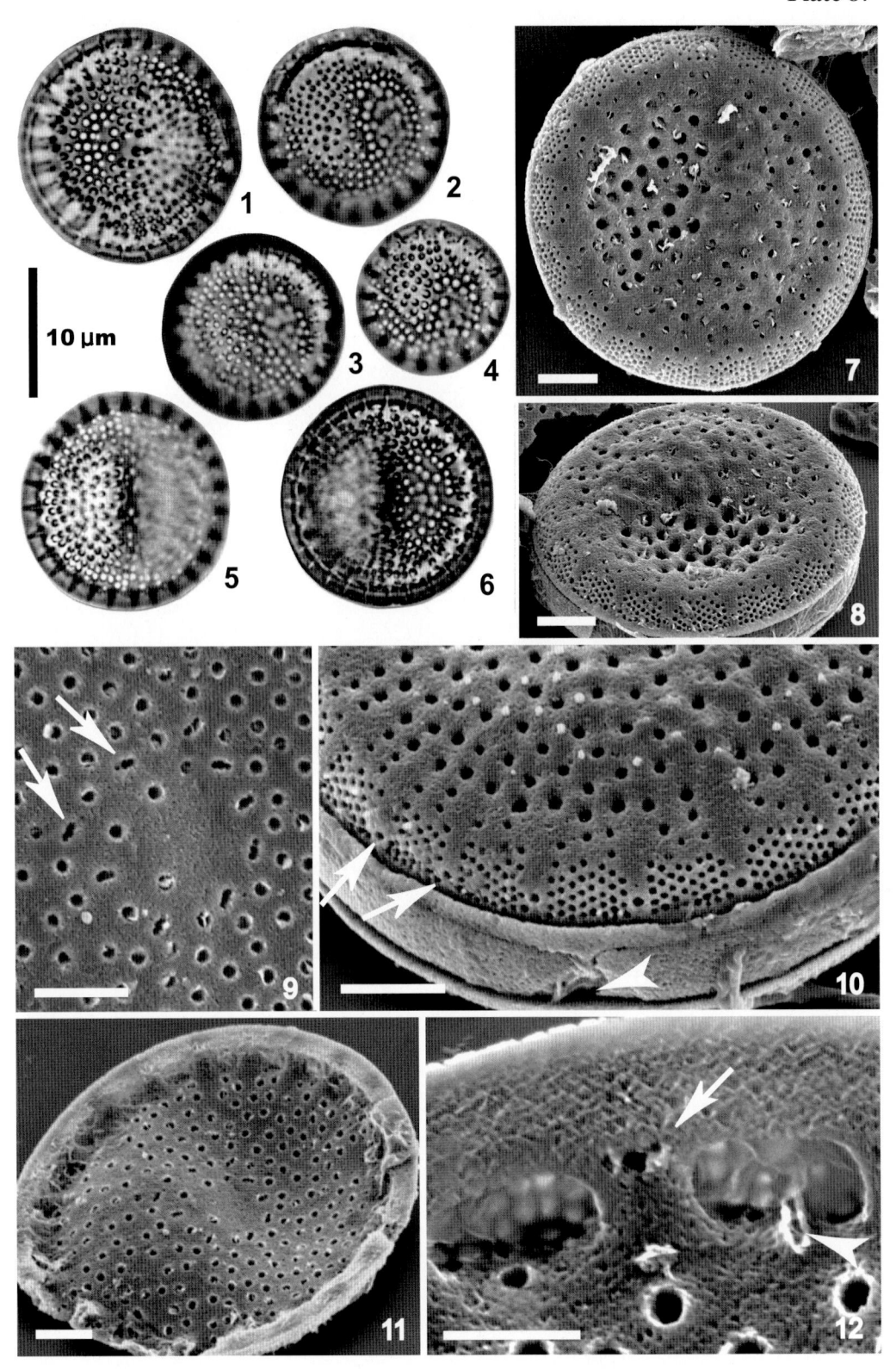

211

Plate 88. Centric diatoms: Thalassiosirales
Pliocaenicus omarensis (Kuptsova) Stachura-Suchoples & Khursevich 2007

基礎異名：*Stephanodiscus omarensis* Kuptsova 1962

文　　献：Stachura-Suchoples, K. & Khursevich, G.K. 2007. On the genus *Pliocaenicus* Round & Håkansson (Bacillariophyceae) from the Northern Hemisphere. *In*: Kusber, W.-H. & Jahn, R.(eds) Proceedings of the 1st Central Europian Diatom Meeting 2007. pp. 155-158. Botanic Garden and Botanical Museum Berlin-Dahlem, Freie Universität Berlin, Berlin-Dahlem.

形　　態：殻は円形で殻面は強く横に波打つ，殻径 8.0-36 μm．殻面中心から単列の点紋列が放射状に配列し，その数は 10 μm に 8-10 本．点紋列を構成する点紋は 10 μm に 8-16 個．殻縁の黒線は 10 μm に約 8 個である．SEM 観察によれば，殻套の胞紋は殻面より細かい列になり 10 μm に 51-66 本である．殻面有基突起は中心部にほぼ円状に分布し，殻套有基突起は 2-3 本ごとの肋に所在する．殻面・殻套有基突起とも付随孔は 3 個である．唇状突起は 1 個で，殻面が突出している側の殻套近くにある．長胞には遠心性覆いがある．殻帯には Ligula-like segment が含まれる．

ノート：Round & Håkansson(1992)で本属へ組み合わせされたが不備があったので Stachura-Suchoples & Khursevich(2007)で正式に *Pliocaenicus* 属へ組み合わせになった．Khursevich & Stachura-Suchoples(2008)は *Pliocaenicus* 属の総括的な研究を行ったが，その中で *P. omarensis* の殻面有基突起は半円状に分布すると記している．日本からは春来層と郡山層から見出されているが，いずれもほぼ円状に分布している．また郡山層産出の個体は殻面の点紋が放射状～不規則に分布している．しかし，他の形質は類似しているので，両層から見出された分類群を本種に同定した．

産　　出：春来層 (Tanaka & Kobayasi 1999)：鮮新世・兵庫県，郡山層 (Tanaka & Nagumo 2009)：後期鮮新世・鹿児島県．

図　　版：春来層標本．

Figs 1-10. *Pliocaenicus omarensis* (Kuptsova) Stachura-Suchoples & Khursevich

LM. Figs 1-4. SEM. Figs 5-10. Materials from Haruki Member (Pliocene), Shin-onsen Town, Hyogo, Japan. Scale bars: Figs 5, 8＝5 μm, Fig. 10＝2 μm, Figs 6-7＝1 μm, Fig. 9＝0.5 μm.

Figs 1-4.　Four different size valves.
Fig. 5.　External view of whole valve.
Fig. 6.　External view of cingulum, ligula-like segment (arrow).
Fig. 7.　Detailed view of valve margin, openings of mantle fultoportulae (arrows).
Fig. 8.　External oblique view of valve.
Fig. 9.　Detailed internal view of margin showing domed cribra, rimoportula (arrowhead) and mantle fultoportulae with three satellite pores (arrows).
Fig. 10.　Internal oblique view of valve, rimoportula (arrowhead).

Plate 88

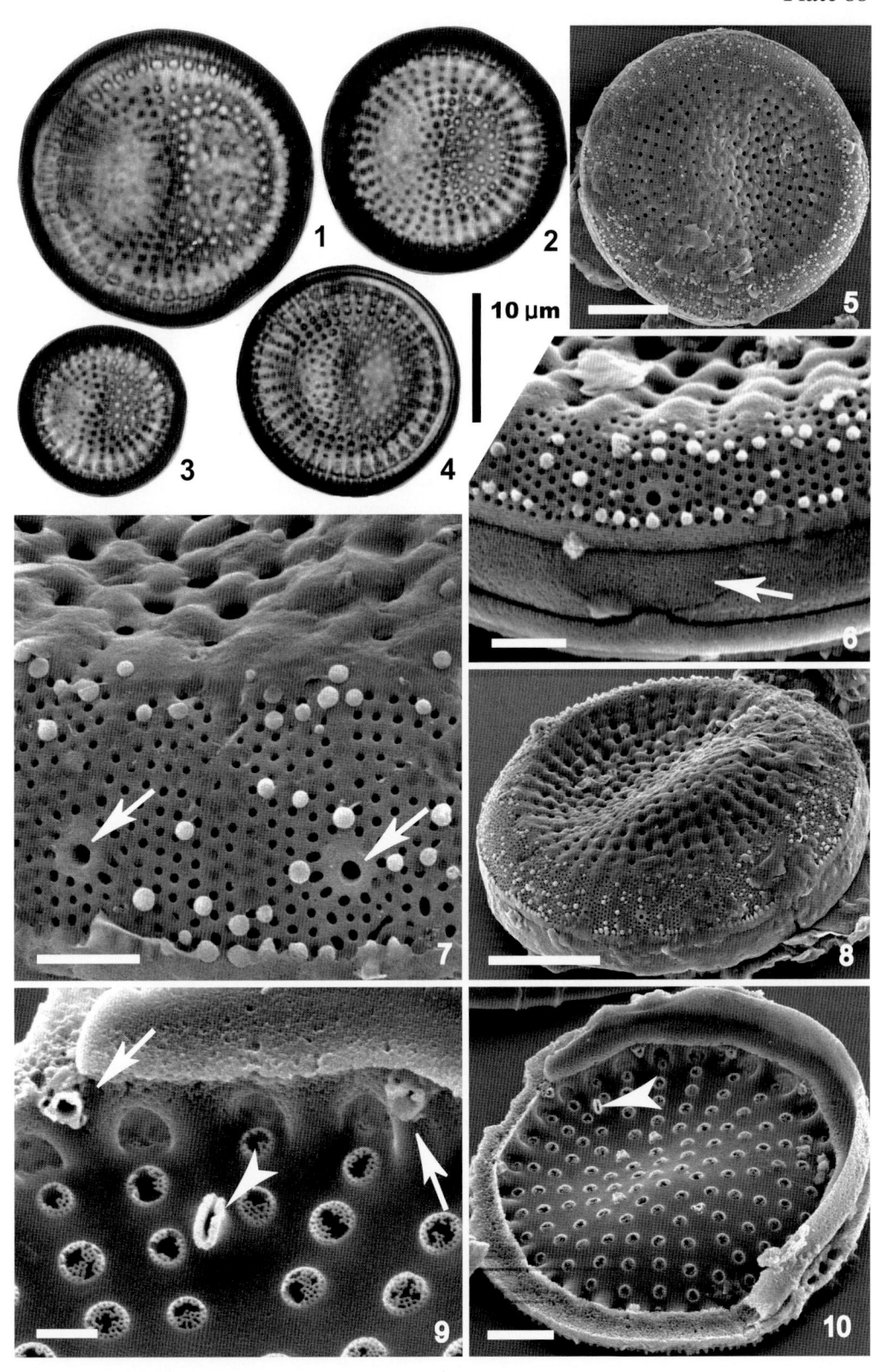

Plate 89. Centric diatoms: Thalassiosirales
Pliocaenicus radiatus H. Tanaka in Houk *et al.* 2014 (in press)

文　献：Houk, V., Klee, R. & Tanaka, H. 2014. Atlas of freshwater centric diatoms with a brief key and descriptions. Part Ⅳ. Stephanodiscaceae B. *In*: Pouličková, A.(ed.): Fottea **14** Supplement.(in press)

形　態：殻は円形で殻面はゆるく横に波打つ，殻径 6.5-22 μm. 殻面中心から単列（まれに 2列）の点紋が放射状に配列し，その数は 10 μm に 3-4 本．点紋列を構成する点紋は 10 μm に 8-16 個である．SEM 観察によれば，殻面有基突起は 5-10 個で，中心部に円状に分布する．殻套有基突起は毎 3-7 本ごとにあるわずか窪んだ太い肋に所在する．殻面・殻套有基突起とも付随孔は 3 個である．唇状突起は 1 個で殻面が突出している側の殻套近くで肋の基部に所在する．長胞には遠心性覆いがある．殻帯には Ligula-like segment が含まれる．

ノート：本種は殻面の点紋列が粗いことが他と最も区別しやすい形質である．類似種との 10 μm における点紋列の殻縁における計数値は，*P. radiatus*(3-4 本)，*P. changbaiense*(ca. 8 本)，*P. jilinensis*(ca. 6 本)，*P. omarensis*(ca. 9 本) である．

産　出：筆ん崎層（Houk *et al*. 2014）：鮮新世・沖縄県．

図　版：筆ん崎層標本．

Figs 1-11.　*Pliocaenicus radiatus* H. Tanaka

LM. Figs 1-4. SEM. Figs 5-11. From the type material, AGU-001, Fudenzaki Formation (Pliocene), Aguni Island, Okinawa, Japan. Scale bars: Fig. 10＝5 μm, Figs 7-8, 11＝2 μm, Figs 5-6, 9＝0.5 μm.

Figs 1-4.　Four different size valves.
Fig. 5.　Detailed view of internal valve margin, mantle fultoportula with three satellite pores (arrow) and rimoportula (arrowhead).
Fig. 6.　Broken valve showing valve margin.
Fig. 7.　Oblique view of whole valve showing tangential undulated central area of valve face.
Fig. 8.　View of whole valve, opening of valve face fultoportula (arrow).
Fig. 9.　External view of valve margin and cingulum showing ligula-like segment (arrow) and opening of mantle fultoportula (arrowhead).
Fig. 10.　Internal view of whole valve, valve face fultoportulae with three satellite pores (arrows).
Fig. 11.　Internal oblique view of whole valve showing undulated valve face.

Plate 89

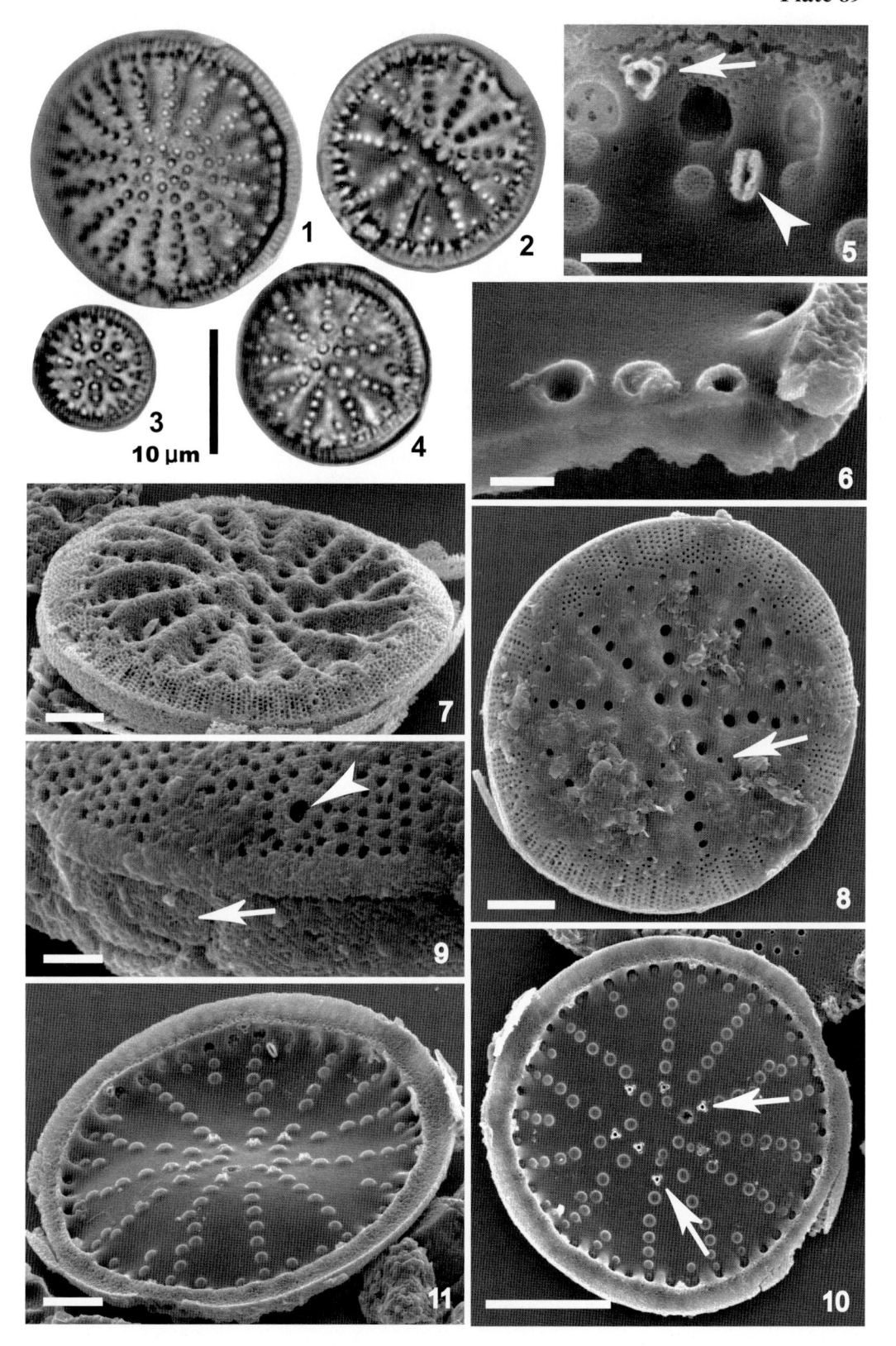

Plate 90. Centric diatoms: Thalassiosirales
Pliocaenicus tanimurae H. Tanaka & Saito-Kato 2011

文　　献：Tanaka, H. & Saito-Kato, M. 2011. *Pliocaenicus tanimurae*, a new Pliocene diatom species from Aguni Island, Okinawa Prefecture, Southwestern Japan. Diatom Research **26**: 155-160.

形　　態：大形の殻の殻面は円形であるが，一般的には楕円形であり，殻面はゆるく横に波打つ．長径 9-27 μm，短径 7-27 μm．殻面中心から単列の点紋列が放射状に配列し，その数は殻縁で 10 μm に約 5 本である．中心付近には黒色（あるいは白色）の細点として，光顕においても円形に分布する殻面有基突起を観察することができる．SEM 観察によれば，殻面有基突起は 4-17 個で，中心部に円状に分布する．殻套有基突起は太い肋（ときどきわずか凹む）にあるが，殻面が突出する側では毎 2-6 本ごとの肋で，凹む側では毎 7-15 本ごとの肋にある．殻面・殻套有基突起とも 3 個の付随孔を伴う．唇状突起は 1 個で，殻面縁辺の殻套近くで肋の基部に所在する．長胞には遠心性覆いがある．殻帯には Ligula-like segment が含まれる．

ノ ー ト：Khursevich & Stachura-Suchoples (2008) によれば *Pliocaenicus* 属は北半球のみから見出されている．最南地点はエチオピアであり，本種の産出地である沖縄県粟国島は 2 番目に南方の地点である．粟国島筆ん崎層からは 3 分類群（*P. omarensis*, *P. radiatus*, *P. tanimurae*）の *Pliocaenicus* 属珪藻を産する．*P. tanimurae* の種小名は元国立科学博物館谷村好洋博士を記念して付けられた．

産　　出：筆ん崎層（Tanaka & Saito-Kato 2011）：鮮新世・沖縄県．

図　　版：筆ん崎層標本．

Figs 1-10. *Pliocaenicus tanimurae* H. Tanaka & Saito-Kato
LM. Figs 1-5. SEM. Figs 6-10. From the type material, AGU-107, Fudenzaki Formation (Pliocene), Aguni Island, Okinawa, Japan. Scale bars: Figs 6-7, 9=2 μm, Figs 8, 10=0.5 μm.

Figs 1-5.　Five different size valves.
Fig. 6.　Oblique view of valve showing undulated valve face.
Fig. 7.　External view of whole valve, opening of valve face fultoportula (arrow).
Fig. 8.　Detailed view of valve margin, opening of mantle fultoportula (arrow).
Fig. 9.　Internal oblique view, rimoportula (arrowhead).
Fig. 10.　Detailed view of mantle fultoportula with three satellite pores.

Plate 90

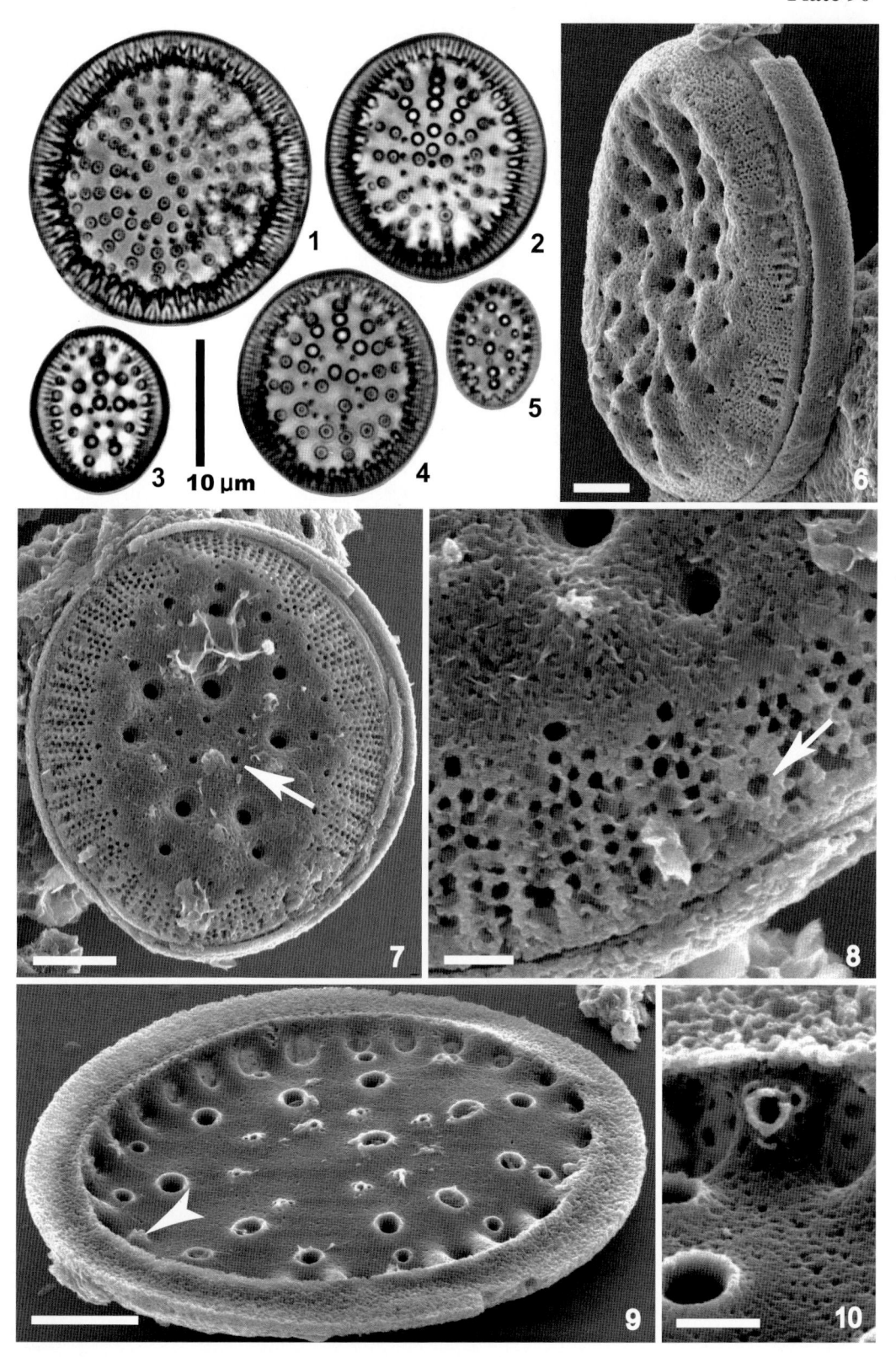

Plate 91. Centric diatoms: Thalassiosirales
Spicaticribra kingstonii Johansen, Kociolek & Lowe 2008

文　　献：Johansen, J., Kociolek, P. & Lowe, R. 2008. *Spicaticribra kingstonii* gen. nov. et sp. nov. (Thalassiosirales, Bacillariophyta) from Great Smoky Mountains National Park, U.S.A. Diatom Research **23**: 367-375.

形　　態：殻は円盤状で殻面はほぼ平ら，殻径 10-26.5 μm，胞紋列は 10 μm に 16-18 本．胞紋列を構成する胞紋は 10 μm に 16-24 個．殻の中心から放射状に胞紋列が配列するが，殻中央の胞紋はいくつかが融合して大きな外側開口を形成しており，また全体的に胞紋の外側開口の大きさは不揃いである．殻套有基突起は 10 μm に約 4 個所在し 3 個の付随孔を有する．唇状突起は殻套中位に 1-3 個存在し，基部が伸張している．有基突起（外側開口は肥厚している），唇状突起とも外管を欠く．

ノ ー ト：本種はネパールの湖成堆積物から見出され記載された *Thalassiosira inlandica* Hayashi (Hayashi *et al.* 2007) およびブラジルから記載された *Thalassiosira rudis* Tremarin, Ludwig, Becker & Torgan (Ludwig *et al.* 2008) の 2 種に形態が類似しているので詳細な検討が必要と思われる．

産　　出：池田湖（田中・南雲 2009）：鹿児島県（現生），福上湖（福地ダム湖）（田中 2010）：沖縄県（現生）．

図　　版：池田湖標本．

Figs 1-11.　*Spicaticribla kingstonii* Johansen, Kociolek & Lowe
LM. Figs 1-5. SEM. Figs 6-11. Material from Lake Ikeda (Recent), Kagoshima, Japan. Scale bars: Figs 6-7＝5 μm, Figs 8, 11＝2 μm, Figs 9-10＝1 μm.

Figs 1-5.　Five different size valves.
Fig. 5.　Three rimoportulae (arrowheads).
Fig. 6.　External view of whole valve.
Fig. 7.　External oblique view of valve, openings of mantle fultoportulae (arrows).
Fig. 8.　Internal view of whole valve, rimoportulae (arrowheads).
Fig. 9.　Detailed view of internal valve margin showing rimoportula (arrowhead) and mantle fultoportulae with three satellite pores (arrows).
Fig. 10.　Detailed view of epicingulum.
Fig. 11.　Detailed view of valve margin, opening of rimoportula (arrowhead) and openings of mantle fultoportulae (arrows).

Plate 91

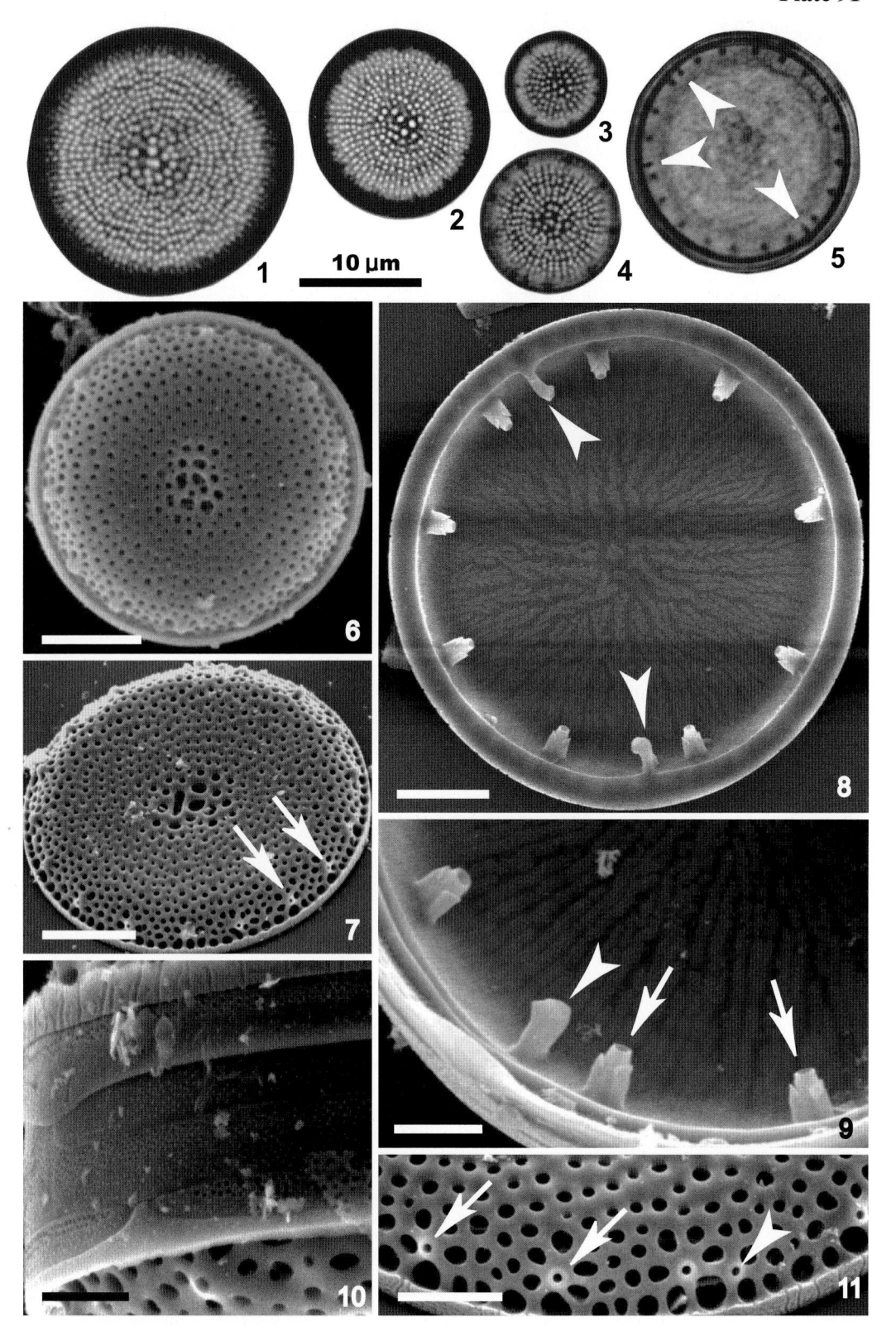

Plates 92-93. Centric diatoms: Thalassiosirales
Stephanodiscus akanensis Tuji, Kawashima, Julius & Stoermer 2003

文　　献：Tuji, A., Kawashima, A., Julius, M.L. & Stoermer, E.F. 2003. *Stephanodiscus akanensis* sp. nov., a new species of extant diatom from Lake Akan, Hokkaido, Japan. Bulletin of the National Science Museum, Ser. B **29**: 1-8.

形　　態：殻は円盤状で同心円状に波打つ，直径13-60 μm，束線は 10 μm に 7-8 本で，殻面縁辺部では 2-3(4)本の胞紋列から構成される．殻面有基突起は 1-2 個で2個の付随孔を伴い殻面縁辺にあるが，殻面中心に所在することもある．殻套有基突起は毎 1-3 本ごとの間束線にあり，3個の付随孔を伴う．唇状突起は(1)2 個で互いに反対側の殻面／殻套境界あるいは針よりわずか下方にある（まれには殻套に近い殻面縁辺の場合もある）．

ノート：河島・小林(1993)では未同定として産出が報告されていたが，Tuji *et al.*(2003)により *Stephanodiscus akanensis* として記載された．

産　　出：阿寒湖（Tuji *et al.* 2003）：北海道（現生）．

図　　版 92-93：阿寒湖標本．

Plate 92, Figs 1-5. *Stephanodiscus akanensis* Tuji, Kawashima, Julius & Stoermer

LM. Figs 1-5. Material from the type locality, Lake Akan (Recent), Kushiro City, Hokkaido, Japan.

Figs 1-5.　Four different size valves.
Figs 4-5.　Same valve at different focal planes.

Plate 92

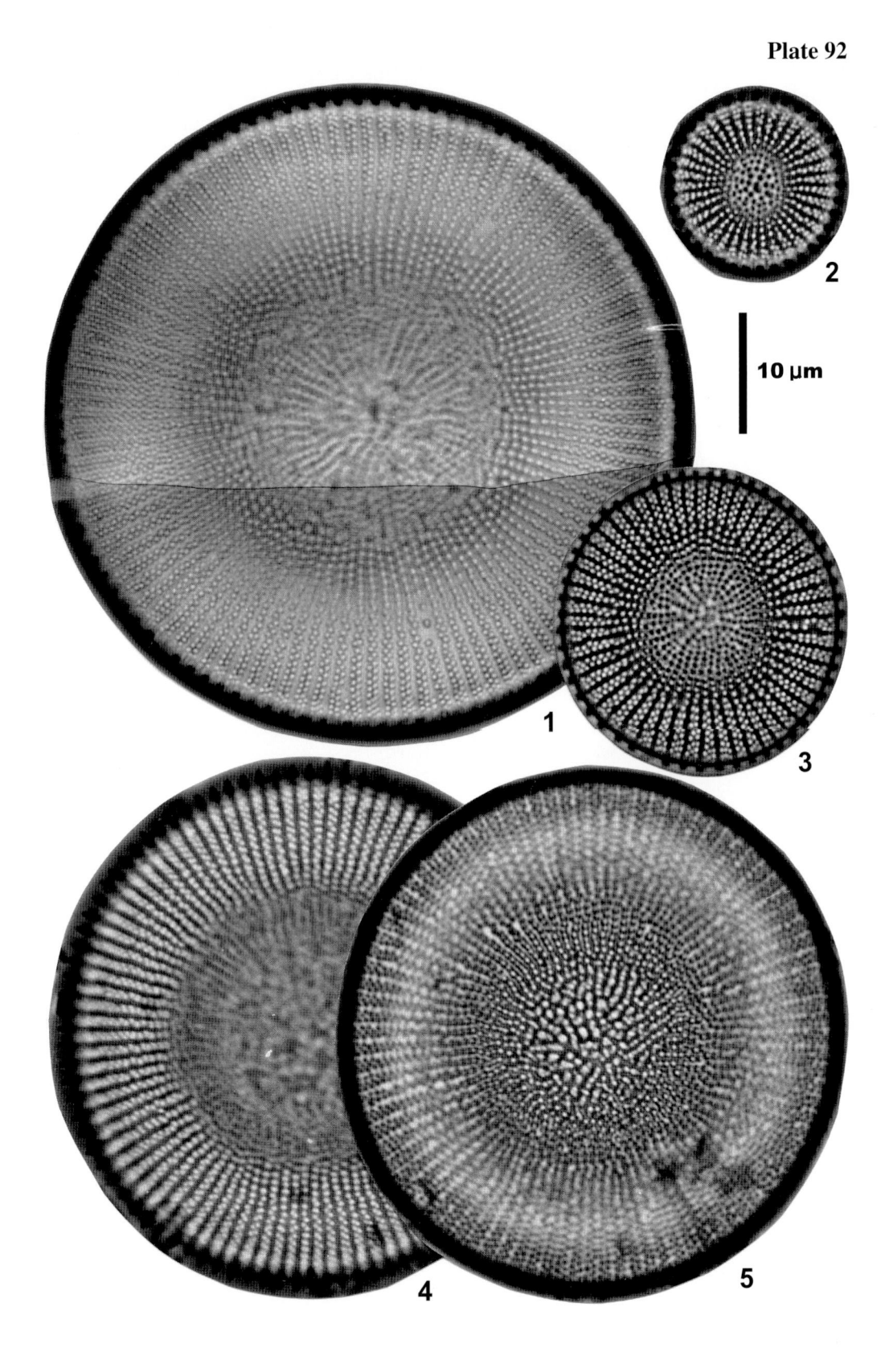

Plate 93, Figs 1–5. *Stephanodiscus akanensis* **Tuji, Kawashima, Julius & Stoermer**

SEM. Figs 1–5. Material from the type locality, Lake Akan (Recent), Kushiro City, Hokkaido, Japan. Scale bars: Fig. 1=5 μm, Figs 3, 5=2 μm, Figs 2, 4=1 μm.

Fig. 1. External oblique view of entire valve.

Fig. 2. Enlarged view of valve margin of Fig. 1 showing outer opening of rimoportula (arrowhead), mantle fultoportulae with short tubes and valve face fultoportula (arrow).

Fig. 3. Internal view showing rimoportula (arrowhead) and valve face fultoportula (arrow).

Fig. 4. Detailed view of valve margin showing rimoportula (arrowhead) and mantle fultoportulae with three satellite pores.

Fig. 5. Internal view showing rimoportulae (arrowheads) and valve face fultoportula (arrow).

Plate 93

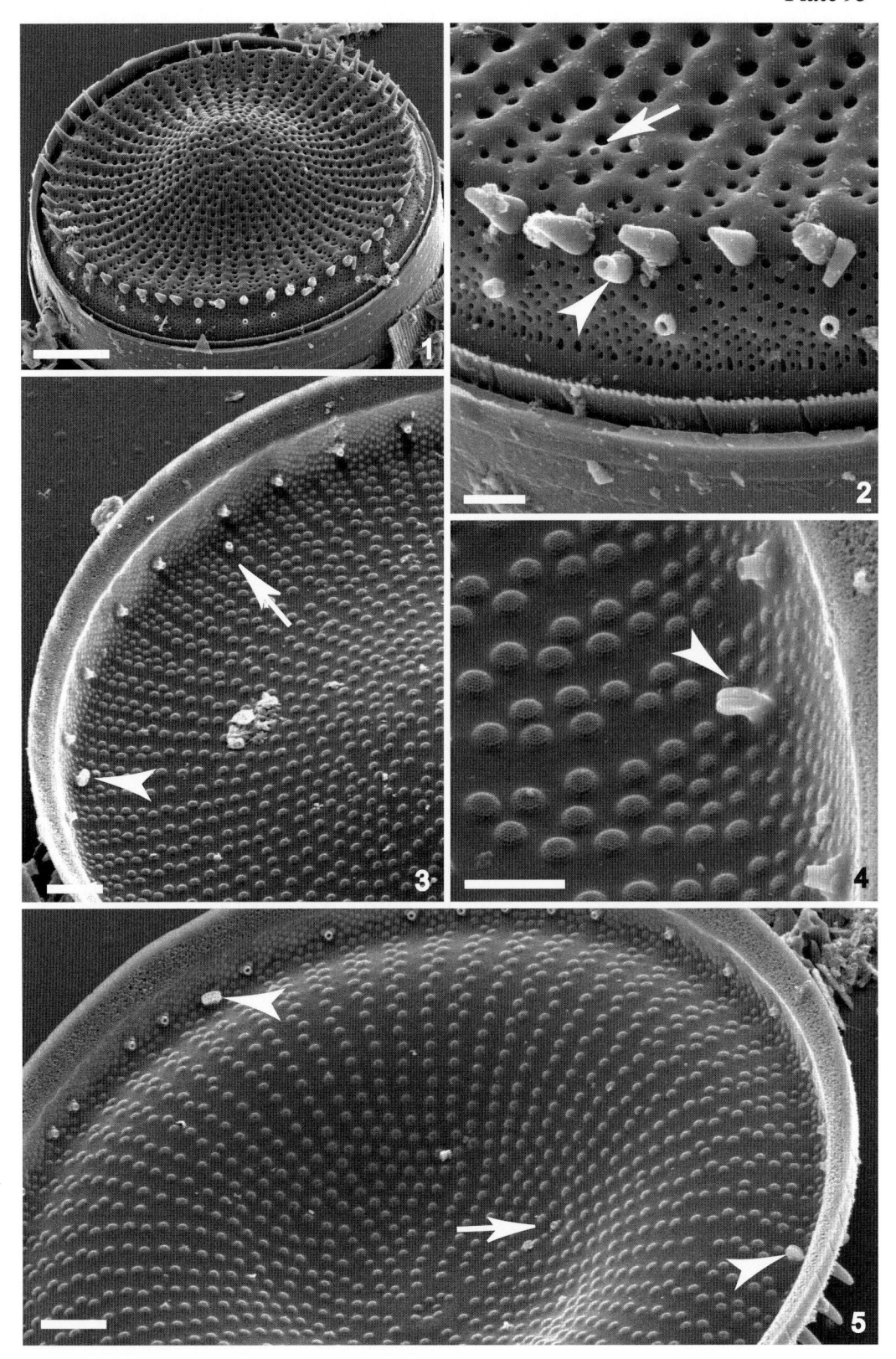

Plate 94. Centric diatoms: Thalassiosirales
Stephanodiscus hashiensis H. Tanaka in Houk *et al.* 2014 (in press)

文　献：Houk, V., Klee, R. & Tanaka, H. 2014. Atlas of freshwater centric diatoms with a brief key and descriptions. Part Ⅳ. Stephanodiscaceae B. *In*: Poulíčková, A.(ed.): Fottea 14 Supplement.(in press)

形　態：殻は小形円盤状で，殻面は同心円状に波打っている．SEM観察では外側表面は微粒子が密に付着したような外観を呈している．図に使用した波志江沼産の個体は殻径7.5-9.5 µm，束線は10 µmに12-14本である．殻面／殻套境界のすべての間束線に針が所在する．殻面中心近くに1個の有基突起があり付随孔は2または3個である．一方，殻套有基突起は2-3本ごとの間条線にあり付随孔は3個である．開口は扇形をした板状物のスリット状の裂け目上位にある．唇状突起は1個であり，開口は針より太い管で針列の中にある．

ノート：*Stephanodiscus vestibulis* Håk., E.C. Ther. & Stoerm. とよく類似しており，著者は同種へ同定したこともある（田中・南雲 2007）．種小名は原産地（波志江沼）に由来する．

産　出：波志江沼（田中・南雲 2007）：群馬県（現生）．

図　版：波志江沼標本．

Figs 1-11. *Stephanodiscus hashiensis* H. Tanaka

LM. Figs 1-5. SEM. Figs 6-11. Figs 1-11. From the type material, HAS-201, Hashie-numa Pond (Recent), Isesaki City, Gunma, Japan. Scale bars: Figs 6, 9=2 µm, Fig. 7=1 µm, Figs 8, 10-11=0.5 µm.

Figs 1-5.　Three different size valves.
Figs 1-2, 4-5.　Two valves at different focal planes.
Fig. 6.　External view of entire valve, opening of valve face fultoportula (arrow).
Fig. 7.　Oblique view of Fig. 6, external tube of rimoportula (arrowhead).
Fig. 8.　Enlarged view of Fig. 7, external opening of mantle fultoportula with sector-shaped flaps (arrow), external tube of rimoportula (arrowhead).
Fig. 9.　Internal oblique view of valve, valve face fultoportula (arrow).
Fig. 10.　Detailed view of valve margin, mantle fultoportula with three satellite pores (arrow), rimoportula (arrowhead).
Fig. 11.　Detailed view of valve margin, opening of mantle fultoportula with sector-shaped flaps (arrow), top of rimoportula (arrowhead).

Plate 94

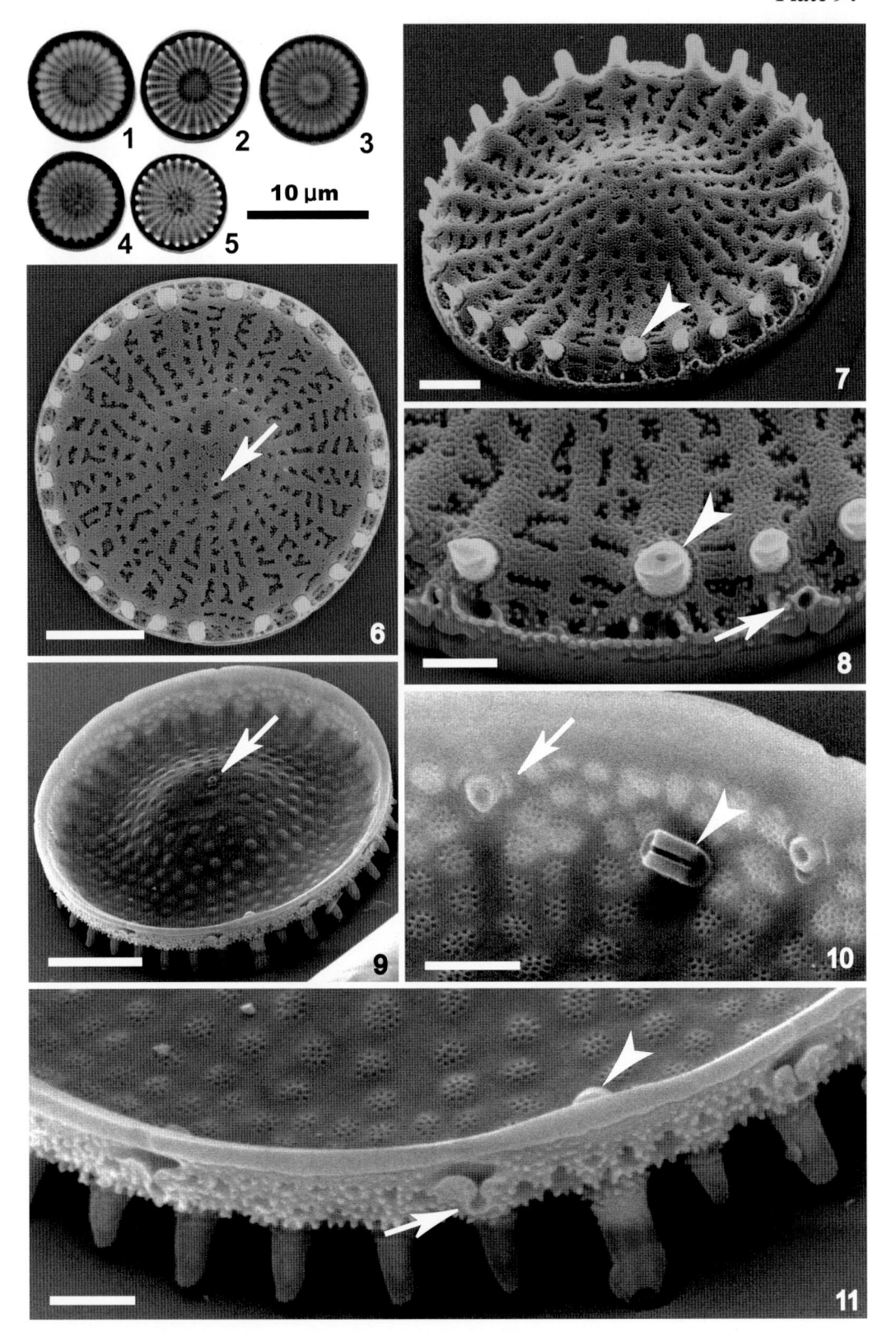

Plates 95-96. Centric diatoms: Thalassiosirales
Stephanodiscus iwatensis H. Tanaka sp. nov.

新　　種：記載文（英文）は8頁参照．
形　　態：殻は小形で円盤形，殻径 11-23 μm，殻面は同心円状に波打っている．束線は殻縁では普通2列の胞紋列から構成され，10 μm に約 10 本．SEM 観察によれば，すべての間束線の殻面／殻套境界には針がある．殻中心近くに1個の殻面有基突起がある．殻面有基突起の殻中心をはさんだ反対側の殻面／殻套境界に唇状突起が所在する．殻套有基突起は 3-5 本ごとの間束線末端にあり，殻面有基突起・殻套有基突起とも付随孔は3個である．殻面有基突起の外側開口は円錐状であるが，殻套有基突起・唇状突起は管である．
ノ ー ト：*Stephanodiscus minutulus*（Kütz.）Round と類似するが，殻面有基突起付随孔が *S. minutulus* は常に2個なのに対し3個であること，殻直径が *S. minutulus* は 2-12 μm（Håkansson 2002）に対して，*S. iwatensis* は 11-23 μm であることから区別される．種小名は原産地（岩手県）に由来する．
産　　出：舛沢層：中新-鮮新世・岩手県．
図　　版 95-96：舛沢層標本．

Plate 95, Figs 1-12. *Stephanodiscus iwatensis* H. Tanaka sp. nov.
LM. Figs 1-7. SEM. Figs 8-12, external views. From the type material, MAS-117, Masuzawa Formation (Mio-Pliocene), Shizukuishi Town, Iwate, Japan. Scale bars: Figs 8-10＝5 μm, Figs 11-12＝1 μm.

Figs 1-2.　Holotype, MPC-25056, same valve at different focal planes.
Figs 3-4.　Same valve at different focal planes.
Figs 1-6.　Four different size valves.
Fig. 7.　Oblique view showing convex center of valve face.
Fig. 8.　View of whole valve, opening of valve face fultoportula (arrow).
Fig. 9.　Oblique view of Fig. 8 showing opening of rimoportula (arrowhead), opening of valve face fultoportula (arrow) and spines.
Fig. 10.　Oblique view of concave valve, rimoportula (arrowhead).
Fig. 11.　Detailed view of valve margin showing opening of rimoportula (arrowhead) and openings of mantle fultoportulae (arrows).
Fig. 12.　Enlarged view of part of Fig. 9, showing conical opening of valve face fultoportula (arrow).

Plate 95

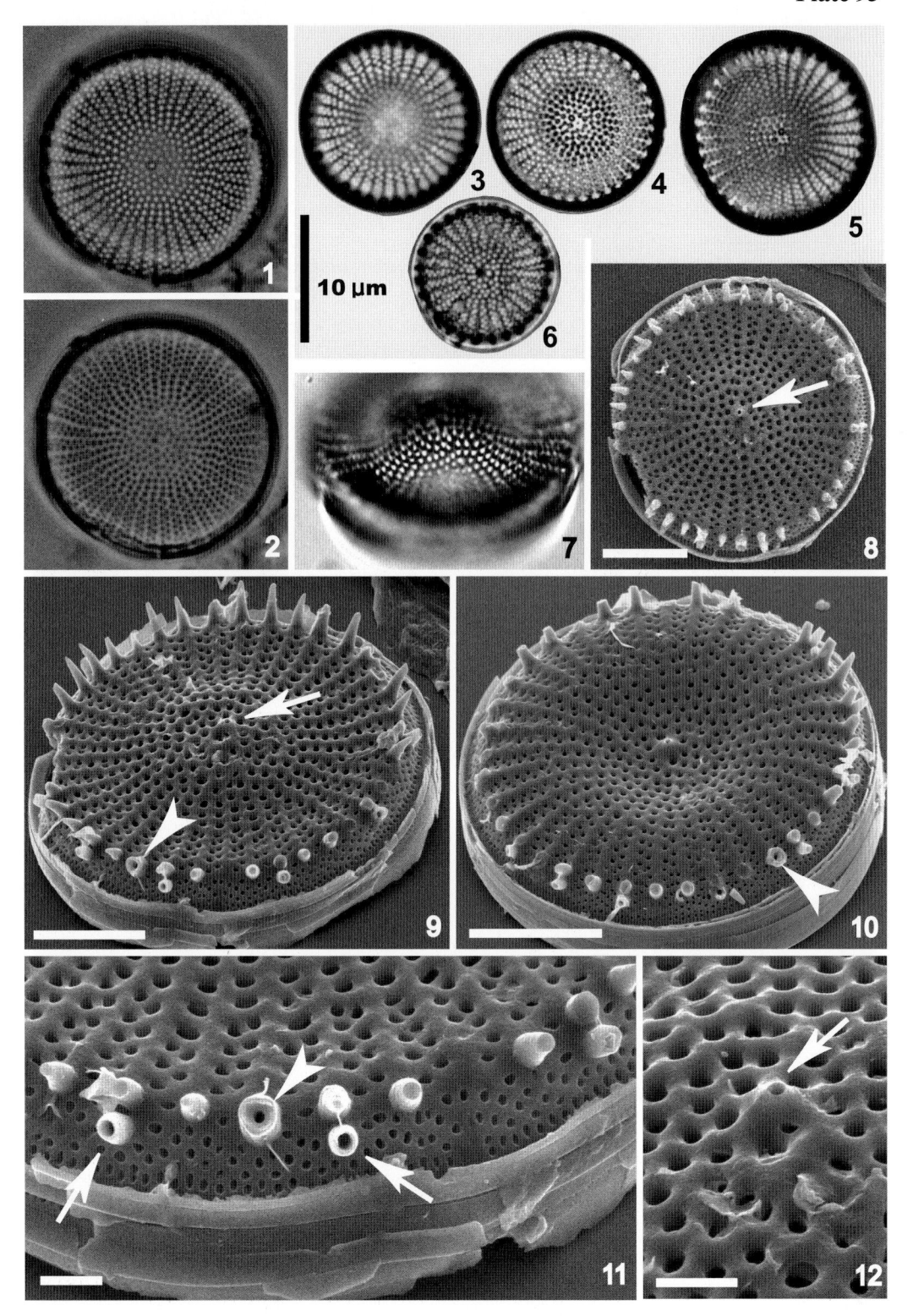

Plate 96, Figs 1-6. *Stephanodiscus iwatensis* **H. Tanaka** sp. nov.
SEM. Figs 1-6. Internal views, from the type material, MAS-117, Masuzawa Formation (Mio-Pliocene), Shizukuishi Town, Iwate, Japan. Scale bars: Figs 1-2=5 μm, Fig. 6=2 μm, Fig. 4=1 μm, Figs 3, 5=0.5 μm.

Fig. 1. View of whole valve, rimoportula (arrowhead), valve face fultoportula (arrow).
Fig. 2. Oblique view of convex valve.
Fig. 3. Enlarged view of Fig. 1 showing valve face fultoportula with three satellite pores and domed cribra.
Fig. 4. Enlarged view of part of Fig. 2 showing valve margin, rimoportula (arrowhead) and two mantle fultoportulae.
Fig. 5. Detailed view of valve margin showing rimoportula and mantle fultoportula with three satellite pores (arrow).
Fig. 6. Oblique view of concave valve of Fig. 1.

Plate 96

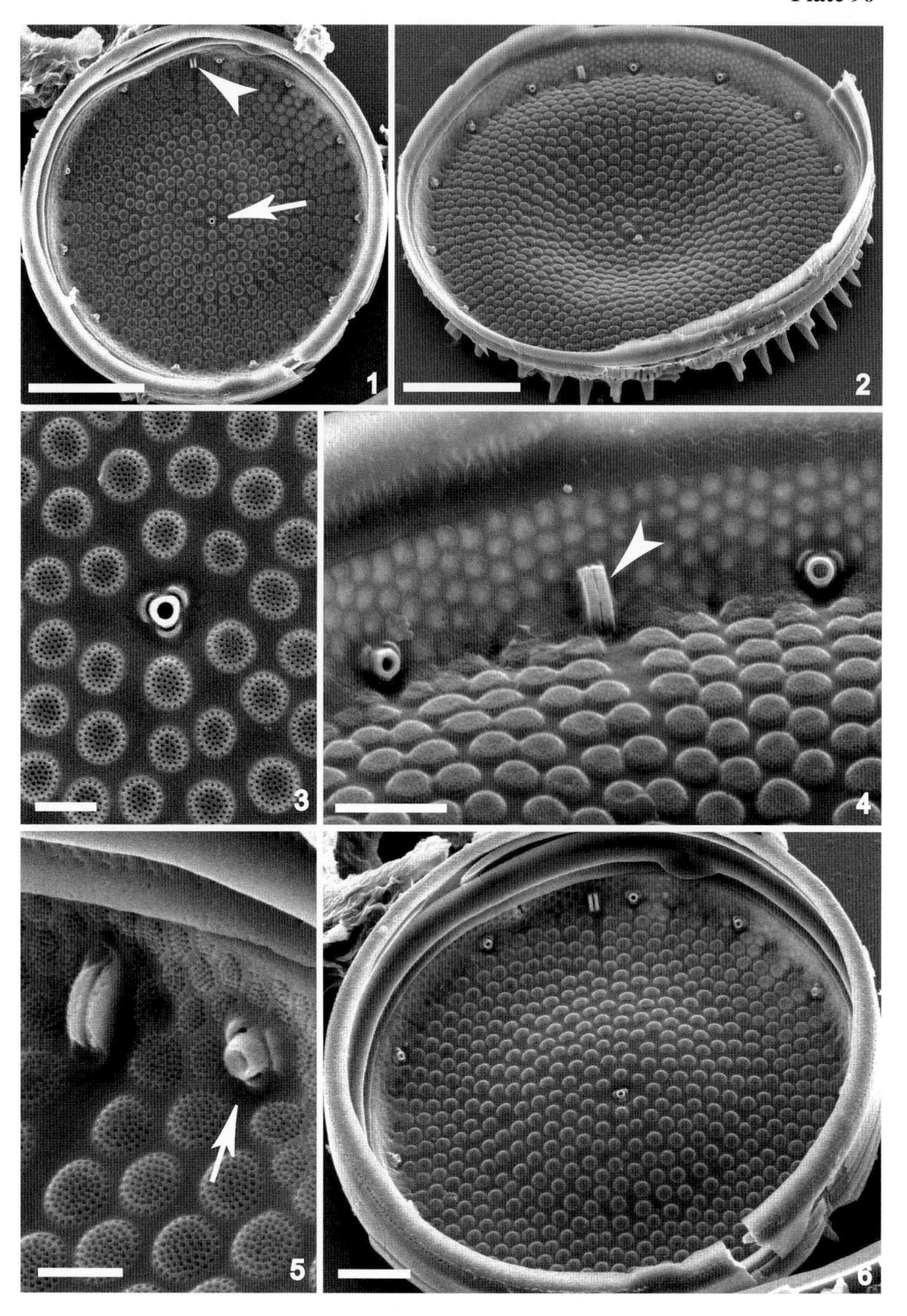

Plates 97-98.　Centric diatoms: Thalassiosirales
Stephanodiscus kobayasii H. Tanaka in Houk *et al.* 2014 (in press)

文　　献：Houk, V., Klee, R. & Tanaka, H. 2014. Atlas of freshwater centric diatoms with a brief key and descriptions. Part Ⅳ. Stephanodiscaceae B. *In*: Pouličková, A.(ed.): Fottea **14** Supplement.(in press)

形　　態：殻は円盤形で殻面はほぼ平ら，殻径 42-95 μm．束線は殻面縁辺部では 5-9 本の胞紋列から構成される．間束線は殻中心から殻套有基突起または唇状突起まで連続する一次間束線と，一次間束線が形成する束線内に所在する殻面有基突起の殻中心側に形成された，短い二次間束線の 2 種類ある．一次間束線の殻面／殻套境界には針が所在する．殻面有基突起は 9-32 個で 2 または 3 個の付随孔を伴い，束線内で二次間束線末端に所在する場合と，間束線を伴わないで存在する場合がある．殻套有基突起は殻套下部にあり，3 個の付随孔を伴う．殻面，殻套有基突起の付随孔の覆いは，ふつう，ゆるいらせん状であるが直線状の場合もある．唇状突起は(1)2 個で 1 個は常に一次間束線末端，他方は束線中に所在する場合が多い．

ノ ー ト：束線・間束線の区別が不明瞭で，本種に類似する有基突起付随孔覆いがらせん状になる分類群は *Thalassiobeckia* Khurs & Fedenya 属がある．*S. kobayasii* は SEM 観察による殻表面の束線・間束線の区別は不明瞭であるが，内面および光顕では明らかであり，付随孔覆いはらせんにならない場合もあるので *Stephanodiscus* に所属が適当である．種小名は元日本珪藻学会会長の小林　弘博士を記念して付けられた．

産　　出：尾本層（Houk *et al.* 2013）：前期更新世・大分県．

図　　版 97-98：尾本層標本．

Plate 97, Figs 1-3.　*Stephanodiscus kobayasii* H. Tanaka
LM. Figs 1-3. From the type materiall, OIT-021, Omoto Formation (Early Pleistocene), Kitsuki City, Oita, Japan.

Figs 1-3.　Three different size valves.

Plate 97

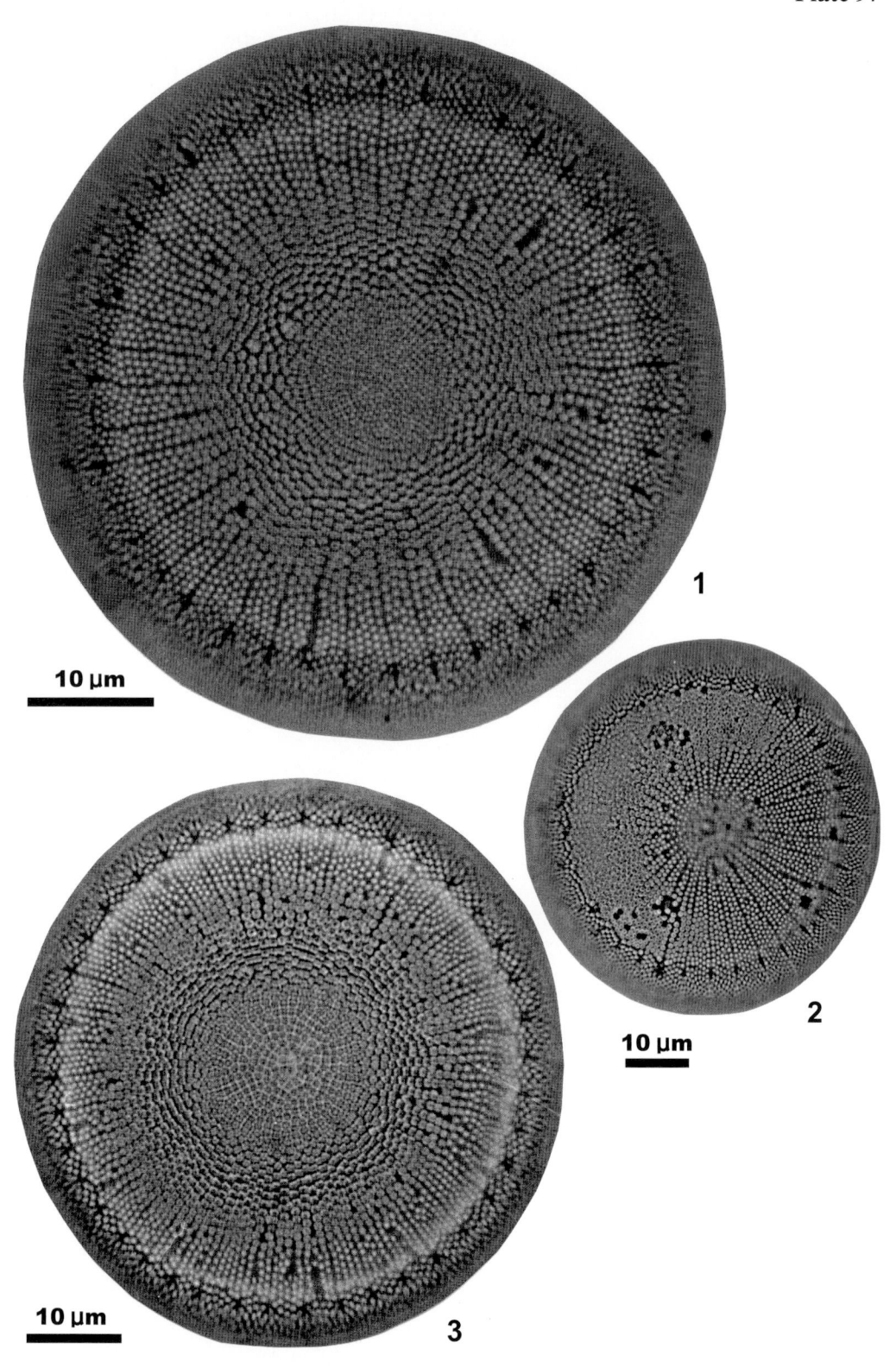

10 µm

1

10 µm

2

10 µm

3

Plate 98, Figs 1-6. *Stephanodiscus kobayasii* H. Tanaka

SEM. Figs 1-6. From the type material, OIT-021, Omoto Formation (Early Pleistocene), Kitsuki City, Oita, Japan. Scale bars: Figs 1, 5-6=10 μm, Fig. 3=5 μm, Figs 2, 4=2 μm.

Fig. 1. External oblique view of whole valve.

Fig. 2. View of valve margin showing opening of rimoportula (arrowhead) on the valve face/mantle boundary in the ring of spines.

Fig. 3. View of mantle showing high mantle, openings of mantle fultoportulae (arrows) and opening of rimoportula slightly under valve face/mantle boundary (arrowhead).

Fig. 4. Detailed internal view of valve margin showing valve face fultoportula with three spiral cowlings (arrow), one mantle fultoportula with three straight cowlings (arrowhead), one mantle fultoportula with spiral cowlings and rimoportula in a fascicle.

Fig. 5. Oblique internal view showing valve face fultoportulae in the fascicles, mantle fultoportulae at the end of interfascicles and one rimoportula located in a fascicle (large arrowhead) and another at end of interfascicle (small arrowhead).

Fig. 6. Oblique internal view showing short interfascicles on valve face with valve face fultoportulae on centrifugal ends.

Plate 98

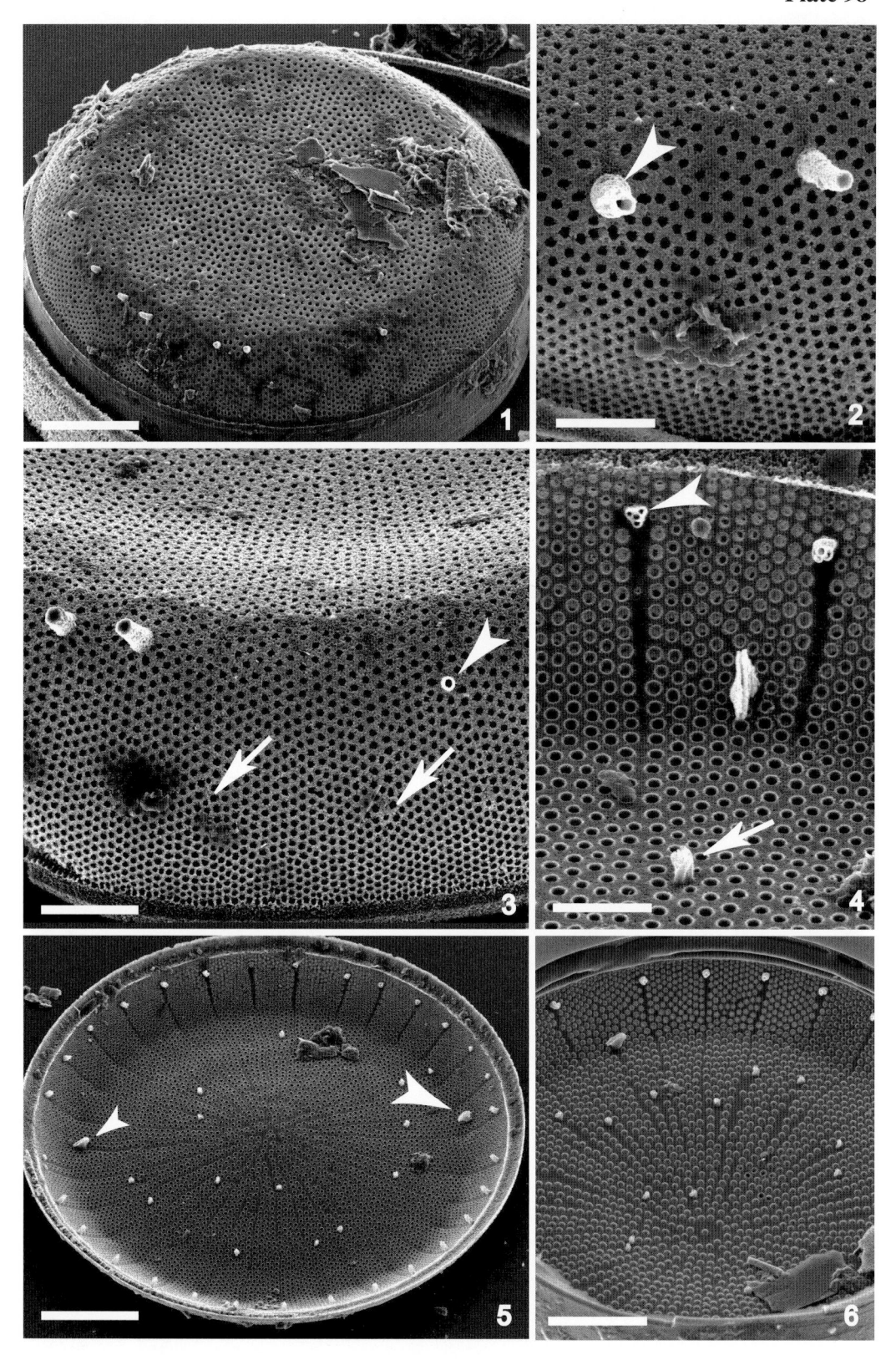

233

Plates 99-100.　Centric diatoms: Thalassiosirales
Stephanodiscus komoroensis H. Tanaka 2000

文　　献：Tanaka, H. 2000. *Stephanodiscus komoroensis* sp. nov., a new Pleistocene diatom from central Japan. Diatom Research 15: 149-157.

形　　態：殻は円盤状で，殻径 26-87 μm，針は存在しない．束線は殻面／殻套境界で 10 μm に(4)5(6)本であるが，光顕あるいは SEM 観察においても束線と間束線の区別が外側では明瞭でないことが多い．内側では常に明瞭である．束線内側末端の 2-4 本ごとに柄の長い殻套有基突起があり，付随孔は常に 3 個で，付随孔の覆いも長く伸張している．殻面有基突起は束線中にあり付随孔は 3(2) 個である．唇状突起は 5-9 個存在し，外管は太く長く先端が三角形になっている．

ノ　ー　ト：特徴的な唇状突起の外管の存在，および針が欠如することは一般的な *Stephanodiscus* 属の形態とは異なっているが，他の形態は *Stephanodiscus* そのものであり，本属に所属する種であることは明らかである．種小名は原産地（小諸市）に由来する．

産　　出：瓜生坂層（Tanaka 2000，田中・南雲 2000）：前期更新世・長野県，小野上層（田中・南雲 2000）：前期更新世・群馬県，八束村珪藻土（田中・南雲 2000）：中期更新世・岡山県，野原層（Tanaka et al.(2004)では松本ら(1984)のとおり津房川層の産出としているが，石塚ら(2005)によって当地の地層は津房川層とは異なることが判明し，野原層と命名された）：前期更新世・大分県，大鷲湖沼性堆積物（田中ら 2011）：鮮新世・岐阜県．

図　　版 99-100：瓜生坂層標本．

Plate 99, Figs 1-6.　*Stephanodiscus komoroensis* H. Tanaka
LM. Figs 1-6. From the type material, OKU-101, Uryuzaka Formation (Early Pleistocene), Komoro City, Nagano, Japan.

Figs 1-2, 3-4, 5-6. Three valves at different focal planes. Outer openings of rimoportulae with triangular tubes (arrowheads).

Plate 99

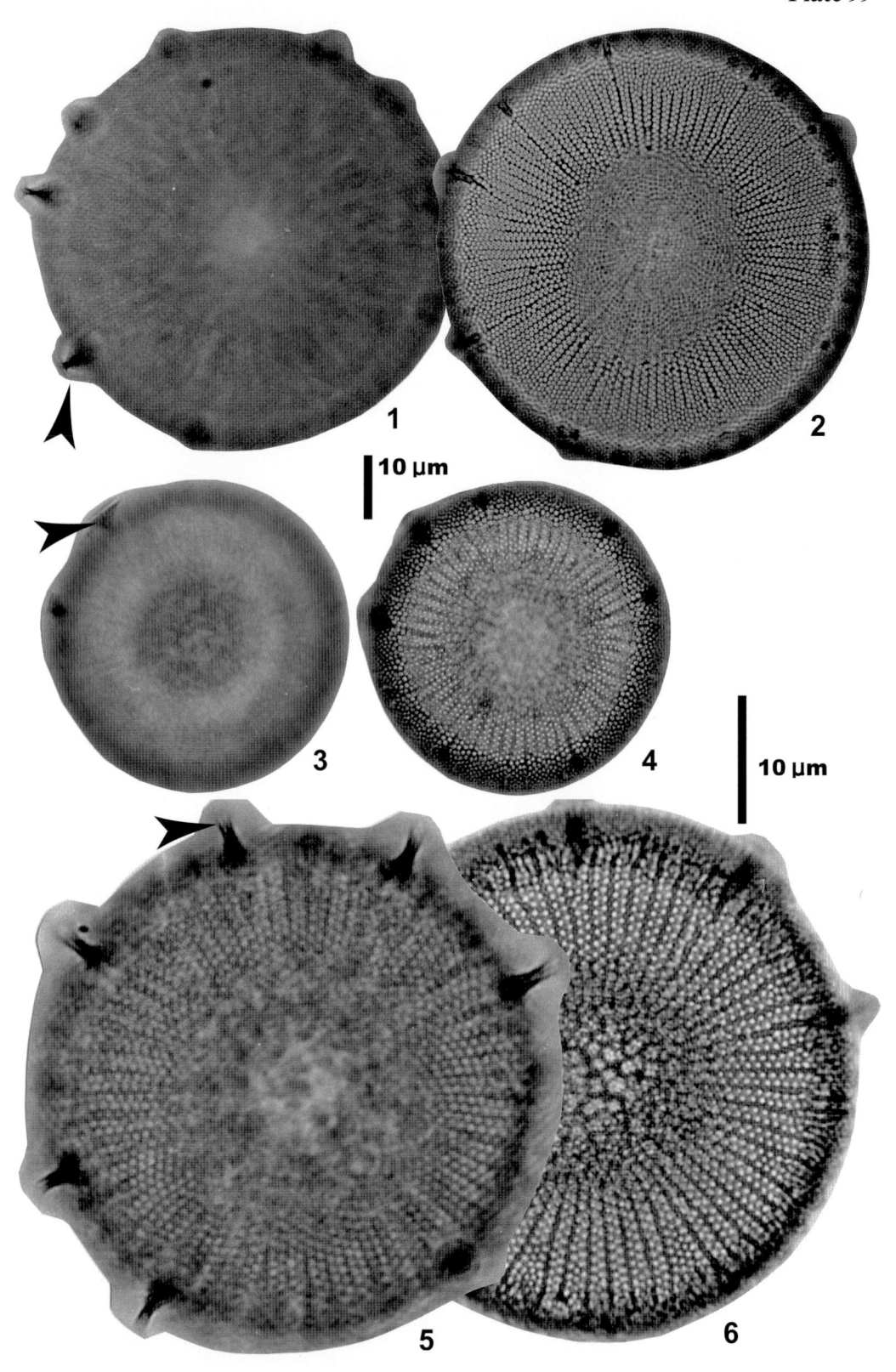

Plate 100, Figs 1-5. *Stephanodiscus komoroensis* **H. Tanaka**

SEM. Figs 1-5. From the type material, OKU-101, Uryuzaka Formation (Early Pleistocene), Komoro City, Nagano, Japan. Scale bars: Fig. 1=10 μm, Figs 2, 4-5=5 μm, Fig. 3=1 μm.

Fig. 1. External view of entire valve.

Fig. 2. Enlarged view of part of Fig. 1 showing thick tubes with triangular tips of outer openings of rimoportulae (arrows).

Fig. 3. Detailed view of internal valve margin showing domed cribra, rimoportula (arrowhead) and mantle fultoportulae with three long cowlings (arrows).

Fig. 4. Oblique view of valve margin, showing two tubes of outer openings of rimoportulae.

Fig. 5. Internal view of broken valve showing rimoportulae (arrowheads), valve face fultoportulae (black arrows) and mantle fultoportulae (white arrows).

Plate 100

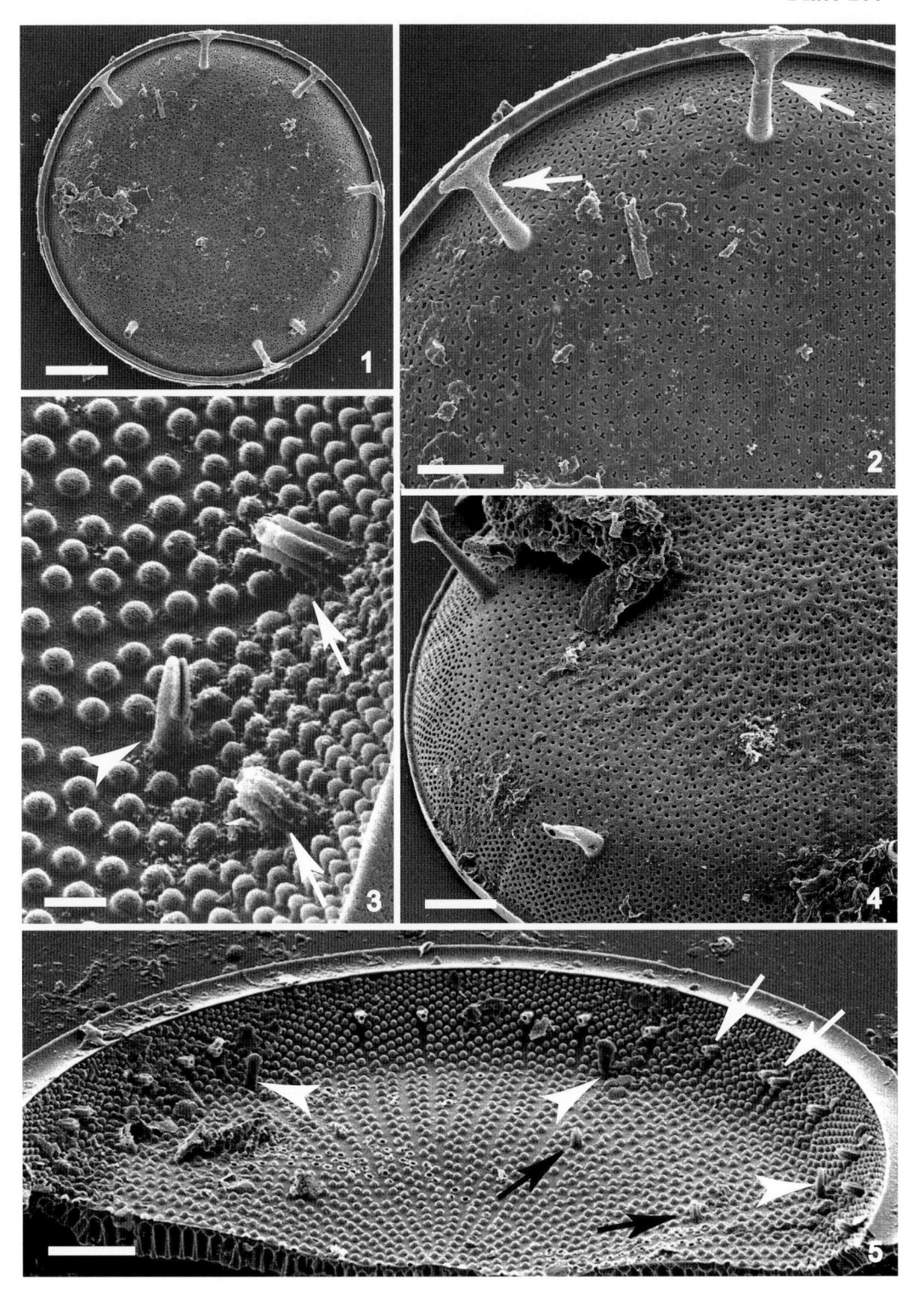

Plates 101-102.   Centric diatoms: Thalassiosirales
Stephanodiscus kusuensis Julius, Tanaka & Curtin
in Julius *et al.* 2006

文　　献：Julius, M.L., Curtin, M. & Tanaka, H. 2006. *Stephanodiscus kusuensis*, sp. nov a new Pleistocene diatom from southern Japan. Phycological Research **54**: 294-301.

形　　態：殻面は円形で同心円状に強く波打つ，殻径 35-49 μm，束線は殻縁で 10 μm に (5) 6-8 本．殻面有基突起は中央が突出する殻では中央の凸部，中央が凹む殻では最も央心よりの円形の凸部に分布する．さらに殻面／殻套境界（この場合針は存在しない），および境界近くの殻面に所在する場合がある．SEM 観察によれば，殻面／殻套境界の針は 2-4 本ごとの間束線にあり，針のすぐ下方には殻套有基突起の開口がある．殻面有基突起の付随孔は 2 (3) 個であるが，殻套有基突起の付随孔は常に 3 個である．唇状突起は 3-4 個で殻面／殻套境界にあり，外側への開口は管を伴う．

ノート：本種の他分類群と識別する特徴は，殻面が同心円状に二重に強く波打つこと，殻面有基突起の配置，有基突起がしばしば殻面／殻套境界（あるいはその近く）に所在し針が欠如することである．

産　　出：野上層（Julius *et al.* 2006，田中ら 2008）：中期更新世・大分県．

図　　版 101-102：野上層標本．

Plate 101, Figs 1-6. *Stephanodiscus kusuensis* Julius, Tanaka & Curtin
LM. Figs 1-6. Material from Nogami Formation (Middle Pleistocene), Migita, Kokonoe Town, Oita, Japan.

Figs 1-2, 3-4, 5-6.   Three valves at different focal planes, valve face fultoportulae (arrows).

Plate 101

Plate 102, Figs 1-6. *Stephanodiscus kusuensis* **Julius, Tanaka & Curtin**
SEM. Figs 1-6. Material from Nogami Formation (Middle Pleistocene), Migita, Kokonoe Town, Oita, Japan. Scale bars: Figs 1-3=10 µm, Fig. 4=2 µm, Figs 5-6=1 µm.

Fig. 1. External view of entire valve.
Fig. 2. Oblique view of Fig. 1.
Fig. 3. Internal view of concave valve showing valve face fultoportulae (arrows) and rimoportulae (arrowheads).
Fig. 4. Internal view of convex valve center showing valve face fultoportulae with two or three satellite pores.
Fig. 5. Detailed view of internal valve margin showing domed cribra, rimoportula and mantle fultoportulae with three satellite pores.
Fig. 6. Detailed view of external oblique convex valve showing openings of fultoportulae: central area of valve face (arrow F1), marginal area of valve face (arrow F2), valve face/mantle boundary (arrow B) and mantle (arrow M).

Plate 102

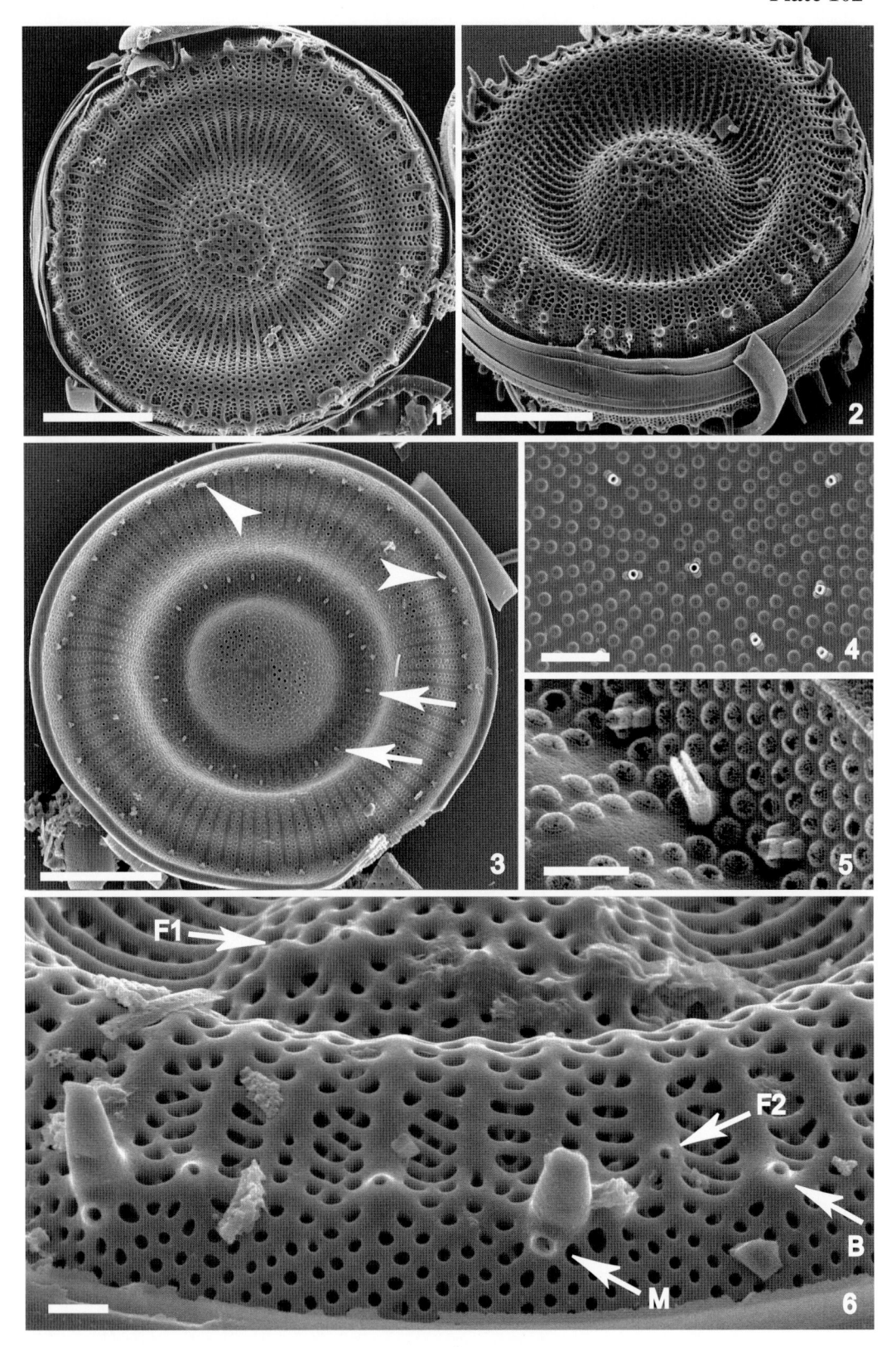

Plates 103-104. Centric diatoms: Thalassiosirales
Stephanodiscus kyushuensis H. Tanaka in Houk *et al.* 2014 (in press)

文　　献：Houk, V., Klee, R. & Tanaka, H. 2014. Atlas of freshwater centric diatoms with a brief key and descriptions. Part Ⅳ. Stephanodiscaceae B. *In*: Poulíčková, A.(ed.): Fottea **14** Supplement.(in press)

形　　態：殻は円盤状，中心部はほぼ平らで急に上がって凸状になるか，急に下がって凹状になる．殻面／殻套境界の針は全間束線にある．殻径 25-45 μm，束線は殻縁で 10 μm に 5-7 本．束線を構成する胞紋はぼ 1 列であるが殻套近くで 2 列になる．殻面の胞紋は形・大きさが変化し，中心付近ではやや楕円形で大きいが，殻套近くでは円形で小さくなる．殻面有基突起は殻中心部と縁辺部に分布するが，数と所在がかなり変化し，殻中心部の凸または凹状部には 1-13 個，殻套寄りの縁辺部には 0-17 個観察された．付随孔は 3 個のものが多いが 2 個のものもしばしば見られた．唇状突起は 1 個で外側への開口は針の列よりもわずか殻端よりの殻套にあり短管である．これよりも下位に殻套有基突起の開口が毎～3 本ごとの間束線末端にある．殻套有基突起の付随孔は常に 3 個である．殻套は浅い．

ノ ー ト：種小名は原産地（九州）に由来する．

産　　出：永野層（Houk *et al.* 2014）：後期鮮新世・鹿児島県．

図　　版 103-104：永野層標本．

Plate 103, Figs 1-6. *Stephanodiscus kyushuensis* H. Tanaka

LM. Figs 1-5. SEM. Fig. 6. From the type material, KAG-204, Nagano Formation (Late Pliocene), Satsuma Town, Kagoshima, Japan. Scale bar: Fig. 6＝5 μm.

Figs 1-5.　Three valves, different sizes.
Figs 2-3, 4-5.　Two valves at different focal planes.
Fig. 6.　External view of whole convex valve.

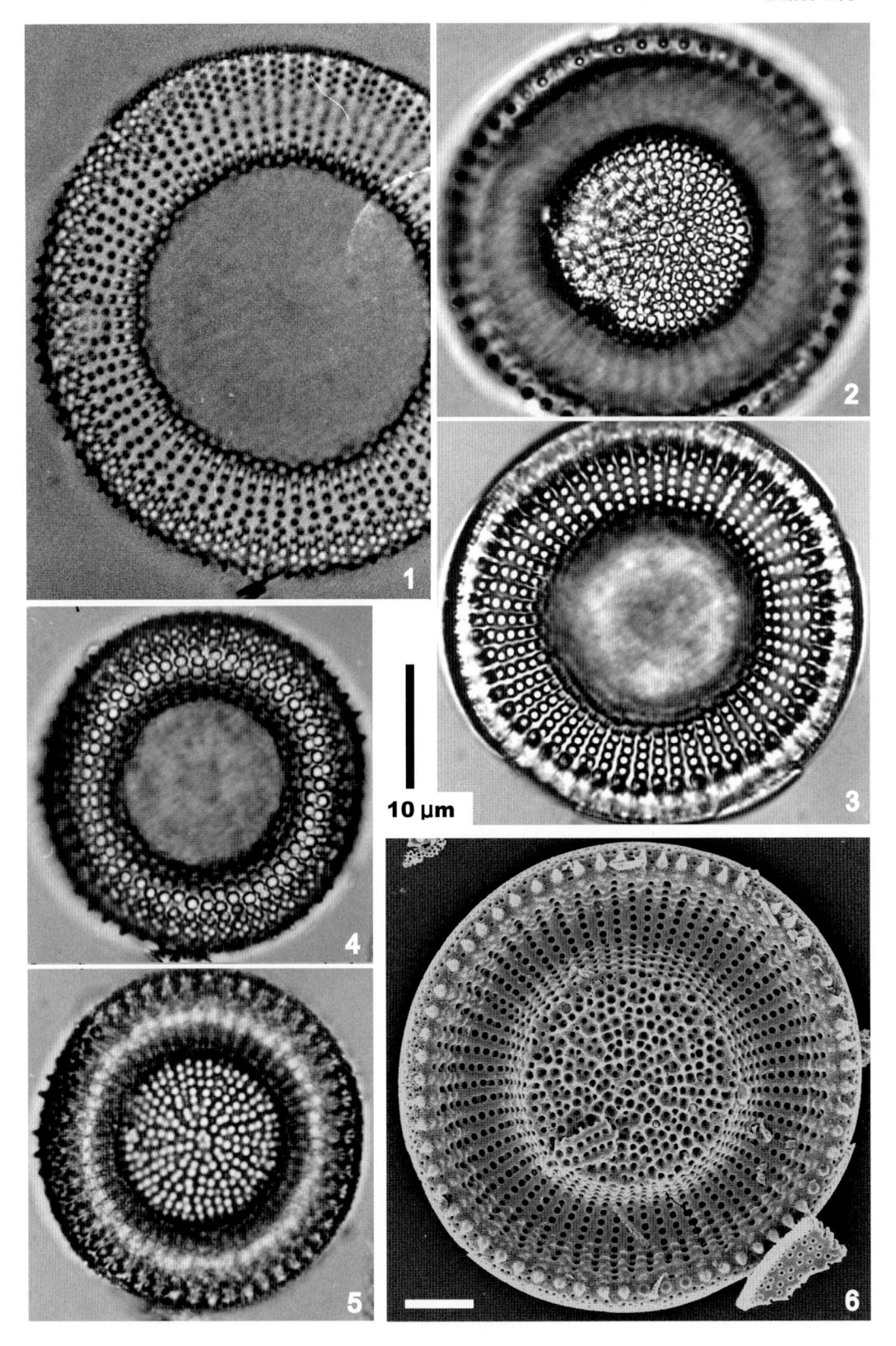

Plate 103

10 µm

Plate 104, Figs 1–6. *Stephanodiscus kyushuensis* **H. Tanaka**

SEM. Figs 1–6. From the type material, KAG-204, Nagano Formation (Late Pliocene), Satsuma Town, Kagoshima, Japan. Scale bars: Figs 1–2, 6=5 µm, Figs 3–5=2 µm.

Fig. 1. Oblique view of Plate 103, Fig. 6 showing sharp angle between convex central zone and marginal zone of valve face and spines located every interfascicle at valve face/mantle boundary.

Fig. 2. External valve margin, opening of rimoportula (arrowhead).

Fig. 3. Internal view of whole concave valve.

Fig. 4. Enlarged view of valve center of Fig. 3 showing remains of valve face fultoportulae with three satellite pores (arrows).

Fig. 5. Internal valve margin, remains of mantle fultoportulae with three satellite pores (arrows).

Fig. 6. Oblique view of Fig. 3 showing large areolae located on the valve center and small areolae on marginal valve face, valve face fultoportulae located on central area (small arrows) and marginal area (large arrows).

Plate 104

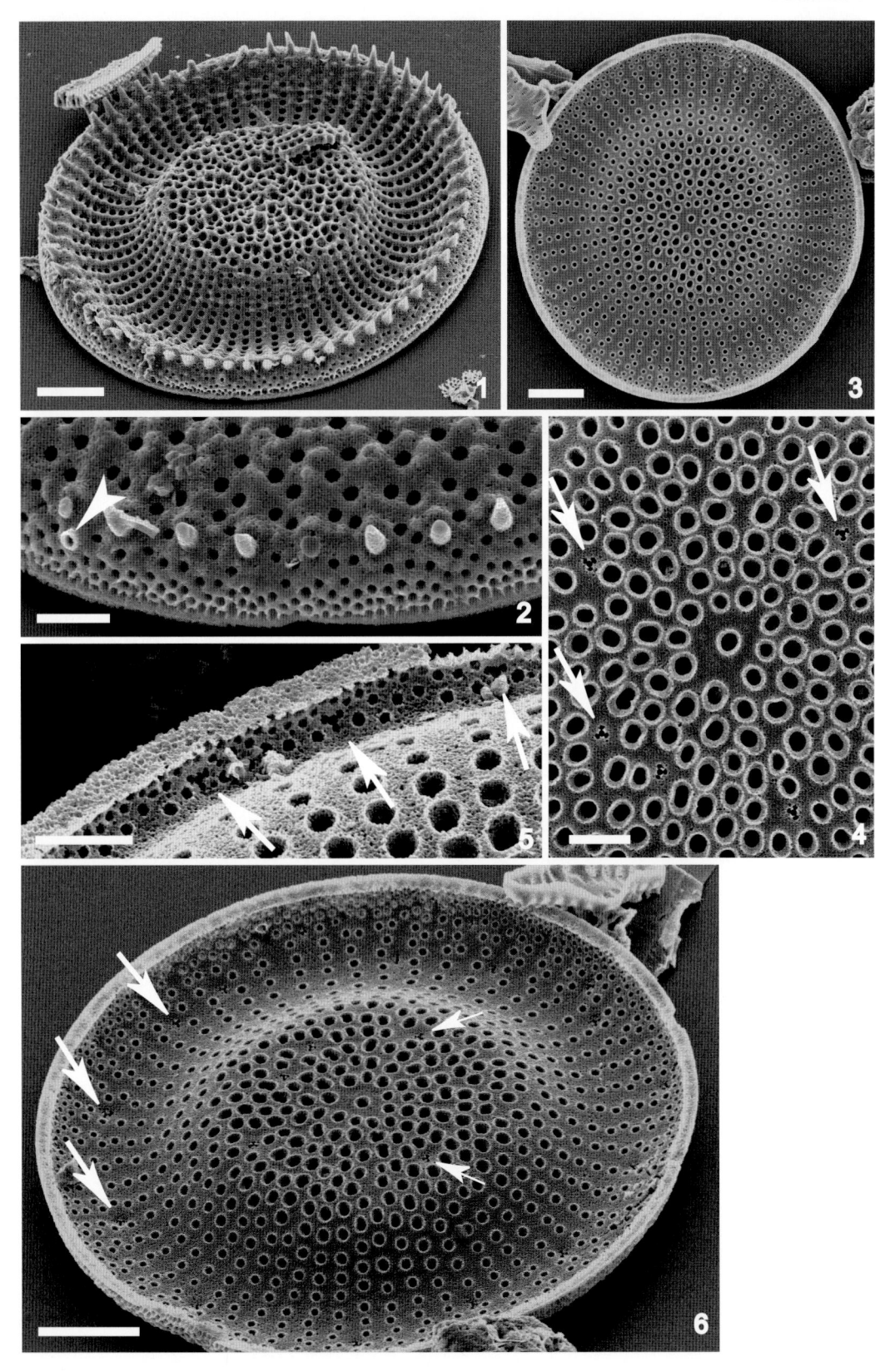

245

Plate 105.　Centric diatoms: Thalassiosirales
Stephanodiscus minutulus (Kützing) Cleve & Möller 1882

基礎異名：*Cyclotella minutula* Kützing 1844
文　　献：Håkansson, H. 2002. A compilation and evaluation of species in the general *Stephanodiscus, Cyclostephanos* and *Cyclotella* with a new genus in the family Stephanodiscaceae. Diatom Research **17**: 1-139.
形　　態：殻は小形円盤状，殻径 7-17 μm，殻面は同心円状に波打っている．束線は殻縁で 10 μm に約 11 本．SEM 観察によれば，すべての間束線の殻面／殻套境界には針がある．凸殻では殻中心近くに 1 個の殻面有基突起があり，凹殻ではそれより殻縁よりに所在する．これら殻面有基突起の殻中心をはさんだ反対側の殻面／殻套境界に唇状突起がある．殻套有基突起は (1)2-3(4) ごとの間束線にある．殻面有基突起は 2 個の付随孔をもつが，殻套有基突起は 3 個である．
ノ　ー　ト：現生の本種は阿寒湖（河島・小林 1993），波志江沼（田中・南雲 2007），琵琶湖（辻・伯耆 2001）等，各地から知られている．
産　　出：塩原湖成層：前期更新世・栃木県．
図　　版：塩原湖成層標本．

Figs 1-12.　*Stephanodiscus minutulus* (Kützing) Cleve & Möller
LM. Figs 1-5. SEM. Figs 6-12. Material from Shiobara Lacustrine Deposit (Early Pleistocene), Nasushiobara City, Tochigi, Japan. Scale bars: Fig. 7＝5 μm, Figs 6, 10-11＝2 μm, Figs 9, 12＝1 μm, Fig. 8＝0.5 μm.

Figs 1-5.　Three different size valves.
Figs 2-3, 4-5.　Two valves at different focal planes.
Fig. 6.　External view of whole convex valve, opening of valve face fultoportula (arrow) and outer-tube of rimoportula (arrowhead).
Fig. 7.　Oblique view of convex valve.
Fig. 8.　Detailed internal view of valve face fultoportula with two satellite pores.
Fig. 9.　Enlarged oblique view of part of Fig. 6 showing outer-tube of rimoportula (arrowhead).
Fig. 10.　Internal view of convex valve, valve face fultoportula (arrow) and rimoportula (arrowhead).
Fig. 11.　Internal oblique view of concave valve, valve face fultoportula (arrow) and rimoportula (arrowhead).
Fig. 12.　Detailed internal oblique view of convex valve showing valve face fultoportula with two satellite pores (arrow), mantle fultoportulae with three satellite pores (dark arrow) and rimoportula (arrowhead).

Plate 105

Plates 106-107. Centric diatoms: Thalassiosirales
Stephanodiscus miyagiensis H. Tanaka & Nagumo 2006

文　　献：Tanaka, H. & Nagumo, T. 2006. *Stephanodiscus miyagiensis*, sp. nov. from Pleistocene sediment in northeastern Japan. Diatom Research **21**: 371-378.

形　　態：殻面は円盤形，殻は同心円状に波打つ，殻径 20-68 μm，束線を構成する点紋列は殻面／殻套境界で(3)4-13(15)本である．間束線はわずかにうねり，殻套有基突起で終るが，各間束線の殻面／殻套境界には針がある．殻套では殻套部のみに伸張する 1-2 本の間束線が，殻面から連続する間束線の間に所在する．SEM 観察によれば，殻面有基突起は 2-4 個で(1)2-3(4)個の付随孔を伴う．殻套有基突起は殻面から連続する間束線および殻套のみに伸張する間束線の末端にあり，付随孔は 3 個である．唇状突起は 1(2)個で殻面／殻套境界にある．

ノ ー ト：種小名は原産地（宮城県）に由来する．

産　　出：鬼首層（Tanaka & Nagumo 2006）：後期更新世・宮城県．

図　　版 106-107：鬼首層標本．

Plate 106, Figs 1-4. *Stephanodiscus miyagiensis* **H. Tanaka & Nagumo**

LM. Figs 1-4. From the type material, ONI-07, Onikobe Formation (Late Pleistocene), Osaki City, Miyagi, Japan.

Figs 1-2, 3-4. Two valves at different focal planes.
Figs 2, 4. Short mantle interfascicles (arrows).

Plate 106

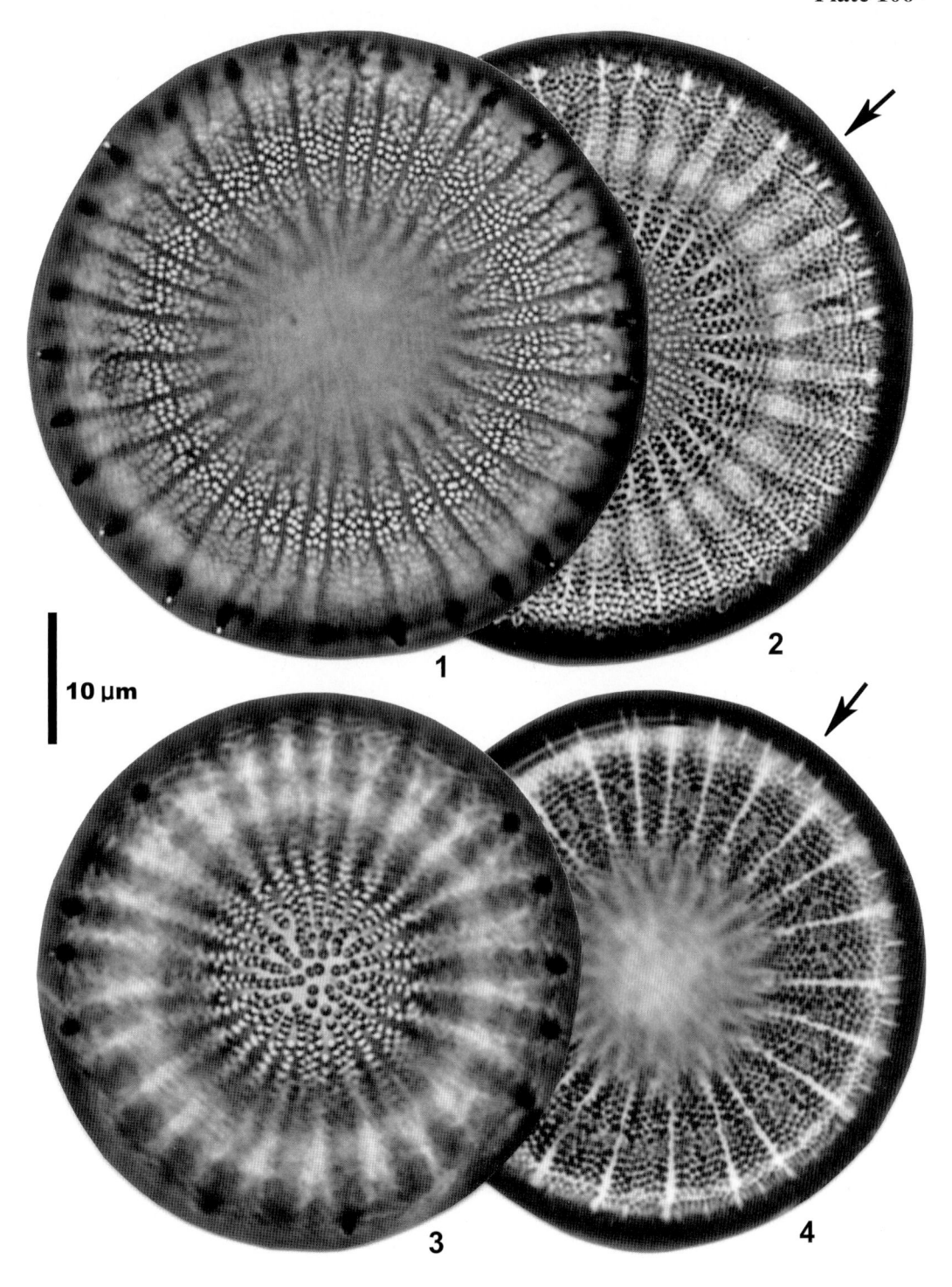

10 μm

1

2

3

4

Plate 107, Figs 1-7. *Stephanodiscus miyagiensis* **H. Tanaka & Nagumo**
SEM. Figs 1-7. From the type material, ONI-07, Onikobe Formation (Late Pleistocene), Osaki City, Miyagi, Japan. Scale bars: Figs 1-2, 4=10 μm, Fig. 3=5 μm, Figs 5, 7=2 μm, Fig. 6=0.5 μm.

Fig. 1. External view of convex valve.
Fig. 2. Oblique view of concave valve.
Fig. 3. Enlarged view of valve margin showing fascicles and interfacicles on valve face, mantle areolae rows, spines and openings of mantle fultoportulae (arrows).
Fig. 4. Internal oblique view of concave valve.
Fig. 5. Detailed view of valve margin showing domed cribra, interfascicles, short mantle interfascicles, rimoportula (arrowhead) and mantle fultoportulae with three satellite pores.
Fig. 6. Detailed view of valve face fultoportulae with two or three satellite pores.
Fig. 7. Detailed view of internal mantle, rimoportula (arrowhead).

Plate 107

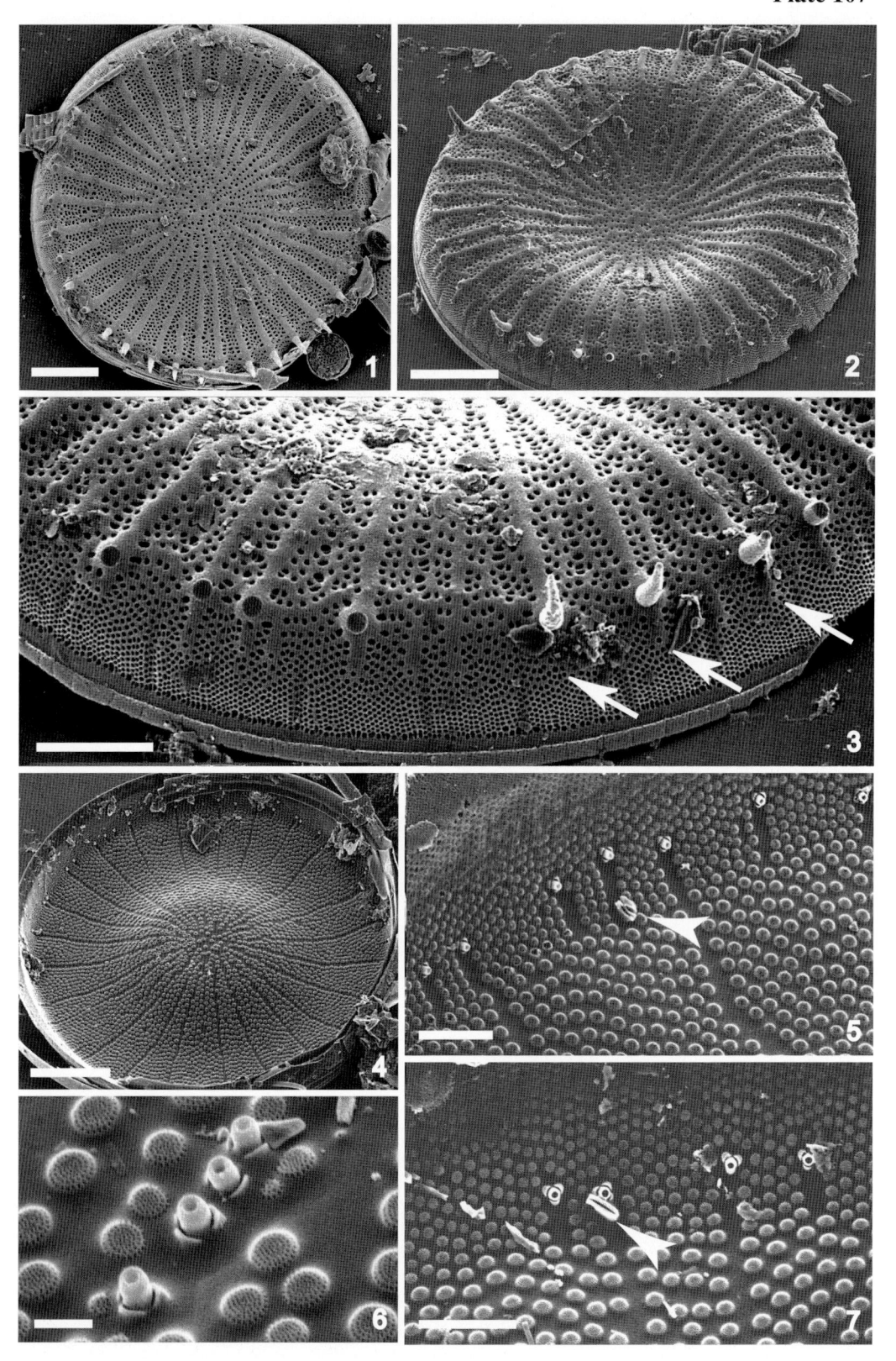

251

Plates 108-109.　Centric diatoms: Thalassiosirales
Stephanodiscus nagumoi H. Tanaka in Houk *et al*. 2014 (in press)

文　献：Houk, V., Klee, R. & Tanaka, H. 2014. Atlas of freshwater centric diatoms with a brief key and descriptions. Part Ⅳ. Stephanodiscaceae B. *In*: Poulíčková, A.(ed.): Fottea **14** Supplement.(in press)

形　態：殻は円盤状で同心円状に波打つ，殻径 30-75 µm，束線は 10 µm に (5)6-7(8) 本で，殻面縁辺部では 2-3(4) 本の胞紋列から構成される．束線を構成する点紋は 10 µm に約 17 個である．殻面有基突起は付随孔が (2)3(4) 個で円状に分布する．殻套有基突起は殻面／殻套境界と殻套下部の 2 箇所に分布し，外面では星形の装飾がある．前者は殻面から連続する間束線末端に所在するが，後者は殻套だけに分布する間束線末端に所在する場合が多い．付随孔は 3 個であるが，極てまれに 4 個の場合もある．唇状突起は 1 個で殻面殻套境界にあり，外管を伴う．

ノート：殻面／殻套境界に有基突起が所在する種類は，他に *Stephanodiscus kusuensis* Julius, H. Tanaka & Curtin と *Stephanodiscus uemurae* H. Tanaka があるが，前者とは殻面同心円状の波打ちが弱いこと，殻面有基突起の分布と付随孔数，後者とは殻のサイズ，殻面有基突起の数の違いで区別できる．種小名は日本歯科大学生物学教室の南雲　保博士を記念して付けられた．

産　出：尾本層（Houk *et al*. 2014）：前期更新世・大分県．

図　版 108-109：尾本層標本．

Plate 108, Figs 1-3.　*Stephanodiscus nagumoi* H. Tanaka

LM. Figs 1-2. SEM. Fig. 3. From the type material, OIT-009, Omoto Formation (Early Pleistocene), Kitsuki City, Oita, Japan. Scale bar: Fig. 3＝5 µm.

Figs 1-2.　Two different size valves.
Fig. 3.　Openings of valve face fultoportulae (arrows).

Plate 108

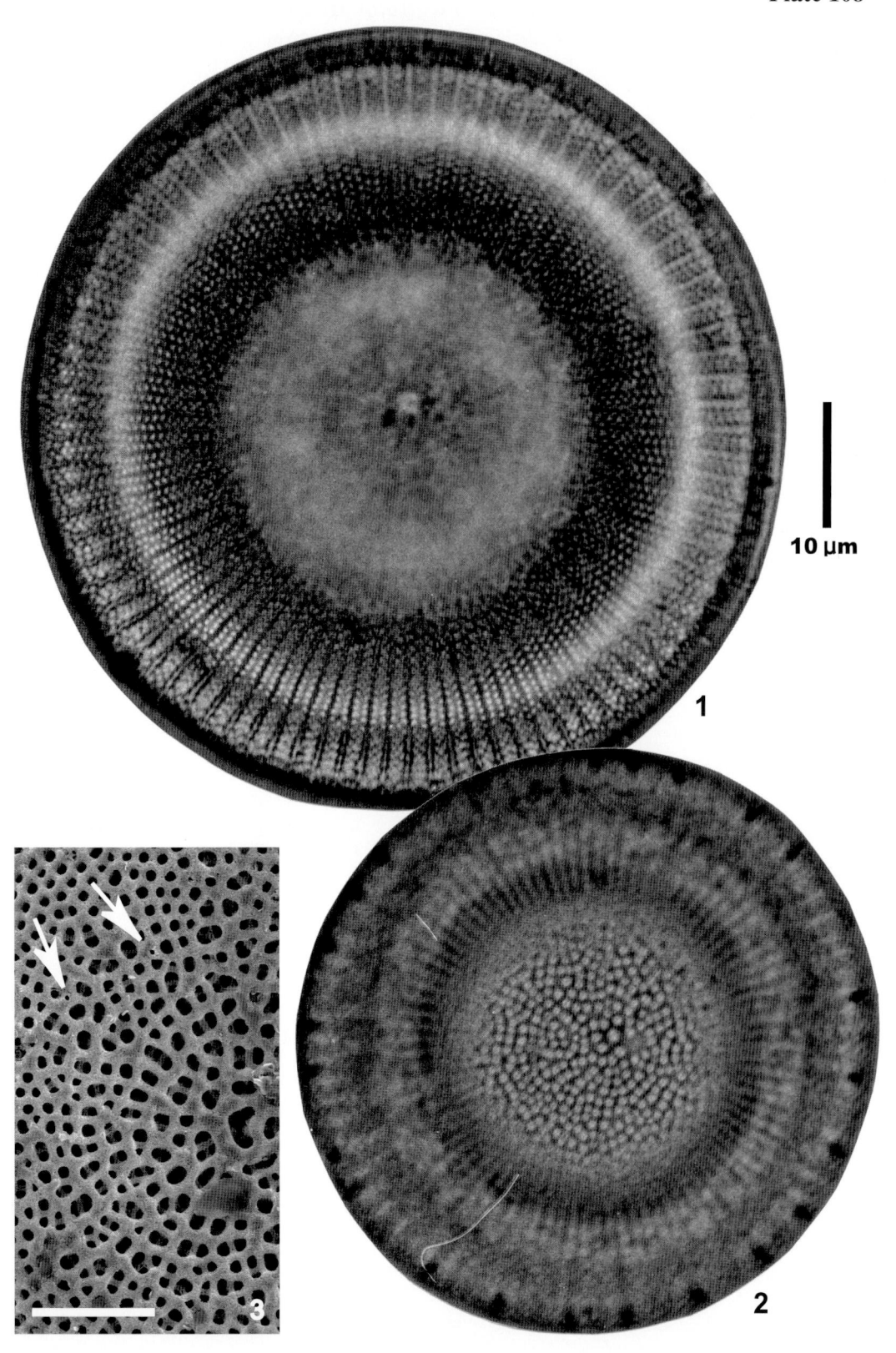

Plate 109, Figs 1-6. *Stephanodiscus nagumoi* **H. Tanaka**

SEM. Figs 1-6. From the type material, OIT-009, Omoto Formation (Early Pleistocene), Kitsuki City, Oita, Japan. Scale bars: Figs 1-2=10 μm, Figs 3-4, 6=2 μm, Fig. 5=1 μm.

Fig. 1. External view of whole valve.
Fig. 2. Oblique view of Fig. 1.
Fig. 3. Enlarged view of part of Fig. 2 showing outer openings of fultoportulae on valve face/mantle boundary (small arrows) and on lower mantle (large arrows), both having small, attached ornamentation.
Fig. 4. Enlarged internal view of valve margin showing rimoportula (arrowhead), fultoportulae on valve face/mantle boundary (small arrows) and on lower mantle (large arrows).
Fig. 5. Detailed view of valve face showing domed cribra and fultoportula with three satellite pores.
Fig. 6. Internal oblique view showing fultoportula near valve center (arrowhead) and fultoportulae on valve face/mantle boundary (small arrow) and on lower mantle (large arrow).

Plate 109

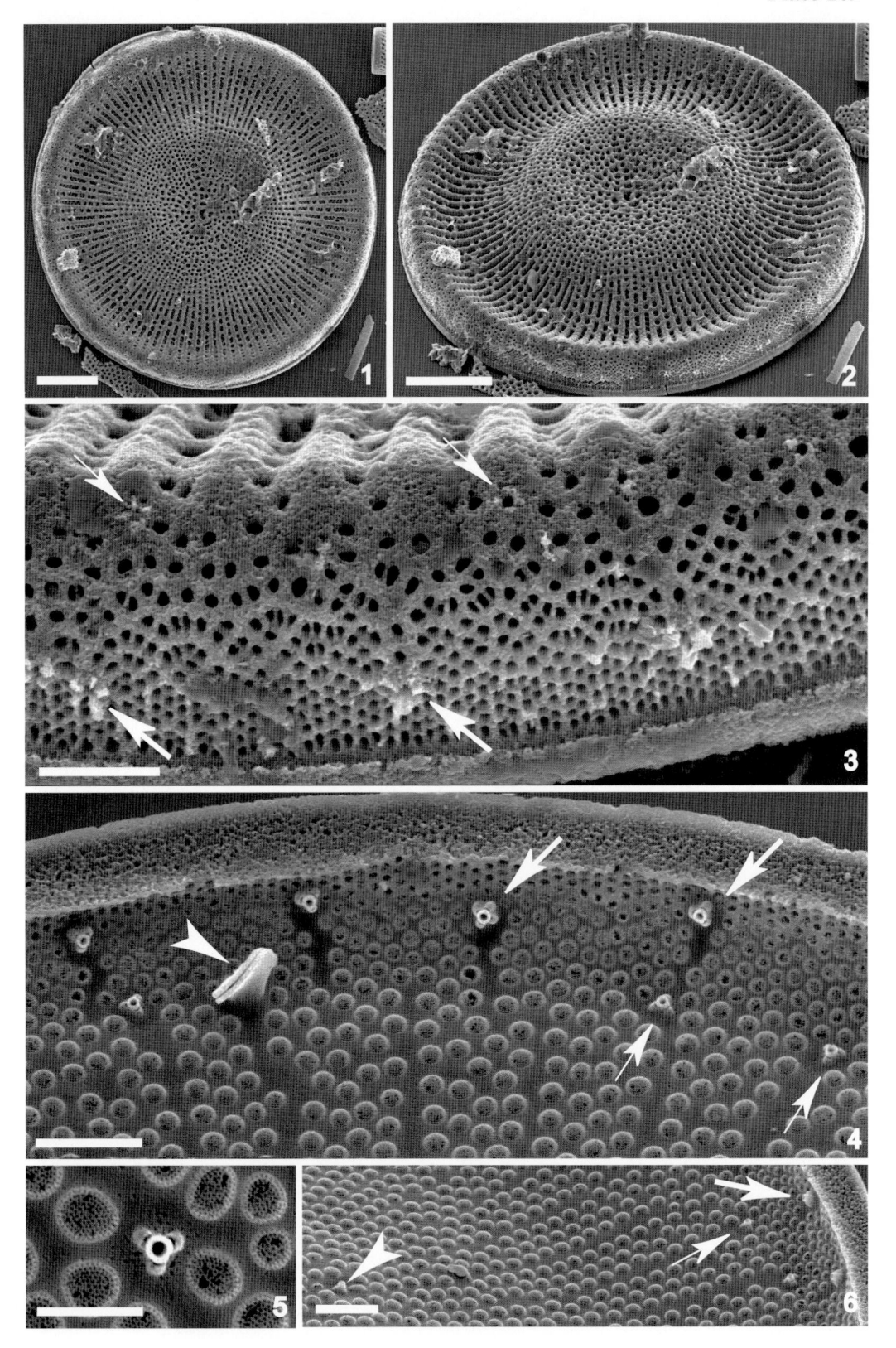

Plates 110-111.　Centric diatoms: Thalassiosirales
Stephanodiscus rotula (Kützing) Hendey 1964

基礎異名：*Cyclotella rotula* Kützing 1844

文　　献：Hendey, N.I. 1964. An introductory account of the smaller algae of British coastal waters. Part V: Bacillariophyceae (Diatoms). 317 pp. 1-45 pls. Fishery Investigations Series IV. HMSO, London.

　　田中宏之・小林　弘 1992. 群馬県, 中之条上部湖成層（中期更新世）の珪藻. 群馬県立歴史博物館紀要 13 号: 17-38.

　　小林　弘・出井雅彦・真山茂樹・南雲　保・長田敬五 2006. 小林弘珪藻図鑑「第1巻」, 531 pp. 内田老鶴圃, 東京.

形　　態：殻面は円盤形で同心円状に波打つ，殻径 27-54 μm，束線は 10 μm に 5-7 本，束線を構成する点紋列は殻面／殻套境界で 2(3) 本である．間束線は殻面／殻套境界または殻套中位に所在する殻套有基突起で終るが，殻套有基突起まで伸びる間束線の殻面／殻套境界には針がある．殻面有基突起は 5-10 個で 2(3) 個の付随孔を伴う．殻套有基突起の付随孔は 3 個である．唇状突起は 1-4 個で殻面／殻套境界にある．

ノ ー ト：*Stephanodiscus niagarae* Ehrenb. と類似しているが，殻面観の殻套部の広さ，唇状突起の外管の位置，殻面有基突起の付随孔の数，殻套有基突起の位置等を総合的に判断するとよい．文献欄に記した図鑑は本書と同じ中之条湖成層の本種を図に用いている．

産　　出：中之条湖成層（田中・小林 1992）：中期更新世・群馬県．

図　　版 110-111：中之条湖成層標本．

Plate 110, Figs 1-5.　*Stephanodiscus rotula* (Kützing) Hendey

LM. Figs 1-5. Material from Nakanojo Lacustrine Deposit (Middle Pleistocene), Nakanojo Town, Gunma, Japan.

Figs 1-5.　Three different size valves.
Figs 1-2, 4-5.　Two valves at different focal planes.
Fig. 3.　Photomontage made by two photographs of same valve.

Plate 110

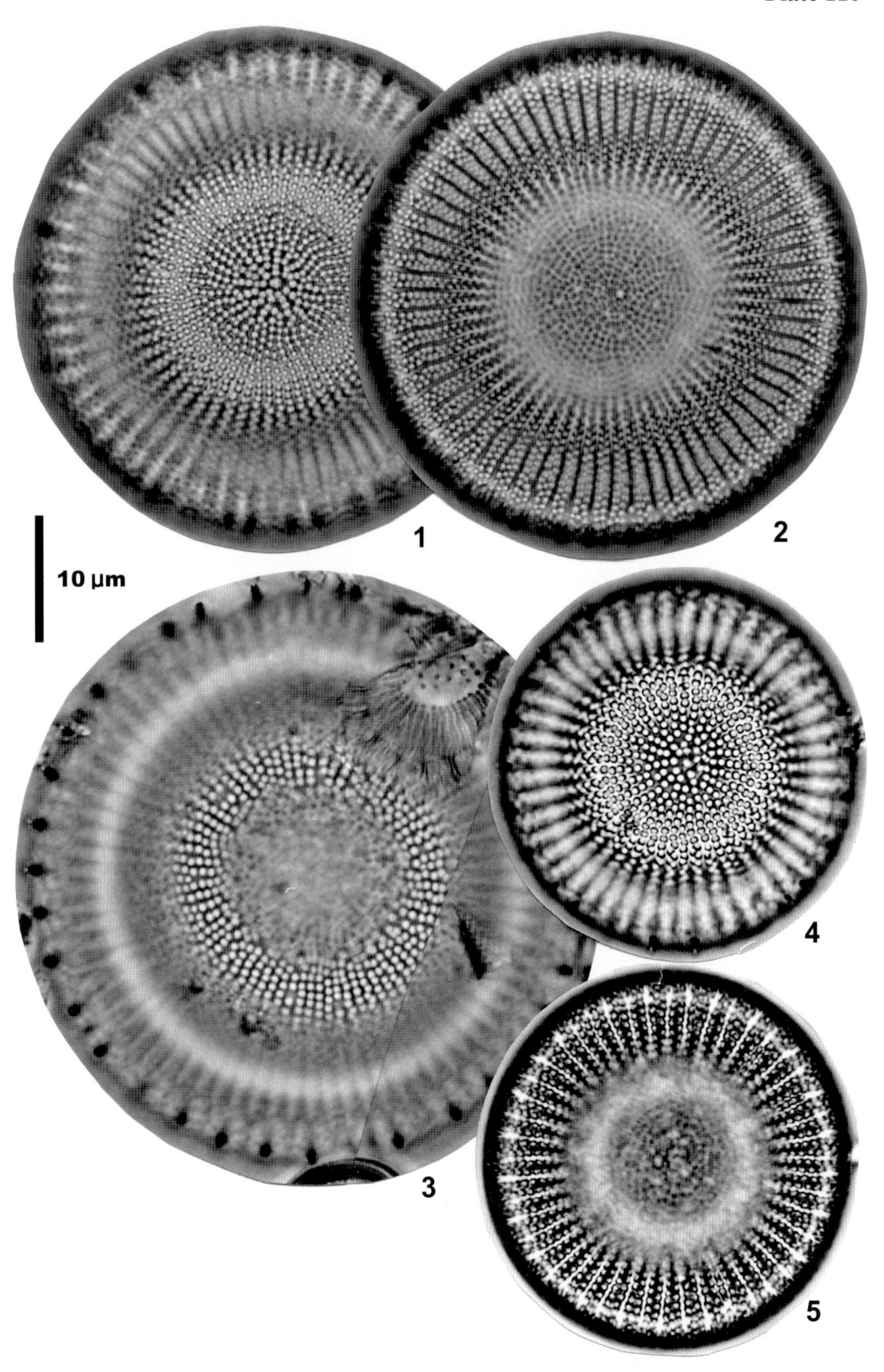

10 μm

1

2

3

4

5

257

Plate 111, Figs 1-5. *Stephanodiscus rotula* **(Kützing) Hendey**
SEM. Figs 1-5. Material from Nakanojo Lacustrine Deposit (Middle Pleistocene), Nakanojo Town, Gunma, Japan. Scale bars: Fig. 1=10 μm, Fig. 3=5 μm, Figs 2, 4-5=2 μm.

Fig. 1. External oblique view of whole valve.

Fig. 2. Enlarged view of part of Fig. 1, opening of mantle fultoportula (arrow) and opening of rimoportula (arrowhead).

Fig. 3. Internal oblique view of whole valve.

Fig. 4. Detailed view of internal valve center showing valve face fultoportulae with two satellite pores (arrows).

Fig. 5. View of internal marginal valve face and mantle, mantle fultoportula with three satellite pores (arrow) and rimoportulae (arrowheads).

Plate 111

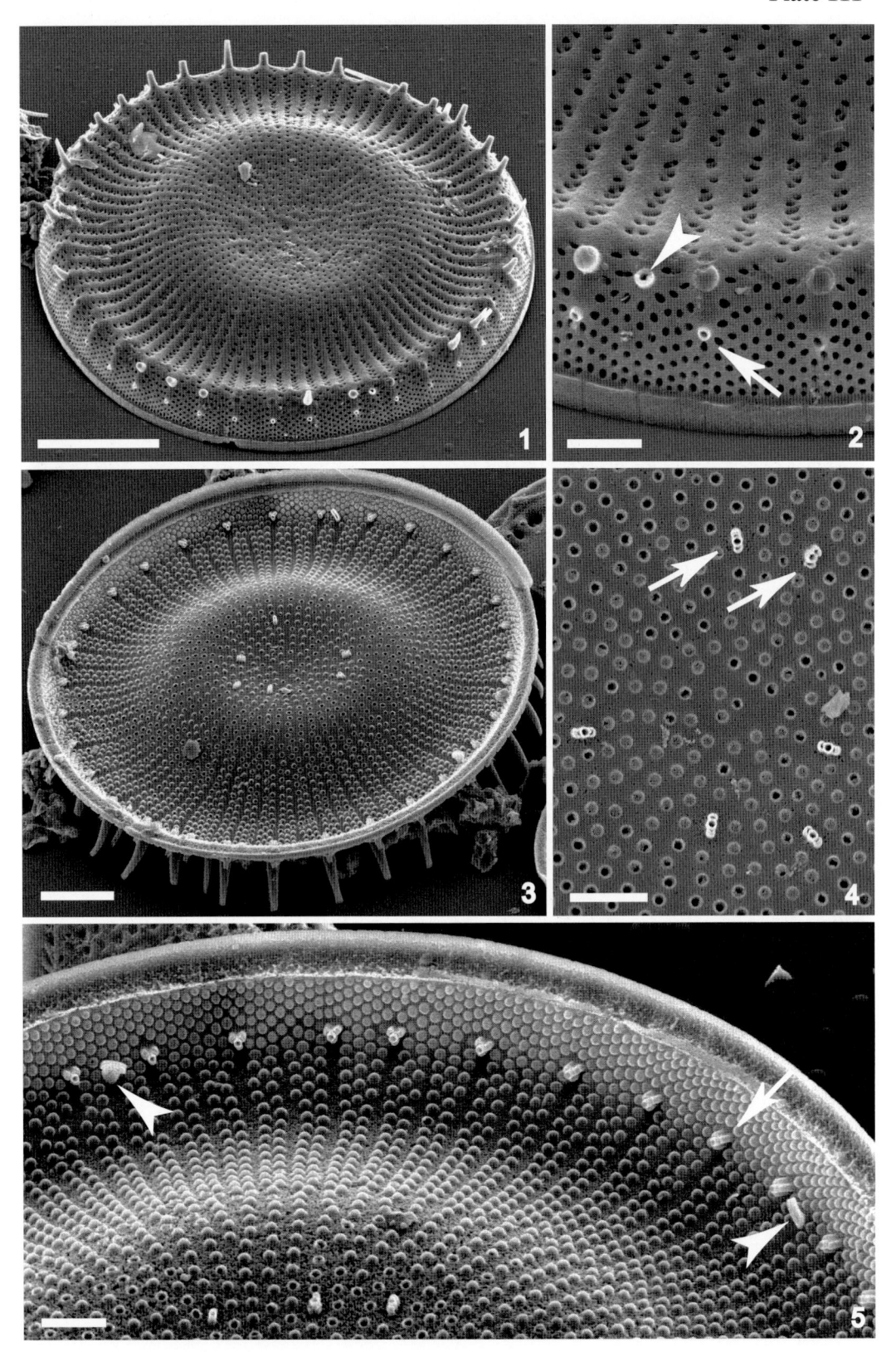

Plates 112-113.　Centric diatoms: Thalassiosirales
Stephanodiscus suzukii Tuji & Kociolek 2000

文　献：Tuji, A. & Kociolek, J.P. 2000. Morphology and taxonomy of *Stephanodiscus suzukii* sp. nov. and *Stephanodiscus pseudosuzukii* sp. nov.(Bacillariophyceae) from Lake Biwa, Japan, and *S. carconensis* from North America. Phycological Research **48**: 231-239.

形　態：殻面は円盤形で，同心円状に波打つ，殻径 14-45 μm，束線は殻面／殻套境界で 10 μm に約 4 本．1 本の束線を構成する点紋列は殻面／殻套境界で 3-5 本．点紋列を構成する点紋は 10 μm に 10-14 個である．間束線は幅広い．ふつう針はない．殻面有基突起が所在する場所は窪んでおり，1(2)個で，3(4)個の付随孔を伴う．殻套有基突起は殻套の低い位置にあり付随孔は 3 個で，外側では短い管と装飾がある．唇状突起は 1 個で殻面縁辺部にあり，太い外管と先端の装飾（しばしば長い板状）がある．

ノート：かつて琵琶湖で *Stephanodiscus carconensis* Grunow に同定されていた分類群である．*Stephanodiscus* 属の他種と比較すると，針がない，間束線の幅が広い，唇状突起の外管の先が T 形である，殻套有基突起が殻端近くに存在し外側に装飾がある，殻面有基突起に付随孔が 4 個のものがある等の特徴がある．

産　出：琵琶湖（Tuji & Kociolek 2000）：滋賀県（現生），水月湖（Kato *et al*. 2003）：福井県（完新世），余呉湖（原口 2001）：滋賀県（現生）．

図　版 112-113：琵琶湖標本．

Plate 112, Figs 1-6.　*Stephanodiscus suzukii* Tuji & Kociolek
LM. Figs 1-6. Material from the type locality, Lake Biwa, Shiga, Japan.

Figs 1-6.　Four different size valves.
Fig. 1.　External tube of rimoportula (arrowhead).
Figs 1-2, 4, 6.　Two valves at different focal planes.

Plate 112

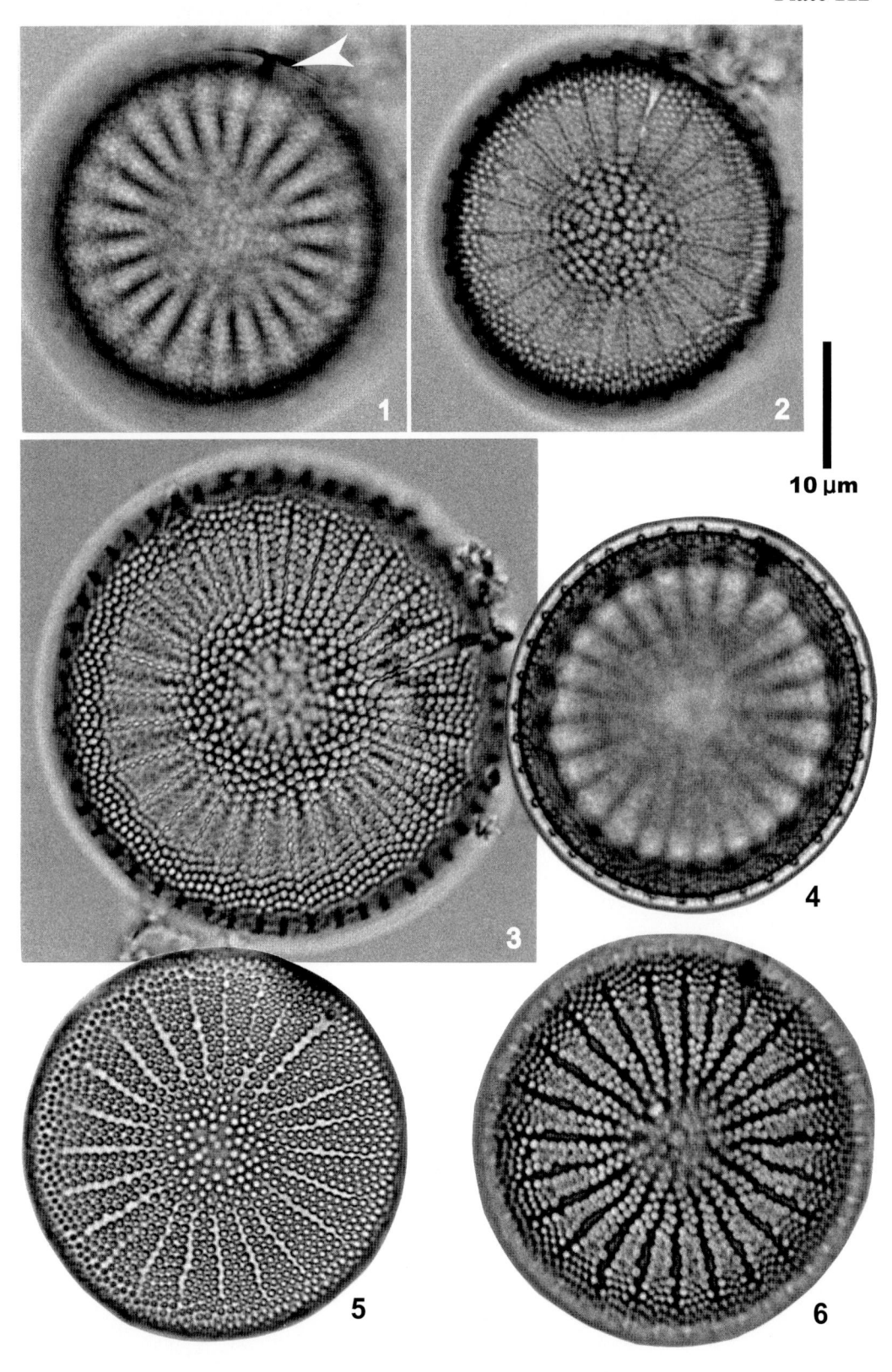

10 µm

Plate 113, Figs 1–6. *Stephanodiscus suzukii* **Tuji & Kociolek**
SEM. Figs 1–6. Materials from the type locality, Lake Biwa, Shiga, Japan. Scale bars: Figs 1–2, 4=5 μm, Fig. 3=2 μm, Fig. 5=1 μm, Fig. 6=0.5 μm.

Fig. 1. External view of whole valve, external tube of rimoportula (arrowhead).
Fig. 2. Oblique view of Fig. 1, external tube of rimoportula (arrowhead).
Fig. 3. Enlarged view of Fig. 2, external tube of rimoportula (arrowhead).
Fig. 4. Internal oblique view of valve, rimoportula (arrowhead) and valve face fultoportula (arrow).
Fig. 5. Enlarged view of Fig. 4, rimoportula (arrowhead) and mantle fultoportula with three satellite pores (arrow).
Fig. 6. Detailed view of valve face fultoportula with three satellite pores in depression (arrow).

Plate 113

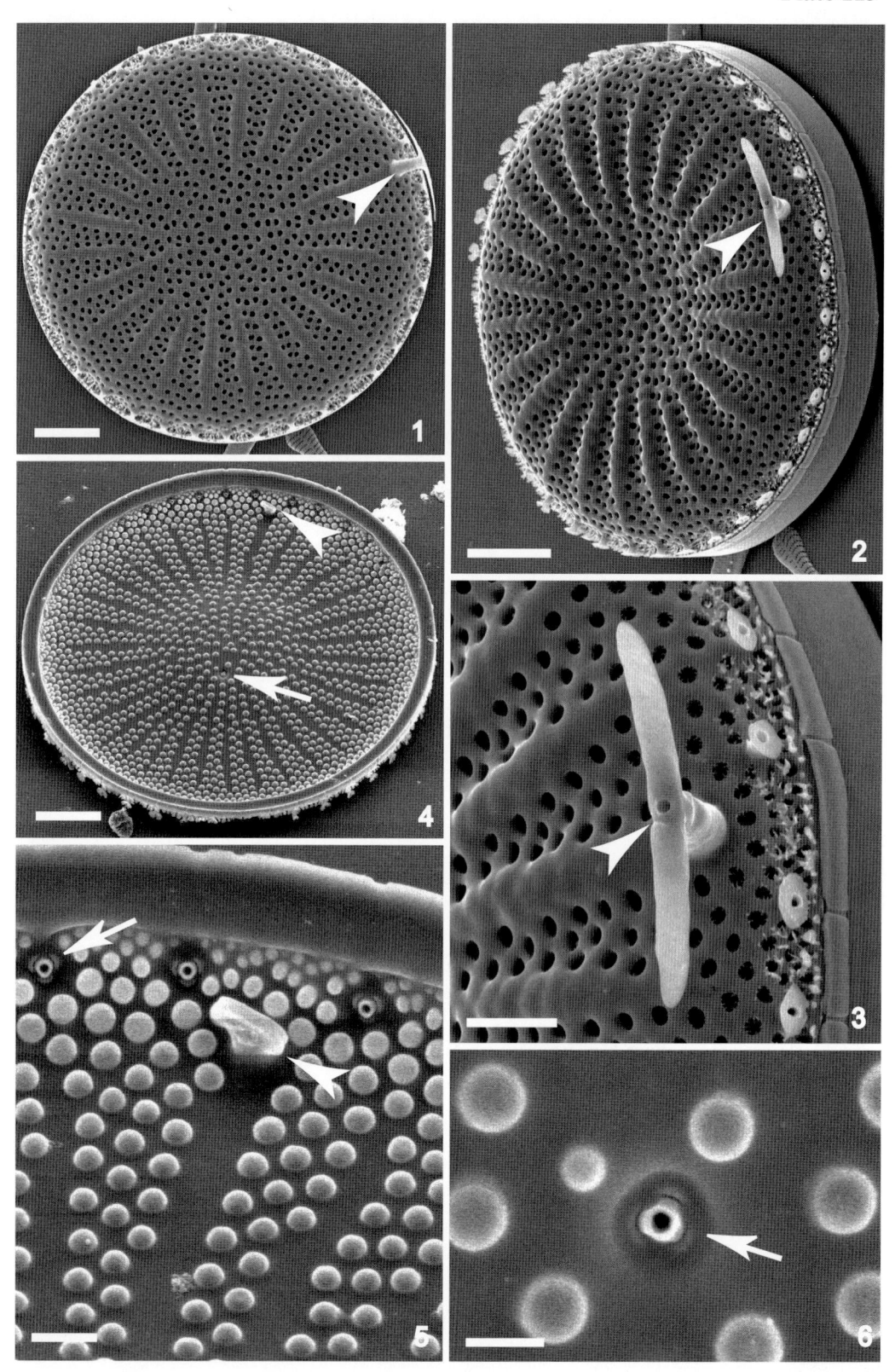

263

Plate 114.　Centric diatoms: Thalassiosirales
Stephanodiscus tenuis Hustedt 1939

文　　献：Hustedt, F. 1939. Die Diatomeenflora des Küstengebietes der Nordseen von Dollart bis zur Elbemündung. Ⅰ. Abhanlungen, naturwissenschaftl. Verein zu Bremen, Bd. **31**: Heft 3, S. 571-677, 123 Textfig.

形　　態：殻は円盤形，殻面はほぼ平らで中心に環紋がある．殻径 9-17 μm．束線は細かい点紋からなり中心から放射状に分布し，10 μm に 8-12 本，束線を構成する胞紋列は殻面／殻套境界で 2-3 列である．間束線は殻面／殻套境界の針，または殻套有基突起で終わる．針はしばしば先端が平らか，わずかであるが V 字状に割れている．殻套は高い．針のすぐ下には 3-5 本ごとに殻套有基突起の開口がある．殻套有基突起は殻套上部に所在し 3 個の付随孔を伴う．殻面有基突起は存在しない．唇状突起は 1 個で，外部への開口は針列の中で 1 個の針と置き換わって所在する．

ノ　ー　ト：本種は Hustedt(1939) によって *Stephanodiscus tenuis* として新種記載されたが，Håkansson & Stoermer(1984) によって *S. hantzschii* の品種に位置づけられた．多くの研究者はこれを引用して本分類群を *S. hantzschii* f. *tenuis* と同定してきたが，Houk *et al.* (2014)の原稿(印刷中)では Hustedt(1939) のとおり種のランクに位置づけているので，本書ではこれに倣った．ただし日本で *S. hantzschii* f. *tenuis* と同定されている分類群は束線が殻面／殻套境界で 2-3 列の胞紋列で構成されているものが多いが，*S. tenuis* は 4-5 列である (Håkansson 2002)．

産　　出：*Stephanodiscus hantzschii* f. *tenuis* として各地から報告がある．琵琶湖（辻・伯耆 2001）：滋賀県（現生），波志江沼（田中・南雲 2007）：群馬県（現生），等．

図　　版：諏訪湖（3-4, 7-9），波志江沼（Figs 1-2, 5-6, 10-11）標本．

Figs 1-11.　*Stephanodiscus tenuis* Hustedt

LM. Figs 1-6. SEM. Figs 7-11. Material, Figs 3-4, 7-9, from Lake Suwa (Recent), Suwa City, Nagano and Figs 1-2, 5-6, 10-11, from Hashie Pond (Recent), Isesaki City, Gunma, Japan. Scale bars: Figs 8-10＝2 μm, Figs 7, 11＝1 μm.

Figs 1-2, 3-4, 5-6.　Three valves at different focal planes.
Fig. 7.　External whole valve view.
Fig. 8.　Oblique view of Fig. 7.
Fig. 9.　Enlarged view of Fig. 8 showing openings of mantle fultoportulae with short tubes (arrows) and external tube of rimoportula (arrowhead).
Fig. 10.　Internal oblique view of valve.
Fig. 11.　Enlarged view of Fig. 10 showing mantle fultoportulae with three satellite pores (arrows) and rimoportula (arrowhead).

Plate 114

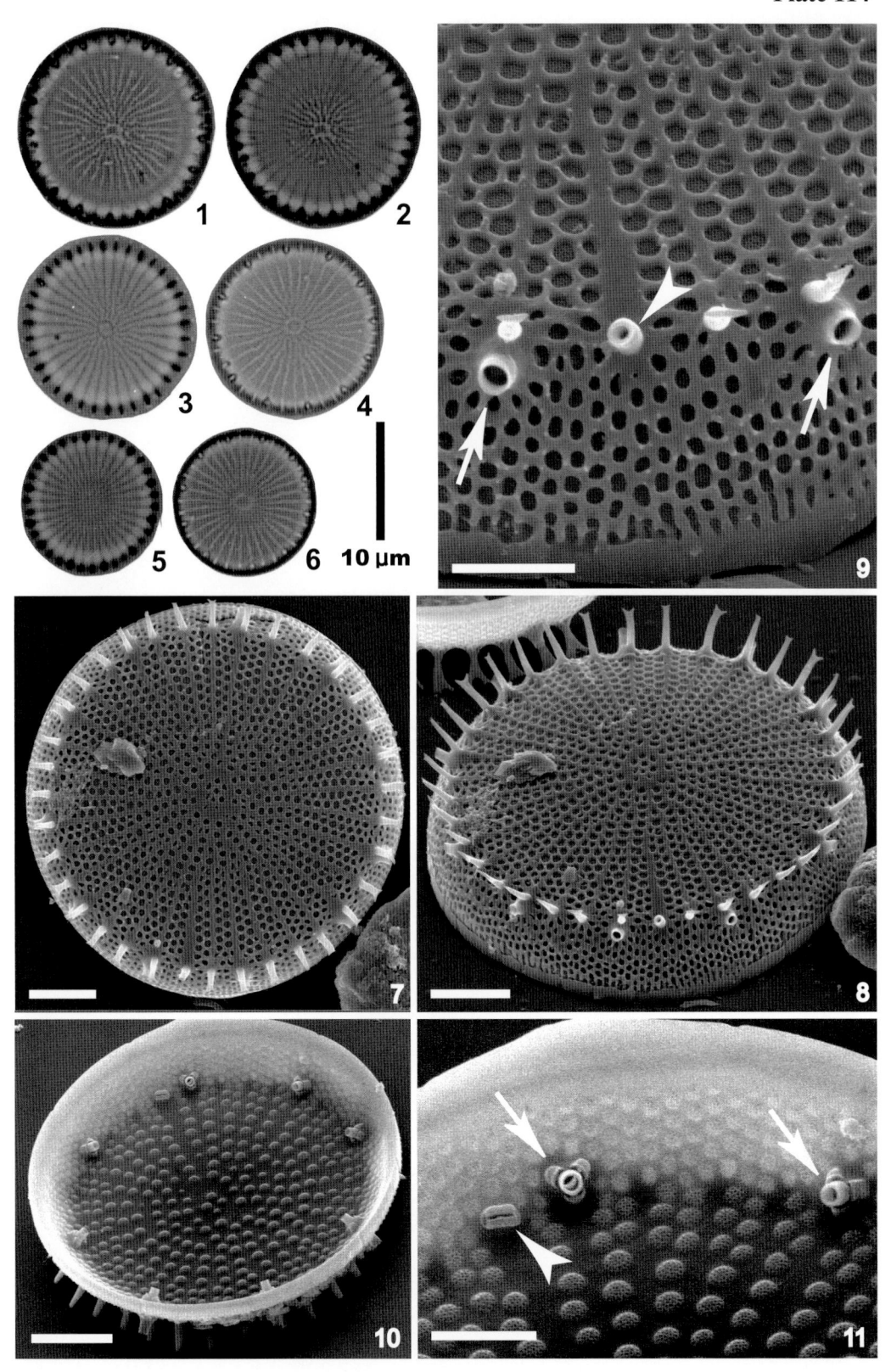

Plate 115.　Centric diatoms: Thalassiosirales
Stephanodiscus uemurae H. Tanaka in Tanaka & Nagumo 2009

文　　献：Tanaka, H. & Nagumo, T. 2009. *Stephanodiscus uemurae*, a new Mio-Pliocene diatom species from the Miyata Formation, Akita Prefecture, northern Honshu, Japan. Diatom **25**: 45-51.

形　　態：殻は円盤形で，同心円状に二重に波打つ．殻径 19-28 μm．束線は殻面／殻套境界で 10 μm に 6-8 本，中心付近では放射状に配列した 1 列の胞紋列だが，殻縁では 2-3 列になる．殻面／殻套境界の針は間束線上にあるが，殻套有基突起の開口が殻面／殻套境界あるいはその近くにある場合は所在しない．唇状突起は 1 個で殻面／殻套境界にあり，針と同じ高さに所在する．まれに，わずか殻の中心寄りに所在する場合もある．殻面有基突起はふつう欠如しているが，まれに存在する場合は 2 個の付随孔を伴う．殻套有基突起の付随孔は 3 個である．

ノ　ー　ト：種小名は本種が見出された宮田層の地質・植物化石研究者である元国立科学博物館植村和彦博士を記念して付けられた．

産　　出：宮田層（Tanaka & Nagumo 2009）：中新-鮮新世・秋田県．

図　　版：宮田層標本．

Figs 1-8.　*Stephanodiscus uemurae* H. Tanaka
LM. Figs 1-3. SEM. Figs 4-8. From the type material, MIY-104, Miyata Formation (Mio-Pliocene), Senboku City, Akita, Japan. Scale bars: Figs 4-6＝5 μm, Fig. 8＝2 μm, Fig. 7＝1 μm.

Figs 1-2.　Same valve at different focal planes.
Fig. 3.　Large valve.
Fig. 4.　External view of convex valve.
Fig. 5.　Oblique view of Fig. 4.
Fig. 6.　Internal oblique view of concave valve, rimoportula (arrowhead).
Fig. 7.　Internal detailed view of valve margin, rimoportula (arrowhead) and mantle fultoportula with three satellite pores (arrow).
Fig. 8.　Enlarged view of part of Fig. 5 showing outer opening of rimoportula (arrowhead) and outer openings of mantle fultoportulae located on or near valve face/mantle boundary (arrows).

Plate 115

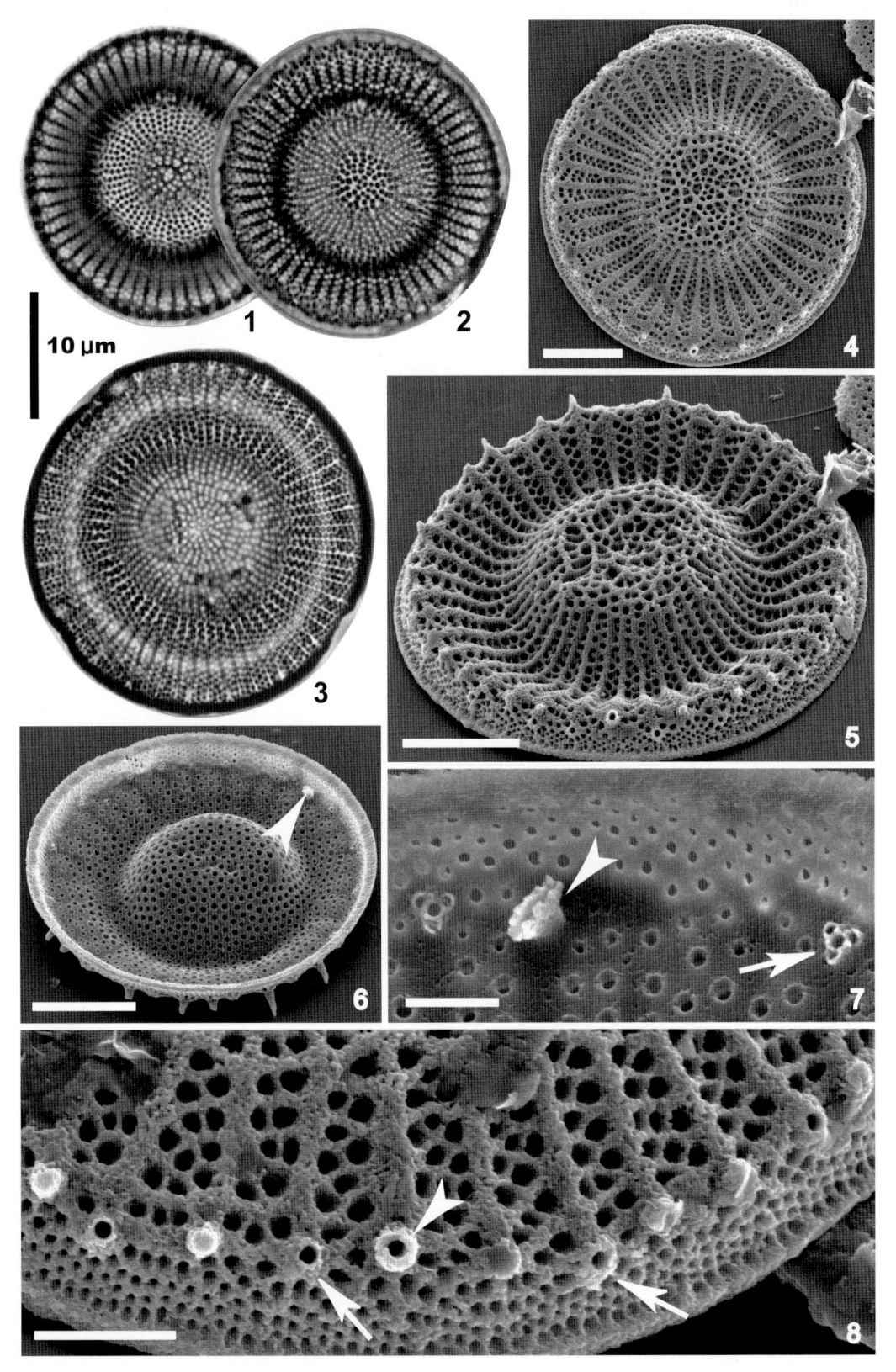

Plate 116.　Centric diatoms: Thalassiosirales
Tertiariopsis costatus H. Tanaka in Tanaka & Nagumo 2012

文　　献：Tanaka, H. & Nagumo, T. 2012. *Tertiariopsis costatus*, a new diatom species from the Mio-Pliocene freshwater sediment of Masuzawa Formation, Iwate Prefecture, Japan. Diatom **28**: 1-6.

形　　態：殻は円盤形で殻面はほぼ平ら，殻径 6-29 μm．殻面には 1 列の点紋列が殻中心から放射状に分布し，殻面／殻套境界で 10 μm に 16-20 本，点紋列を構成する点紋は 10 μm に 12-18 個である．殻套では胞紋列を区分して細い筋があるが，この筋の内側は肥厚して細い肋を形成し，唇状突起（肋がない場合もある）または殻套有基突起から殻端間に配列している．内側では殻面・殻套有基突起とも 3 個の付随孔をもつ．唇状突起は 1 個で殻套に所在する．

産　　出：舛沢層（Tanaka & Nagumo 2012）：中新-鮮新世・岩手県．

図　　版：舛沢層標本．

Figs 1-12.　*Tertiariopsis costatus* H. Tanaka

LM. Figs 1-5. SEM. Figs 6-12. From the type material, MAS-108, Masuzawa Formation (Mio-Pliocene), Shizukuishi Town, Iwate, Japan. Scale bars: Figs 6, 8-10＝2 μm, Figs 7, 12＝1 μm, Fig. 11＝0.5 μm.

Figs 1-5.　Five different size valves.
Fig. 6.　External view of whole valve.
Fig. 7.　Enlarged oblique view of Fig. 6 showing opening of mantle fultoportula (arrow) and opening of rimoportula (arrowhead).
Fig. 8.　Internal view of whole valve, valve face fultoportula with three satellite pores (small arrow), mantle fultoportula (large arrow) and rimoportula (arrowhead).
Fig. 9.　Detailed view of cingulum showing ligula-like segment (arrow).
Fig. 10.　Oblique view of Fig. 8 showing valve face fultoportula (small arrow), mantle fultoportula (large arrow) and rimoportula (arrowhead).
Fig. 11.　Detailed view of rimoportula (arrowhead) located on the edge of a thin costa (arrow) and covered by marginal lamina.
Fig. 12.　Detailed view of mantle fultoportulae with three satellite pores (arrows) located on the edge of thin costae and covered by marginal lamina.

Plate 116

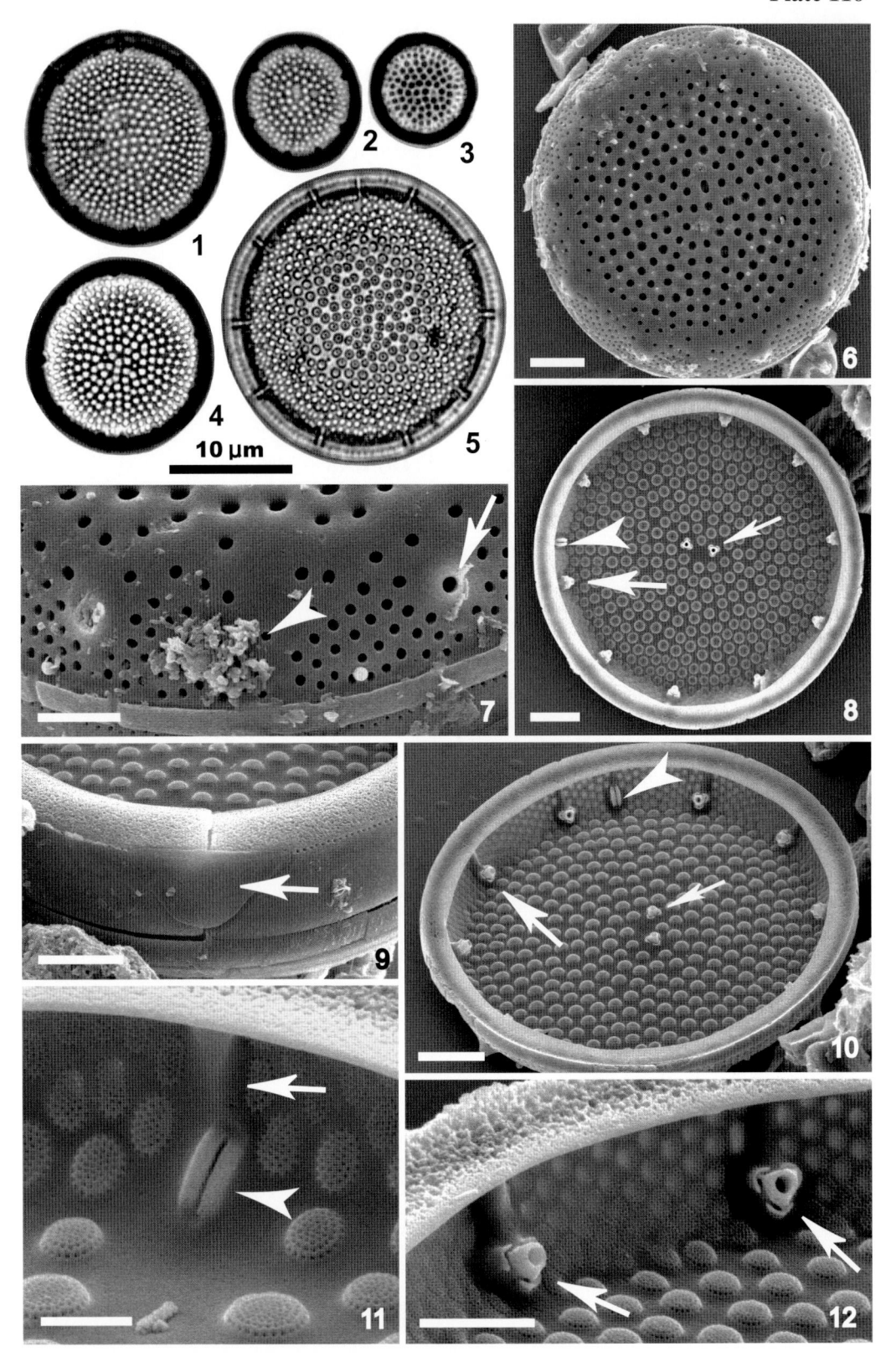

Plate 117.　Centric diatoms: Thalassiosirales
Tertiariopsis nipponicus H. Tanaka in Tanaka *et al.* 2010

文　　献：Tanaka, H., Nagumo, T. & Kashima, K. 2010. *Tertiariopsis nipponicus* sp. nov., from a Pliocene freshwater deposit in southwestern Japan. Diatom Research **25**: 175-183.

形　　態：殻は小形円盤状で，殻径 4.5-14.0 μm，殻の中心付近は凸または凹状であるが，しばしば凸または凹状の中心，および点紋の射出中心が殻の中心とは一致しない．中心付近には1個の殻面有基突起が所在し，付随孔は2個である．殻套有基突起の付随孔は3個である．唇状突起は1個で殻套にある．

ノート：本種は *Cyclostephanos* の特徴ももつが，全体的な形態を合わせ考えると Khursevich *et al.*(2002)によって設立された *Tertiariopsis* に，より類似している．

産　　出：津房川層（Tanaka *et al.* 2010）：後期鮮新世・大分県．

図　　版：津房川層標本．

Figs 1-15.　*Tertiariopsis nipponicus* H. Tanaka

LM. Figs 1-9. SEM. Figs 10-15. From the type material, OIT-327, Tsubusagawa Formation (Late Pliocene), Kamiuchikawano, Usa City, Oita, Japan. Scale bars: Figs 10-13=2 μm, Figs 14-15=0.5 μm.

Figs 1-9.　Seven different size valves.
Figs 1-2, 6-7.　Two valves at different focal planes.
Fig. 10.　External view of concave valve.
Fig. 11.　Oblique view of a frustule, antiligula (arrow).
Fig. 12.　Oblique view of convex valve showing valve pattern center not at geometric valve center.
Fig. 13.　Internal oblique view of convex valve, valve face fultoportula (arrow).
Fig. 14.　Enlarged view of valve showing a valve face fultoportula with two satellite pores and a mantle fultoportula with three satellite pores (arrow).
Fig. 15.　Enlarged view of a rimoportula (arrowhead) and mantle fultoportula with three satellite pores, both covered by marginal lamina.

Plate 117

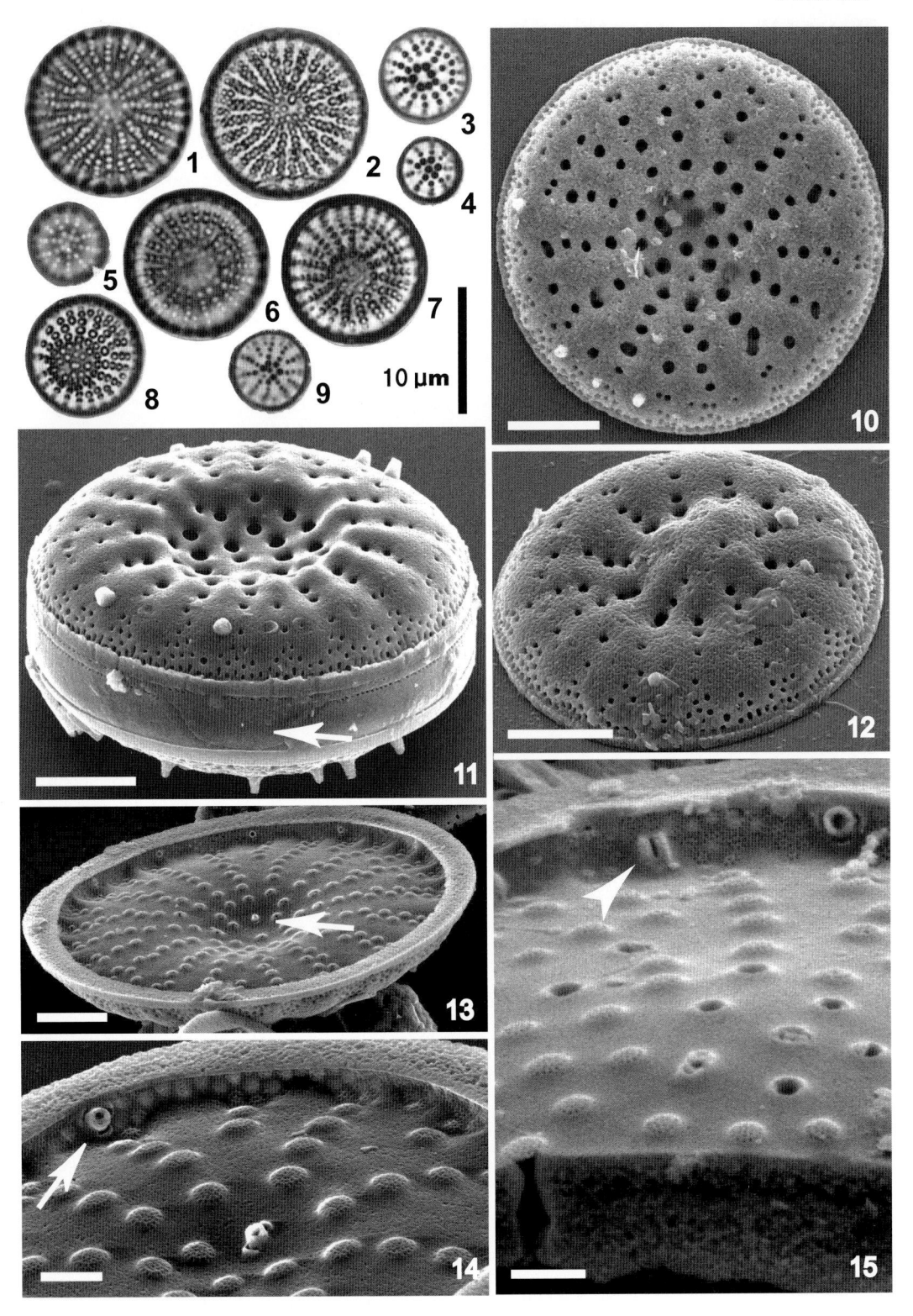

271

Plates 118-119. Centric diatoms: Thalassiosirales
Tertiarius agunensis H. Tanaka sp. nov.

新　　種：記載文（英文）は9頁参照．
形　　態：殻は円盤形，殻面は平らで，点紋が散在する中心域と条線が分布する縁辺域に分けられる．殻径 9-24 μm，殻面中心域には2個の付随孔をもつ殻面有基突起が円状に分布する．条線は 10 μm に 10-18 本で殻面から殻套へ連続し殻端まで達する．間条線の多くは殻面または殻套で分岐する．殻套有基突起は各太肋に所在し2個の付随孔を横に伴う．唇状突起は1個で長胞の中に所在する．殻帯は ligula-like segment を含む．
ノ　ー　ト：本種は *Cyclotella* 属の特徴ももつが，唇状突起が長胞中に所在することから *Tertiarius* に所属するのが適当である．種小名は原産地（粟国島）に由来する．
産　　出：筆ん崎層：鮮新世・沖縄県．
図　版 118-119：筆ん崎層標本．

Plate 118, Figs 1-10.　*Tertiarius agunensis* H. Tanaka sp. nov.
LM. Figs 1-5. SEM. Figs 6-10, external views. From the type material, AGU-001, Fudenzaki Formation (Pliocene), Aguni Island, Okinawa, Japan. Scale bars: Figs 6-8＝2 μm, Figs 9-10＝1 μm.

Figs 1-2.　Holotype, MPC-25057, housed in National Museum of Nature and Science, Tokyo. Same valve shown at different focal planes.
Figs 3-5.　Three different size valves.
Fig. 6.　Whole view of valve.
Fig. 7.　Enlarged view of central area of Fig. 6, openings of valve face fultoportulae (arrowheads).
Fig. 8.　Oblique view of Fig. 6 showing flat valve face.
Fig. 9.　Detailed view of cingulum showing band structure, VC: valvocopula, S: second band, T: third band, ligula-like segment (arrow).
Fig. 10.　Mantle areolae rows and openings of mantle fultoportulae (arrows).

Plate 118

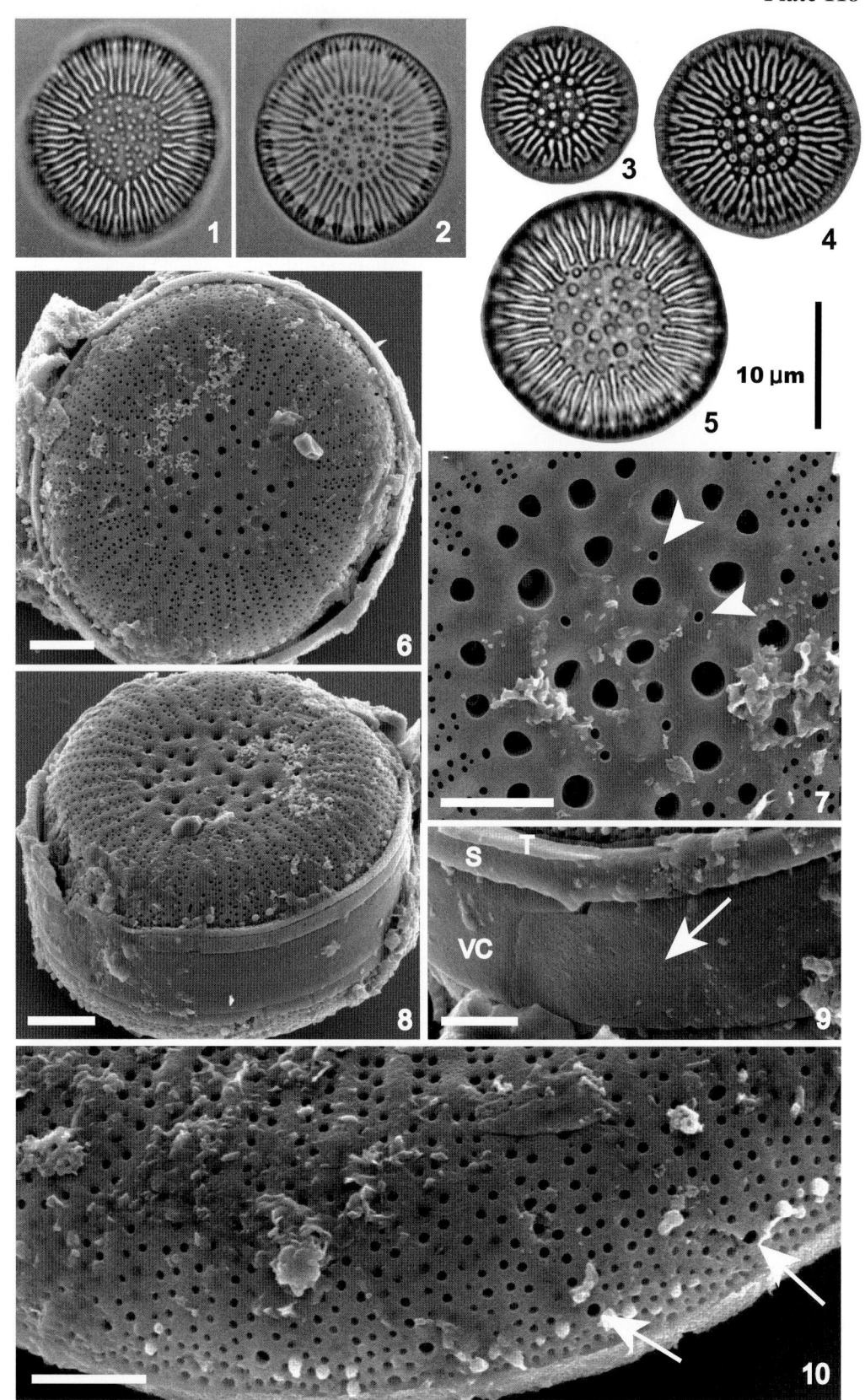

273

Plate 119, Figs 1-6. *Tertiarius agunensis* **H. Tanaka** sp. nov.

SEM. Figs 1-6, internal views. From the type material, AGU-001, Fudenzaki Formation (Pliocene), Aguni Island, Okinawa, Japan. Scale bars: Figs 1-2=2 μm, Figs 3, 5=1 μm, Figs 4, 6=0.5 μm.

Fig. 1. Whole view of valve.
Fig. 2. Oblique view of Fig. 1, rimoportula (arrowhead).
Fig. 3. Detailed view of valve center, valve face fultoportula with two satellite pores (arrow).
Fig. 4. Enlarged view of Fig. 2 showing rimoportula inside alveolus (arrowhead).
Fig. 5. Broken valve showing alveolus with centrifugal and centripetal roofing over as well as mantle fultoportula with two satellite pores (arrow).
Fig. 6. Detailed view showing rimoportula inside alveolus (arrowhead).

Plate 119

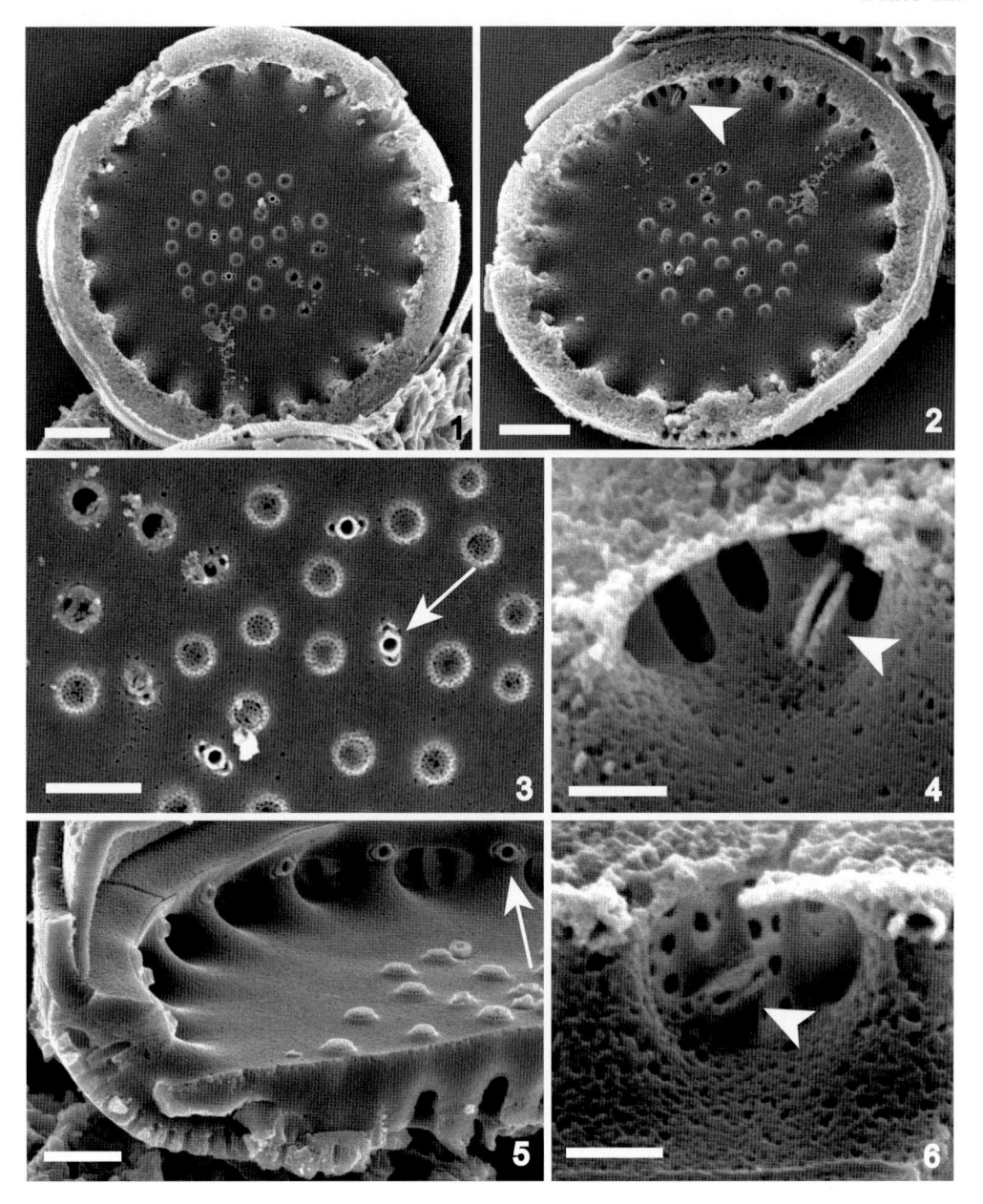

Plate 120.   Centric diatoms: Thalassiosirales
Thalassiocyclus pankensis H. Tanaka & Nagumo 2008

文　　献：Tanaka, H. & Nagumo, T. 2008. *Thalassiocyclus pankensis* sp. nov., a new diatom from the Panke Swamp, northern Japan(Bacillariophyta). Phycological Research **56**: 83-88.

形　　態：殻は円盤状，殻径 4-16 μm，束線は 10 μm に 6-8 本．殻面中心域は強く横方向に波打ち，その凸部に 1-3 個，凹部に 1 個の有基突起が所在する．殻面有基突起は 2-4 個，殻套有基突起は 4 個の付随孔を伴う．殻中心から胞紋列が放射状に分布するが，中心域では不明瞭であり，縁辺域では明瞭である．殻面縁辺域の間束線は外側へ強く突出し，束線は幅広く凹み，束線と間束線が凹凸する．唇状突起は 1 個で殻面／殻套境界にある．

ノート：本種はアメリカのサンフランシスコ湾から記載された単種属であった *Thalassiocyclus* 属の第 2 の種であるとともに，日本から最初に記載された本属の分類群である．種小名は原産地（パンケ沼）に由来する．

産　　出：パンケ沼（Tanaka & Nagumo 2008）：北海道（現生）．

図　　版：パンケ沼（タイプ試料）標本．

Figs 1-12.   *Thalassiocyclus pankensis* H. Tanaka & Nagumo
LM. Figs 1-6. SEM. Figs 7-12. From the type material, PAN-101, Panke Swamp (Recent), Horonobe Town, Hokkaido, Japan. Scale bars: Figs 7-9＝2 μm, Fig. 11＝1 μm, Figs 10, 12 ＝0.5 μm.

Figs 1-6.   Four different size valves.
Figs 2-3, 5-6.   Two valves at different focal planes.
Fig. 7.   External view of entire valve, openings of two valve face fultoportulae (arrows).
Fig. 8.   Oblique view of valve showing strongly undulated central area.
Fig. 9.   Internal oblique view of valve, rimoportula (arrowhead).
Fig. 10.   Detailed view of internal valve margin, rimoportula (arrowhead).
Fig. 11.   Detailed view of internal valve center showing three valve face fultoportulae with three satellite pores.
Fig. 12.   Detailed view of valve margin showing spatulaed spines and outer openings of mantle fultoportulae (arrows).

Plate 120

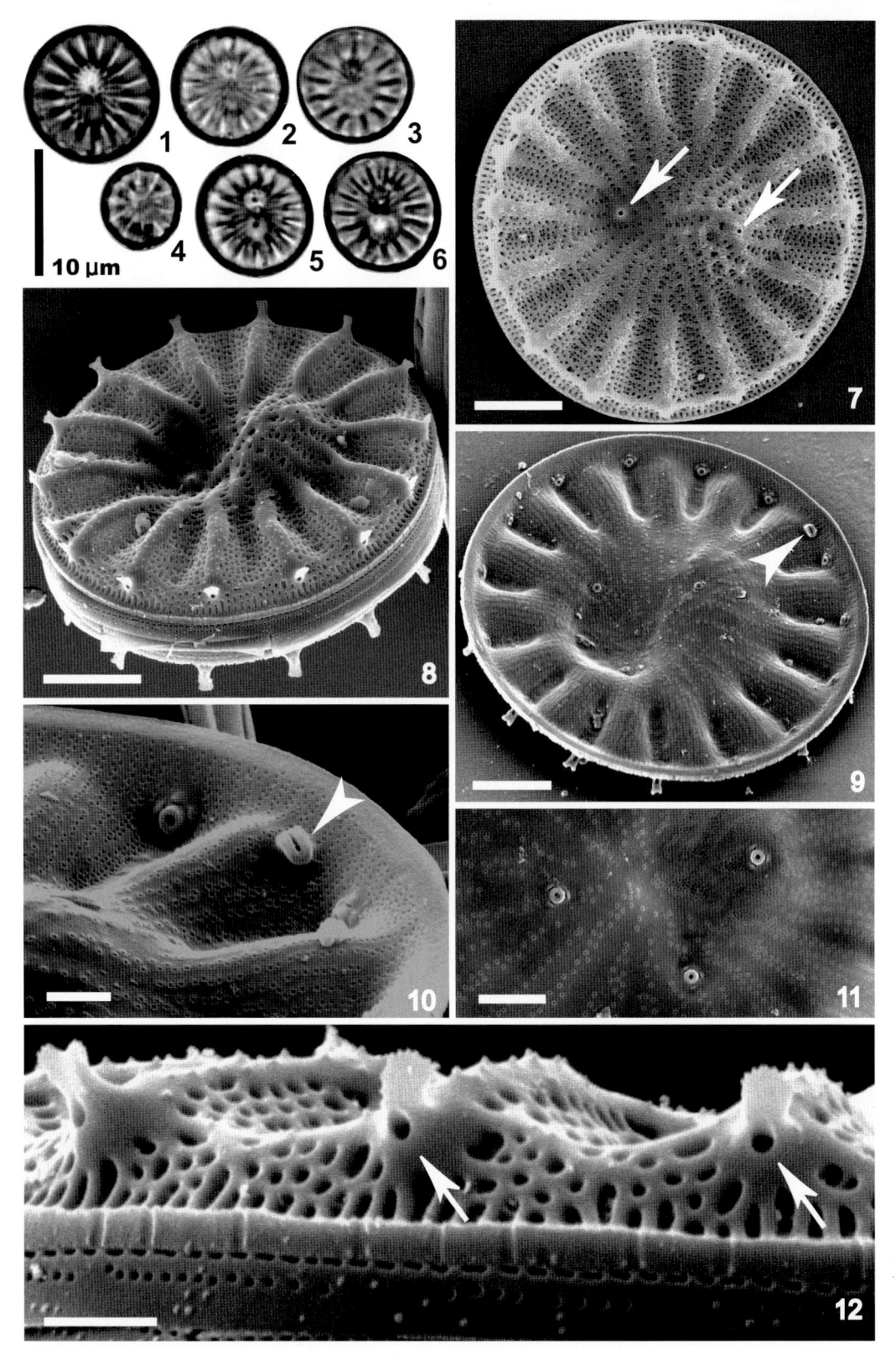

Plate 121.　Centric diatoms: Thalassiosirales
Thalassiosira lacustris (Grunow) Hasle
in Hasle & Fryxell 1977

基礎異名：*Coscinodiscus lacustris* Grunow in Cleve & Grunow 1880
文　　献：Hasle, G.R. & Fryxell, G.A. 1977. The genus *Thalassiosira*: some species with a linear areolae array. Nava Hedwigia, Beiheft **54**: 15-66.
形　　態：殻は円盤状，図に使用した稲城層の本種は，殻径 35-60 μm，殻面は強く横に波打つ．殻面の点紋列は 10 μm に約 12 本，点紋列を構成する点紋は 10 μm に約 12 個である．SEM 観察では殻面／殻套境界に中空の針が分布し，この下側にある短い管は殻套有基突起の開口である．殻面の突出している側にやや偏って 10 個程度の殻面有基突起が分布する．殻面有基突起・殻套有基突起とも付随孔は 4 個である．唇状突起は殻面／殻套境界にある．
ノ　ー　ト：本種は *Thalassiosira bramaputurae* とされたことがあったが，小林ら（2006）に記してあるとおり，標記の *T. lacustris* と同定するのが適当である．原口ら（1998）によると適応性は広塩性(低濃度から高濃度まで広い範囲の塩分濃度に適応して出現する種類・著者注)とされている．
産　　出：稲城層：前期更新世・東京都，荒川低地（南雲・安藤 1984）：完新世・埼玉県，琵琶湖（辻・伯耆 2001）：滋賀県（現生），波志江沼（田中・南雲 2007）：群馬県（現生）等．
図　　版：稲城層標本．

Figs 1-8.　*Thalassiosira lacustris* (Grunow) Hasle
LM. Figs 1-2. SEM. Figs 3-8. Material from Inagi Formation (Early Pleistocene), Fuchu City, Tokyo. Scale bars: Figs 4, 6-7＝5 μm, Fig. 8＝2 μm, Fig. 5＝1 μm, Fig. 3＝0.5 μm.

Figs 1-2.　Two different size valves.
Fig. 3.　Internal detailed view of a valve face fultoportula with four satellite pores.
Fig. 4.　Internal oblique view of valve, valve face fultoportulae (arrows), rimoportula (arrowhead).
Fig. 5.　Detailed view of internal valve margin, mantle fultoportula (arrow), rimoportula (arrowhead).
Fig. 6.　External view of entire valve, opening of valve face fultoportula (arrow).
Fig. 7.　Oblique view of Fig. 6.
Fig. 8.　Detailed view of mantle showing openings of mantle fultoportulae (arrows) and two broken spines (arrowheads).

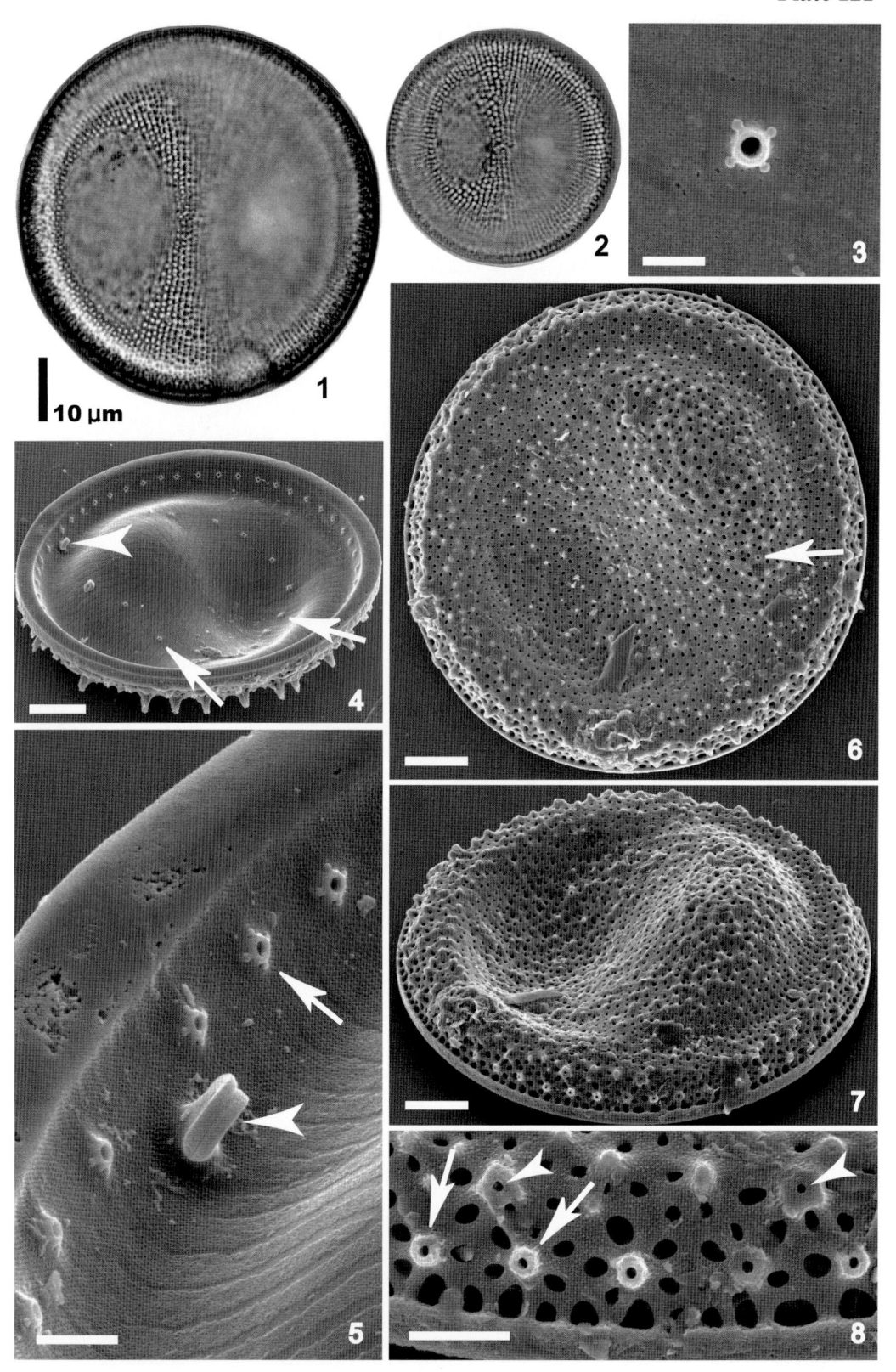

Plate 121

Plate 122.　Centric diatoms: Triceratiales
Pleurosira laevis (Ehrenberg) Compère 1982

基礎異名：*Biddulphia laevis* Ehrenberg 1843
文　　献：Hendey, I. 1964. An introductory account of the smaller algae of British coastal waters. 317 pp, 65 pls. Her Majesty's Stationery Office, London.
形　　態：大形で殻面は円形～わずか楕円形，長径 65-83 μm，短径 58-66 μm．殻面は平らで，中心付近からほぼ放射状に小点紋列が分布する．この小点紋列は殻套を下って殻端まで届く．両端に大きな眼域がある．眼域を結ぶ線の両側殻面に 1 個あるいは 2 個の唇状突起があり，1 殻には計 2-3 個の唇状突起を有する．殻套は比較的高い．
ノ ー ト：本種は Ehrenberg(1843)により *Biddulphia laevis* として原記載されたが，Compère(1982)によって *Pleurosira* 属へ組み合わせになった．Hendey(1964)は本種（*Biddulphia laevis*）の形態を，数値を交えて詳しく記している．加藤ら(1977)は秋田県八郎潟調整池から本種を見出し，塩分濃度に対する適応性は広塩性と記している．
産　　出：稲城層：前期更新世・東京都，八郎潟調整池（加藤ら 1977）：秋田県（現生），琵琶湖（辻・伯耆 2001）：滋賀県（現生），池田湖（田中・南雲 2009）：鹿児島県（現生）．
図　　版：稲城層標本．

Figs 1-7.　*Pleurosira laevis* (Ehrenberg) Compère
LM. Figs 1-2. SEM. Figs 3-7. From Inagi Formation (Early Pleistocene), Fuchu City, Tokyo, Japan. Scale bars: Figs 3, 6＝10 μm, Fig. 7＝5 μm, Figs 4-5＝2 μm.

Figs 1-2.　Valve view of two different valves.
Fig. 1.　Three rimoportulae (arrowheads).
Fig. 3.　External oblique view of whole valve, ocelli (arrows).
Fig. 4.　Detailed view of ocellus.
Fig. 5.　External opening of rimoportula (arrowhead).
Fig. 6.　Internal oblique view of whole valve.
Fig. 7.　Enlarged view of internal valve face showing three rimoportulae (arrowheads).

Plate 122

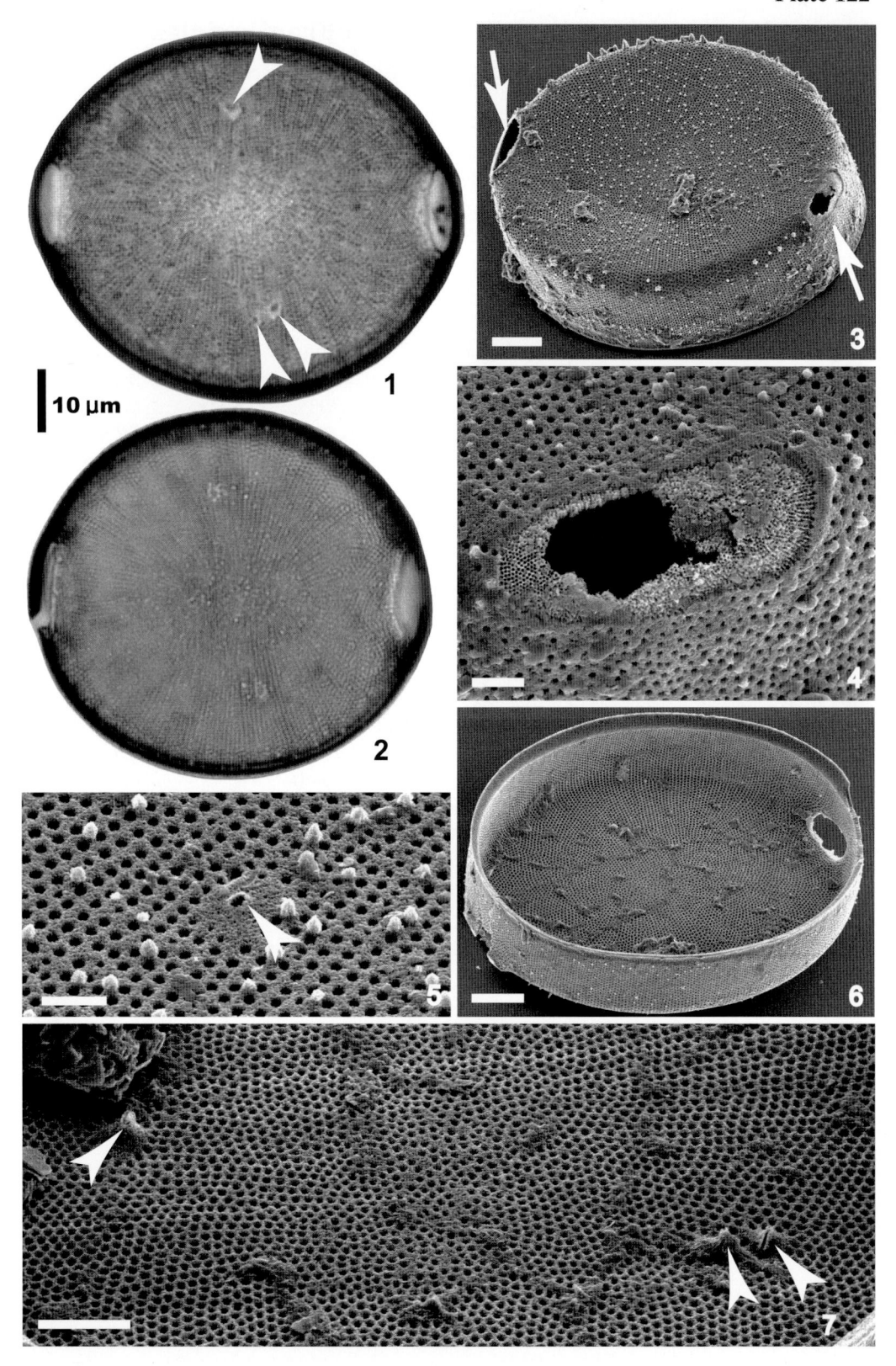

Plate 123. Araphid, pennate diatoms: Fragilariales
Diatoma anceps (Ehrenberg) Kirchner 1878

基礎異名：*Fragilaria*？*anceps* Ehrenberg 1843
文　　献：Krammer, K. & Lange-Bertalot, H. 1991.　Bacillariophyceae. Teil 3. Centrales, Fragilariaceae, Eunotiaceae. 576 pp. Süβwasserflora von Mitteleuropa, Bd. **2/3**, Begründet von A. Pascher. Gustav Fischer Verlag, Stuttgart, Jena.
形　　態：殻はやや短い線形，両端は頭状に突出する，殻長 27-67 µm，殻幅約 7 µm．横走肋は 10 µm に約 3 本，条線は 10 µm に 12-18 本．唇状突起は 1 個で，片方の殻端突出部の殻端近くに所在する．
ノ ー ト：殻面観で両端が頭状に突出するのが特徴であり，太櫓層の個体（中村層の個体も同様である）は突出がやや少ないが本種に同定できる．Krammer & Lange-Bertalot(1991) の *Diatoma anceps* の図（Fig. 102：5），Fricke(1906)の図（Fig. 52）に類似している．*Meridion* へ所属したこともあるが（Williams 1985），殻形が上下対称であることから，Krammer & Lange-Bertalot(1991)のとおり *Diatoma* へ所属するのが適当であろう．
産　　出：太櫓層（瀬棚の珪藻土）：前期中新世・北海道，中村層：前期中新世・岐阜県．
図　　版：太櫓層標本．

Figs 1-8. *Diatoma anceps* (Ehrenberg) Kirchner
LM. Figs 1-2. SEM. Figs 3-8. Materials from Futoro Formation "diatomite of Setana" (Early Miocene), Setana Town, Hokkaido, Japan. Scale bars: Fig. 5＝5 µm, Figs 3, 6＝2 µm, Figs 4, 7-8＝1 µm.

Figs 1-2.   Two different size valves.
Fig. 3.   External view of valve, opening of rimoportula (arrowhead).
Fig. 4.   Enlarged oblique view of valve apex of Fig. 3, opening of rimoportula (arrowhead).
Fig. 5.   Internal view of whole valve, rimoportula (arrowhead).
Fig. 6.   Oblique view of Fig. 5, rimoportula (arrowhead).
Fig. 7.   Enlarged view of spines.
Fig. 8.   Enlarged view of Fig. 5, rimoportula (arrowhead).

Plate 123

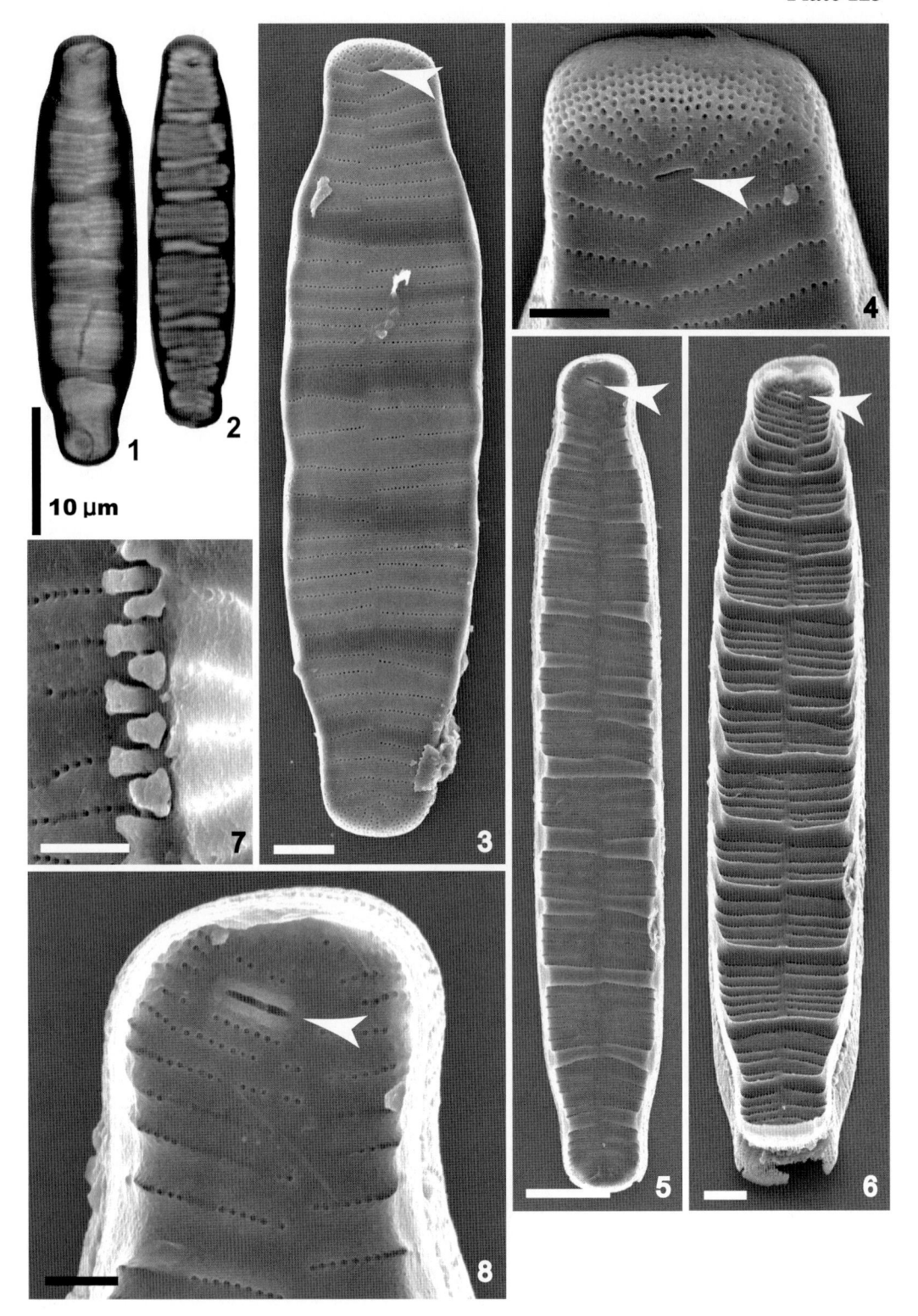

10 μm

Plate 124.   Araphid, pennate diatoms: Fragilariales
Diatoma ehrenbergii Kützing 1844

文　献：Krammer, K. & Lange-Bertalot, H. 1991. Bacillariophyceae. Teil 3. Centrales, Fragilariaceae, Eunotiaceae. 576 pp. Süßwasserflora von Mitteleuropa, Bd. **2/3**, Begründet von A. Pascher. Gustav Fischer Verlag, Stuttgart, Jena.

　　　Fukushima, H., Ko-Bayashi, T., Terao, K. & Yoshitake, S. 1988. Morphological variability of *Diatoma vulgare* Bory var. *grande*. *In*: Round, F.E.(ed.) Proceedings of the 9th International Diatom Symposium. pp. 377-389. Biopress, Bristol.

形　態：殻は皮針形～中央が膨らむ線形，殻端は頭状に突出する，殻長 38-52 μm，殻幅 7-10 μm．横走肋は 10 μm に 7-9 本．唇状突起は 1 個で，一般的に殻端近くの殻面／殻套境界付近に所在するが，殻面に存在する場合も観察された．両殻端に殻端小孔域がある．

ノート：Krammer & Lange-Bertalot(1991)で本種の異名とされている *D. vulgaris* var. *grande* は，Fukushima *et al.*(1988) によると *D. vulgaris* var. *capitata*, *D. vulgaris* var. *producta*, *D. vulgaris* var. *linearis* と形態が連続し，それら 3 変種は var. *grande* の異名である．本書では Fukushima *et al.*(1988) で var. *grande* の異名と報告された分類群も合わせて *D. ehrenbergii* と考えている．

産　出：沼田湖成層：中期更新世・群馬県，瓜生坂層（*D. vulgaris* var. *producta* として，窪田ら 1976）：鮮新世・長野県．

図　版：沼田湖成層標本．

Figs 1-5.   *Diatoma ehrenbergii* Kützing

LM. Figs 1-2. SEM. Figs 3-5. Materials from Numata Lacustrine Deposit (Middle Pleistocene), Numata City, Gunma, Japan. Scale bars: Figs 5, 7＝5 μm, Fig. 6＝2 μm.

Figs 1-2.   Valve view of two different size valves.
Fig. 3.   External view of whole valve, opening of rimoportula (arrowhead).
Fig. 4.   Enlarged view of an apex of Fig. 3, opening of rimoportula (arrowhead) and apical pore field (arrow).
Fig. 5.   Internal view of whole valve, rimoportula (arrowhead).

Diatoma vulgaris Bory 1824

文　献：Krammer, K. & Lange-Bertalot, H. 1991. Bacillariophyceae. Teil 3. Centrales, Fragilariaceae, Eunotiaceae. 576 pp. Süßwasserflora von Mitteleuropa, Bd. **2/3**, Begründet von A. Pascher. Gustav Fischer Verlag, Stuttgart, Jena.

形　態：殻は皮針形～広楕円形，殻端は広円形，殻長 29-55 μm，殻幅 10-13 μm．横走肋は 10 μm に約 6 本．唇状突起は 1 個で，殻端近くの殻面に所在する．

ノート：*Diatoma ehrenbergii* の類似種なので，現生個体の写真であるが参考に掲げた．

図　版：神流川標本．

Figs 6-7.   *Diatoma vulgaris* Bory

LM. Fig. 6. SEM. Fig. 7. Materials from Kanna River (Recent), Gunma, Japan. Scale bar: Fig. 2＝5 μm.

Fig. 6.   Whole view of valve, rimoportula (arrowhead).
Fig. 7.   Internal view of whole valve, rimoportula (arrowhead).

Plate 124

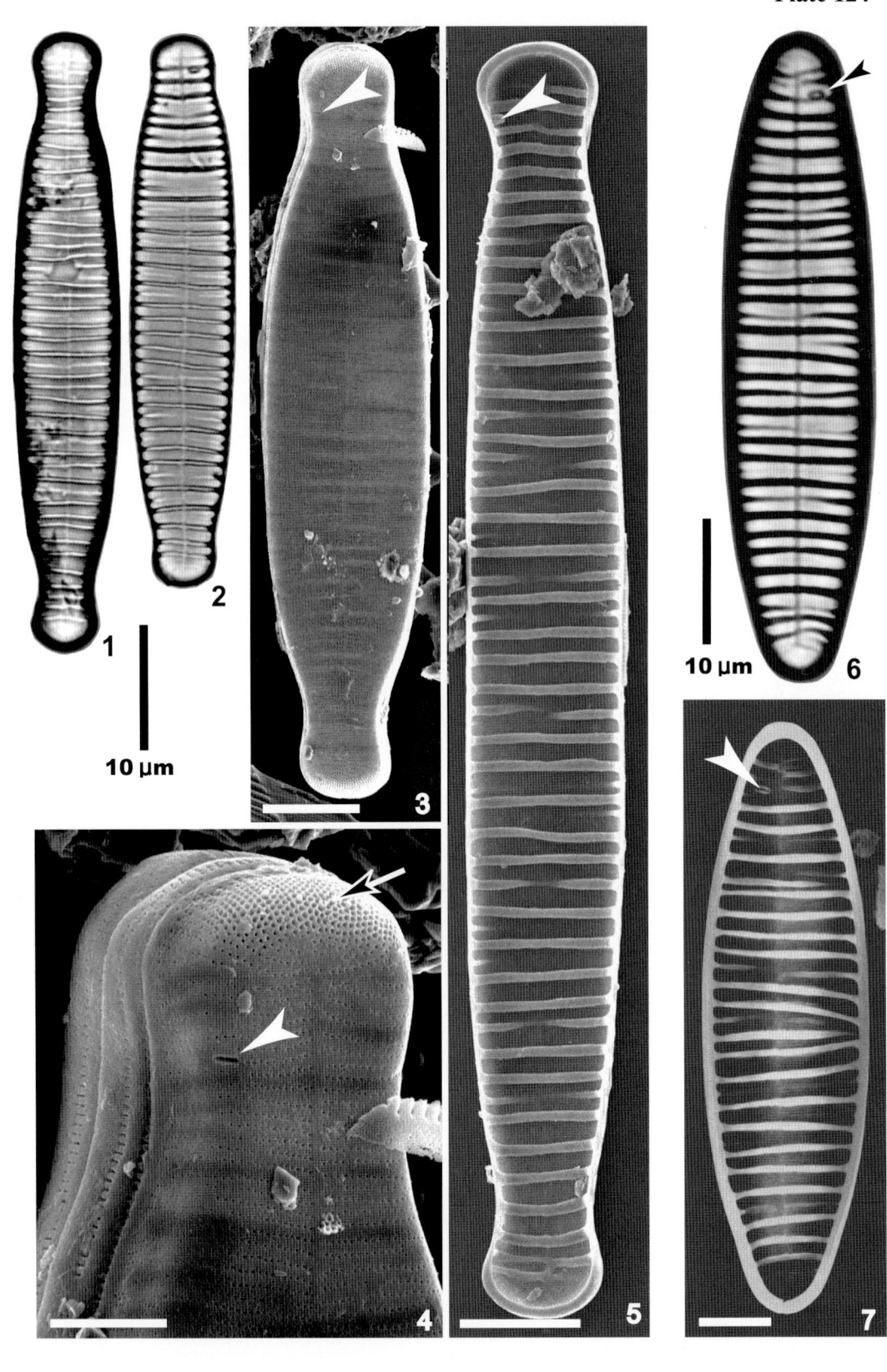

285

Plate 125.　Araphid, pennate diatoms: Fragilariales
Diatoma hyemalis (Roth) Heiberg 1863

基礎異名：*Conferva hyemalis* Roth 1800
文　　献：Hustedt, F. 1930. Bacillariophyta. *In*: Pascher, A.(ed.) Die Süsswasser-Flora Mitteleuropas. Heft 10. 466 pp. Gustav Fischer, Jena.
形　　態：殻は皮針形～狭楕円形，殻端はややかさび形の広円形，殻長 33-50 µm，殻幅 8.5-9.5 µm．横走肋は 10 µm に約 4 本，殻全体の数は 6-15 本．唇状突起は 1-2 個で，殻端近くの殻面に所在する．
ノ ー ト：類似する *Diatoma mesodon*(Ehrenb.)Kütz. とは，横走肋の数によって，*Diatoma vulgaris* Bory とは横走肋の密度によって区別できる．
産　　出：鬼首層（Ichikawa 1955）：後期更新世・宮城県，沼田湖成層：中期更新世・群馬県，瓜生坂層（窪田ら 1976）：鮮新世・長野県，奄芸層群（根来・後藤 1981）：鮮新世・三重県．
図　　版：沼田湖成層標本．

Figs 1-5.　*Diatoma hyemalis* (Roth) Heiberg
LM. Figs 1-2. SEM. Figs 3-5. Material from Numata Lacustrine Deposit (Middle Pleistocene), Numata City, Gunma, Japan. Scale bars: Figs 3-5=2 µm.

Figs 1-2.　Two different size valves.
Figs 1, 3-4 and Figs 2, 5, same valves LM and SEM photographs.
Fig. 3.　Whole view of valve, openings of rimoportulae (arrowheads).
Fig. 4.　Enlarged oblique view of Fig. 3 showing openings of rimoportulae (arrowheads), granules of valve face/mantle boundary and mantle.
Fig. 5.　Internal oblique view showing rimoportula (arrowhead) and transapical costa (arrow).

Diatoma mesodon (Ehrenberg) Kützing 1844

基礎異名：*Fragilaria mesodon* Ehrenberg 1839
文　　献：Kützing, F.T. 1844. Die kieselschaligen Bacillarien oder Diatomeen. Nordhausen, W. Kohne. 152 pp, pls 1-30.
形　　態：沼田湖成層から産出した本種は，殻が広皮針形～狭楕円形，殻長 10-18 µm，殻幅 5.5-6.5 µm．横走肋は 2-4(5)本／殻．唇状突起は 1 個で，殻端近くの殻面に所在する．
ノ ー ト：*Diatoma hiemalis* var. *mesodon*(Ehrenb.)Grunow と同定されてきた分類群であるが独立した種として扱われるようになってきたので，本書でもそれに倣った．*Diatoma hyemalis*(Roth)Heiberg とは殻の横断肋の数の違いで区別できる．図版の SEM 写真（Fig. 9）は現生個体標本で示す．
産　　出：野殿層（中島・南雲 1999）：中期更新世・群馬県，沼田湖成層：中期更新世・群馬県，瓜生坂層（窪田ら 1976）：鮮新世・長野県．
図　　版：沼田湖成層（Fig. 6），利根川源流（現生）（Figs 7-9）標本．

Figs 6-9.　*Diatoma mesodon* (Ehrenberg) Kützing
LM. Figs 6-8. SEM. Fig. 9. Fig. 6. Material from Numata Lacustrine Deposit (Middle Pleistocene), Numata City, Gunma and Figs 7-9 material from the uppermost part of the Tone River (Recent), Gunma, Japan. Scale bar: Fig. 9=2 µm.

Figs 6-7.　View of two different size valves.
Fig. 8.　Girdle view of a frustule.
Fig. 9.　Internal view of a valve showing rimoportula (arrowhead) and transapical costa (arrow).

Plate 125

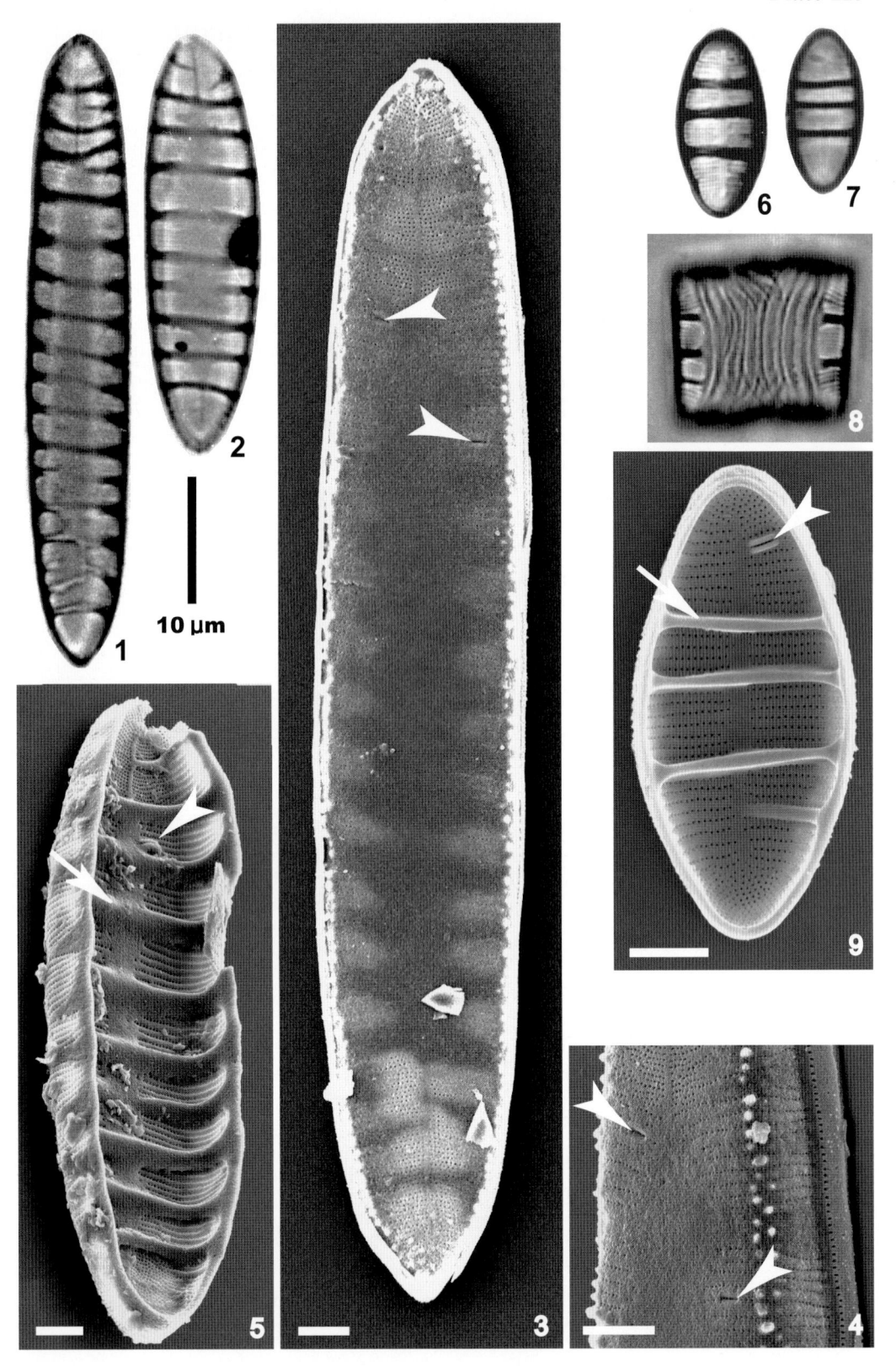

Plate 126.　Araphid, pennate diatoms: Fragilariales
Fragilaria neoproducta Lange-Bertalot 1993

文　　献：Lange-Bertalot, H. 1993. 85 neue taxa und über 100 weitere neu difinierte taxa ergänzerd zur süßwasserflora von Mitteleuropa. **2/1-4**. Bibliotheca Diatomologica **27**: pp. 1-164, 134 pls.

形　　態：殻は皮針形〜短い線形，殻長 13-44 μm，殻幅 6.5-7 μm，先端は広いくちばし形．左右の条線は互い違いに配列し 10 μm に約 16 本．SEM 観察では殻面の条線は殻套へ続き，各間条線の殻面／殻套境界には針があり，軸域は狭い．唇状突起はない．

ノート：類似している *Fragilariforma virescens* とは *F. neoproducta* の方が軸域が広いこと，唇状突起を欠くことから区別できる．

産　　出：津森層（田中ら 2005）：中期更新世・熊本県，大鷲湖沼性堆積物（田中ら 2011）：鮮新世・岐阜県．

図　　版：大鷲湖沼性堆積物標本．

Figs 1-7.　*Fragilaria neoproducta* Lange-Bertalot
LM. Figs 1-3. SEM. Figs 4-7. Material from lacustrine deposit of Owashi (Pliocene), Gujo City, Gifu, Japan. Scale bars: Figs 4-6=2 μm, Fig. 7=1 μm.

Figs 1-3.　Three different size valves.
Fig. 4.　External view of whole valve.
Fig. 5.　Oblique view of Fig. 4.
Fig. 6.　Internal oblique view of whole valve.
Fig. 7.　Detailed oblique view of terminal area, ocellulimbus (arrow).

Plate 126

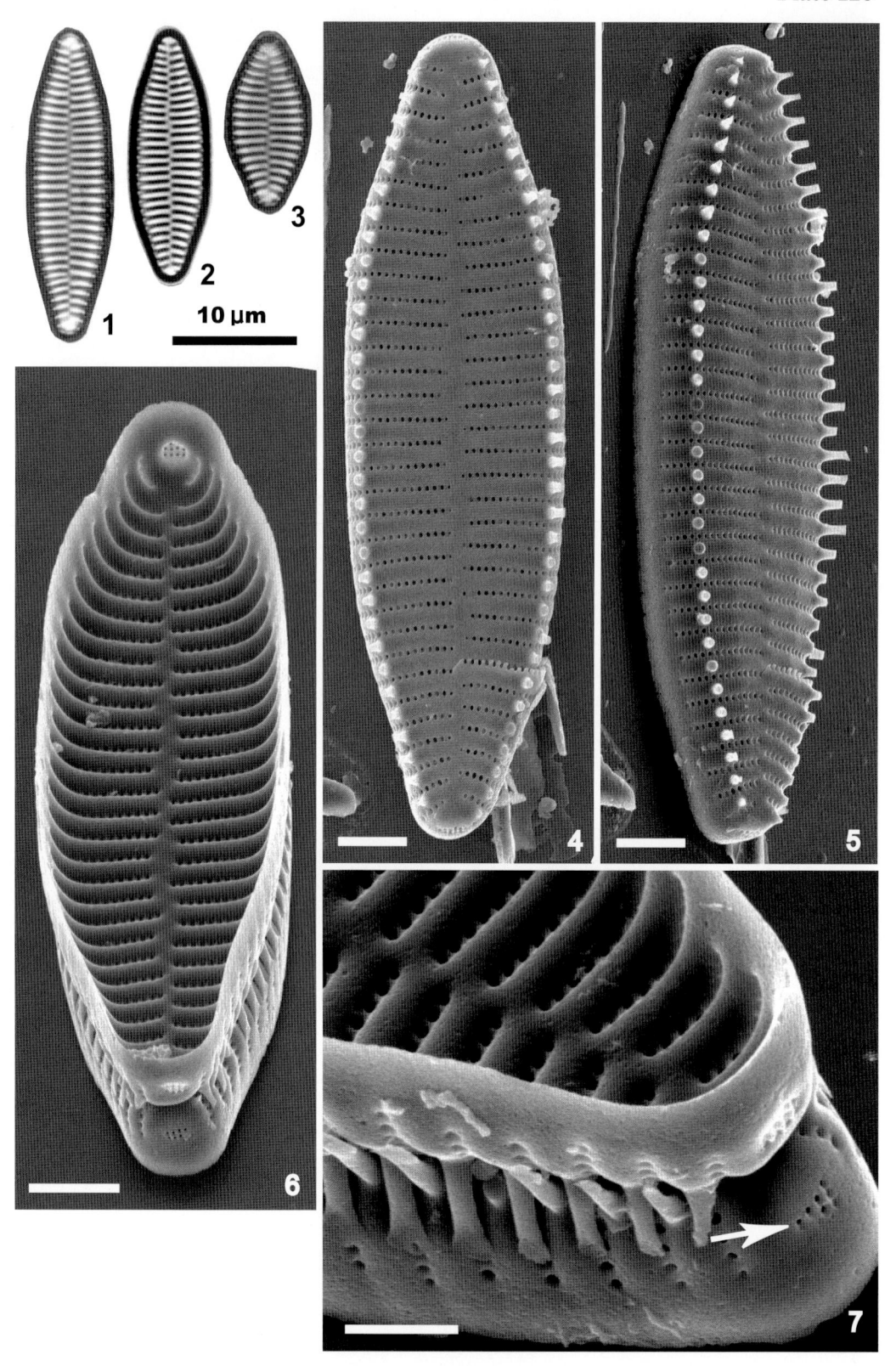

Plate 127.　Araphid, pennate diatoms: Fragilariales
Fragilaria vaucheriae (Kützing) Petersen 1938

基礎異名：*Exilaria vaucheriae* Kützing 1833
文　　献：小林　弘・出井雅彦・真山茂樹・南雲　保・長田敬五　2006．小林弘珪藻図鑑「第1巻」，内田老鶴圃，東京．
形　　態：殻は線状皮針形で両殻端は頭状に突出する．殻長 27-41 μm，殻幅 3-4 μm．条線は平行で，左右の条線は互い違いに配列し 10 μm に約 11-14 本．中心域は縁まで無紋で片側がわずかに膨らむが，SEM 観察では膨らまない側に短い条線が観察されることが多い．唇状突起は片側の殻端付近に 1 個存在する．
ノ ー ト：本種は現在までにいろいろな分類がされてきた種類の一つなので，最新の図鑑を文献として記した．
産　　出：野殿層（中島・南雲 1999）：更新世・群馬県，大鷲湖沼性堆積物：鮮新世・岐阜県．
図　　版：大鷲湖沼性堆積物標本．

Figs 1-6.　*Fragilaria vaucheriae* (Kützing) Petersen
LM. Figs 1-2. SEM. Figs 3-6. Material from lacustrine deposit of Owashi (Pliocene), Gujo City, Gifu, Japan. Scale bars: Figs 3-5＝2 μm, Fig. 6＝1 μm.

Figs 1-2.　Two different size valves.
Fig. 3.　External view of whole valve.
Fig. 4.　Internal view of whole valve.
Fig. 5.　Oblique view of Fig. 4.
Fig. 6.　Detailed view of terminal area of Fig. 5, rimoportula (arrowhead).

Plate 127

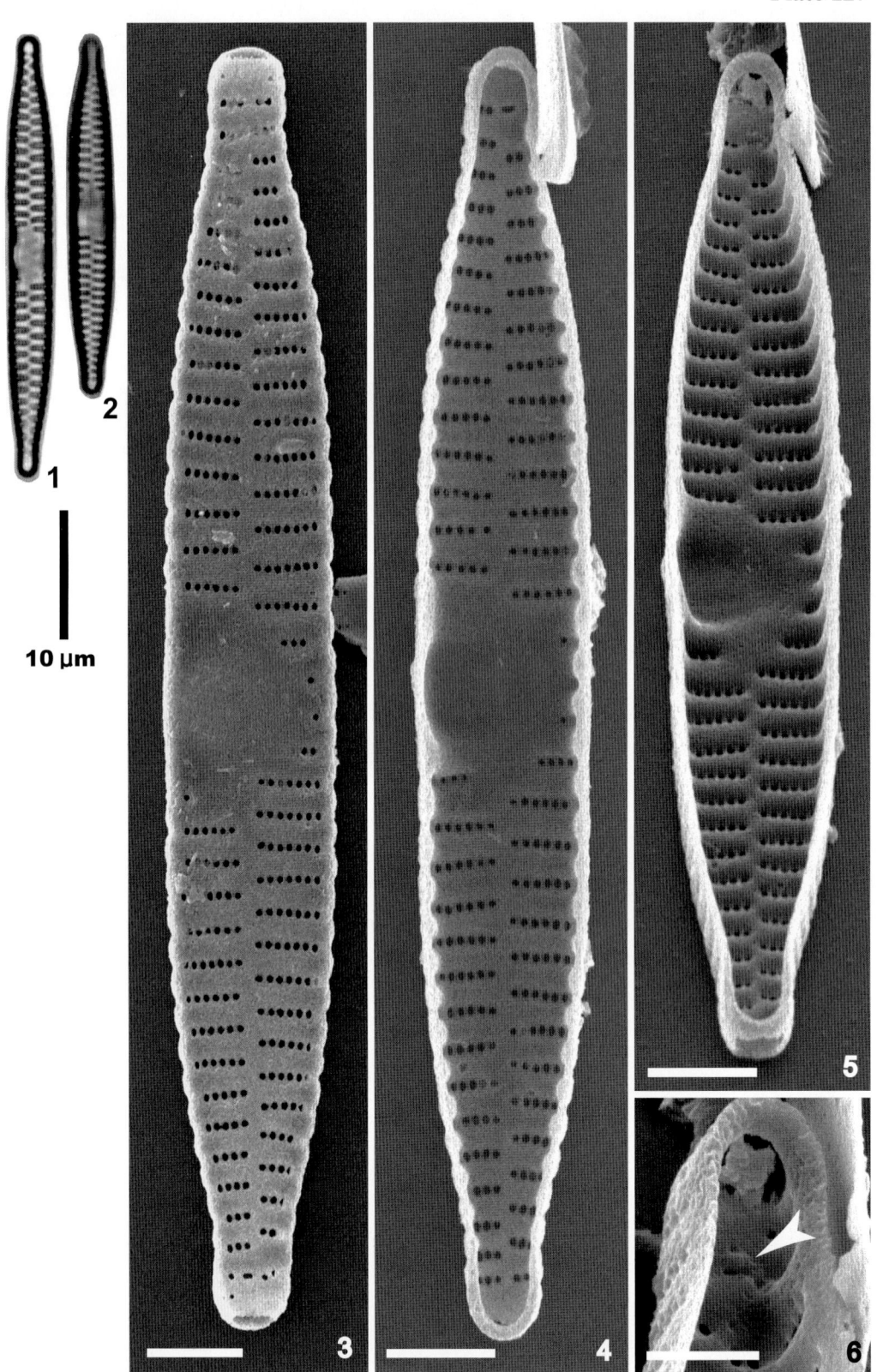

Plate 128. Araphid, pennate diatoms: Fragilariales
Fragilariforma fossilis (Pantocsek) H. Tanaka
comb. nov. et stat. nov.

基礎異名：*Diatoma anceps* var. *fossilis* Pantocsek 1905
組み合わせ・地位変更：17頁参照（英文）．
文　　献：Pantocsek, J. 1905. Beiträge zur Kenntnis der Fossilen Bacillarien Ungarns. 3 Teil. 118 pp. 1-42 pls. W. Junk, Berlin.
　　奥野春雄 1959. 北海道瀬棚町の珪藻土について(4)．植物研究雑誌 **34**: 272-277.
形　　態：殻は棒状で，殻面は平らであり，両殻端は頭状に突出する．殻長 59-83 μm，殻幅 6-10 μm．条線は単列の胞紋列からなり，全体的に平行であるが，多くの条線はわずかであるが不規則に向きが変化する．また条線の間隔も不規則で，計測場所により数がかなり異なる，10 μm に約 7-14 本．軸域は非常に狭い．殻面／殻套境界には肥厚または結合針がある．唇状突起，殻端小孔域は認められなかった．
ノ　ー　ト：Pantocsek(1905)は本分類群を *Diatoma anceps* var. *fossilis* Pant. として記載したが，横肋線がないこと等から奥野(1959)は *Fragilaria bicaptata* A. Mayer(=*Fragilariforma bicapitata*(A. Mayer)Williams & Round) に同定した．しかし SEM で形態を観察すると *F. bicaptata* とも異なっており，独立した種とするのが適当と思われる．
産　　出：太櫓層（瀬棚の珪藻土）：前期中新世・北海道．
図　　版：太櫓層標本．

Figs 1-6. *Fragilariforma fossilis* (Pantocsek) H. Tanaka
comb. nov. et stat. nov.

LM. Figs 1-3. SEM. Figs 4-6. Material from Futoro Formation "diatomite of Setana" (Early Miocene), Setana Town, Hokkaido, Japan. Scale bars: Figs 4-6＝5 μm.

Fig. 1.　Neotype specimen.
Figs 1-3.　Three different size valves.
Fig. 4.　External view of whole valve.
Fig. 5.　Internal view of whole valve.
Fig. 6.　Oblique view of part of Fig. 4 showing frustule of one valve with linking spines (arrow) and the other with ridge between valve face and mantle (arrowhead).

Plate 128

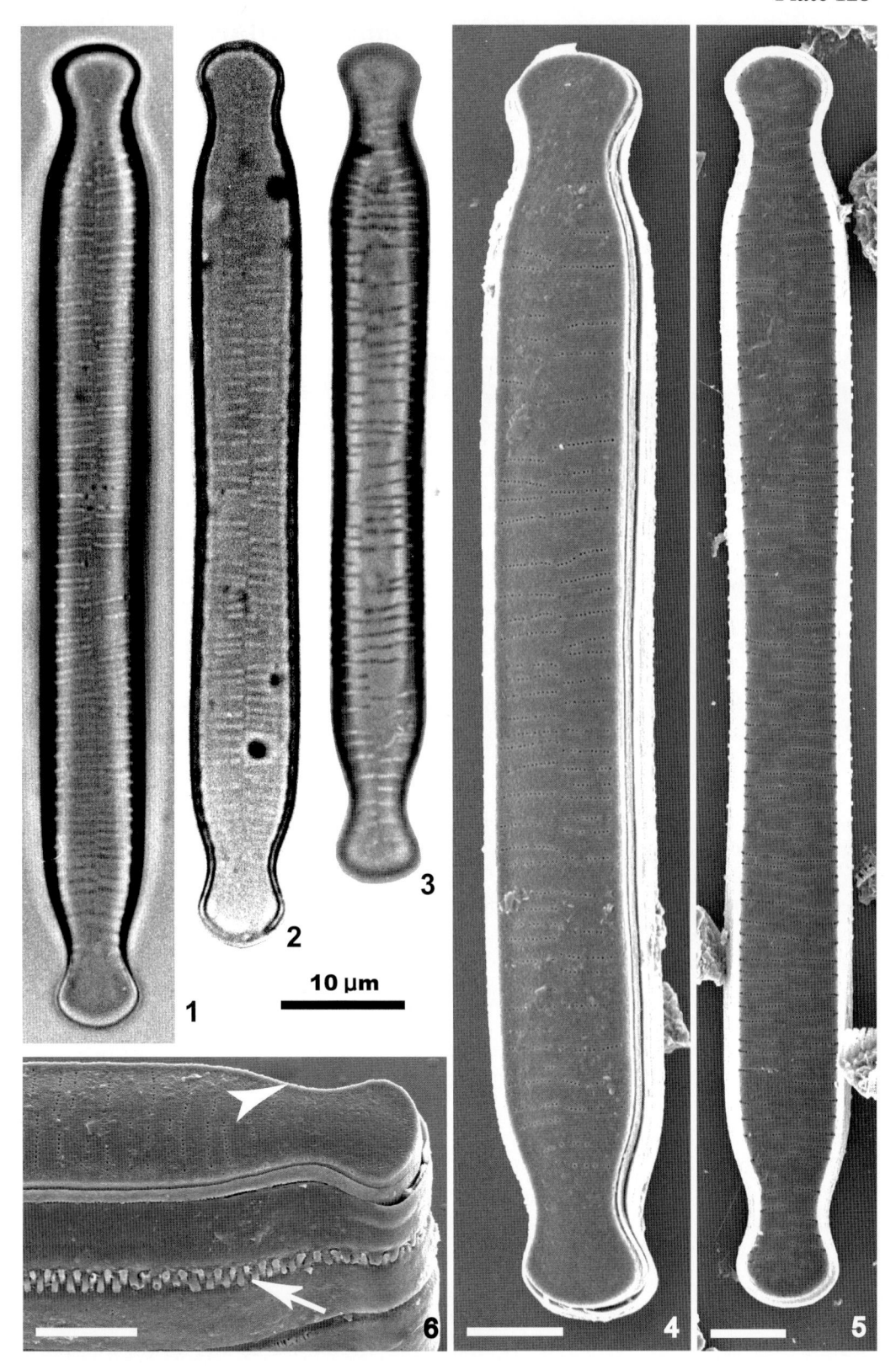

10 μm

293

Plate 129.　Araphid, pennate diatoms: Fragilariales
Fragilariforma kamczatica (Lupikina) H. Tanaka
comb. nov. et stat. nov.

基礎異名：*Fragilaria nitzschioides* var. *kamczatica* Lupikina
組み合わせ・地位変更：17頁参照（英文）．
文　　献：Lupikina, E.G. 1965. Diatomeae novae et curiosae e stratis ermanicis partis Kamczatkae occidentalis. Novitates Systematicae Plantarum non Vascularium. Tom **2**: 15-22.
形　　態：殻は棒状で，殻端がやや膨らみ殻面は平らである．殻長58-108 µm，殻幅7-8.5 µm．条線は殻端を除き平行で，単列の胞紋列からなる．10 µm に約18本．軸域は非常に狭い．ふつう殻面／殻套境界の間条線上には連続〜5列ごとに針がある．殻端の片方は小孔域が殻面まで広がっているが，他方は殻套内に分布する．唇状突起は殻面に所在するが，殻端小孔域近くに所在する場合，殻套小孔域側に所在する場合，中心付近に所在する場合等さまざまであった．
ノ　ー　ト：Lupikina(1965)は本分類群を*Fragilaria nitzschioides* var. *kamczatica* Lupik. として記載したが，*Fragilaria nitzschioides* var. *nitzschioides* はすでに*Fragilariforma*属へ組み合わせになっている．および var. *kamczatica* は独立した種とするのが適当と思われる．
産　　出：太櫓層（瀬棚の珪藻土）：前期中新世・北海道．
図　　版：太櫓層標本．

Figs 1-9.　*Fragilariforma kamczatica* (Lupikina) H. Tanaka
comb. nov. et stat. nov.
LM. Figs 1-2. SEM. Figs 3-9. Material from Futoro Formation "diatomite of Setana" (Early Miocene), Setana Town, Hokkaido, Japan. Scale bars: Figs 3-5, 8-9＝5 µm, Figs 6-7＝2 µm.

Figs 1, 3-4.　Same valve, LM and SEM photographs.
Figs 1-2.　Two different size valves.
Fig. 3.　External view of whole valve.
Fig. 4.　Oblique view of part of Fig. 3 showing spines and areolae rows.
Fig. 5.　Internal view of whole valve, rimoportula (arrowhead).
Fig. 6.　Enlarged view of an apex of Fig. 5.
Fig. 7.　Enlarged view of opposite apex of Fig. 5 showing rimoportula (arrowhead) and apical pore field (arrow).
Fig. 8.　External view of an apex showing opening of rimoportula (arrowhead).
Fig. 9.　Opposite apex of Fig. 8 valve showing apical pore field (arrow).

Plate 129

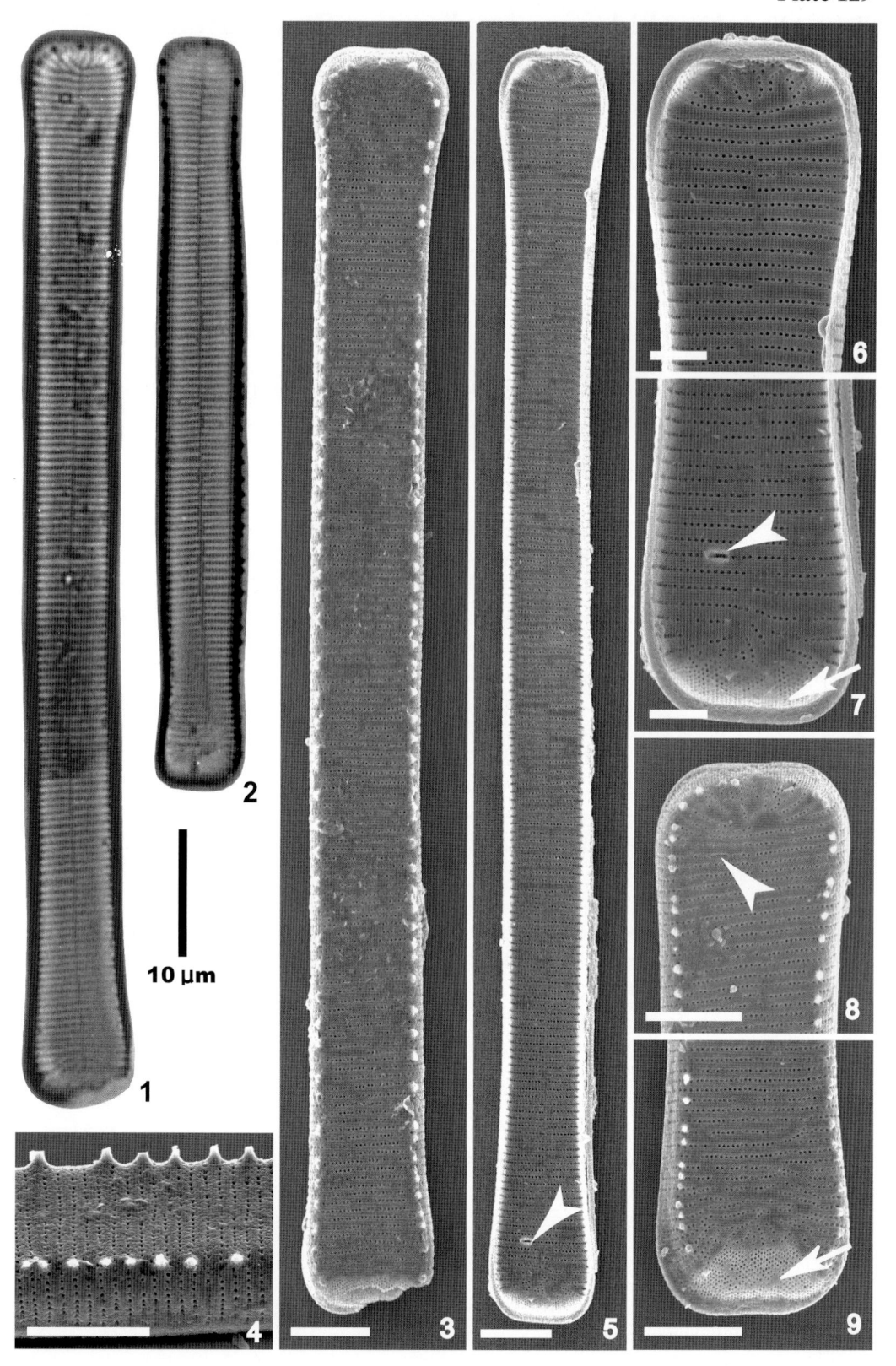

10 μm

Plate 130. Araphid, pennate diatoms: Fragilariales

Hannaea arcus (Ehrenberg) R.M. Patrick var. *arcus* 1961
基礎異名：*Navicula arcus* Ehrenberg 1838

Hannaea arcus var. *hattoriana* (F. Meister) Ohtsuka 2002
基礎異名：*Ceratoneis arcus* var. *hattoriana* F. Meister 1914

Hannaea arcus var. *recta* (Cleve) M. Idei 2006
基礎異名：*Fragilaria arcus* var. *recta* Cleve 1898

文　献：Meister, F. 1914. Beiträge zur Bacillaria ceenflora Japan. Archiv für Hydrobiologie undPlanktonkunde, Bd. **9**: 226-232.

　　　Ohtsuka, T. 2002. Checklist and illustration of diatoms in the Hii River. Diatom **18**: 23-56.

　　　小林　弘・出井雅彦・真山茂樹・南雲　保・長田敬五 2006. 小林弘珪藻図鑑「第1巻」，pp. 531. 内田老鶴圃，東京.

形　態：この3分類群は殻の形は異なるが他は類似しているので，ここでは基本種（var. *arcus*）について記す．殻は線状で両殻端に向かい殻幅は徐々に小さくなり殻端は頭状に突出する．殻長39-75 μm, 殻幅約6 μm. 条線は平行で，左右の条線は互い違いに配列し10 μmに約14本．中軸は狭く，中心域は腹側が殻縁まで無紋で，背側は条線が短くなり広くなる．中心域腹側殻縁は膨らむ．SEM観察によると殻面の条線と殻套の条線は連続せず，間の殻面／殻套境界に針が存在する．唇状突起は1個で片方の殻端の頭状に突出した部分にある（var. *hattoriana* では，まれに両殻端に所在する場合も見られた）．両殻端に殻套眼域がある．

ノート：3分類群とも沼田湖成層から産出した，一見類似しているが，殻面観において var. *arcus* は殻が腹側へくの字形に曲がること，var. *hattoriana* は腹側の殻縁は直線的であるが中軸・背側は腹側へ曲がること，var. *recta* は中軸が直線的で両殻縁は中軸側へ曲がることに注意して識別するとよい．var. *hattoriana* は日本（横浜）から記載された分類群である．

産　出：沼田湖成層：中期更新世・群馬県.
図　版：沼田湖成層標本.

Figs 1-4. *Hannaea arcus* (Ehrenberg) R. M. Patrick var. *arcus*
Figs 5-6. *Hannaea arcus* var. *hattoriana* (F. Meister) Ohtsuka
Figs 7-8. *Hannaea arcus* var. *recta* (Cleve) M. Idei

LM. Figs 1-2, 5, 7. SEM. Figs 3-4, 6, 8. Material from Numata lacustrine deposit (Middle Pleistocene), Numata City, Gunma, Japan. Scale bars: Fig. 4＝1 μm, Figs 3, 8＝2 μm, Fig. 6＝5 μm.

Figs 1-2. Two different size valves of *H. arcus* var. *arcus*.
Fig. 3. External view of whole valve, opening of rimoportula (arrowhead).
Fig. 4. Enlarged oblique view of an apex of Fig. 3, opening of rimoportula (arrowhead), ocellulimbus (arrow).
Fig. 5. Whole valve view of *H. arcus* var. *hattoriana*.
Fig. 6. Internal view of valve, rimoportula (arrowhead).
Fig. 7. Whole valve view of *H. arcus* var. *recta*.
Fig. 8. Internal view of valve.

Plate 130

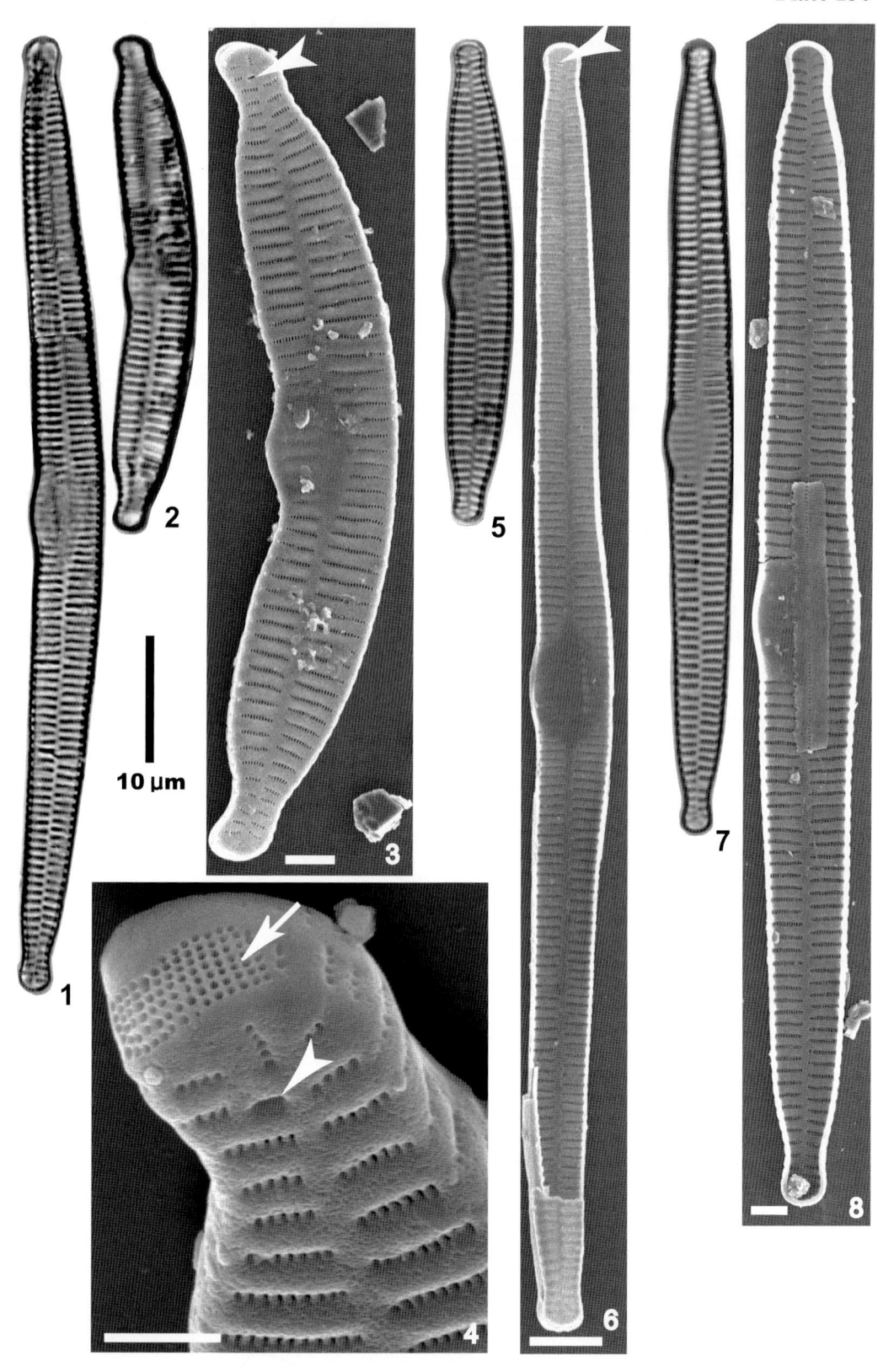

297

Plate 131.　Araphid, pennate diatoms: Fragilariales
Meridion circulare var. *constrictum* (Ralfs) Van Heurck 1880

基礎異名：*Meridion constrictum* Ralfs 1843

文　　献：Van Heurck, H. 1880-1885. Synopsis des Diatomées de Belgique. Atlas: pls. 132, text 235 pp. 3 pls. Ducaju et Cie. Anvers.

　　河島綾子・小林　弘 1995. 阿寒湖の珪藻(3. 羽状類―広義の *Fragilaria* を除く無縦溝類). 自然環境科学研究 **8**: 35-49.

形　　態：殻はくさび形で，頭部は頭状に突出し足部は細くなる．殻長 30-42 μm，殻幅 8-10 μm．横走肋は 10 μm に約 3 本，条線は 10 μm に約 15 本．唇状突起は 1 個で，頭状突出部近くの殻面に所在する．足部に殻端小孔域がある．

ノート：殻端の片方が頭状に突出するのが特徴であるが，頭状の程度はさまざまで突出の付け根がほとんど凹まない個体も観察できた．本変種を独立の種とする考えもあるが（Williams 1985），本書では河島・小林(1995)に倣って *M. circulare* の変種とする．

産　　出：太櫓層（瀬棚の珪藻土）：前期中新世・北海道，野殿層（中島・南雲 1999）：中期更新世・群馬県，中村層：前期中新世・岐阜県，琵琶湖堆積物（Mori 1975）：更新世・滋賀県，奄芸層群（根来・後藤 1981）：鮮新世・三重県，俣水層（田中・鹿島 2007）：前期更新世・大分県．

図　　版：太櫓層標本．

Figs 1-6.　*Meridion circulare* var. *constrictum* (Ralfs) Van Heurck

LM. Figs 1-2. SEM. Figs 3-6. Materials from Futoro Formation "diatomite of Setana" (Early Miocene), Setana Town, Hokkaido, Japan. Scale bars: Figs 3, 5＝2 μm, Figs 4, 6＝1 μm.

Figs 1-2.　Two different size valves.
Fig. 3.　External view of whole valve, opening of rimoportula (arrowhead).
Fig. 4.　Enlarged oblique view of foot pole of Fig. 3, apical pore field (arrow).
Fig. 5.　Internal oblique view of whole valve, rimoportula (arrowhead).
Fig. 6.　Enlarged view of Fig. 5, rimoportula (arrowhead).

Plate 131

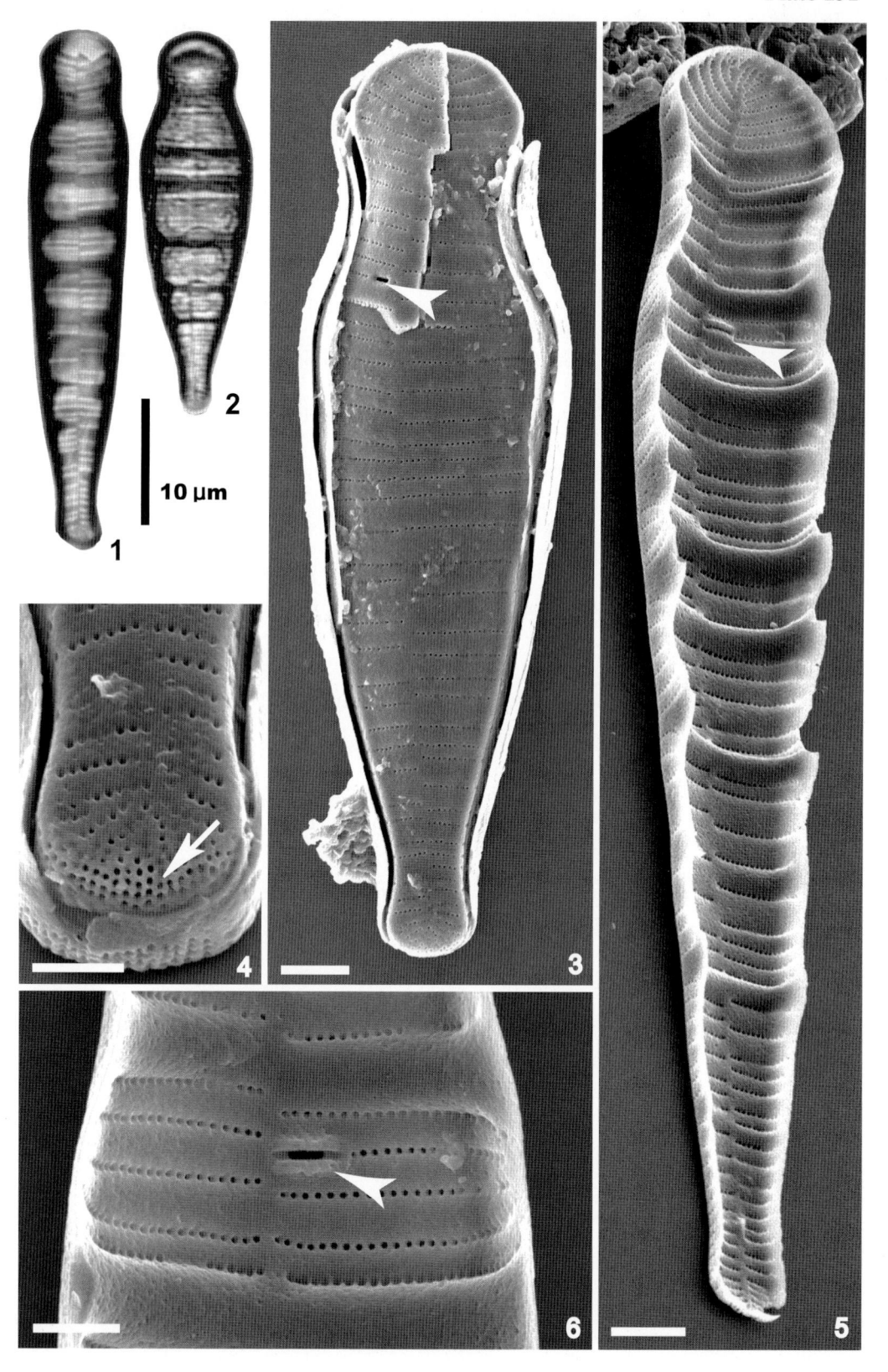

Plate 132. Araphid, pennate diatoms: Fragilariales
Pseudostaurosira brevistriata var. *nipponica* (Skvortsov) H. Kobayasi in Mayama *et al.* 2002

基礎異名：*Fragilaria brevistriata* var. *nipponica* Skvortsov 1936

文　献：Skvortsov, B.W. 1936. Diatoms from Kizaki Lake, Honshu Island, Nippon. Philippine Journal of Science **61**(1): 9-73, 16 pls.

　　　Mayama, S., Idei, M., Osada, K. & Nagumo, T. 2002. Nomenclatural changes for 20 diatom taxa occurring in Japan. Diatom **18**: 89-91.

形　態：殻は中央でゆるやかにくびれる，殻長 13-17.5 μm，殻幅約 4 μm．条線は単列で短く殻縁のみに所在し，軸域が広く，10 μm に 14-15 本で，殻面では 1(2)個の胞紋のみからなり殻套へ連続するが，殻套では 1 個である．胞紋は楕円～円形である．連結針は各間束線の殻面／殻套境界にある．

ノート：Skvortsov(1936)により木崎湖から *Fragilaria brevistriata* var. *nipponica* として原記載された．化石としても日本の各地から報告がある．

産　出：長井川湖沼性堆積物（南雲ら 1998）：中期更新世・群馬県，大戸湖沼性堆積物：中期更新世・群馬県，大鷲湖沼性堆積物（田中ら 2011）：鮮新世・岐阜，俣水層（田中・鹿島 2007）：前期更新世・大分県．

図　版：大戸湖沼性堆積物標本．

Figs 1-9. *Pseudostaurosira brevistriata* var. *nipponica* (Skvortsov) H. Kobayasi

LM. Figs 1-3. SEM. Figs 4-9. Material from lacustrine deposit of Odo (Middle Pleistocene), Higashiagatsuma Town, Gunma, Japan. Scale bars: Figs 4-7=2 μm, Figs 8-9=0.5 μm.

Figs 1-3.　Three different size valves.
Fig. 4.　External view of whole valve.
Fig. 5.　Oblique view of Fig. 4.
Fig. 6.　Internal view of whole valve.
Fig. 7.　Oblique view of Fig. 6.
Fig. 8.　Detailed view of an apex showing ocellulimbus (arrow) and spines.
Fig. 9.　Oblique view of part of inner side and spines.

Plate 132

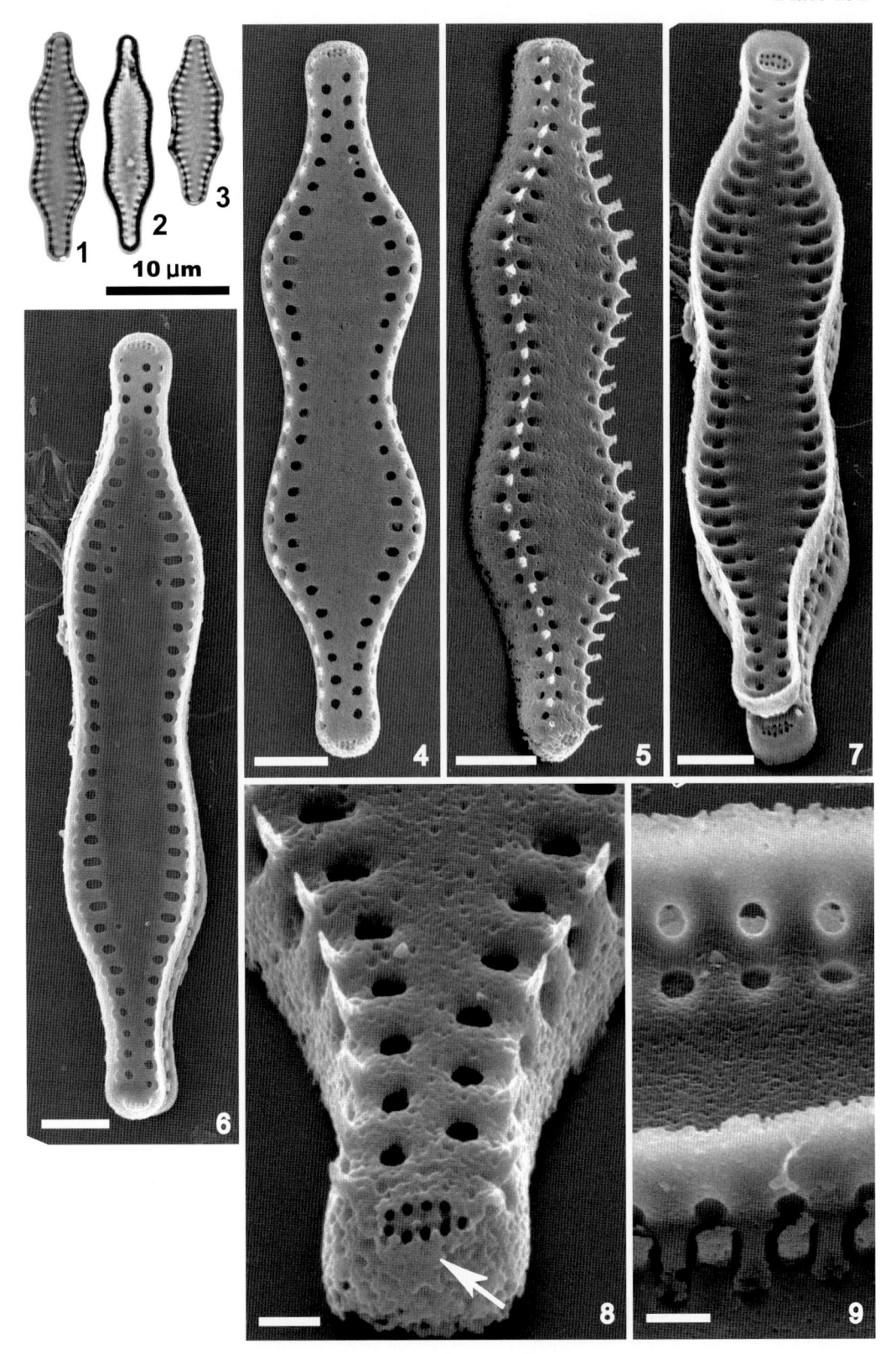

Plate 133.   Araphid, pennate diatoms: Fragilariales
Staurosira construens Ehrenberg var. *construens* 1987

文　　献：Williams, D.M. & Round, F.E. 1987. Revision of the genus *Fragilaria*. Diatom Research **2**: 267-288.
　　出井雅彦・南雲　保 1995. 無縦溝珪藻 *Fragilaria* 属（狭義の）とその近縁属. 藻類 **43**: 227-239.

形　　態：殻は小形で，中央で左右に張り出し，殻面観では十字形をしている．殻長 6-17 µm，殻幅 6-10 µm．条線は平行～わずか放射状で 10 µm に約 14 本．胞紋は細長の楕円形である．殻端には殻套眼域がある．殻面／殻套境界には針があり，他の殻と結合して群体を形成する．

ノ ー ト：かつて *Fragilaria* 属に一括されていた分類群であるが，Williams & Round (1987) によって整理・細分された．この細分された属と形態については出井・南雲 (1995) が参考になる．

産　　出：瓜生坂層（窪田　1976）：更新世・長野県，大鷲湖沼性堆積物（田中ら 2011）：鮮新世・岐阜県，俣水層（田中・鹿島 2007）：前期更新世・大分県，津森層（田中ら 2005）：中期更新世・熊本県．その他各地から産出報告がある．

図　　版：津森層（Figs 1, 3-7, 9），大鷲湖沼性堆積物（Figs 2, 8）標本．

Figs 1-9.   *Staurosira construens* Ehrenberg var. *construens*
Figs 1-3. LM. Figs 4-9. SEM. Materials, Figs 1, 3-7, 9, from Tsumori Formation (Middle Pleistocene), Mashiki Town, Kumamoto and Figs 2, 8, from Owashi lacustrine deposit (Pliocene), Gujo City, Gifu, Japan. Scale bars: Figs 4-5, 9=2 µm, Fig. 8=1 µm, Figs 6-7=0.5 µm.

Figs 1-3.　Three different size valves.
Fig. 4.　External view of whole valve.
Fig. 5.　Oblique view of Fig. 4, ocellulimbus (arrow).
Fig. 6.　Detailed external view of two areolae rows of Fig. 4.
Fig. 7.　Detailed internal view of two areolae rows.
Fig. 8.　Internal oblique view of whole valve.
Fig. 9.　Oblique view of sibling valves.

Plate 133

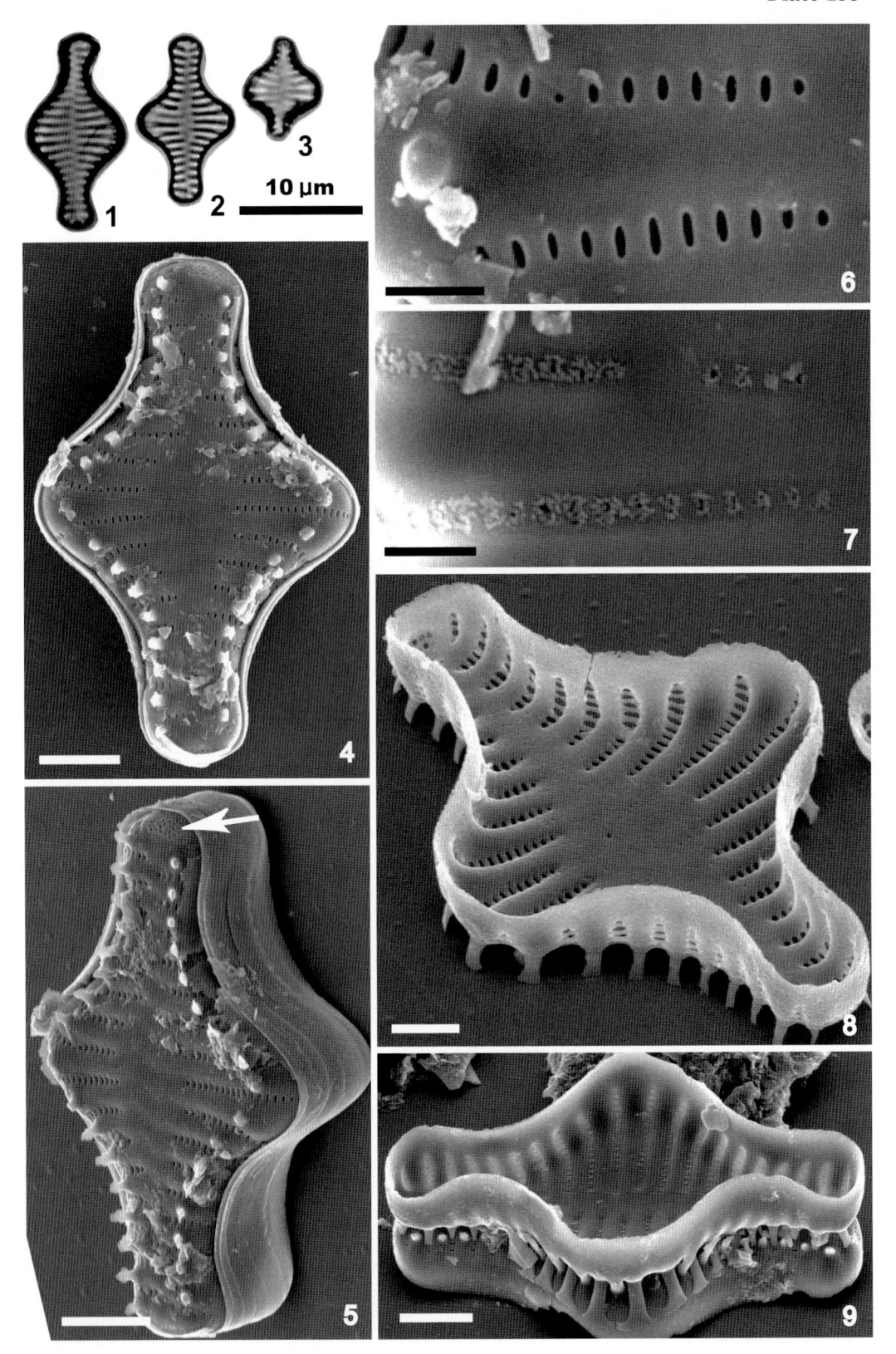

Plate 134. Araphid, pennate diatoms: Fragilariales
Staurosira construens var. *binodis* (Ehrenberg) Hamilton in Hamilton *et al.* 1992

基礎異名：*Fragilaria binodis* Ehrenberg 1854

文　　献：Hamilton, P.B., Poulin, M., Charles, D.F. & Angell, M. 1992. Americanarum Diatomarum Exsiccata: CANA, Voucher slides from eight acidic lakes in northeastern North America. Diatom Research **7**: 25-36.

形　　態：殻面は両殻端方向へ細いが突出し，横方向へ2回膨らむ（中央部でくびれる），小形の殻では片側がくびれない場合がある．殻長 13-23 μm，殻幅 7-9 μm．条線は 10 μm に約 13 本で，平行～わずか放射状である．

産　　出：横子珪藻土：前期更新世・群馬県，和村珪藻土（上村・小林 1983）：第三紀・長野県，由布院珪藻土（Okuno 1952）：鮮新世または更新世・大分県，入小野珪藻土（Okuno 1952）：更新世・大分県．

図　　版：横子珪藻土標本．

Figs 1-7. *Staurosira construens* var. *binodis* (Ehrenberg) Hamilton

LM. Figs 1-4. SEM. Figs 5-7. Material from diatomite of Yokogo (Early Pleistocene), Numata City, Gunma, Japan. Scale bars: Figs 5-6=2 μm, Fig. 7=1 μm.

Figs 1-4.　Four different size valves.
Fig. 5.　External view of whole valve.
Fig. 6.　Oblique view of sibling valves showing internal view of valve and linking spines.
Fig. 7.　Enlarged view of valve apices of sibling valves showing ocellulimbi (arrows).

Plate 134

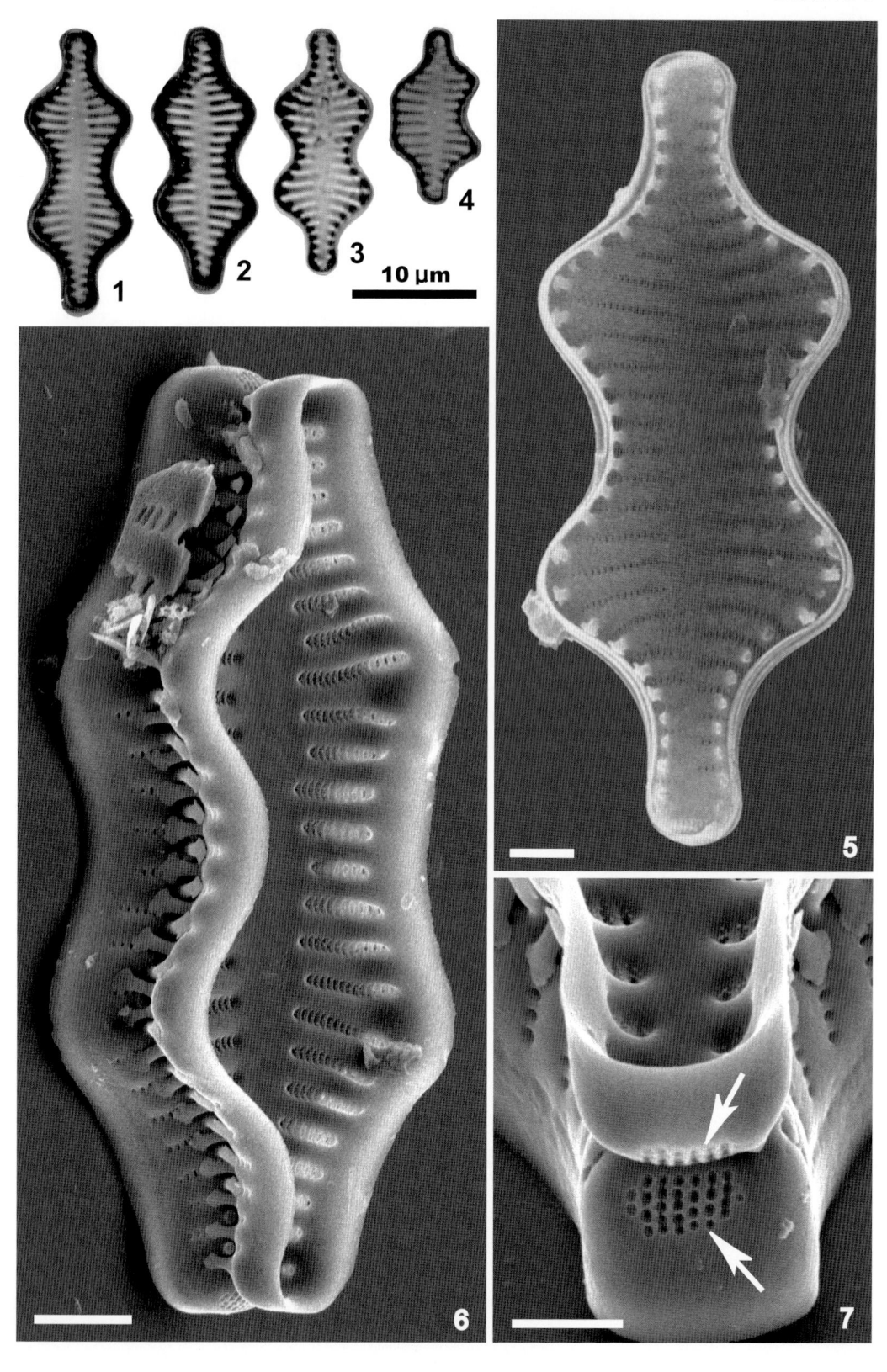

Plate 135.　Araphid, pennate diatoms: Fragilariales
Staurosira construens var. *triundulata* (H. Reichelt) H. Kobayasi in Mayama *et al.* 2002

基礎異名：*Fragilaria construens* var. *triundulata* H. Reichelt 1899

文　　献：Mayama, S., Idei, M., Osada, K. & Nagumo, T. 2002. Nomenclatural changes for 20 diatom taxa occurring in Japan. Diatom **18**: 89-91.

形　　態：殻は殻面が3回横方向へ張り出す棒状～線形で，中央の張り出しが最も大きい．両殻端はくちばし形である．殻長 15-35 μm, 殻幅 4.5 μm. 条線は平行で 10 μm に 14-16 本である．

ノ ー ト：本種は *Fragilaria* 属へ所属していたが，Mayama *et al.*(2002)により *Staurosira* へ組み合わせになった．

産　　出：野殿層（中島・南雲 1999）：中期更新世・群馬県，伊賀層（田中・松岡 1985）：鮮新世・三重県，佐山層（Negoro 1981）：鮮新世・滋賀県．

図　　版：伊賀層標本．

Figs 1-8.　*Staurosira construens* var. *triundulata* (H. Reichelt) H. Kobayasi

LM. Figs 1-3. SEM. Figs 4-8. Material from Iga Formation (Pliocene), Kobiwako Group, Iga City, Mie, Japan. Scale bars: Figs 4-6＝2 μm, Figs 7-8＝1 μm.

Figs 1-3.　Three different size valves.
Fig. 4.　External view of whole valve.
Fig. 5.　Oblique view of Fig. 4.
Fig. 6.　Oblique internal view of valve.
Fig. 7.　Enlarged view of valve center of Fig. 4.
Fig. 8.　Detailed internal view of valve center.

Plate 135

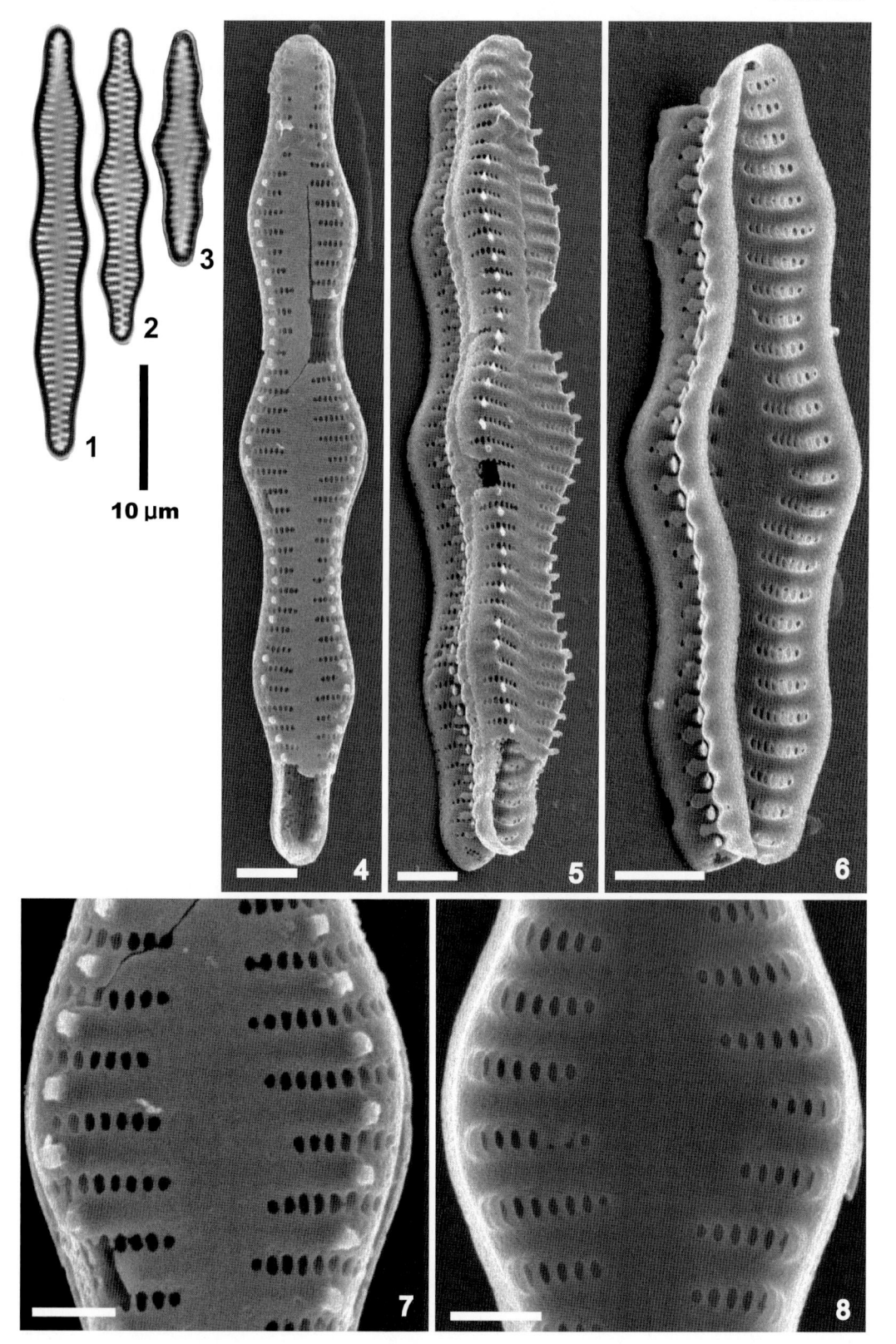

Plate 136.　Araphid, pennate diatoms: Fragilariales
Staurosirella lapponica (Grunow) D.M. Williams & Round 1987

基礎異名：*Fragilaria lapponica* Grunow 1881
文　　献：Williams, D.M. & Round, F.E. 1987. Revision of the Genus *Fragilaria*. Diatom Research **2**: 267-288.
形　　態：殻面は長楕円形で殻端は広円形，軸域は広い，殻長 9.5-19 μm，殻幅 5.5-7 μm．条線は太く 10 μm に約 6 本で，SEM 観察によると縦長の胞紋から構成されている．
ノ ー ト：本種は *Fragilaria* 属へ所属していたが，Williams & Round (1987) により *Staurosirella* 属へ組み合わせになった．
産　　出：小松沢層：鮮新世・北海道，鬼首層（Ichikawa 1955）：後期更新世・宮城県，野殿層（中島・南雲 1999）：中期更新世・群馬県，倉渕湖沼性堆積物（南雲ら 1998）：中期更新世・群馬県，大戸湖沼性堆積物（田中 1991）：中期更新世・群馬県，三徳層（Tanaka & Nagumo 2006）：後期中新世・鳥取県，嬉野珪藻土（Okuno 1952）：新第三紀・佐賀県．
図　　版：小松沢層標本．

Figs 1-9.　*Staurosirella lapponica* (Grunow) D.M. Williams & Round
LM. Figs 1-3. SEM. Figs 4-9. Materials from Komatsuzawa Formation (Pliocene), Rubeshibe, Kitami City, Hokkaido, Japan. Scale bars: Figs 4-6, 8＝2 μm, Figs 7, 9＝0.5 μm.

Figs 1-3.　Three different size valves.
Fig. 4.　External view of whole valve.
Fig. 5.　Oblique view of Fig. 4.
Fig. 6.　Oblique internal view of valve.
Fig. 7.　Enlarged internal view of Fig. 6 showing striae with slit-like areolae.
Fig. 8.　Oblique view of Fig. 4.
Fig. 9.　Enlarged external view of Fig. 4 showing striae with slit-like areolae.

Plate 136

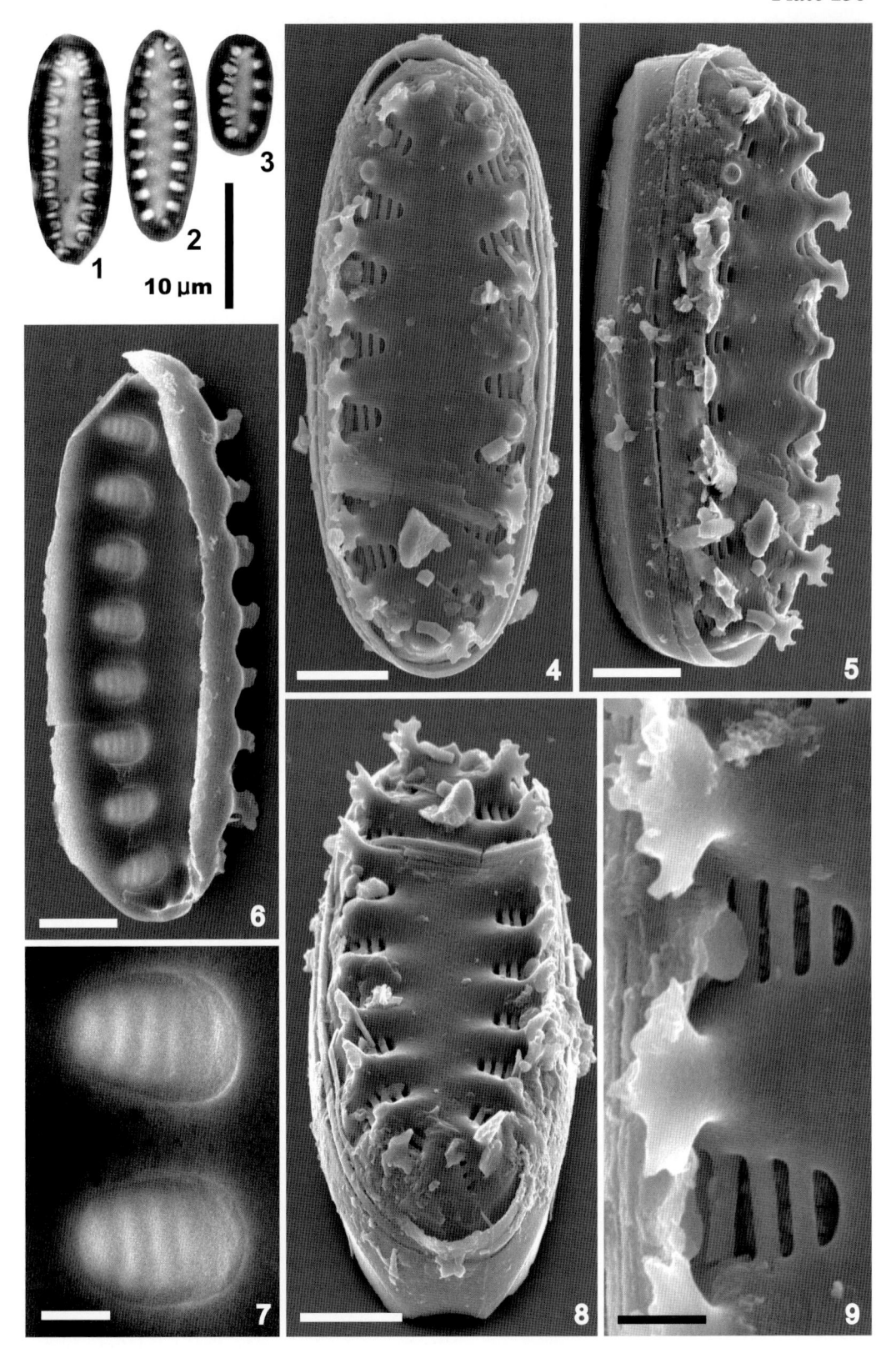

Plate 137.   Araphid, pennate diatoms: Fragilariales
Ulnaria capitata (Ehrenberg) Compère 2001

基礎異名：*Synedra capitata* Ehrenberg 1836
文　　献：Compère, P. 2001. *Ulnaria* (Kützing) Compère, a new genus name for *Fragilaria* subgen. *Alterasynedra* Lange-Bertalot with comments on the typification of *Synedra* Ehrenberg. pp. 97-101. *In*: Jahn, R., Kociolek, J.P., Witkowski, A. & Compère, P.(eds) Lange-Bertalot-Festschrift. 633 pp. A.R.G. Gantner, Ruggel.
形　　態：殻は細いが長く大形で，両端が三角形状に膨らむ．殻長 260-319 μm，殻幅は中央部で 5-6 μm，先端の膨らんでいる部分で約 8 μm，条線は 10 μm に約 8 本であった．先端の唇状突起の所在は光顕でも認めることができる．殻套眼域がある．
ノ ー ト：従来 *Synedra capitata* と同定されてきた分類群であるが，Compère(2001)により *Ulnaria* 属へ組み合わせになった．図版に使用した標本は人形峠層高清水部層産出である．藤田(1973)によれば人形峠層の最上位の部層であり，人形峠層は赤城ら(1984)によると後期中新世～鮮新世の堆積物なので，高清水部層は鮮新世であろう．
産　　出：塩原湖成層（Akutsu 1964）：前期更新世・栃木県，人形峠層（田中ら 2008）：後期中新世～鮮新世・岡山-鳥取県境，嬉野珪藻土（Okuno 1964）：新第三紀・佐賀県．
図　　版：人形峠層標本．

Figs 1-4.   *Ulnaria capitata* (Ehrenberg) Compère
LM. Fig. 1. SEM. Figs 2-4. Material from Ningyo-toge Formation (Late Miocene-Pliocene), boundary of Okayama and Tottori Prefectures, Japan. Scale bars: Fig. 2＝5 μm, Figs 3-4 ＝2 μm.

Fig. 1.   Whole valve view, photomontage, divided into three.
Fig. 2.   External oblique view of part of valve, opening of rimoportula (arrowhead).
Fig. 3.   Oblique apex view of valve, opening of rimoportula (arrowhead) and ocellulimbus (arrow).
Fig. 4.   External view of valve surface.

Plate 137

Plate 138.   Araphid, pennate diatoms: Tabellariales
Tabellaria fenestrata (Lyngbye) Kützing 1844

基礎異名：*Diatoma fenestratum* Lyngbye 1819
文　　献：Kützing, F.T. 1844. Die kieselschaligen Bacillarien order Diatomeen. Nordhausen, W. Kohne. 152 pp, pls 1-30.
形　　態：殻は殻端と中央が膨らんだ線形，殻長 50-94 μm，殻幅約 6.5 μm．条線は平行で 10 μm に 16-22 本．唇状突起は中央の膨れた部分にあり，光学顕微鏡でも観察できる．両殻端には小孔域がある．殻帯は隔壁を有する．
ノ　ー　ト：化石として産出する場合は殻が分離している場合がほとんどなので，類似する *Tabellaria flocculosa* とは軸域が中央で広がるか（*T. fenestrata* は広がらない）が識別点として有効である．
産　　出：鬼首層（Ichikawa 1955）：後期更新世・宮城県，大鷲湖沼性堆積物（田中ら 2011）：鮮新世・岐阜県，野上層（Okuno 1952）：中期更新世・大分県．
図　　版：大鷲湖沼性堆積物標本．

Figs 1-6. *Tabellaria fenestrata* (Lyngbye) Kützing
LM. Figs 1-2. SEM. Figs 3-6. Materials from lacustrine deposit of Owashi (Pliocene), Gujo City, Gifu, Japan. Scale bars: Figs 3, 5-6＝5 μm, Fig. 4＝2 μm.

Figs 1-2.　Two different size valves, rimoportula (arrowhead).
Fig. 3.　External view of valve, an apical pore field (arrow).
Fig. 4.　Enlarged view of valve center of Fig. 3, opening of rimoportula (arrowhead).
Fig. 5.　Internal oblique view of valve, rimoportula (arrowhead).
Fig. 6.　Band with septum (arrow).

Plate 138

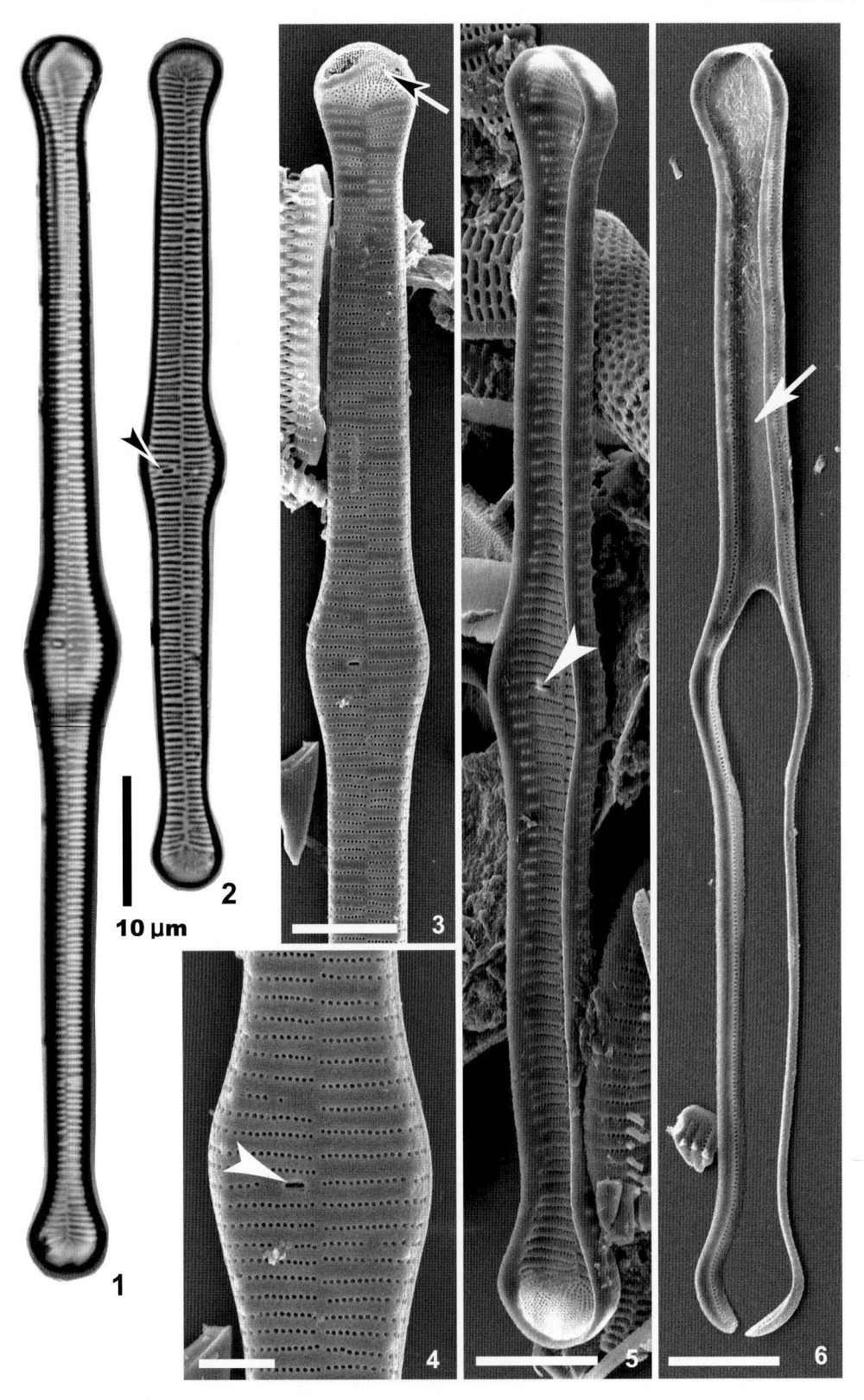

313

Plates 139-140.   Araphid, pennate diatoms: Tabellariales
Tabellaria japonica H. Tanaka sp. nov.

新　　種：記載文（英文）は 10 頁参照．
形　　態：殻は両殻端が弱く頭状〜へら状に突出した幅が狭い皮針形〜線形，殻長 39-90 µm，殻幅 4-5.5 µm．条線間隔は不規則であるが平行で 10 µm に 10-18 本．唇状突起は中央近くの条線中にある．両殻端には小孔域がある．殻帯は隔壁を有する．
ノート：殻中央部の膨らみがないことが類似する *Tabellaria fenestrata*（Lyngbye）Kütz. や *T. flocculosa*（Roth）Kütz. との区別点である．
産　　出：紫竹層：前期中新世・福島県．
図　　版 139-140：紫竹層標本．

Plate 139, Figs 1-9.   *Tabellaria japonica* H. Tanaka sp. nov.
LM. Figs 1-6. SEM. Figs 7-9. From type material, SCH-102, Shichiku Formation (Early Miocene), Iwaki City, Fukushima, Japan. Scale bars: Figs 7-9=2 µm.

Figs 1-2.   Holotype, MPC-25058, same valve at different focal planes.
Figs 3-5.   Three different size valves.
Fig. 6.   Band with septum.
Fig. 7.   External view of whole valve.
Fig. 8.   Band with septum (arrow).
Fig. 9.   External oblique view of valve apices showing apical pore fields (arrows).

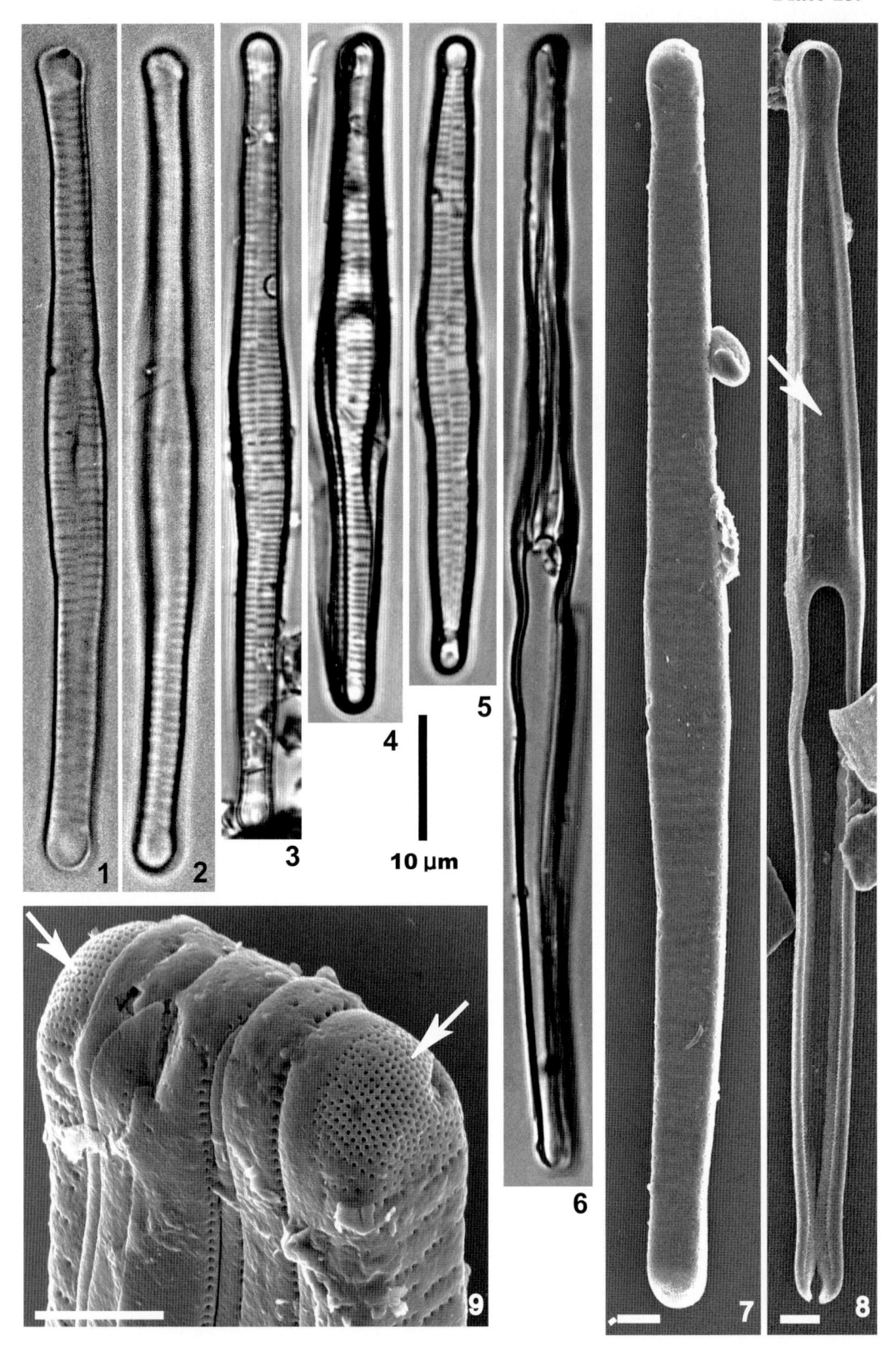

Plate 139

10 μm

315

Plate 140, Figs 1–5. *Tabellaria japonica* **H. Tanaka** sp. nov.

SEM. Figs 1–5. From type material, SCH-102, Shichiku Formation (Early Miocene), Iwaki City, Fukushima, Japan. Scale bars: Figs 1–2＝2 µm, Figs 3–4＝1 µm, Fig. 5＝0.5 µm.

Fig. 1. External oblique view of valve.
Fig. 2. Internal oblique view of valve.
Fig. 3. Detailed view of valve center showing opening of rimoportula (arrowhead).
Fig. 4. Enlarged view of Fig. 2 showing rimoportula (arrowhead).
Fig. 5. Detailed internal oblique view of valve apex showing apical pore field (arrow).

Plate 140

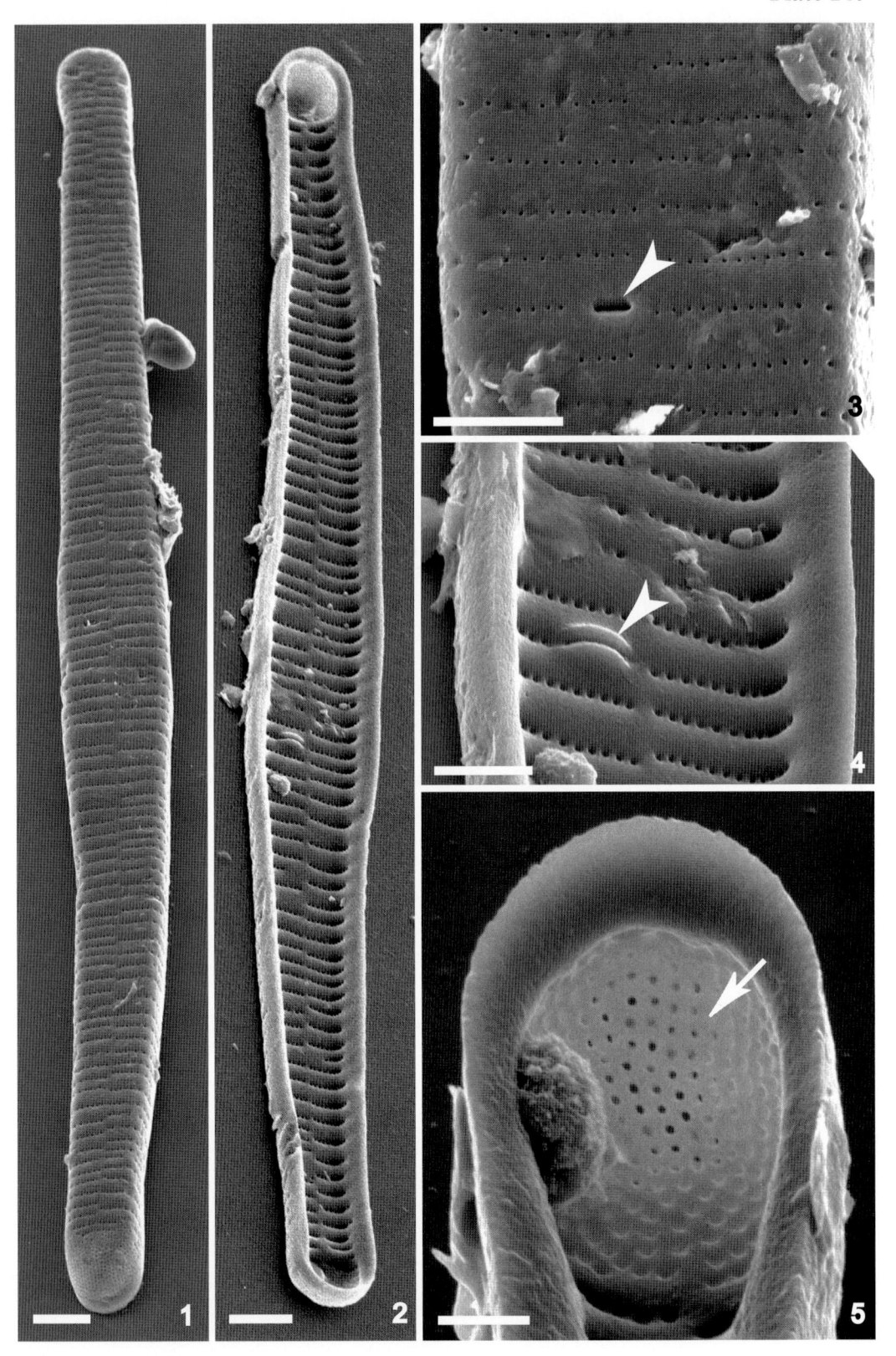

Plate 141. Araphid, pennate diatoms: Tabellariales
Tetracyclus castellum (Ehrenberg) Grunow 1862

基礎異名：*Biblarim castellum* Ehrenberg 1843
文　　献：Zimmermann, C., Poulin, M. & Pienitz, R. 2010. Diatoms of North America: The Pliocene-Pleistocene freshwater flora of Bylot Island, Nunavut, Canadian High Arctic. 407 pp. *In*: Lange-Bertalot, H.(ed.) Iconographia Diatomologica **21**: A.R.G. Gantner Verlag, Ruggell.
形　　態：殻は八方に突出部があり先端は丸みを帯びている．殻長 24-30 µm，殻幅 22-26 µm．肋線はふつう殻の左右に連続しているものは 4-5 本，途中で終了してしまうものは両側とも 3-5 本である．
ノ ー ト：筆者は人形峠層（高清水部層）の珪藻群集を調査した折には *T. stellare* var. *eximius* と同定したが（田中ら 2008），Zimmermann *et al.*(2010)によれば，*T. stellare* var. *eximius* は *T. castellum* の異名であるので，これに従った．
産　　出：人形峠層（田中ら 2008）：後期中新世～鮮新世・岡山-鳥取県境．
図　　版：人形峠層標本．

Figs 1-6. *Tetracyclus castellum* (Ehrenberg) Grunow
LM. Figs 1-3. SEM. Figs 4-6. Material from Ningyo-toge Formation (Late Miocene-Pliocene), boundary of Okayama and Tottori Prefectures, Japan. Scale bars: Fig. 4=5 µm, Figs 5-6=2 µm.

Figs 1-2.　Two different valves.
Fig. 3.　Band with septa.
Fig. 4.　External oblique view of whole valve, opening of rimoportula (arrowhead).
Fig. 5.　Internal oblique view of whole valve, rimoportula (arrowhead).
Fig. 6.　Detailed view of external valve margin showing opening of rimoportula (arrowhead).

Plate 141

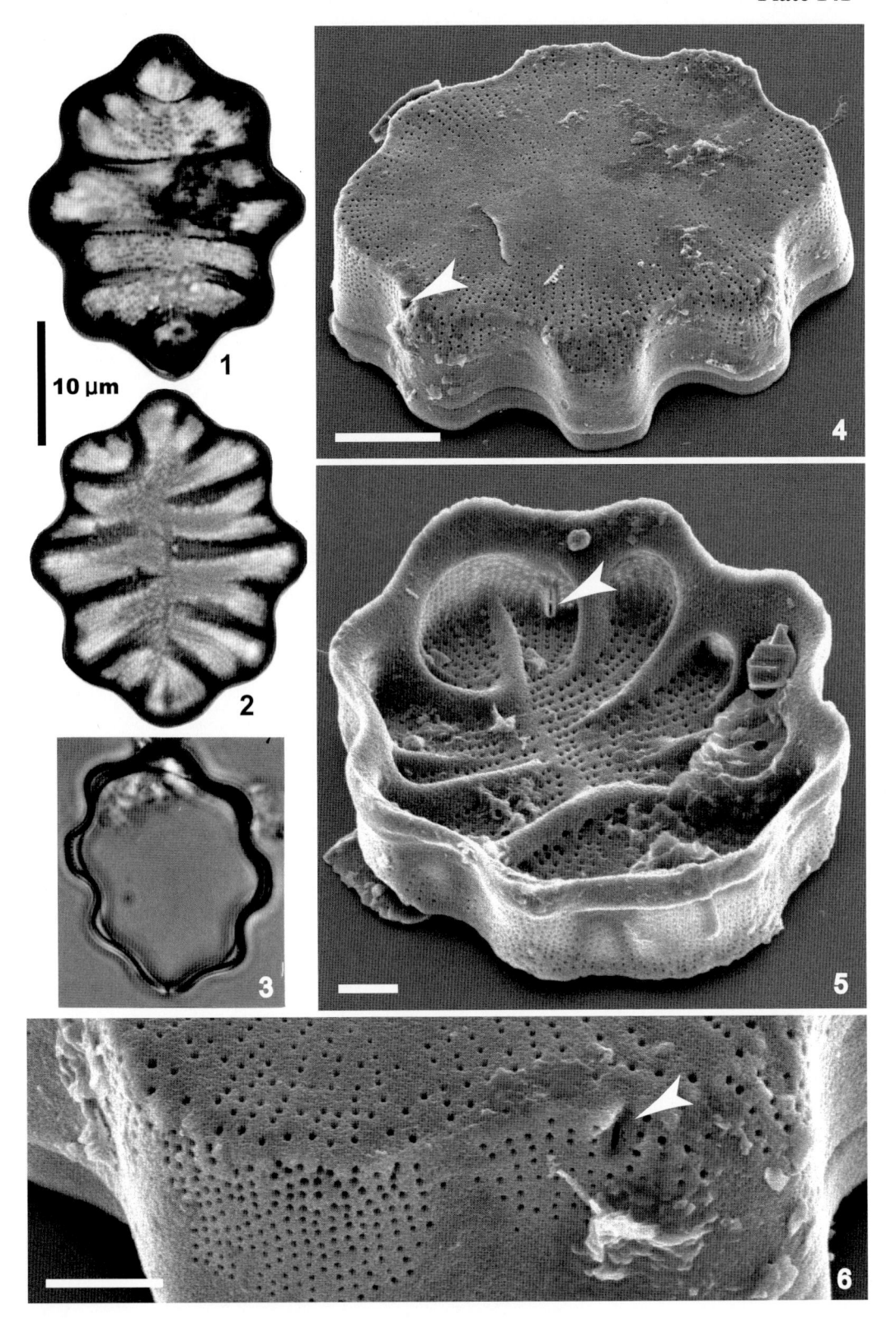

Plate 142. Araphid, pennate diatoms: Tabellariales
Tetracyclus cruciformis Andrews 1970

文　　献：Andrews, G.W. 1970. Late Miocene nonmarine diatoms form the Kilgore area, Cherry County, Nebraska: U.S. Geologocal Survey Professional Paper 683-A: 24, 3 pls.

形　　態：殻は十字形で四方向に強く突出し，先端は広円形である．殻長は上下，左右とも34-76 μm．肋は 10 μm に 2(3) 本で，中心を通るものは，ふつう両殻縁まで連続しない．

ノ　ー　ト：類似した，殻が十字形の種類は他に *Tetracyclus floriformis* があるが，この種名はAndrews(1970)によると命名規約上の不備があるので無効である．太櫓層から見出した本種は中心部に唇状突起が観察されるものとないものが見られた．

産　　出：太櫓層（瀬棚の珪藻土）：前期中新世・北海道．

図　　版：太櫓層標本．

Figs 1-6. *Tetracyclus cruciformis* Andrews
LM. Fig. 1. SEM. Figs 2-6. Material from Futoro Formation "diatomite of Setana" (Early Miocene), Setana Town, Hokkaido, Japan. Scale bars: Figs 2-3＝10 μm, Figs 4-6＝5 μm. Figs 1-4, 6, same valve.

Fig. 1.　Valve view.
Fig. 2.　External whole valve view.
Fig. 3.　Internal oblique view of whole valve, note: no rimoportulae on center.
Fig. 4.　Part of enlarged view of Fig. 3.
Fig. 5.　Internal oblique view of valve center, rimoportula (arrowhead).
Fig. 6.　Enlarged oblique view of part of Fig. 2.

Plate 142

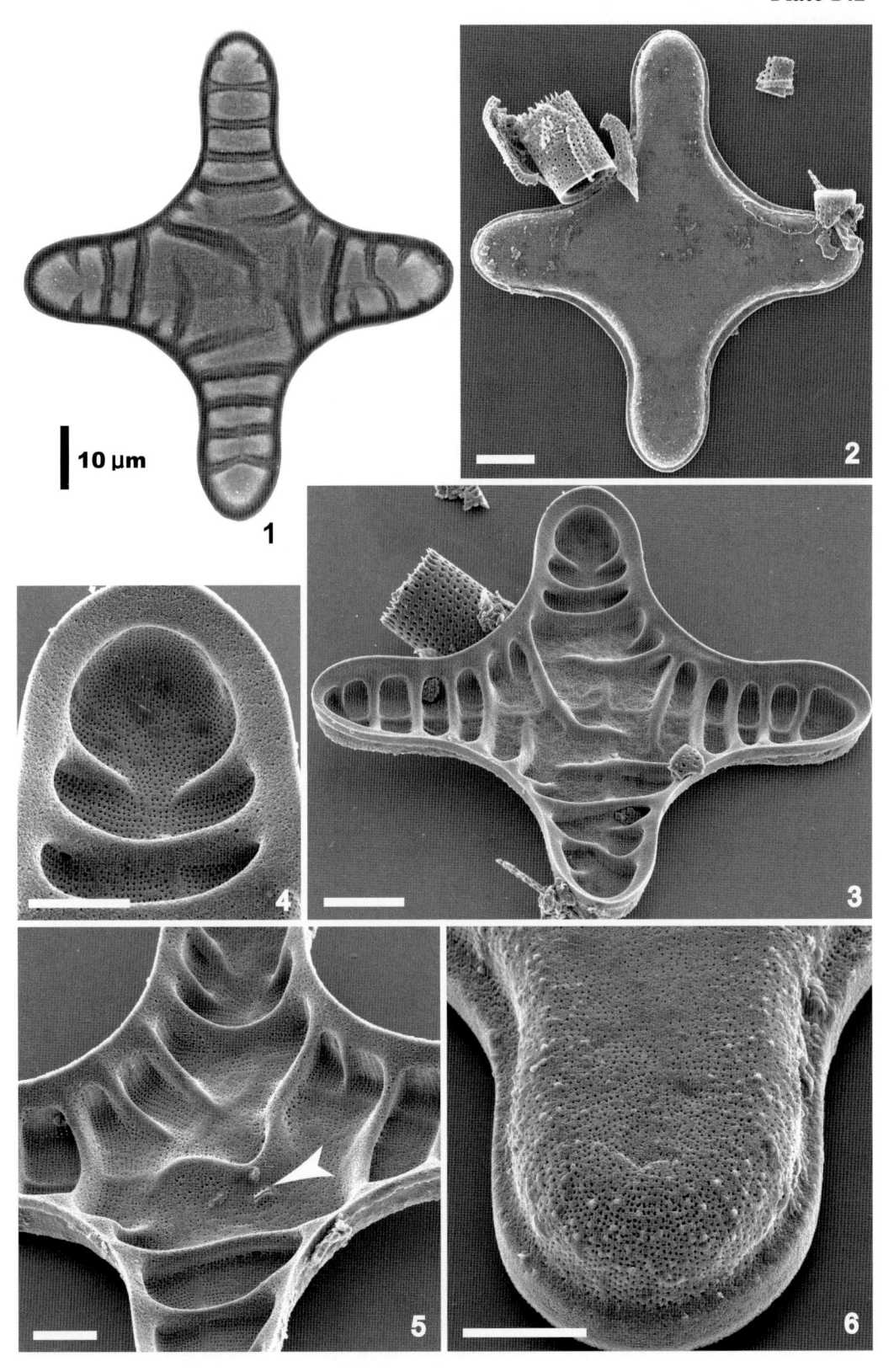

Plate 143. Araphid, pennate diatoms: Tabellariales
Tetracyclus ellipticus (Ehrenberg) Grunow var. *ellipticus* 1862

基礎異名：*Biblarim ellipticum* Ehrenberg 1843
文　　献：Hustedt, F. 1912. *In*: A. Schmidt's Atlas der Diatomaceen-Kunde(1874-1959), pls 280/9-15, 281/24. O.R. Reisland, Leipzig.
形　　態：殻は楕円形．殻長 13-49 μm，殻幅 9-18 μm．肋線は殻の左右に連続しているものは 2-6 本，途中で終了してしまうものは両側とも 0-2 本である．唇状突起は 1(2)個．
ノ ー ト：殻が楕円形の個体を本分類群へ同定した．
産　　出：太櫓層（瀬棚の珪藻土）（奥野 1959）：前期中新世・北海道，紫竹層：前期中新世・福島県．
図　　版：紫竹層標本．

Figs 1-7. *Tetracyclus ellipticus* (Ehrenberg) Grunow var. *ellipticus*
LM. Figs 1-3. SEM. Figs 4-7. Materials from Shichiku Formation (Early Miocene), Iwaki City, Fukushima, Japan. Scale bars: Figs 4-5＝5 μm, Figs 6-7＝2 μm.

Figs 1-3.　Three different size valves.
Fig. 4.　External view of whole valve, opening of rimoportula (arrowhead).
Fig. 5.　Band with septum.
Fig. 6.　Internal view of valve, rimoportula (arrowhead).
Fig. 7.　Enlarged oblique view of Fig. 4, opening of rimoportula (arrowhead).

Plate 143

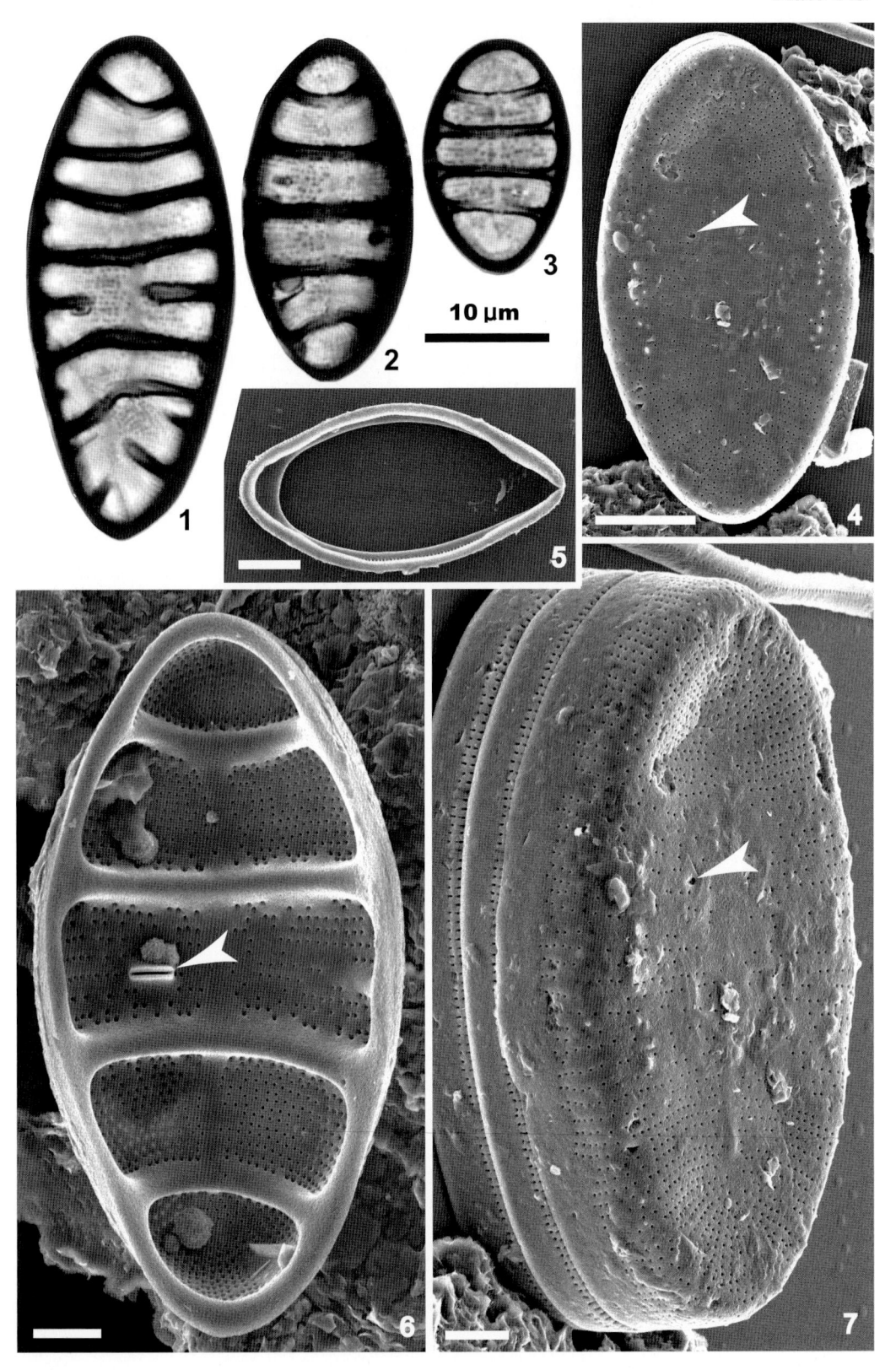

10 μm

Plate 144.   Araphid, pennate diatoms: Tabellariales
Tetracyclus ellipticus var. *constricta* Hustedt
in Schmidt *et al.* 1912

文　　献：Hustedt, F. 1912. *In*: A. Schmidt's Atlas der Diatomaceen-Kunde (1874-1959), pl. 281/9. O.R. Reisland, Leipzig.

形　　態：殻は線形，殻面は丸みを帯びている．殻面観において中心殻縁が短軸方向に凹み，殻端は広円形である．殻長69-102 μm，殻幅17-22 μm．肋線はほとんどが殻の左右に連続しており10 μmに約2本，条線は10 μmに約20本である．唇状突起は1個であった．

ノ ー ト：殻は比較的大形で殻面は丸みを帯び，殻面観において中心が短軸方向に凹むことが特徴で，Williams (1990) のとおり初生殻と思われる．太櫓層から産出する他の栄養細胞の初生殻と思われるが，特定が難しいのでHustedt (1912) で記述された分類群名で記載する．

産　　出：太櫓層（瀬棚の珪藻土）：前期中新世・北海道.

図　　版：太櫓層標本.

Figs 1-4.   *Tetracyclus ellipticus* var. *constricta* Hustedt
LM. Figs 1-2. SEM. Figs 3-4. Material from Futoro Formation "diatomite of Setana" (Early Miocene), Setana Town, Hokkaido, Japan. Scale bars: Figs 3-4=5 μm.

Figs 1-2.   Same valve at different focal planes, rimoportula (arrowhead).
Fig. 3.   External view of whole valve.
Fig. 4.   Internal view showing rimoportula (arrowhead).

Plate 144

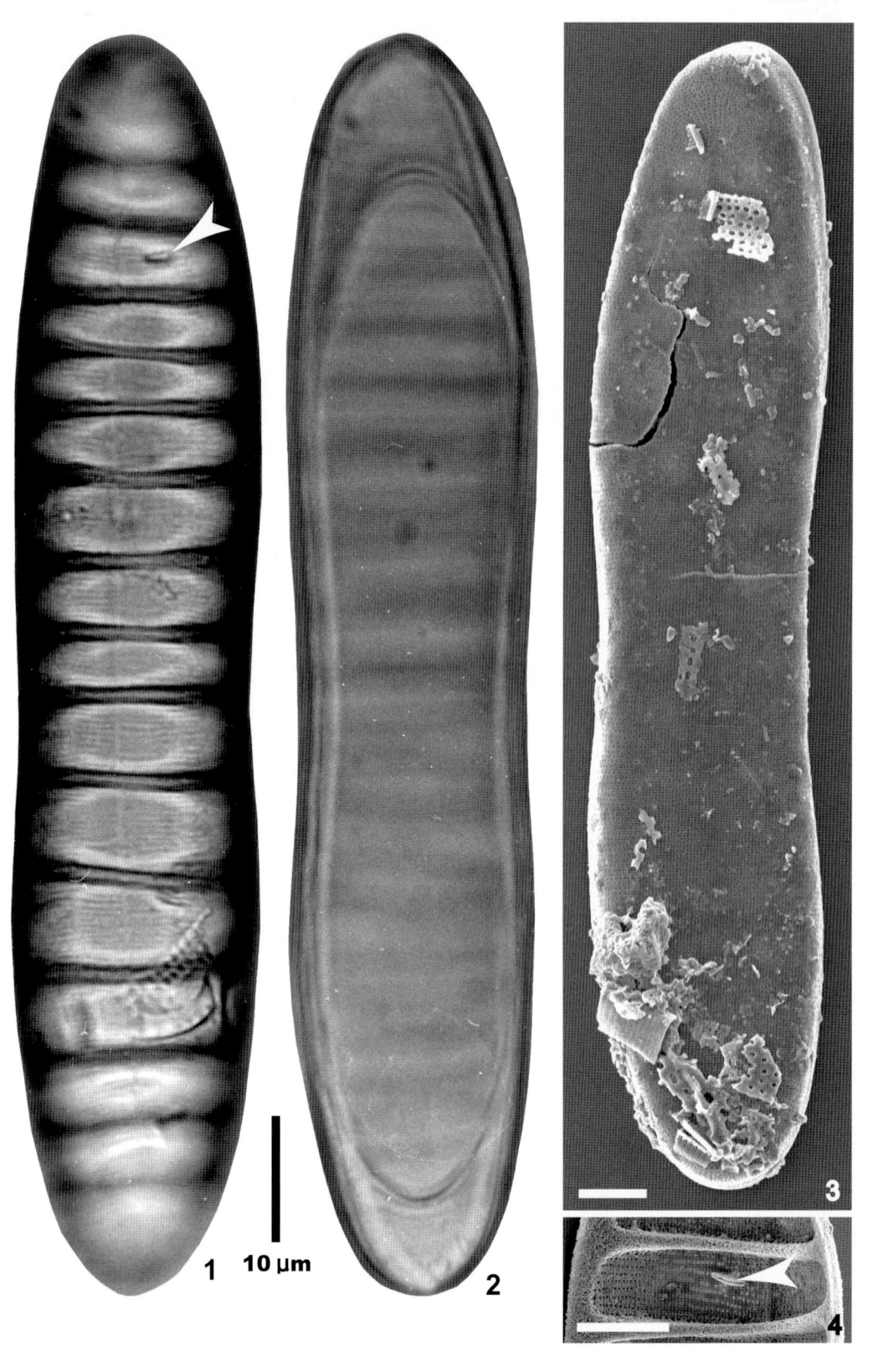

1 10 µm

325

Plate 145.   Araphid, pennate diatoms: Tabellariales
Tetracyclus ellipticus var. *lancea* f. *subrostrata* Hustedt
in A. Schmidt *et al.* 1912

文　　献：Hustedt, F. 1912. *In*: A. Schmidt's Atlas der Diatomaceen-Kunde（1874-1959），
　　pl. 281/17-18. O.R. Reisland, Leipzig.
形　　態：殻は楕円形であるが長軸方向の先端がわずかくちばし状である．殻長 14-36 μm，
　　殻幅 10-20 μm．肋線は殻の左右に連続しているものは 2-4 本，途中で終了してしまうもの
　　は両側とも 0-2 本である．唇状突起は殻面に 1 個．
ノ ー ト：殻の長軸方向の先端がややくちばし状であるのが本分類群の特徴である．
産　　出：紫竹層：前期中新世・福島県．
図　　版：紫竹層標本．

Figs 1-8. *Tetracyclus ellipticus* var. *lancea* f. *subrostrata* **Hustedt**
LM. Figs 1-3. SEM. Figs 4-8. Material from Shichiku Formation (Early Miocene), Iwaki
　　City, Fukushima, Japan. Scale bars: Figs 4-5, 7＝5 μm, Figs 6, 8＝2 μm.

Figs 1-3.　　Three different size valves.
Fig. 4.　　External view of whole valve.
Fig. 5.　　Enlarged view of valve center of Fig. 4.
Fig. 6.　　Enlarged oblique view of an apex of Fig. 4.
Fig. 7.　　Internal view of small valve, rimoportula (arrowhead).
Fig. 8.　　Enlarged oblique view of part of Fig. 7 showing rimoportula.

Plate 145

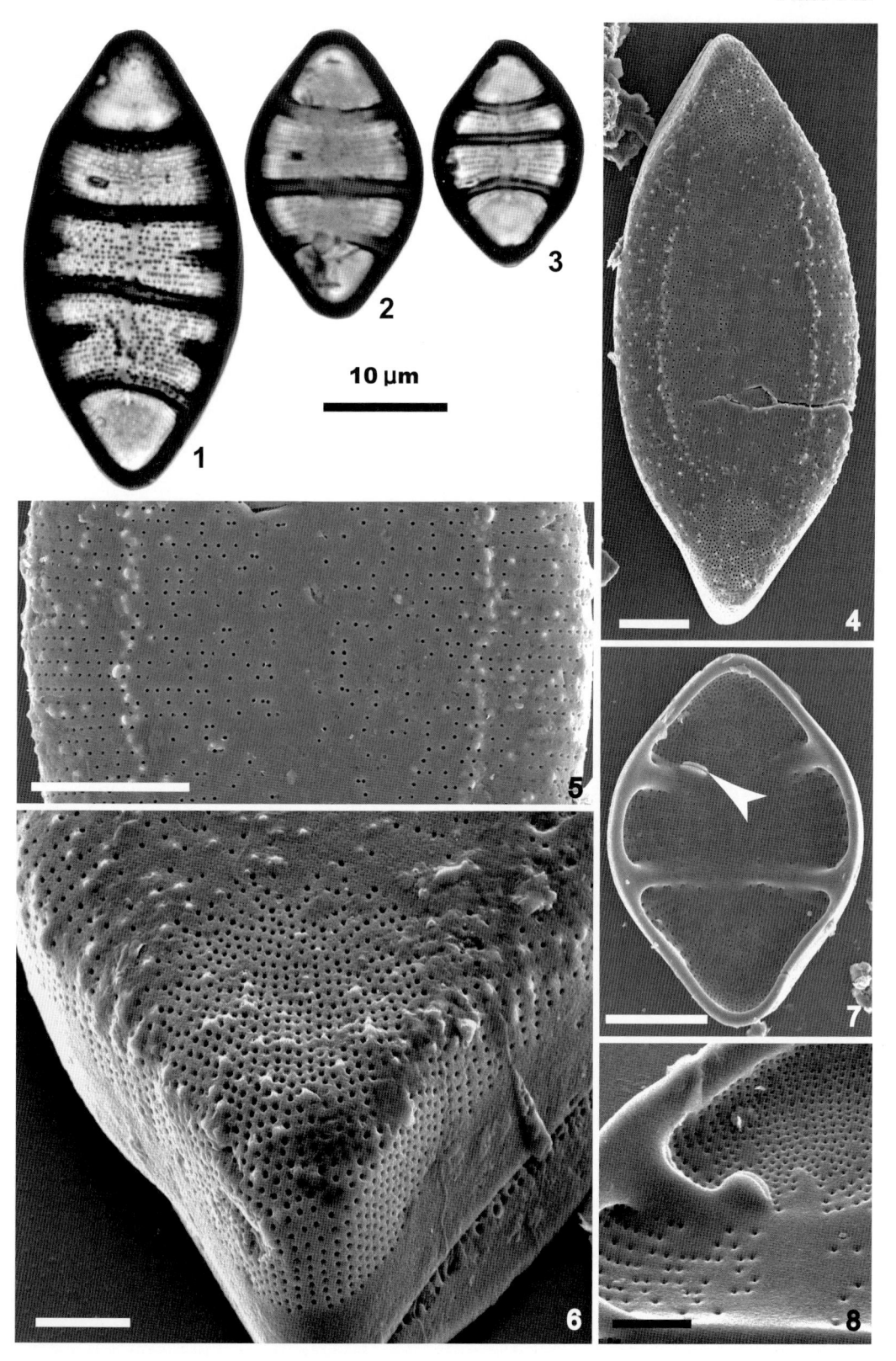

10 μm

Plate 146. Araphid, pennate diatoms: Tabellariales
Tetracyclus ellipticus var. *latissima* f. *minor* Hustedt in A. Schmidt *et al.* 1912

文　　献：Hustedt, F. 1912. *In*: A. Schmidt's Atlas der Diatomaceen-Kunde (1874-1959), pl. 281/22. O.R. Reisland, Leipzig.
　　Simonsen, R. 1987. Atlas and catalogue of the diatom types of Friedrich Hustedt. Vol. **1**: Catalogue. pp. 1-525. Vol. **2**: Atlas, Taf. 1-395. Vol. **3**: Atlas, Taf. 396-772. J. Cramer Berlin/Stuttgart.

形　　態：殻は小形で円形からやや楕円形．殻長 17-20 μm，殻幅 16-20 μm．肋線は殻の左右に連続しているものは 1-3 本，途中で終了してしまうものは両側とも約 2 本である．唇状突起は 1 個で殻面に所在する．

ノ ー ト：殻が小形なので上記文献の図から本分類群へ同定したが，瀬棚産の分類群は Hustedt (1912) の図よりも放射肋の数が多い．放射肋の形態だけからは *Tetracyclus clypeus* (Ehrenb.) Li により近いと思われるが，Williams (1989) によれば *T. clypeus* は唇状突起が殻套に所在するので，これには同定しづらい．

産　　出：太櫓層（瀬棚の珪藻土）：前期中新世・北海道．

図　　版：太櫓層標本．

Figs 1-7. *Tetracyclus ellipticus* var. *latissima* f. *minor* Hustedt
LM. Fig. 1. SEM. Figs 2-7. Material from Futoro Formation "diatomite of Setana" (Early Miocene), Setana Town, Hokkaido, Japan. Scale bars: Figs 2, 6=5 μm, Figs 4, 7=2 μm, Figs 3, 5=1 μm.

Fig. 1.　Valve view.
Fig. 2.　External view of whole valve.
Fig. 3.　Enlarged view of part of Fig. 2 showing opening of rimoportula (arrowhead).
Fig. 4.　Enlarged oblique view of Fig. 2, opening of rimoportula (arrowhead).
Fig. 5.　Detailed view of part of Fig. 6 showing rimoportua.
Fig. 6.　Internal view of whole valve.
Fig. 7.　Enlarged oblique view of Fig. 6.

Plate 146

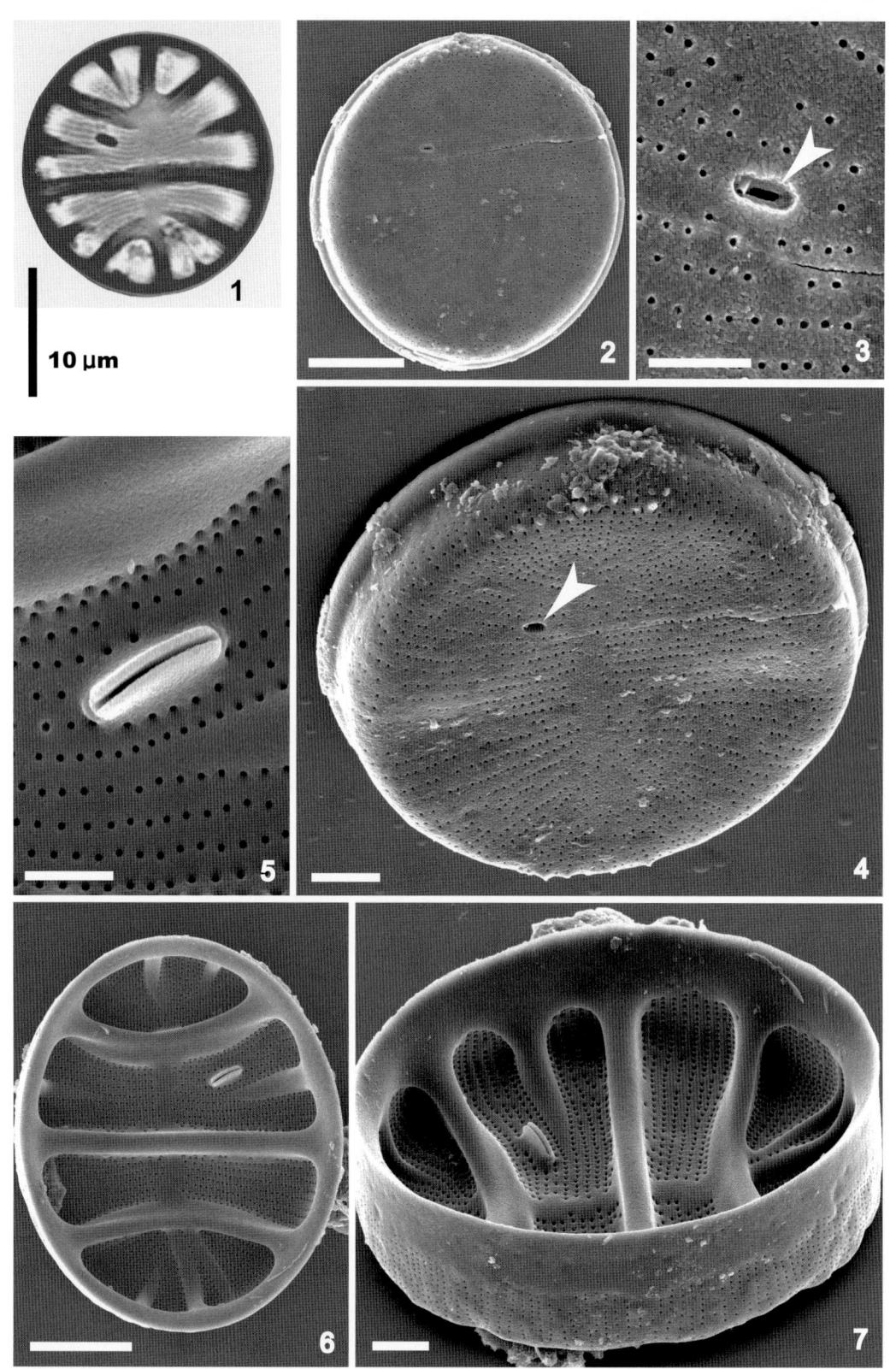

10 µm

329

Plate 147.　Araphid, pennate diatoms: Tabellariales
Tetracyclus emarginatus (Ehrenberg) W. Smith 1856

基礎異名：*Biblarim emarginatum* Ehrenberg 1854
文　　献：Smith, W. 1856. A synopsis of the British Diatomaceae. Vol. 2. 107 p, pls 32-60. Supplementary pls 61-62, pls A-E. John van Voorst, London.
形　　態：殻は縦に長い十字形であるが，中央が凹むので横に2回膨らむ形状になる．殻長26-77 μm, 殻幅21-40 μm. 肋線は10 μmに約3本，途中で終了してしまう肋も見受けられる．唇状突起は1個で条線中にあることが多く，頭部から中央部までの広い範囲に所在する．
ノ ー ト：横に膨らむ突出の程度はさまざまであるが，本書に用いた宮田層の個体は突出程度の大きいものである．唇状突起の所在はWilliams(1987)の記述のとおりさまざまであった．
産　　出：宮田層：中新-鮮新世・秋田県，鬼首層（Ichikawa 1955）：後期更新世・宮城県，大戸湖成堆積物（田中 1991）：中期更新世・群馬県，大鷲湖沼性堆積物（田中ら 2011）：鮮新世・岐阜県，琵琶湖底堆積物（Mori 1975）：更新世・滋賀県，奄芸層群（根来・後藤 1981）：鮮新世・三重県，人形峠層：後期中新世～鮮新世・岡山-鳥取県境，西瀬珪藻土（Okuno 1952）：更新世・熊本県．
図　　版：宮田層標本．

Figs 1-4.　*Tetracyclus emarginatus* (Ehrenberg) W. Smith
LM. Figs 1-2. SEM. Figs 3-4. Materials from Miyata Formation (Mio-Pliocene), Senboku City, Akita, Japan. Scale bars: Figs 3-4=10 μm.

Fig. 1.　Valve view.
Fig. 2.　Band with septa.
Fig. 3.　External oblique view of whole valve.
Fig. 4.　Internal view of whole valve, rimoportula (arrowhead).

Plate 147

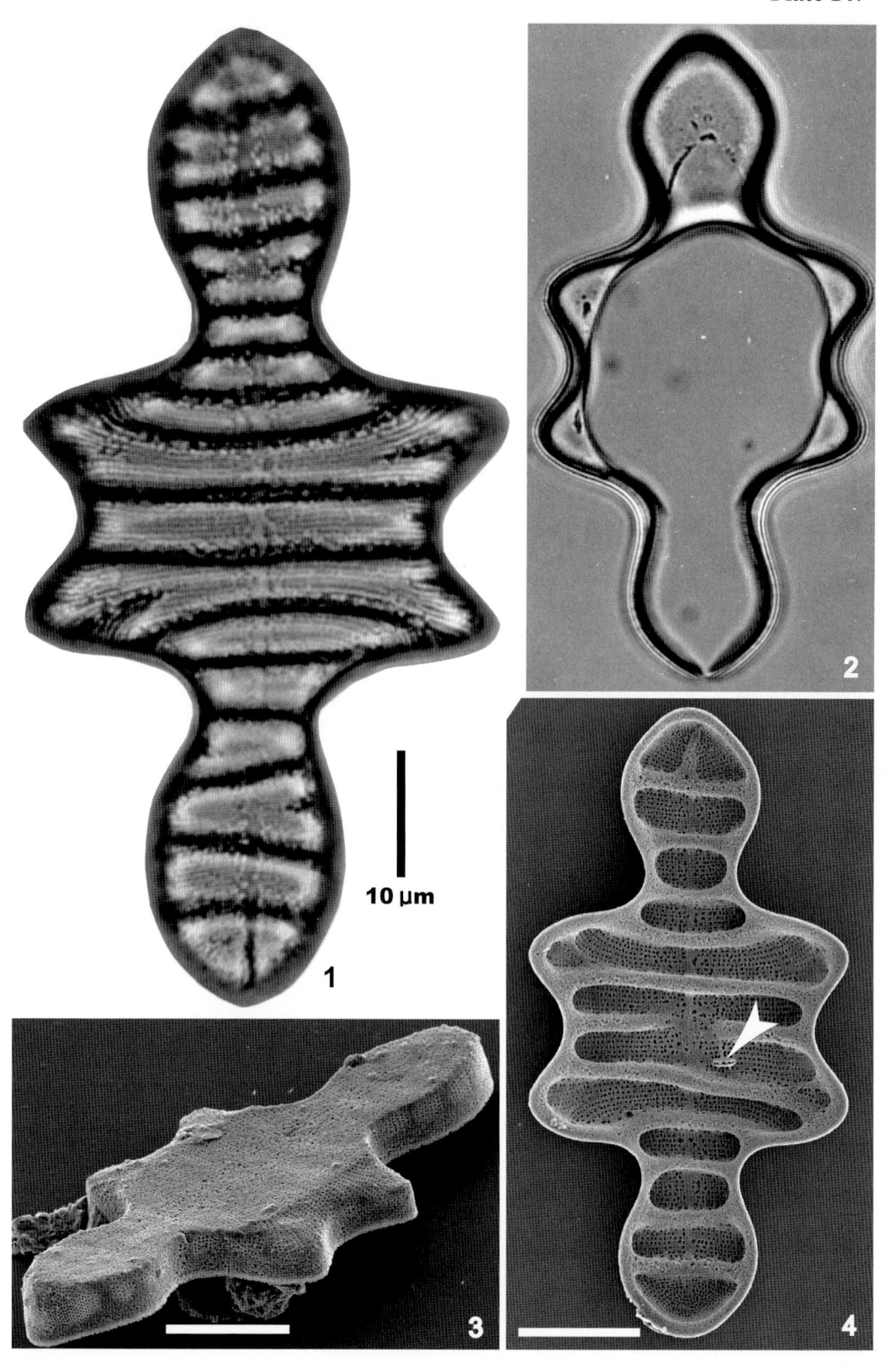

10 µm

Plate 148.　Araphid, pennate diatoms: Tabellariales
Tetracyclus glans (Ehrenberg) Mills 1935

基礎異名：*Navicula* ? *glans* Ehrenberg 1838
文　　献：Williams, D.M. 1987. Observations on the genus *Tetracyclus* Ralfs (Bacillariophyta) I. Valve and girdle structure of the extant species. British Phycological Journal **22**: 383-399.
形　　態：殻は十字形で長軸方向の突出部はややくさび形であることが多いが，他は広円形である．殻長22-53 μm，殻幅14.5-29 μm．肋線は殻の左右に連続しているもの5-14本，連続しないものは左右とも約2本が見られた．肋線間は細かい点紋から構成される条線が走る．唇状突起は観察できなかった．
ノート：Williams(1987)によれば，唇状突起について "No rimoportula can be seen" と記してある．尾瀬沼から見出された本分類群は *Tetracyclus lacustris* Ralfs にも類似するがSEMを使用していろいろ角度を変えて観察したが，唇状突起の存在が確認できなかったので本種に同定した．
産　　出：尾瀬沼（田中・中島 1983）：完新世・群馬県．
図　　版：尾瀬沼標本．

Figs 1-6.　*Tetracyclus glans* (Ehrenberg) Mills
LM. Figs 1-3. SEM. Figs 4-6. Material from Lake Oze-numa (Recent), boundary of Gunma and Fukushima, Japan. Scale bars: Figs 4-6＝5 μm.

Figs 1-3.　Three different size valves.
Fig. 3.　Internal view showing septum (arrow).
Fig. 4.　External view of whole valve.
Fig. 5.　Oblique view of Fig. 4.
Fig. 6.　Internal oblique view of valve.

Plate 148

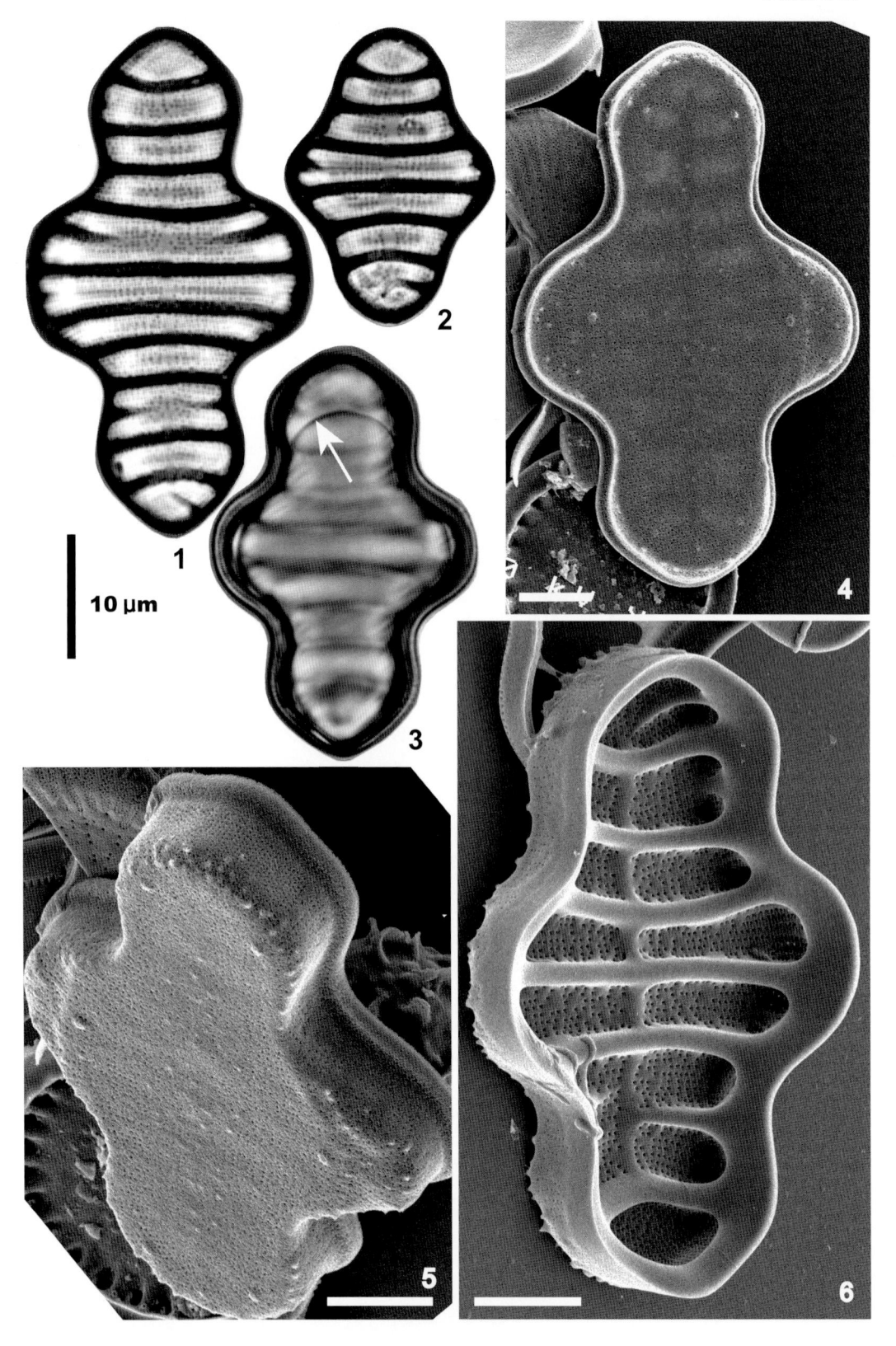

10 μm

333

Plate 149. Araphid, pennate diatoms: Tabellariales
Tetracyclus lacustris Ralfs 1843

文　　献：Ralfs, J. 1843.　On the Diatomaceae. Annals and Magazine of Natural History **12**: 104-111.

形　　態：殻は中央が幅広く膨らむ十字形で突出部は広円形である．殻長 45-60 µm，殻幅 16-28 µm．肋線は殻の左右に連続しているもの 7-11 本，連続しないものは左右とも 1-2 本が見られた．肋線間は細かい点紋から構成される条線が走る．唇状突起は 1-2 個で殻面にある．

ノ　ー　ト：類似種に *Tetracyclus stella* (Ehr.) Hérib. があるが，*T. lacustris* の発表が早い．*T. glans* とは唇状突起が明瞭であることから区別できる．

産　　出：太櫓層（瀬棚の珪藻土）（奥野 1959）：前期中新世・北海道，奄芸層群（根来・後藤 1981）：鮮新世・三重県．

図　　版：太櫓層標本．

Figs 1-4.　*Tetracyclus lacustris* Ralfs
LM. Fig. 1. SEM. Figs 2-4. Material from Futoro Formation "diatomite of Setana" (Early Miocene), Setana Town, Hokkaido, Japan. Scale bars: Fig. 3＝10 µm, Figs 2, 3＝5 µm.

Fig. 1.　Valve view, rimoportula (arrowhead).
Figs 2-4.　Same valve, external and internal SEM views.
Fig. 2.　External view of whole valve, openings of rimoportulae (arrowheads).
Fig. 3.　Internal view, opening of rimoportulae (arrowheads).
Fig. 4.　Enlarged view of rimoportulae of Fig. 3.

Plate 149

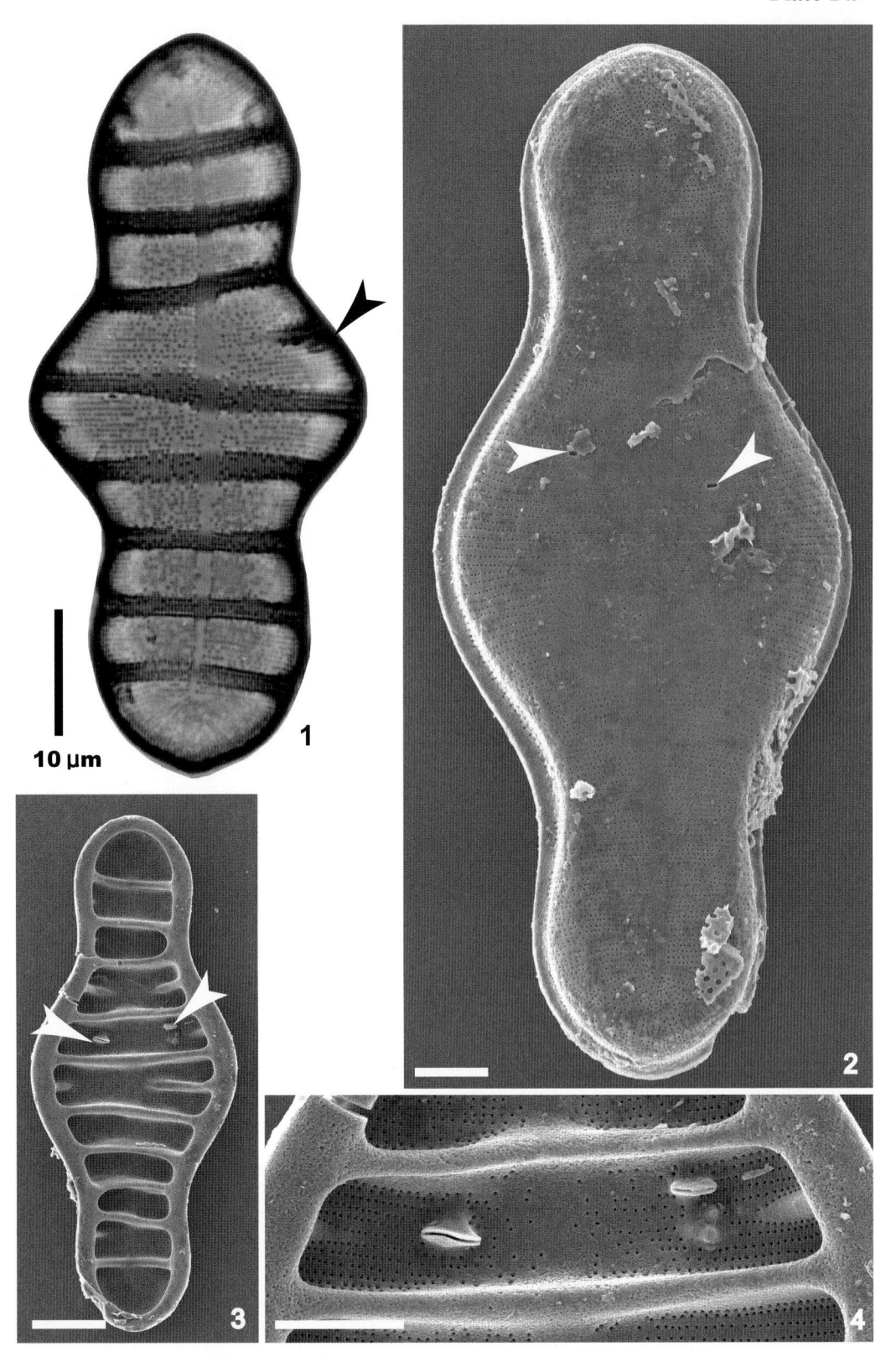

10 μm

Plate 150. Monoraphid, pennate diatoms: Achnanthales
Achnanthes exigua var. *angustirostrata* (Krasske) Lange-Bertalot in Lange-Bertalot & Krammer 1989

基礎異名：*Achnanthes exigua* var. *heterovalvata* f. *angustirostrata* Krasske 1939

文　献：Lange-Bertalot, H. & Krammer, K. 1989. *Achnanthes* eine monographie der gattung, mit definition der gattung *Cocconeis* und nachträgen zu den Naviculaceae. Bibliotheca Diatomologioca Bd. **9**: 1-393.

形　態：被殻は縦溝殻と無縦溝殻から構成され，楕円形．両殻端はくちばし状に突出する．殻長 13-17 μm，殻幅 7-8.5 μm，縦溝殻の条線は 10 μm に約 28 本であるが，無縦溝殻では 20-24 本であった．外側縦溝の両殻端末端は反対方向に曲がって終了するが，中心ではわずかであるが同じ方向へ曲がって終了する．縦溝殻の中心域は横に広がり殻縁に達する．無縦溝殻では，片側は殻縁に達するが他方の広がりはわずかで殻縁には達しない．中心域の内側は肥厚している．軸域も肥厚しており，無縦溝殻では中央で広く殻端に近づくにつれ狭くなる．

ノート：本変種は殻の両側がやや膨らみ楕円形を呈するので，本変種よりも両側が平行的であり，すでに *Achnanthidium* 属へ組み合わせになっている基本種の *A. exiguum* var. *exiguum* と区別できる．本変種は *Achnanthidium* へ未だ組み合わせになっていないので *Achnanthes* のまま掲載する．筆者らは大鷲湖成堆積物の調査を行った際は本変種を *Achnanthidium* sp. とした（田中ら 2011）．

産　出：横子珪藻土：前期更新世・群馬県，大鷲湖沼性堆積物（田中ら 2011）：鮮新世・岐阜県．

図　版：横子珪藻土（Figs 1-2, 7, 9），大鷲湖沼性堆積物（Figs 3-6, 8, 10-12）標本．

Figs 1-12. *Achnanthes exigua* var. *angustirostrata* (Krasske) Lange-Bertalot

LM. Figs 1-6. SEM. Figs 7-12. Figs 1-2, 7, 9, material from diatomite of Yokogo (Early Pleistocene), Numata City, Gunma, Japan, Figs 3-6, 8, 10-12, material from lacustrine deposit of Owashi (Pliocene), Gujo City, Gifu, Japan. Scale bars: Fig. 9＝1 μm, Figs 7-8, 10-12＝2 μm.

Figs 1-2, 3-4.　Two frustules, raphid valves (Figs 1, 3) and araphid valves (Figs 2, 4).
Fig. 5.　Raphid valve, showing extended central area.
Fig. 6.　Araphid valve, showing almost no extended central area on one side.
Fig. 7.　External view of whole raphid valve.
Fig. 8.　Internal oblique view of raphid valve.
Fig. 9.　Enlarged view of center of Fig. 7, showing central raphe endings (arrows).
Fig. 10.　Internal view of araphid valve.
Fig. 11.　Oblique view of Fig. 10.
Fig. 12.　External view of araphid valve.

Plate 150

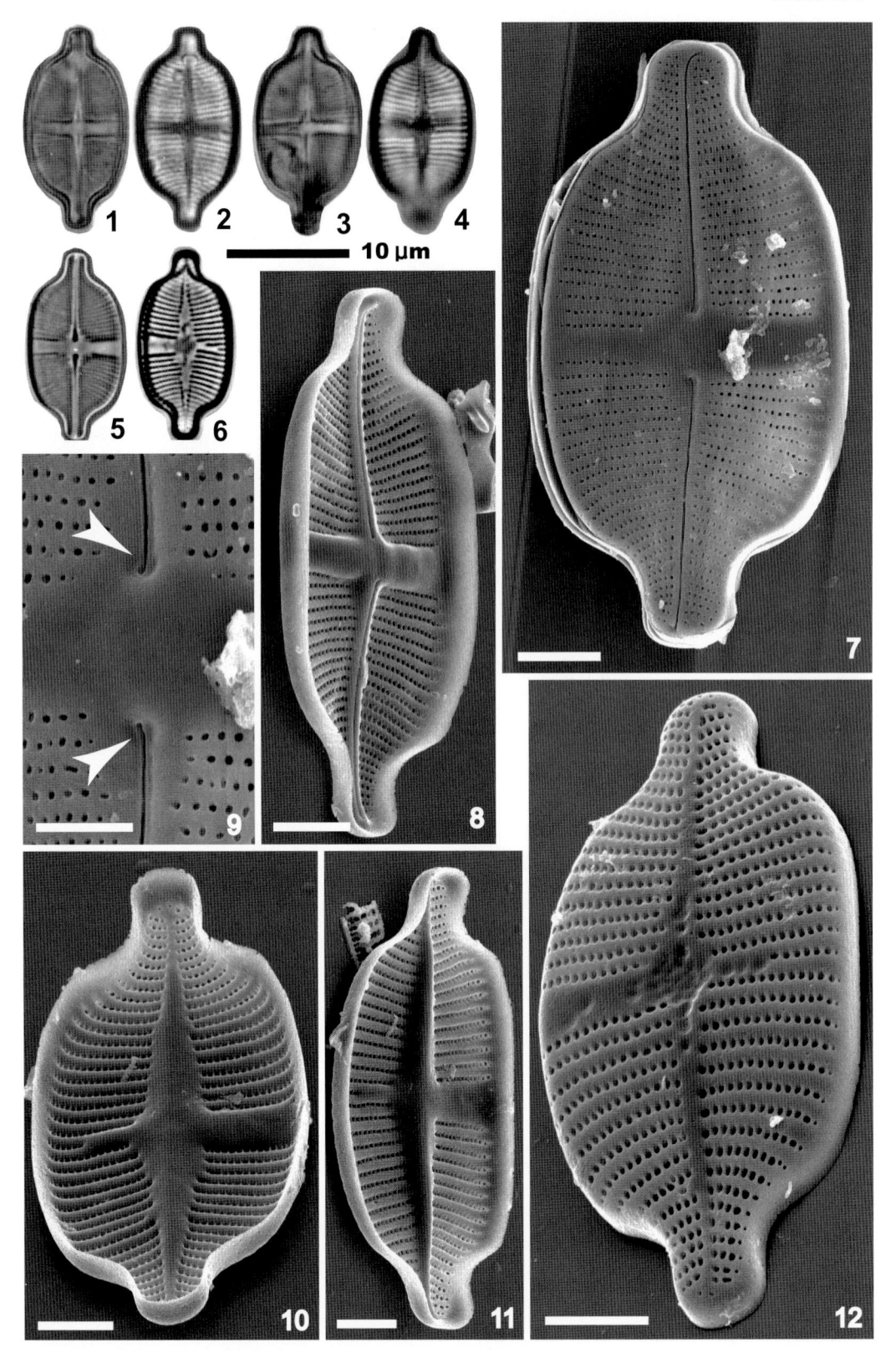

Plate 151.　Monoraphid, pennate diatoms: Achnanthales
Achnanthes obliqua (Gregory) Hustedt 1924

基礎異名：*Stauroneis oblique* Gregory 1856

文　　献：Hustedt, F. 1937.　*In*: A. Schmidt's Atlas der Diatomaceen-Kunde(1874-1959), pl. 414, Figs 1-5. O.R. Reisland, Leipzig.

形　　態：殻は広皮針形で，殻長 25.5-33 μm，殻幅 12-14 μm，条線は 10 μm に 16-20 本であった．殻面観において，縦溝殻・無縦溝殻とも軸域はわずかに長軸に斜行するが，一般に無縦溝殻の斜行は非常にわずかである．斜行に伴って縦溝殻の軸域に所在する縦溝も長軸に斜行する．中心域はやや表側では凹み，内側では膨出する．

ノ ー ト：Figs 1-2, 6 は同じ個体である．最初に SEM で Fig. 6 を撮影し，その後 LM 用のスライドに移して Figs 1-2 を撮影した．現生では八郎潟調整池から見出されている（加藤ら 1977）．

産　　出：三徳層（Tanaka & Nagumo 2006）：後期中新世・鳥取県，堅田層（根来 1981）：更新世・滋賀県，琵琶湖底堆積物（Mori 1975）：更新世・滋賀県，人吉層：後期鮮新世・熊本県．

図　　版：人吉層標本．

Figs 1-6.　*Achnanthes obliqua* (Gregory) Hustedt
LM. Figs 1-2. SEM. Figs 3-6. Material from Hitoyoshi Formation (Late Pliocene), Hitoyoshi City, Kumamoto, Japan. Scale bars: Figs 4, 6＝5 μm, Fig. 3＝2 μm, Fig. 5＝1 μm.

Figs 1-2.　A frustule, raphid valve (Fig. 1) and araphid valve (Fig. 2).

Fig. 3.　External oblique view of whole araphid valve, wide central area on one side (arrow).

Fig. 4.　Internal oblique view of raphid valve, helictoglossa (arrowhead).

Fig. 5.　Enlarged internal view of an apex with helictoglossa (arrowhead).

Fig. 6.　Internal oblique view of araphid valve (same valve as Fig. 2), wide central area on one side (arrow).

Plate 151

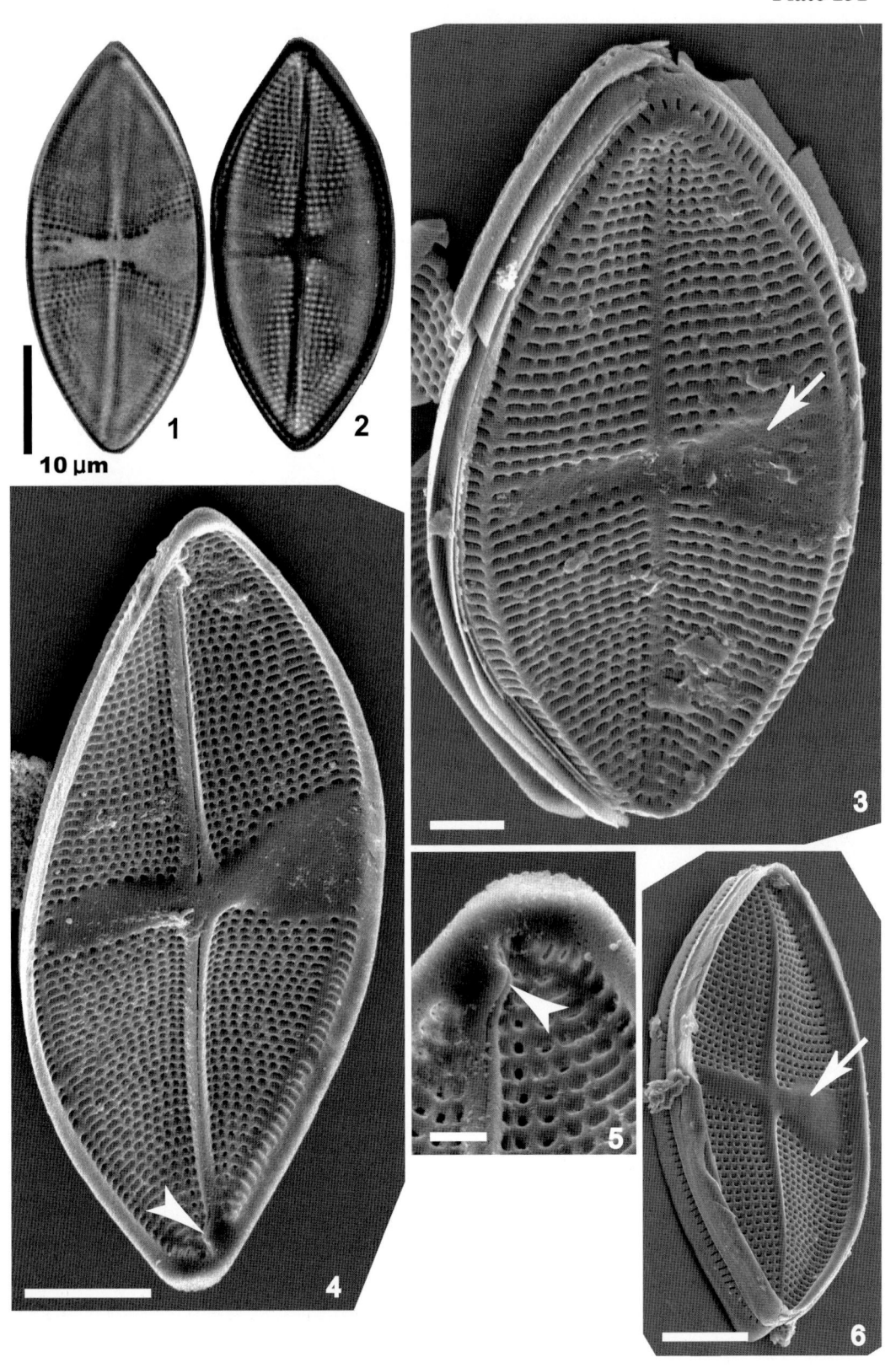

Plates 152-153. Monoraphid, pennate diatoms: Achnanthales
Achnanthes okunoi (Hustedt) H. Tanaka comb. nov.

基礎異名：*Navicula okunoi* Hustedt 1966
組み合わせ変更：16 頁参照（英文）.
文　　献：Hustedt, F. 1961-1966.　Die Kieselalgen Deutschlands, Österreichs und der Schweiz unter Berücksichtigung der übrigen Länder Europas sowie der angrenzenden Meeresgebiete.　*In*: Rabenhorst, L.(ed.) Kryptogamen-Flora **III**. pp. 1-816. Leipzip.
形　　態：殻はひし形で，殻長 71-85 μm，殻幅 31-39 μm，条線は殻端に近くなるほど放射が強くなる放射状で 10 μm に 16-18 本であった．殻面観において，殻中央部は長軸に沿って強く突出している．殻の片側中央縁辺部には無紋域がある．帯面観では殻が外側に反っていることがわかる．片方の殻には縦溝と殻端内側には蝸牛舌があり，他方には外側軸域にすじ模様が見られるが，内部まで貫通しておらず，縦溝・蝸牛舌はない．
ノ ー ト：本種は Hustedt(1966)によって *Navicula* 属として記載されたが，SEM 観察の結果，被殻の片方の殻に縦溝がない（縦溝の位置にすじ摸様があるが，これは表面のみの模様である）ことが判明した．本種は *Achnanthes* 属へ所属するのが適当であると思われる．
産　　出：太櫓層（瀬棚の珪藻土）（Hustedt 1961-1966）：前期中新世・北海道．
図　　版 152-153：太櫓層標本．

Plate 152, Figs 1-5.　*Achnanthes okunoi* (Hustedt) H. Tanaka comb. nov.
LM. Figs 1-2. SEM. Figs 3-5. Material from Futoro Formation "diatomite of Setana" (Early Miocene), Setana Town, Hokkaido, Japan. Scale bars: Figs 3-4=10 μm, Fig. 5=2 μm.

Figs 1-2.　Same valve at different focal planes.
Fig. 3.　External view of whole raphid valve, raphe (arrowhead), no areolae zone (arrow).
Fig. 4.　Oblique view of Fig. 3.
Fig. 5.　Enlarged view of valve center of Fig. 3.

Plate 152

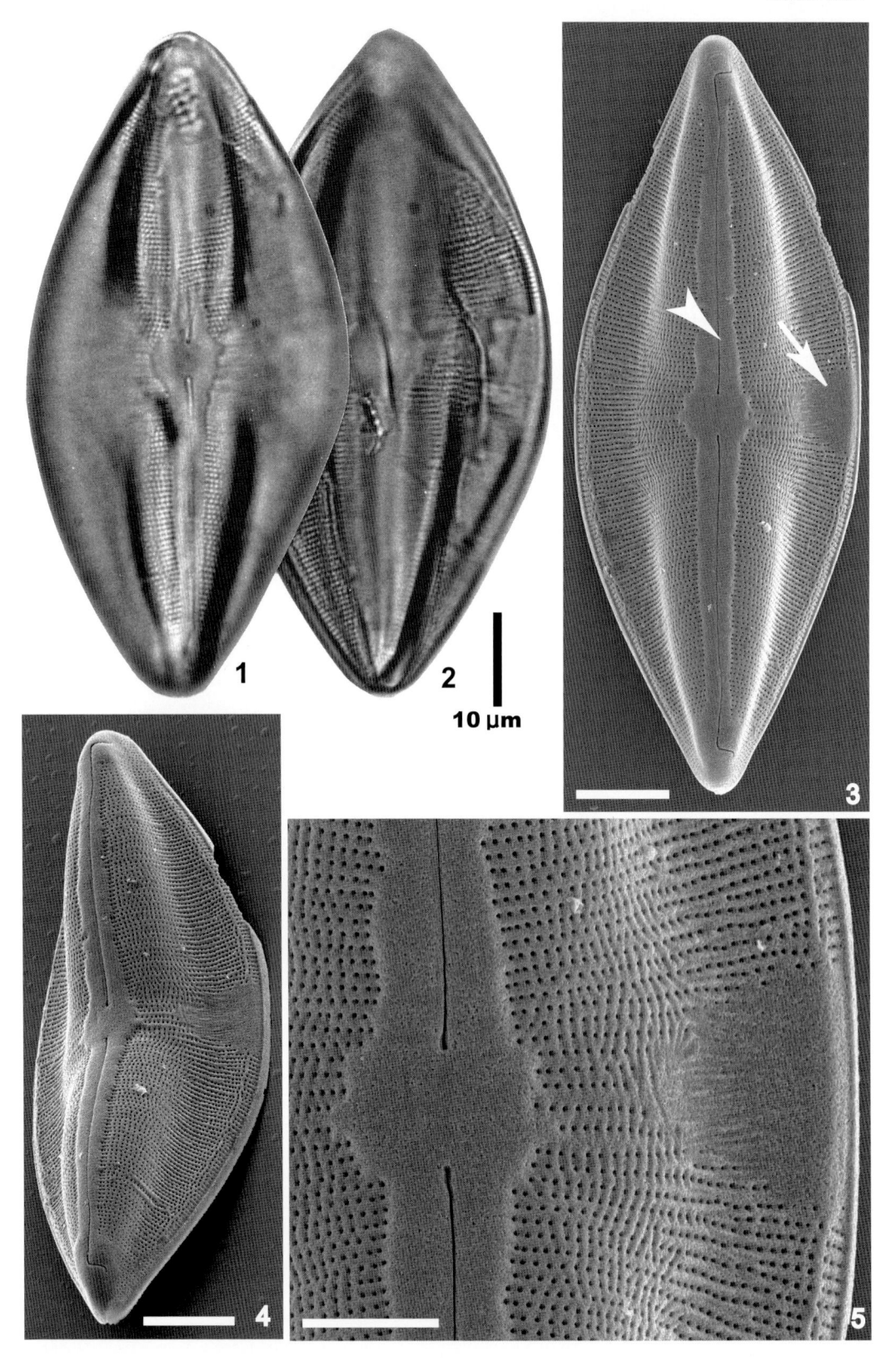

10 μm

Plate 153, Figs 1-5. *Achnanthes okunoi* **(Hustedt) H. Tanaka** comb. nov.
SEM. Figs 1-5. Material from Futoro Formation "diatomite of Setana" (Early Miocene), Setana Town, Hokkaido, Japan. Scale bars: Figs 1, 3-4=10 µm, Fig. 2=2 µm, Fig. 5=1 µm.

Fig. 1. Internal view of whole araphid valve.
Fig. 2. Enlarged view of an apex of Fig. 1 without helictoglossa.
Fig. 3. Internal view of whole raphid valve showing helictoglossae on both apices (arrows).
Fig. 4. Broken frustule showing raphid valve over araphid valve, arrowhead indicates edge of broken axial area.
Fig. 5. Enlarged view of broken axial area of Fig. 4 showing raphe-like line on surface, but without penetration (arrowhead).

Plate 153

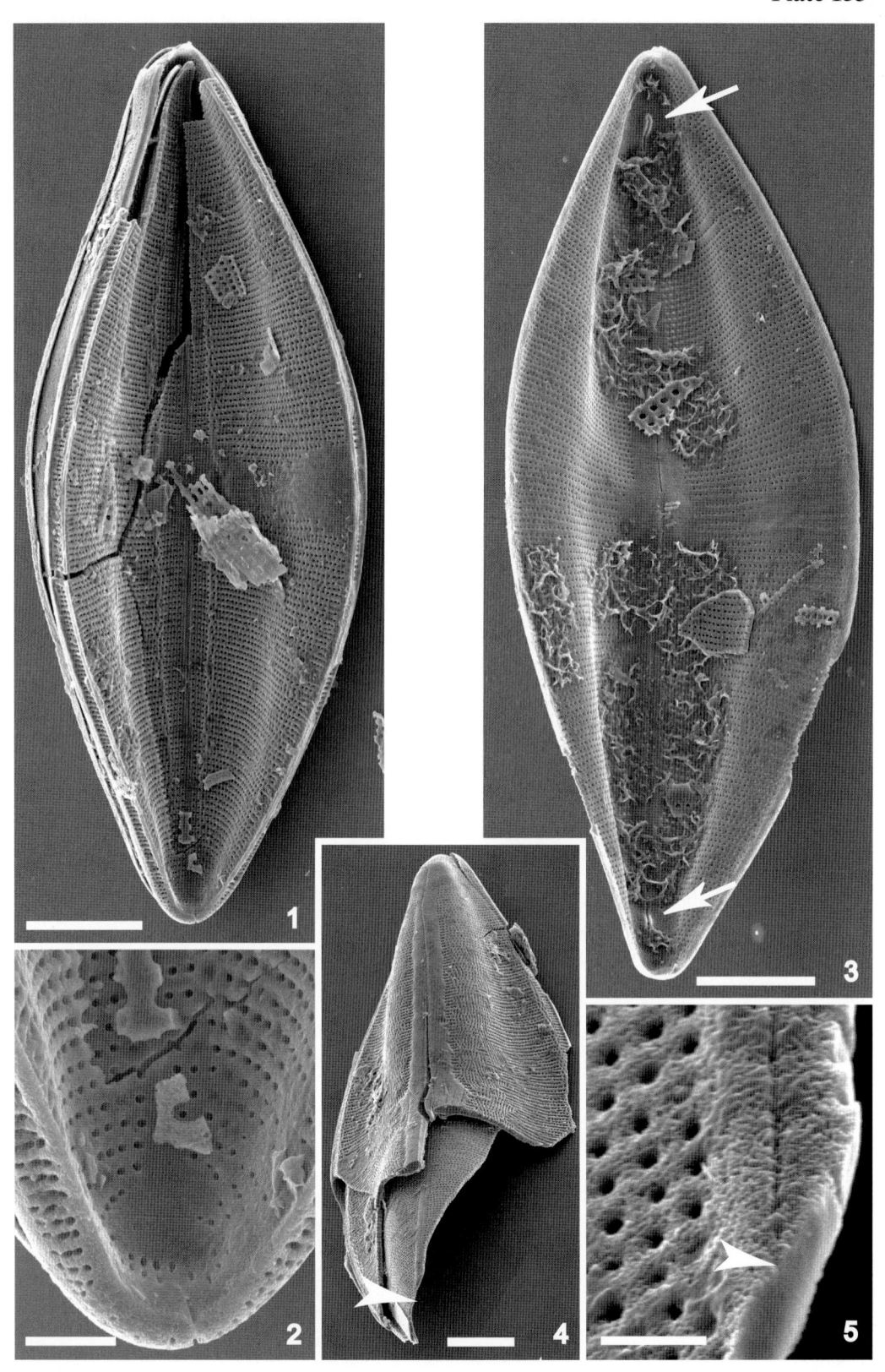

343

Plate 154.   Monoraphid, pennate diatoms: Achnanthales
Cocconeis jimboites VanLandingham 1968

基礎異名：*Surirella jimboi* Pantocsek 1905
文　　献：Pantocsek, J. 1905.　Beiträge zur Kenntnis der Fossilen Bacillarien Ungarns. 3 Teil. pp. 1-118, pls 1-42. W. Junk, Berlin.
　　奥野春雄 1959．北海道瀬棚町の珪藻土について(4)．植物研究雑誌 **34**: 272-277．
　　VanLandingham, S.L. 1968.　Calalogue of the fossil and recent genera and species of diatoms and their synonyms. Part Ⅱ. pp. 1-1086. J. Cramer.
形　　態：殻は楕円形，被殻の片方に縦溝があるが他方にはなく，殻の構造も異なっている．殻長 19.5-43 μm，殻幅 15-25.5 μm．縦溝殻の条線は 10 μm に約 16 本で 1 列の胞紋列から構成される．間条線は殻縁で枝分かれをし，短い条線が生じていることが多い．胞紋の外側開口は円形，殻面／殻套境界には肥厚した無紋域があり，殻面の胞紋は殻套へ連続しない．無縦溝殻は間条線の内側が肥厚して肋を形成しており，その数は 10 μm に 6-8 本である．条線を構成する胞紋の開口は円形で，中央では 4 列，殻縁では 5 列程度の胞紋列から構成されるが，両外側の開口が大きい．
ノ　ート：本種は *Surirella jimboi* Pant. として原記載されたが (Pantocsek 1905)，奥野 (1959) により被殻が縦溝殻と無縦溝殻から構成されていることが見出され，*Cocconeis* 属へ組み換えられた．ところが *Cocconeis jimboi* はすでに存在していたので，VanLandingham (1968) により，*C. jimboites* として記載された．現在のところ瀬棚町の珪藻土（太櫓層）からしか産出報告がない．
産　　出：太櫓層（瀬棚の珪藻土）（奥野 1959）：前期中新世・北海道．
図　　版：太櫓層標本．

Figs 1-7.　*Cocconeis jimboites* VanLandingham
LM. Figs 1-2. SEM. Figs 3-7. Material from Futoro Formation "diatomite of Setana" (Early Miocene), Setana Town, Hokkaido, Japan. Scale bars: Figs 3, 5-6＝5 μm, Fig. 4＝2 μm, Fig. 5＝1 μm.

Figs 1-2.　A frustule, raphid valve (Fig. 1), raphe (arrowhead) and araphid valve (Fig. 2).
Fig. 3.　Internal oblique view of araphid valve, costae (arrow).
Fig. 4.　External view of raphid valve center and areolae rows.
Fig. 5.　External view of araphid valve center areolae rows.
Fig. 6.　External oblique view of raphid valve, raphe (arrowhead).
Fig. 7.　External oblique view of araphid valve.

Plate 154

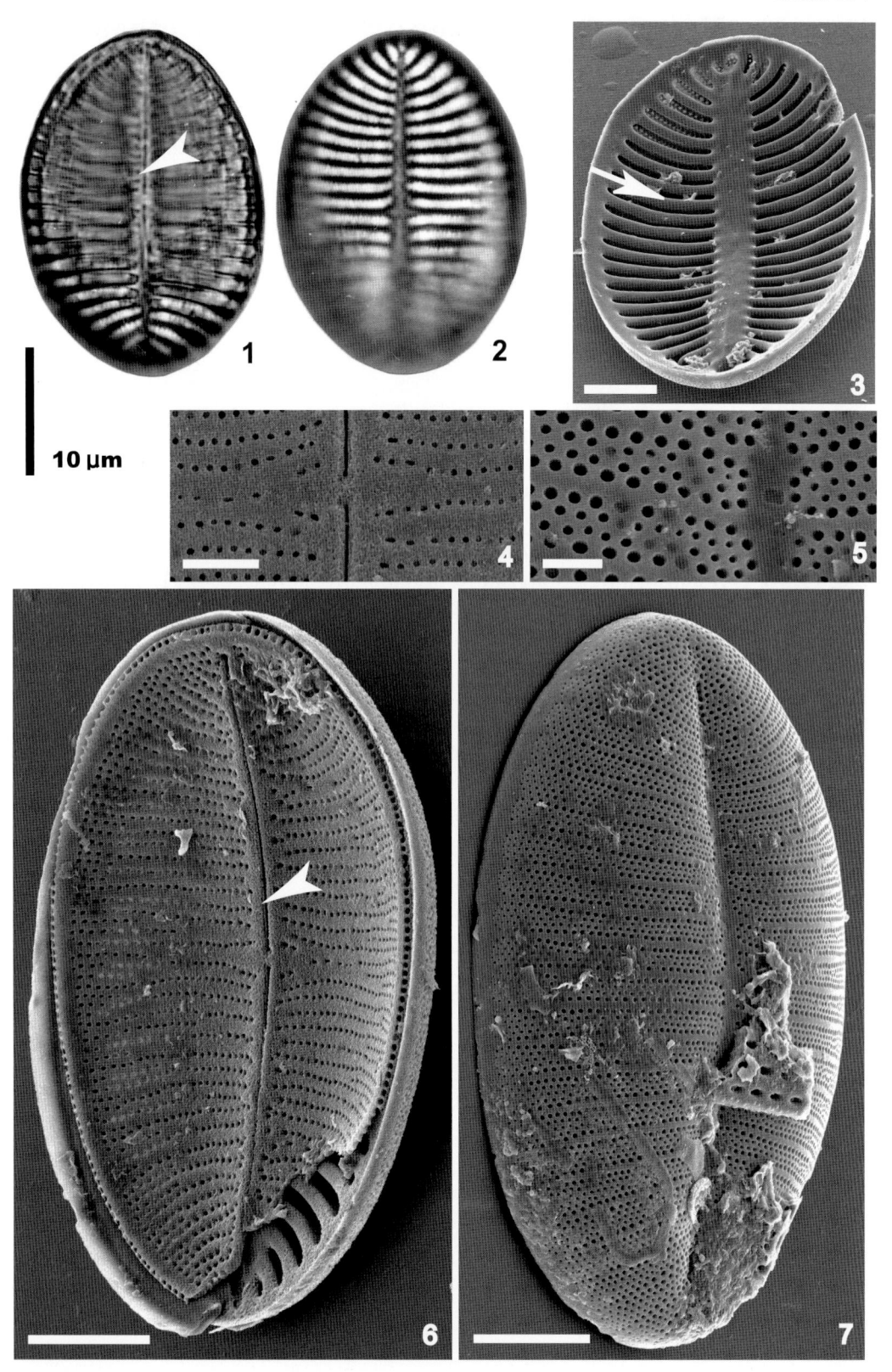

10 µm

Plate 155. Monoraphid, pennate diatoms: Achnanthales
Cocconeis placentula Ehrenberg var. *placentula* 1838

文　献：Krammer, K. & Lange-Bertalot, H. 1991. Bacillariophyceae. Teil 4. Achnanthaceae, Kritische Ergänzungen zu Navicula (Lineolatae) und Gomphonema. 437 pp. Süsswasserflora von Mitteleuropa Bd. **2/4**, Gustav Fischer Verlag, Stuttgart, Jena.

形　態：殻は楕円形，被殻の片方に縦溝があるが他方にはなく，殻面の構造も異なっている．殻長 15-41 μm，殻幅 12-29 μm，条線は無縦溝殻で 10 μm に約 24 本であった．縦溝殻では外中心裂溝は互いに反対方向へわずか曲がり終了する．胞紋の形は円形で殻面／殻套境界に肥厚した無紋域があり，殻面の胞紋は殻套へ連続しない．無縦溝殻では胞紋は横長形で殻套へ連続するが，殻内側では外側よりも横方向が短くなる．

ノート：本種には多くの変種が記載されているが，縦溝殻ではそれらを区別できないこと，無縦溝殻では区別はできるが変異は連続的であることが多いことから，変種を区別しない研究者もいる（小林・吉田 1984）．

産　出：小野上層（田中・小林 1995）：前期更新世・群馬県，横子珪藻土：前期更新世・群馬県，津森層（田中ら 2005）：中期更新世・熊本県．

図　版：横子珪藻土標本．

Figs 1-5. *Cocconeis placentula* Ehrenberg var. *placentula*
LM. Figs 1-2. SEM. Figs 3-5. Material from diatomite of Yokogo (Early Pleistocene), Numata City, Gunma, Japan. Scale bars: Figs 3, 5＝5 μm, Fig. 4＝2 μm.

Figs 1, 3. 　Raphid valves.
Figs 2, 4-5.　Araphid valves.
Figs 1-2.　Two different size valves.
Fig. 3.　External view of whole raphid valve.
Fig. 4.　Enlarged internal view of araphid valve center.
Fig. 5.　Internal oblique view of araphid valve.

Plate 155

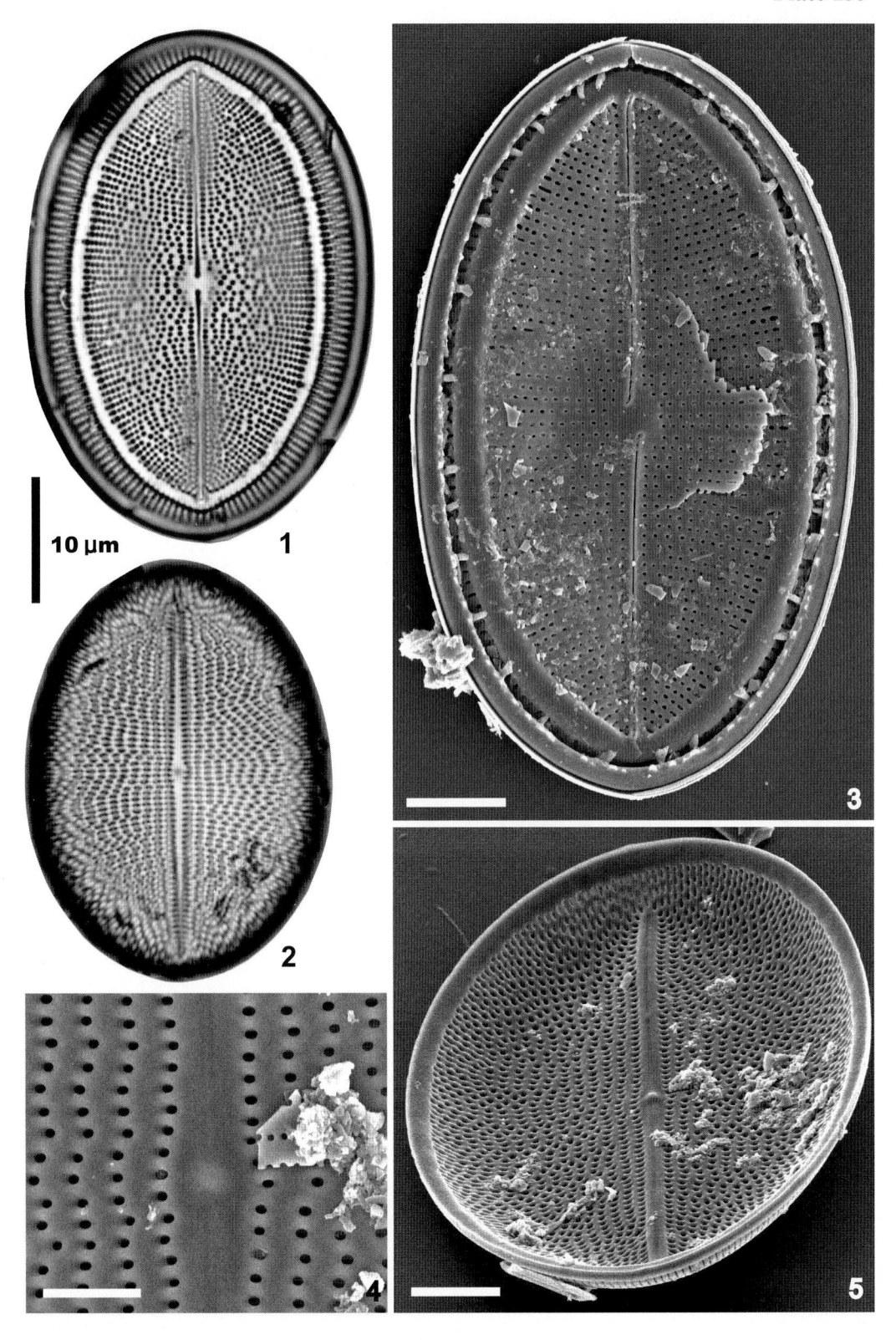

Plate 156. Monoraphid, pennate diatoms: Achnanthales
Cocconeis placentula var. *lineata* (Ehrenberg) Van Heurck 1880-1885

基礎異名：*Cocconeis lineata* Ehrenberg 1843
文　　献：Van Heurck, H. 1880-1885. Synopsis des diatomées de Belgique. Atlas: 132 pls. Text: 235 pp. 3 pls. Ducaju et Cie. Anvers.
形　　態：殻は楕円形，被殻の片方に縦溝があるが他方にはなく，殻面の構造も異なっている．殻長 13-35 μm，殻幅 7-19 μm．縦溝殻の条線は 10 μm に 20-26 本で，胞紋の形は円形，殻面／殻套境界に肥厚した無紋域があり，殻面の胞紋は殻套へ連続しないが，無縦溝殻の条線は 10 μm に 20-28 本で，胞紋は横長形で殻套へ連続する，胞紋密度は 10 μm に 10-16 個．
ノ ー ト：本分類群と基本種（var. *placentula*）の区別は難しいが，無縦溝殻の胞紋密度の比較（var. *lineata* の方がやや粗い）が参考になる．および条線密度，内側開口の形，縦走条線の数等を勘案するとよい．
産　　出：大鷲湖沼性堆積物：鮮新世・岐阜県，人形峠層（田中ら 2008）：後期中新世～鮮新世・岡山-鳥取県境．
図　　版：大鷲湖沼性堆積物標本．

Figs 1-7. *Cocconeis placentula* var. *lineata* (Ehrenberg) Van Heurck
LM. Figs 1-2. SEM. Figs 3-7. Material from lacustrine deposit of Owashi (Pliocene), Gujo City, Gifu, Japan. Scale bars: Fig. 5＝5 μm, Figs 3-4, 6-7＝2 μm.

Figs 1, 3-5.　Raphid valves.
Figs 2, 6-7.　Araphid valves.
Figs 1-2.　Two different size valves.
Fig. 3.　External view of whole raphid valve, raphe (arrowhead).
Fig. 4.　Oblique view of Fig. 3.
Fig. 5.　Internal oblique view of raphid valve.
Fig. 6.　External oblique view of araphid valve.
Fig. 7.　Internal oblique view of araphid valve.

Plate 156

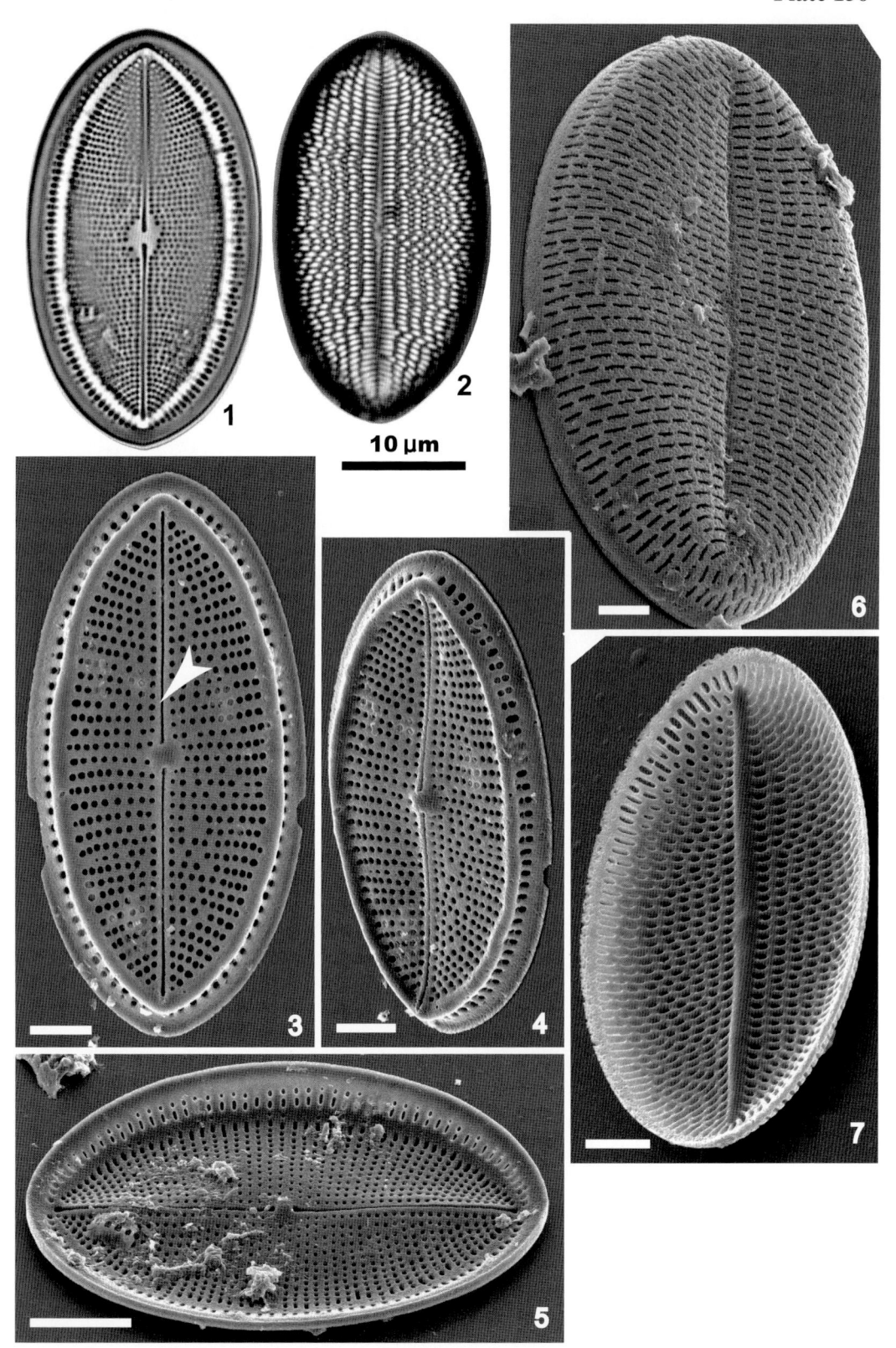

10 µm

Plate 157.　Biraphid, pennate diatoms: Cymbellales
Cymbella cymbiformis C. Agardh 1830

文　　献：Metzeltin, D., Lange-Bertalot, H. & Nergui, S. 2009.　Diatoms in Mongolia. 686 pp. Iconographia Diatomologica **20**: A.R.G. Gantner Verlag K.G.

形　　態：殻は比較的細長く殻面は三日月形，殻長 75-90 μm，殻幅約 16 μm，条線は 10 μm に約 9-11 本，条線を構成する点紋は 10 μm に約 20 個，遊離点は 1-2 個，両殻端には殻端小孔域がある．外裂溝は殻端と中心付近では腹側へ寄るが，その間は背側へ寄っている．極裂は背側へ曲がり，外中心裂溝は腹側へ曲がる．

ノ　ー　ト：中之条湖成層産の個体は Krammer(2002)の本種の写真と比較すると殻端がより広円形である．しかしそれよりも最新の出版物である Metzeltin *et al.*(2009)の写真とは似ている．類似している *Cymbella cistula* とは，軸域がほぼ同じ幅で，中心孔付近でもほとんど広がらない，中心孔付近の縦溝の左右への屈曲が大きいことから区別できる．

産　　出：中之条湖成層（田中・小林 1992）：中期更新世・群馬県，奄芸層群（根来・後藤 1981）：鮮新世・三重県．

図　　版：中之条湖成層標本．

Figs 1-6.　*Cymbella cymbiformis* C. Agardh
LM. Figs 1-2. SEM. Figs 3-6. Material from Nakanojo Lacustrine Deposit (Middle Pleistocene), Nakanojo Town, Gunma, Japan. Scale bars: Figs 4-5＝10 μm, Figs 3, 6＝5 μm.

Figs 1-2.　Two different valves.
Fig. 2.　Photomontage.
Fig. 3.　External view of whole valve.
Fig. 4.　Oblique view of Fig. 3.
Fig. 5.　Internal oblique view of whole valve.
Fig. 6.　Enlarged internal view of central area.

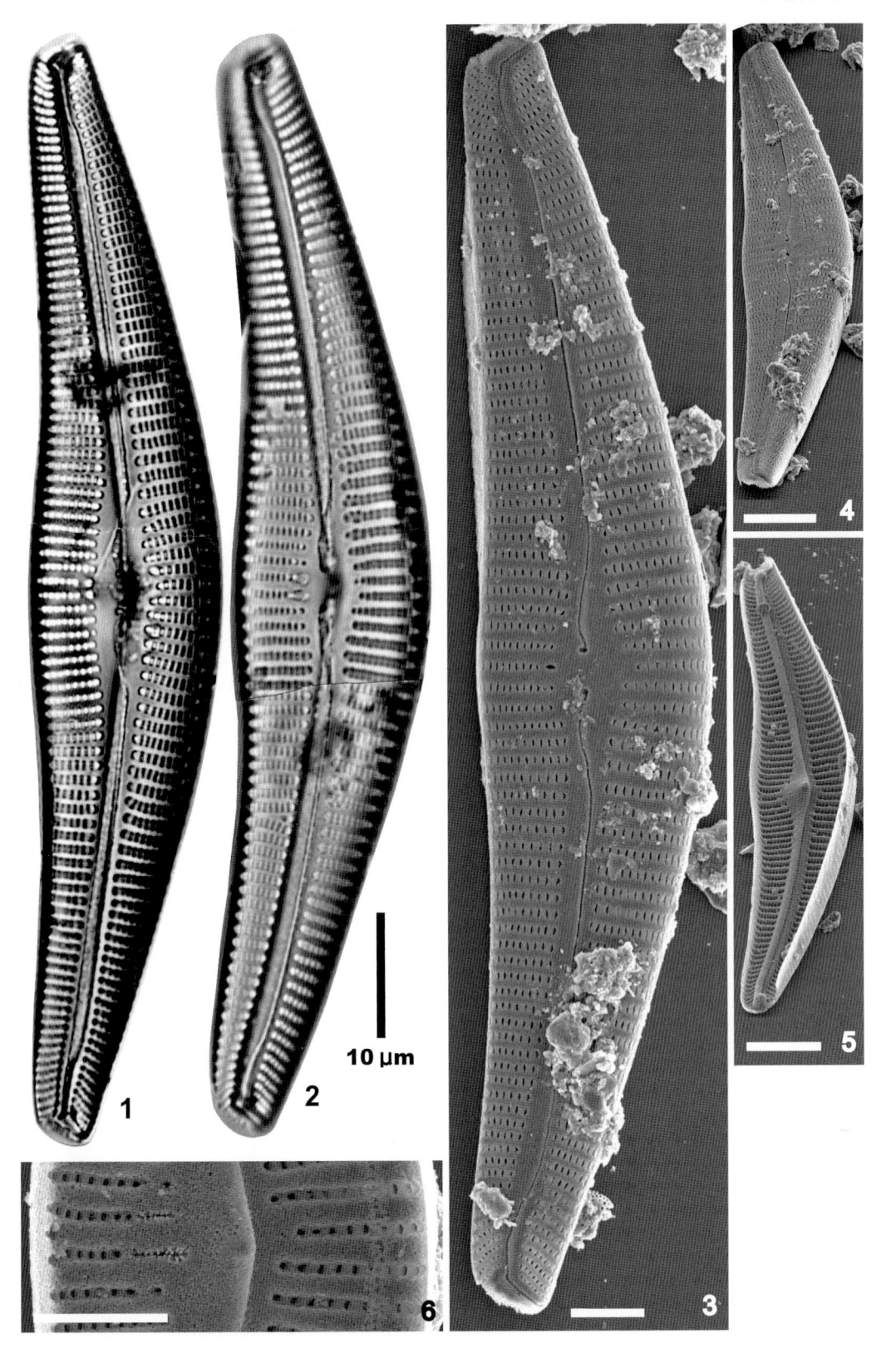

Plate 157

10 μm

Plate 158. Biraphid, pennate diatoms: Cymbellales
Cymbella neoleptoceros Krammer 2002

文　献：Krammer, K. 2002. *Cymbella. In*: Lange-Bertalot, H. (ed.) Diatoms of Europe. Diatoms of the European inland waters and comparable habitats **3**: 1-584 pp. A.R.G. Gantner Verlag K.G.

形　態：殻面は半皮針形，殻長 75-90 μm，殻幅約 16 μm，条線は 10 μm に約 9-11 本，条線を構成する点紋は 10 μm に約 20 個，遊離点は 1-2 個，両殻端には殻端小孔域がある．極裂は背側へ曲がり，外中心裂溝は腹側へ曲がる．

ノート：*Cymbella subleptoceros* (Ehrenb.) Kütz. にも類似しているが，条線の幅の太いこと，点紋の密度により識別できる．

産　出：大鷲湖沼性堆積物（田中ら 2011）：鮮新世・岐阜県．

図　版：大鷲湖沼性堆積物標本．

Figs 1-7. *Cymbella neoleptoceros* Krammer

LM. Figs 1-2. SEM. Figs 3-7. Material from lacustrine deposit of Owashi (Pliocene), Gujo City, Gifu, Japan. Scale bars: Figs 3-5＝2 μm, Figs 6-7＝1 μm.

Figs 1-2.　Two different size valves.
Fig. 3.　External view of whole valve.
Fig. 4.　Internal oblique view of whole valve.
Fig. 5.　Oblique view of valve center of Fig. 3.
Fig. 6.　Detailed oblique view of an apex of Fig. 3, apical pore field (arrow).
Fig. 7.　Detailed internal oblique view of an apex, helictoglossa (arrowhead).

Plate 158

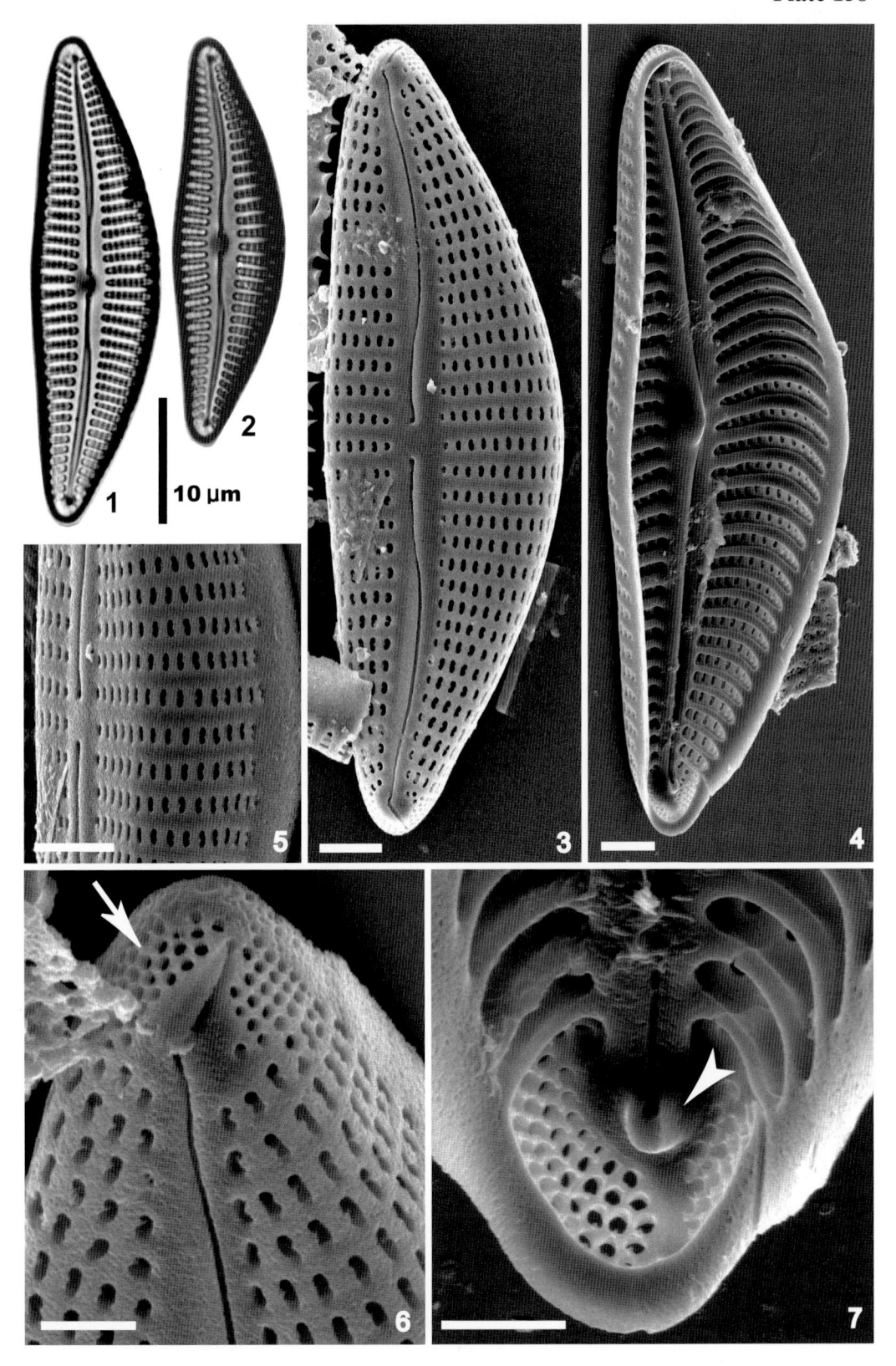

Plates 159-160. Biraphid, pennate diatoms: Cymbellales
Cymbella ocellata H. Tanaka sp. nov.

新　　種：記載文（英文）は 10 頁参照.
形　　態：殻は半円形で，殻端は小さく先細りに突出する，殻長 26-32 μm，殻幅 11-14 μm. 中心域・軸域は広いが殻端では狭くなる．両殻端には殻端小孔域がある．縦溝はほぼ軸域の中心を通る．極裂は背側に曲がり外中心裂溝は腹側に曲がる．遊離点は存在しない．条線はほぼ平行であるが，殻端ではわずか放射状になることが多く，中心付近で 10 μm に 8-10 本，殻端では 10 μm に 10-14 本である．
ノ ー ト：鹿児島市北部，八重山南方に分布する郡山層から見出されたものである．
産　　出：郡山層：後期鮮新世・鹿児島県.
図　　版 159-160：郡山層（タイプ試料）標本.

Plate 159, Figs 1-10. *Cymbella ocellata* **H. Tanaka** sp. nov.
LM. Figs 1-6. SEM. Figs 7-10, internal views. From the type material, KAG-506, Koriyama Formation (Late Pliocene), Kagoshima City, Kagoshima, Japan. Scale bars: Figs 7-8＝2 μm, Figs 9-10＝0.5 μm.

Figs 1-2.　Holotype, MPC-25059, National Science Museum, Tokyo, Japan. Same valve shown at different focal planes.
Figs 3-6.　Four different size valves.
Fig. 7.　Internal view of whole valve.
Fig. 8.　Oblique view of Fig. 7, apical pore field (arrow).
Fig. 9.　Enlarged view of Fig. 8 showing apical pore field (arrow) and helictoglossa (arrowhead).
Fig. 10.　Areolae. Note: no silica struts between areolae.

Plate 159

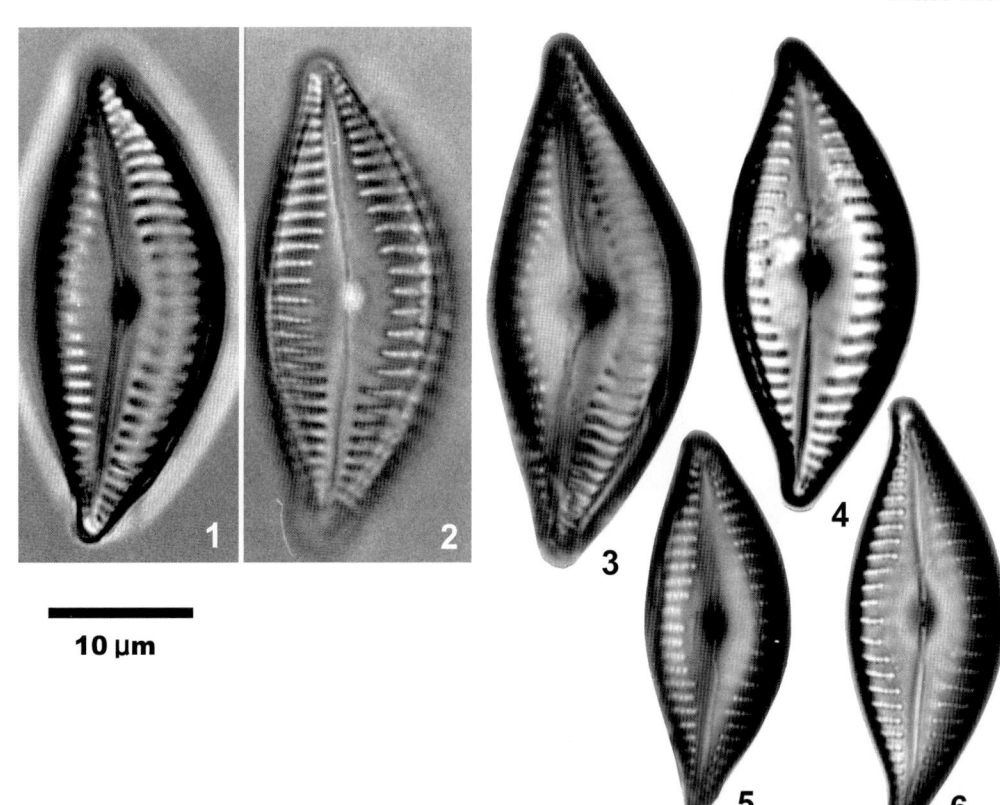

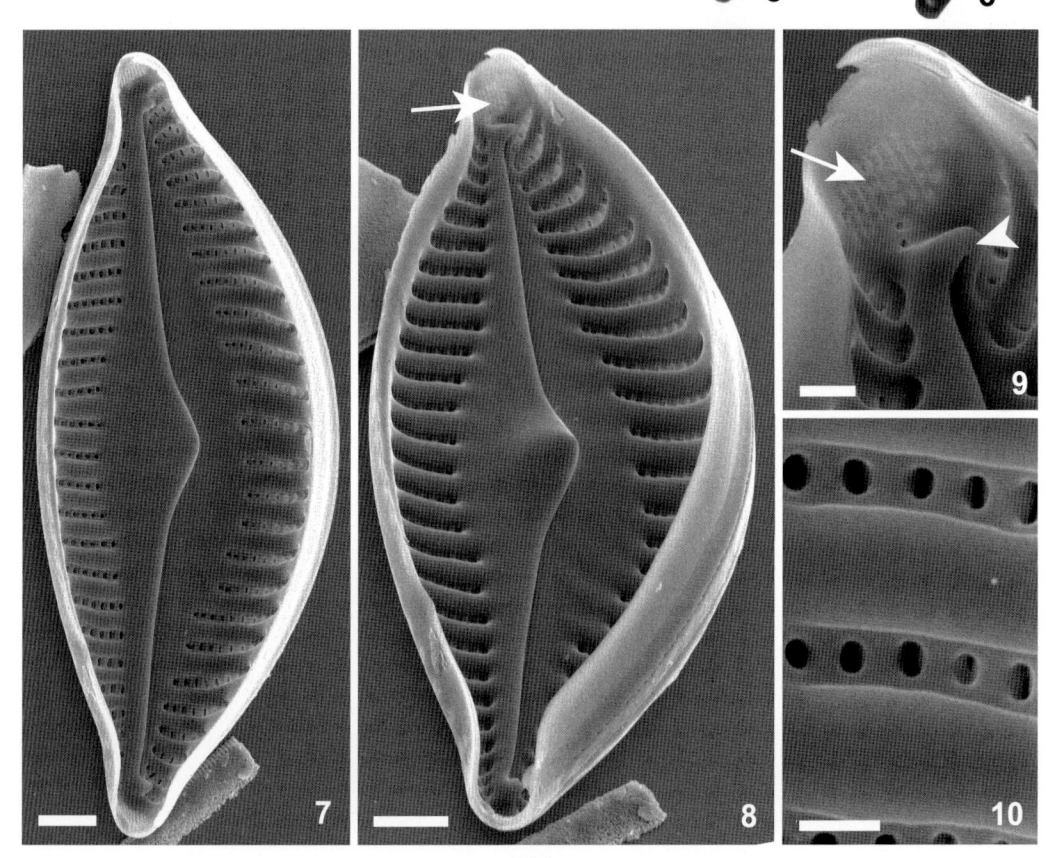

10 µm

Plate 160, Figs 1-4. *Cymbella ocellata* **H. Tanaka** sp. nov.

SEM. Figs 1-4, external views. From the type material, KAG-506, Koriyama Formation (Late Pliocene), Kagoshima City, Kagoshima, Japan. Scale bars: Figs 1-2=2 μm, Fig. 3= 1 μm, Fig. 4=0.5 μm.

Fig. 1. External view of whole valve, terminal fissure (arrow), central fissure (arrowhead).
Fig. 2. Oblique view of Fig. 1.
Fig. 3. Valve edge with apical pore field (arrow).
Fig. 4. Areolae. Note: slit-shaped foramina.

Plate 160

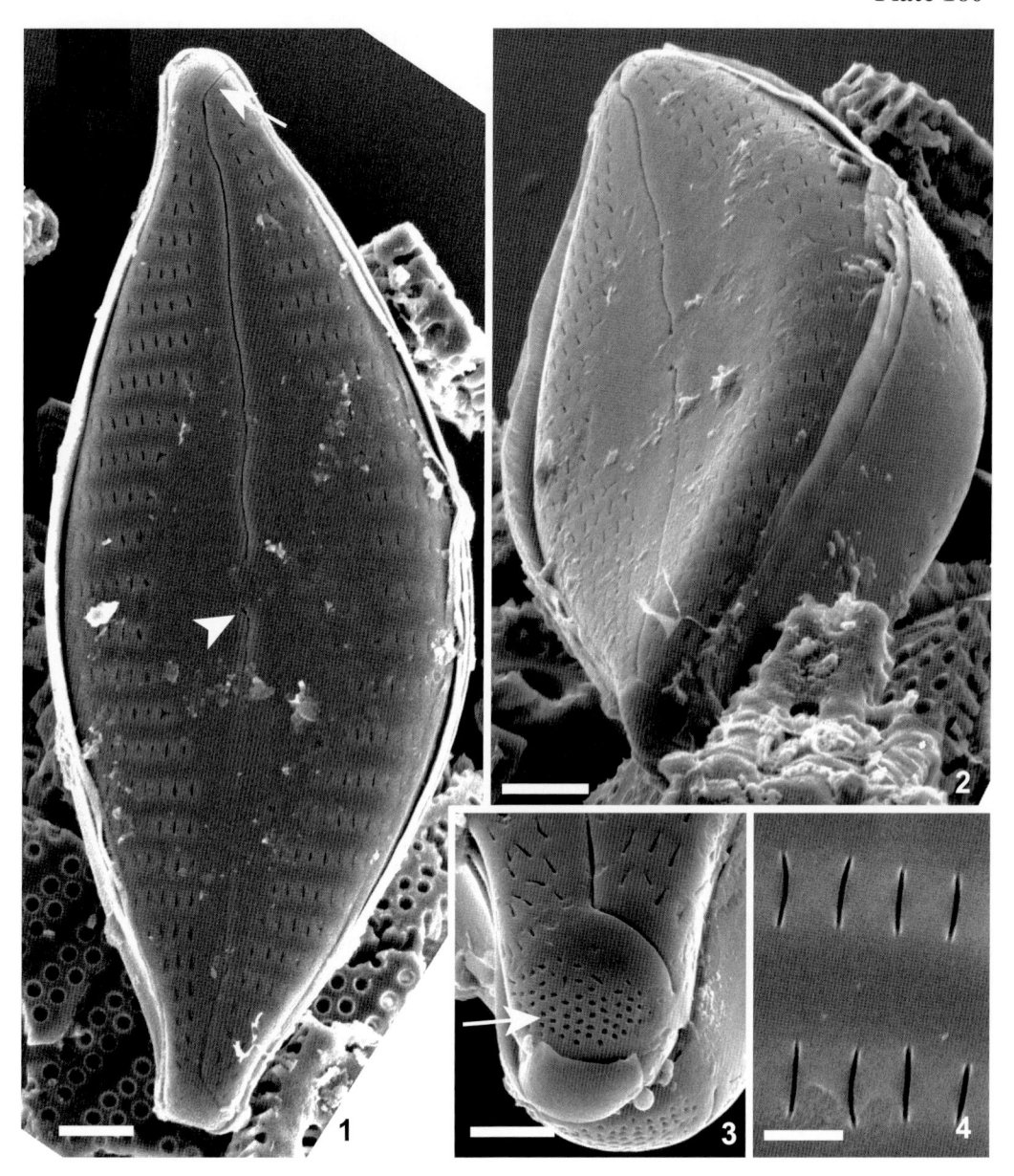

Plates 161-162.   Biraphid, pennate diatoms: Cymbellales
Cymbella okunoi H. Tanaka sp. nov.

新　　種：記載文（英文）は 12 頁参照.

形　　態：殻は強く背側が膨らみ三角形をしている．腹側はほぼ直線であるが中央がわずか膨らむ．殻中央と殻端の間はややくびれ，殻端は比較的小さい．殻長 43-64 μm，殻幅 15-17 μm．条線は腹側中央で 10 μm に 7-8 本，条線を構成する点紋は 10 μm に 16-20 個．胞紋の外側開口は細長いスリット状であるが，X や Y 字状等の場合もある．縦溝の外側表面は中心近くで大きく腹側に振れてから中心孔で終了し，殻端では背側へ曲がる．遊離点は 2-3 個，両端には殻端小孔域がある．殻内側の殻套周辺では胞紋の境に細い珪酸質の細壁が見られた．

ノ ー ト：Okuno(1952)で，筆者と同じ野上層から採取し *Cymbella cistula* var. *insignis* Meister として報告されている分類群は同じ分類群と考えられる．しかし *C. cistula* var. *insignis* の原記載の図を見ると野上層から見出した分類群は当種へ同定しづらく，他に同定できる種が見当たらないので，最初に本分類群を見出した奥野春雄博士を記念した種小名の新種として記載する．

産　　出：野上層（Okuno 1952）：中期更新世・大分県．

図　　版 161-162：野上層（タイプ試料）標本．

Plate 161, Figs 1-5.   *Cymbella okunoi* H. Tanaka sp. nov.
LM. Figs 1-4. SEM. Fig. 5. From the type material, OIT-229, Nogami Formation (Middle Pleistocene), Kokonoe Town, Oita, Japan. Scale bar: Fig. 5＝2 μm.

Figs 1-2.　Holotype, MPC-25060, National Science Museum, Tokyo, Japan. Same valve shown at different focal planes.

Figs 1-4.　Three different size valves.

Fig. 5.　External view of valve center of Plate 162, Fig. 1 showing expanded dorsal side, foramina slits, wave roll external raphe branches near center and two stigmata (arrows).

Plate 161

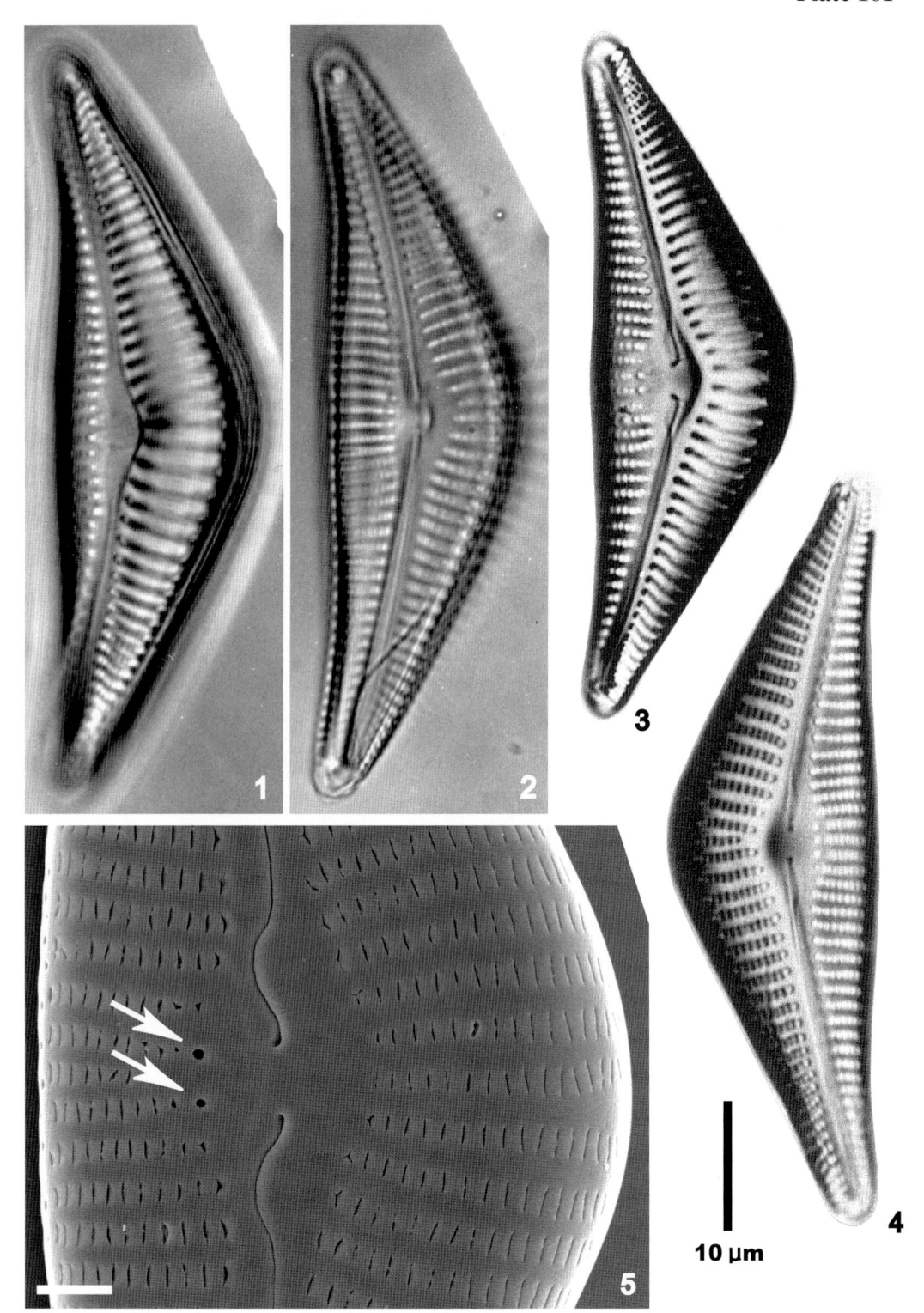

10 µm

Plate 162, Figs 1-7. *Cymbella okunoi* **H. Tanaka** sp. nov.

SEM. Figs 1-7. From the type material, OIT-229, Nogami Formation (Middle Pleistocene), Kokonoe Town, Oita, Japan. Scale bars: Figs 1-2, 4=5 μm, Figs 5, 7=2 μm, Fig. 3=1 μm, Fig. 6=0.5 μm.

Fig. 1. External view of whole valve.
Fig. 2. Oblique view of Fig. 1 showing expanded dorsal side.
Fig. 3. Enlarged view of an apex of Fig. 2, terminal fissure (arrowhead) and apical pore field (arrow).
Fig. 4. Internal oblique view of whole valve.
Fig. 5. Enlarged view of an apex of Fig. 4, helictoglossa (arrowhead) and apical pore field (arrow).
Fig. 6. Enlarged view of areolae rows showing remains of silica struts.
Fig. 7. Enlarged view of Fig. 4 showing three stigmata (arrow).

Plate 162

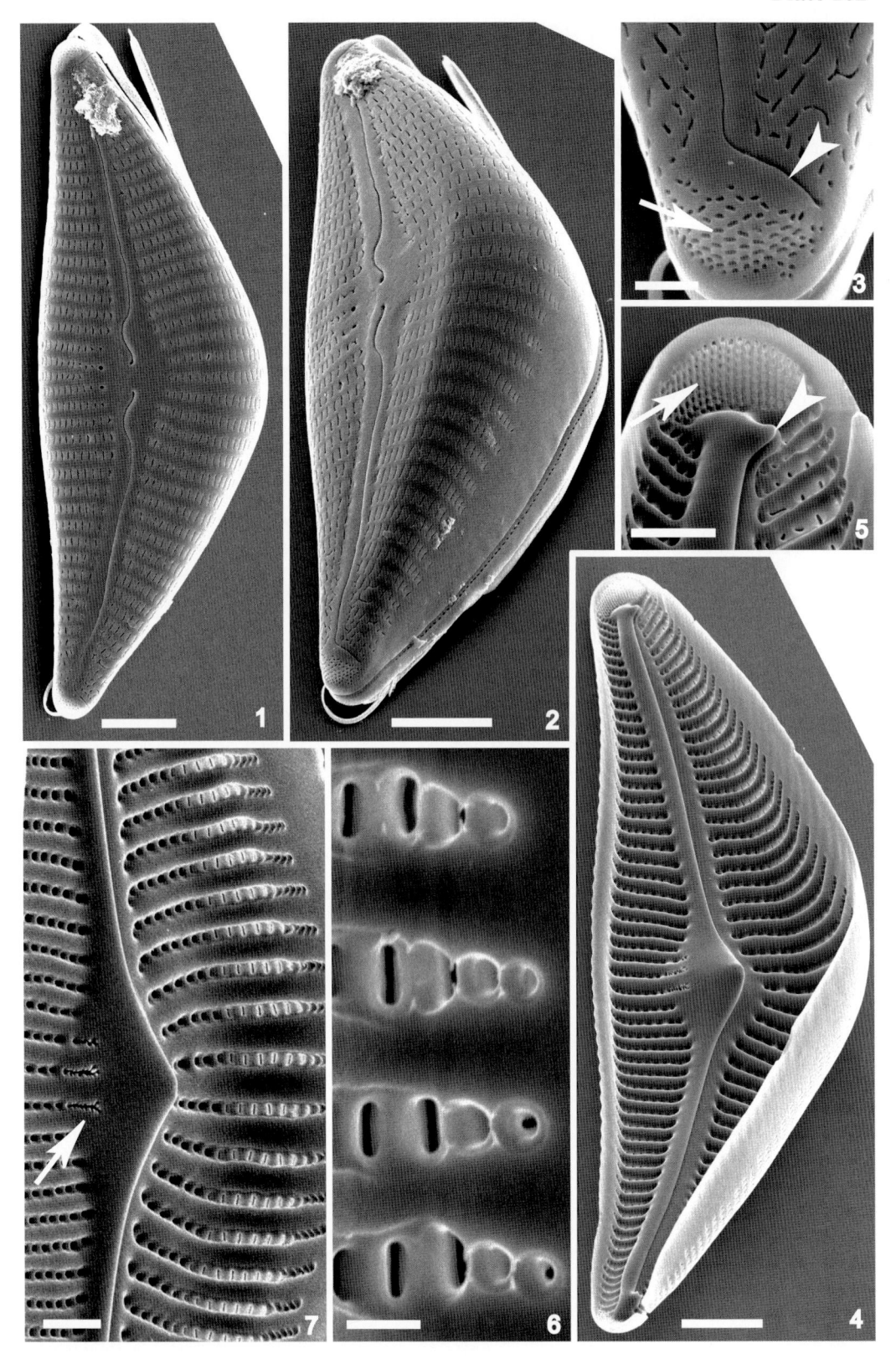

Plate 163.　Biraphid, pennate diatoms: Cymbellales
Cymbella orientalis Lee in Lee *et al.* 1993

文　　献：Lee, J.H., Gotoh, T. & Chung, J. 1993. *Cymbella orientalis* sp. nov., a freshwater diatom from the Far East. Diatom Research **8**: 99-108.

形　　態：殻は背側が腹側よりもやや膨らむ．殻長 17-33 μm, 殻幅約 6 μm. 条線は平行〜わずか放射状で，腹側は連続するが背側は中心付近で途切れる, 10 μm に約 12 本．SEM 観察によると縦溝は中心近くの外側では曲がるが，内側では直線であり，殻端では背側に曲がる．

ノ ー ト：Lee *et al.*(1993)は広く極東地域から見出している．筆者はかって南西諸島喜界島の上嘉鉄湧泉池，および雁股の泉から見出し *Cymbella subzewarensis* Foget と報告した分類群である（田中 1989）．

産　　出：中国・朝鮮・南西諸島・台湾（Lee *et al.* 1993)：(現生），上嘉鉄湧泉池（田中 1989）：鹿児島県（現生）．

図　　版：上嘉鉄湧泉池（喜界島）標本.

Figs 1-9.　*Cymbella orientalis* Lee

LM. Figs 1-4. SEM. Figs 5-9. Material from Kamikatetsu spring (recent), Kikai Island, Kagoshima, Japan. Scale bars: Figs 5-6＝2 μm, Figs 7-9＝1 μm.

Figs 1-4.　Four different size valves.
Fig. 5.　External view of whole valve.
Fig. 6.　Internal oblique view of whole valve.
Fig. 7.　View of an apex showing terminal fissure (arrowhead) and apical pore field.
Fig. 8.　View of an apex showing short band.
Fig. 9.　Detailed external view of central area.

Plate 163

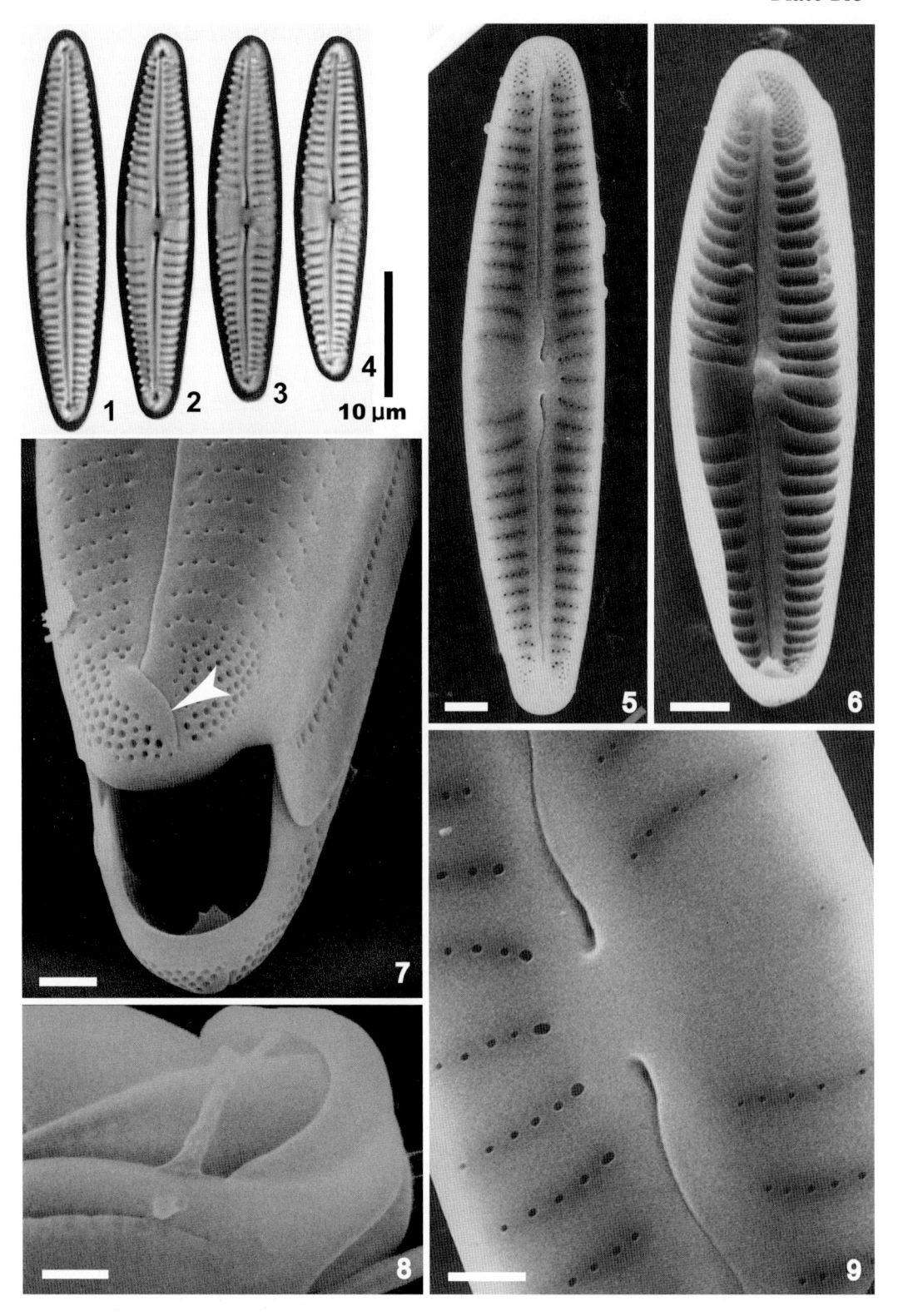

Plate 164. Biraphid, pennate diatoms: Cymbellales
Cymbella peraspera Krammer 2002

文　　献：Krammer, K. 2002.　*Cymbella*. *In*: Lange-Bertalot, H.(ed.) Diatoms of Europe. Diatoms of the European inland waters and comparable habitats **3**: 1-584 pp. A.R.G. Gantner Verlag K.G.

形　　態：殻は大形でゆるく弓なりになり左右不対称，殻長 215-302 μm，殻幅 42-45 μm. 腹側はほぼ直線で背側が膨らむ．軸域は広い．条線は放射状で 10 μm に 7-10 本．遊離点は多い．殻端小孔域は広い．

ノ ー ト：大形の *Cymbella* 属の中では *C. amelieana* Vijver & Lange-Bert. に殻形，条線密度等類似するが，中心域の外中心裂溝が *C. amelieana* は釣針状なのに対して，本分類群はやや腹側に傾くが直線状のままで終了している．また *C. aspera* に類似するが，大形の *C. aspera* 類似種に対して Krammer(2002)は *C. peraspera* として新種を記載している．鬼首層の本分類群は *C. aspera* か *C. peraspera* のどちらかに同定できると考えられるが，Krammer(2002)では *C. aspera* の殻長が 110-200 μm，*C. peraspera* の殻長は (130)154-320 μm と記されており，鬼首層の分類群は *C. peraspera* の範囲にある．

産　　出：鬼首層：後期更新世・宮城県．

図　　版：鬼首層標本．

Figs 1-4.　*Cymbella peraspera* **Krammer**
LM. Fig. 1. SEM. Figs 2-4. Material from Onikobe Formation (Late Pleistocene), Osaki City, Miyagi, Japan. Scale bars: Figs 2, 4＝20 μm, Fig. 3＝10 μm.

Figs 1-4.　Same valve, LM and SEM photographs.
Fig. 1.　Whole valve view.
Fig. 2.　External view of whole valve.
Fig. 3.　Enlarged view of valve center of Fig. 2, central pore (arrowhead).
Fig. 4.　Oblique internal view of whole valve, helictoglossa (arrow).

Plate 164

20 μm

Plate 165. Biraphid, pennate diatoms: Cymbellales
Cymbella proxima Reimer in Patrick & Reimer 1975

文　献：Patrick, R. & Reimer, C.W. 1975. The diatoms of exclusions of Alaska and Hawaii. Vol. **2**, Part 1. 213 pp. Monographs of the Academy of Natural Sciences of Philadelphia No. 13.

形　態：殻は左右不対称で，腹側はわずか凹んだカーブを描くが中心部はやや膨らむ．殻長 90 µm，殻幅 20 µm．軸域は狭い．中心域腹側に数個の遊離点がある．条線は 10 µm に約 7 本で，中心部でやや放射状，殻端では放射がきつくなる．条線を構成する点紋は 10 µm に約 15 個．縦溝は外裂溝・内裂溝ともほぼ直線であるが，外裂溝は殻端で背側に曲がる．殻端小孔域がある．

産　出：津森層：中期更新世・熊本県．

図　版：津森層標本．

Figs 1-5. *Cymbella proxima* Reimer
LM. Fig. 1. SEM. Figs 2-5. Material from Tsumori Formation (Middle Pleistocene), Mashiki Town, Kumamoto, Japan. Scale bars: Figs 2, 4-5＝5 µm, Fig. 3＝2 µm.

Fig. 1.　　Whole valve view.
Figs 2-5.　Same valve, external and internal SEM views.
Fig. 2.　　External view of whole valve.
Fig. 3.　　Oblique view of an apex.
Fig. 4.　　Internal view of valve center.
Fig. 5.　　Internal oblique view of valve.

Plate 165

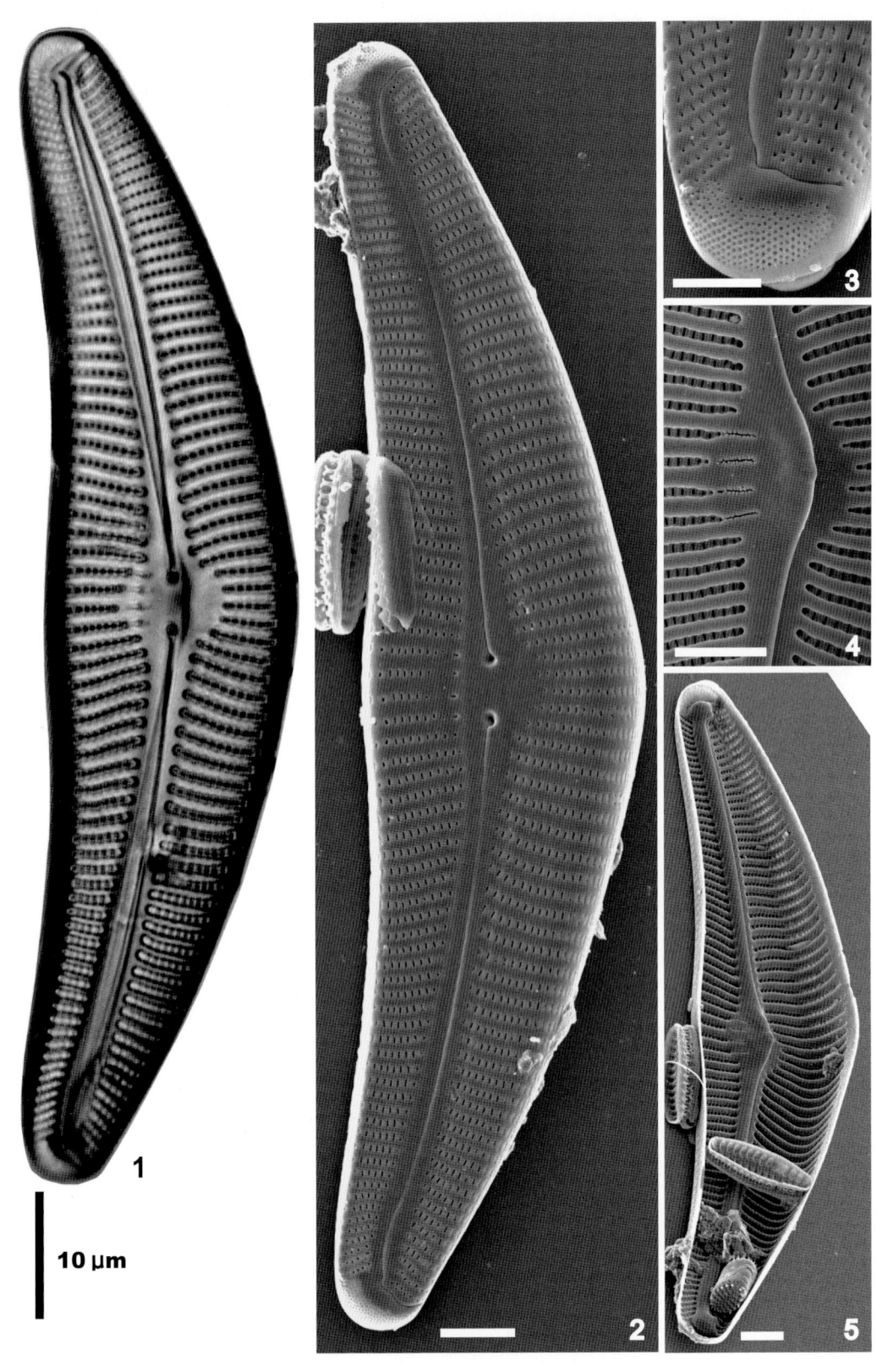

10 μm

Plate 166.   Biraphid, pennate diatoms: Cymbellales
Cymbella stuxbergii var. *robusta* Okuno 1959

文　　献：奥野春雄 1959. 北海道瀬棚町の珪藻土について(5). 植物研究雑誌 **34**: 353-360.
形　　態：殻は左右不対称で，腹側はやや凹んだカーブを描く．殻長 84-210 μm，殻幅 28-50 μm．中心域腹側に条線の先端から離れて 5-10 個程度の遊離点がある．条線は 10 μm に 5-8 本，放射状ないし平行であるがゆるくカーブすることもある．条線を構成する点紋は 10 μm に 8-14(16) 個．胞紋の外側開口はスリット状である．縦溝は殻中心側では背側にカーブし中心孔で終了する．極裂は背側に曲がる．殻端小孔域がある．
ノ　ー　ト：本種は奥野(1959)によって瀬棚町に分布する珪藻土（太櫓層）から新変種として記載された．広円形の殻端，遊離点が条線から離れて存在すること，条線密度が比較的疎であることがおもな特徴である．
産　　出：太櫓層（瀬棚の珪藻土）（奥野 1959）：前期中新世・北海道．
図　　版：太櫓層標本．

Figs 1-5.   *Cymbella stuxbergii* var. *robusta* Okuno
LM. Figs 1-2. SEM. Figs 3-5. Material from Futoro Formation "diatomite of Setana" (Early Miocene), Setana Town, Hokkaido, Japan. Scale bars: Figs 3, 5＝5 μm, Fig. 4＝2 μm.

Figs 1-2.   Two different size valves.
Fig. 3.   External whole valve view, stigmata (arrow).
Fig. 4.   Internal view of an apex.
Fig. 5.   Internal view of valve center.

Plate 166

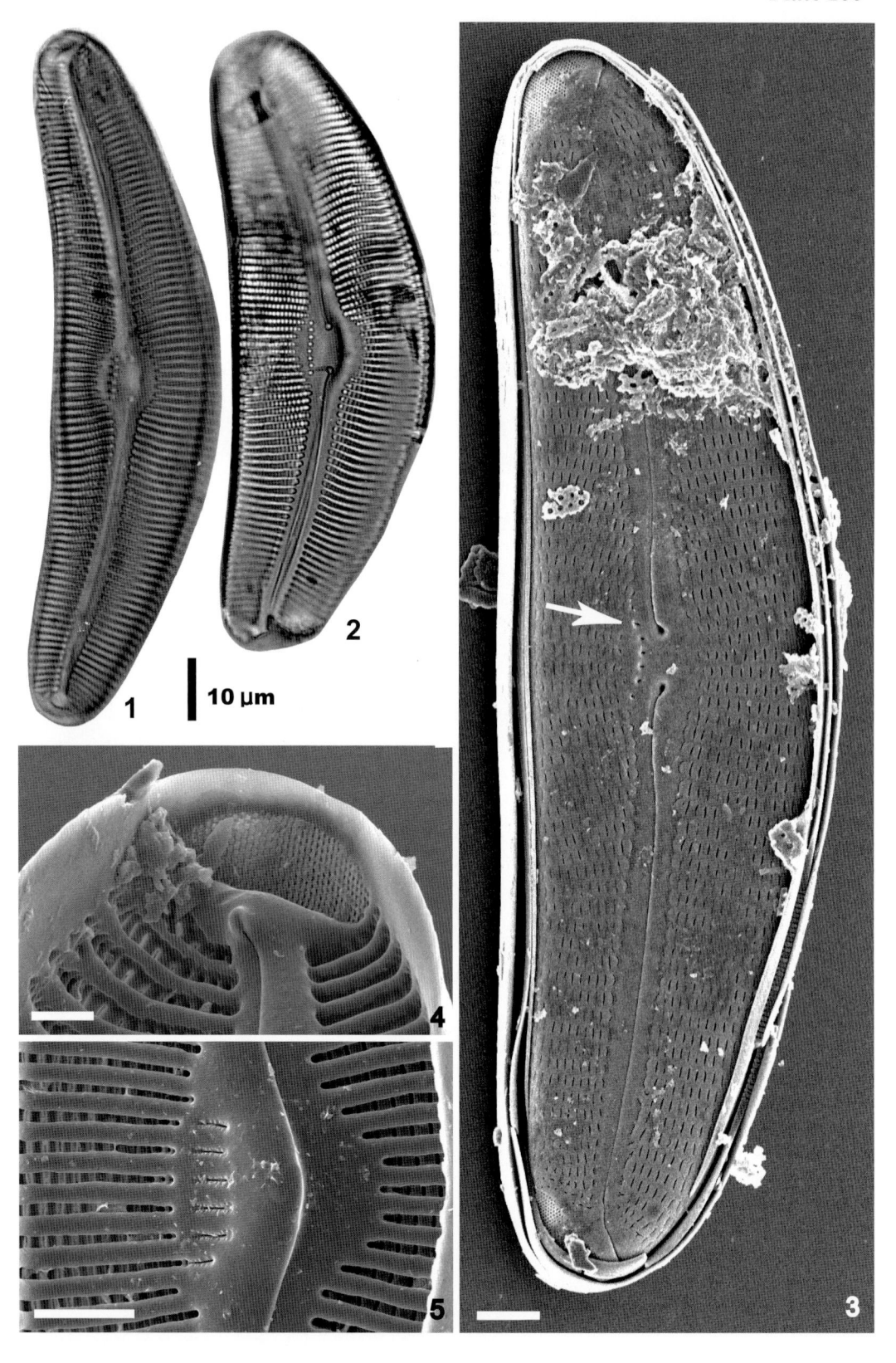

Plates 167-169.　Biraphid, pennate diatoms: Cymbellales
Cymbella tsumurae H. Tanaka sp. nov.

新　　種：記載文（英文）は 13 頁参照.
形　　態：殻は大形で左右不対称．殻端は幅広で台形的である．腹側はわずか凹んだカーブを描くが中心部はやや膨らむ．殻長 123-188 μm，殻幅 26-30 μm．軸域は比較的広い．中心域腹側条線先端に最大 14 個の遊離点がある．条線は中心部でやや放射状であるが，ほぼ軸域に直角に配列し，中心部で 10 μm に 8-10 本，殻端では約 12 本になる．条線を構成する点紋は 10 μm に 10-12 個．縦溝はほぼ軸域の中心を通り，外中心裂溝はわずか腹側へ曲がるが，極裂は背側へ曲がる．殻端小孔域がある．胞紋の外側開口はふつうスリット状であるがしばしばアルファベットの X, Y 字形等になる．内側では各胞紋間には細いすじ状の仕切りがある．
ノート：種小名は，横子珪藻土について筆者へ教示された津村孝平博士を記念して付けられた．
産　　出：横子珪藻土：前期更新世・群馬県.
図　　版 167-169：横子珪藻土（タイプ試料）標本.

Plate 167, Figs 1-5.　*Cymbella tsumurae* H. Tanaka sp. nov.
LM. Figs 1-5. From the type material, YOK-101, diatomite of Yokogo (Early Pleistocene), Numata City, Gunma, Japan.

Figs 1-4.　Holotype.
Figs 1-2.　Same frustule, epitheca (Fig. 1) and hypotheca (Fig. 2).
Figs 3-4.　Enlarged view of two apices.
Fig. 5.　Different size and well preserved valve, photomontage.

Plate 167

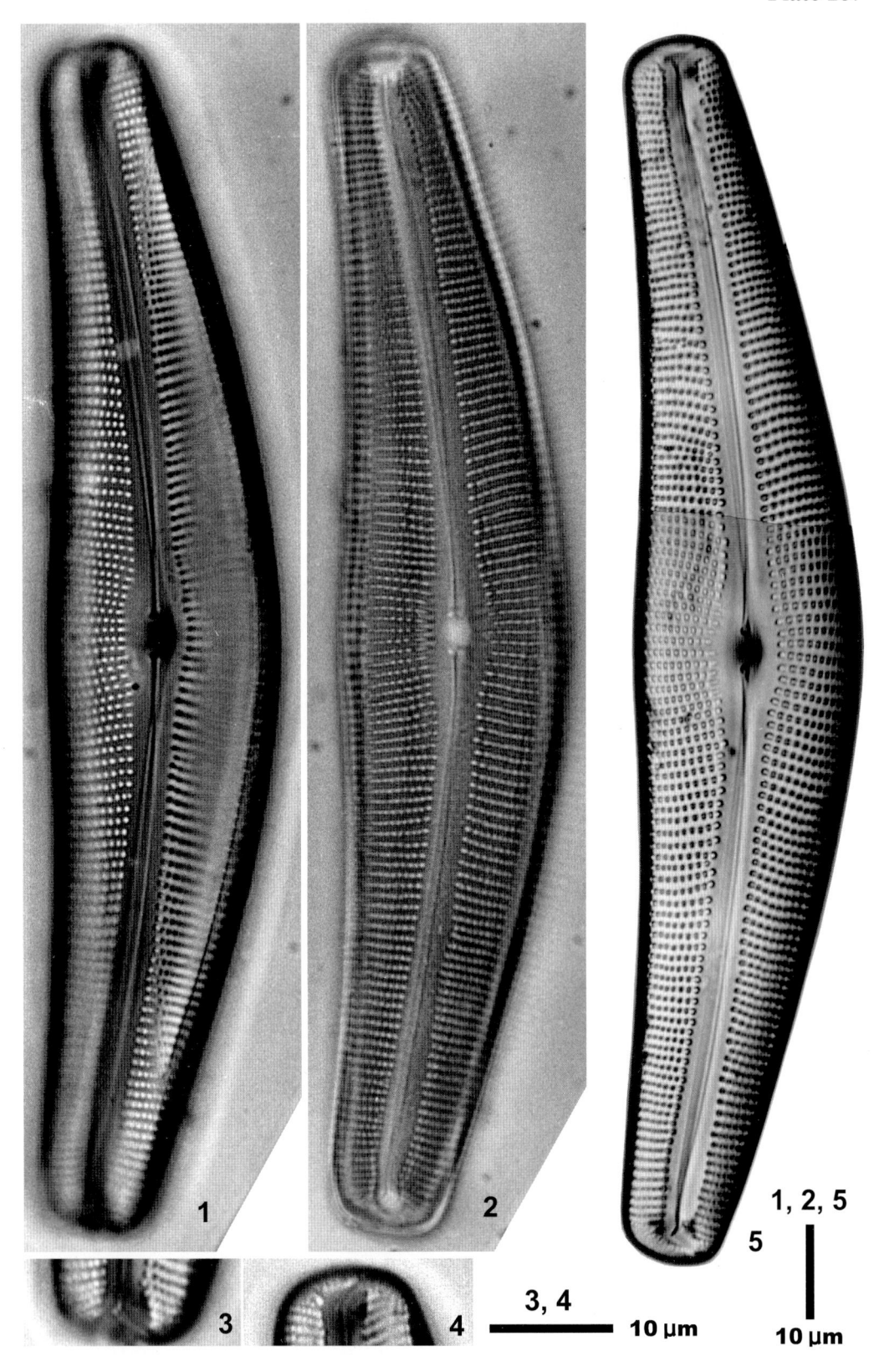

1, 2, 5

10 µm

3, 4

10 µm

371

Plate 168, Figs 1–5. *Cymbella tsumurae* **H. Tanaka** sp. nov.

SEM. Figs 1–5, external views. From the type material, YOK-101, diatomite of Yokogo (Early Pleistocene), Numata City, Gunma, Japan. Scale bars: Fig. 1=10 μm, Figs 3–4=5 μm, Fig. 2=2 μm, Fig. 5=1 μm.

Fig. 1. Whole valve view.
Fig. 2. Enlarged oblique view of an apex of Fig. 1, apical pore field (arrow).
Fig. 3. Enlarged view of valve center of Fig. 1, stigmata (arrows).
Fig. 4. Oblique view of Fig. 1 showing central pores (arrows) and mantle areolation.
Fig. 5. Detailed view of foramina, in rare cases, X or Y shaped.

Plate 168

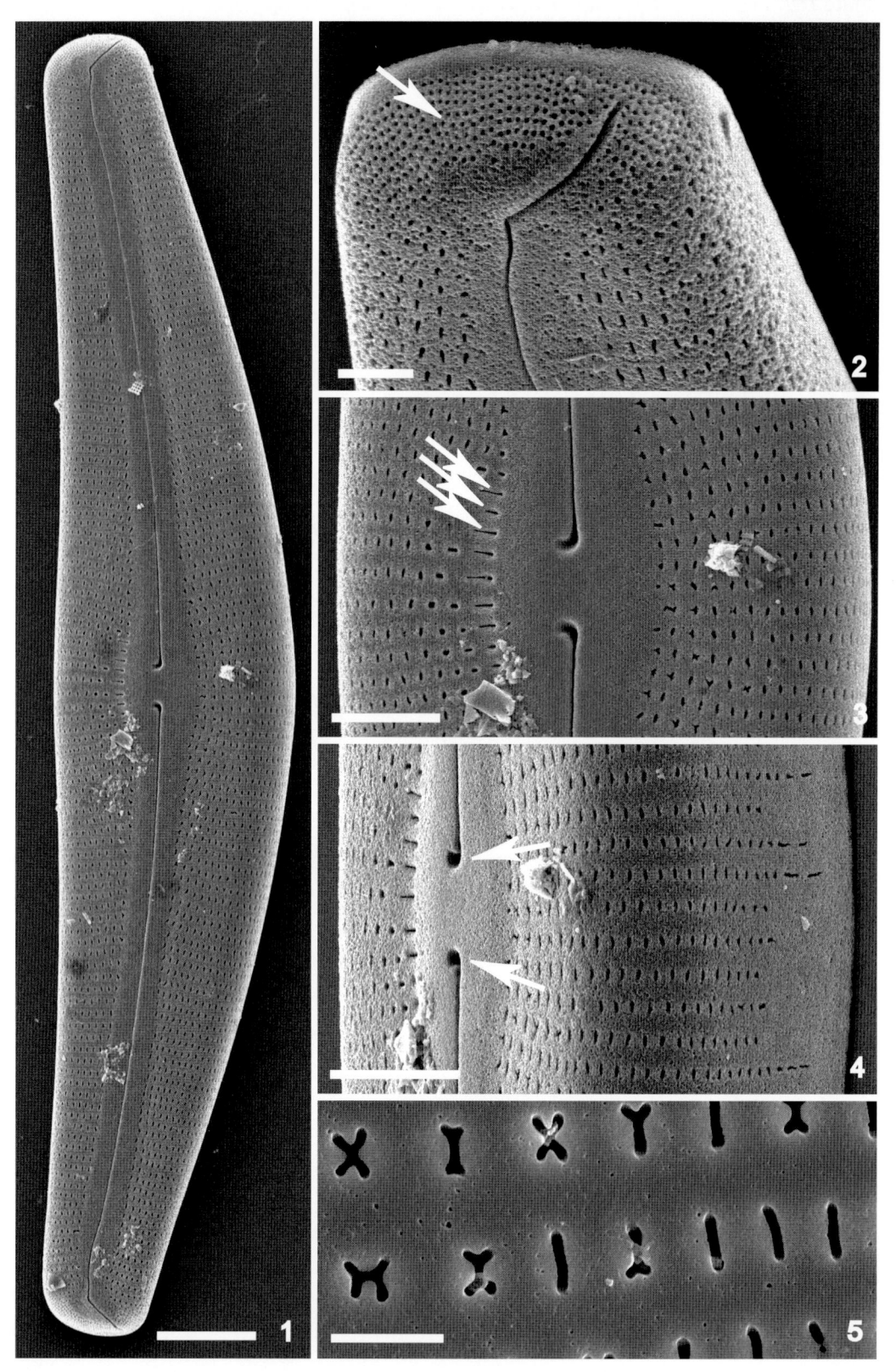

Plate 169, Figs 1-4. *Cymbella tsumurae* **H. Tanaka** sp. nov.

SEM. Figs 1-4, internal views. From the type material, YOK-101, diatomite of Yokogo (Early Pleistocene), Numata City, Gunma, Japan. Scale bars: Fig. 1=10 μm, Fig. 3=5 μm, Fig. 2 =2 μm, Fig. 5=1 μm.

Fig. 1. Whole valve view.
Fig. 2. Enlarged oblique view of an apex of Fig. 1, apical pore field (arrow).
Fig. 3. Enlarged view of valve center of Fig. 1.
Fig. 4. Detailed view of areolae separated by silica struts.

Plate 169

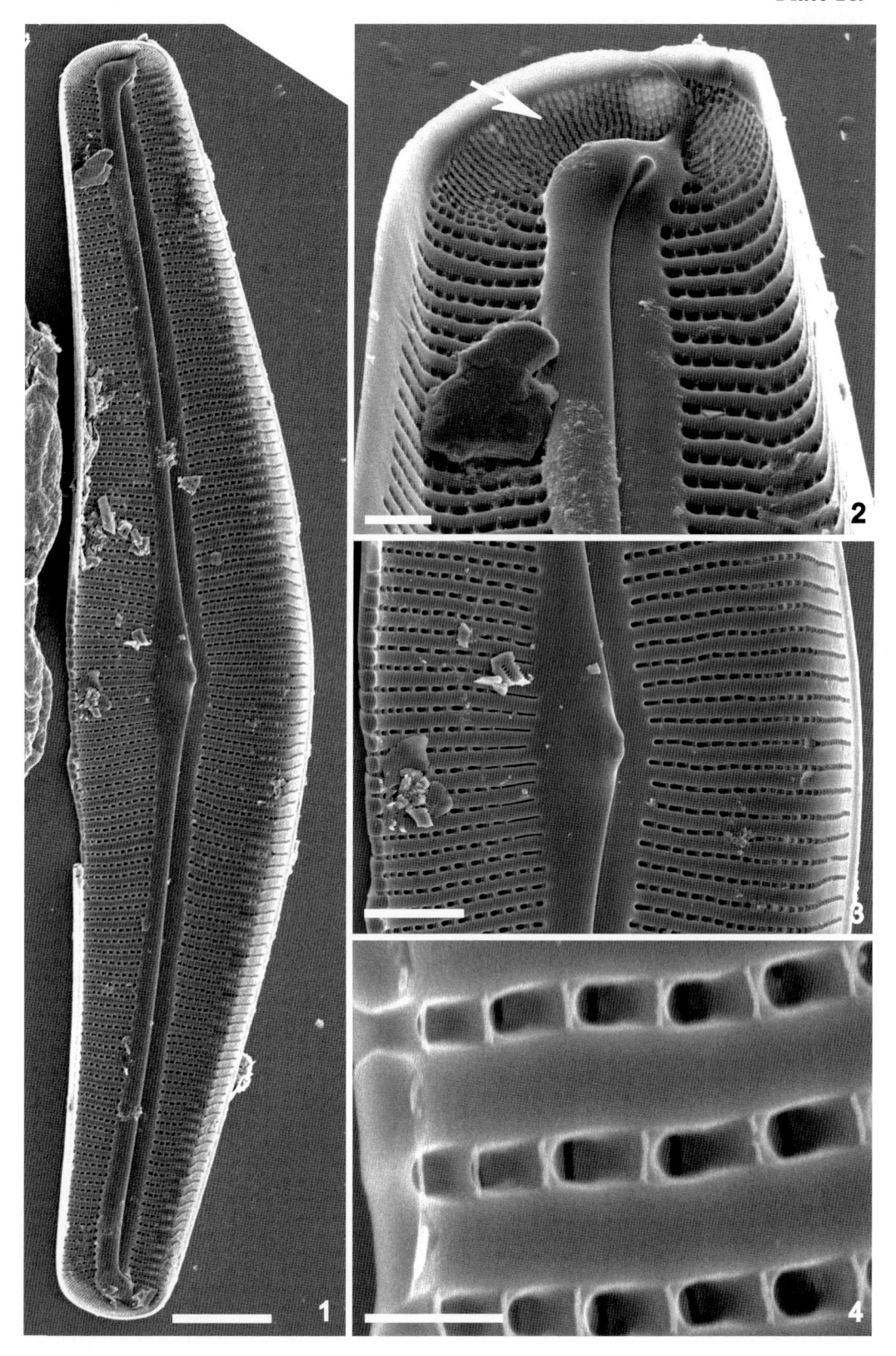

Plate 170. Biraphid, pennate diatoms: Cymbellales
Cymbopleura apiculata Krammer 2003

文　　献：Krammer, K. 2003. *Cymbopleura, Delicata, Navicymbula, Gomphocymbellopsis, Afrocymbella. In*: Lange-Bertalot, H.(ed.) Diatoms of Europe. Diatoms of the European inland waters and comparable habitats **4**: 1-530 pp. 164 pls. A.R.G. Gantner Verlag K.G.

形　　態：殻面は左右不対称の皮針形で，両殻端はくちばし状に突出する．殻長 75-97 μm，殻幅 23-27 μm，条線は 10 μm に 7（中心付近）-14 本，条線を構成する点紋は 10 μm に約 24 個，遊離点はない．極裂は背側に曲がり，外中心裂溝は腹側へ釣針状に曲がって終わる．内側での各胞紋間は薄い仕切りで境される．

ノート：本種は *Cymbopleura cuspidata*（Kütz.）Krammer に似るが，Krammer(2003)によると *C. cuspidata* の殻長は 28-61 μm で小さい．および横子珪藻土の個体は殻端が頭状の個体を含まないことから *C. subcuspidata* と区別できる．

産　　出：横子珪藻土：前期更新世・群馬県．

図　　版：横子珪藻土標本．

Figs 1-4. *Cymbopleura apiculata* Krammer
LM. Fig. 1. SEM. Figs 2-4. Material from diatomite of Yokogo (Early Pleistocene), Numata City, Gunma, Japan. Scale bars: Fig. 2=5 μm, Fig. 3=2 μm, Fig. 4=1 μm.

Fig. 1.　Whole valve view.
Fig. 2.　External view of whole valve.
Fig. 3.　Oblique view of an apex of Fig. 2.
Fig. 4.　Internal detailed view of areolae separated by silica struts.

Plate 170

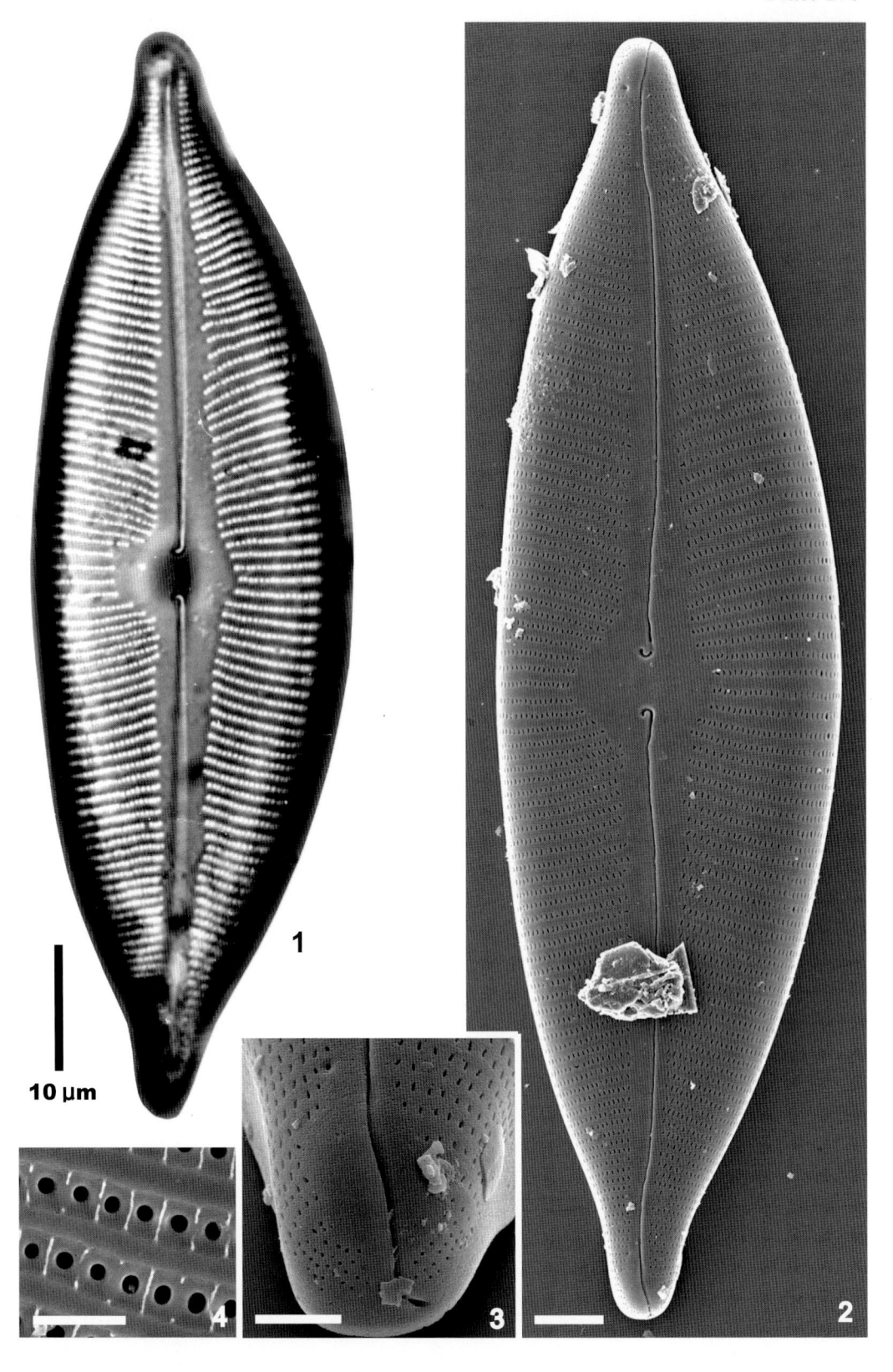

10 µm

Plates 171-172. Biraphid, pennate diatoms: Cymbellales
Cymbopleura inaequalis (Ehrenberg) Krammer 2003

基礎異名：*Navicula inaequalis* Ehrenberg 1836

文　献：Krammer, K. 2003. *Cymbopleura, Delicata, Navicymbula, Gomphocymbellopsis, Afrocymbella. In*: Lange-Bertalot, H.(ed.) Diatoms of Europe. Diatoms of the European inland waters and comparable habitats **4**: 1-530 pp. 164 pls. A.R.G. Gantner Verlag K.G.

　　Kützing, F.T. 1844. Die kieselschaligen Bacillarien oder Diatomeen. Nordhausen, W. Kohne. 152 pp, pls 1-30.

　　河島綾子・真山茂樹 2001. 阿寒湖の珪藻（8. 羽状類-縦溝類：*Cymbella, Encyonema, Gomphoneis, Gomphonema, Gomphosphenia, Reimeria*）. 自然環境科学研究 **14**: 89-109.

形　態：殻は左右不対称、殻長 72-91 μm、殻幅 22-35 μm、条線は腹側で 10 μm に 9-11 本. 遊離点・殻端小孔域は存在しない.

ノート：*Cymbopleura*(Krammer)Krammer 属は Krammer(1999)により正当化されている．*Cymbella inaequalis*(Ehrenb.)Rabenh. は *Cymbopleura* 属へ組み合わせになっており，類似種の *Cymbella ehrenbergii* Kütz. は本種の異名とされる（Krammer 2003）．しかし逆に *C. inaequalis* を *C. ehrenbergii* の異名とする見解もあり，その経過と *C. ehrenbergii* の形態の検討が河島・真山(2001)によって述べられている.

産　出：和村珪藻土（Skvortzov 1937）：新第三紀・長野県，中之条湖成層：中期更新世・群馬県，津森層：中期更新世・熊本県.

図　版 171-172：津森層標本.

Plate 171, Figs 1-2. *Cymbopleura inaequalis* (Ehrenberg) Krammer
LM. Figs 1-2. Material from Tsumori Formation (Middle Pleistocene), Mashiki Town, Kumamoto, Japan.

Figs 1-2. Two different valves.

Plate 171

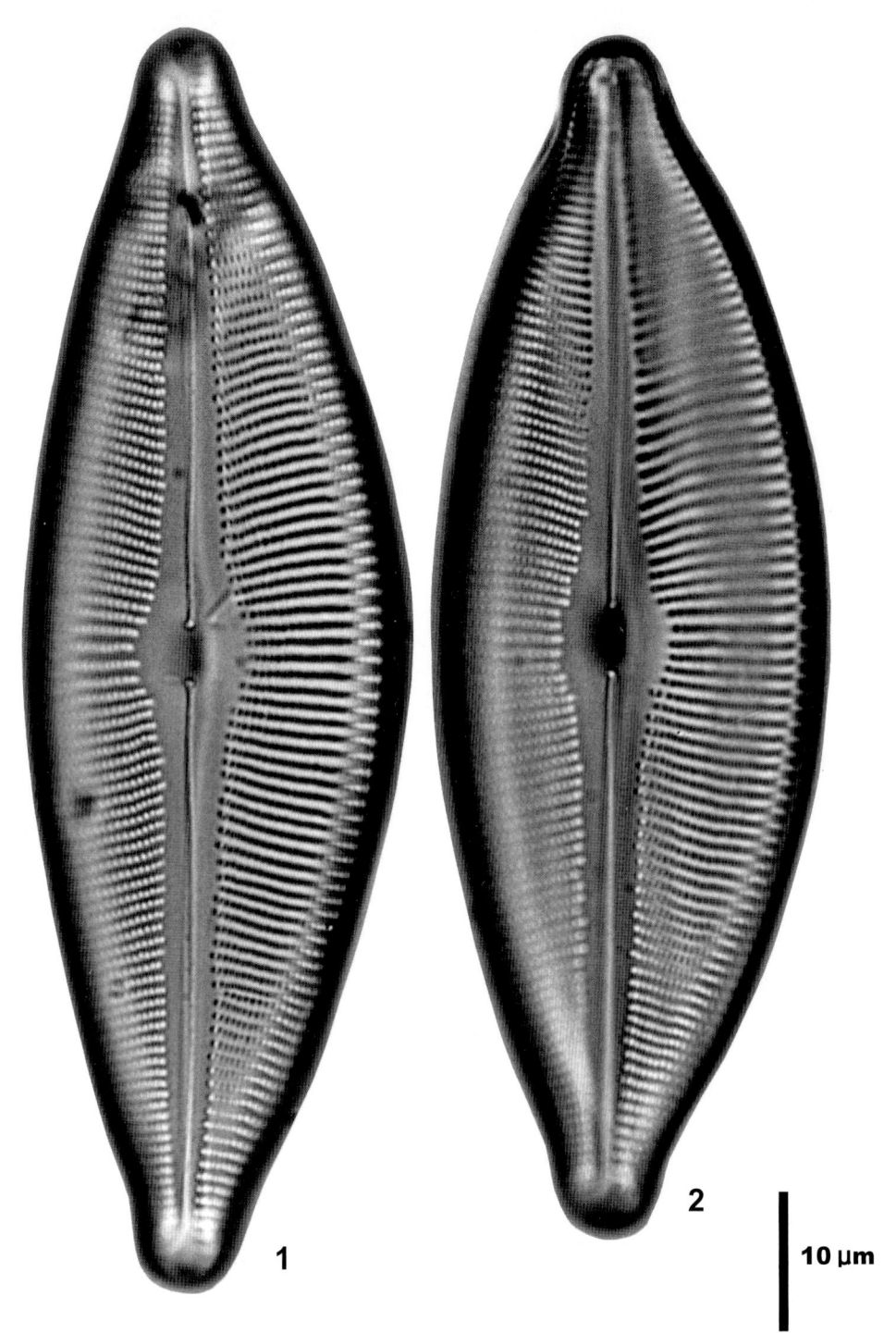

10 μm

Plate 172, Figs 1-4. *Cymbopleura inaequalis* **(Ehrenberg) Krammer**
SEM. Figs 1-4. Material from Tsumori Formation (Middle Pleistocene), Mashiki Town, Kumamoto, Japan. Scale bars: Figs 1-2=5 μm, Figs 3-4=0.5 μm.

Fig. 1. External oblique view of whole valve.
Fig. 2. Internal view of valve.
Fig. 3. Enlarged view of part of Fig. 1, showing outer openings of areolae.
Fig. 4. Enlarged view of part of Fig. 2, showing areolae separated by silica struts.

Plate 172

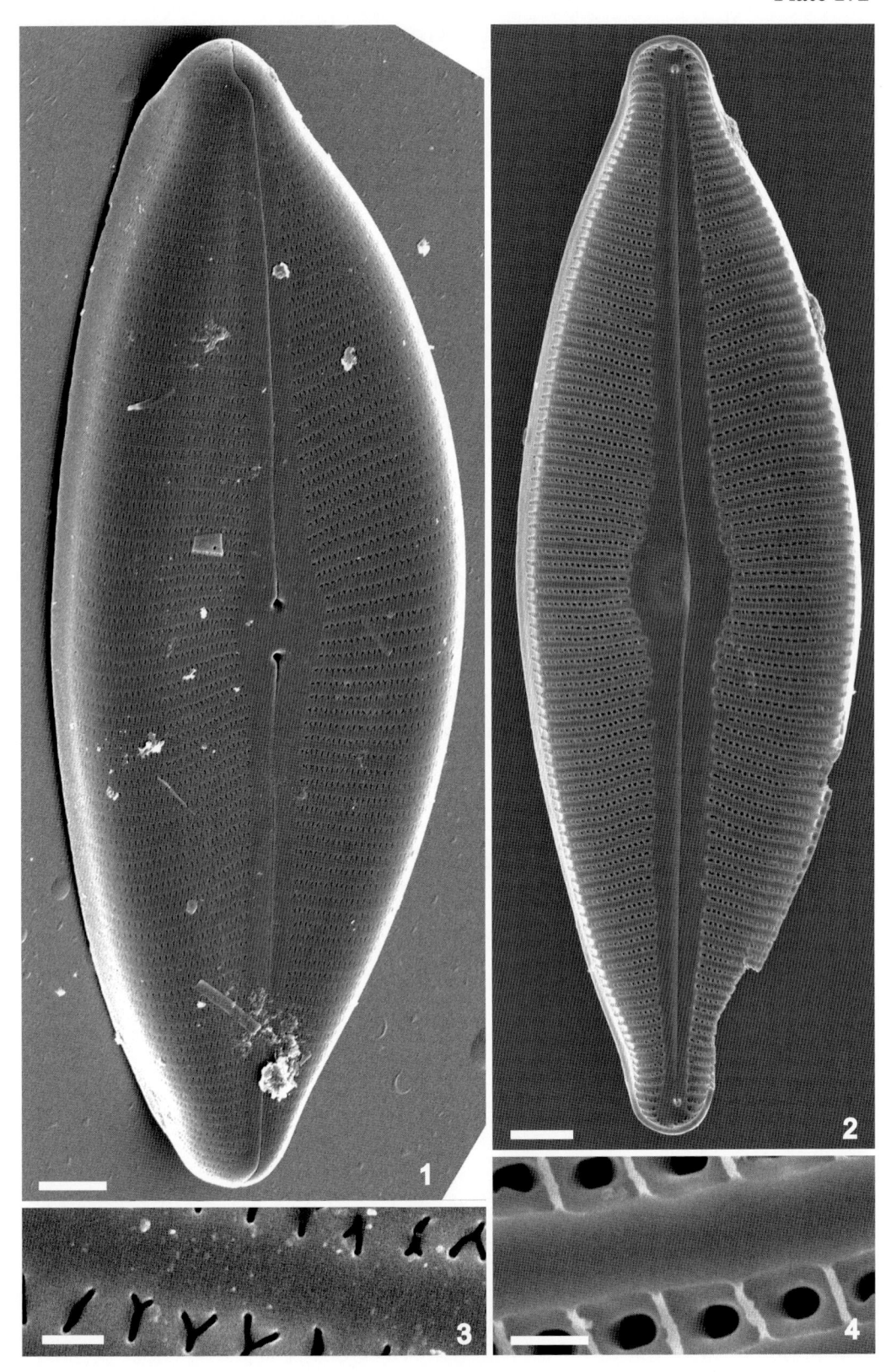

381

Plate 173.　Biraphid, pennate diatoms: Cymbellales
Cymbopleura naviculiformis (Auerswald) Krammer 2003

基礎異名：*Cymbella naviculiformis* Auerswald 1863
文　　献：Krammer, K. 2003. *Cymbopleura, Delicata, Navicymbula, Gomphocymbellopsis, Afrocymbella. In*: Lange-Bertalot, H.(ed.) Diatoms of Europe. Diatoms of the European inland waters and comparable habitats **4**: 1-530 pp. 164 pls. A.R.G. Gantner Verlag K.G.
　　　Hustedt, F. 1930.　Bacillariophyta (Diatomeae). *In*: Pascher, A.(ed.) Die Süsswasser-Flora Mitteleuropas. 1-466 pp. Jena, Gustav Fischer.
形　　態：殻面は左右不対称の広皮針形，殻端は頭状に突出する．殻長 33.5-44.5 μm，殻幅 8.5-13.5 μm，条線はゆるい放射状で，腹側 10 μm に 12-16 本，背側 10 μm に 10-14 本で腹側のほうがやや密であった．中心域はほぼ円形，縦溝の中心孔はわずか腹側に曲がり，両殻端の極裂は背側に曲がる．遊離点・殻端小孔域はない．
ノ　ー　ト：Atlas(A. Schmidt's)の図（A. Schmidt 1885, Plate 9）では *Cymbella naviculiformis* として中心域がない図が描かれているが，同じ Atlas の Plate 377 (Hustedt 1944) および Hustedt(1930, 1938)の図では横子珪藻土の本分類群程度にあり，よく似ている．Krammer(2003)により *Cymbopleura* 属へ組み合わせになった．
産　　出：横子珪藻土：前期更新世・群馬県，和村珪藻土（Skvortzov 1937）：新第三紀・長野県．
図　　版：横子珪藻土標本．

Figs 1-6.　*Cymbopleura naviculiformis* (Auerswald) Krammer
LM. Figs 1-2. SEM. Figs 3-6. Material from diatomite of Yokogo (Early Pleistocene), Numata City, Gunma, Japan. Scale bars: Figs 3-5＝5 μm, Fig. 6＝1 μm.

Figs 1-2.　Two different size valves.
Fig. 3.　External view of whole valve.
Fig. 4.　Oblique view of Fig. 3.
Fig. 5.　Internal view of whole valve.
Fig. 6.　Enlarged view of central area of Fig. 5.

Plate 173

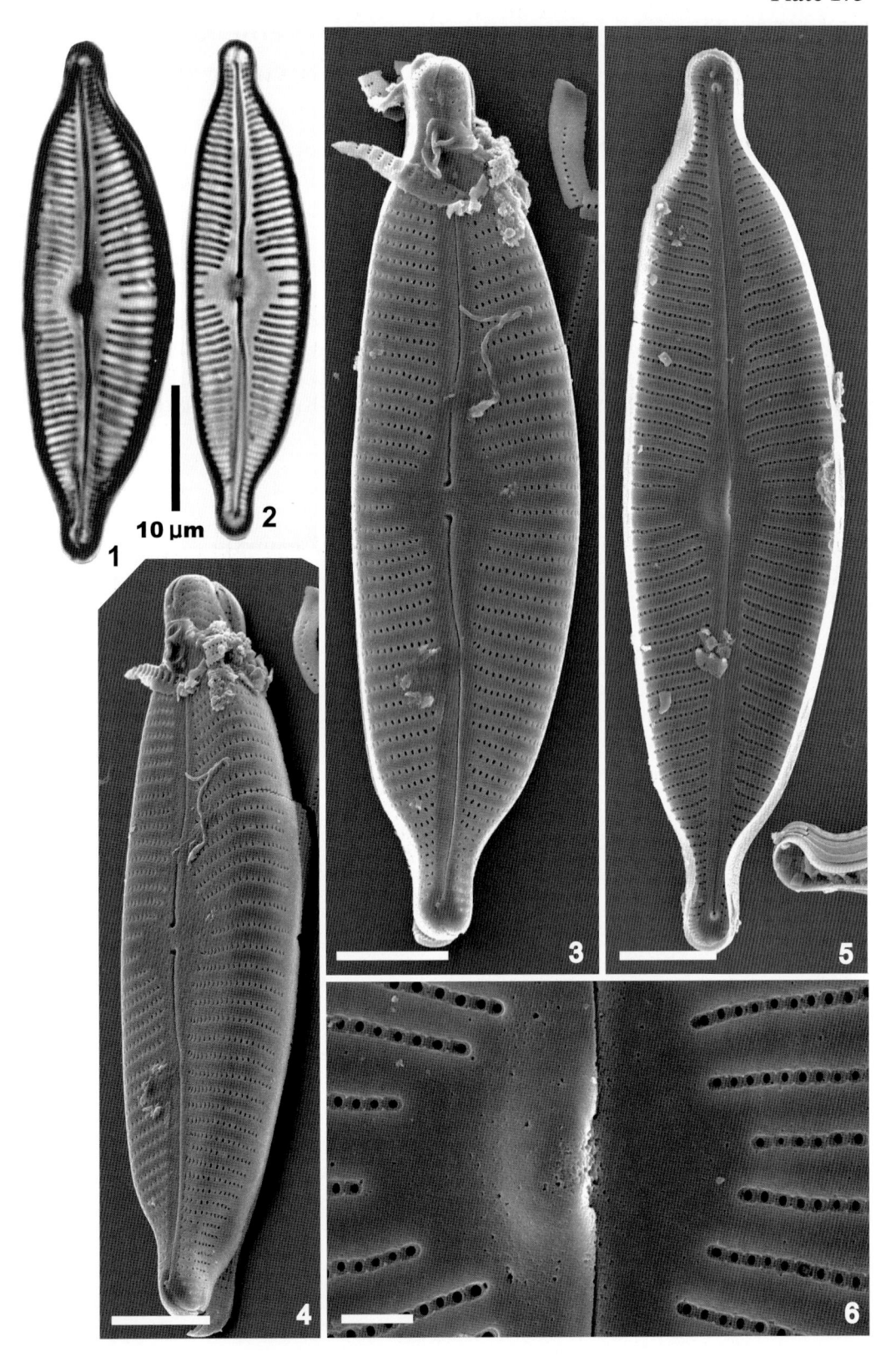

Plate 174.　Biraphid, pennate diatoms: Cymbellales
Cymbopleura subaequalis (Grunow) Krammer 2003

基礎異名：*Cymbella subaequalis* Grunow 1880
文　　献：Krammer, K. 2003. *Cymbopleura, Delicata, Navicymbula, Gomphocymbellopsis, Afrocymbella. In*: Lange-Bertalot, H.(ed.) Diatoms of Europe. Diatoms of the European inland waters and comparable habitats **4**: 1-530 pp. 164 pls. A.R.G. Gantner Verlag K.G.
形　　態：殻面はわずか左右不対称，殻端は広円形．殻長 31-51 μm，殻幅 9-10 μm．条線は放射状で，中心付近の腹側では 10 μm に 14-15 本，背側は 10 μm に 12-14 本であった．両殻端の軸域は狭いが中心に近づくにつれ広くなる．中心域の広がりは少ない，外裂溝は中心孔と極裂の間では背側に偏り，極裂は背側に曲がる．遊離点・殻端小孔域はない．
ノート：*Cymbella obtusa* Grunow と同定されたことも多いが，原口ら(1998)によると *C. obtusa* は *Cymbella aequalis* W. Sm の異名であり，*C. aequalis* は *Cymbella subaequalis* Grunow に同定されるべきものである．*C. subaequalis* は Krammer(2003)により *Cymbopleura* 属が設立されると共に同属へ組み合わせになった．
産　　出：横子珪藻土：前期更新世・群馬県．
図　　版：横子珪藻土標本．

Figs 1-7.　*Cymbopleura subaequalis* (Grunow) Krammer
LM. Figs 1-3. SEM. Figs 4-7. Material from diatomite of Yokogo (Early Pleistocene), Numata City, Gunma, Japan. Scale bars: Figs 4-5, 7＝2 μm, Fig. 6＝1 μm.

Figs 1-3.　Three different size valves.
Fig. 4.　External view of whole valve.
Fig. 5.　Oblique view of Fig. 4.
Fig. 6.　Enlarged oblique view of valve center of Fig. 4.
Fig. 7.　Internal oblique view of whole valve.

Plate 174

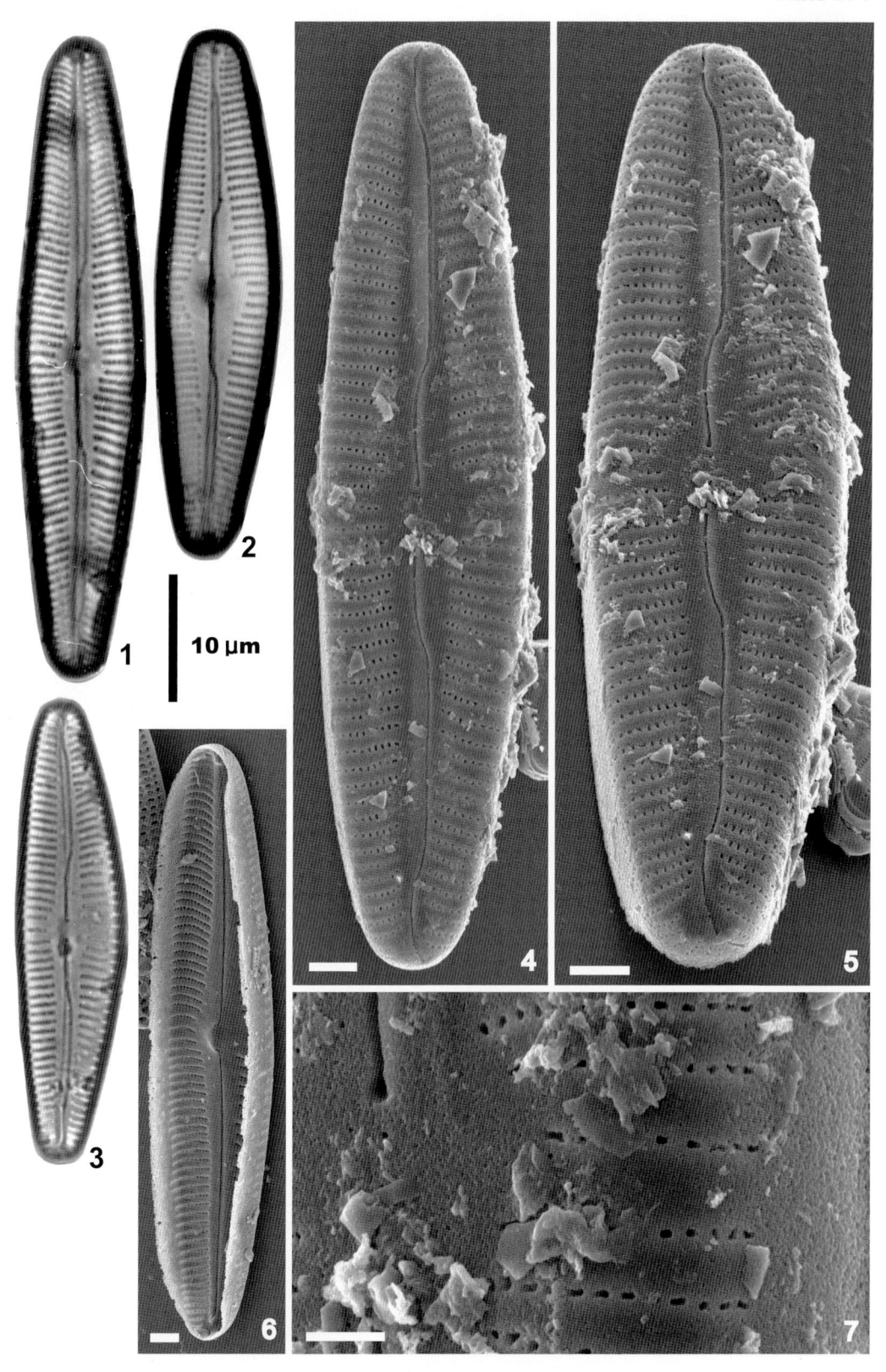

Plate 175.   Biraphid, pennate diatoms: Cymbellales
Didymosphenia curvata (Skvortsov & Meyer) Metzeltin & Lange-Bertalot 1995

基礎異名：*Didymosphenia geminata* var. *curvata* Skvortsov & Meyer 1928

文　　献：Metzeltin, D. & Lange-Bertalot, H. 1995.　Kritische wertung der taxa in *Didymosphenia* (Bacillariophyceae). Nova Hedwigia **60**: 381-405.

形　　態：殻は上下不対称で，わずかに左または右に曲がる．殻長 145-170 μm，殻幅約 44 μm，条線は 10 μm に約 15 本，条線を構成する点紋は 10 μm に 18-22 個．条線は中央では放射状であるが頭部に向かって平行～反放射になり，足部に向かっては徐々に平行的になるが急に放射方向にきつくなる．遊離点は 2-5 個．SEM 観察によると頭部（まれにはさらに広く）の殻面／殻套境界付近に線状の肥厚が認められる．

ノ ー ト：Metzeltin & Lange-Bertalot(1995)の図に殻形は一致する（微細構造は完全には一致しないが変異の範囲と考えられる）．*D. lineata* Skabichevskij にも類似するが殻形の曲がりが異なる．

産　　出：稲城層：前期更新世・東京都，野上層：中期更新世・大分県．

図　　版：稲城層標本．

Figs 1-7.   *Didymosphenia curvata* (Skvortsov & Meyer) Metzeltin & Lange-Bertalot

LM. Figs 1-3. SEM. Figs 4-7. Material from Inagi Formation (Early Pleistocene), Fuchu City, Tokyo, Japan. Scale bars: Fig. 4＝10 μm, Figs 5-6＝5 μm, Fig. 7＝0.5 μm.

Figs 1-2.   Two different size valves.
Figs 2-3.   Same valve at different focal planes.
Fig. 4.   External view of whole valve.
Fig. 5.   Detailed view of areolae with volae.
Fig. 6.   Oblique view of head pole of Fig. 4, ridge (arrow).
Fig. 7.   Internal view of foot pole, helictoglossa (arrowhead).

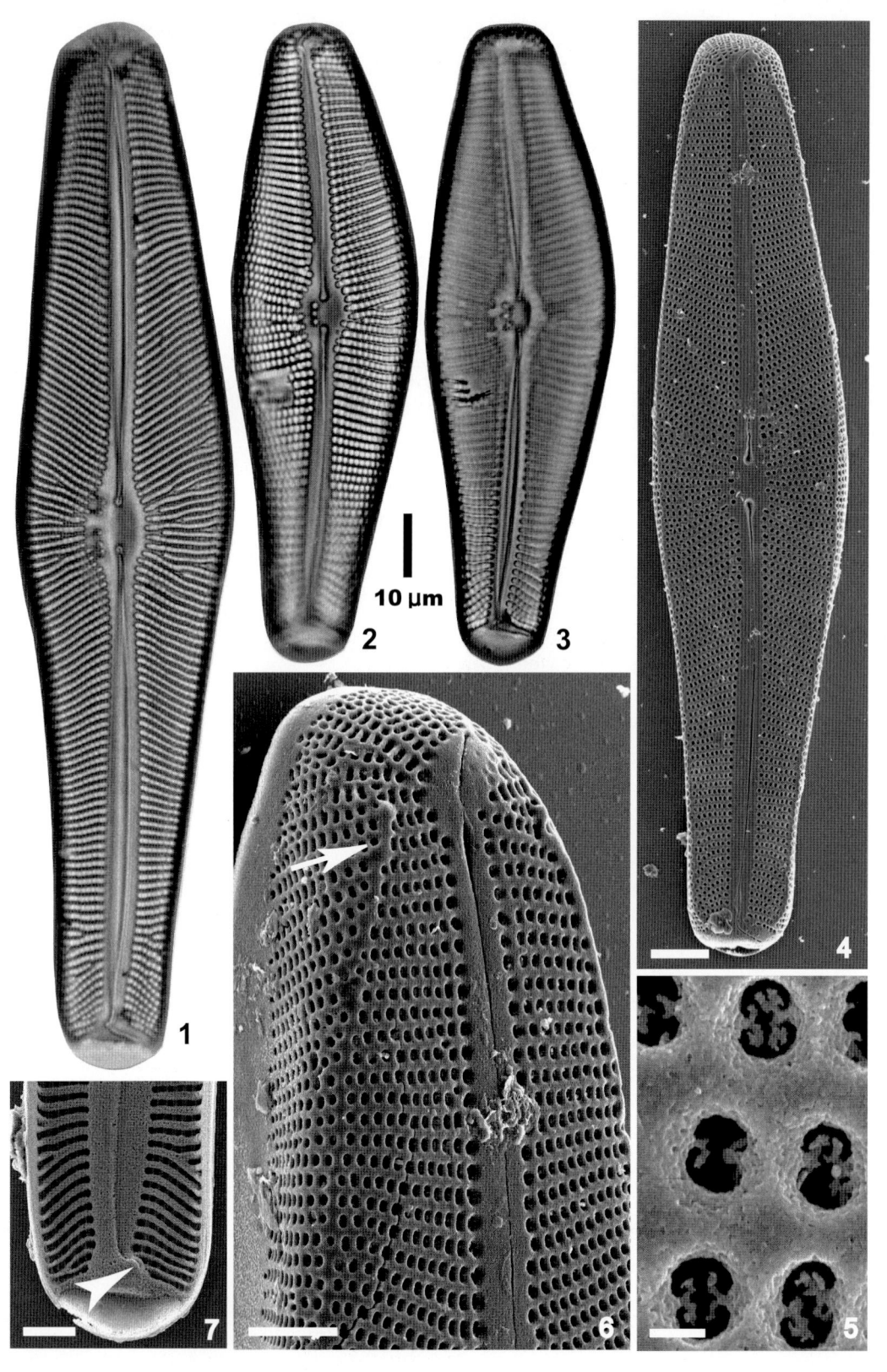

Plate 175

Plate 176.   Biraphid, pennate diatoms: Cymbellales
Didymosphenia fossils Horikawa & Okuno in Okuno 1944

文　　献：奥野春雄 1944. 日本珪藻土鉱床の植物分類学的研究(第2報). 植物学雑誌 **58**: 8-14.

形　　態：殻は上下不対称でくさび形，殻面観では長軸に対して右または左へわずか曲がっていることも多い．頭部の中央が凹む，殻長 125-174 µm，殻幅 33-40 µm，遊離点は1個．条線は中央では放射状であるが，頭部に向かって並行〜反放射になり，足部に向かっては徐々に平行になるが急に放射方向にきつくなる．条線は 10 µm に約7本．条線を構成する点紋は 10 µm に約9個．頭部の殻面／殻套境界はわずか肥厚が認められることがあるが，ないことが多い．小刺が頭部に認められる．

ノート：奥野(1944)により岡山県八束村珪藻土から記載された種であり，頭部が凹むことにより他種と明瞭に区別されるが，凹みの程度は様々である．

産　　出：小野上層（田中・小林 1995）：前期更新世・群馬県，大鷲湖沼性堆積物（田中ら 2011）：鮮新世・岐阜県，奄芸層群（根来・後藤 1981）：鮮新世・三重県，濃尾平野の更新統（Mori 1986）：更新世・愛知県，八束村珪藻土（奥野 1944）：更新世・岡山県．

図　　版：大鷲湖沼性堆積物標本.

Figs 1-5.   *Didymosphenia fossilis* Horikawa & Okuno
LM. Figs 1-2. SEM. 3-5. Material from lacustrine deposit of Owashi (Pliocene), Gujo City, Gifu, Japan. Scale bars: Figs 3-4＝10 µm, Fig. 5＝1 µm.

Figs 1-2.   Two different size valves.
Fig. 3.   External oblique view of whole valve.
Fig. 4.   Internal oblique view of whole valve.
Fig. 5.   Enlarged oblique view of part of Fig. 3 showing external openings of areolae.

Plate 176

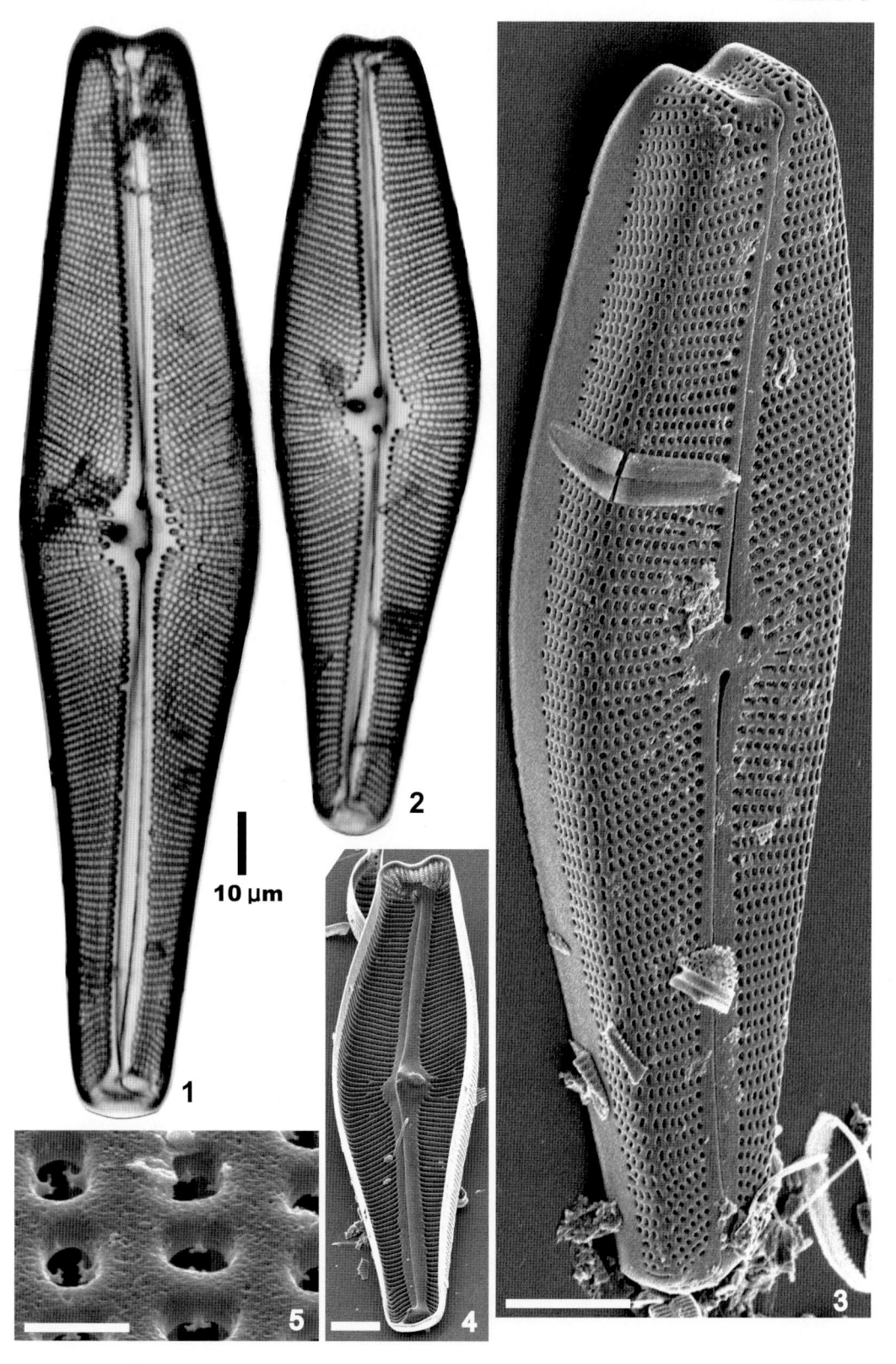

389

Plate 177. Biraphid, pennate diatoms: Cymbellales
Didymosphenia geminata (Lyngbye) M. Schmidt in A. Schmidt *et al.* 1899

基礎異名：*Echinella geminata* Lyngbye 1819

文　　献：Schmidt, M. 1899. *In*: A. Schmidt's Atlas der Diatomaceen-Kunde (1874-1959). Taf. 214. O.R. Reisland, Leipzig.

形　　態：殻は上下不対称でくさび形，頭部は大きく頭状に突出し，中央は膨らむ．殻長 79-127 μm，殻幅 30-42 μm，遊離点は 1-3 個．条線は中央では放射状であるが両殻端では弱放射～平行（まれに弱反放射）になる．条線は中央殻縁で 10 μm に約 10 本．条線を構成する点紋は 10 μm に約 12 個．殻左右の殻面／殻套境界は肥厚し，頭部の肥厚末端には刺が認められる．足部には殻端小孔域がある．外側の胞紋開口はすり鉢状である．

ノ　ー　ト：日本からは，現生では北海道から報告があるが（辻・Nergui 2008），化石としての産出報告は今回が初めてである．

産　　出：沼田湖成層：中期更新世・群馬県．

図　　版：沼田湖成層標本．

Figs 1-6. *Didymosphenia geminata* (Lyngbye) M. Schmidt

LM. Fig. 1. SEM. Figs 2-6. Material from Numata Lacustrine Deposit (Middle Pleistocene), Numata City, Gunma, Japan. Scale bars: Figs 1-3＝10 μm, Figs 4, 6＝5 μm, Fig. 5＝0.5 μm.

Fig. 1.　Whole valve view, photomontage.
Fig. 2.　External view of whole valve.
Fig. 3.　Internal oblique view of whole valve.
Fig. 4.　Oblique view of head pole of Fig. 2, spine (arrow).
Fig. 5.　Detailed view of foramina with rotae.
Fig. 6.　Enlarged view of foot pole of Fig. 3, helictoglossa (arrowhead).

Plate 177

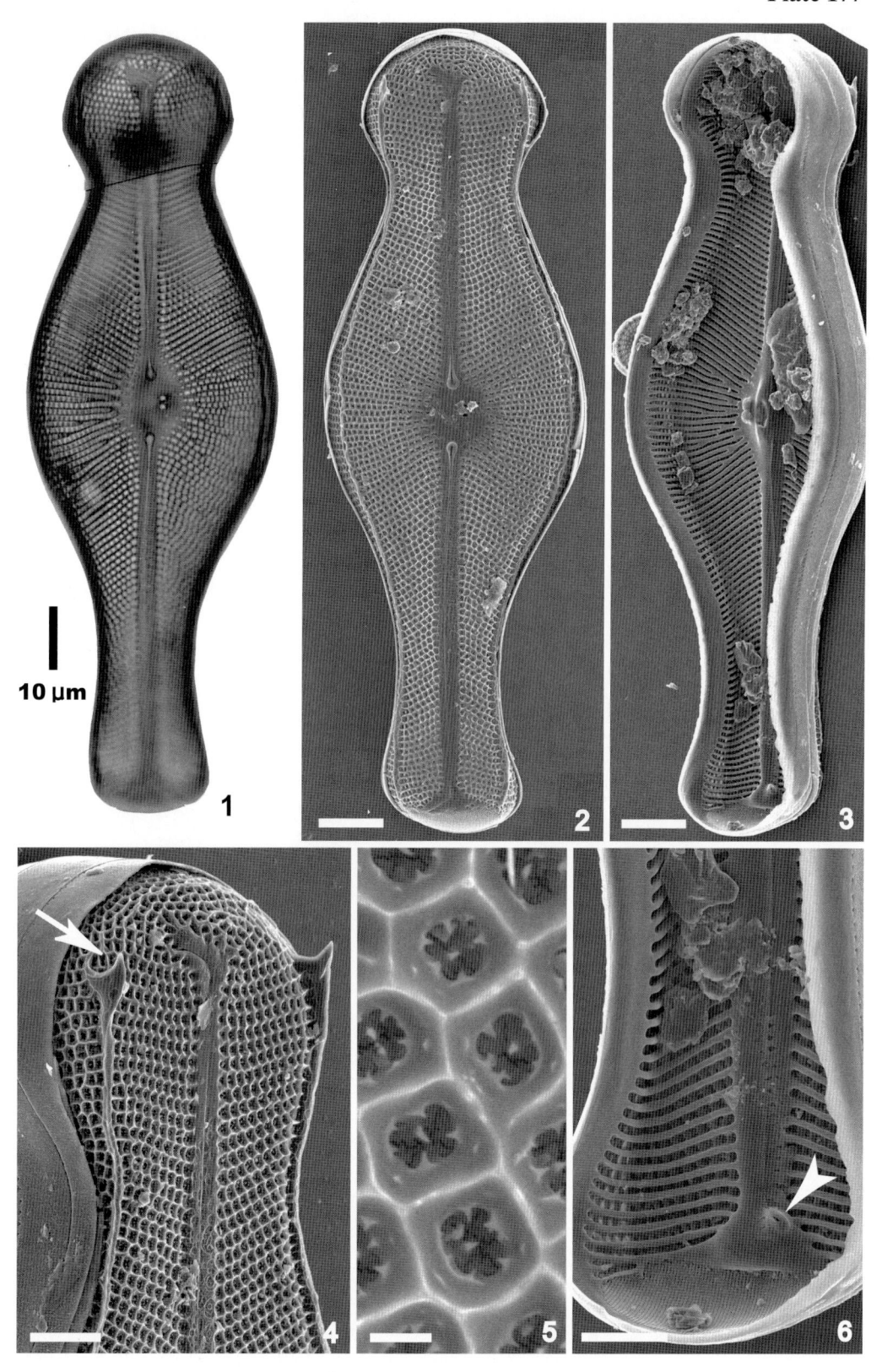

391

Plates 178-179.　Biraphid, pennate diatoms: Cymbellales
Didymosphenia nipponica H. Tanaka sp. nov.

新　　種：記載文（英文）は 14 頁参照.

文　　献：田中宏之・鹿島　薫 2007. 大分県杵築市西俣水に分布する珪藻土から見出された前期更新世淡水生珪藻群集. 地学研究 **56**: 137-145.

形　　態：殻は上下不対称で幅は比較的狭く，頭部は幅の広いくさび形で終る．殻長 93-191 μm，殻幅 26-38 μm．条線は 10 μm に約 8 本で，条線を構成する点紋は 10 μm に 7-12 個．条線は中央では放射状であるが頭部では（平行から）反放射になり，足部に向かっては平行になるが急に放射方向にきつくなる．軸域は中心域近くでやや狭くなる．遊離点は 2-5 個．殻面／殻套境界に刺・肥厚は認められない．殻端小孔域がある．

ノート：田中・鹿島(2007)では *Didymosphenia* sp. として報告したが，その後の検討で新種であることが判明したものである．頭部が幅の広いくさび形であることが他種との有用な識別点である．

産　　出：俣水層（田中・鹿島 2007）：前期更新世・大分県.

図　　版 178-179：俣水層（タイプ試料）標本.

Plate 178, Figs 1-5.　*Didymosphenia nipponica* H. Tanaka sp. nov.

LM. Figs 1-2. SEM. Figs 3-5. From the type material, OIT-104, diatomite of Matamizu Formation (Early Pleistocene), Kitsuki City, Oita, Japan. Scale bars: Fig. 3＝10 μm, Fig. 4＝5 μm, Fig. 5＝1 μm.

Fig. 1.　Photomontage using two photographs.
Figs 1-2.　Two different size valves.
Fig. 2.　Holotype.
Fig. 3.　External view of whole valve.
Fig. 4.　Enlarged oblique view of foot pole of Fig. 3 showing apical pore field.
Fig. 5.　Detailed view of external areolae with volae.

Plate 178

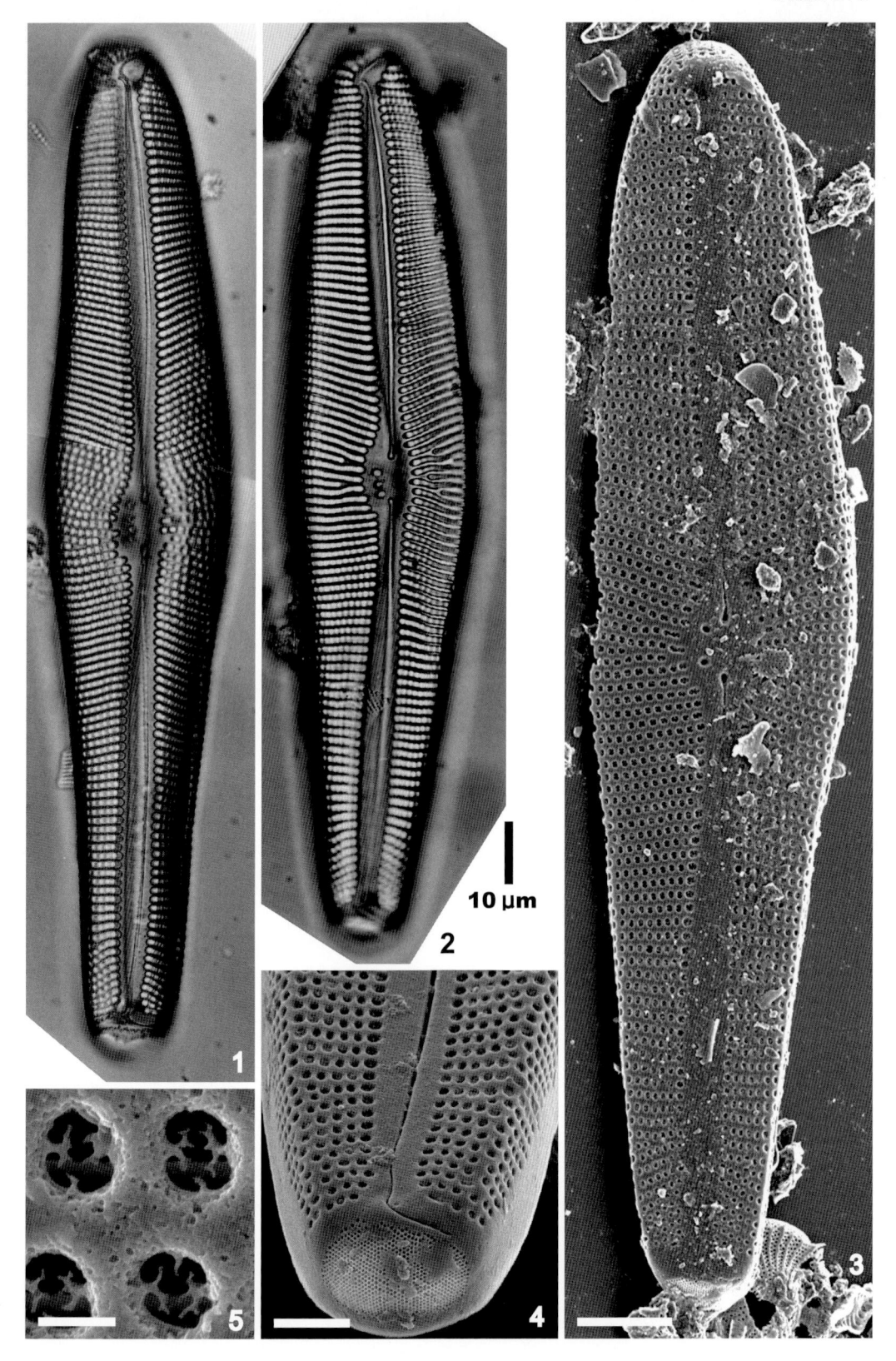

Plate 179, Figs 1–4. *Didymosphenia nipponica* **H. Tanaka** sp. nov.
SEM. Figs 1–4. From the type material, OIT-104, diatomite of Matamizu Formation (Early Pleistocene), Kitsuki City, Oita, Japan. Scale bars: Figs 1, 3=10 μm, Figs 2, 4=2 μm.

Fig. 1. Internal view of whole valve.
Fig. 2. Enlarged oblique view of head pole of Fig. 1.
Fig. 3. Oblique view of Fig. 1.
Fig. 4. Oblique view of valve center, stigma (arrowhead).

Plate 179

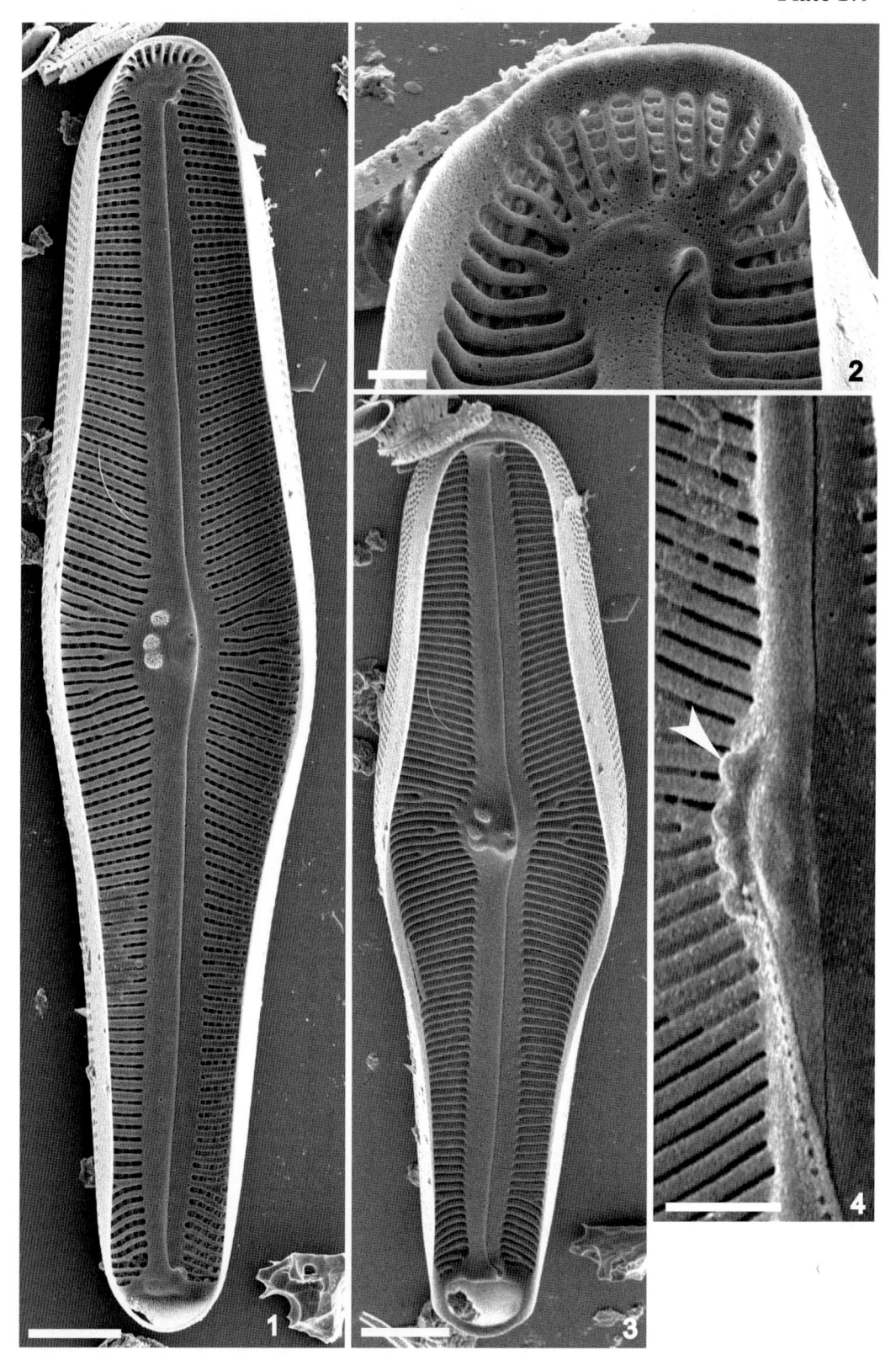

Plate 180. Biraphid, pennate diatoms: Cymbellales
Encyonema geisslerae Krammer 1997

文　　献：Krammer, K. 1997. Taxonomische probleme bei ähnlichen syntopischen sippen am beispiel einer *Encyonema*-assoziation (Bacillariophyceae) aus dem Kivu-See, Zaire. Nava Hedwigia **65**: 131-146.

形　　態：殻は左右不対称で，背側に大きく膨れるが腹側は中央で弱く膨らむ．殻長 41-57 μm，殻幅 14-15 μm．条線は太く 10 μm に腹側で約 6 本，背側で約 7 本，条線を構成する点紋は 10 μm に約 18 個，腹側は中心で放射状，殻端で反放射状になる．背側ではほぼ全域が放射状である．縦溝は中心では背側に曲がるが，両殻端では腹側に曲がる．遊離点，殻端小孔域は存在しない．

産　　出：稲城層：前期更新世・東京都．

図　　版：稲城層標本．

Figs 1-7. *Encyonema geisslerae* **Krammer**
LM. Figs 1-3. SEM. Figs 4-7. Material from Inagi Formation (Early Pleistocene), Fuchu City, Tokyo, Japan. Scale bars: Figs 4, 6＝5 μm, Fig. 7＝2 μm, Fig. 5＝1 μm.

Figs 1-3.　Three different size valves.
Fig. 4.　External view of whole valve.
Fig. 5.　Enlarged internal view of an apex showing helictoglossa (arrowhead).
Fig. 6.　Internal oblique view of whole valve.
Fig. 7.　Enlarged view of valve center of Fig. 6 showing internal central fissures.

Plate 180

Plate 181.　Biraphid, pennate diatoms: Cymbellales
Encyonema vulgare Krammer 1997

文　　献：Krammer, K. 1997. Die cymbelloiden diatomeen. Eine monographic der weltweit bekannten taxa. Teil 1. Allgemeines und *Encyonema* part. Bibliotheca Diatomologica **36**: 1-382.

形　　態：殻は左右不対称で，腹側中心はやや膨らみ背側は大きく膨らむ．殻長 32-58 μm，殻幅 10-15 μm．条線はわずかに放射状で，中心付近で 10 μm に約 8 本，殻端では約 10 本である．条線を構成する点紋は 10 μm に約 22 個．外裂溝は腹側近くを走り，極裂は腹側へ曲がるが，内裂溝はほぼ直線で中央末端は背側へ曲がる．遊離点は 1 個である．内側で胞紋を境する小壁，および殻端小孔域は存在しない．

ノ ー ト：本種は Krammer(1997)で新種記載された．大鷲湖沼性堆積物の分類群とは，殻形が Krammer(1997)の Taf. 42；Figs 1, 2 が似ているが，他の図はこれよりやや細長い．しかし大鷲湖沼性堆積物からの個体も図版に使用した物より細長い個体がある．Krammer(1997)によれば，*Cymbella turgida* は異名である．

産　　出：大鷲湖沼性堆積物：鮮新世・岐阜県．*C. turgida* としては，嬉野珪藻土（Skvortzov 1937）：新第三紀・佐賀県等から報告がある．

図　　版：大鷲湖沼性堆積物標本．

Figs 1-7.　*Encyonema vulgare* Krammer

LM. Figs 1-2. SEM. Figs 3-7. Figs 1-7. Material from lacustrine deposit of Owashi (Pliocene), Gujo City, Gifu, Japan. Scale bars: Figs 3-4＝5 μm, Figs 5-7＝1 μm.

Figs 1-2.　Two different size valves.

Fig. 3.　External view of whole valve.

Fig. 4.　Internal view of whole valve.

Fig. 5.　Enlarged oblique view of an apex of Fig. 3 showing terminal fissure (arrow).

Fig. 6.　Enlarged view of central area of Fig. 4 showing internal central fissures and stigma (arrowhead).

Fig. 7.　Enlarged view of central area of Fig. 3 showing central pores of raphes and stigma (arrowhead).

Plate 181

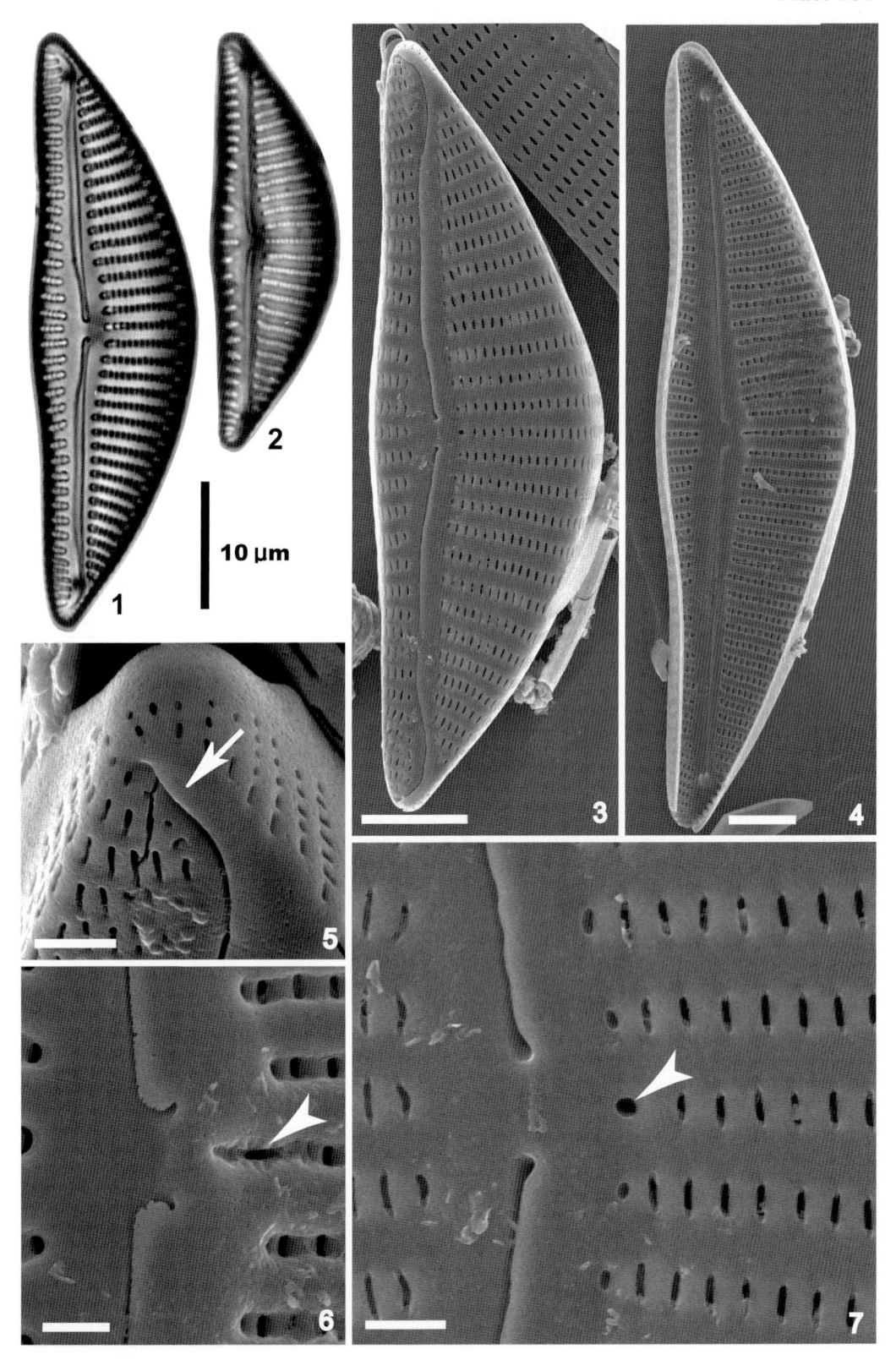

Plate 182.   Biraphid, pennate diatoms: Cymbellales
Gomphoneis okunoi Tuji 2005

文　　献：Tuji, A. 2005. Taxonomy of the *Gomphoneis tetrastigmata* complex. Bulletin of the National Science Museum, Tokyo, Ser. B. **31**: 89-108.
　　　　渡辺仁治・浅井一視・大塚泰介・辻　彰洋・伯耆晶子 2005. 淡水珪藻生態図鑑. 666 pp. 内田老鶴圃, 東京.

形　　態：殻はくさび形で頭端は広円状．中央部から足部にかけて細くなる．殻長 22-24 μm, 殻幅 7.5-9 μm, 殻長／殻幅は 2.6-3.0 であった．LM 観察では，縦溝は軸域の中心を通り直線的に走るが，SEM 観察によると外裂溝はわずか横に波打っている．内中心裂溝は上下の縦溝枝が同じ方向へ曲がって終了する．遊離点は（3）4個である．条線は全体的にやや放射状であるが，殻端で平行的になる殻もある．条線数は 10 μm に 13-14 本で, 2 列の胞紋列から構成されている．足部には殻端小孔域がある．

ノ ー ト：類似している *Gomphoneis pseudokunoi* Tuji とは殻の長軸方向と短軸方向の長さの比で比較するとわかりやすい．図版の Fig. 2 と Figs 3-4 は同じ殻であるが LM 用のスライドにする際足部が破損した（LM の殻の向きは SEM と左右が逆）．

産　　出：沼田湖成層：中期更新世・群馬県．
図　　版：沼田湖成層標本．

Figs 1-7.   *Gomphoneis okunoi* Tuji

LM. Figs 1-2. SEM. Figs 3-7. Material from Numata Lacustrine Deposit (Middle Pleistocene), Numata City, Gunma, Japan. Scale bars: Figs 3-5＝2 μm, Figs 6-7＝1 μm.

Figs 1, 5-7.   Same valve, LM and SEM views.
Figs 2-4.   Same valve, LM (broken) and SEM views.
Fig. 3.   External view of whole valve.
Fig. 4.   Oblique view of Fig. 3.
Fig. 5.   Internal oblique view of whole valve.
Fig. 6.   Enlarged view of valve center of Fig. 5 showing four stigmata and striae consisting of two areolae rows, stigma (arrowhead).
Fig. 7.   Enlarged oblique view of head pole of Fig. 5, helictoglossa (arrow).

Plate 182

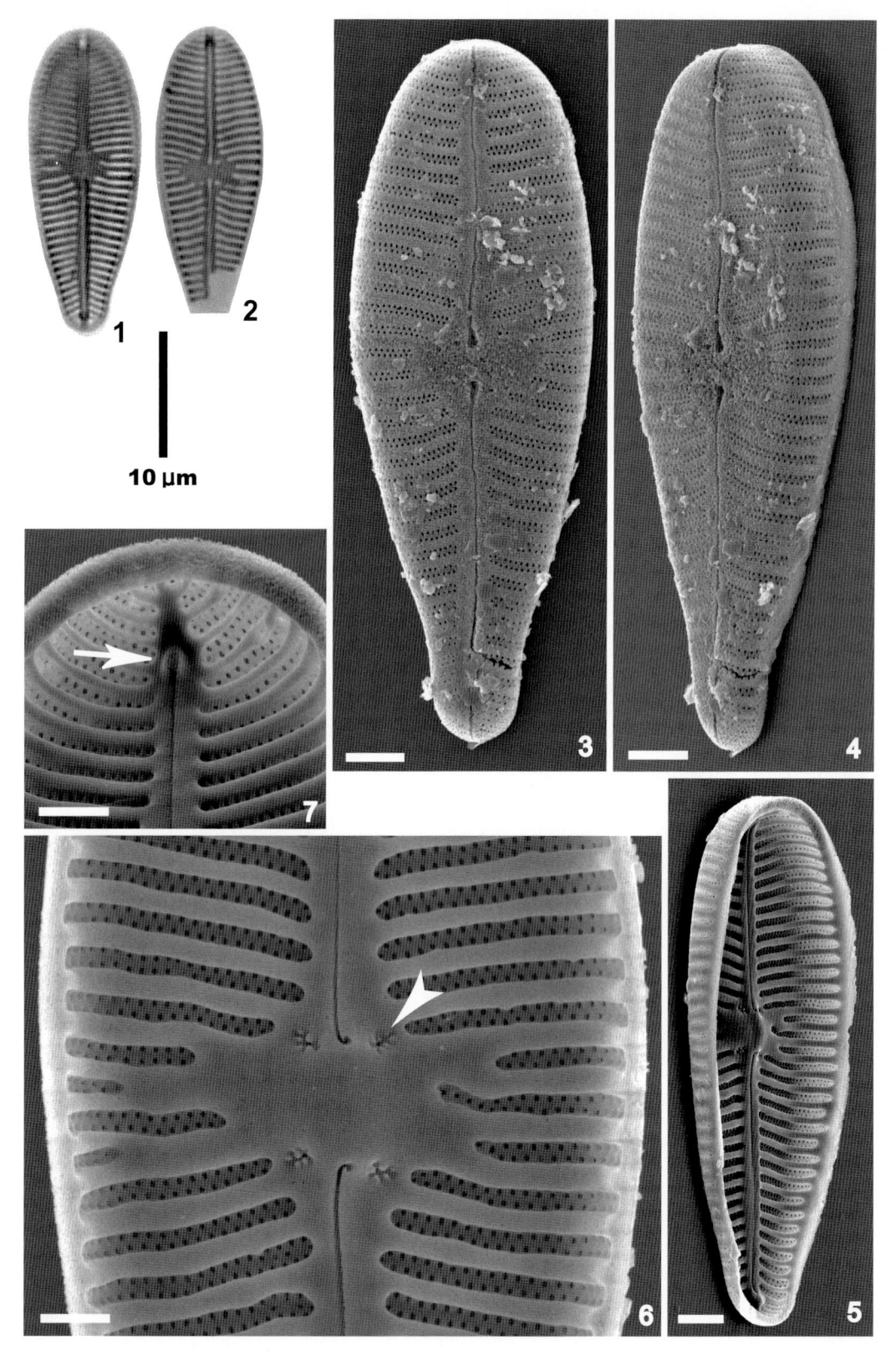

Plate 183.   Biraphid, pennate diatoms: Cymbellales
Gomphoneis tumida (Skvortsov) Kociolek & Stoermer 1988

基礎異名：*Gomphonema quadripunctatum* var. *genuina* f. *tumida* Skvortzov 1928
文　　献：Kociolek, J.P. & Stoermer, E.F. 1988.　Taxonomy and systematic position of the *Gomphoneis quadripunctata* species complex. Diatom Research **3**: 95-108.
形　　態：殻はくさび形で頭端は広円状．中央部から足部にかけて細くなる．殻長 33-50 μm，殻幅 10-14.5 μm．中心域は円状．外中心孔はやや幅が広がる程度である．内中心裂溝は上下の縦溝枝が同じ方向へ曲がって終了する．遊離点は 4-6 個である．条線は全体的に放射状で，条線数 10 μm に 11-12 本，2 列の胞紋列から構成されている．足部には殻端小孔域がある．
ノ ー ト：類似している *Gomphoneis okunoi* Tuji とは中心域の形，条線の密度が異なり，*Gomphoneis hastata*（Wislouch）Kociolek & Stoermer とは条線の密度，殻中央部の膨らみが異なる．Kociolek & Stoermer(1988)で示されている *G. tumida* とは，殻頭部における条線の向きが沼田湖成層産の個体は放射状であることが異なるが，中心域の形，条線の密度（10 μm に 10-14 本），4-6 個の遊離点などの形態が沼田湖成層産の本種と一致した．
産　　出：沼田湖成層：中期更新世・群馬県．
図　　版：沼田湖成層標本．

Figs 1-7.   *Gomphoneis tumida* (Skvortsov) Kociolek & Stoermer
LM. Figs 1-2. SEM. Figs 3-6. Material from Numata Lacustrine Deposit (Middle Pleistocene), Numata City, Gunma, Japan. Scale bars: Figs 3-4＝5 μm, Figs 5-6＝2 μm.

Figs 1-2.   Two different size valves.
Fig. 3.   Internal oblique view of whole valve.
Fig. 4.   External view of whole valve (broken).
Fig. 5.   Enlarged central area of Fig. 4 showing central raphe endings, stigmata and striae consisting two areolae rows, stigmata (arrowheads).
Fig. 6.   Enlarged view of valve center of Fig. 3 showing six stigmata and striae consisting of two areolae rows, stigmata (arrowheads).

Plate 183

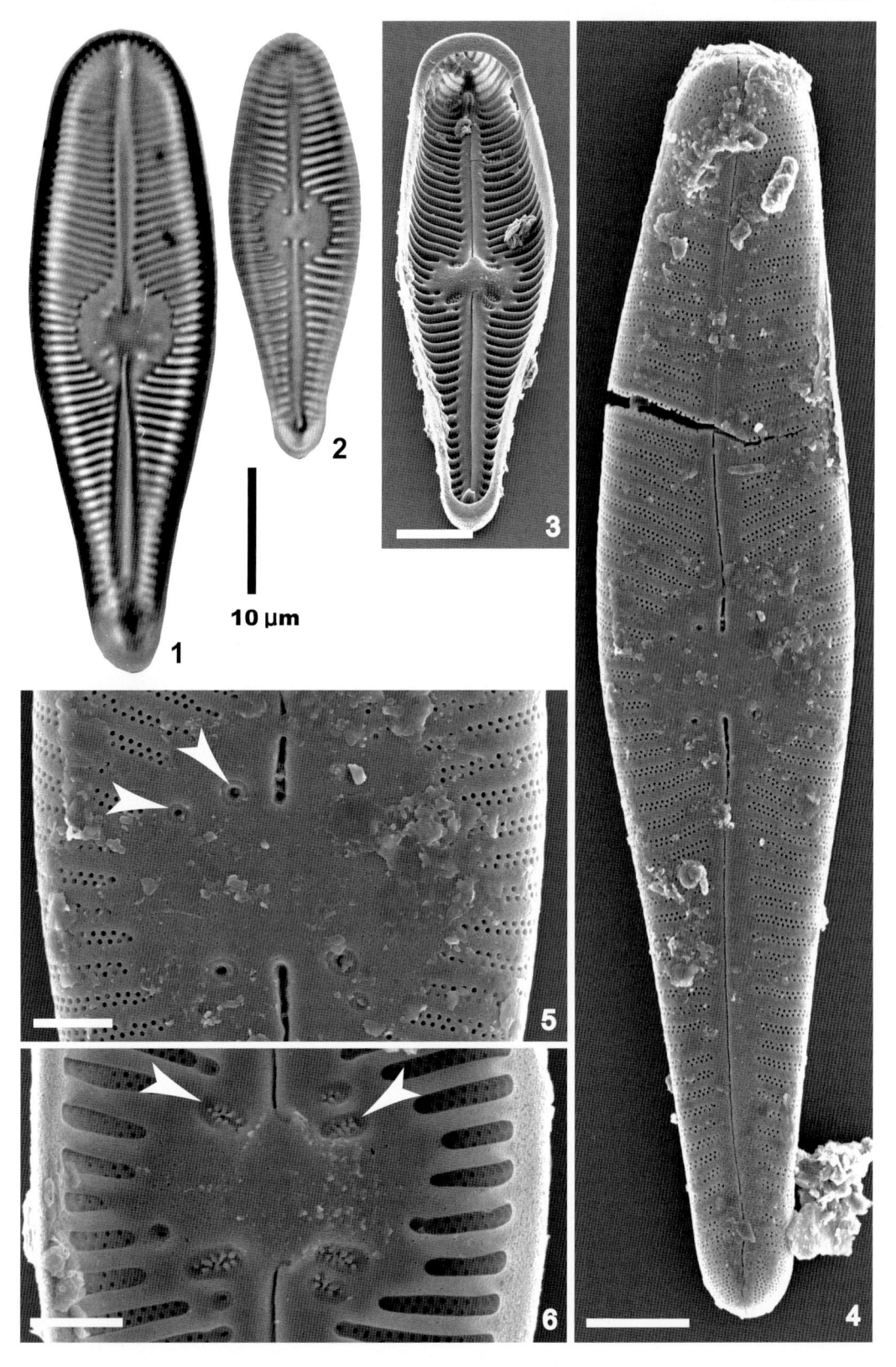

403

Plate 184.　Biraphid, pennate diatoms: Cymbellales
Gomphonema augur var. *gautieri* Van Heurck 1885

文　　献：Schmidt, M. & Fricke, F. 1902. *In*: A. Schmidt's Atlas der Diatomaceen-Kunde (1874-1959), pls 233-240. O.R. Reisland, Leipzig.

形　　態：殻はくさび形で頭端は細いくちばし状に突出する．頭部と中央部の間が凹み，中央部から足部にかけて細くなる．殻長68-76 µm，殻幅15-17 µm，縦溝の外裂溝は軸域を遊離点とは反対側に偏って走るが，内側では中央を走る．遊離点は1個である．条線は中央では平行であるが，両端へ向かうに従い放射状になり，頭部では突出した先端を中心とした弧状になることが多い．条線数10 µmに約10本，条線を構成する点紋は10 µmに約20個である．足部には殻端小孔域がある．

ノ ー ト：頭部と中央部の間が凹むのが本変種の特徴であるが，阿寒湖の分類群（現生）は凹むものから凹まないものまで変化が連続する（河島・真山 2001）．横子珪藻土の分類群はすべて凹む．

産　　出：夕張堆積物（Okuno 1952）：鮮新世・北海道，横子珪藻土：前期更新世・群馬県，奄芸層群（根来・後藤 1981）：鮮新世・三重県，嬉野珪藻土（Skvortzov 1937）：新第三紀・佐賀県．

図　　版：横子珪藻土標本．

Figs 1-5.　*Gomphonema augur* var. *gautieri* Van Heurck

LM. Fig. 1. SEM. Figs 2-5. Material from diatomite of Yokogo (Early Pleistocene), Numata City, Gunma, Japan. Scale bars: Fig. 5＝10 µm, Figs 2-3＝5 µm, Fig. 4＝0.5 µm.

Fig. 1.　Whole valve view.
Fig. 2.　External view of whole valve, stigma (arrowhead).
Fig. 3.　Internal view of whole valve, stigma (arrowhead).
Fig. 4.　External detailed view of rotae of external areolae openings.
Fig. 5.　External oblique view of whole valve.

Plate 184

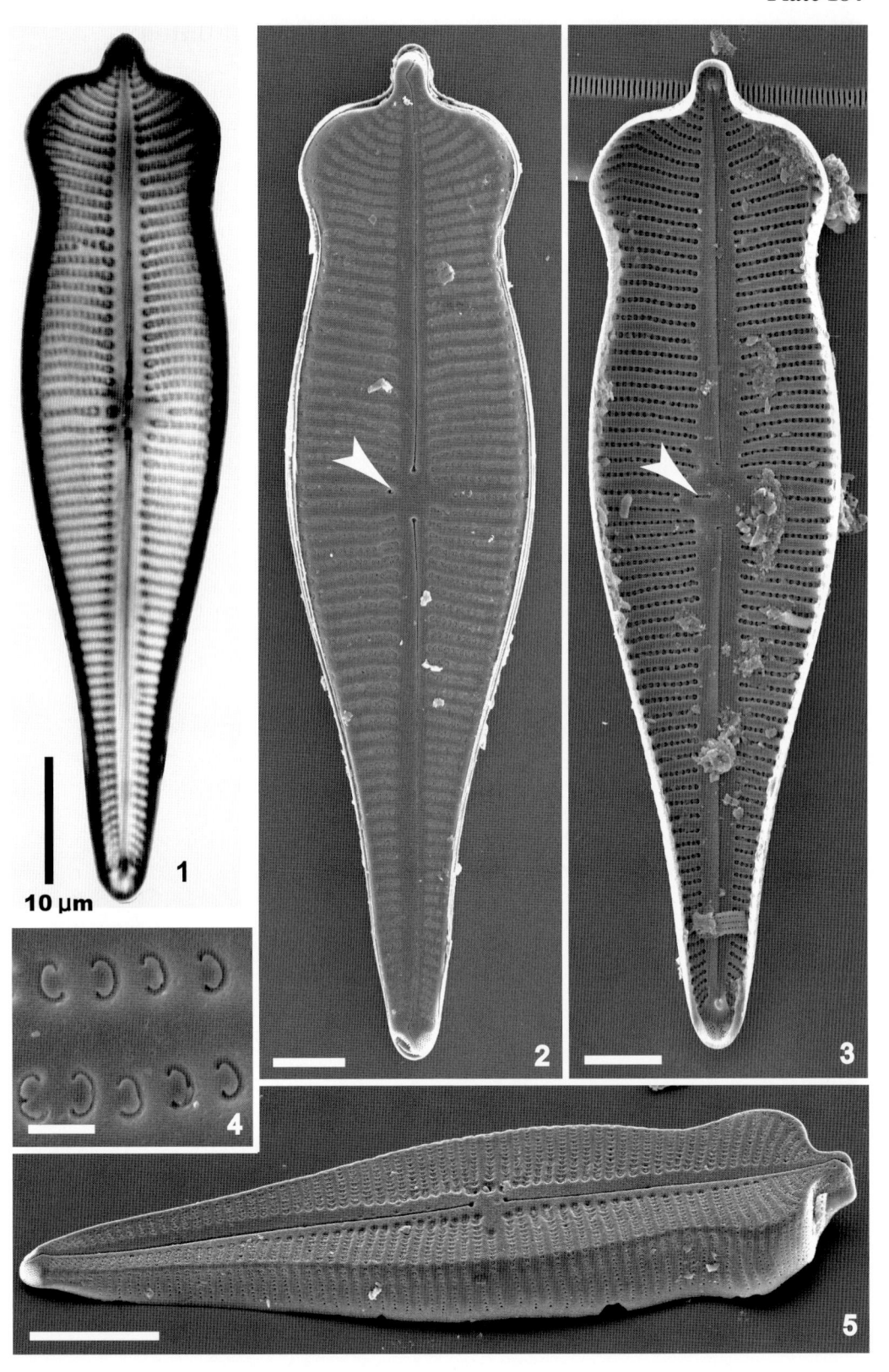

Plate 185.　Biraphid, pennate diatoms: Cymbellales
Gomphonema biceps F. Meister 1934

文　　献：Meister, F. 1934.　Seltene und neue Kieselalgen. I. Berichte der Schweizerischen Botanischen Gesellschaft (Zürich). Bd. 44. S. 87-106.

形　　態：殻はわずかくさび形の皮針形で両殻端は頭状に突出する．殻長 27-32 μm，殻幅 6.5-7.5 μm，縦溝の外裂溝はカーブし，中央末端は極裂とは反対側（遊離点側）に曲がり，中心孔で終了するが，内裂溝は直線で内中心裂溝は遊離点側へ釣針状に曲がる．遊離点は 1 個．条線は放射状で，10 μm に約 13 本，単列の胞紋から構成されている．軸域はやや広く，中心域は軸域よりわずか広い．足部に殻端小孔域がある．

ノ ー ト：Meister(1934)は写真 1 点を図示しているが，この図と比較すると桐生川産の本種は頭部・足部の突出・伸長の程度が少なく頭部はやや小形である．しかし既存の種の中では最も類似しているので本種に同定した．Meister(1934)の記述では殻長 30 μm，殻幅 75 μm，条線は 10 μm に 12 本とあり，桐生産の本種と計測値はよい一致を示した．今までのところ化石からの報告は見当たらない．

産　　出：斐伊川（Ohtsuka 2002）：鳥取県（現生），那賀川（小藤ら 2004）：徳島県（現生），桐生川：群馬県（現生）．

図　　版：桐生川標本．

Figs 1-8.　*Gomphonema biceps* F. Meister

LM. Figs 1-2. SEM. Figs 3-8. Material from Kiryu River (Recent), Kiryu City, Gunma, Japan. Scale bars: Figs 3-5＝2 μm, Fig. 6＝1 μm, Figs 7-8＝0.5 μm.

Figs 1-2.　Two different size valves.
Fig. 3.　External view of whole valve, stigma (arrowhead), apical pore field (arrow).
Fig. 4.　Oblique view of Fig. 3.
Fig. 5.　Enlarged oblique view of foot-pole of Fig. 3 showing apical pore field (arrow).
Fig. 6.　Detailed view of an areolae row of Fig. 3.
Fig. 7.　Internal view of whole valve, stigma (arrowhead).
Fig. 8.　Detailed view of an areolae row of Fig. 7.

Plate 185

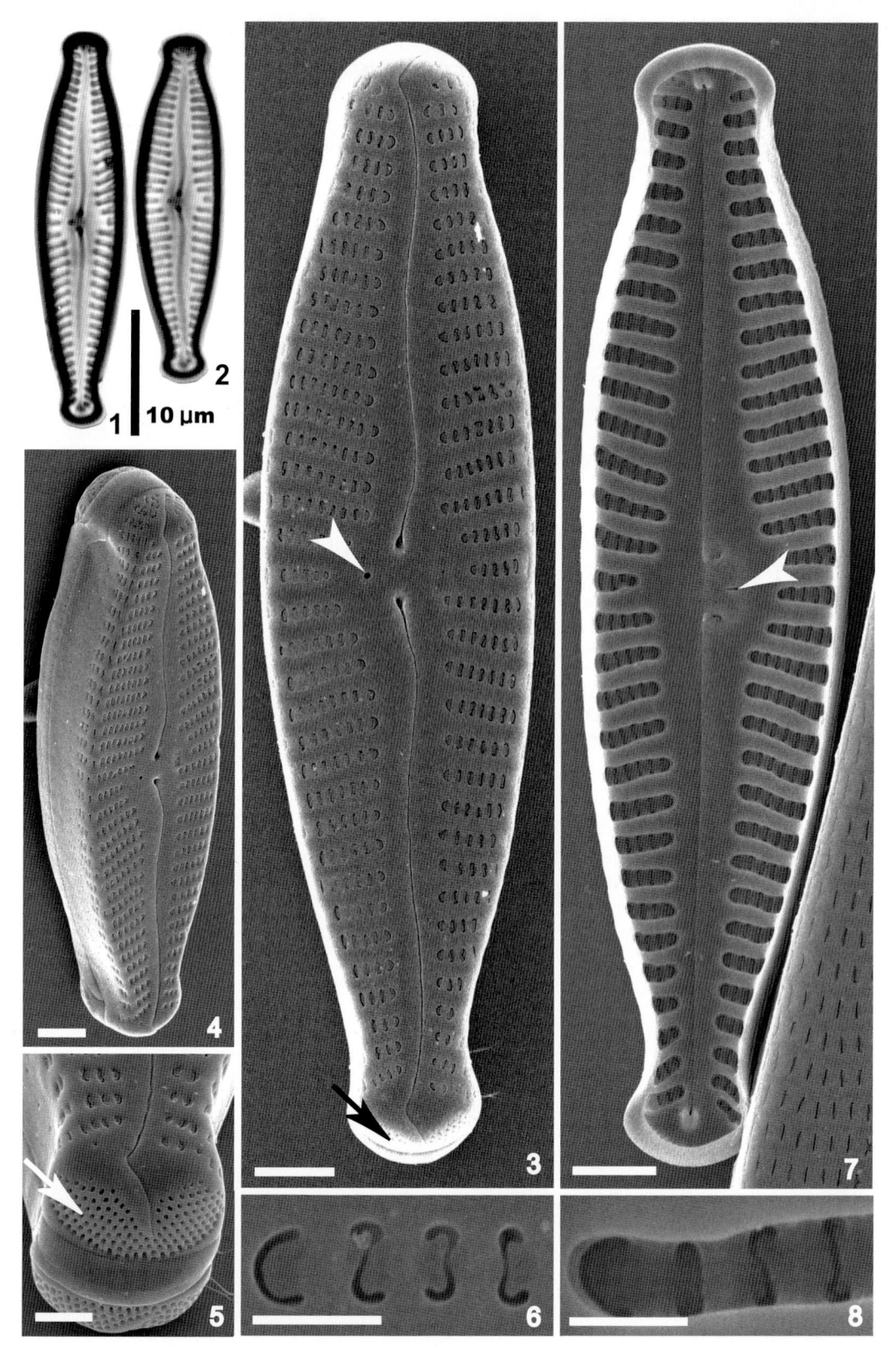

Plate 186.　Biraphid, pennate diatoms: Cymbellales
Gomphonema coronatum Ehrenberg 1840

文　　献：河島綾子・真山茂樹　2004．阿寒湖の珪藻（11．羽状類-縦溝類：*Campylodiscus, Cymatopleura, Surirella* and additional 10 taxa）．自然環境科学研究 **17**: 1-21.

形　　態：殻は中央と頭部が膨れ，頭部は3方向へ突出する．殻長 50-57 μm，殻幅は頭部で 13-14 μm，中央で約 13 μm．縦溝の外側中心域末端はやや遊離点側へ偏り直線的に終了するが，内側の中心裂溝は釣針状に曲がる．遊離点は1個である．条線は中央では放射状で，10 μm に 9-12 本，首部および頭部では平行～弱い放射状，頭部末端では放射状，足部末端では強い放射状になる．単列の胞紋から構成されているが，殻縁近くで2列になる場合も観察された．

ノート：従来 *G. acuminatum*（たとえば Ichikawa 1951）または *G. acuminatum* var. *coronata*（たとえば Okuno 1952）と同定されてきたが，河島・真山(2004)によると *G. acuminatum* とは別種であることがすでに報告されているので，本書ではそれに従った．

産　　出：横子珪藻土：前期更新世・群馬県，卯辰山層（Ichikawa 1951）：鮮新世または更新世・石川県，野上層（Okuno 1952）：中期更新世・大分県．

図　　版：横子珪藻土標本．

Figs 1-8.　*Gomphonema coronatum* Ehrenberg

LM. Figs 1-2. SEM. Figs 3-8. Material from diatomite of Yokogo (Early Pleistocene), Numata City, Gunma, Japan. Scale bars: Figs 3-4＝5 μm, Figs 6-7＝2 μm, Fig. 5＝1 μm, Fig. 8＝0.5 μm.

Figs 1-2.　Two different size valves.
Figs 2, 4-5, 8.　Same valve, LM and internal SEM views.
Fig. 3.　External view of whole valve.
Fig. 4.　Internal view of whole valve.
Fig. 5.　Detailed oblique view of head-pole of Fig. 4, helictoglossa (arrowhead).
Fig. 6.　Enlarged oblique view of head-pole of Fig. 3.
Fig. 7.　Enlarged oblique view of foot-pole of Fig. 3, apical pore field (arrow).
Fig. 8.　Detailed view of areolae rows of Fig. 4.

Plate 186

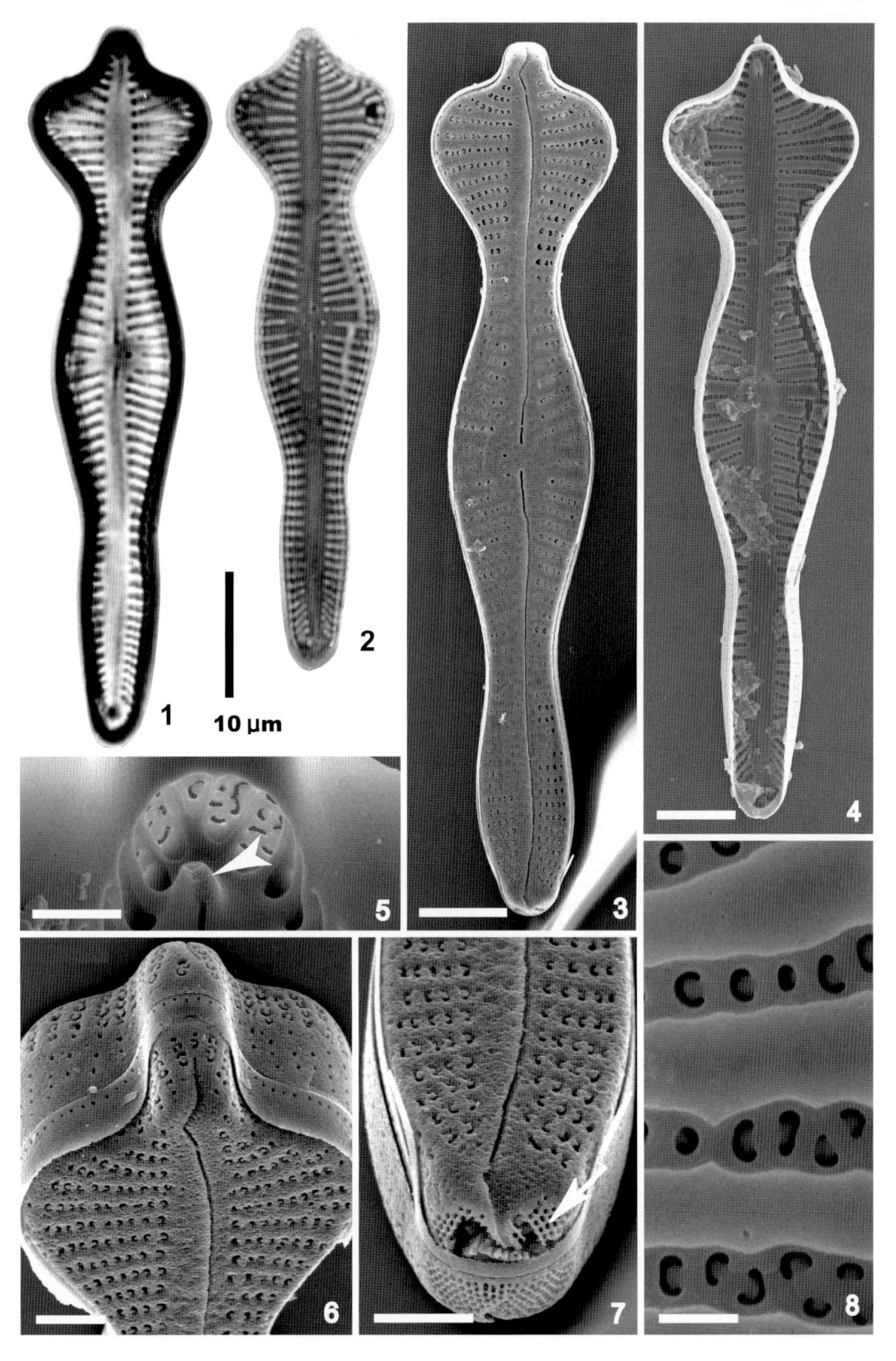

Plate 187.　Biraphid, pennate diatoms: Cymbellales
Gomphonema nipponicum Skvortsov 1936

文　　献：Skvortsow, B.W. 1936.　Diatoms from Kizaki Lake, Honshu Island. The Philippine Journal of Science **61**: 9-73, pls 1-16.
　　　Lee, J.H., Gotoh, T. & Chung, J. 1992.　A study of diatom species *Gomphonema vibrio* Ehr. var. *subcapitatum*（Mayer）Lee, comb. nov. The Korean Journal of Phycology **7**: 79-87.

形　　態：殻は上下不対称で頭部が頭状に突き出る．殻長 49-92 μm，殻幅 8-12 μm，遊離点の反対側の中心域は殻縁まで広がっている．条線は放射状で中央では疎になり 10 μm に 8-10 本，殻端では密になる 10 μm に約 14 本．SEM 観察では縦溝の外中心裂溝は直線で終わるが，内中心裂溝は釣針状に遊離点側に曲がる．

ノ　ー　ト：木崎湖から Skvortsow（1936）により原記載された種である．Lee *et al.*（1992）は本種を *G. vibrio* var. *subcapitatum*（Mayer）J.H. Lee の異名とした．筆者は津森層から産出した本種を Lee *et al.*（1992）に従って *G. vibrio* var. *subcapitatum*（Mayer）J.H. Lee と同定したが（田中ら 2005），後に出版された小林弘珪藻図鑑（小林ら 2006）は *G. nipponicum* の種名を使用しているので，これに揃えた．文献欄には Lee *et al.*（1992）も記したので参考にしてもらいたい．

産　　出：中之条湖成層（田中・小林 1992）：中期更新世・群馬県，津森層：中期更新世・熊本県．

図　　版：津森層標本．

Figs 1-7.　*Gomphonema nipponicum* Skvortsov
LM. Figs 1-3. SEM. Figs 4-7. Material from Tsumori Formation (Middle Pleistocene), Mashiki Town, Kumamoto, Japan. Scale bars: Figs 4-5＝5 μm, Figs 6-7＝2 μm.

Figs 1-3.　Three different size valves.
Fig. 4.　External oblique view of whole valve.
Fig. 5.　Internal oblique view of whole valve.
Fig. 6.　Enlarged oblique view of head-pole of Fig. 5, helictoglossa (arrowhead).
Fig. 7.　Enlarged view of valve center of Fig. 4, stigma (arrow).

Plate 187

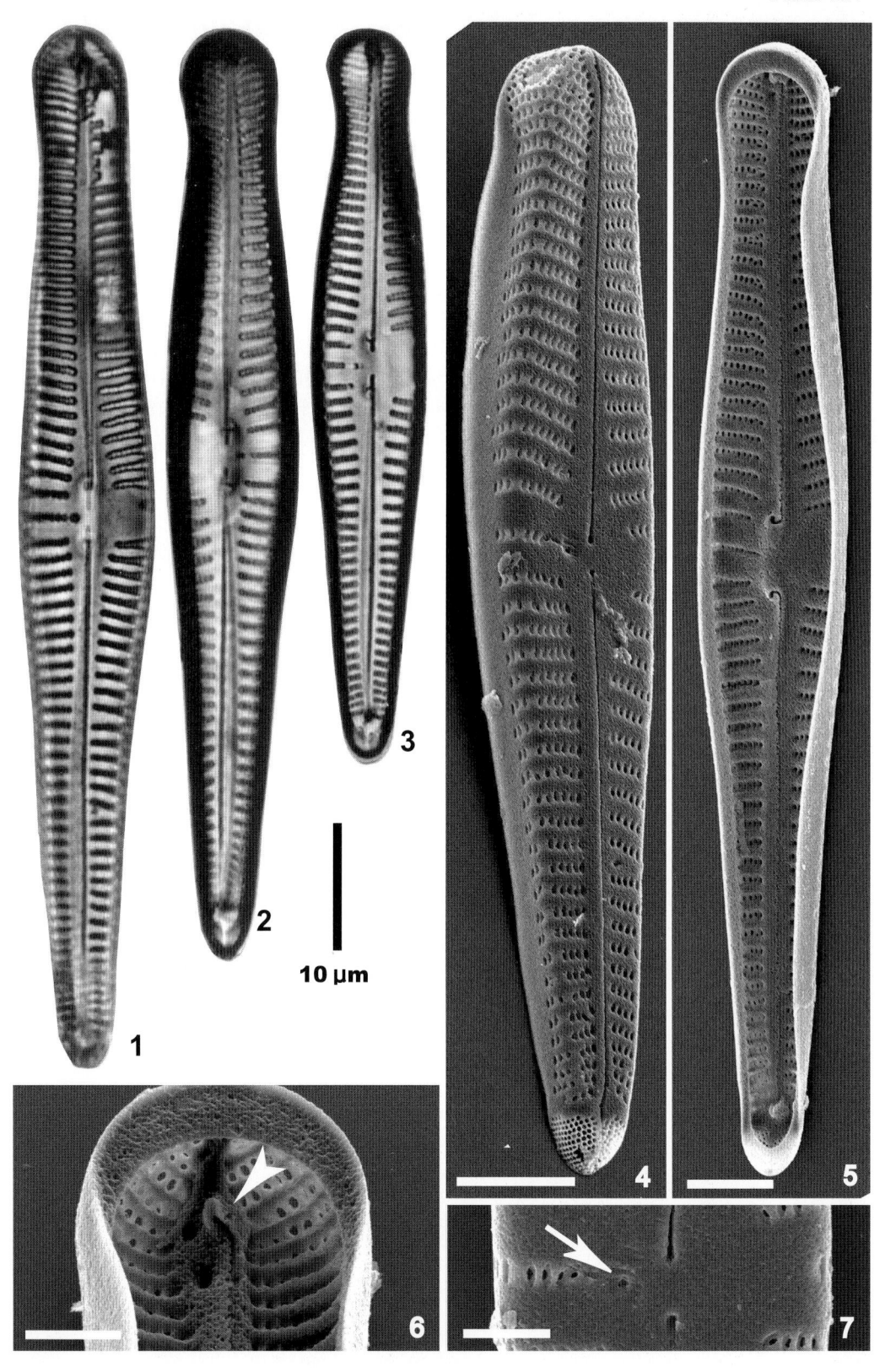

411

Plate 188. Biraphid, pennate diatoms: Cymbellales
Gomphonema truncatum Ehrenberg 1832

文　献：Patrick, R. & Reimer, C.W. 1975. The diatoms of the United States 2 (1). Monographs of the Academy of Natural Sciences of Philadelphia, no. 13. Philadelphia.

形　態：殻は上下不対称で頭部と中央部が膨らむ．殻長 27-41 μm，殻幅 12-14 μm，条線は放射状で 10 μm に 11-13 本．条線を構成する胞紋は軸域寄りでは 1 列で殻縁側では 2 列になる．遊離点は 1 個である．SEM 観察では縦溝の外側中心裂溝は中心孔で終わり，内側中心裂溝は極裂とは反対側の遊離点側に曲がる．

ノート：Patrick & Reimer (1975) により類似している *Gomphonema constrictum* Ehrenb. は異名であり *G. truncatum* が正名であることが報告されている．

産　出：大鷲湖沼性堆積物（田中ら 2011）：鮮新世・岐阜県，和村珪藻土（上山・小林 1983）：新第三紀・長野県．*G. constrictum* としては，鬼首層（Ichikawa 1955）：後期更新世・宮城県，琵琶湖堆積物（Mori 1975）：更新世・滋賀県，濃尾平野沖積層（森 1981）：完新世・愛知県．

図　版：大鷲湖沼性堆積物標本．

Figs 1-7. *Gomphonema truncatum* Ehrenberg
LM. Figs 1-2. SEM. Figs 3-7. Material from lacustrine deposit of Owashi (Pliocene), Gujo City, Gifu, Japan. Scale bars: Figs 3-4＝5 μm, Figs 5-7＝2 μm.

Figs 1-2.　Two different size valves.
Fig. 3.　External view of whole valve.
Fig. 4.　Internal view of whole valve.
Fig. 5.　Oblique view of Fig 3.
Fig. 6.　Enlarged view of valve center of Fig. 3, stigma (arrowhead).
Fig. 7.　Enlarged view of valve center of Fig. 4, stigma (arrowhead).

Plate 188

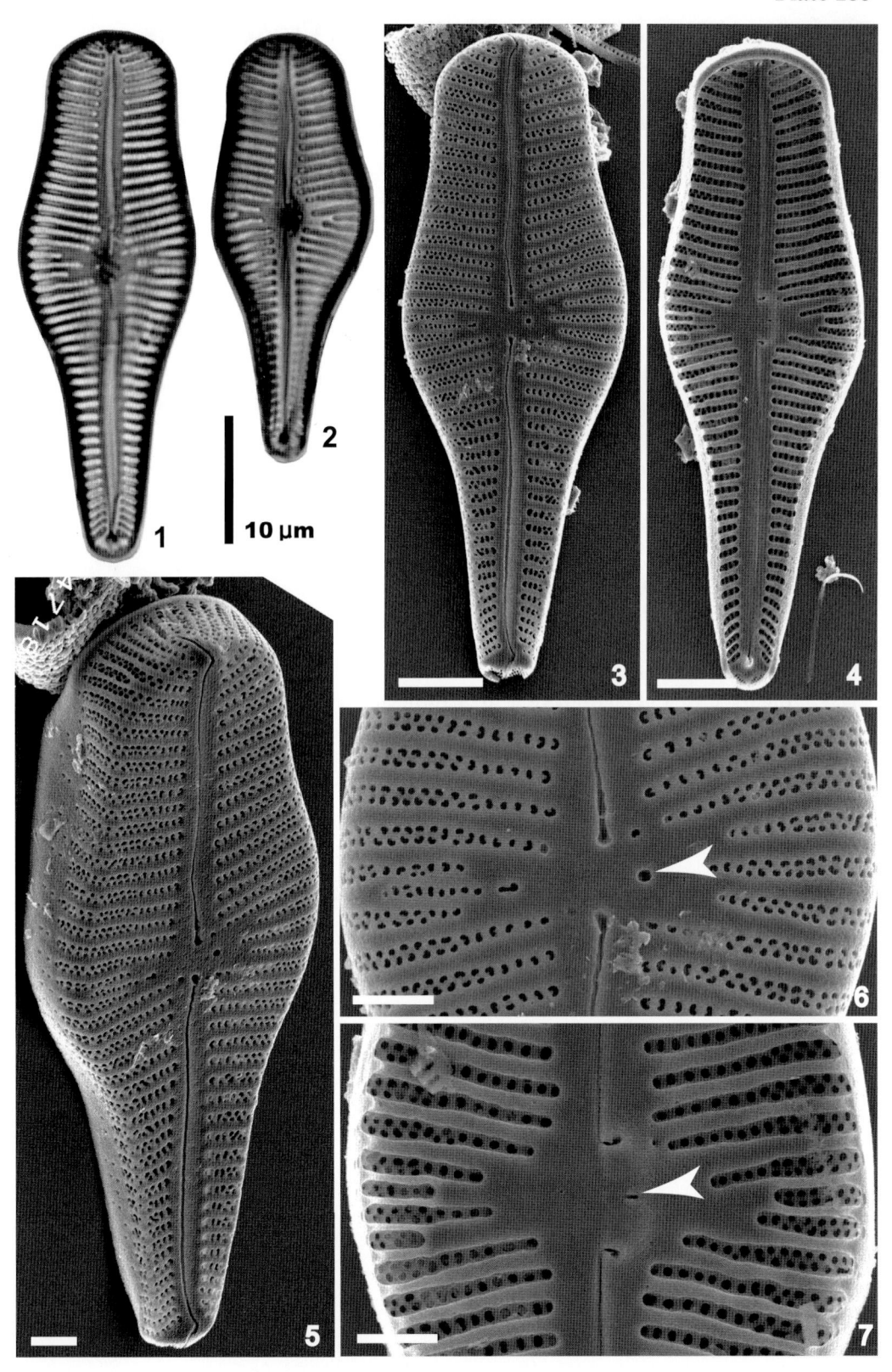

Plate 189.　Biraphid, pennate diatoms: Cymbellales
Gomphonema vastum Hustedt 1927

文　　献：Hustedt, F. 1927. Bacillariales aus dem Aokiko-see in Japan. Archiv für Hydrobiologie und Plankonkunde **18**: 155-172.

形　　態：殻は上下不対称で頭部が頭状に突き出る．殻長 25-44 μm，殻幅 5-7 μm，遊離点は1個で軸域・中心域とも広く，縦に細長い菱形を呈する．条線はわずか放射状で 10 μm に約 16 本，SEM 観察では縦溝の外側中心裂溝は直線で終わるが，内側中心裂溝はほぼ直角に遊離点側へ曲がる．また，両極裂は外側で遊離点とは反対方向へ曲がる．足部に殻端小孔域がある．

ノ ー ト：青木湖から Hustedt(1927)により原記載された分類群である．化石としては田中ら(2011)による報告がある．

産　　出：大鷲湖沼性堆積物（田中ら 2011）：鮮新世・岐阜県．

図　　版：大鷲湖沼性堆積物標本．

Figs 1-7.　*Gomphonema vastum* Hustedt
LM. Figs 1-3. SEM. Figs 4-7. Material from lacustrine deposit of Owashi (Pliocene), Gujo City, Gifu, Japan. Scale bars: Figs 4-5＝2 μm, Figs 6-7＝1 μm.

Figs 1-3.　Three different size valves.
Fig. 4.　External view of whole valve, stigma (arrow) and apical pore field (arrowhead).
Fig. 5.　Internal view of whole valve, stigma (arrow).
Fig. 6.　Enlarged view of head-pole of Fig. 4.
Fig. 7.　Enlarged view of apical pore field (arrowhead) of Fig. 4.

Plate 189

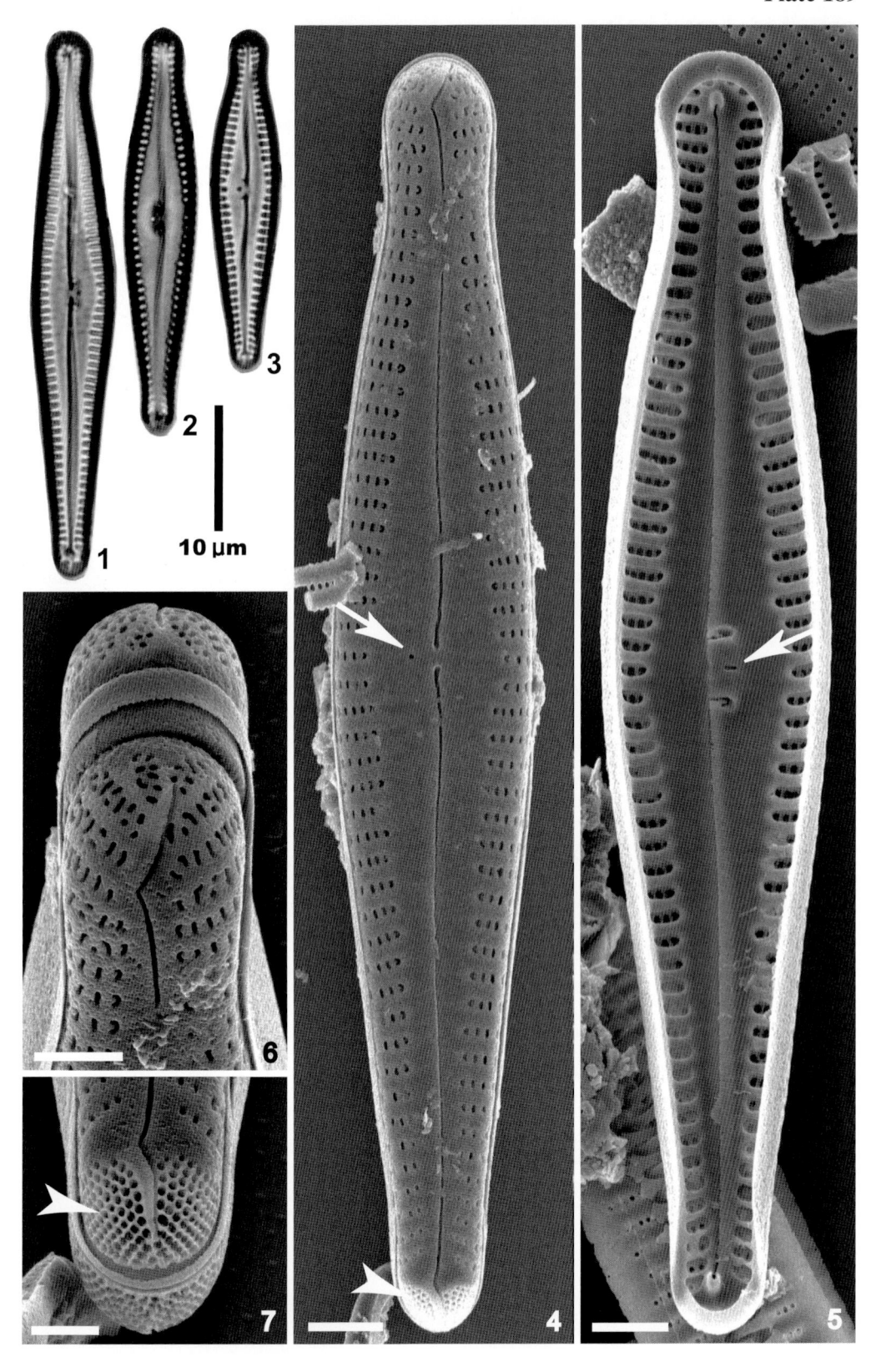

Plate 190. Biraphid, pennate diatoms: Cymbellales
Gomphopleura frickei Reichelt 1904

文　　献：Fricke, F. 1904. *In*: A. Schmidt's Atlas der Diatomaceen-Kunde (1874-1959), pl. 247. O.R. Reisland, Leipzig.
　　奥野春雄　1959．北海道瀬棚町の珪藻土について(5)．植物研究雑誌 **34**: 353-360.

形　　態：殻は上下不対称で頭部が広くちばし状に長く突き出る．殻長 184-232 μm，殻幅約 28 μm，縦溝は短く上下とも殻の長さの4分の1程度である．縦溝が所在する部分は軸域が広いが他は狭くなる．条線は中央では平行，縦溝が所在する付近ではわずか放射状で 10 μm に約8本であった．SEM 観察によると，胞紋の外側開口は弧状で，頭部には太い1本の刺がある．殻面／殻套境界は肥厚して隆起している．内側では足部の蝸牛舌はやや突出して唇状突起様である．

ノ ー ト：瀬棚町に分布する珪藻土（太櫓層）から記載された種である（Fricke 1904）．現在のところ他の地域からの産出報告は見当たらない稀産種である．邦人では奥野(1959)の報告がある．

産　　出：太櫓層（瀬棚の珪藻土）（奥野　1959）：前期中新世・北海道．

図　　版：太櫓層標本．

Figs 1-5. *Gomphopleura frickei* Reichelt
LM. Fig. 1. SEM. Figs 2-5. Material from Futoro Formation "diatomite of Setana" (Early Miocene), Setana Town, Hokkaido, Japan. Scale bars: Fig. 2＝10 μm, Fig. 3＝5 μm, Figs 4-5＝2 μm.

Figs 1-4.　Same valve, LM and SEM views.
Fig. 1.　Whole valve view.
Fig. 2.　External view of whole valve.
Fig. 3.　Enlarged oblique view of head-pole of Fig. 2 showing apical spine (arrow).
Fig. 4.　Enlarged oblique view of foot-pole of Fig. 2 showing ridge boundary of valve face and mantle (arrow).
Fig. 5.　Internal oblique view of foot pole, helictoglossa (arrowhead).

Plate 190

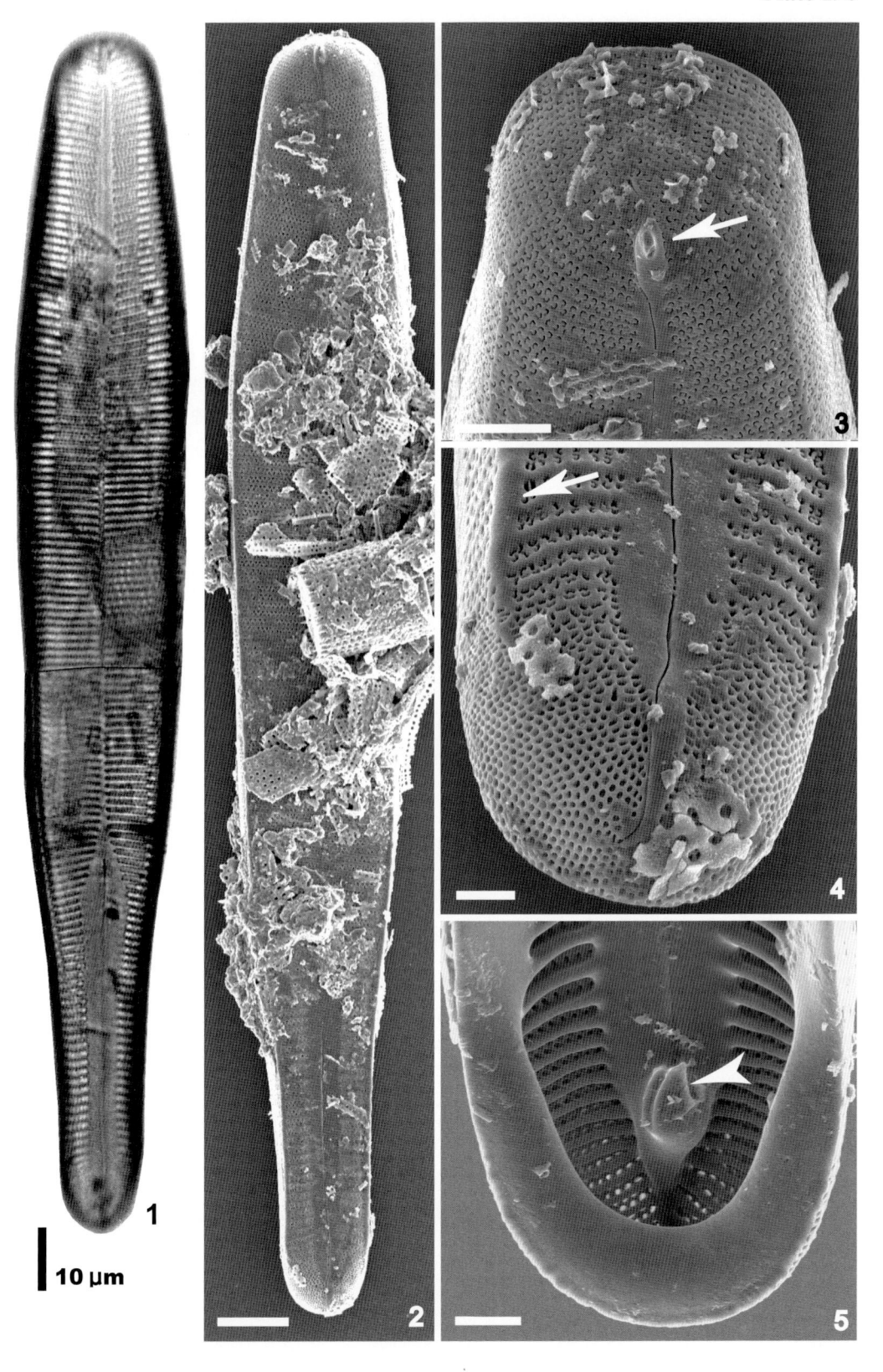

Plate 191.　Biraphid, pennate diatoms: Cymbellales
Gomphosphenia grovei var. *lingulata* (Hustedt) Lange-Bertalot 1995

基礎異名：*Gomphonema lingulatum* Hustedt 1927

文　献：Hustedt, F. 1927. Bacillariales aus dem Aokiko-see in Japan. Archiv für Hydrobiologie und Plankonkunde **18**: 155-172.

Lange-Bertalot, H. 1995. *Gomphosphenia paradoxa* nov. spec. et nov. gen. und Vorschlag zur Lösung taxonomischer probleme infloge eines veränderten Gattungskonzepts von *Gomphonema* (Bacillariophyceae). Nova Hedwigia **60**: 241-252.

形　態：殻は上下不対称で頭部が膨らみ，足部は細くなるがわずか頭状になる．殻長19-29 µm，殻幅約7.5 µm．縦溝は直線で，LMでは中心裂溝間に丸形の影が見られるが，この間の珪酸壁が肥厚しているためである．外中心裂溝は直線のまま終了するが，内中心裂溝は両側へ分岐し釣り針状になる．殻面の条線は殻面の縁にあり，10 µmに約15本，わずか放射状である．軸域は広い．殻套の胞紋と縁との間に内部まで貫通していない凹みが存在する．殻端小孔域を持たない．

ノート：青木湖からHustedt(1927)により*Gomphonema lingulatum* Hust.として原記載されたが，Lange-Bertalot(1995)により本属へ組み合わせになった．*Gonphonema yatukaensis* Horik. & Okunoは本種の異名である（Tuji 2004）．

産　出：中之条湖成層（田中・小林 1992）：中期更新世・群馬県，琵琶湖底堆積物（Mori 1975）：更新世・滋賀県，八束珪藻土（奥野 1944）：更新世・岡山県，嬉野珪藻土（Skvortzov 1937）：新第三紀・佐賀，俣水層（田中・鹿島 2007）：前期更新世・大分県，吹上浜（泥炭層）：完新世・鹿児島県．

図　版：吹上浜（泥炭層）(1, 4-7図), 中之条湖成層 (2-3図) 標本．

Figs 1-7.　*Gomphosphenia grovei* var. *lingulata* (Hustedt) Lange-Bertalot

LM. Figs 1-3. SEM. Figs 4-7. Materials, Figs 1, 4-7, from peat at Fukiagehama (Holocene), Kagoshima, Japan and Figs 2-3, Nakanojo Lacustrine Deposit (Middle Pleistocene), Nakanojo Town, Gunma, Japan. Scale bars: Figs 4-6＝2 µm, Fig. 7＝1 µm.

Figs 1-3.　Three different size valves.
Fig. 4.　External view of whole valve.
Fig. 5.　Oblique view of Fig. 4.
Fig. 6.　Internal oblique view of whole valve.
Fig. 7.　Enlarged oblique view of valve center of Fig. 6, internal central fissure (arrowhead).

Plate 191

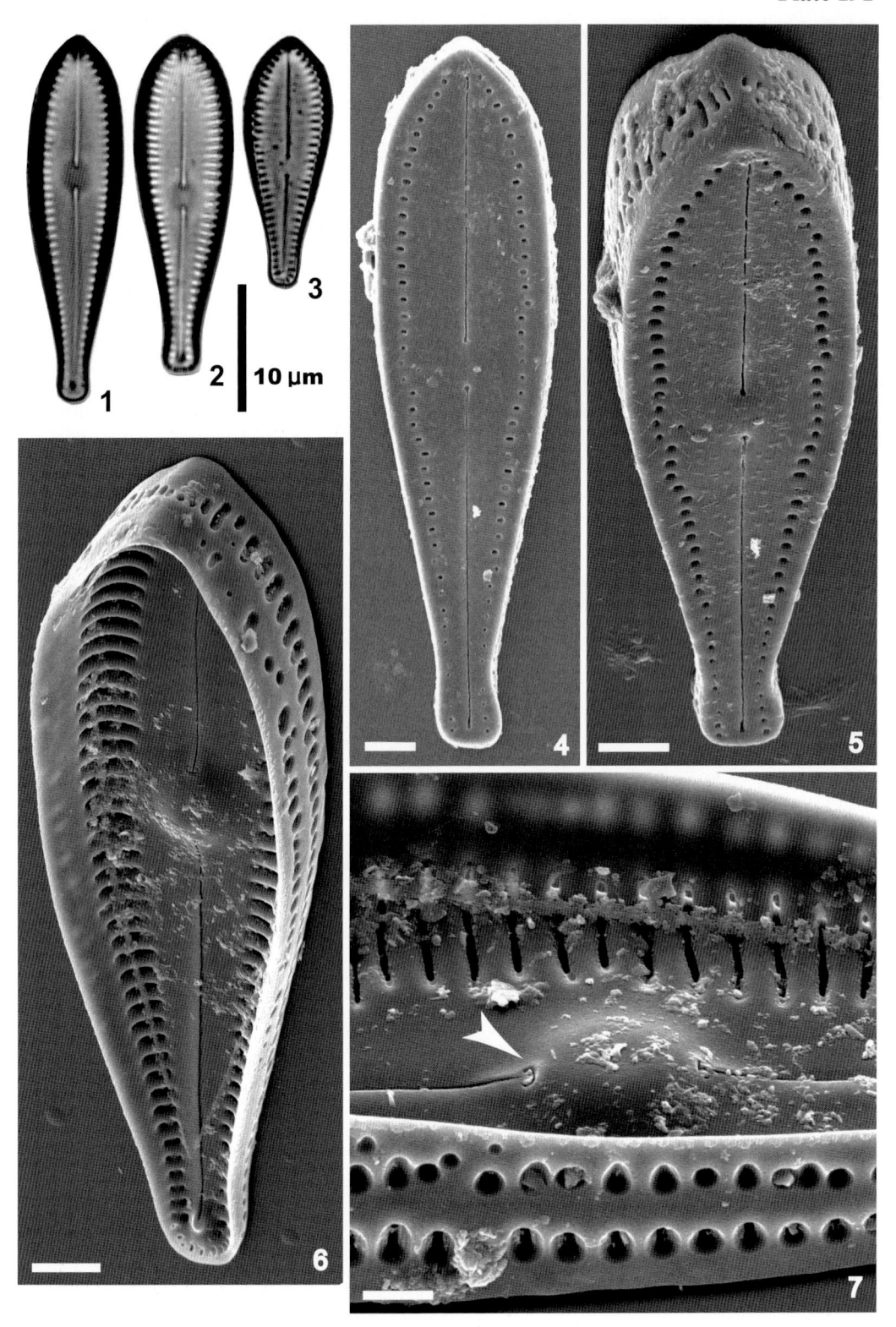

Plates 192-193. Biraphid, pennate diatoms: Cymbellales
Oricymba cunealjaponica H. Tanaka sp. nov.

新　　種：記載文（英文）は 15 頁参照.
形　　態：殻はわずか左右不対称. 殻端は急に細くなりくさび形になる. 中心部は背側・腹側ともやや膨らむ. 中心域は腹側にわずか広くなり, 1 個の遊離点がある. 殻長 38-67 μm, 殻幅 10-13 μm. 軸域は比較的広い. 条線はわずかに放射状であり両端では中心付近より放射が強くなる, 中心付近で 10 μm に 7-8 本, 条線を構成する点紋は 10 μm に 20-24 個であった. 縦溝は外裂溝が軸域の中心より背側へ偏るが中心付近と殻端では腹側へ寄り, 殻端では極裂が背側に曲がって終了する. SEM 観察によると両殻端に殻端小孔域がある. 胞紋の外側開口はスリット状である. 殻面／殻套境界は肥厚し 1 本の隆起線を形成している. 遊離点は外側では 1 個であるが, 内側の開口は 2 個になる.
ノ ー ト：種小名は, 殻の両端がくさび形であることとタイプ地に因む. Jüttner *et al.* (2010) は *Oricymba* 属を, 新組み合わせ（1 種）と新種（3 種）とともに設立した. *O. cunealjaponica* は殻の両端が急に細くなりくさび形であることにより, 既存の本属の他の分類群と区別できる.
産　　出：津森層：中期更新世・熊本県.
図　　版 192-193：津森層標本.

Plate 192, Figs 1-8. *Oricymba cunealjaponica* H. Tanaka sp. nov.

LM. Figs 1-4. SEM. Figs 5-8. From the type material, KUM-001, Tsumori Formation (Middle Pleistocene), Mashiki Town, Kumamoto, Japan. Scale bars: Figs 5, 7＝5 μm, Fig. 8＝2 μm, Fig. 6＝0.5 μm.

Figs 1-4.　Three different size valves.
Figs 2-3.　Holotype, same valve at different focal planes.
Fig. 5.　External whole valve view.
Fig. 6.　Detailed view of external areolae openings.
Fig. 7.　Oblique view of Fig. 5.
Fig. 8.　Enlarged view of central part of Fig. 7 showing edge between valve face and mantle (arrow) and a stigma (arrowhead).

Plate 192

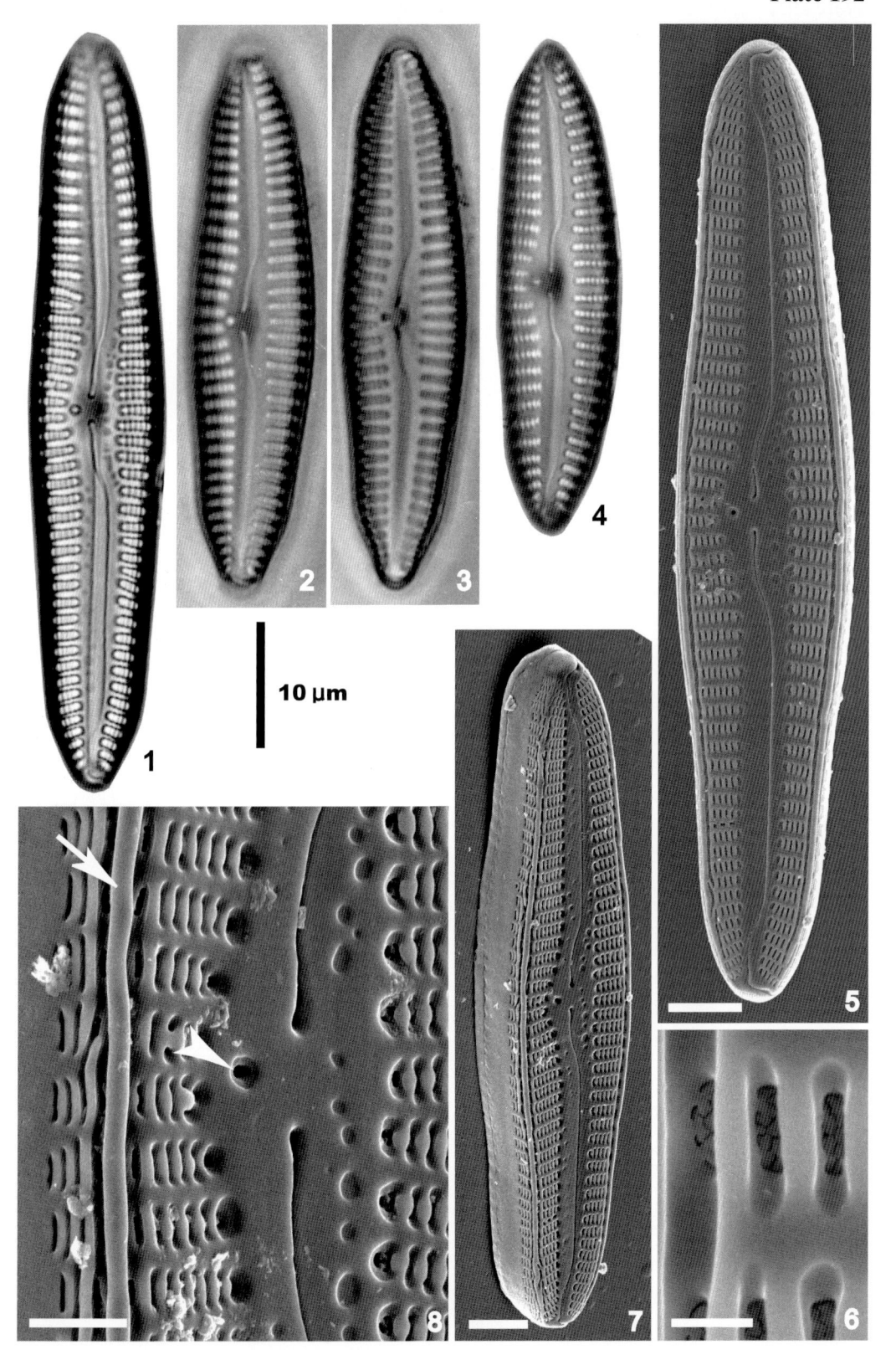

Plate 193, Figs 1–5. *Oricymba cunealjaponica* **H. Tanaka** sp. nov.

SEM. Figs 1–5. From the type material, KUM-001, Tsumori Formation (Middle Pleistocene), Mashiki Town, Kumamoto, Japan. Scale bars: Fig. 1=5 μm, Fig. 2=2 μm, Figs 3, 5 =1 μm, Fig. 4=0.5 μm.

Fig. 1. Internal whole valve view.
Fig. 2. Enlarged view of valve center of Fig. 1, stigmata (arrows).
Fig. 3. Enlarged oblique view of an apex of Fig. 1 showing inner apical pore field.
Fig. 4. Detailed view of internal areolae.
Fig. 5. Oblique view of an apex of Plate 192, Fig. 5, showing outer apical pore field and terminal fissure.

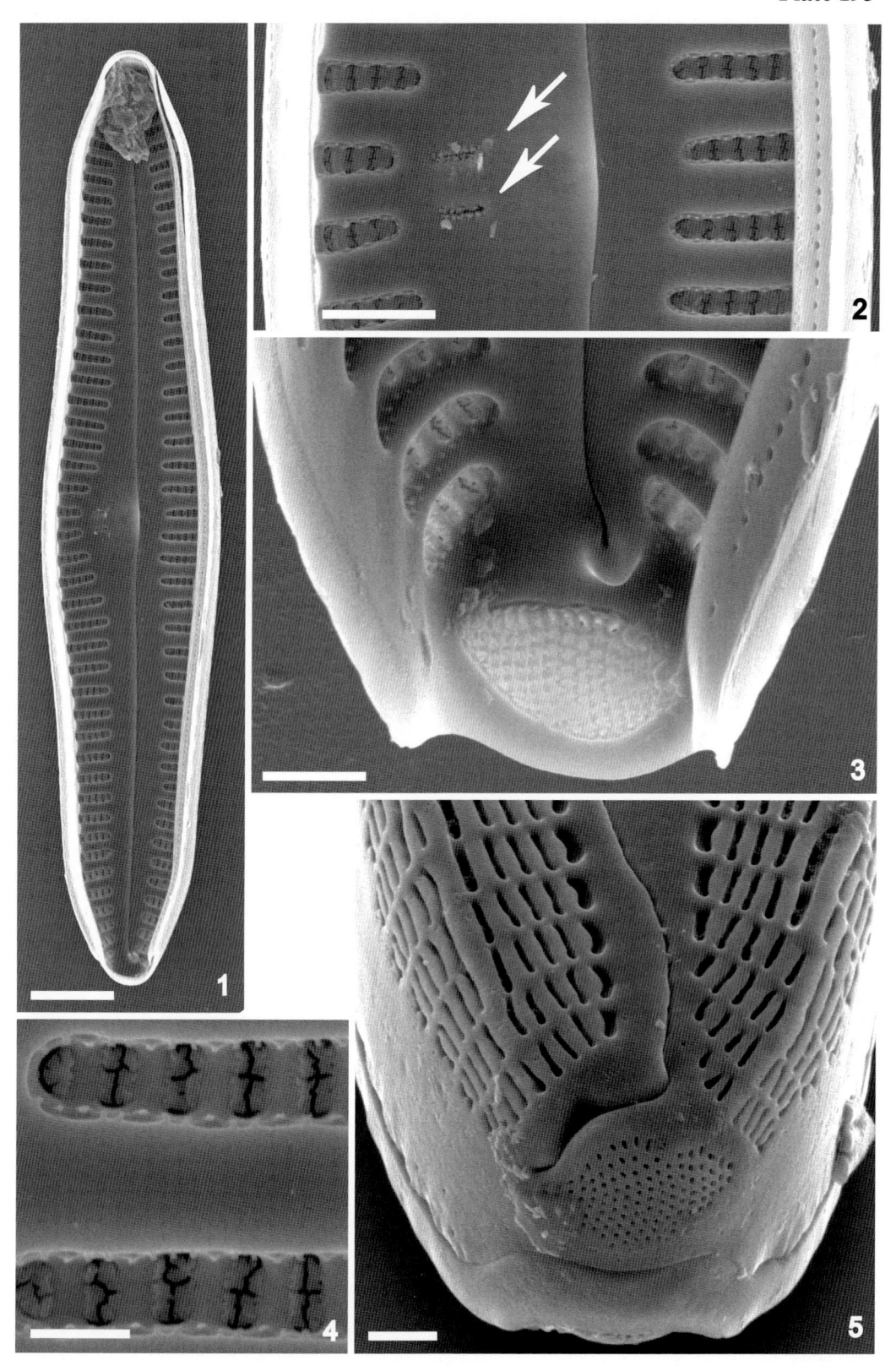

Plate 193

Plate 194.　Biraphid, pennate diatoms: Eunotiales
Actinella brasiliensis Grunow 1881

文　献：Grunow, A. 1881. in Van Heurck, H.(1880-1885). Synopsis des diatomées de Belgique. Atlas: 132 pls. Text: 235 pp. 3 pls. Ducaju et Cie. Anvers.

形　態：殻はわずかに湾曲するが背側と腹側の殻の間隔は先端部を除くとほぼ同じ，片方（頭部）の殻端は頭状に膨れるが，他方（足部）は鈍形～くさび形である．殻長 100-174 μm，殻幅は頭部で 9-13 μm，中央で 7-9 μm，条線は 10 μm に 9-13 本．唇状突起は頭部または足部の片方に認められた．

ノート：本種は頭部中央が刺状に突出している形態を示すが（田中・松岡 1985, Okuno 1964），産出欄に記した紫竹層のものは多くは突出していない．しかし少ないが突出する殻も認められたことと，Moiseeva(1971)では頭部が丸いものを本種に同定しているので，紫竹層産の分類群も *A. brasiliensis* に同定した．

産　出：太櫓層（瀬棚の珪藻土）：前期中新世・北海道，紫竹層：前期中新世・福島県，伊賀層（田中・松岡 1985）：鮮新世・三重県，奄芸層群（根来・後藤 1981）：鮮新世・三重県，嬉野珪藻土（Okuno 1964）：新第三紀・佐賀県．

図　版：太櫓層標本．

Figs 1-7.　*Actinella brasiliensis* Grunow
LM. Fig. 1. SEM. Figs 2-7. Material from Futoro Formation "diatomite of Setana" (Early Miocene), Setana Town, Hokkaido, Japan. Scale bars: Figs 2-3, 5-6＝5 μm, Figs 4, 7＝2 μm.

Fig. 1.　Whole valve view, photomontage.
Fig. 2.　External view of whole valve.
Fig. 3.　Oblique view of Fig. 2.
Fig. 4.　Enlarged view of head pole of Fig. 3 showing opening of rimoportula (arrowhead).
Fig. 5.　Internal oblique view of whole valve.
Fig. 6.　Enlarged oblique view of of head pole of Fig. 5, rimoportula (arrowhead).
Fig. 7.　Enlarged oblique view of foot pole of Fig. 5.

Plate 194

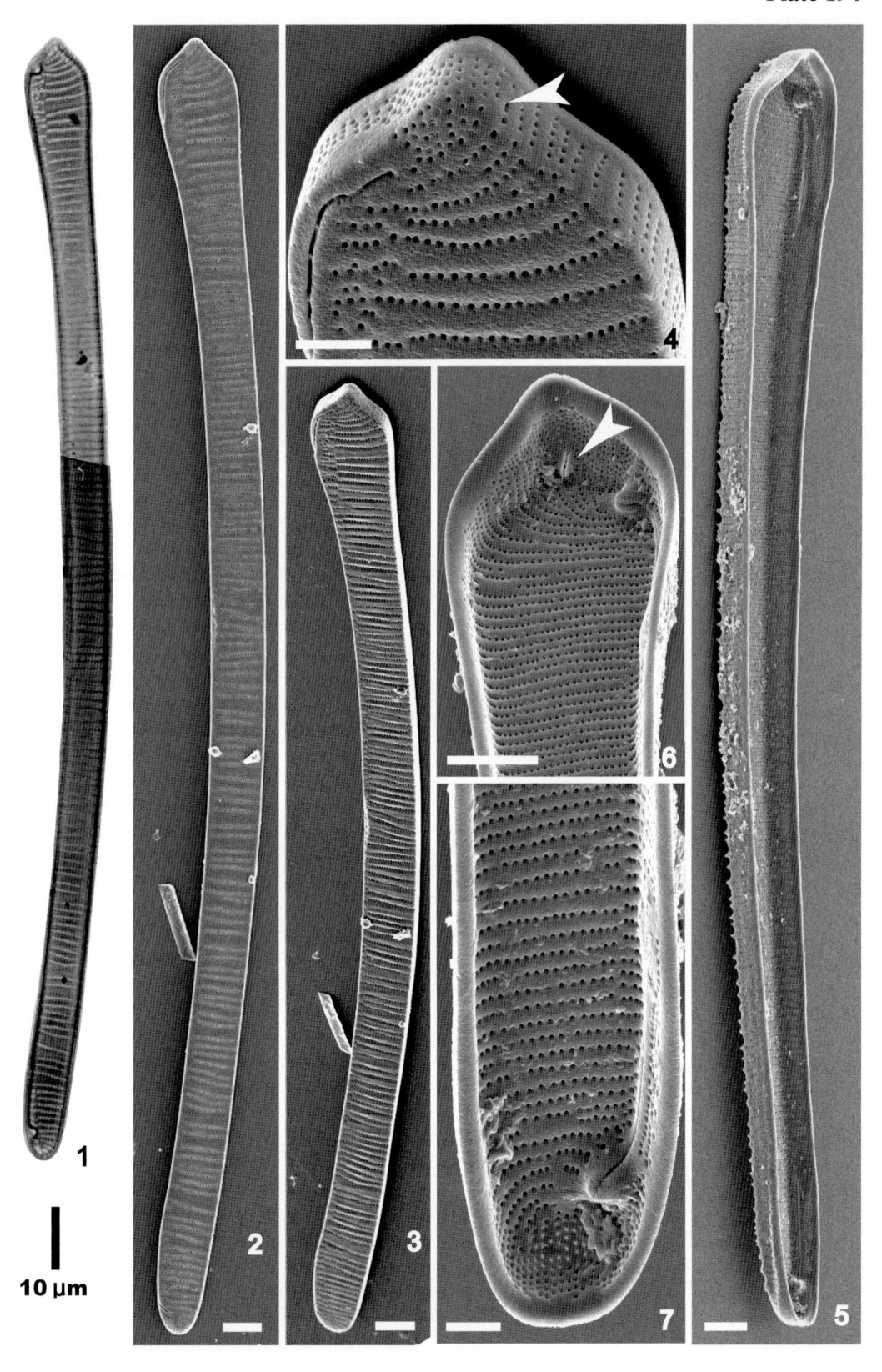

10 μm

Plate 195.　Biraphid, pennate diatoms: Eunotiales
Eunotia arcus Ehrenberg 1838

文　　献：Mayama, S. & Kobayasi, H. 1991. Observations of *Eunotia arcus* Ehr., type species of the genus *Eunotia* (Bacillariophyceae). The Japanese Journal of Phycology **39**: 131-141.

形　　態：殻はわずかに湾曲するが背側と腹側の殻の間隔は平行，両殻端は長軸方向へ頭状に突き出る．唇状突起は片方の殻端のみにある．殻長 40-58 μm，殻幅約 7.5 μm，条線は 10 μm に約 11 本．極裂は殻面に達する．

ノ ー ト：Mayama & Kobayasi(1991)は同地基準標本の詳細を報告している．それと比較すると，中之条湖成層産の個体はやや殻端が丸く背側のくびれが少ないが，本種に含められるものと思われる．

産　　出：中之条湖成層（田中・小林 1992）：中期更新世・群馬県．

図　　版：中之条湖成層標本．

Figs 1-6.　*Eunotia arcus* Ehrenberg

LM. Figs 1-2. SEM. Figs 3-6. Material from Nakanojo Lacustrine Deposit (Middle Pleistocene), Nakanojo Town, Gunma, Japan. Scale bars: Figs 3, 6=5 μm, Figs 4-5=2 μm.

Figs 1-2.　*Eunotia arcus* Ehrenb. copies from Bulletin of the Gunma Prefectural Museum of History, No. 13, 1992. Plate 4. Figs 18-19. Whole valve view, rimoportula (arrowhead).

Fig. 3.　External view of whole valve.

Fig. 4.　Oblique view of an apex of Fig. 3, opening of rimoportula (arrowhead).

Fig. 5.　Internal oblique view of whole valve, rimoportula (arrowhead).

Fig. 6.　Part of oblique view of Fig. 5, outer fissure of raphe (arrow).

Plate 195

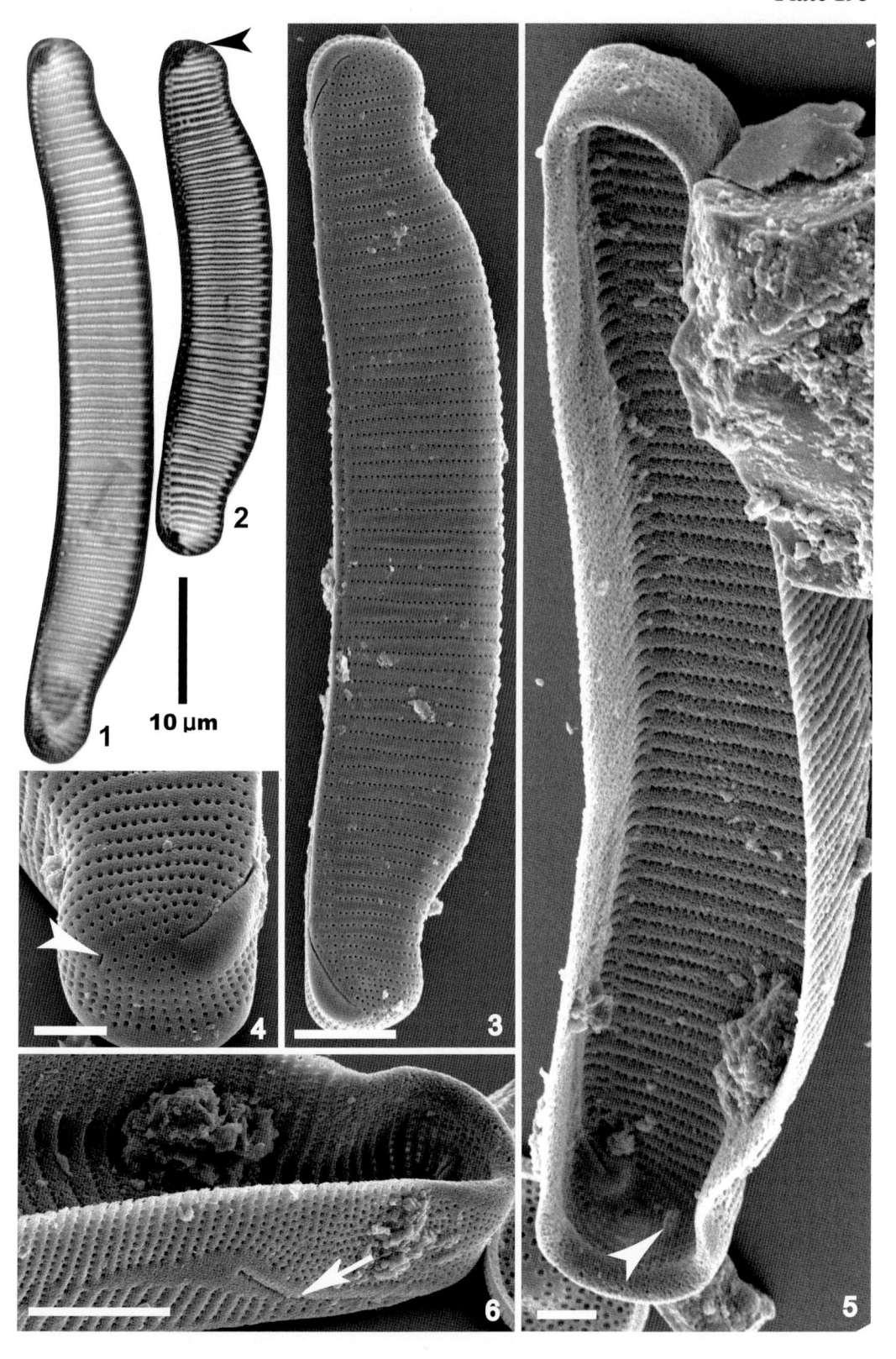

Plate 196. Biraphid, pennate diatoms: Eunotiales
Eunotia biareofera f. *linearis* H. Kobayasi
in Kobayashi & Ando 1975

文　　献：Kobayasi, H. & Ando, K. 1975. Diatoms from Hozoji-numa, Jizoin-numa and Nakashinden-numa ponds in Hanyu City, Saitama Prefecture. Bulletin of Tokyo Gakugei University, Ser. 4. **27**: 178-204.

形　　態：殻はわずかであるがカーブし腹側は凹み，背側は膨れる．殻長 115-143 μm，殻幅 9-10 μm，条線は 10 μm に 14-17 本，条線を構成する点紋は 10 μm に約 30 個．殻面／殻套境界には小針がある．SEM 観察によると両殻端にはそれぞれ唇状突起が 1 個所在する．

ノート：紫竹層産の本分類群は標記の *Eunotia biareofera* f. *linearis* または *Eunotia goretskyi* Churs. に類似するが，前者より大形で，後者より条線密度が高い．殻の大きさは，条線密度よりも生育場所による変異が大きいので前者に同定した．

産　　出：紫竹層：前期中新世・福島県．

図　　版：紫竹層標本．

Figs 1-6. ***Eunotia biareofera* f. *linearis* H. Kobayasi**
LM. Fig. 1. SEM. Figs 2-6. Material from Shichiku Formation (Early Miocene), Iwaki City, Fukushima, Japan. Scale bars: Figs 5-6＝5 μm, Figs 2-4＝2 μm.

Fig. 1.　Whole valve view.
Fig. 2.　Internal oblique view of an apex of Fig. 5 showing rimoportula (arrowhead) and helictoglossa (arrow).
Fig. 3.　Enlarged oblique view of part of Fig. 6.
Fig. 4.　Internal oblique view of opposite apex of Fig. 2, rimoportula (arrowhead) and helictoglossa (arrow).
Fig. 5.　Internal view of whole valve.
Fig. 6.　Oblique view of Fig. 5.

Plate 196

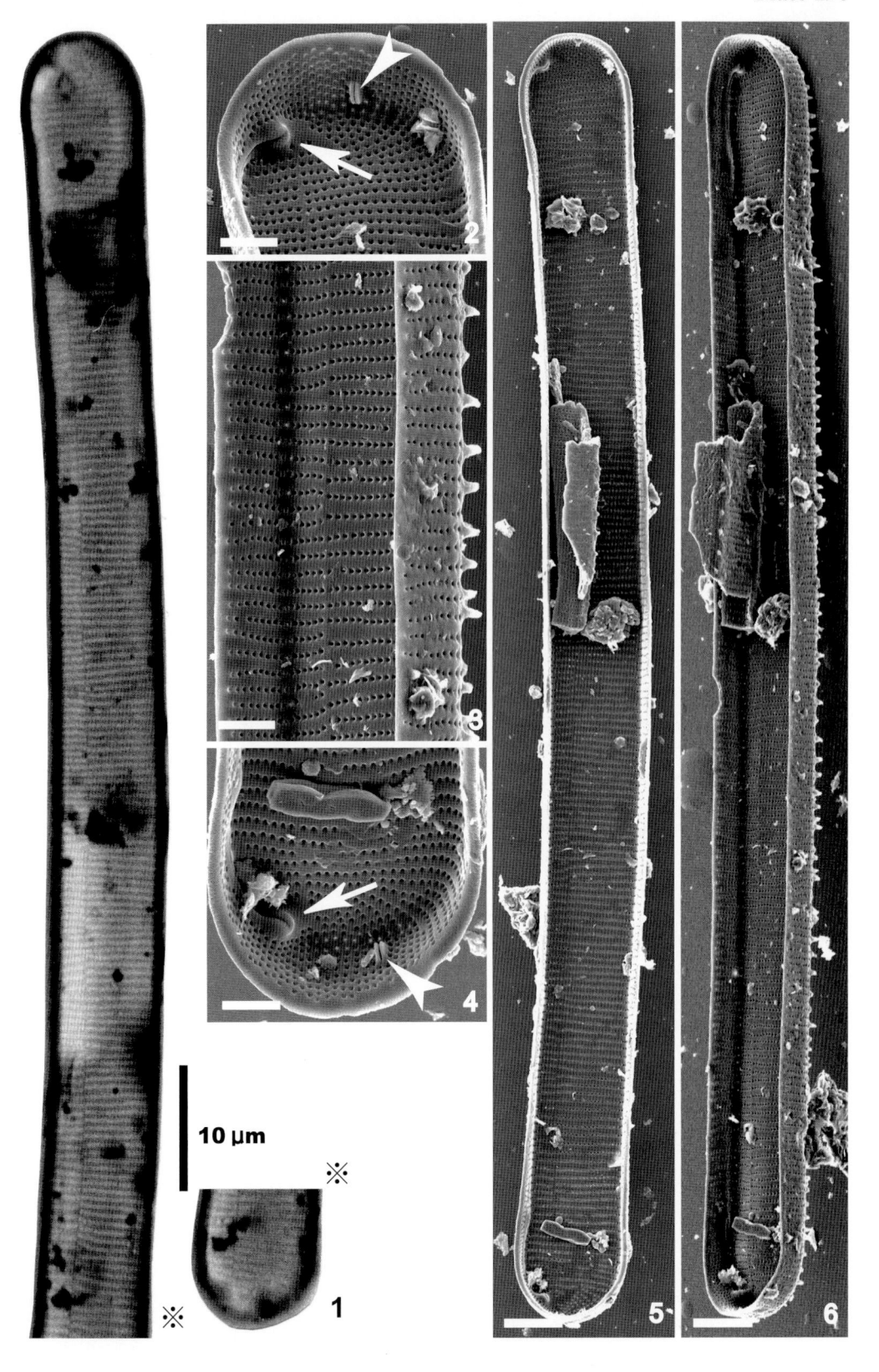

429

Plate 197. Biraphid, pennate diatoms: Eunotiales
Eunotia bidens Ehrenberg 1843

文　献：Ehrenberg, C.G. 1843. Verbreitung und Einfluss des mikroskopischen Lebens in Süd- und Nord-America. Abhandlungen der Königlichen Akademie der Wissenschaften zu Berlin(1841): Teil 1: 291-445.

形　態：殻は幅広くゆるい弓形で，腹側がわずか凹み，背側は膨れるが中央で凹む．殻長 50-68 μm，殻幅 12-13.5 μm，条線は粗く中心部腹側で 10 μm に約 11 本であった．両殻端は幅広く上下に突出するが，先端は平らで背側にやや傾いている．縦溝殻端末端は殻面まで届き，片方の殻端には唇状突起が所在する．

ノート：本分類群は（Cleve & Grunow 1880）に従って *E. praerupta* の変種に格付けられることが一般的であったが，近年原記載と同じく独立した種として扱う研究者が増えてきたのでそれに倣った（Mayama 2001, Metzeltin *et al.* 2009）．

産　出：太櫓層（瀬棚の珪藻土）：前期中新世・北海道，奄芸層群（根来・後藤 1981）：鮮新世・三重県．

図　版：太櫓層標本．

Figs 1-5. *Eunotia bidens* **Ehrenberg**
LM. Fig. 1-2. SEM. Figs 3-6. Material from Futoro Formation "diatomite of Setana" (Early Miocene), Setana Town, Hokkaido, Japan. Scale bars: Figs 3, 5＝5 μm, Figs 4, 6＝2 μm.

Figs 1-2.　Two different size valves.
Figs 2-4.　Same valve, LM and SEM photographs.
Fig. 3.　External view of whole valve.
Fig. 4.　Oblique view of an apex of Fig. 3 showing opening of rimoportula (arrowhead).
Fig. 5.　Internal oblique view of whole valve.
Fig. 6.　Oblique view of an apex of Fig. 5 showing rimoportula (arrowhead).

Plate 197

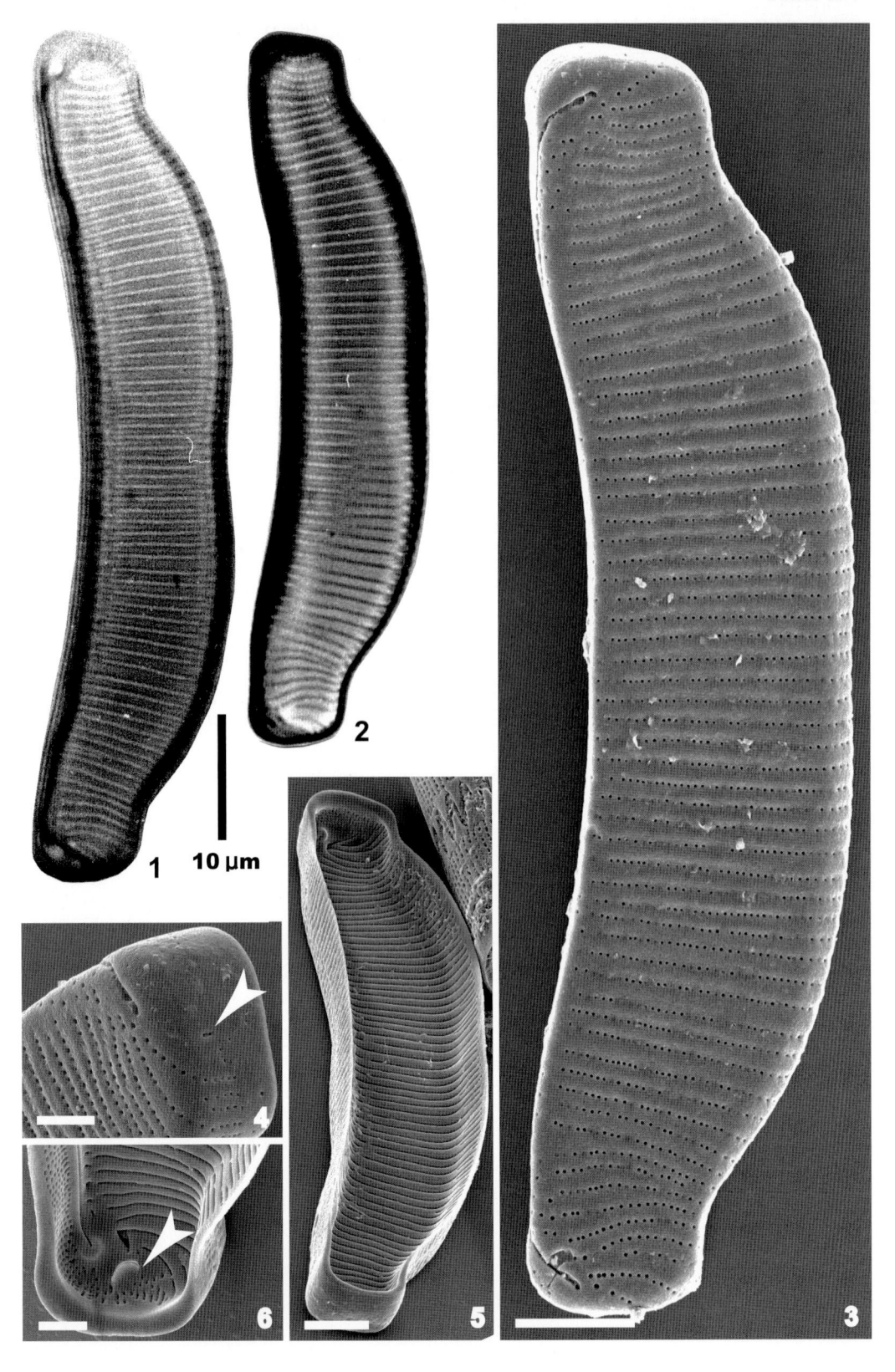

10 μm

Plate 198.　Biraphid, pennate diatoms: Eunotiales
Eunotia clevei Grunow 1878

文　　献：Hustedt, F. 1913. *In*: A. Schmidt's Atlas der Diatomaceen-Kunde(1874-1959), pl. 290, Figs 1-4. O.R. Reisland, Leipzig.

形　　態：殻は大形で腹側が大きく凹み，背側は膨れる．殻長96-256 μm，殻幅23-34 μm，条線は中心部腹側で10 μmに13-14本，条線を構成する点紋は10 μmに約16個であった．縦溝殻端末端は殻面まで届き，両殻端には唇状突起が所在する．被殻の厚さは背側で厚く，腹側でそれよりも薄い．

ノ ー ト：瀬棚町の珪藻土（太櫓層）の標本から Hustedt(1913)が記している図は比較的大形であるが，奥野(1959)は小形の個体（殻長約60 μm）しか見出せなかったと記している．筆者は大形の個体を多く見出すことができた．

産　　出：太櫓層（瀬棚の珪藻土）（奥野 1959）：前期中新世・北海道，釣懸層（秦・長谷川 1970）：中期中新世・北海道，豊川村の珪藻土（津村 1967）：年代不明・石川県．

図　　版：太櫓層標本．

Figs 1-5.　*Eunotia clevei* Grunow

LM. Fig. 1. SEM. Figs 2-5. Material from Futoro Formation "diatomite of Setana" (Early Miocene), Setana Town, Hokkaido, Japan. Scale bars: Figs 2, 5=10 μm, Figs 3-4=2 μm.

Fig. 1.　Whole valve view.

Fig. 2.　External view of whole valve, openings of rimoportulae (arrowheads).

Fig. 3.　External enlarged view of an apex showing opening of rimoportula (arrowhead) and raphe.

Fig. 4.　Internal view of an apex showing helictoglossa (arrow) and rimoportula (arrowhead).

Fig. 5.　Oblique view of whole frustule showing expanded dorsal side.

Plate 198

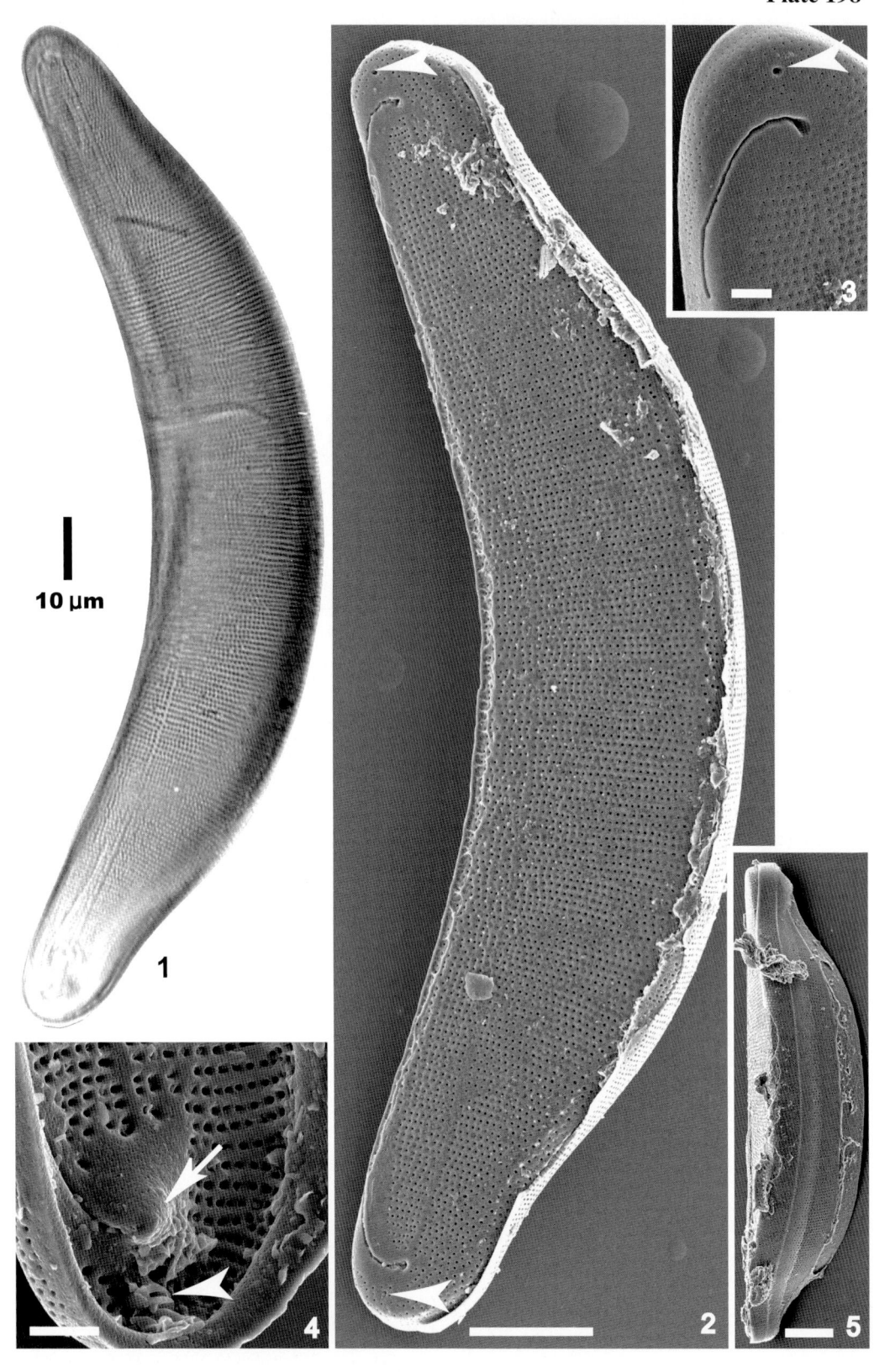

10 μm

Plate 199.　Biraphid, pennate diatoms: Eunotiales
Eunotia diadema Ehrenberg 1837

文　　献：Kobayasi, H., Ando, K. & Nagumo, T. 1981. On some endemic species of the genus *Eunotia* in Japan. *In*: Ross, R.(ed.) Proceedings of the 6th Symposium on recent and fossil diatoms. pp. 93-114. Koenigstein, Otto Koeltz.

形　　態：殻は腹側が凹み，背側は波打ちながら膨れる．殻長 44-50 µm，殻幅 17-19 µm，条線は中心部腹側で 10 µm に 11-12 本，条線を構成する点紋は 10 µm に 20-22 個であった．極節は殻端近くにある．片方の殻端には唇状突起が蝸牛舌と向かい合った殻面に近い殻套に所在し，外部への開口は細長く胞紋開口に似た大きさで，小さなスリットとして点紋列中に所在する．

ノ ー ト：本種は背側の波打ちが目立った特徴であるが，波打ちの回数の 6 回は *Eunotia serra* Ehrenberg にも（Brant & Furey 2011），また *E. diadema* にも（Mayama & Kobayasi 1990）存在する．しかし横子珪藻土産の分類群は 6 回未満の個体も観察できたことと極節の位置が *E. diadema* に類似している．

産　　出：横子珪藻土：前期更新世・群馬県．

図　　版：横子珪藻土標本．

Figs 1-7.　*Eunotia diadema* Ehrenberg

LM. Figs 1-2. SEM. Figs 3-7. Material from diatomite of Yokogo (Early Pleistocene), Numata City, Gunma, Japan. Scale bars: Figs 3, 7＝5 µm, Figs 4-6＝2 µm.

Figs 1-2.　Two different size valves.
Fig. 3.　External view of whole valve.
Fig. 4.　Internal oblique view of an apex, helictoglossa (arrow).
Fig. 5.　Same valve as Fig. 4, internal oblique view of opposit apex, helictoglossa (arrow) and rimoportula (arrowhead).
Fig. 6.　Enlarged view of an apex of Fig. 3, terminal fissure (arrow) and opening of rimoportula (arrowhead).
Fig. 7.　Oblique view of Fig. 3 showing undulated dorsal side.

Plate 199

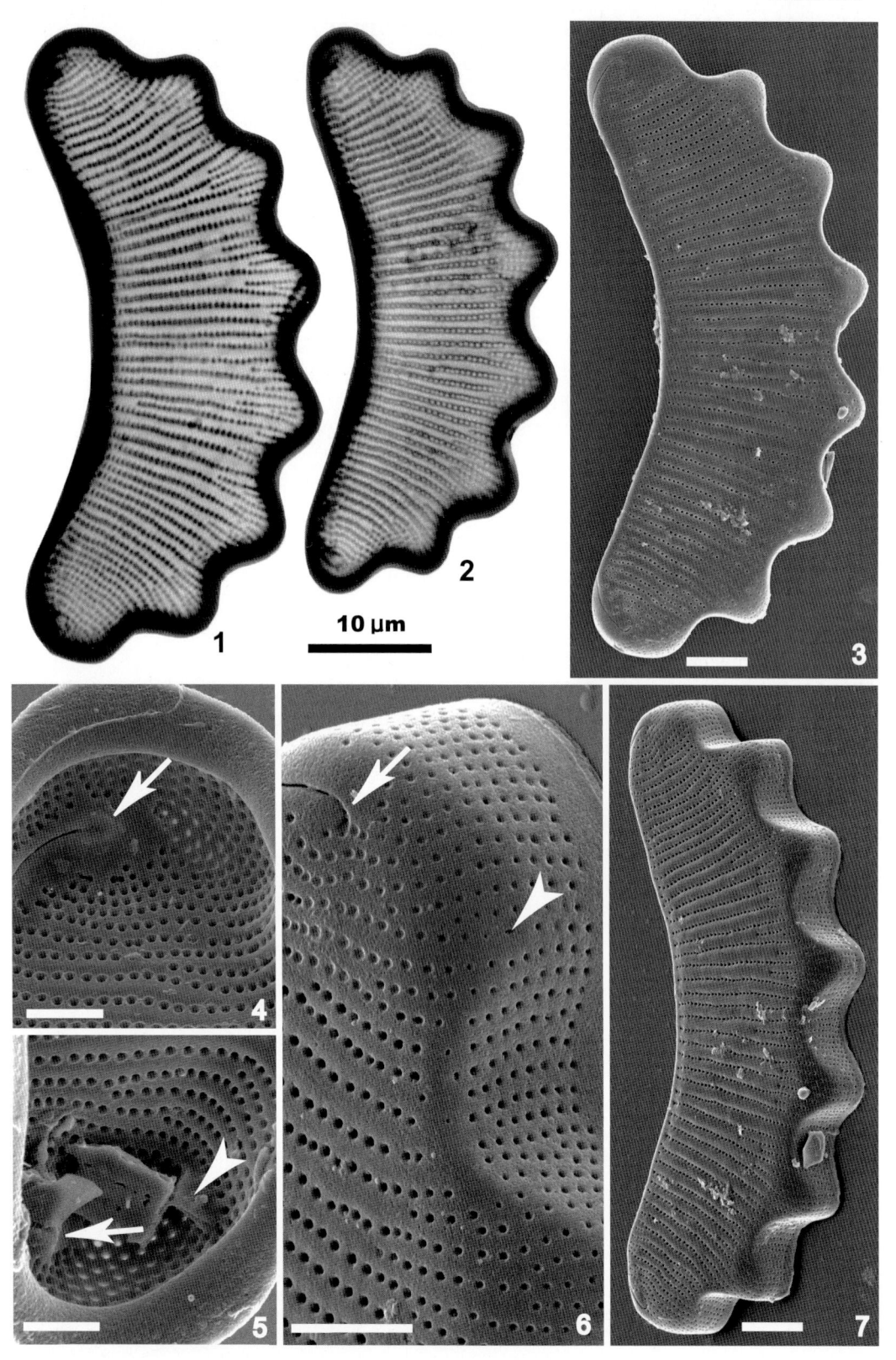

Plate 200.　Biraphid, pennate diatoms: Eunotiales
Eunotia duplicoraphis H. Kobayasi, Ando & Nagumo 1981

文　　献：Kobayasi, H., Ando, K. & Nagumo, T. 1981. On some endemic species of the genus *Eunotia* in Japan. *In*: Ross, R.(ed.) Proceedings of the 6th Symposium on recent and fossil diatoms. pp. 93-114. Koenigstein, Otto Koeltz.

形　　態：殻はわずかであるが腹側は凹み，背側は膨れる．殻長 21-45 μm，殻幅 3.5-6.5 μm，条線は中心部で 10 μm に 11-17 本，殻端では多少密になる．極節は明瞭で，極裂は殻面に達する．唇状突起は片方の殻端に所在する．

ノ ー ト：Kobayasi *et al.*(1981)によると本種は片方の殻端に唇状突起が存在する．中之条の個体も片方の殻端に唇状突起が存在し，計測値もほぼ一致した．蝸牛舌の形態はやや異なるところもあるが，LM 像が非常によく似ていることと，類似する *Eunotia sudetica* var. *emycephala*(A. Cl.)A. Berg. の微細構造が明瞭でないので本種に同定した．*E. duplicoraphis* の原産地は新潟県の牛池である．

産　　出：中之条湖成層（田中・小林　1992）：中期更新世・群馬県．

図　　版：中之条湖成層標本．

Figs 1-9.　*Eunotia duplicoraphis* H. Kobayasi, Ando & Nagumo
LM. Figs 1-4. SEM. Figs 5-9. Material from Nakanojo Lacustrine Deposit (Middle Pleistocene), Nakanojo Town, Gunma, Japan. Scale bars: Figs 5-6=2 μm, Figs 7-9=1 μm.

Figs 1-4.　Four different size valves.
Fig. 5.　Internal oblique view of whole valve, rimoportula (arrowhead).
Fig. 6.　External oblique view of whole valve, outer fissure of raphe (arrow).
Fig. 7.　Enlarged view of an apex of Fig. 5 showing helictoglossa (arrow).
Fig. 8.　Enlarged view of opposite apex of Fig. 5 showing helictoglossa and rimoportula (arrowhead).
Fig. 9.　Enlarged view of an apex of upper side of Fig. 6 showing opening of rimoportula (arrowhead).

Plate 200

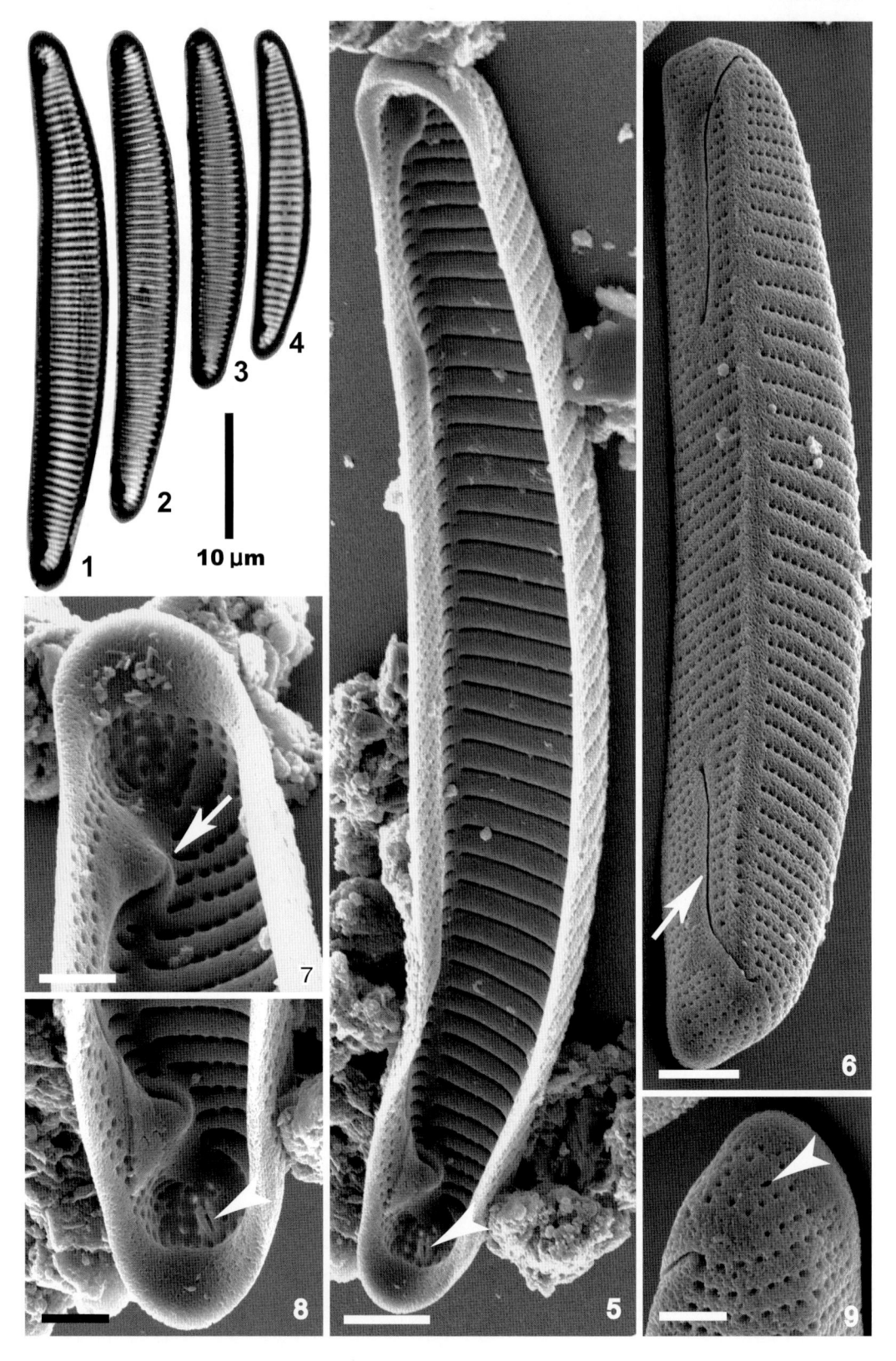

Plate 201.　Biraphid, pennate diatoms: Eunotiales
Eunotia epithemioides Hustedt in A. Schmidt *et al.* 1913

文　　献：Hustedt, F. 1913. *In*: A. Schmidt's Atlas der Diatomaceen-Kunde(1874-1959), pl. 287, Figs 16-19. O.R. Reisland, Leipzig.

形　　態：殻の腹側はわずか凹み，背側はわずか膨れる．両端はやや頭状になり殻幅は多少小さくなる．殻長 91 μm，殻幅 17 μm，条線は 10 μm に約 8 本だが，1-3 本の条線が束状になっており計測箇所によって異なる．

ノ ー ト：原記載の図 Hustedt(1913)では，条線が 2-8 本ずつ条線間隔が密になり束状になっている．人吉層から見出された分類群は条線が 1 本の場合もあり，また多くて 3 本であるが，Hustedt(1938)で示された図では 1 本の場合もあり，光顕での形態がほぼ一致した．稀産種で，日本において現生では奈良県上北山村から見出されている（渡辺ら 2005）．

産　　出：人吉層：後期鮮新世・熊本県．

図　　版：人吉層標本．

Figs 1-4.　*Eunotia epithemioides* Hustedt

LM. Fig. 1. SEM. Figs 2-4. Material form Hitoyoshi Formation (Late Pliocene), Hitoyoshi City, Kumamoto, Japan. Scale bars: Figs 2-4＝5 μm.

Fig. 1.　Whole valve view.
Fig. 2.　External view of whole valve.
Fig. 3.　Enlarged oblique view of an apex showing outer fissure of raphe.
Fig. 4.　Oblique view of Fig. 2.

Plate 201

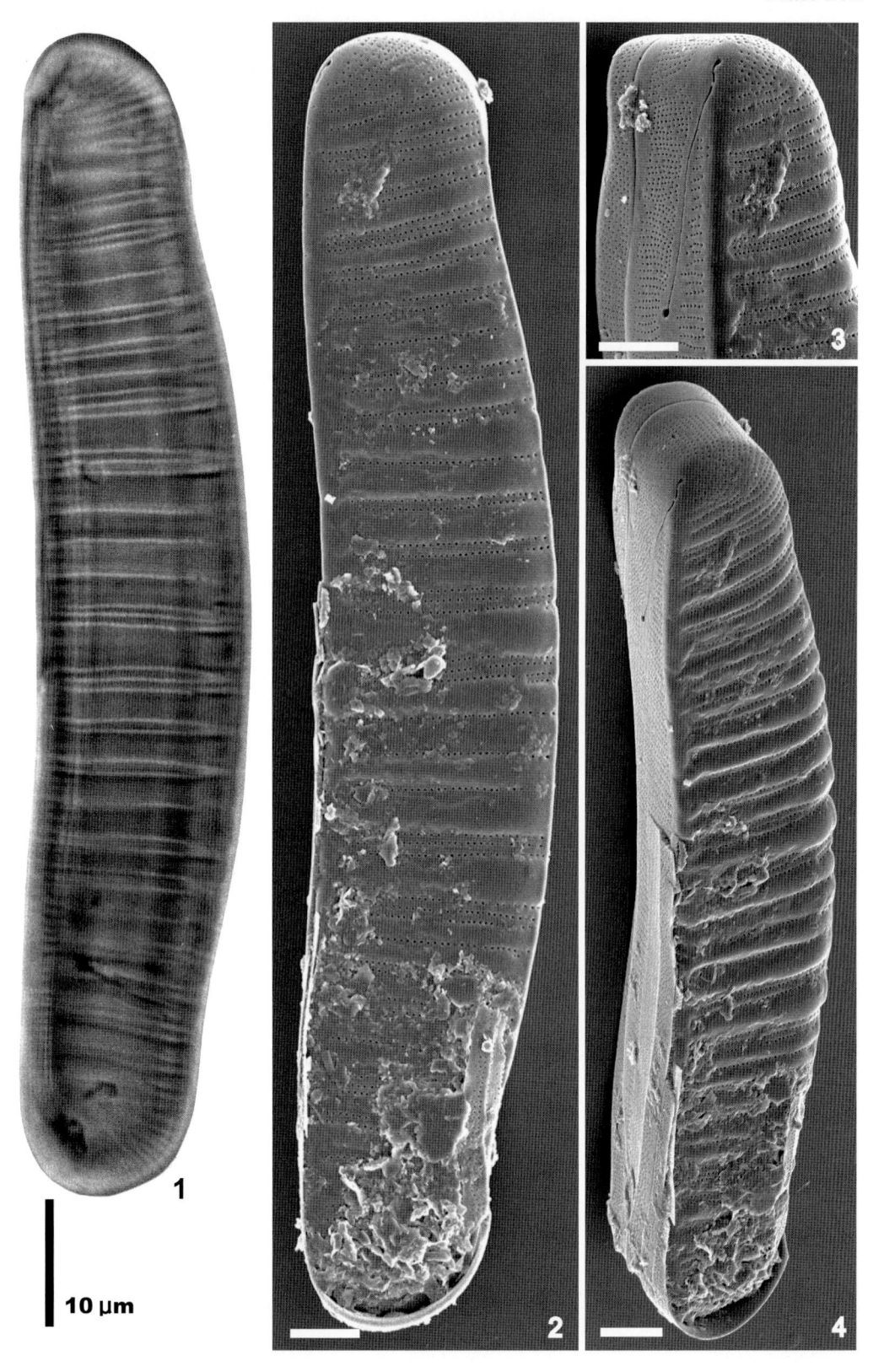

10 µm

Plate 202. Biraphid, pennate diatoms: Eunotiales
Eunotia formica Ehrenberg 1843

文　献：Hustedt, F. 1911. *In*: A. Schmidt's Atlas der Diatomaceen-Kunde (1874-1959), pl. 271, Figs 3-5. O.R. Reisland, Leipzig.

形　態：殻は大形で湾曲は少ない．腹側の中央部がやや膨れるが，背側も腹側より少ないがまれに膨れることがある．殻端はおおよそ三角形状で，膨れて頭状になることが多い．殻長70-152 μm，殻幅11-14 μm，条線は10 μmに8-10本である．極節は腹側で殻端近くにある．SEM観察では蝸牛舌は殻端近くの腹側で殻面／殻套境界にあるが，極裂は殻面先端を通り背側まで回り込む．

ノート：殻中央の膨らみは文献欄に記したHustedt(1911)の図（Plate 271, Figs 3-5）では明瞭であるが，同じA. Schmidt's AtlasのHustedt(1913)の図（Plate 291, Figs 4, 5）ではほとんど膨れていない．この膨らみの程度は変異が大きいと思われる．

産　出：宮田層：中新-鮮新世・秋田県，紫竹層：前期中新世・福島県．

図　版：宮田層標本．

Figs 1-5. *Eunotia formica* Ehrenberg
LM. Figs 1-2. SEM. Figs 3-5. Material from Miyata Formation (Mio-Pliocene), Senboku City, Akita, Japan. Scale bars: Fig. 4=10 μm, Figs 3, 5=5 μm.

Figs 1, 3-5.　Same valve, LM and SEM photographs.
Figs 1-2.　Two different size valves. Whole valve views.
Fig. 3.　External view of an apex, raphe (arrowhead).
Fig. 4.　External view of whole valve.
Fig. 5.　Enlarged view of valve center of Fig. 4.

Plate 202

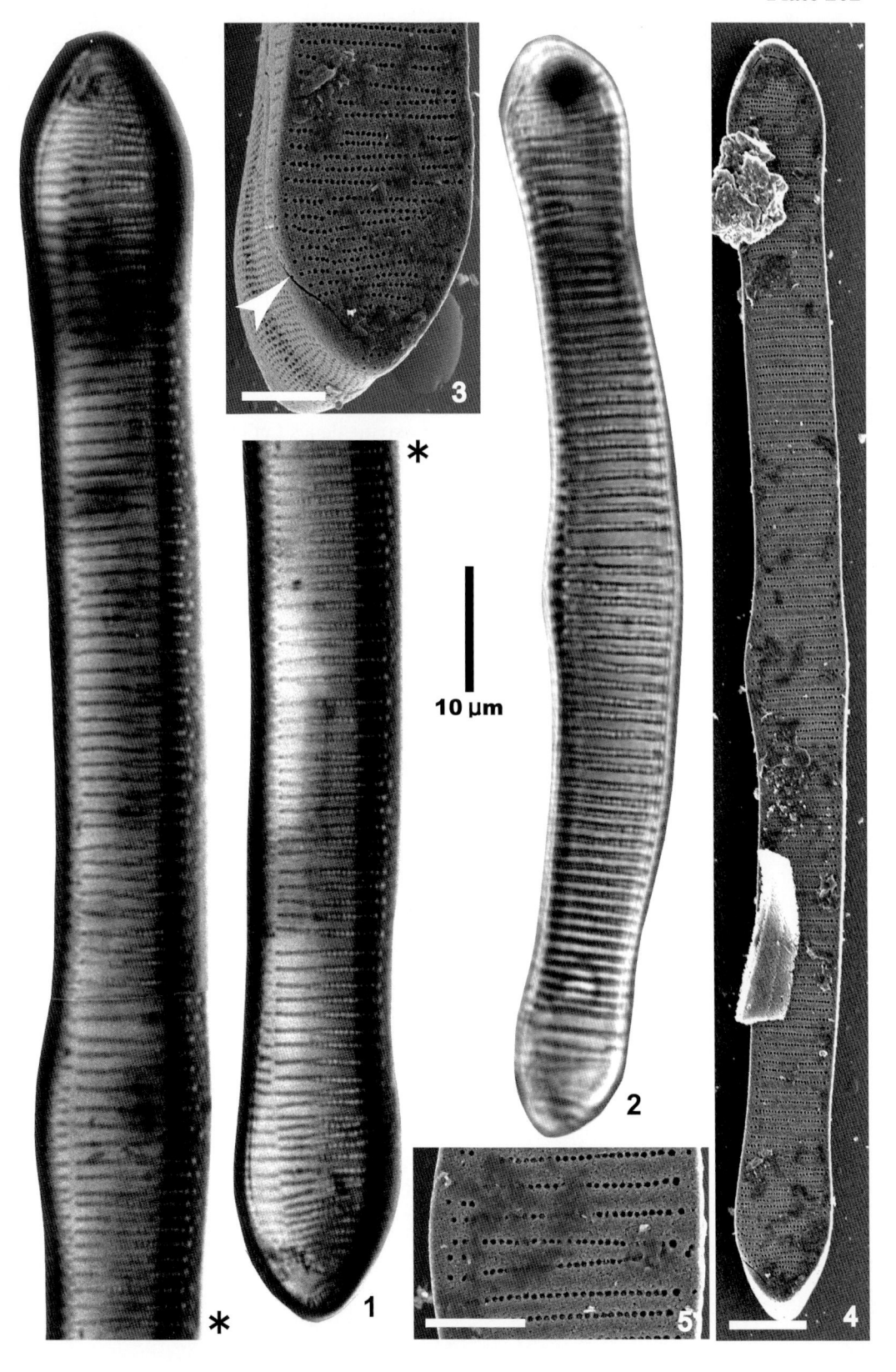

10 μm

Plate 203.   Biraphid, pennate diatoms: Eunotiales
Eunotia incisa W. Gregory 1854

文　　献：Gregory, W. 1854. Notice of the new forms and varieties of known forms occurrings in the diatomaceous earth of Mull; with remarks on the classification of the Diatomaceae. Quarterly Journal of Microscopical Sciences vol. **2**: 90-100, pl. 4.

形　　態：殻は小形で細長く，腹側はほぼ直線であるが背側は弓状にカーブする．殻長 23-39 μm，殻幅 4.5-5.5 μm，条線は中心付近で 10 μm に 13-17 本であるが，殻端では密になる．極節は腹側で殻端からかなり中央よりにある．SEM 観察によると縦溝は殻套部に分布しており，唇状突起は片方の殻端のみに見られた．

産　　出：大鷲湖沼性堆積物：鮮新世・岐阜県．

図　　版：大鷲湖沼性堆積物標本．

Figs 1-10.   *Eunotia incisa* W. Gregory
LM. Figs 1-3. SEM. Figs 4-10. Material from lacustrine deposit of Owashi (Pliocene), Gujo City, Gifu, Japan. Scale bars: Figs 4-7＝2 μm, Figs 8-10＝1 μm.

Figs 1-3.   Three different size valves.
Fig. 4.   External view of whole valve, opening of rimoportula (arrowhead).
Fig. 5.   Oblique view of Fig. 4, opening of rimoportula (arrowhead).
Fig. 6.   Internal oblique view of whole valve.
Fig. 7.   An apex of Fig. 4 showing opening of rimoportula (arrowhead) and outer fissure of raphe (arrow).
Fig. 8.   Opposite side of Fig. 7. Note: no opening of rimoportula.
Fig. 9.   Internal apex of Fig. 6, helictoglossa (arrow). Note: no rimoportulae.
Fig. 10.   Opposite side apex of Fig. 6 showing rimoportula (arrowhead) and helictoglossa (arrow).

Plate 203

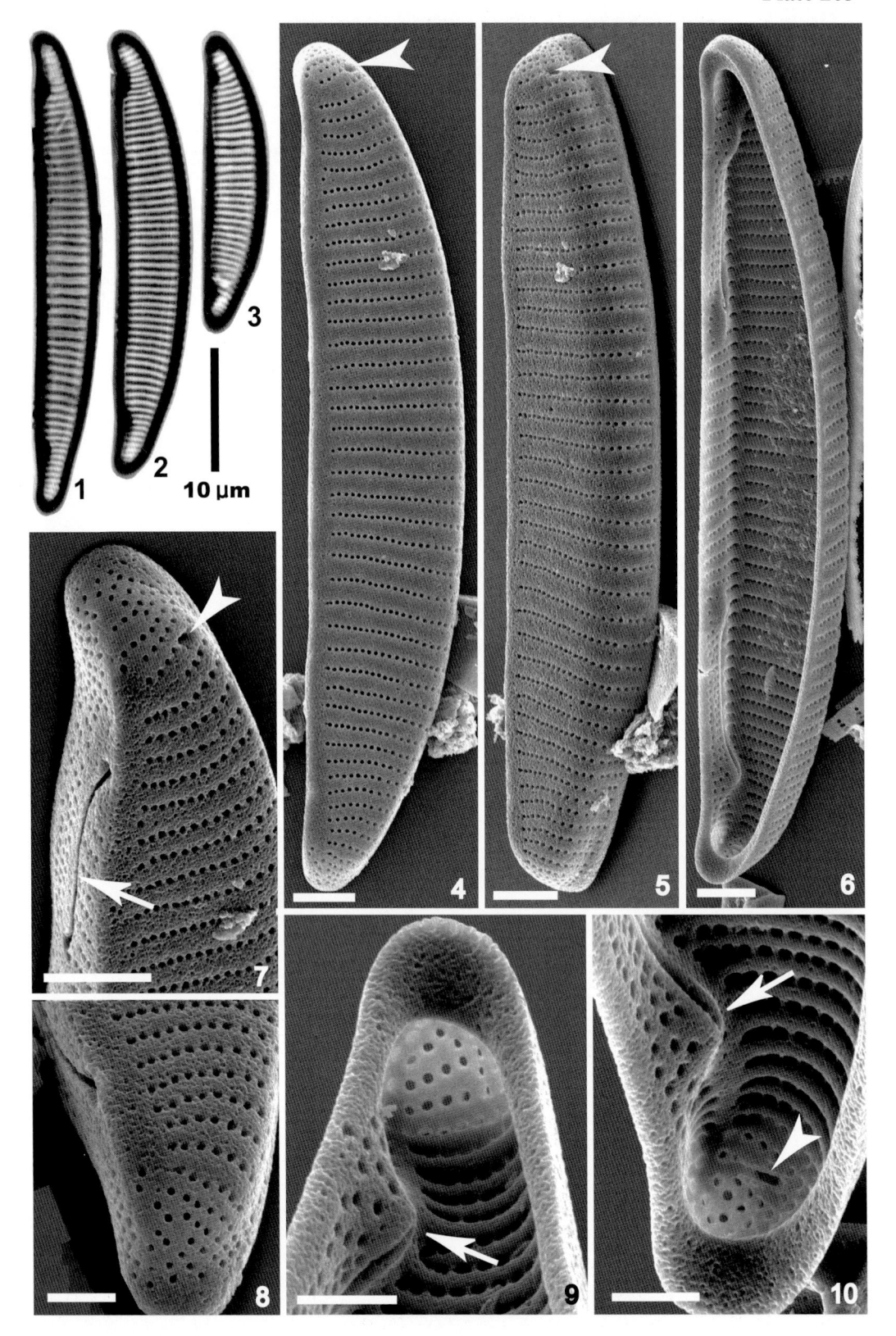

10 μm

Plate 204.   Biraphid, pennate diatoms: Eunotiales
Eunotia monodon var. *tropica* (Hustedt) Hustedt 1933

基礎異名：*Eunotia tropica* Hustedt 1927
文　　献：Hustedt, F. 1933.　*In*：A. Schmidt's Atlas der Diatomaceen-Kunde(1874-1959), pl. 381, Figs 3-6. O.R. Reisland, Leipzig.
形　　態：殻はゆるくカーブし背側に4-6回突出する．殻端はやや腹側に偏ってくさび形．殻長66-113 μm，殻幅14-15 μm，条線は10 μmに9-13本，条線を構成する点紋は10 μmに約28個．外裂溝は殻面に達する．唇状突起は片方の殻端にのみ存在する．
ノ　ー　ト：本分類群は熱帯性のものらしいとのことであるが（小林・原口 1969），今のところ横子珪藻土が日本で産出が記録された最も北の地点である．
産　　出：横子珪藻土：前期更新世・群馬県，琵琶湖底堆積物（Mori 1975）：更新世・滋賀県，有井珪藻土（Okuno 1952）：完新世・三重県．
図　　版：横子珪藻土標本．

Figs 1-5.　*Eunotia monodon* var. *tropica* (Hustedt) Hustedt
LM. Figs 1-2. SEM. Figs 3-5. Material from diatomite of Yokogo (Early Pleistocene), Numata City, Gunma, Japan. Scale bars: Fig. 3＝5 μm, Figs 4-5＝2 μm.

Figs 1-2.　Two different size valves.
Fig. 3.　External view of whole valve.
Fig. 4.　Internal view of an apex. Note: no rimoportulae.
Fig. 5.　Opposite apex of Fig. 4, rimoportula (arrowhead).

Plate 204

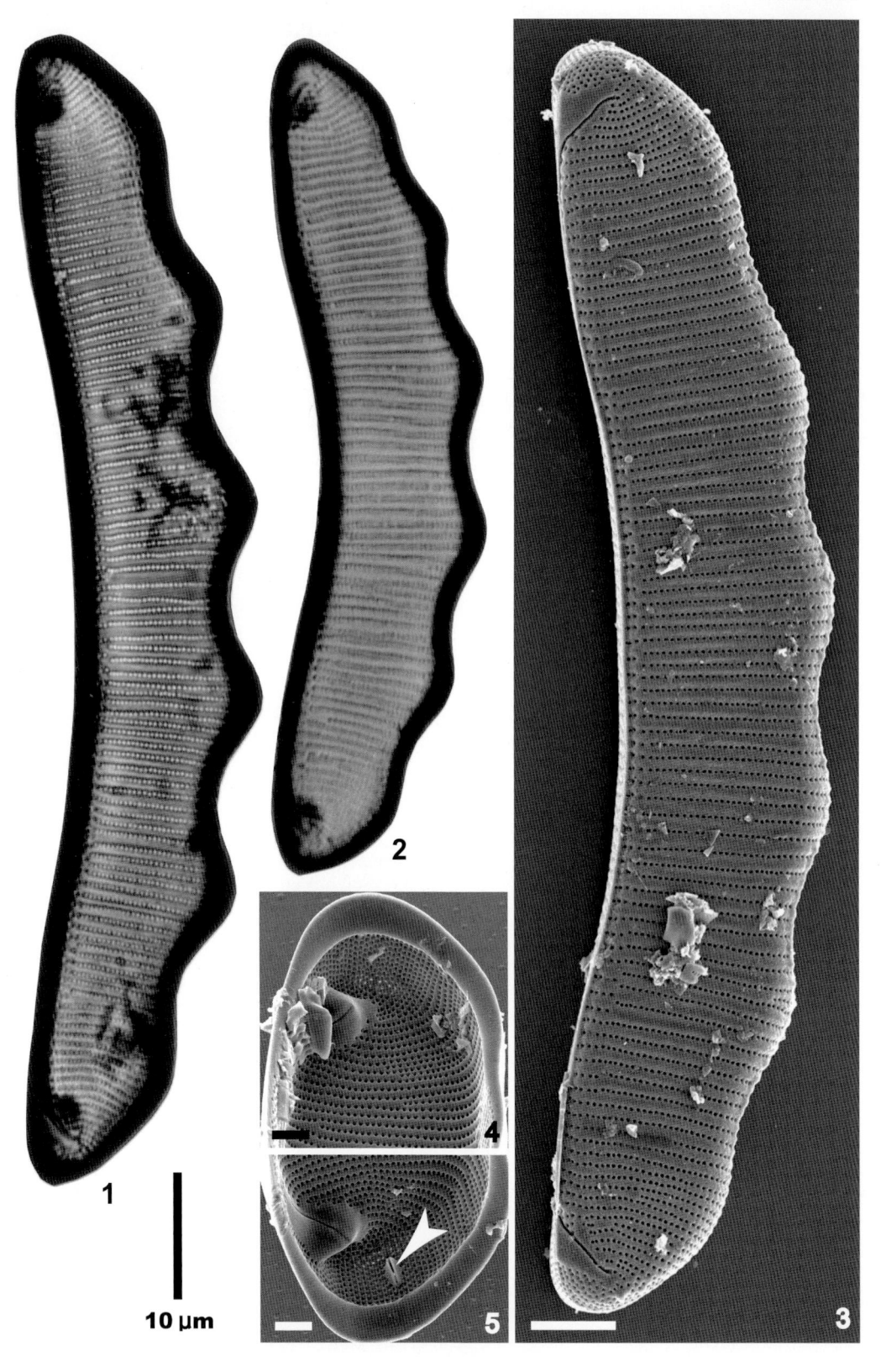

10 μm

Plate 205. Biraphid, pennate diatoms: Eunotiales
Eunotia nipponica **Skvortsov** 1938

文　献：Skvortsov, B.V. 1938.　Diatoms collected by Mr. Yoshikazu Okada in Nippon. The Journal of Japanese Botany **14**: 52-65.

　　Kobayasi, H., Ando, K. & Nagumo, T. 1981.　On some endemic species of the genus *Eunotia* in Japan. *In*: Ross, R.(ed.) Proceedings of the 6th Symposium on recent and fossil diatoms. pp. 93-114. Koenigstein, Otto Koeltz.

形　態：殻は太い線状で弓形に曲がる．殻面観における腹側・背側の両殻縁はほぼ平行，殻端ではわずか背側にカーブの向きが変わる．殻面／殻套境界には針がある．両殻端に唇状突起が認められる．殻長 52-99 µm，殻幅 11-12 µm，条線は両殻端を除き平行で 10 µm に約 17-20 本．外裂溝は殻面に達する．

ノート：原記載は Skvortsov(1938)により鎌が池（長野県・八島ガ原湿原）の試料を使用して行われた．微細構造については Kobayasi *et al.*(1981)による報告がある．本種は弱酸性のミズゴケ池沼に広く分布する種類（長田・南雲 1983）なので，産出水域の環境を推定するための有用な種と思われる．

産　出：嬬恋湖成層：中期更新世・群馬県．

図　版：嬬恋湖成層標本．

Figs 1-7.　*Eunotia nipponica* Skvortsov

LM. Figs 1-2. SEM. Figs 3-7. Material from Tsumagoi Lake Deposit (Middle Pleistocene), Tsumagoi Village, Gunma, Japan. Scale bars: Figs 3, 6＝5 µm, Figs 4-5, 7＝2 µm.

Figs 1-2.　Two different size valves.
Fig. 3.　Internal view of whole valve, helictoglossa (arrow).
Fig. 4.　Enlarged oblique view of an apex of Fig. 3, rimoportula (arrowhead).
Fig. 5.　Enlarged oblique view of opposite side apex of Fig. 3, helictoglossa (arrow) and rimoportula (arrowhead).
Fig. 6.　External view of whole valve.
Fig. 7.　Enlarged oblique view of an apex of Fig. 6, outer fissure of raphe (arrow) and opening of rimoportula (arrowhead).

Plate 205

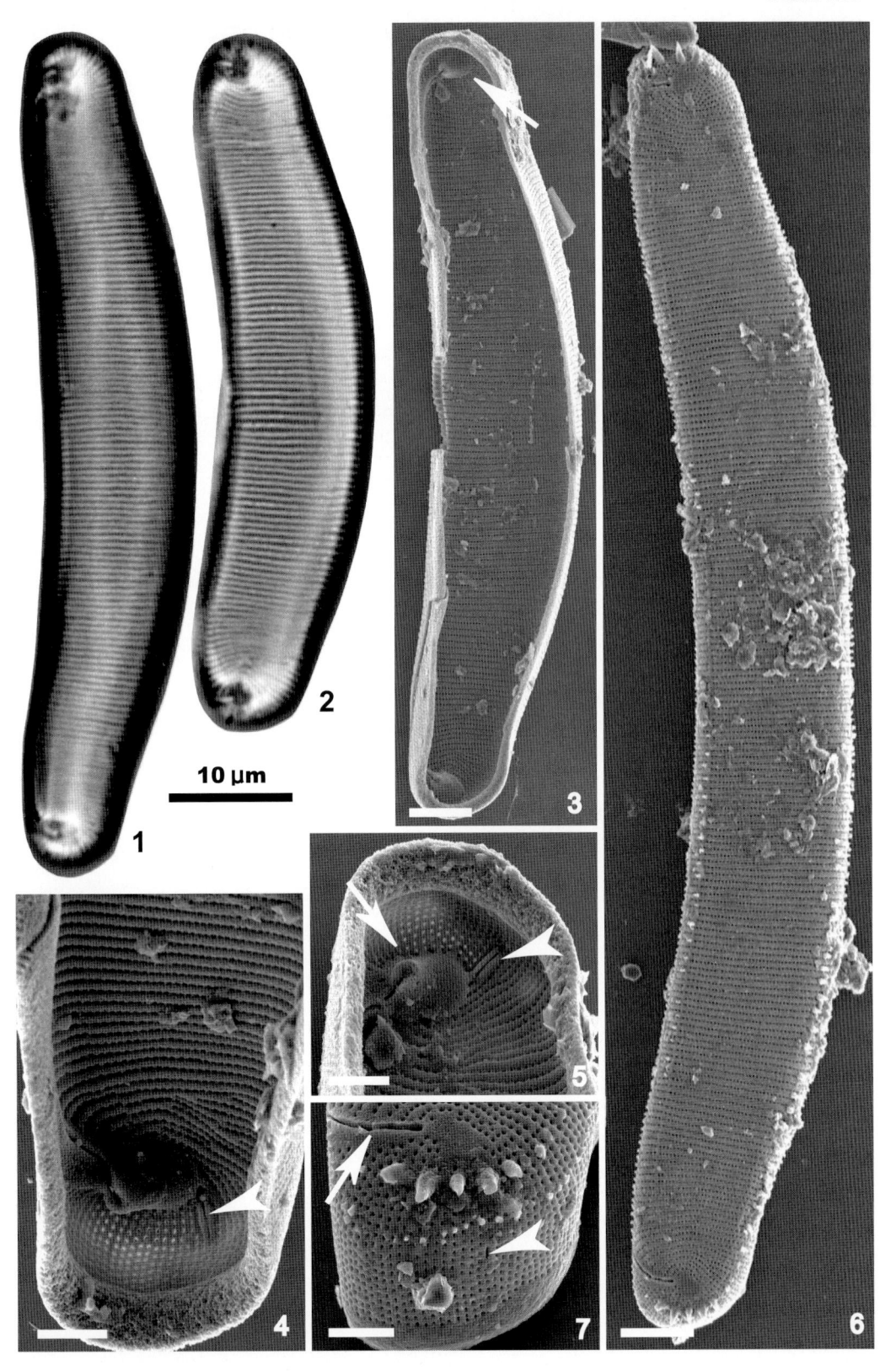

10 μm

Plate 206.   Biraphid, pennate diatoms: Eunotiales
Eunotia serra Ehrenberg 1837

文　献：Krammer, K. & Lange-Bertalot, H. 1991.　Bacillariophyceae. Teil 3. Bacillariaceae, Centrales, Fragilariaceae, Eunotiaceae. 576 pp. Süßwasserflora von Mitteleuropa, Bd. 2/3, Begründet von A. Pascher. Gustav Fischer Verlag, Stuttgart.

形　態：殻は弓形に曲がり，太い線状で，背側が6-20回波打つ．殻端外側には針がある．殻長54-168 μm，殻幅12-17 μm，長い個体の殻幅は中央部が殻端部よりもわずか細い傾向が見られた．条線は背側に広がる放射状で，両端ではやや密になり，10 μm に 12-16 本．唇状突起は片方の殻端内側に 1 個所在する．

ノ ー ト：化石としては報告が見当たらないが，類似種が産出していることと，SEM写真が既存の図鑑にはないので比較のため収録する．尾瀬沼の西に所在する小沼（群馬県片品村）からの個体で図を構成した．類似する *E. diadema* とは背側の波打ち回数が 6 回以上であることにより区別できる．

　　　古くは *Eunotia robusta* Ralfs と同定されていた．

産　出：各地の高地の湿原等．

図　版：尾瀬地域（小沼：片品村，群馬県）・現生標本．

Figs 1-6.　*Eunotia serra* Ehrenberg

LM. Fig. 1. SEM. Figs 2-6. Material from Ko-numa (Recent), Oze, Katashina Village, Gunma, Japan. Scale bars: Figs 2-3＝10 μm, Fig. 6＝5 μm, Figs 4-5＝2 μm.

Fig. 1.　Whole valve view.

Fig. 2.　External oblique view of whole valve.

Fig. 3.　Internal whole valve view, rimoportula (arrowhead) and helictoglossae (arrows).

Figs 4-5.　Enlarged view of valve apices of Fig. 3 showing helictoglossae (arrows) and rimoportula (arrowhead). Note: only one rimoportula.

Fig. 6.　Areolae rows of Fig. 2.

Plate 206

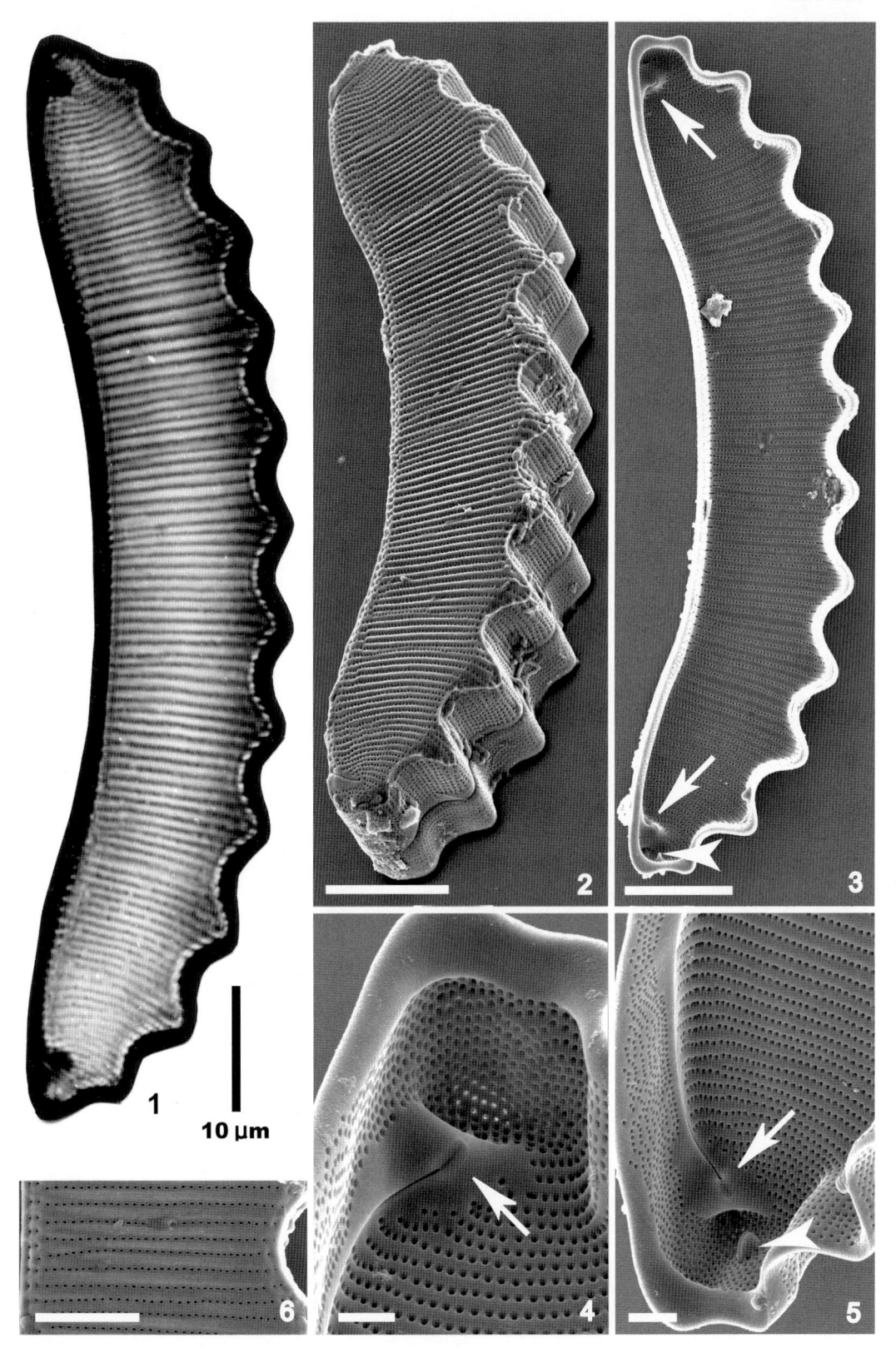

10 μm

Plate 207.　Biraphid, pennate diatoms: Naviculales
Aneumastus tusculus (Ehrenberg) D.G. Mann & Stickle
in Round *et al.* 1990

基礎異名：*Navicula tuscula* Ehrenberg 1840

文　　献：Round, F.E., Crawford, R.M. & Mann, D.G. 1990. The diatoms, Biology & morphology of the genera. 747 pp. Cambridge University Press, Cambridge.

形　　態：殻形は広皮針形で先端はくちばし状，条線は放射状に配列する．中心域は狭い．殻長 27-73 μm，殻幅 12-26 μm，条線は 10 μm に 10-13 本であった．縦溝の外裂溝は通常は直線であるが，わずかであるがうねっている殻も観察できた（Fig. 1）．殻面の条線を構成する胞紋は，縦溝側では大形で 1 個の胞紋列であるが縁辺では小形で 2 列になる．また最も縦溝に近い 1 個は光学顕微鏡観察においてやや横長の形状を示すことがある．殻面／殻套境界は肥厚しており，殻面と殻套の胞紋は連続しない．殻套の胞紋列は殻面縁辺部より粗い．

ノート：本分類群は従来 *Navicula tuscula* とされていたが，Round *et al.*(1990) により本属が設立され，そのタイプ種となった．図に使用した尾本層の個体は Round *et al.*(1990) と同じく殻面／殻套境界が肥厚して殻面の条線と殻套の条線は連続しないが，人吉層の個体はほとんど肥厚がなく，肥厚の程度は変化する．

産　　出：八束村珪藻土（奥野 1943）：更新世・岡山県，琵琶湖底堆積物（Mori 1975）：更新世・滋賀県，宮島層（Akutsu 1964）：前期更新世・栃木県，以上は *Navicula tuscula* として．*A. tusculus* としては，現生であるが河島・真山(1997)による阿寒湖からの報告がある．筆者は尾本層：前期更新世・大分県，人吉層：後期鮮新世・熊本県から見出している．

図　　版：尾本層標本．

Figs 1-7.　*Aneumastus tusculus* (Ehrenberg) D.G. Mann & Stickle
LM. Figs 1-2. SEM. Figs 3-7. Material from Omoto Formation (Early Pleistocene), Kitsuki City, Oita, Japan. Scale bars: Fig. 7＝5 μm, Figs 3-5＝2 μm, Fig. 6＝1 μm.

Figs 1-2.　Two different size valves.
Fig. 3.　Part of external oblique view showing thick valve face/mantle boundary (arrow).
Fig. 4.　External oblique view of an apex.
Fig. 5.　Internal oblique view of whole valve.
Fig. 6.　Enlarged view of internal valve margin.
Fig. 7.　External view of whole valve.

Plate 207

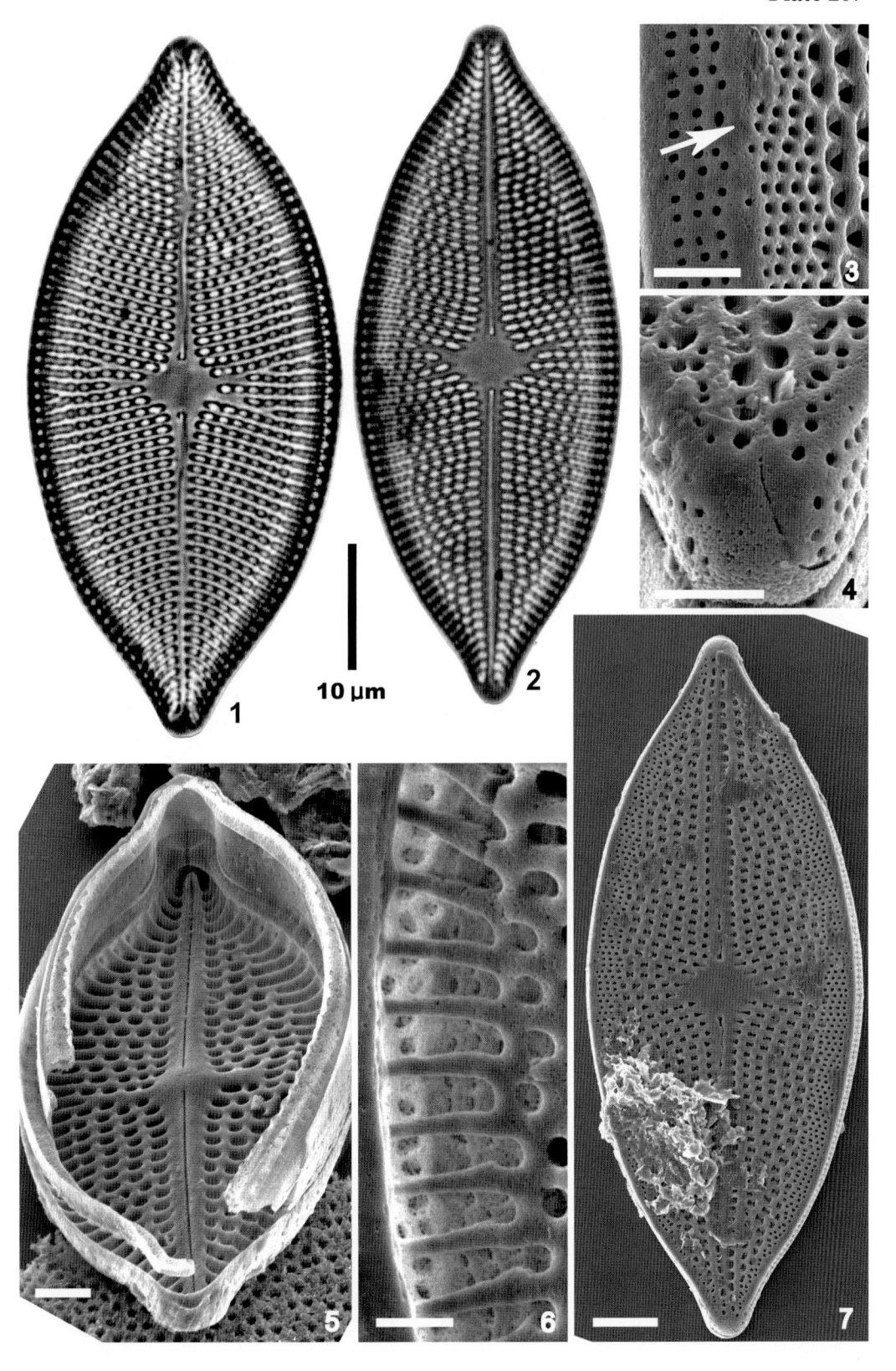

Plate 208. Biraphid, pennate diatoms: Naviculales
Caloneis schumanniana (Grunow) Cleve 1894

基礎異名：*Navicula schumanniana* Grunow 1880-1885
文　　献：Cleve, P.T. 1894. Synopsis of the Naviculoid Diatoms. Kongliga Svenska Vetenskaps-Akademiens Handlingar, Bd. **26**: 1-194, pls 1-5.
形　　態：殻は太い線形で中央部が膨らみ，中心域の両端には半月状の模様がある．条線は中心部でほぼ平行～わずか放射状であるが，両殻端では明瞭な放射状に配列する．津森層・人吉層から見出された本種は殻長 51-122 µm，殻幅 12-18 µm，条線は 10 µm に約 16 本であった．1 本の条線は 5 列の胞紋列から構成されている．
ノ ー ト：原口ら(1998)によれば *Caloneis limosa* (Kütz.) Patrick は本種の異名である．渡辺ら(2005)では *C. schumanniana* が *C. limosa* の異名であるとしている．本書では原口ら (1998)に基づいている．
産　　出：津森層：中期更新世・熊本県，人吉層：後期鮮新世・熊本県．*C. limosa* としては長野県和村珪藻土（上山・小林 1893）から報告がある．
図　　版：津森層標本．

Figs 1-6. *Caloneis schumanniana* (Grunow) Cleve
LM. Figs 1-2. SEM. Figs 3-6. Material from Tsumori Formation (Middle Pleistocene), Mashiki Town, Kumamoto, Japan. Scale bars: Figs 3, 5＝5 µm, Fig. 4＝1 µm, Fig. 6＝0.5 µm.

Fig. 1 (LM) and Figs 3-6 (SEM) are same valve.
Figs 1-2.　Two different size valves.
Fig. 3.　External view of whole valve.
Fig. 4.　Enlarged view of part of Fig. 3 showing striae consisting of five areolae rows.
Fig. 5.　Internal oblique view of whole valve.
Fig. 6.　Enlarged view of part of Fig. 5 showing alveolus openings.

Plate 208

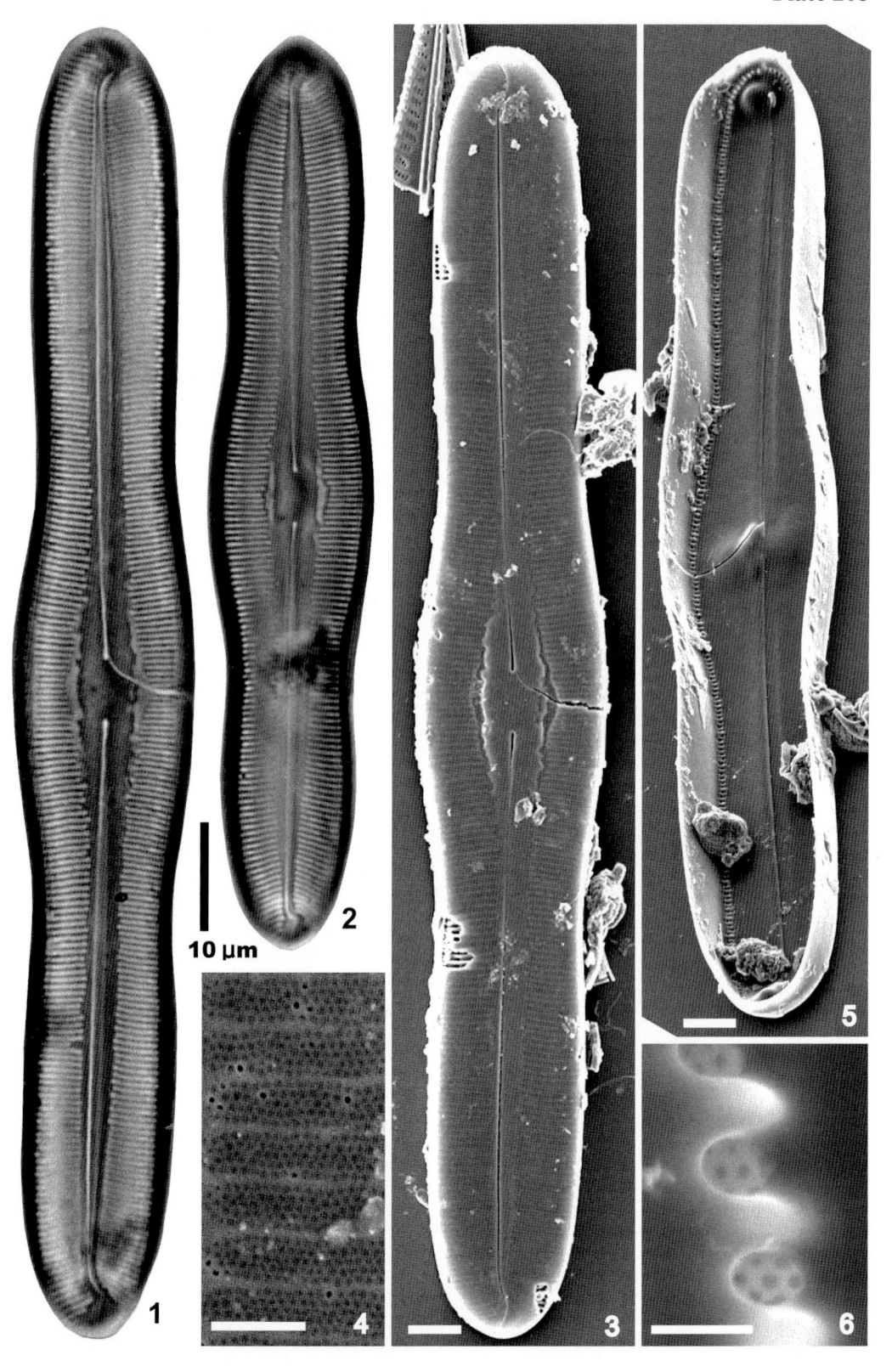

10 μm

Plate 209. Biraphid, pennate diatoms: Naviculales
Cavinula pseudoscutiformis (Hustedt) D.G. Mann & Stickle in Round *et al.* 1990

基礎異名：*Navicula pseudoscutiformis* Hustedt 1930

文　　献：Round, F.E., Crawford, R.M. & Mann, D.G. 1990. The diatoms. Biology & morphology of the genera. 747 pp. Cambridge University Press, Cambridge.

形　　態：殻はほとんど円形に近い楕円形で，殻長 10.5-15 μm，殻幅 9-12.5 μm．条線は強く放射し 10 μm に 20-26 本，条線を構成する点紋は 10 μm に 20-24 個である．軸域は殻外側ではわずか両殻端に向かって狭くなり，内側では強く肥厚している．

産　　出：太櫓層（瀬棚の珪藻土）（奥野 1959）：前期中新世・北海道，瓜生坂層（窪田ら 1976）：前期更新世・長野県，大鷲湖沼性堆積物：鮮新世・岐阜県，奄芸層群（根来・後藤 1981）：鮮新世・三重県．

図　　版：大鷲湖沼性堆積物標本．

Figs 1-7. *Cavinula pseudoscutiformis* (Hustedt) D.G. Mann & Stickle

LM. Figs 1-2. SEM. Figs 3-7. Material from lacustrine deposit of Owashi (Pliocene), Gujo City, Gifu, Japan. Scale bars: Figs 3-6＝2 μm, Fig. 7＝0.5 μm.

Figs 1-2.　Two different size valves.
Fig. 3.　External view of whole valve.
Fig. 4.　Oblique view of Fig. 3.
Fig. 5.　Internal view of whole valve.
Fig. 6.　Enlarged oblique view of Fig. 5.
Fig. 7.　Enlarged view of terminal area of Fig. 5, helictoglossa (arrowhead).

Plate 209

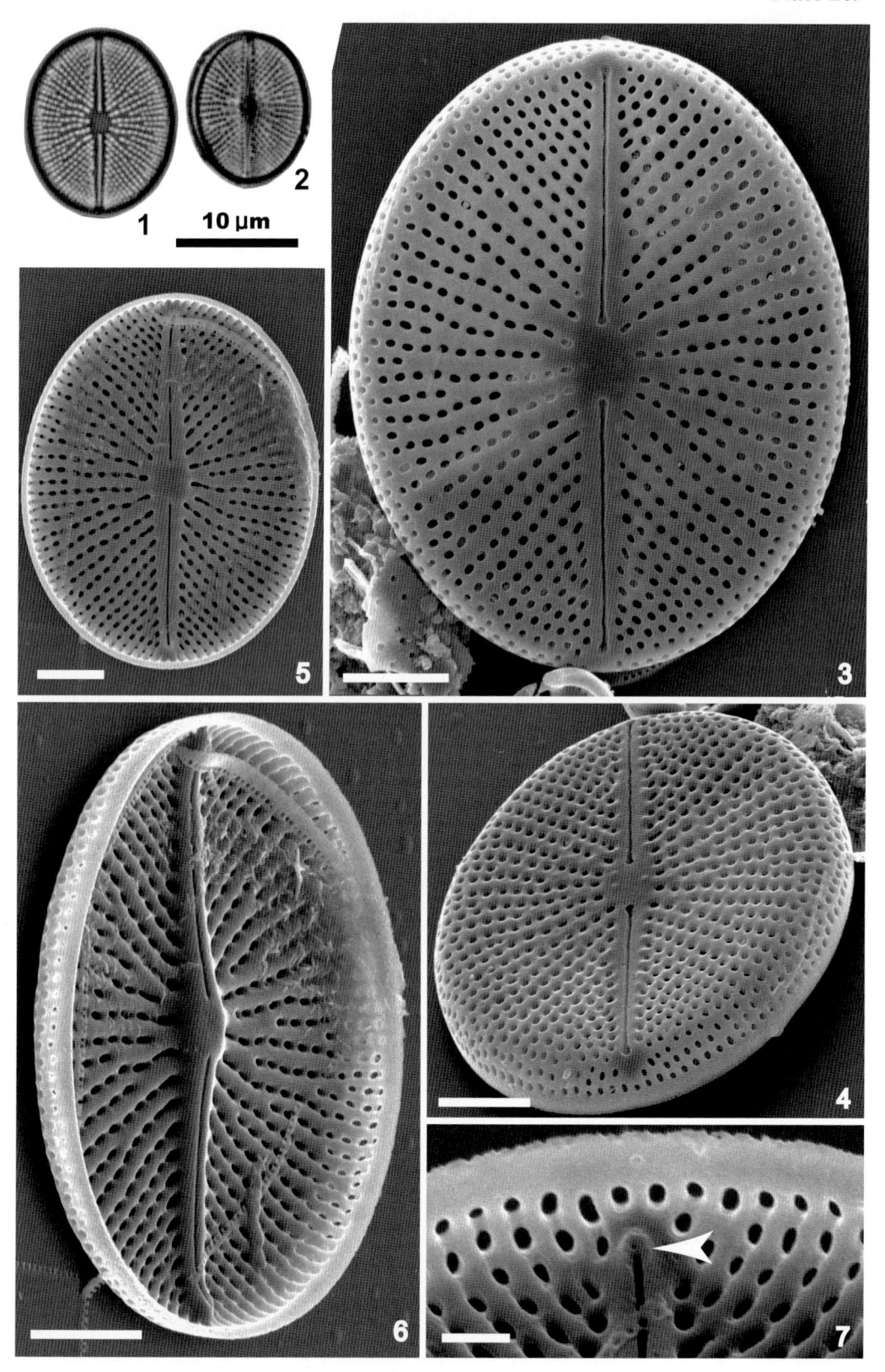

Plate 210.   Biraphid, pennate diatoms: Naviculales
Craticula ambigua (Ehrenberg) D.G. Mann
in Round *et al.* 1990

基礎異名：*Navicula ambigua* Ehrenberg 1843

文　　献：Round, F.E., Crawford, R.M. & Mann, D.G. 1990. The diatoms. Biology & morphology of the genera. 747 pp. Cambridge University Press, Cambridge.

形　　態：殻は皮針形，殻端はくちばし状で長く突き出る，殻長 53-72 μm，殻幅 15-17 μm．条線は短軸に平行で 10 μm に約 18 本，殻外側に縦の肥厚線があるので LM 像では殻面条線域が格子模様を呈する．条線を構成する点紋は 10 μm に約 30 個．軸域は狭い．外中心裂溝はわずかに曲がり中心孔で終了する．

産　　出：奄美大島泥染公園の泥田：鹿児島県（現生）．

図　　版：泥染公園の泥田（奄美大島）標本．

Figs 1-5.　*Craticula ambigua* (Ehrenberg) D.G. Mann

LM. Fig. 1. SEM. Figs 2-5. Material from"dorota"of Dorozome-Park (Recent), Amami-Oshima, Kagoshima, Japan. Scale bars: Figs 2, 4＝5 μm, Fig. 5＝2 μm, Fig. 3＝1 μm.

Fig. 1.　Whole valve view.
Fig. 2.　External view of whole valve.
Fig. 3.　Enlarged view of an apex of Fig. 2, terminal fissure (arrowhead).
Fig. 4.　Internal oblique view of whole valve.
Fig. 5.　View of valve center of Fig. 4, raphe (arrowhead).

Plate 210

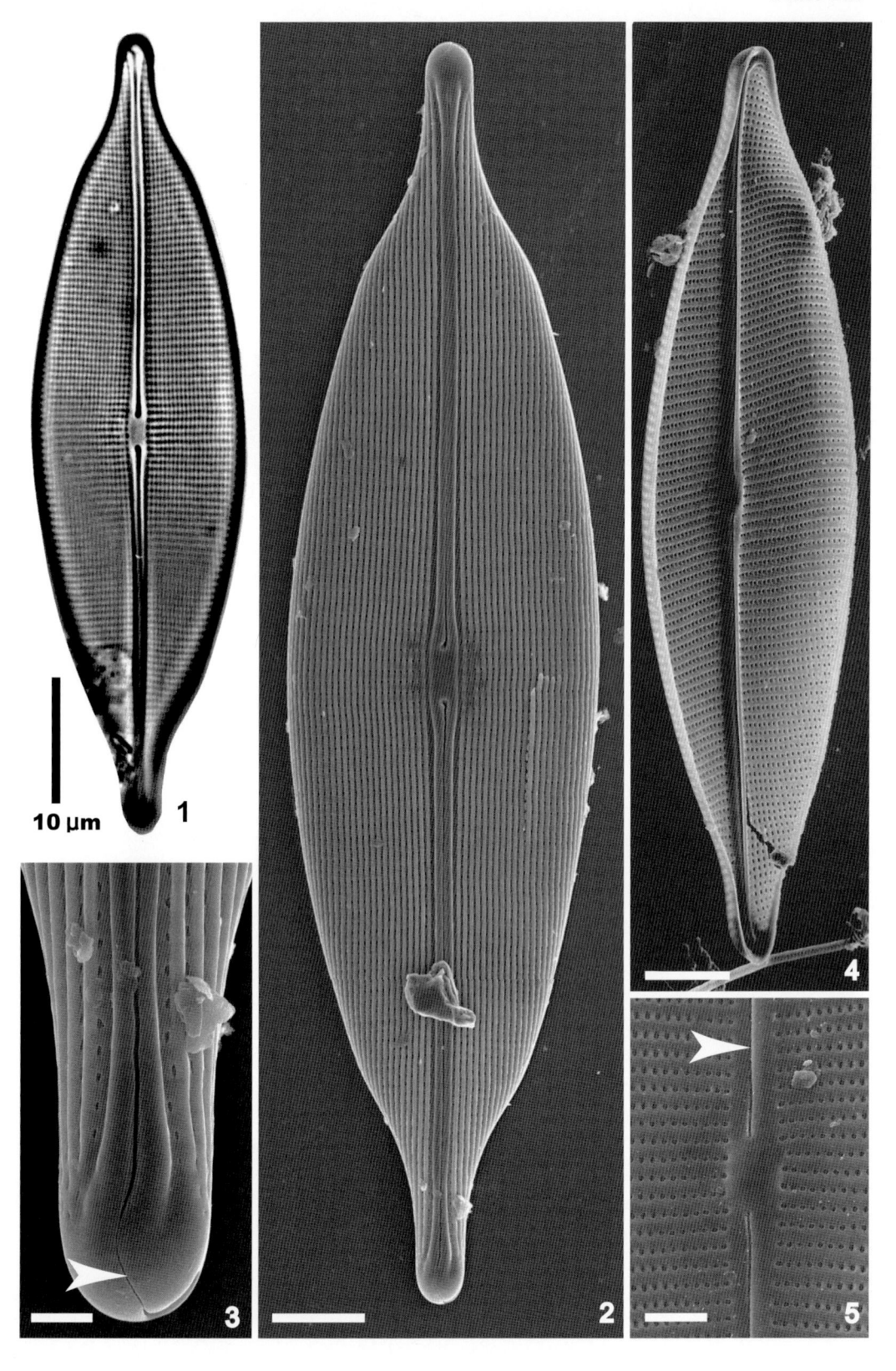

10 µm

Plate 211. Biraphid, pennate diatoms: Naviculales
Craticula cuspidata (Kützing) D.G. Mann in Round *et al.* 1990

基礎異名：*Frustulia cuspidata* Kützing 1833

文　献：Round, F.E., Crawford, R.M. & Mann, D.G. 1990. The diatoms. Biology & morphology of the genera. 747 pp. Cambridge University Press, Cambridge.

形　態：殻は大形で菱形に近い皮針形，殻端はわずかくちばし形になることもある．殻長 96-110 μm，殻幅 20-27 μm．条線は短軸に平行で 10 μm に約 14 本，殻外側表面に縦の肥厚線があるので条線域が格子模様を呈する．条線を構成する点紋は 10 μm に約 28 個．軸域は狭い．外中心裂溝は横に曲がる．

産　出：横子珪藻土：前期更新世・群馬県，和村珪藻土（Skvortzov 1937）：新第三紀・長野県，由布院珪藻土（Okuno 1952）：鮮新世～更新世・大分県．

図　版：横子珪藻土標本．

Figs 1-5. *Craticula cuspidata* (Kützing) D.G. Mann
LM. Fig. 1. SEM. Figs 2-5. Material from diatomite of Yokogo (Early Pleistocene), Numata City, Gunma, Japan. Scale bars: Fig. 4＝10 μm, Figs 2-3＝2 μm, Fig. 5＝0.5 μm.

Fig. 1. Whole valve view.
Fig. 2. Enlarged view of an apex of Fig. 4.
Fig. 3. Enlarged view of valve center of Fig. 4.
Fig. 4. External view of whole valve.
Fig. 5. Detailed internal view of areolae.

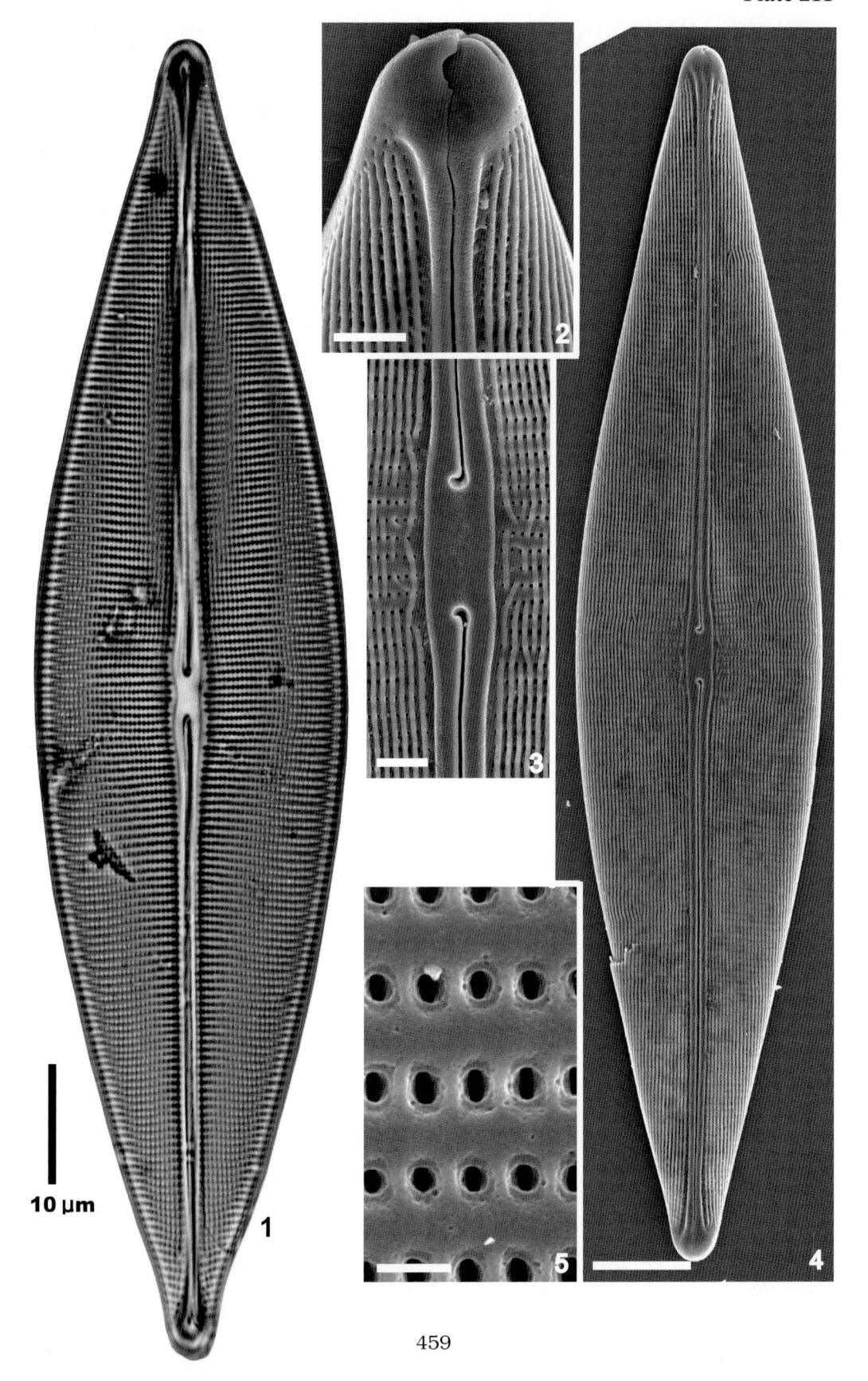

Plate 211

Plate 212.   Biraphid, pennate diatoms: Naviculales
Diadesmis confervacea Kützing 1844

文　　献：Kützing, F.T. 1844.　Die kieselschaligen Bacillarien order Diatomeen. 152 pp. pls 1-30. Nordhausen.

　　　　福島　博・小林艶子・寺尾公子　1984.　羽状ケイ藻 *Navicula confervacea* (Kütz.) Grunow の分類学的検討(1)．日本水処理生物学会誌 **20**: 20-33.

形　　態：殻は先端がやや突出した楕円形で，中心域は広い．殻長 14-19 μm，殻幅 6.5-8 μm，条線は放射状で 10 μm に約 20 本．条線を構成する点紋は横長形である．殻面／殻套境界は肥厚して針様の突出物があり，殻面の条線は殻套へ連続しない．殻套は縦長の 1 個の胞紋が分布している．

ノート：本種は *Diadesmis confervacea* として Kützing(1844) により原記載されたが，Grunow(1880) によって *Navicula* へ組み合わせになり，現在は最初の所属に位置づけられている．形態については和文で詳しい研究があるので文献欄に記した．その中で本種は熱帯性とされている（福島ら 1984）．Skvortzow(1936) が木崎湖から記載した *N. confervacea* f. *nipponica* は f. *confervacea* と同じである（福島ら 1984）．

産　　出：津房川層：後期鮮新世・大分県，大正池：粟国島・沖縄県（現生）．現生としては多くの地域から報告がある．

図　　版：大正池（Figs 1-3, 5-8），津房川層（Fig. 4）標本．

Figs 1-8.   *Diadesmis confervacea* Kützing

LM. Figs 1-4. SEM. Figs 5-8. Material Figs 1-3, 5-8 from Taisho-ike Pond (Recent), Aguni Island, Okinawa and Fig. 4 from Tsubusagawa Formation (Late Pliocene), Kitsuki, Oita, Japan. Scale bars: Figs 5-8＝2 μm.

Figs 1-4.   Four different size valves.
Fig. 5.   Internal view of whole valve.
Fig. 6.   Oblique view of Fig. 5.
Fig. 7.   External view of whole valve.
Fig. 8.   Oblique view of Fig. 7.

Plate 212

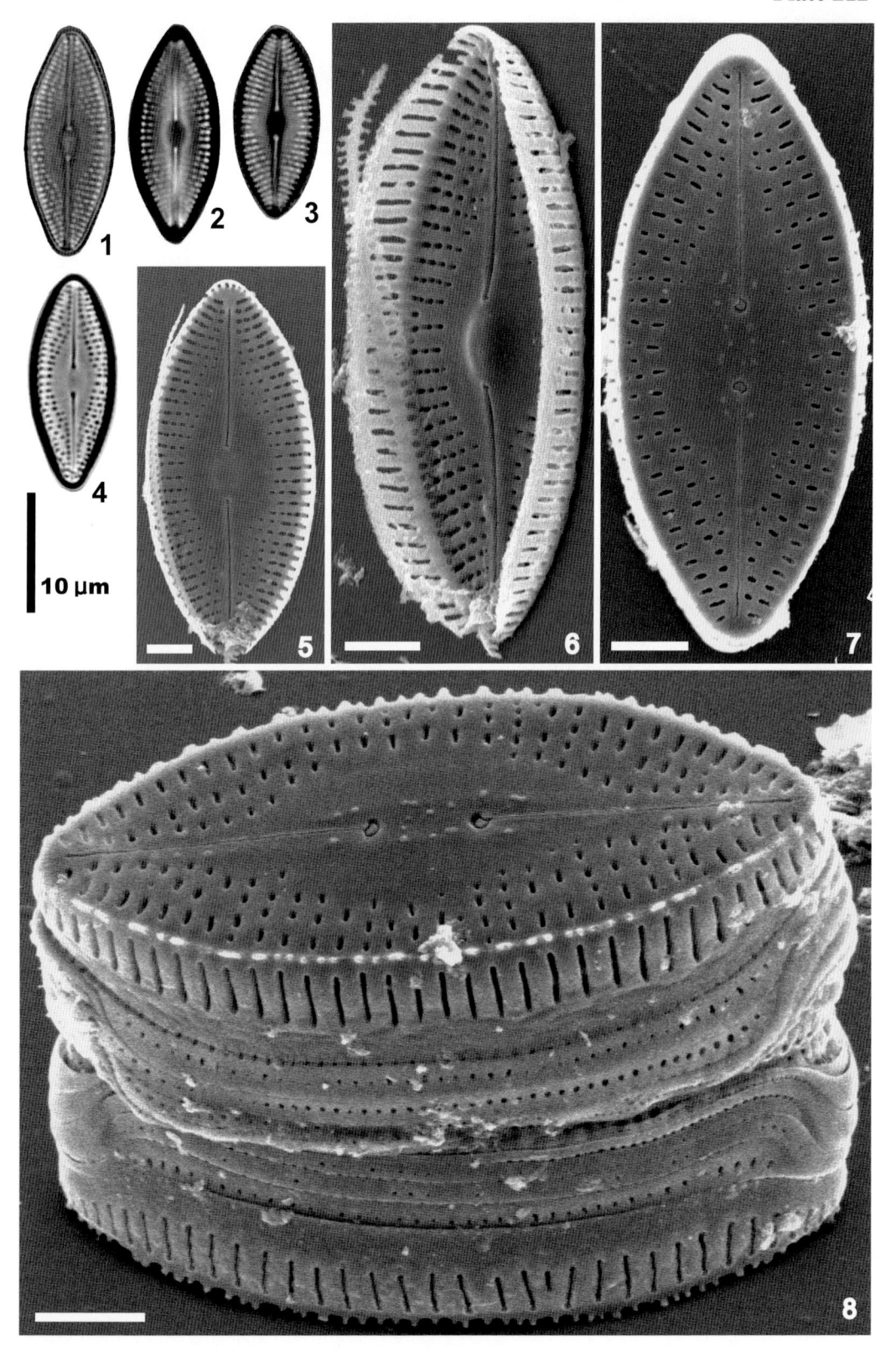

Plate 213.   Biraphid, pennate diatoms: Naviculales
Diploneis elliptica (Kützing) Cleve 1894

基礎異名：*Navicula elliptica* Kützing 1844
文　　献：Kützing, F.T. 1844.　Die kieselschaligen Bacillarien order Diatomeen. 152 pp. pls 1-30. Nordhausen.
形　　態：殻は楕円形，殻長 30-44 μm，殻幅 16-23 μm．条線は放射状で，１列の大形の胞紋列から構成され，10 μm に約 8 本．胞紋はほぼ四角形で 10 μm に 10-12 個である．中心域は比較的広い．縦溝の外側中心末端はわずか横へ曲がるが，内側では直線のまま終了する．縦溝の両側に所在する縦走管は狭く，中央では横に広がる．
産　　出：宮田層：中新-鮮新世・秋田県，鬼首層（Ichikawa 1955）：後期更新世・宮城県，野殿層（中島・南雲 1999）：中期更新世・群馬県，瓜生坂層（田中・南雲 2000）：前期更新世・長野県，投入堂層（赤木ら 1984）：後期中新世・鳥取県．
図　　版：宮田層標本．

Figs 1-6. *Diploneis elliptica* (Kützing) Cleve
LM. Figs 1-2. SEM. Figs 3-6. Material from Miyata Formation (Mio-Pliocene), Senboku City, Akita, Japan. Scale bars: Figs 3, 5＝5 μm, Figs 4, 6＝1 μm.

Figs 1-2.　Two different size valves.
Fig. 3.　Internal view of a valve.
Fig. 4.　Internal view of central raphe ending.
Fig. 5.　External view of whole valve.
Fig. 6.　External view of central raphe ending (arrowhead).

Plate 213

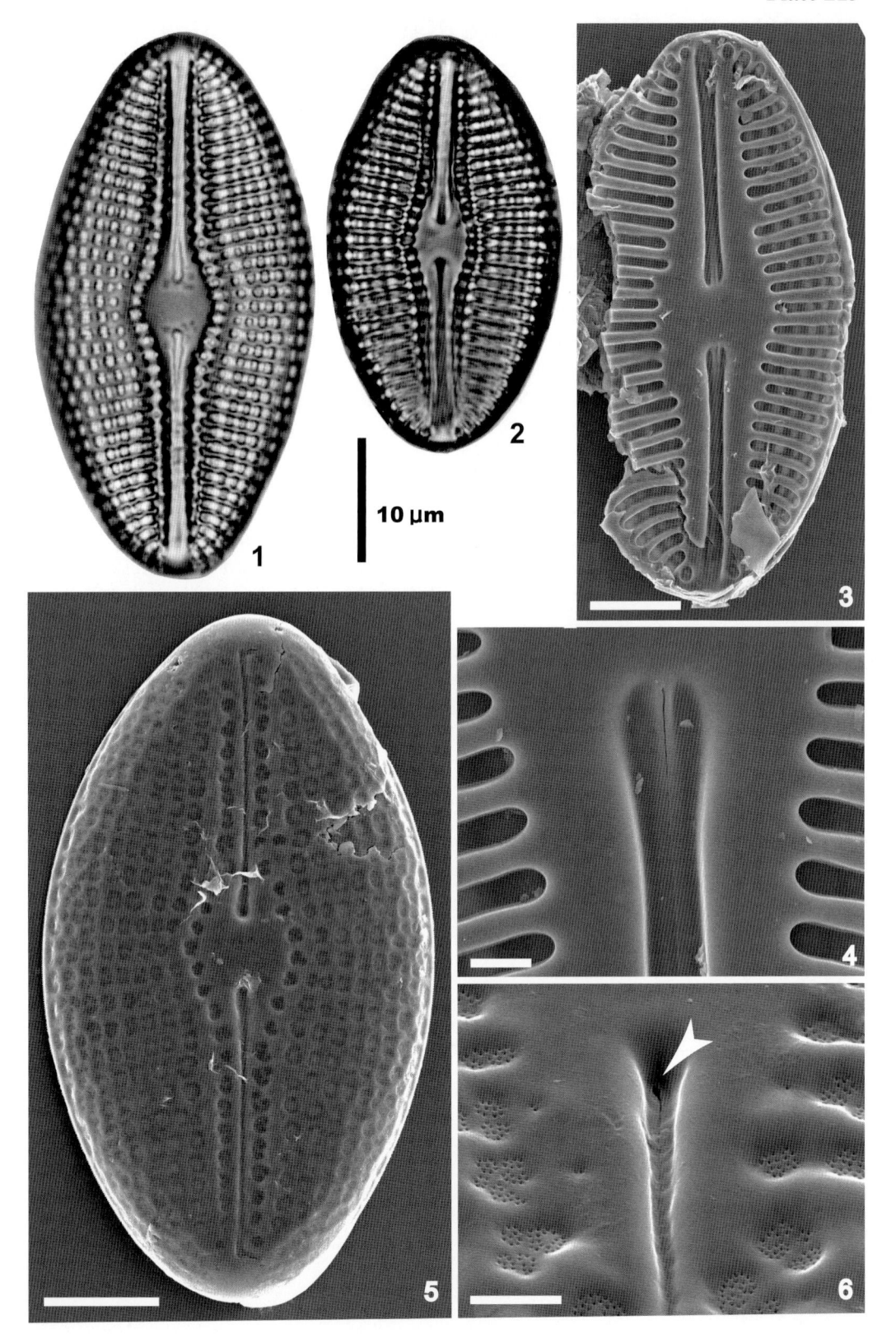

Plate 214.　Biraphid, pennate diatoms: Naviculales
Diploneis finnica (Ehrenberg) Cleve 1891

基礎異名：*Cocconeis finnica* Ehrenberg 1838

文　　献：Cleve, P.T. 1891. The diatoms of Finland. Acta Societatis pro fauna et flora fennica. **8**: 1-68, pls 1-3.

　　Idei, M. & Kobayasi, H. 1989. The fine structure of *Diploneis finnica* with special reference to the marginal openings. Diatom Research **4**: 25-37.

形　　態：殻は楕円形，殻長 45-74 μm，殻幅 25-37 μm．殻面の条線は 2 列の胞紋列から構成される．条線の途中に四角い穴が存在する際立った特徴がある（ない殻もある）．縦走管の幅は広い．外裂溝は両殻端で同方向へ曲がる．

ノート：Idei & Kobayasi(1989)で形態について詳しい報告がある．本種は北半球の極地周辺に多産する（安藤ら 1971）ので，気候を指示する有用な種になる可能性がある．

産　　出：中之条湖成層（田中・小林 1992）：中期更新世・群馬県，人吉層：後期鮮新世・熊本県，仙女ヶ池（安藤ら 1971）：埼玉県（現生）．

図　　版：中之条湖成層（Figs 1, 5），人吉層（Figs 2-4）標本．

Figs 1-5.　*Diploneis finnica* (Ehrenberg) Cleve

LM. Fig. 1. SEM. Figs 2-5. Materials: Figs 1, 5, from Nakanojo Lacustrine Deposit (Middle Pleistocene), Nakanojo Town, Gunma and Figs 2-4, from Hitoyoshi Formation (Late Pliocene), Hitoyoshi City, Kumamoto, Japan. Scale bars: Figs 2, 5＝5 μm, Figs 3-4＝2 μm.

Fig. 1.　Whole valve view, square opening (arrow).
Fig. 2.　External view of whole valve, square opening (arrow).
Fig. 3.　Enlarged view of part of Fig. 2, square opening (arrow).
Fig. 4.　Enlarged oblique view of an apex of Fig. 2, terminal fissure (arrowhead).
Fig. 5.　Internal oblique view of whole valve.

Plate 214

Plate 215.　Biraphid, pennate diatoms: Naviculales
Diploneis ovalis (Hilse) Cleve 1891

基礎異名：*Pinnularia ovalis* Hilse 1861

文　献：Krammer, K. & Lange-Bertalot, H. 1986. Bacillariophyceae. 876 pp. *In*：Ettl, H., Gerloff, J., Heynig, H. & Mollenhauer, D.(eds) Süsswasserflora von Mitteleuropa. Teil 1. Bd. **2/1**. G. Fischer Verlag, Stuttgart & New York.

形　態：殻は楕円形．殻長 19-28 μm，殻幅 11-14 μm．条線は 10 μm に 10-12 本．縦走管は狭く，中央では横に広がる．外裂溝は両殻端で同方向へ曲がる．条線を構成する胞紋は外側表面を師板で覆われるが，師板は細棒状の肥厚部が横に発達していることが多いので，LM 観察では胞紋の上下に（ピントにより）短黒線を観察でき，胞紋が上下に並んでいるように見えることもある．

ノート：各地から産出報告がある種である．胞紋の上下に短黒線を観察できることは，殻形が類似する *D. subovalis* との有効な識別点である．

産　出：野殿層（中島・南雲 1999）：中期更新世・群馬県，瓜生坂層（田中・南雲 2000）：前期更新世・長野県，赤川層（Akutsu 1964）：更新世・栃木県，田染村珪藻土（Okuno 1952）：鮮新世または更新世・大分県，吹上浜（泥炭層）：完新世・鹿児島県．

図　版：吹上浜（泥炭層）標本．

Figs 1-7.　*Diploneis ovalis* (Hilse) Cleve

LM. Figs 1-3. SEM. Figs 4-7. Material from peat deposit of Fukiagehama (Holocene), Kagoshima, Japan. Scale bars: Figs 4-5＝5 μm, Fig. 6＝1 μm, Fig. 7＝0.5 μm.

Figs 1-3.　Three different size valves.
Fig. 4.　Internal view of whole valve.
Fig. 5.　External view of whole valve.
Fig. 6.　Detailed external view of areolae.
Fig. 7.　Detailed internal view of areolae.

Plate 215

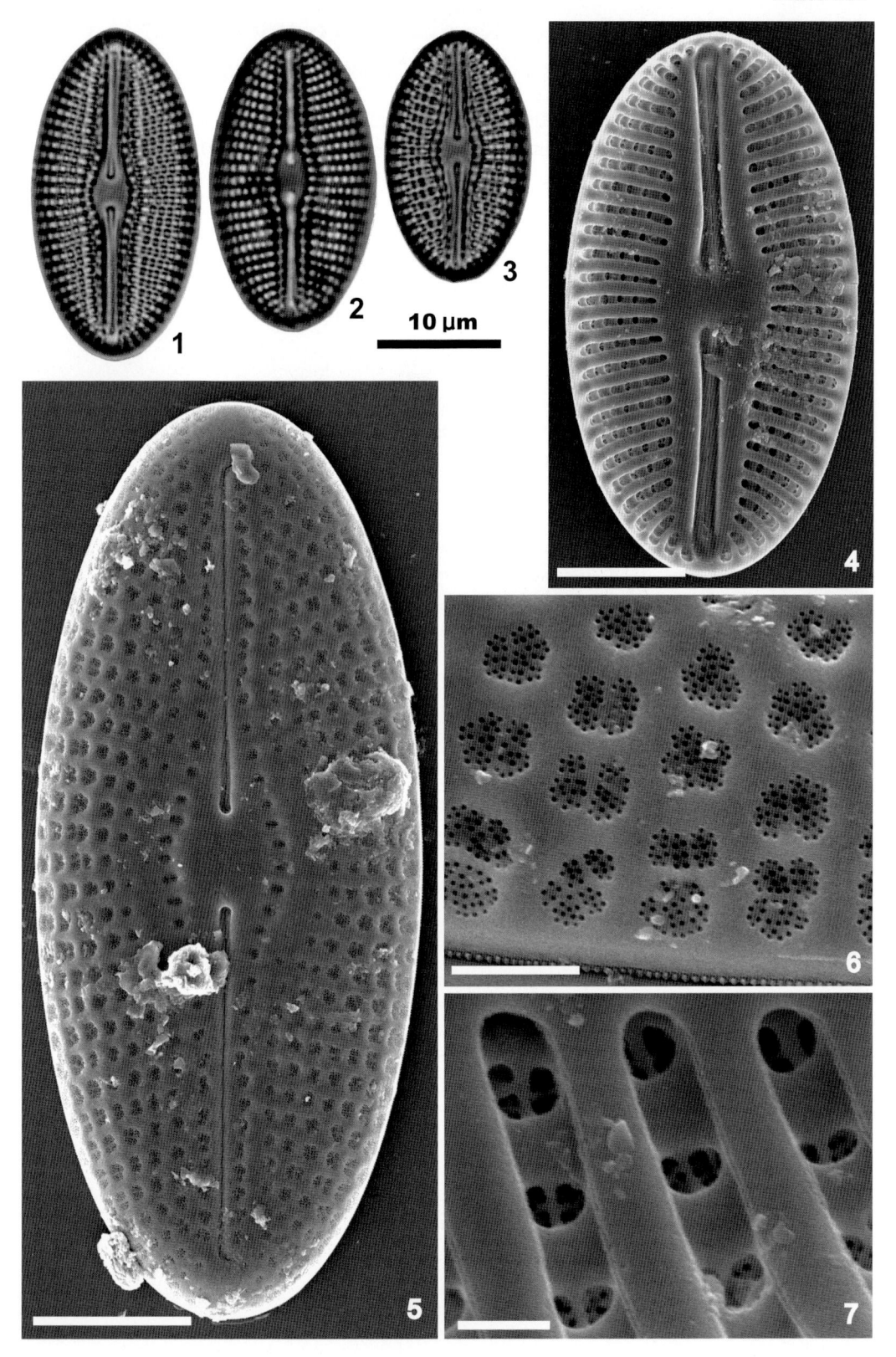

10 µm

Plate 216.　Biraphid, pennate diatoms: Naviculales
Diploneis smithii var. *rhombica* **Mereschkowsky** 1902

文　　献：Krammer, K. & Lange-Bertalot, H. 1986. Bacillariophyceae. 876 pp. *In*：Ettl, H., Gerloff, J., Heynig, H. & Mollenhauer, D.(eds) Süsswasserflora von Mitteleuropa. Teil 1. Bd. 2/1. G. Fischer Verlag, Stuttgart & New York.

形　　態：殻は丸みを帯びた菱形，殻長 65 μm，殻幅 37 μm．中心域および中心域に近い軸域は広い．条線は放射状で互い違いに配列した 2 列の点紋列から構成される．10 μm に約 9 本．縦溝枝の両末端は同方向へ曲がる．殻端の縦走管は狭いが，中央では広くなる．

ノ　ー　ト：稲城層からの完全な殻は 1 個体のみの産出であったので，その計測値を記してある．

産　　出：稲城層：前期更新世・東京都，東京低地（Hasegawa 1975）：完新世・東京都，濃尾平野沖積層（森 1981）：完新世・岐阜県．

図　　版：稲城層標本．

Figs 1-5.　*Diploneis smithii* var. *rhombica* **Mereschkowsky**
LM. Figs 1-2. SEM. Figs 3-5. Material from Inagi Formation (Early Pleistocene), Fuchu City, Tokyo, Japan. Scale bars: Fig. 3＝10 μm, Fig. 5＝5 μm, Fig. 4＝2 μm.

Figs 1-2.　Same valve at different focal planes.
Fig. 3.　External view of whole valve.
Fig. 4.　Enlarged oblique view of an apex of Fig. 3.
Fig. 5.　Internal view of valve center.

Plate 216

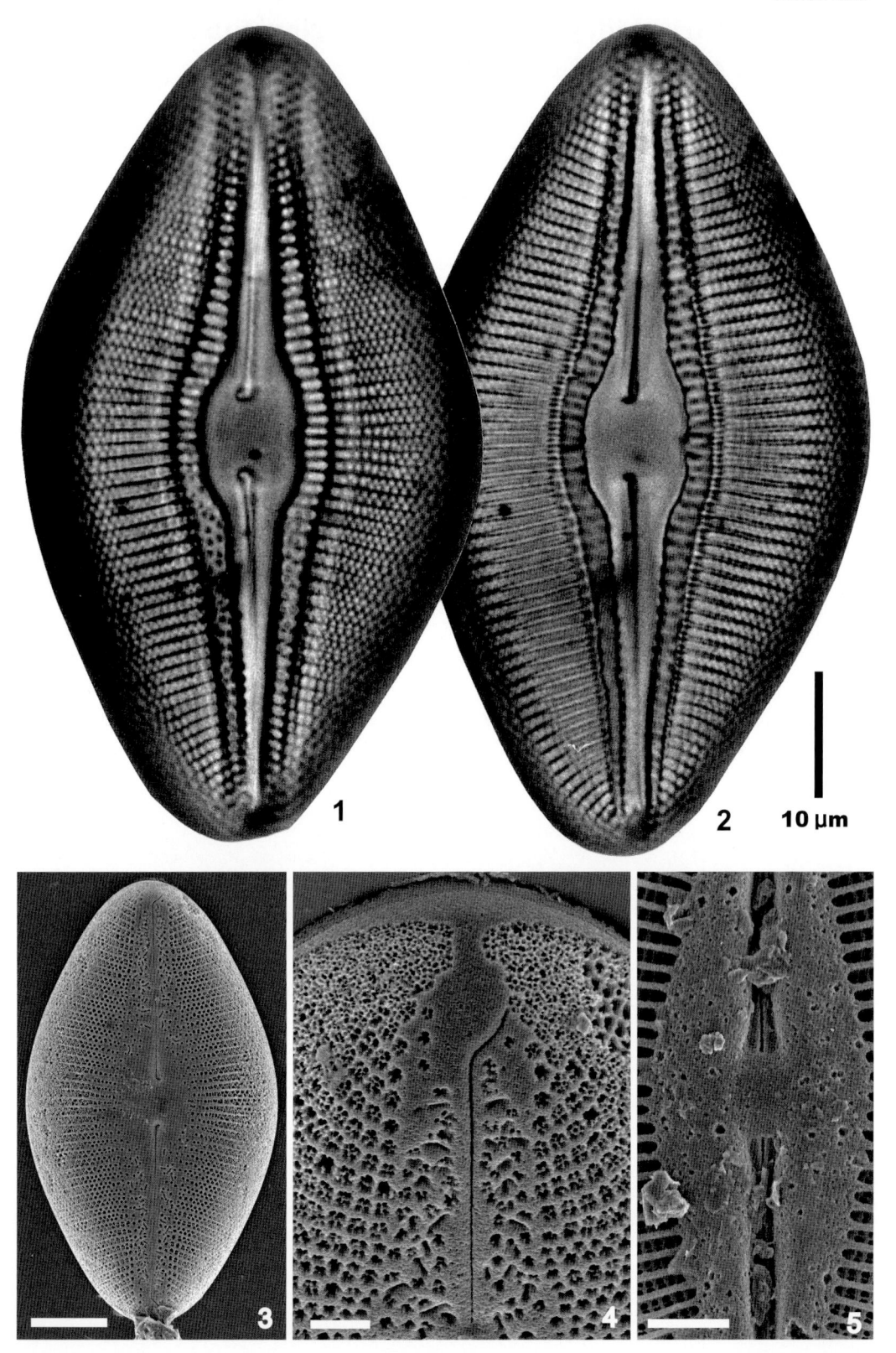

10 µm

Plate 217.   Biraphid, pennate diatoms: Naviculales
Diploneis subovalis Cleve 1894

文　　献：Cleve, P.T. 1894. Synopsis of the naviculoid diatoms. Kongliga Svenska Vetenskaps-Akademiens Handlingar **26**: 1-194, pls 1-5.

形　　態：殻は楕円形，殻長 16-24 μm，殻幅 11-14 μm．条線は 10 μm に 8-9 本で，互い違いに配列する 2 列の胞紋列から構成される．縦走管は幅が狭い．外裂溝は両殻端で同方向へ曲がり，内裂溝もわずかであるが外裂溝と同方向へ両端とも曲がる．

ノ ー ト：類似する *Diploneis ovalis* とは条線における胞紋の配列で区別でき，*Dploneis smithii* とは殻の大きさで区別できる．

産　　出：三徳層（Tanaka & Nagumo 2006）：後期中新世・鳥取県，吹上浜（泥炭層）：完新世・鹿児島県．

図　　版：吹上浜（泥炭層）標本．

Figs 1-9.　*Diploneis subovalis* Cleve
LM. Figs 1-3. SEM. Figs 4-9. Material from peat deposit of Fukiagehama (Holocene), Kagoshima, Japan. Scale bars: Fig. 4＝5 μm, Figs 6-7＝2 μm, Fig. 5＝1 μm, Figs 8-9＝0.5 μm.

Figs 1-3.　Three different size valves.
Fig. 4.　External view of whole valve.
Fig. 5.　Detailed view of a stria of Fig. 4.
Fig. 6.　Oblique view of Fig. 4 showing central and terminal fissures.
Fig. 7.　Internal view of whole valve.
Fig. 8.　Detailed view of Fig. 7 showing central raphe ending.
Fig. 9.　Detailed view of Fig. 7 showing terminal raphe ending.

Plate 217

Plate 218.   Biraphid, pennate diatoms: Naviculales
Frustulia rhomboides var. *amphipleuroides* (Grunow) De Toni 1891

基礎異名：*Navicula (Vanbeurckia) rhomboides* var. *amphipleuroides* Grunow 1880

文　　献：De Toni, G. B. 1891.　Sylloge Algarum omnium hucusque cognitarum vol. **2**. Bacillarieae. Sectio 1. Raphideae, 1-490 pp. Typis Seminarii.

形　　態：殻は大形で，幅の狭い皮針形に近い細長い菱形．殻端は広円形でわずか突き出ることがある．殻長 155-197 μm，殻幅 20-29 μm，条線は縦・横とも平行で，直角に交差する，縦条線は 10 μm に 14-18 本，横条線は 22-24 本．縦溝の外裂溝は軸域の片側にゆるくカーブし，中央および殻端とも末端は短いが左右に分かれて T 字形に終了する．内側では縦溝の両側軸域は肥厚しており，殻端では細く殻端へ向かって伸長した肥厚部がある．殻の長さに比較し縦溝の長さが短いので，両縦溝枝の中央末端はかなり離れている．

ノ ー ト：図に示したのは宮田層産の個体であるが，紫竹層産の本種は宮田層の個体よりかなり小形（殻長 50-104 μm）であった．本変種を独立した種とする見解もあるが，Krammer & Lange-Bertalot(1986)のとおり *F. rhomboides* の変種とした．

産　　出：宮田層：中新-鮮新世・秋田県，紫竹層：前期中新世・福島県．

図　　版：宮田層標本．

Figs 1-3. *Frustulia rhomboides* var. *amphipleuroides* (Grunow) De Toni
LM. Fig. 1. SEM. Figs 2-3. Material from Miyata Formation (Mio-Pliocene), Senboku City, Akita, Japan. Scale bars: Figs 2-3＝5 μm.

Fig. 1.　Whole valve view. Photomontage made by three photographs of same valve.
Fig. 2.　External view of whole valve, central pores (arrowheads).
Fig. 3.　Internal view of valve.

Plate 218

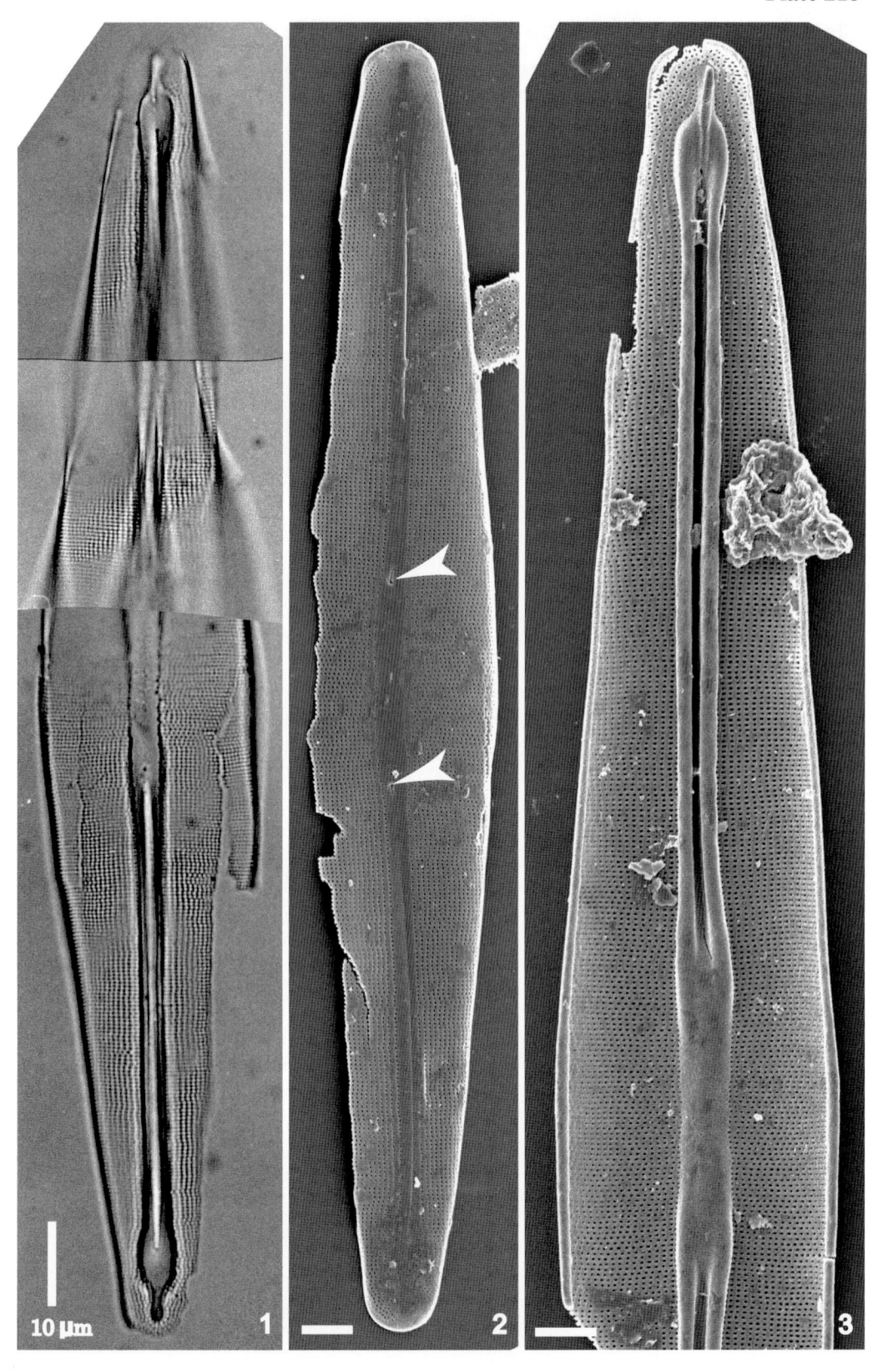

Plate 219. Biraphid, pennate diatoms: Naviculales
Frustulia rhomboides var. *saxonica* (Rabenhorst) De Toni 1891

基礎異名：*Frustulia saxonica* Rabenhorst 1851
文　　献：De Toni, G.B. 1891. Sylloge Algarum omnium hucusque cognitarum, vol. **2**. Bacillarieae. Sectio 1. Raphideae, 1-490 pp. Typis Seminarii.
形　　態：殻は菱形で，殻端はわずか突き出るようになることが多い．殻長 29-79 μm，殻幅 7.5-16 μm，条線は縦・横とも平行で，直角に交差する，10 μm に約 35 本．外側の両縦溝末端は縦溝が左右に分かれて T 字形に終了する．
ノ ー ト：紫竹層産出の本種が強く変形・破損あるいは汚れているので，現生の個体を示す．
産　　出：紫竹層：前期中新世・福島県．現生では各地の湿原等から産出している．
図　　版：池の岳の池塘（新潟県）標本（現生）．

Figs 1-6. *Frustulia rhomboides* var. *saxonica* (Rabenhorst) De Toni
LM. Figs 1-2. SEM. Figs 3-6. Material from a bog of Mt. Ikenodake (Recent), Niigata, Japan. Scale bars: Figs 3-4＝5 μm, Figs 5-6＝2 μm.

Figs 1-2.　Two different size valves.
Fig. 3.　External view of whole valve.
Fig. 4.　Internal view of whole valve.
Fig. 5.　Enlarged view of an apex of Fig. 3.
Fig. 6.　Enlarged view of valve center of Fig. 3, central fissure (arrowhead).

Plate 219

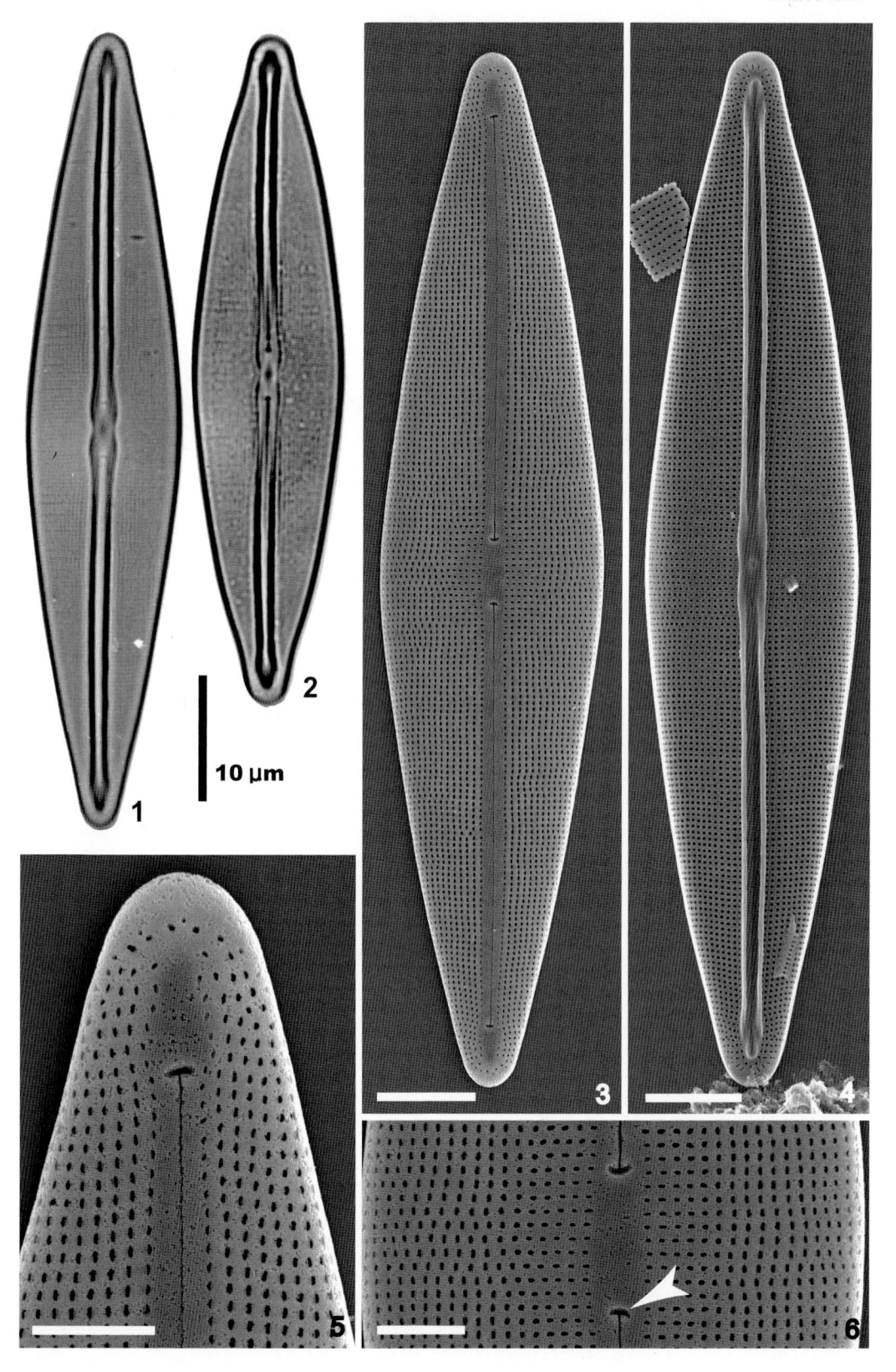

10 µm

Plate 220. Biraphid, pennate diatoms: Naviculales
Gyrosigma spencerii (Quekett) Griffith & Henfrey 1856

基礎異名：*Navicula spencerii* Quekett 1848

文　献：Patrick, R. & Reimer, C.W. 1966. The diatoms of the United States. 1. 688 pp. Monographs of the Academy of Natural Sciences of Philadelphia no. 13.

形　態：殻はS字形で縦・横に条線が分布する．殻中央が幅広く，両端に向かって細くなる．殻長126-164 μm，殻幅15.5-23 μm，横条線10 μmに約20本，縦条線約24本．外・内表面の縦溝裂は軸域の片側に偏って走り，外中心裂溝は横に曲がる．内側中心域の左右には弧状の肥厚部がある．

ノート：条線の密度は，類似する *Gyrosigma acuminatum*（Kütz.）Rabh. が縦横同数で10 μmに18本（原口ら 1998）．一方 *G. spencerii* の縦条線は10 μmに22-24本，横条線は10 μmに18-20本（原口ら 1998）であり，*G. spencerii* の方が密である．津森層からの分類群は *G. spencerii* に計測値が一致するので本種に同定した．加藤ら（1977）によれば Patrick & Reimer（1966）により *Gyrosigma kuetzingii*（Grunow）Cleve は本種の異名であることが報告されている．

産　出：津森層：中期更新世・熊本県．

図　版：津森層標本．

Figs 1-6. *Gyrosigma spencerii* (Quekett) Griffith & Henfrey

LM. Figs 1-2. SEM. Figs 3-6. Material from Tsumori Formation (Middle Pleistocene), Mashiki Town, Kumamoto, Japan. Scale bars: Fig. 3＝10 μm, Figs 4＝5 μm, Figs 5-6＝2 μm.

Figs 1-2.　Same valve at different focal planes.
Figs 3-6.　Same valve shown with external and internal SEM views.
Fig. 3.　Internal view of whole valve.
Fig. 4.　Enlarged view of external central area showing central fissures (arrowheads).
Fig. 5.　Internal oblique view of an apex, helictoglossa (arrowhead).
Fig. 6.　Detailed view of internal central area showing central raphe endings (arrowheads).

Plate 220

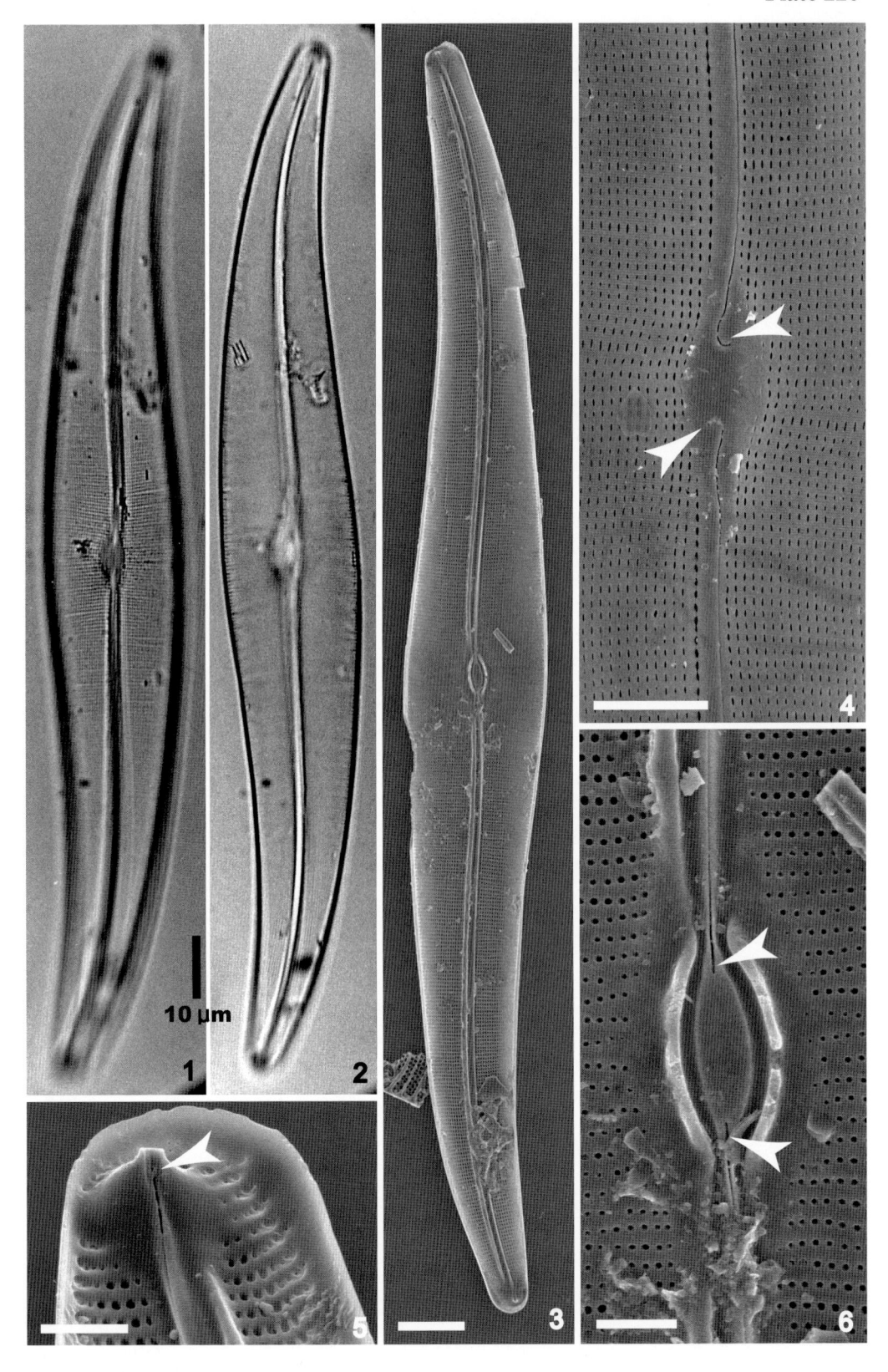

Plate 221. Biraphid, pennate diatoms: Naviculales
Navicula americana Ehrenberg 1843

文　　献：Hustedt, F. 1930. Bacillariophyta. 466 pp. *In*: Pascher, A. (ed.) Die Süsswasser-Flora Mitteleuropas. 10. Gustav Fischer, Jena.

形　　態：殻は大形で殻面観における左右の殻縁は平行，殻端は広円形．軸域は広く，外中心裂溝は同方向の横に曲がり中心孔で終る．条線はわずか放射状．殻長 67-92 μm, 殻幅 15-24 μm, 条線は 10 μm に 18-20 本．

ノート：殻の形，軸域が広いことは本種の有用な識別点である．

産　　出：奄芸層群（根来・後藤 1981）：鮮新世・三重県，野上層（Okuno 1952）：中期更新世・大分県，人吉層：後期鮮新世・熊本県．

図　　版：人吉層標本．

Figs 1-5. *Navicula americana* **Ehrenberg**
LM. Fig. 1. SEM. Figs 2-5. Material from Hitoyoshi Formation (Late Pliocene), Hitoyoshi City, Kumamoto, Japan. Scale bars: Figs 2-3=5 μm, Fig. 4=2 μm, Fig. 5=0.5 μm.

Fig. 1.　Whole valve view of a large valve.
Fig. 2.　External view of whole valve.
Fig. 3.　Oblique view of Fig. 2.
Fig. 4.　Enlarged oblique view of central area of Fig. 2, showing raphe ending in central pores (arrowheads).
Fig. 5.　Detailed internal view of areolae rows.

Plate 221

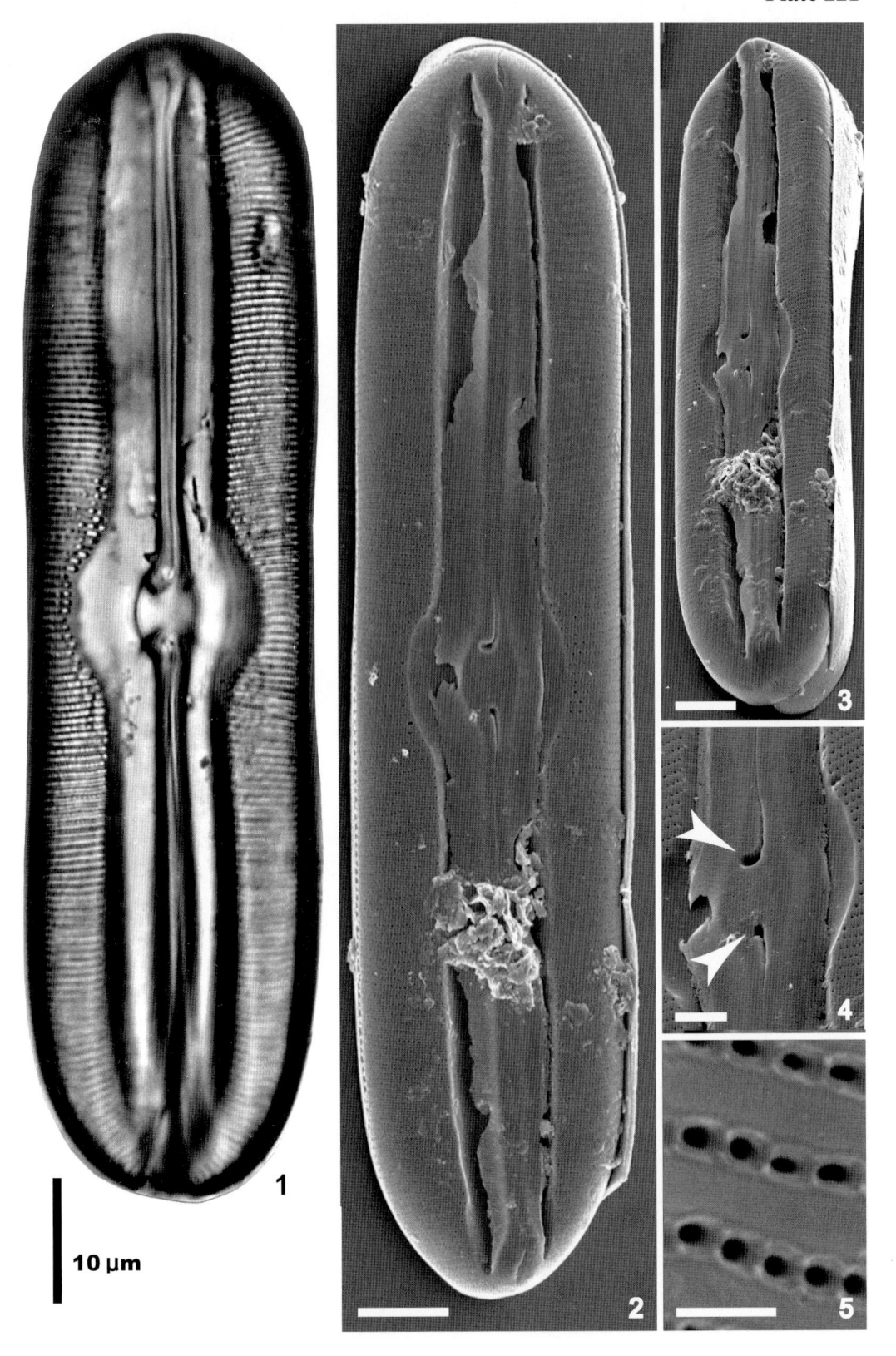

Plate 222. Biraphid, pennate diatoms: Naviculales
Navicula anthracis Cleve & Brun in Brun & Tempére 1889

文　献：Brun, J. & Tempére, J. 1889. Diatomées fossiles du Japon, espèces marines & nouvelles des calcaires argileux de Sendaï & de Yedo. Mémoires de la Société de Physique et D'Histoire Naturelle de Genève **30**: 1-75, pls 1-9.

形　態：殻は菱形，殻長 55-68 μm，殻幅 13-16 μm．殻面観において殻の片側の条線は途中で長軸方向に長く途切れてしまい，無紋域が細長く所在する．条線は 10 μm に約 6 本で，中心では放射状，殻端では平行に配列するようになる．胞紋は縦長のスリット状．縦溝は軸域の片側に偏って走る．

ノート：Brun & Tempére(1889)は江戸と仙台の化石から見出している．その後，他の地域からの産出報告がなく，ここに掲げた向山層（旧八木山層）が 2 番目の産出報告になる．なお，津村(1973)によれば江戸はエゾの可能性がある．

産　出：江戸と仙台の化石（Brun & Tempére 1889），向山層：鮮新世・宮城県．

図　版：向山層標本．

Figs 1-5.　*Navicula anthracis* Cleve & Brun

LM. Figs 1-2. SEM. Figs 3-5. Material from Mukaiyama Formation (Pliocene), Sendai City, Miyagi, Japan. Scale bars: Figs 3-4＝5 μm, Fig. 5＝2 μm.

Figs 1-2.　Two different valves.
Fig. 1.　Hyaline zone, disrupted areolae rows (arrowhead).
Fig. 3.　Internal oblique view of whole valve, disrupted areolae rows (arrowhead).
Fig. 4.　External view of whole valve, disrupted areolae rows (arrowhead).
Fig. 5.　Detailed view of internal valve center.

Plate 222

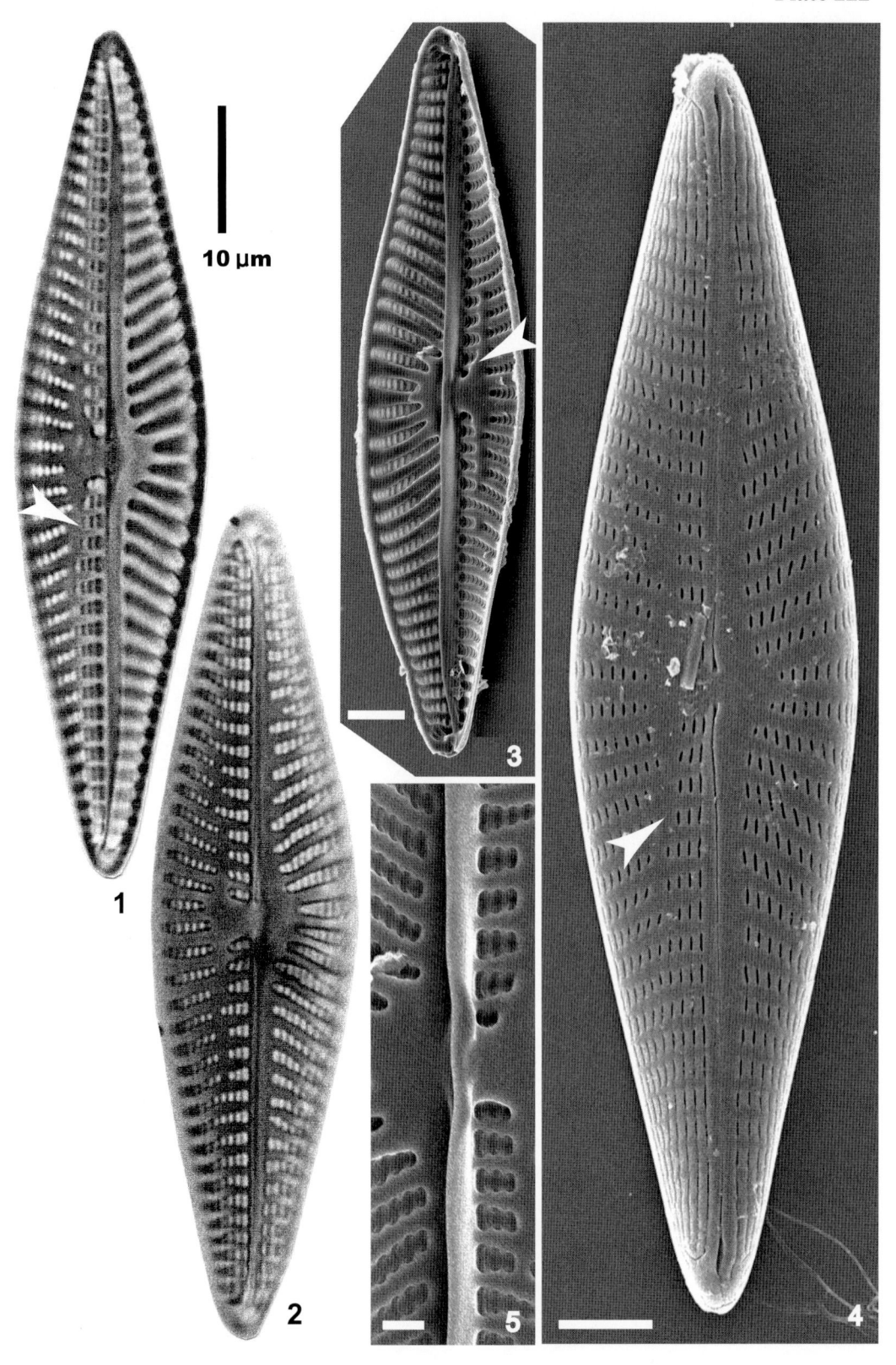

10 μm

Plate 223.   Biraphid, pennate diatoms: Naviculales
Navicula cari Ehrenberg 1836

文　　献：Krammer, K. & Lange-Bertalot, H. 1986.　Bacillariophyceae. Teil 1. Naviculaceae. 876 pp. *In*: Ettl, H., Gerloff, J., Heynig, H. & Mollenhauer, D.(eds) Süsswasserflora von Mitteleuropa, Bd. **2/1**. G. Fischer, Stuttgart & New York.

形　　態：殻は皮針形，中心域は横に広がりほぼ四角形である，殻長 22-39 μm，殻幅 6.5-8.5 μm．条線は 10 μm に 12-14 本，中央部で放射状，殻端でわずか反放射状になる．条線を構成する胞紋は縦長のスリット状．縦溝枝は殻中心外面ではわずか横に曲がり中心孔で終るが，内側ではそのままで終了する．極裂は中心と同方向に曲がる．縦溝の外面の裂溝は軸域の中心を通るが，内面は軸域の片側に寄っている．

産　　出：白沢層：後期中新世・宮城県，三徳層（Tanaka & Nagumo 2006）：後期中新世・鳥取県．

図　　版：白沢層（1-2, 4 図），三徳層（3 図）標本．

Figs 1-4.   *Navicula cari* Ehrenberg
LM. Fig. 1. SEM. Figs 2-4. Materials, Figs 1-2, 4, from Shirasawa Formation (Late Miocene), Sendai City, Miyagi and Fig. 3 from Mitoku Formation (Late Miocene), Misasa Town, Tottori, Japan. Scale bars: Figs 2-3＝2 μm, Fig. 4＝1 m.

Fig. 1.　Whole valve view.
Fig. 2.　External slightly oblique view of whole valve.
Fig. 3.　Internal oblique view of whole valve, helictoglossa (arrow) and inner fissure of raphe (arrowhead).
Fig. 4.　Enlarged view of central area of Fig. 2, central pore (arrowhead).

Plate 223

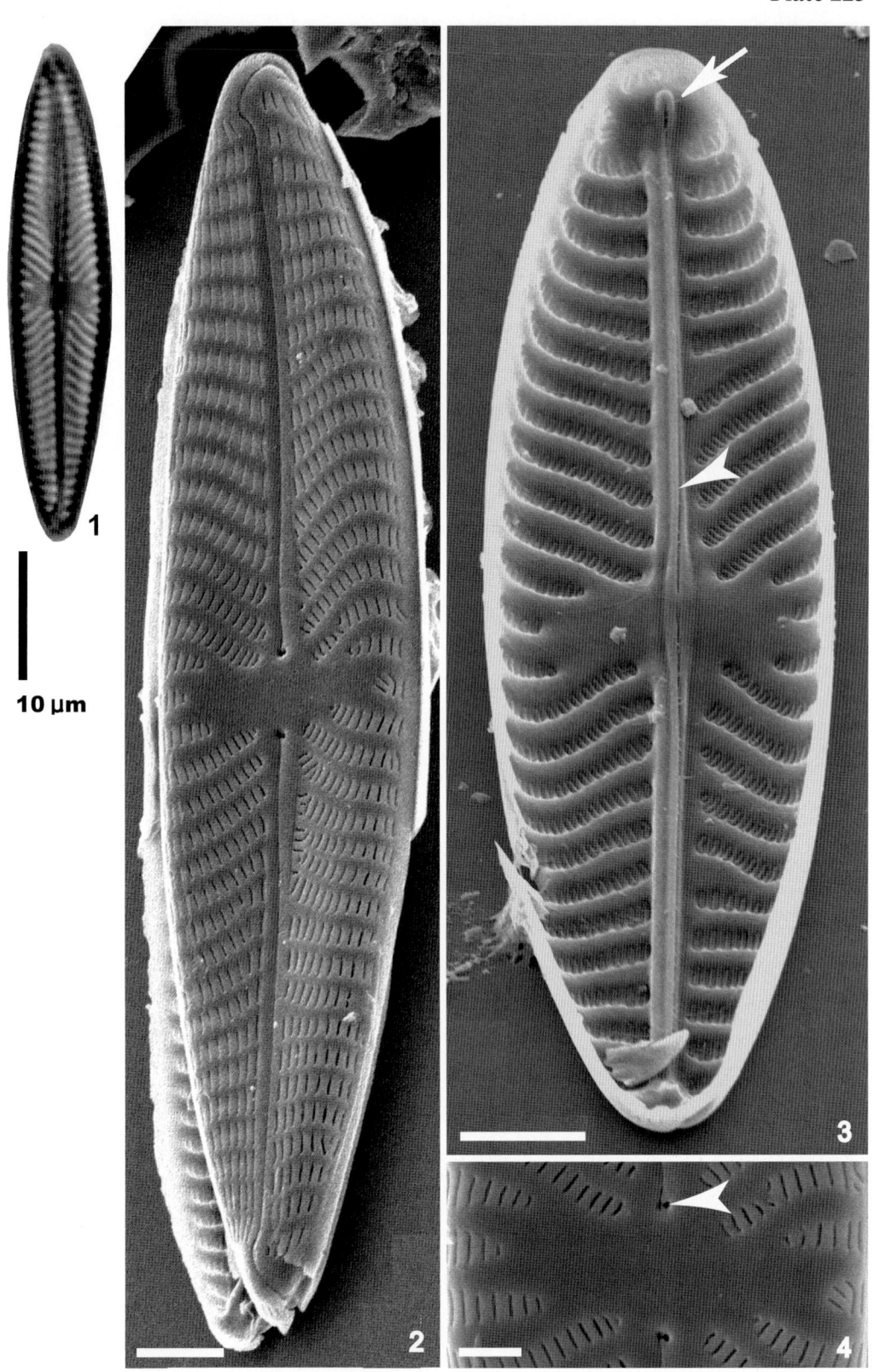

Plate 224. Biraphid, pennate diatoms: Naviculales
Navicula hasta **Pantocsek** 1892

文　　献：Pantocsek, J. 1905.　Beiträge zur Kenntnis der Fossilen Bacillarien Ungarns. Teil 3. 118 pp. pls 1-42. W. Junk, Berlin.

形　　態：殻は皮針形で，殻端に近くなると内側へわずかであるが凹むように細長く突出する．殻長 66-95 μm，殻幅 13-18 μm，条線は放射状，中心では密度が粗になり，10 μm に 6-8 本．胞紋は縦長のスリット状．

ノ ー ト：Pantocsek(1905)によると殻長 63-76 μm，殻幅 14-18 μm，条線は 10 μm に 8-9 本である．Skvortsow(1936)は *Navicula hasta* の殻形をしているが小形のものを琵琶湖から見出し *N. hasta* var. *gracilis* Skvortzov とした．この計測値は殻長 51 μm，殻幅 10 μm，条線は 10 μm に 9 本と記している．

産　　出：中之条湖成層（田中・小林 1992）：中期更新世・群馬県，瓜生坂層（田中・南雲 2000）：前期更新世・長野県，堅田層（根来 1981）：更新世・滋賀県，野上層（Okuno 1952）：中期更新世・大分県，津森層：中期更新世・熊本県，人吉層：後期鮮新世・熊本県．

図　　版：人吉層（1,3-4 図），津森層標本（2 図）．

Figs 1-4. *Navicula hasta* **Pantocsek**

LM. Fig. 1. SEM. Figs 2-4. Materials, Figs 1, 3-4, from Hitoyoshi Formation (Late Pliocene), Hitoyoshi City, Kumamoto and Fig. 2, from Tsumori Formation (Middle Pleistocene), Mashiki Town, Kumamoto, Japan. Scale bars: Figs 2-3＝5 μm, Fig. 4＝0.5 μm.

Fig. 1.　Whole valve view.
Fig. 2.　External view of whole valve.
Fig. 3.　Internal oblique view of whole valve.
Fig. 4.　Detailed internal view of areolae with veluma.

Plate 224

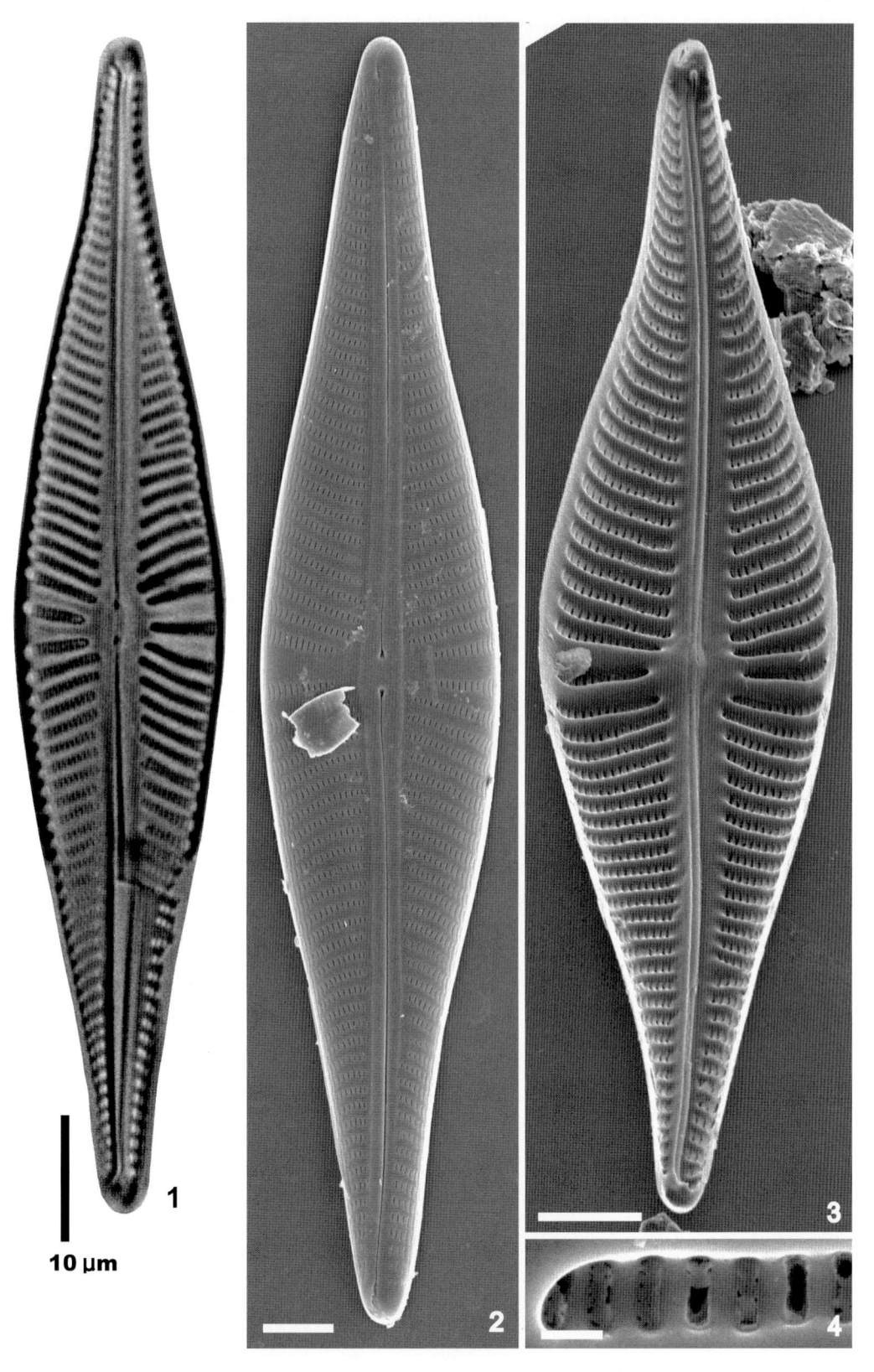

10 µm

Plate 225.   Biraphid, pennate diatoms: Naviculales
Navicula radiosa Kützing 1844

文　　献：Kützing, F.T. 1844. Die kieselschaligen Bacillarien oder Diatomeen. Nordhsusen, W. Kohne. 152 pp. pls 1-30.

形　　態：殻は皮針形，殻端は幅広のくさび形，軸域は狭いが中心域はやや広くなる，殻長66.5-108 μm，殻幅10-13 μm．条線は10 μmに約11本で，中央部では放射状であるが，殻端では反放射状になる．SEM観察では外裂溝中央末端は中心孔で終了し，胞紋は縦長のスリット状をしており，軸域の内側では片側に縦溝に平行した肋が存在することが観察できる．内裂溝末端（殻端）は蝸牛舌で終了する．

ノート：文献欄に記したのは原記載論文であるが，その図（pl. 4, fig. 23）は条線が殻端まで放射状に描いてあるように見える．当時の顕微鏡では殻端の条線がよく観察できなかったと推測できるし，現在の研究者は殻端の条線配列は反放射のものを本種に同定しているので（たとえば，渡辺ら 2005），それに倣った．

産　　出：鬼首層（Ichikawa 1955）：後期更新世・宮城県，奄芸層群（根来・後藤 1981）：鮮新世・三重県，津森層：中期更新世・熊本県．

図　　版：津森層標本.

Figs 1-5.　*Navicula radiosa* Kützing

LM. Figs 1-2. SEM. Figs 3-5. Material from Tsumori Formation (Middle Pleistocene), Mashiki Town, Kumamoto, Japan. Scale bars: Figs 3-4＝5 μm, Fig. 5＝2 μm.

Figs 1-2.　Two different size valves.
Fig. 3.　External view of whole valve.
Fig. 4.　Internal oblique view of valve, helictoglossa (arrowhead).
Fig. 5.　Enlarged view of internal central area, axial costa (arrow).

Plate 225

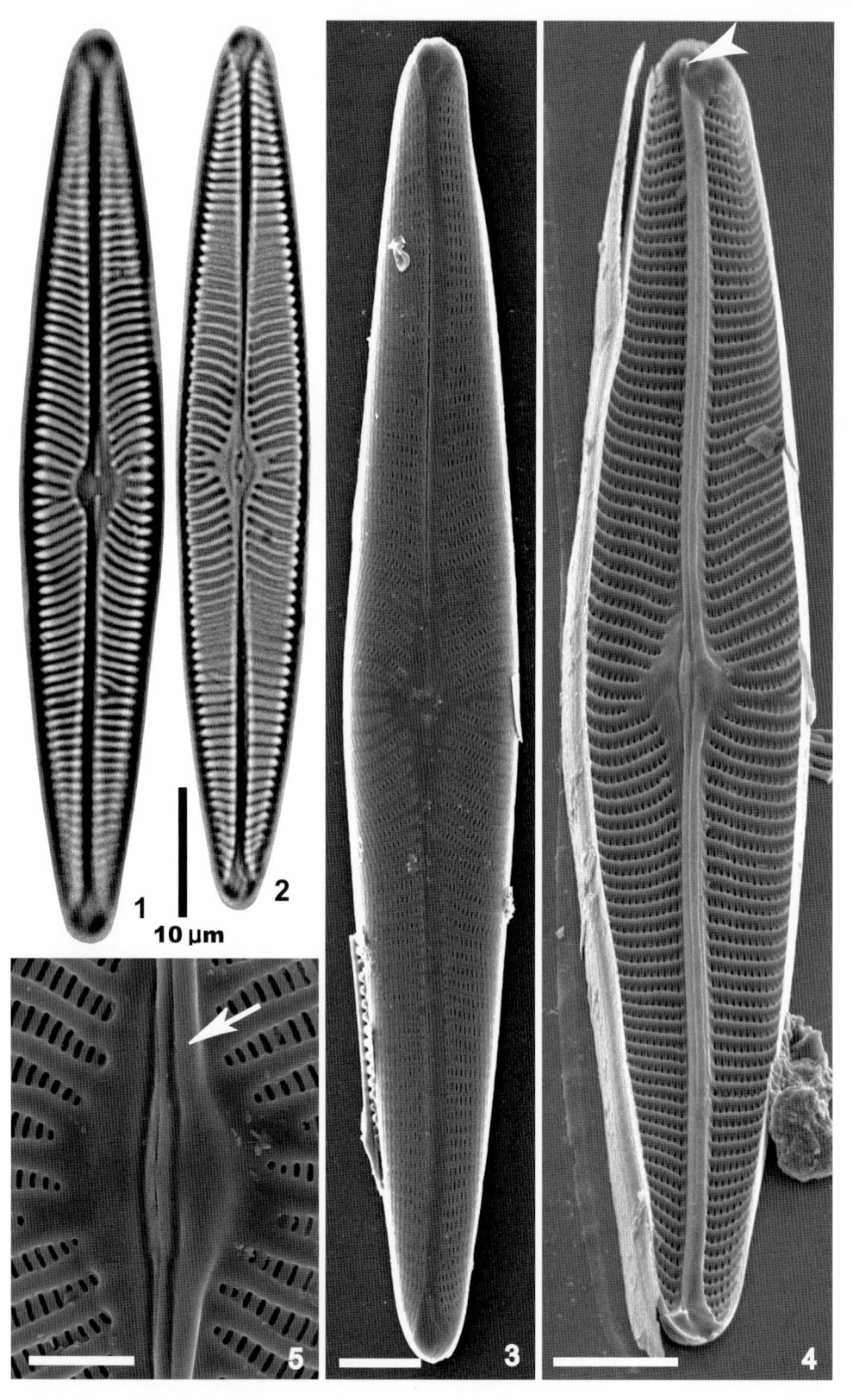

Plate 226. Biraphid, pennate diatoms: Naviculales
Navicula reinhardtii (Grunow) Grunow
in Cleve & Müller 1877

基礎異名：*Stauroneis reinhardtii* Grunow 1860

文　献：Krammer, K. & Lange-Bertalot, H. 1986. Bacillariophyceae. Teil 1. Naviculaceae. 876 pp. *In*: Ettl, H., Gerloff, J., Heynig, H. & Mollenhauer, D.(eds) Süsswasserflora von Mitteleuropa, Bd. **2/1**. G. Fischer, Stuttgart & New York.

　　河島綾子・真山茂樹 1998. 阿寒湖の珪藻（6. 羽状類-縦溝類：*Cavinula, Diadesmis, Geissleria, Hippodonta, Navicula, Placoneis*）. 自然科学環境研究 **11**: 23-41.

形　態：殻は広皮針形～楕円形で殻端は広円形. 殻長 48-79 μm, 殻幅 14-19 μm. 条線は 10 μm に 7-8 本. 条線は太く放射状であるが, 殻端では平行に近くなる. 条線を構成する胞紋は縦長のスリット状である. 極裂は両殻端で反対方向に曲がる.

ノート：河島・真山(1998)は写真が大きくて見やすい. 加藤ら(1977)によると適応性は真アルカリ性または好アルカリ性, 富栄養性, 弱-中腐水性または混腐水性とされている.

産　出：嬉野珪藻土（Skvortzov 1937）：新第三紀・佐賀県, 瓜生坂層（窪田ら 1976）：前期更新世・長野県, 琵琶湖底堆積物（Mori 1975）：滋賀県, 中山香珪藻土（Okuno 1952）：更新世・大分県, 尾本層：前期更新世・大分県.

図　版：尾本層標本.

Figs 1-5.　*Navicula reinhardtii* Grunow
LM. Figs 1-2. SEM. Figs 3-5. Material from Omoto Formation (Early Pleistocene), Kitsuki City, Oita, Japan. Scale bars: Figs 3-4＝5 μm, Fig. 5＝2 μm.

Figs 1-2.　Two different size valves.
Figs 1, 3-5.　Same valve with LM and SEM, external and internal, photographs.
Fig. 3.　External view of whole valve.
Fig. 4.　Oblique view of Fig. 3.
Fig. 5.　Internal view of a terminal area, helictoglossa (arrow).

Plate 226

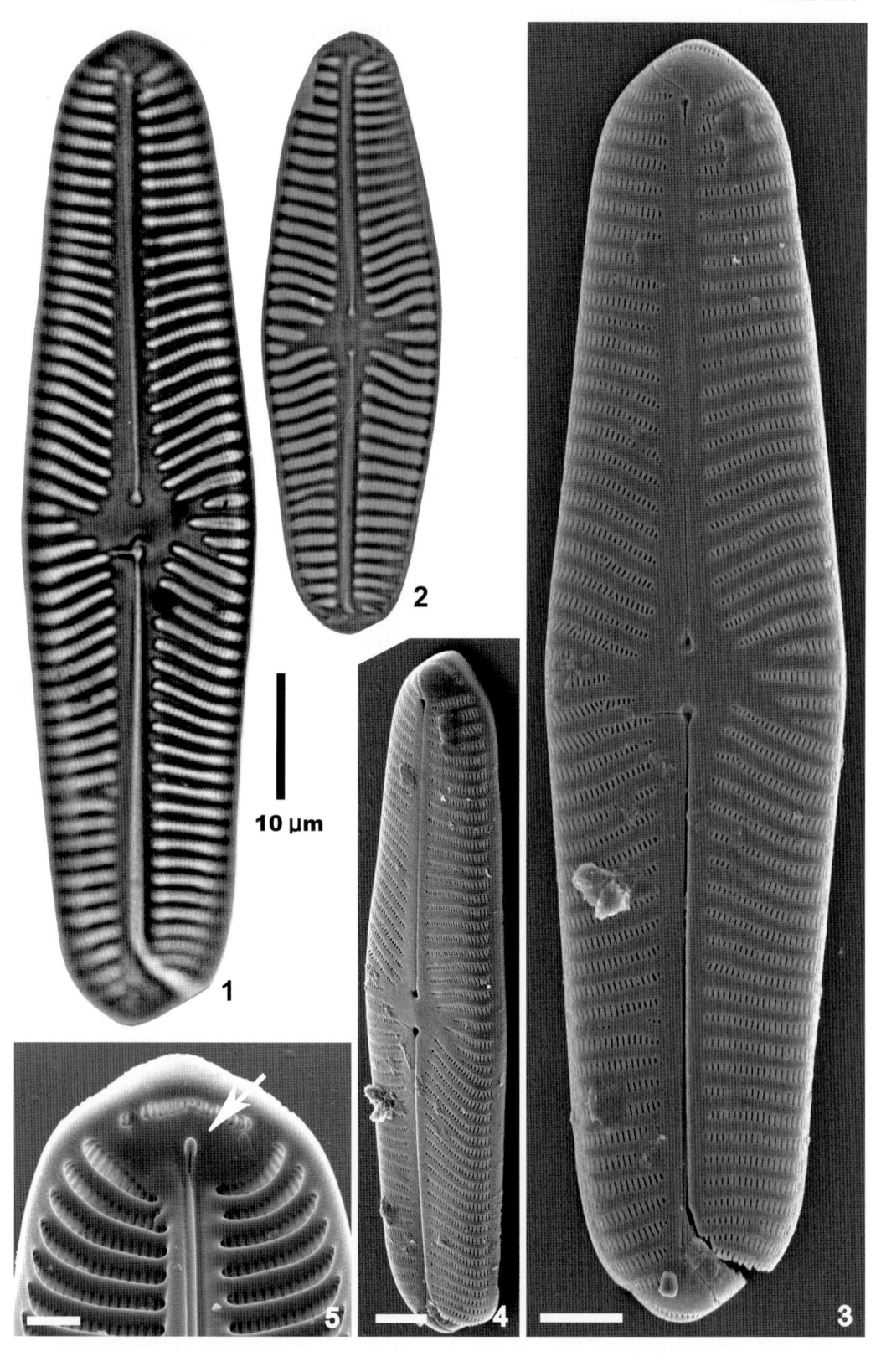

Plate 227. Biraphid, pennate diatoms: Naviculales
Navicula tanakae Fukushima, Ts. Kobayashi & Yoshitake 2002

文　　献：福島　博・小林艶子・吉武佐紀子 2002. 温泉産新種珪藻, *Navicula tanakae* Fukush., Ts. Kobay. & Yoshit. nov. sp. について．Diatom **18**: 13-21.

形　　態：殻は小形でほぼ皮針形，殻端はわずかくちばし状～頭状である．殻長 9-24 µm, 殻幅 3-5 µm. 条線は 10 µm に 13-18 本で，平行～弱い放射状，殻端ではわずか反放射状になる場合も見られた．中心域は小さい．

ノ ー ト：田中・中島(1985)で *Navicula* sp. と報告している分類群である．

産　　出：磯部鉱泉（福島ら 2002）：群馬県（現生）．

図　　版：磯部鉱泉（原産地）標本．

Figs 1-10. *Navicula tanakae* Fukushima, Ts. Kobayashi & Yoshitake
LM. Figs 1-5. SEM. Figs 6-10. Material from Isobe Spa (type locality), Annaka City, Gunma, Japan. Scale bars: Figs 6-8＝2 µm, Figs 9-10＝0.5 µm.

Figs 1-5.　Five different size valves.
Fig. 6.　External view of whole valve.
Fig. 7.　Internal view of whole valve, helictoglossa (arrowhead).
Fig. 8.　Oblique view of Fig. 6.
Fig. 9.　Enlarged view of a terminal area of Fig. 6, terminal fissure (arrowhead).
Fig. 10.　Detailed view of a terminal area of Fig. 7, helictoglossa (arrowhead).

Plate 227

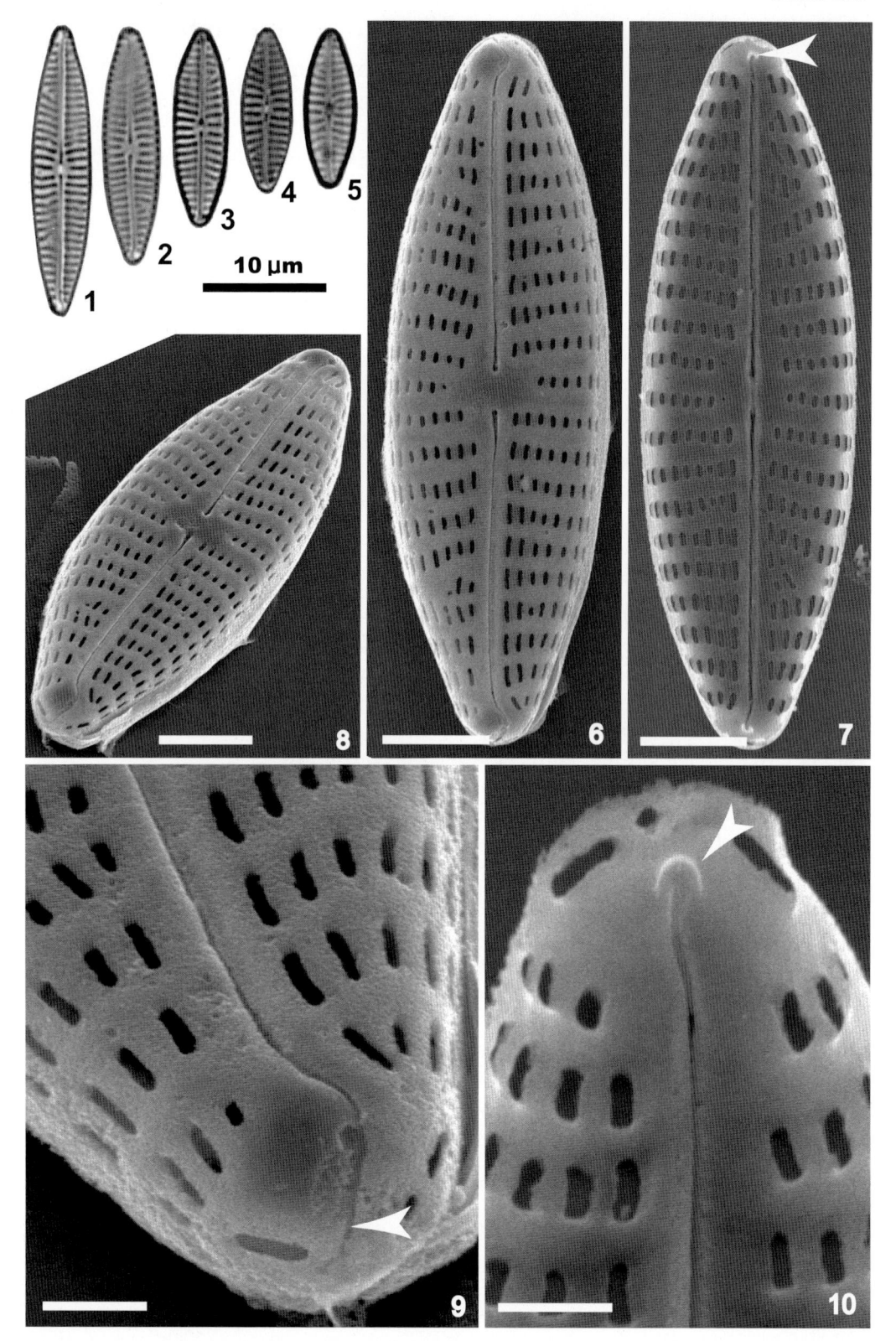

Plate 228. Biraphid, pennate diatoms: Naviculales
Neidium ampliatum (Ehrenberg) Krammer
in Krammer & Lange-Bertalot 1985

文　　献：Ehrenberg, C.G. 1854. Mikrogeologie. Das Erden und Felsen schaffende Wirken des unsichtbar Kleinen selbstständigen Lebens auf der Erde. Leopold Voss, Leipzig. Texte, 374 pp. Atlas, 40 Taf.

　　　　Krammer, H. & Lange-Bertalot, H. 1985. Naviculaceae. Bibliotheca Diatomologica **9**: 1-230.

形　　態：殻は一般的に殻端が突出した皮針形であるが，突出の程度はさまざまである．殻長 64-119 μm，殻幅 15-26 μm．条線は 10 μm に約 20 本で点紋列からなり，全体的に斜行するが殻端では反放射になる．縦溝は中央では相対する縦溝枝が反対方向に曲がって釣針状に終了し，極裂は 2 分岐する．内側の殻面／殻套境界には縦走管が 1 本ある．

ノート：本種は従来 *Neidium iridis* var. *ampliatum* (Ehrenb.) Cl. と同定されてきたが，Krammer & Lange-bertalot (1985) によって種に格付けされた．図に示した津森層の個体は本種の特徴である殻端の突出がやや弱い．

産　　出：奄芸層群（根来・後藤 1981）：鮮新世・三重県，有馬珪藻土（Okuno 1952）：完新世・三重県，津森層：中期更新世・熊本県．

図　　版：津森層標本．

Figs 1-5. ***Neidium ampliatum*** **(Ehrenberg) Krammer**
LM. Figs 1-2. SEM. Figs 3-5. Material from Tsumori Formation (Middle Pleistocene), Mashiki Town, Kumamoto, Japan. Scale bars: Fig. 3＝5 μm, Figs 4-5＝2 μm.

Figs 1-2.　Two different size valves.
Fig. 3.　External view of whole valve, central fissure (arrowhead).
Fig. 4.　Enlarged oblique view of an apex of Fig. 3 showing branched terminal fissure (arrowhead).
Fig. 5.　Internal view of valve apex showing helictoglossa (arrowhead) and longitudinal canal (arrow).

Plate 228

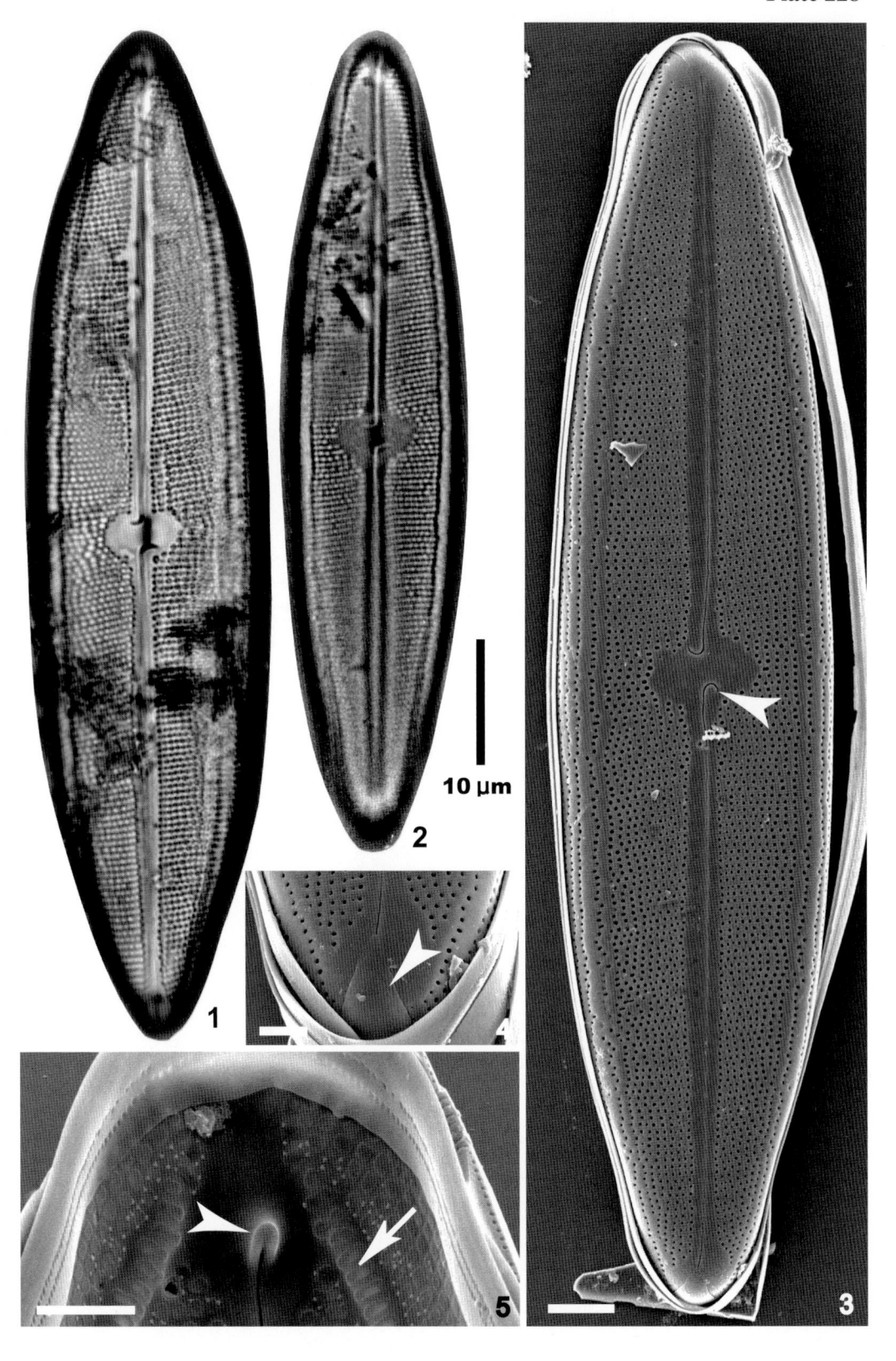

Plate 229.　Biraphid, pennate diatoms: Naviculales
Neidium gracile Hustedt 1938

文　　献：Hustedt, F. 1938. Systematische und ökologische Untersuchungen über die Diatomeen-Flora von Java, Bali und Sumatra nach dem Material der Deutschen Limnologischen Sunda-Expedition. Teil Ⅰ. Systematischer Teil. Archiv für Hydrobiologie, Supplement **15**: 131-506, Taf. 9-28.

形　　態：殻は殻端がくちばし状に突出し，殻縁は3回波打つ．軸域は狭い．条線は斜行するが，殻端では反放射になることがある．奄美大島泥染公園から産出した個体はすべてがほぼ同様な大きさで，殻長約58 μm，殻幅約13 μm．条線は10 μmに約18本．条線を構成する点紋は10 μmに約20個である．縦溝は中央では相対する縦溝枝が反対方向に曲がって終了し，極裂は2分岐する．内側の殻面／殻套境界には縦走管が1本ある．殻の内側に多数の小腎臓形突出物が存在する．

ノート：光顕観察による殻の形，特に殻端の突出の形状から本種に同定した．

産　　出：仙女ヶ池（安藤ら 1971）：埼玉県（現生），奄美大島泥染公園の泥田：鹿児島県（現生）．

図　　版：泥染公園の泥田（奄美大島）標本．

Figs 1-7.　*Neidium gracile* Hustedt
LM. Figs 1-2. SEM. Figs 3-7. Material from "dorota" of Dorozome Park (Recent), Amami-Oshima, Kagoshima, Japan. Scale bars: Figs 3-4＝5 μm, Figs 5, 7＝2 μm, Fig. 6＝0.5 μm.

Figs 1-2.　Two different valves.
Fig. 3.　External view of whole valve.
Fig. 4.　Internal oblique view of whole valve.
Fig. 5.　Enlarged oblique view of an apex of Fig. 3.
Fig. 6.　Detailed view of Fig. 4, renilimbi (arrowheads).
Fig. 7.　Enlarged view of central area of Fig. 3.

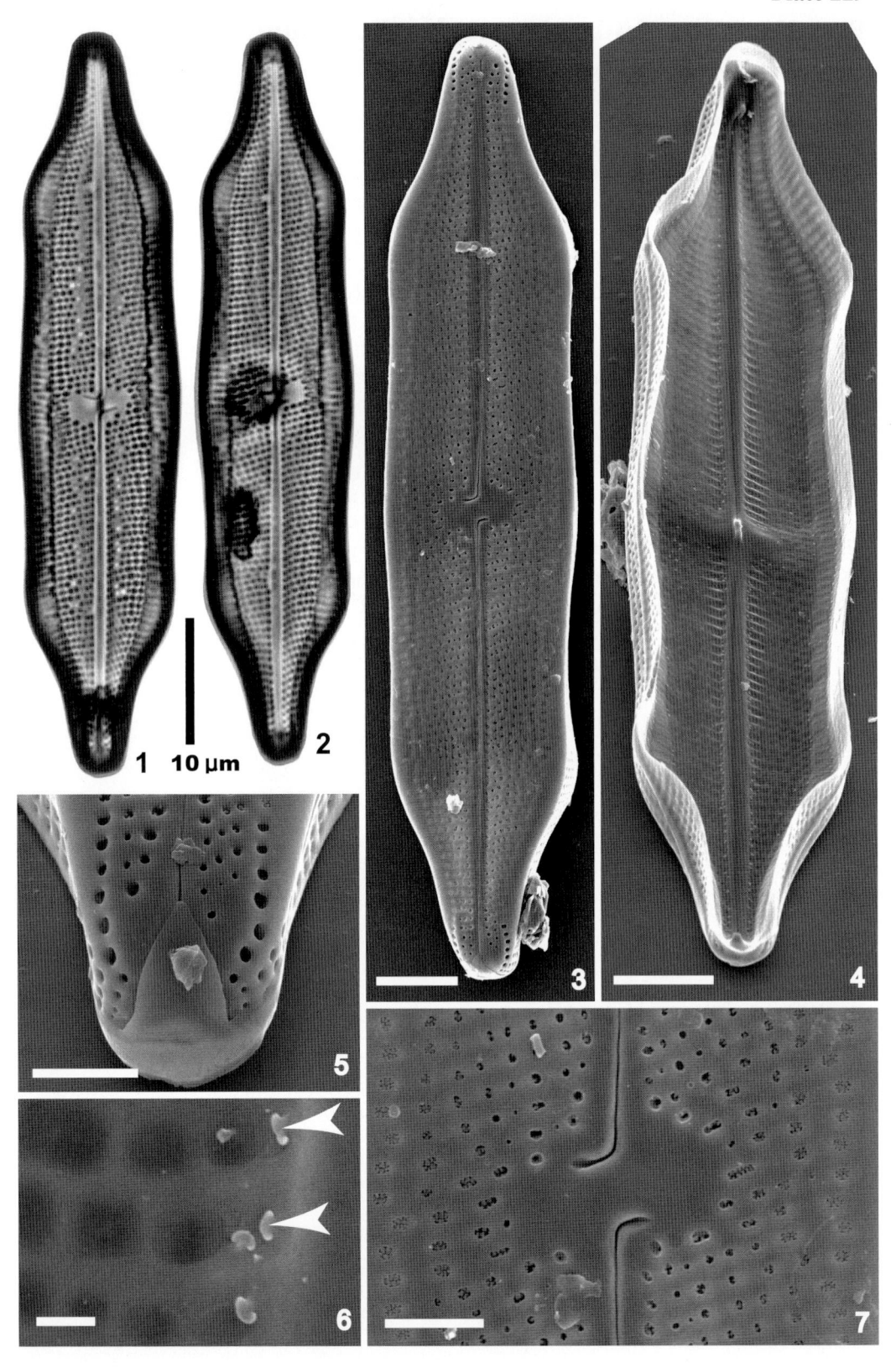

Plate 229

Plate 230. Biraphid, pennate diatoms: Naviculales
Pinnularia episcopalis Cleve 1891

文　　献：Krammer, K. 1992. *Pinnularia* eine monographie der europäeischen taxa. Bibliotheca Diatomologica, Bd. **26**: 1-353.

形　　態：殻は大形で太い線形，両端はややくさび形的な広円形である．殻長 190-280 μm，殻幅 38-44 μm．条線は 10 μm に約 7 本で，中心付近では放射状であるが殻端では反放射になる．軸域は広く，中心域は横に伸びて殻縁に達する．条線の先端（軸域との境）の外側は窪んでおり，窪みは中心域の周りの点紋へ連続する．内側長胞の開口は外側条線域と同程度長い．両端の極裂は縦溝の中心末端とは反対方向に曲がる．

ノ ー ト：類似種としては *Pinnularia imperatrix* Mills があるが，この原記載の図は 2 枚あり，片方はよく似ている．しかし他方は両殻端が膨らんで沼田湖成層産の本種には同定しづらい．*P. episcopalis* は中心域に点紋がない図が多いが（たとえば Hustedt 1930），Krammer（1992, 2000）の写真と比較すると，すべてよく類似しており本種へ同定できると考えられる．Krammer（2000）に化石からよく産すると記されている．

産　　出：沼田湖成層：中期更新世・群馬県．

図　　版：沼田湖成層標本．

Figs 1-3. *Pinnularia episcopalis* Cleve
LM. Fig. 1. SEM. Figs 2-3. Material from Numata Lacustrine Deposit (Middle Pleistocene), Numata City, Gunma, Japan. Scale bars: Figs 2-3＝10 μm.

Fig. 1.　Whole valve view, punctae on central area (arrow).
Fig. 2.　External view of whole valve.
Fig. 3.　Internal oblique view of valve.

Plate 230

10 µm

Plate 231. Biraphid, pennate diatoms: Naviculales
Pinnularia esoxiformis Fusey 1951

文　　献：河島綾子・真山茂樹 2000. 阿寒湖の珪藻（7. 羽状類-縦溝類：*Caloneis, Pinnularia*）. 自然環境科学研究 **13**: 67-83.

形　　態：殻は太い線形で両端は広円状，殻端ではわずか細くなる．殻長 32-86 μm，殻幅 11.5-14.5 μm，条線は 10 μm に 9-11 本．中心域の横帯は通常存在するが，ない場合もしばしばある．軸域は比較的広いが殻端では狭くなる．両端の極裂は縦溝の中心末端とは反対方向に曲がる．

ノ　ー　ト：本分類群は，田中・中島(1985)では別種として報告されたが，河島・真山(2000)で *P. esoxiformis* であると指摘された．他の研究者（たとえば，Krammer 1992）が同定している本種と比較すると，元温泉小屋産の殻の殻端はより広円状の個体も含まれるが，河島・真山（2000）の見解に従ってよいと考えている．

産　　出：元温泉小屋温泉（田中・中島 1985）：福島県（現生），梨木温泉（田中・中島 1985）：群馬県（現生）．

図　　版：元温泉小屋温泉標本.

Figs 1-6. *Pinnularia esoxiformis* Fusey

LM. Figs 1-2. SEM. Figs 3-6. Material from mineral spring of Moto-onsengoya (Recent), Hinoemata Village, Fukushima, Japan. Scale bars: Figs 3-4＝5 μm, Fig. 5＝2 μm, Fig. 6 ＝0.5 μm.

Figs 1-2.　Two different size valves.
Fig. 3.　External view of whole valve.
Fig. 4.　Internal oblique view of valve.
Fig. 5.　Enlarged internal view of a terminal area, helictoglossa (arrow).
Fig. 6.　Enlarged internal view of part of alveoli.

Plate 231

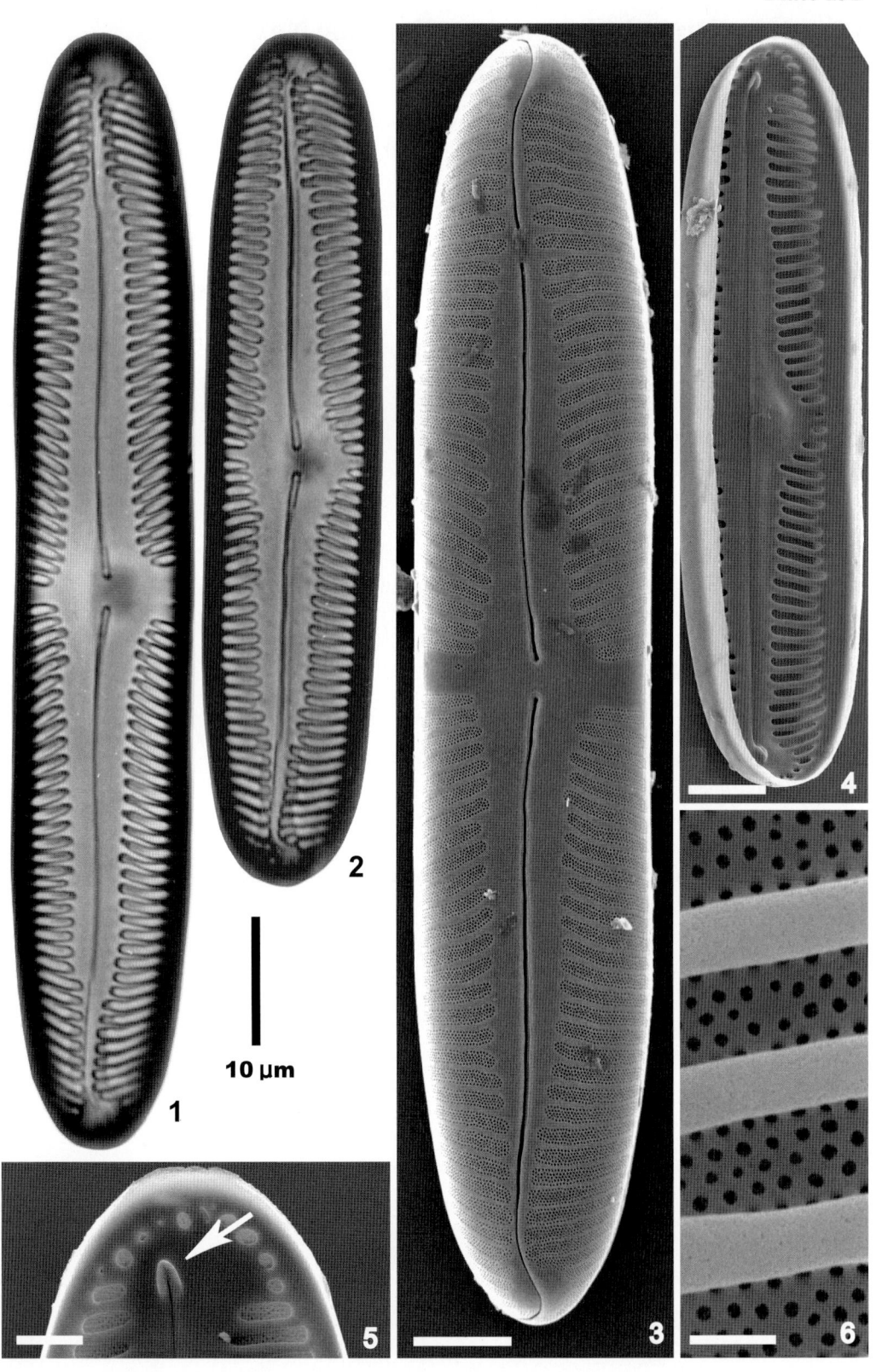

Plate 232.　Biraphid, pennate diatoms: Naviculales
Pinnularia higoensis Okuno 1955

文　　献：Okuno, H. 1955. Electron-microscopic fine structure of fossil diatoms Ⅲ. Transactions and Proceedings of the Palaeontological Society of Japan, New Series, No. 19： 53-58, pls 8-9.

形　　態：殻は大形で細長い，両殻端と中央がやや膨らむ．殻長 244 μm，殻幅 23 μm，軸域は広いが殻端では狭くなる．縦溝は軸域の片側に偏って走る．条線は 10 μm に約 11 本，中央ではわずか放射状，殻端では平行～わずか反放射になる．

ノ　ー　ト：原記載論文では試料を採取した場所は熊本県西瀬村と記されているが，Okuno (1952) で判断すると，現在の人吉市鹿目である．筆者が当地から採取した試料からは細長い *Pinnularia* 属珪藻は本種のみであり，原記載の図とも形態が一致した．Okuno(1955) では *Pinnularia tabellaria* Ehrenb. と比較を行っているが，*P. tabellaria* とは異なる分類群であることは筆者の調査でも確認できた．

産　　出：人吉層（Okuno 1955）：後期鮮新世・熊本県．

図　　版：人吉層標本．

Figs 1-7.　*Pinnularia higoensis* Okuno
LM. Fig. 1. SEM. Figs 2-7. Material from Hitoyoshi Formation (Late Pliocene), Hitoyoshi City, Kumamoto, Japan. Scale bars: Figs 2-3＝10 μm, Figs 4, 7＝5 μm, Fig. 5＝2 μm, Fig. 6＝0.5 μm.

Figs 1-7.　Same valve LM and SEM photographs.

Fig. 1.　Whole valve view. Photomontage made by four photographs of same valve.

Fig. 2.　External view of whole valve.

Fig. 3.　Internal view of whole valve, helictoglossae (arrowheads).

Fig. 4.　Enlarged oblique view of an apex of Fig 2.

Fig. 5.　Enlarged view of central area of Fig. 2.

Fig. 6.　Enlarged internal view of part of Fig. 3 showing alveoli.

Fig. 7.　Enlarged internal oblique view of an apex of Fig. 3, helictoglossa (arrowhead).

Plate 232

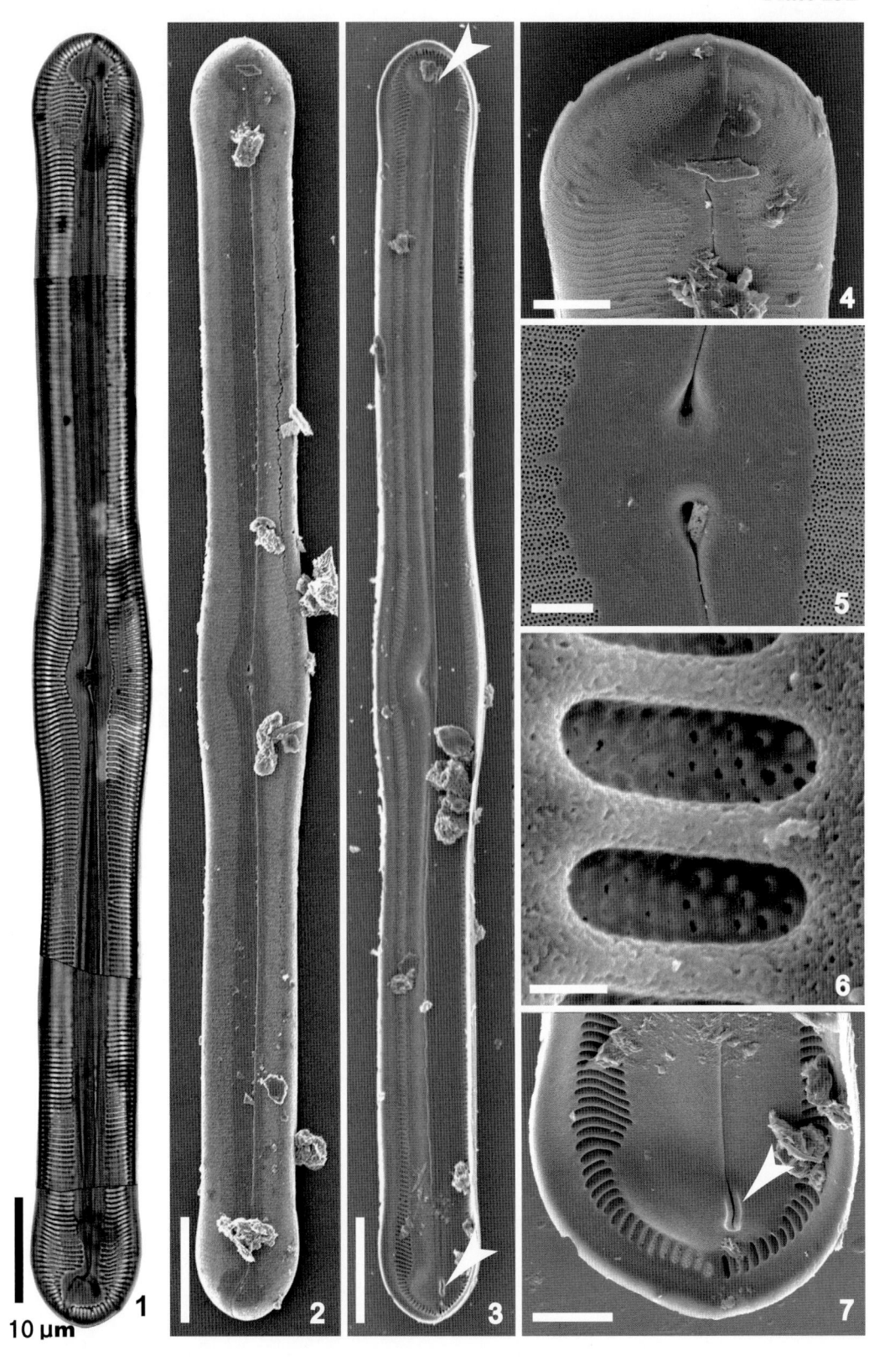

10 μm

Plate 233.　Biraphid, pennate diatoms: Naviculales
Pinnularia lignitica Cleve 1895

文　　献：Cleve, P.T. 1895. Synopsis of the naviculoid diatoms. Konglica Svenska Vetenskaps-Akademiens Handlingar, Bd. **27**, pp 1-219. pls 1-4(Part 2).

形　　態：殻は広皮針形，殻端は広円形，中心外裂溝はわずかカーブしてロート状に広がり，極裂はロート状に幅が広がりながら中心とは反対方向に曲がる．殻長約79 μm，殻幅約19 μm，条線は10 μmに約12本．1本の条線は約5列の胞紋列から形成され，幅が広い．

ノ　ー　ト：本種は仙台産の亜炭試料からCleve(1895)が記載した種である．この試料は秋葉(2008)により「仙台産亜炭」と名付けられている．九州からも見出されている (Skvortzov 1937)．

産　　出：仙台産亜炭（Cleve 1895）：宮城県，嬉野珪藻土（Skvortzov 1937）：新第三紀・佐賀県，堅田層（根来 1981）：更新世・滋賀県，津森層：中期更新世・熊本県．

図　　版：津森層標本．

Figs 1-5.　*Pinnularia lignitica* Cleve
LM. Fig. 1. SEM. Figs 2-5. Material from Tsumori Formation (Middle Pleistocene), Mashiki Town, Kumamoto, Japan. Scale bars: Figs 2-3＝5 μm, Figs 4-5＝1 μm.

Fig. 1.　Whole valve view.
Fig. 2.　External whole valve view, central pore (arrow) and terminal fissure (arrowhead).
Fig. 3.　Internal oblique view of whole valve.
Fig. 4.　Detailed view of alveoli.
Fig. 5.　Enlarged view of striae of Fig. 2 showing one stria made up of five areolae rows.

Plate 233

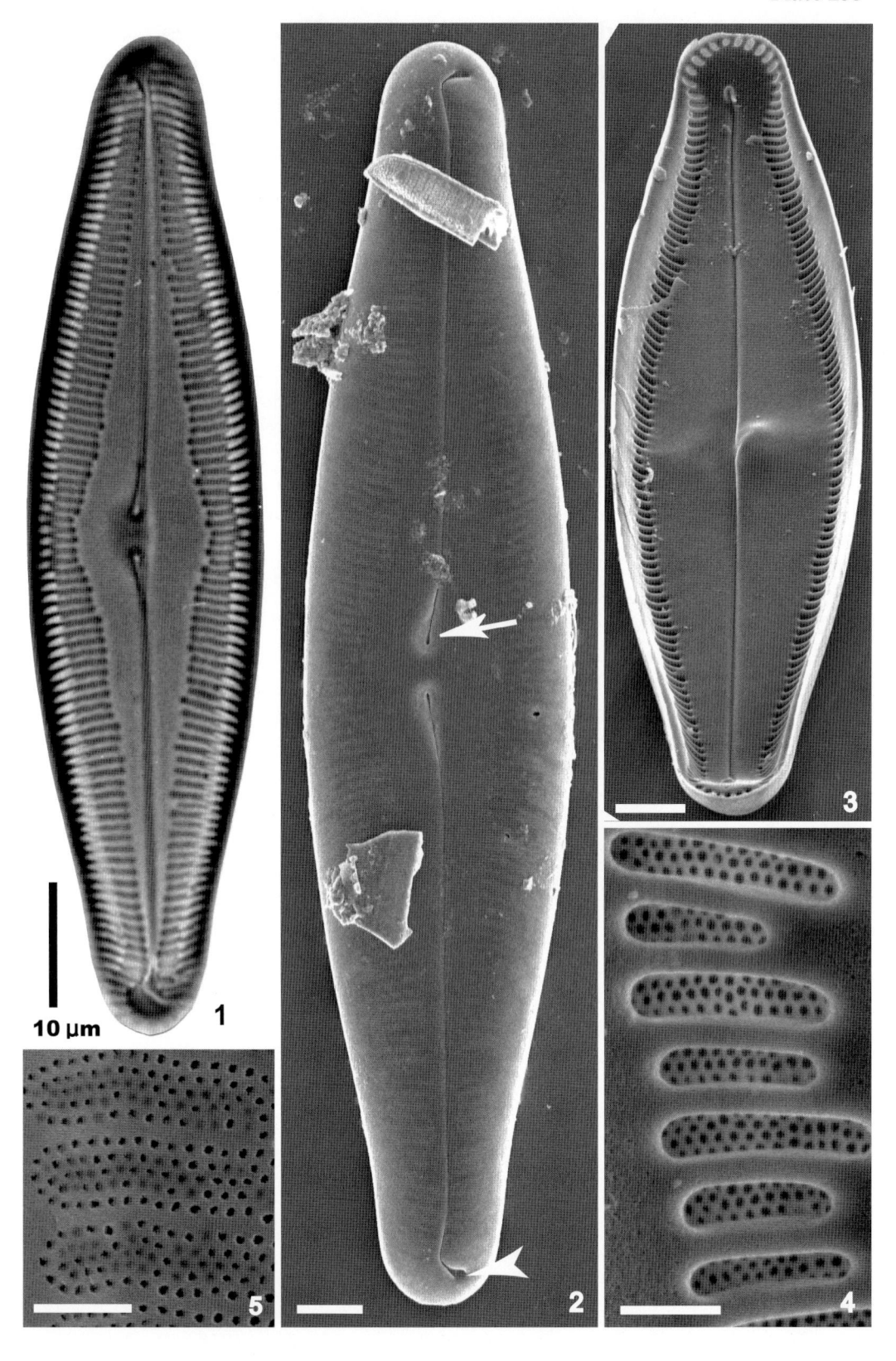

Plate 234.　Biraphid, pennate diatoms: Naviculales
Pinnularia macilenta (Ehrenberg) Ehrenberg 1843

文　　献：Hustedt, F. 1914. *In*: A. Schmidt's Atlas der Diatomaceen-Kunde (1874-1959), pl. 310, O.R. Reisland, Leipzig.

　　Krammer, K. 1992. *Pinnularia* eine monographic der europäischen taxa. Bibliotheca Diatomologica, Bd. **26**, pp. 1-353. J. Cramer, Berlin/Stuttgart.

形　　態：殻は大形で線形，殻縁は平行，殻端は広円形．殻長135-211 μm，殻幅20-28 μm，条線はほぼ平行で，10 μmに約7本．外側殻面／殻套境界には他の殻と接続するための針があるが，片側のみで他方は肥厚である．中心域は狭く中心裂溝が曲がる側の条線がわずかに短くなって形成されている．

ノート：針により他の殻と横に接続して群体を形成する*Pinnularia*属珪藻はいくつか知られているが，日本で化石から報告のあるのは本種のみである．

産　　出：沼田湖成層：中期更新世・群馬県，由布院珪藻土（Okuno 1952）：鮮新世～更新世・大分県，人吉層：後期鮮新世・熊本県．

図　　版：人吉層標本．

Figs 1-5.　*Pinnularia macilenta* (Ehrenberg) Ehrenberg
LM. Fig. 1. SEM. Figs 2-5. Materials from Hitoyoshi Formation (Late Pliocene), Hitoyoshi City, Kumamoto, Japan. Scale bars: Fig. 2＝10 μm, Figs 3, 5＝5 μm, Fig. 4＝2 μm.

Fig. 1.　Whole valve view.
Fig. 2.　External oblique view of valve, spines located at valve face/mantle boundary (arrow) and ridge (arrowhead).
Fig. 3.　Internal view of terminal area, helictoglossa (arrowhead).
Fig. 4.　Enlarged view of part of Fig. 2 showing spines on interstriae (arrow) and striae.
Fig. 5.　Oblique view of part of Fig. 2 showing terminal fissure, spines (arrow) and ridge of valve face/mantle boundary (arrowhead).

Plate 234

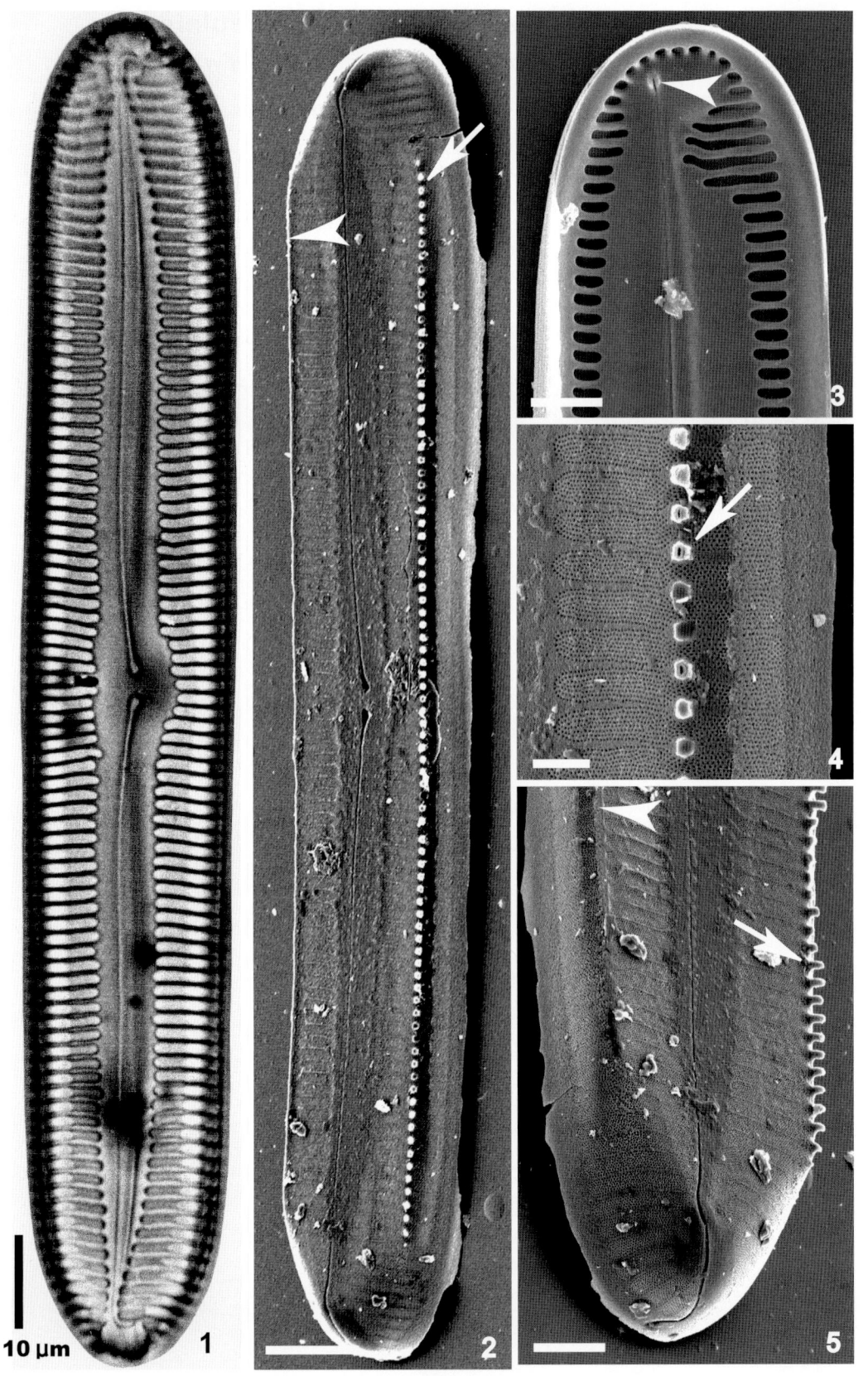

10 μm

Plate 235.　Biraphid, pennate diatoms: Naviculales
Pinnularia rivularis **Hustedt** in A. Schmidt *et al.* 1934

文　　献：Hustedt, F. 1934. *In*: A. Schmidt's Atlas der Diatomaceen-Kunde(1874-1959), pl. 392, O.R. Reisland, Leipzig.
　　Simonsen, R. 1987. Atlas and catalogue of the diatom types of Friedrich Hustedt. 3 vols. Crammer, Berlin & Stuttgart.

形　　態：殻は太い線形であるが，中央部と殻端でわずか幅が広くなることがある．殻縁はほぼ平行，殻端は広円形．殻長 63-83 μm，殻幅約 11 μm，条線は 10 μm に約 9 本で，中心付近では放射状であるが殻端ではわずか反放射になる．縦溝は中心付近でわずかカーブし中心孔で終了するが，両殻端の極裂は中心のカーブとは反対方向に曲がって終了する．

ノ ー ト：Hustedt(1934)によりインドネシアから記載された種である．筆者は奄美大島泥染公園の泥田から見出した．

産　　出：和名湖（安藤 1969）：埼玉県（現生），奄美大島泥染公園の泥田（南雲・田中 2006）：鹿児島県（現生）．

図　　版：泥染公園の泥田（奄美大島）標本．

Figs 1-6.　*Pinnularia rivularis* Hustedt
LM. Figs 1-2. SEM. Figs 3-6. Material from "dorota" of Dorozome Park (Recent), Amami-Oshima, Kagoshima, Japan. Scale bars: Figs 3-4＝5 μm, Fig. 6＝2 μm, Fig. 5＝0.5 μm.

Figs 1-2.　Two different size valves.
Fig. 3.　External oblique view of whole valve.
Fig. 4.　Internal view of whole valve.
Fig. 5.　Enlarged view of Fig. 4 showing alveoli.
Fig. 6.　Oblique external view of an apex.

Plate 235

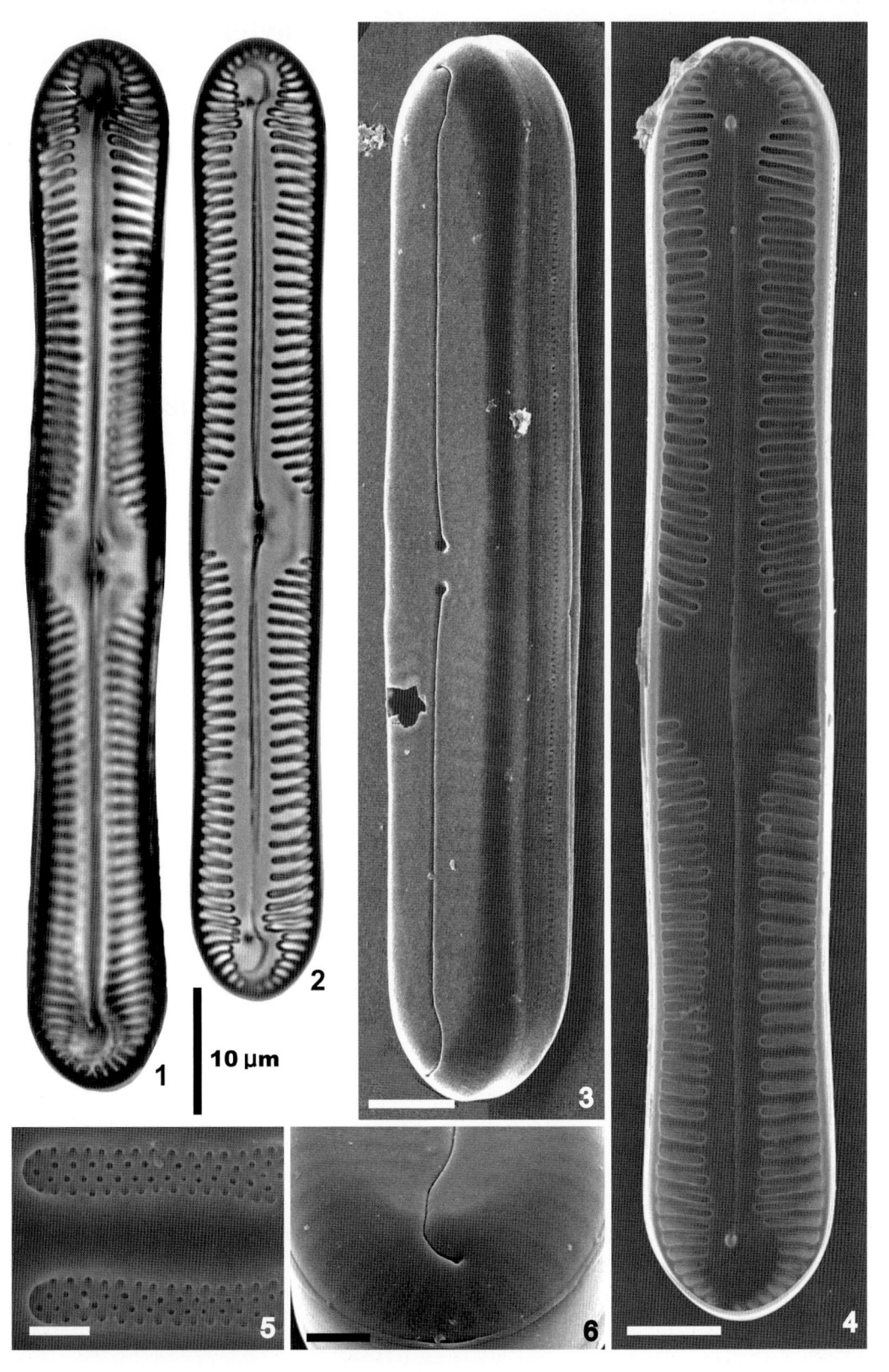

10 µm

Plate 236. Biraphid, pennate diatoms: Naviculales
Pinnularia senjoensis H. Kobayasi in Kobayasi & Ando 1977

文　　献：安藤一男・原口和夫・小林　弘 1971. 埼玉県仙女ヶ池のケイソウ. 秩父自然科学博物館研究報告 16 号: 57-79.
　　Kobayasi, H. & Ando, K. 1977. Diatoms from irrigation ponds in Musashikyuryo-shinrin park, Saitama Prefecture. Bulletin of Tokyo Gakugei University, ser. 4. **29**: 231-263.

形　　態：殻は中央部と両殻端部が特徴的に膨らむが，中央部のほうが殻端部より膨らみが大きい．殻長 54-66 μm，殻幅は中央で約 12.5 μm，条線は 10 μm に約 11 本で，中心付近では放射状，殻端では強い反放射になる．極域，中心域は広い．縦溝の外裂溝は中心付近でわずかカーブして終了する．両極裂は同方向に曲がる．

ノ ー ト：本種の原記載地は埼玉県仙女ヶ池であるが，当池からは筆者が奄美大島泥染公園から見出した，*Pinnularia rivularis* も産出しており，環境の類似性が考えられる．化石からの産出も期待できると考えている．本種は当初，安藤ら(1971)により裸名で報告され，Kobayasi & Ando(1977)で正式に記載された．

産　　出：仙女ヶ池（安藤ら 1971，Kobayasi & Ando 1977）：埼玉県（現生），奄美大島泥染公園の泥田（南雲・田中 2006）：鹿児島県（現生）．

図　　版：泥染公園の泥田（奄美大島）標本．

Figs 1-6. *Pinnularia senjoensis* H. Kobayasi
LM. Figs 1-2. SEM. Figs 3-6. Material from "dorota" of Dorozome Park (Recent), Amami-Oshima, Kagoshima, Japan. Scale bars: Figs 3-4＝5 μm, Figs 5-6＝2 μm.

Figs 1-2.　Two different size valves.
Fig. 3.　External view of whole valve.
Fig. 4.　Internal oblique view of whole valve.
Fig. 5.　Enlarged oblique view of an apex, terminal fissurae (arrows).
Fig. 6.　Enlarged view of internal valve center.

Plate 236

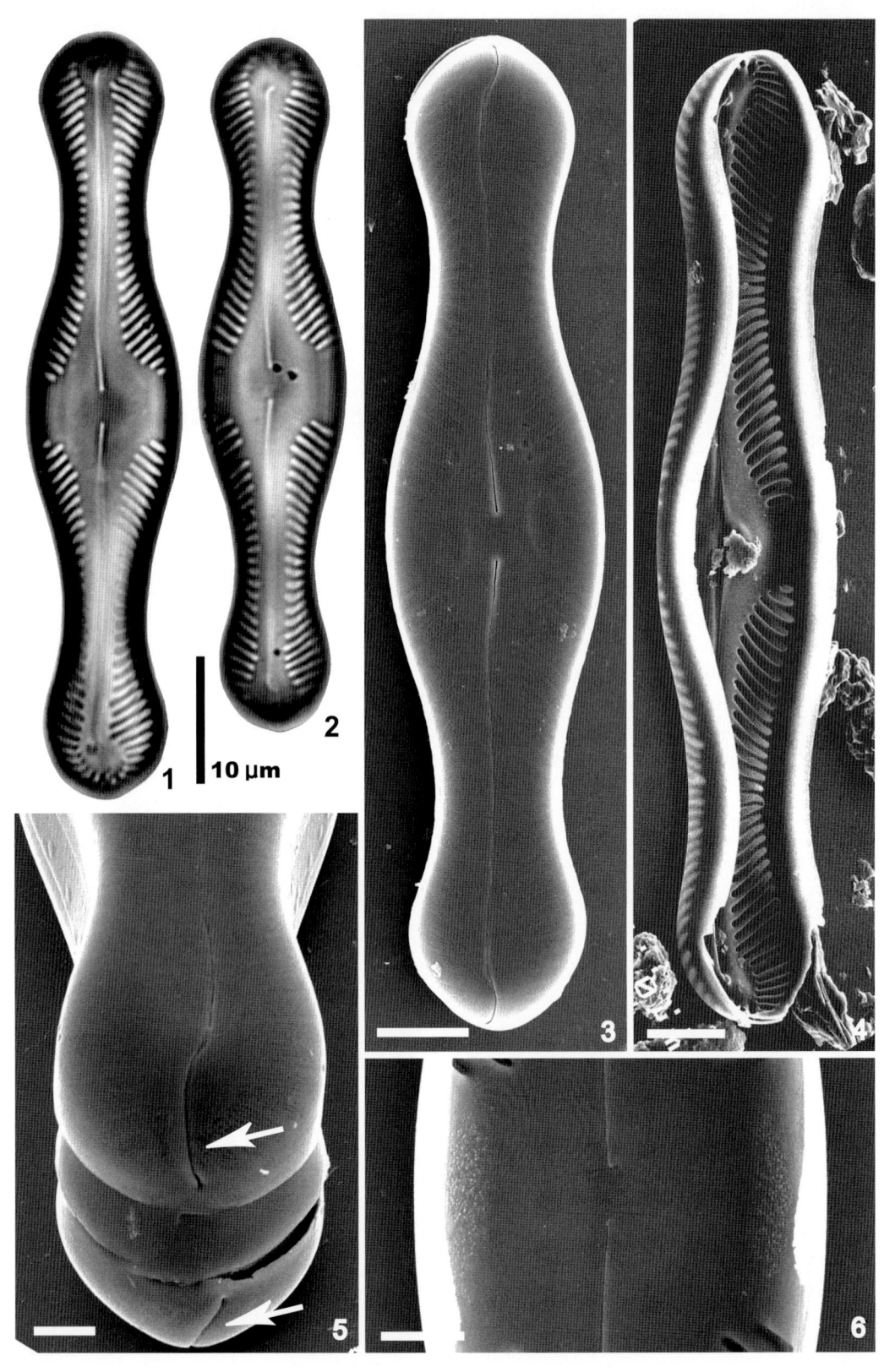

10 μm

Plate 237.　Biraphid, pennate diatoms: Naviculales
Pinnularia subgibba var. *lanceolata* Gaiser & Johansen 2000

文　　献：Gaiser, E.E. & Johansen, J. 2000. Freshwater diatoms from Carolina Bays and other isolated wetlands on the Atlantic costal plain of South Carolina, U.S.A., with descriptions of seven taxa new to science. Diatom Research **15**: 75-130.

形　　態：殻は線状披針形で殻端はわずか頭状になる．殻長 60-98 μm, 殻幅 9-11.5 μm, 条線は 10 μm に約 12 本で，中央では放射状であるが殻端では反放射になる．中心域は横に伸び殻縁に達するが，左右の広さは通常異なり，途切れる条線の幅も左右で異なる．縦溝は外側中央では片側へわずか曲がりながら中心孔で終了する．軸域は広く殻幅の 3 分の 1 程度であるが殻端では狭くなる．

ノ ー ト：タイプ地はアメリカ合衆国南カロライナ州で，葉片・浮遊藻類・アオウキクサ・ヒンジモに付着して普通に見出され，水域の pH は 5.2 と記されている（Gaiser & Johansen 2000）．

産　　出：嬬恋湖成層：中期更新世・群馬県．

図　　版：嬬恋湖成層標本．

Figs 1-5.　*Pinnularia subgibba* var. *lanceolata* Gaiser & Johansen
LM. Figs 1-2. SEM. Figs 3-5. Material from Tsumagoi Lacustrine Deposit (Middle Pleistocene), Tsumagoi Village, Gunma, Japan. Scale bars: Figs 2, 4＝5 μm, Fig. 5＝2 μm.

Figs 1-2.　Two different size valves.
Fig. 3.　External view of whole valve.
Fig. 4.　Internal oblique view of whole valve.
Fig. 5.　Enlarged oblique view of central area of Fig. 3.

Plate 237

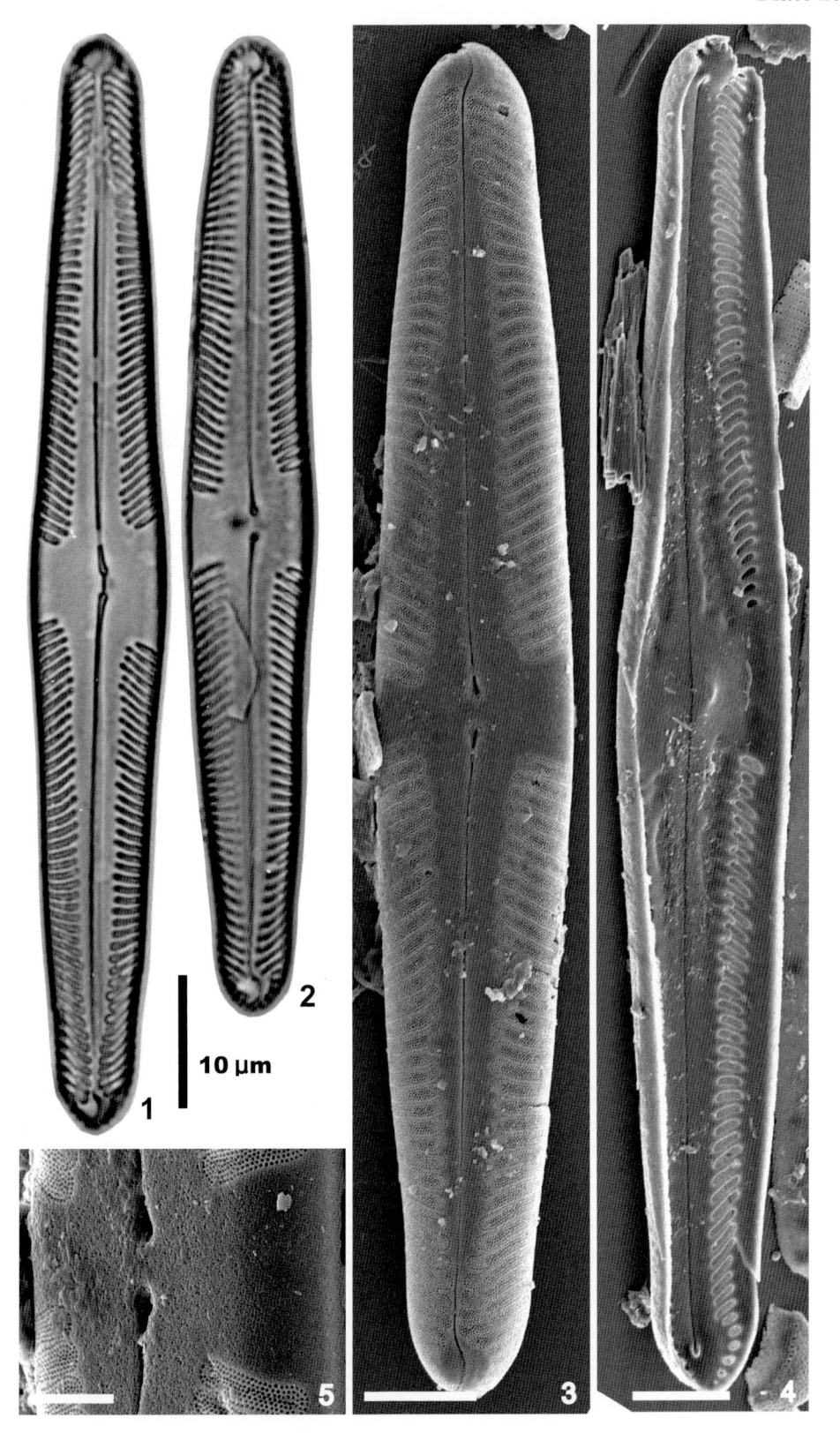

Plate 238.   Biraphid, pennate diatoms: Naviculales
***Plagiotropis lepidoptera* var. *proboscidea* (Cleve) Reimer**
in Patrick & Reimer 1975

基礎異名：*Tropidoneis lepidoptera* var. *proboscidea* Cleve 1894
文　　献：Patrick, R. & Reimer, C.W. 1975. The diatoms of the United States, exclusive of Alaska and Hawaii. Vol. **2**, Part 1. pp. 213. Monographs of the Academy of Natural Sciences of Philadelphia, No. 13.
形　　態：殻は皮針形で殻端はくちばし状に突出する．縦溝が通る軸域が隆起しているので，ふつう傾いて見える．殻長 67-82.5 μm．条線は 10 μm に約 18-19 本，条線を構成する胞紋は 10 μm に約 24 個であった．
ノ ー ト：Patrick & Reimer(1975)によれば，本種のタイプ地は汽水域であるがアメリカ合衆国では淡水域から見出された．
産　　出：奄美大島泥染公園の泥田（南雲・田中 2006）：鹿児島県（現生）．
図　　版：泥染公園の泥田（奄美大島）標本．

Figs 1-4. *Plagiotropis lepidoptera* var. *proboscidea* (Cleve) Reimer
LM. Figs 1-2. SEM. Figs 3-4. Material from "dorota" of Dorozome Park (Recent), Amami-Oshima, Kagoshima, Japan. Scale bars: Fig. 3=5 μm, Fig. 4=2 μm.

Figs 1-2.   Two different size valves.
Fig. 3.   External view of whole valve.
Fig. 4.   Internal detailed view of central area.

Plate 238

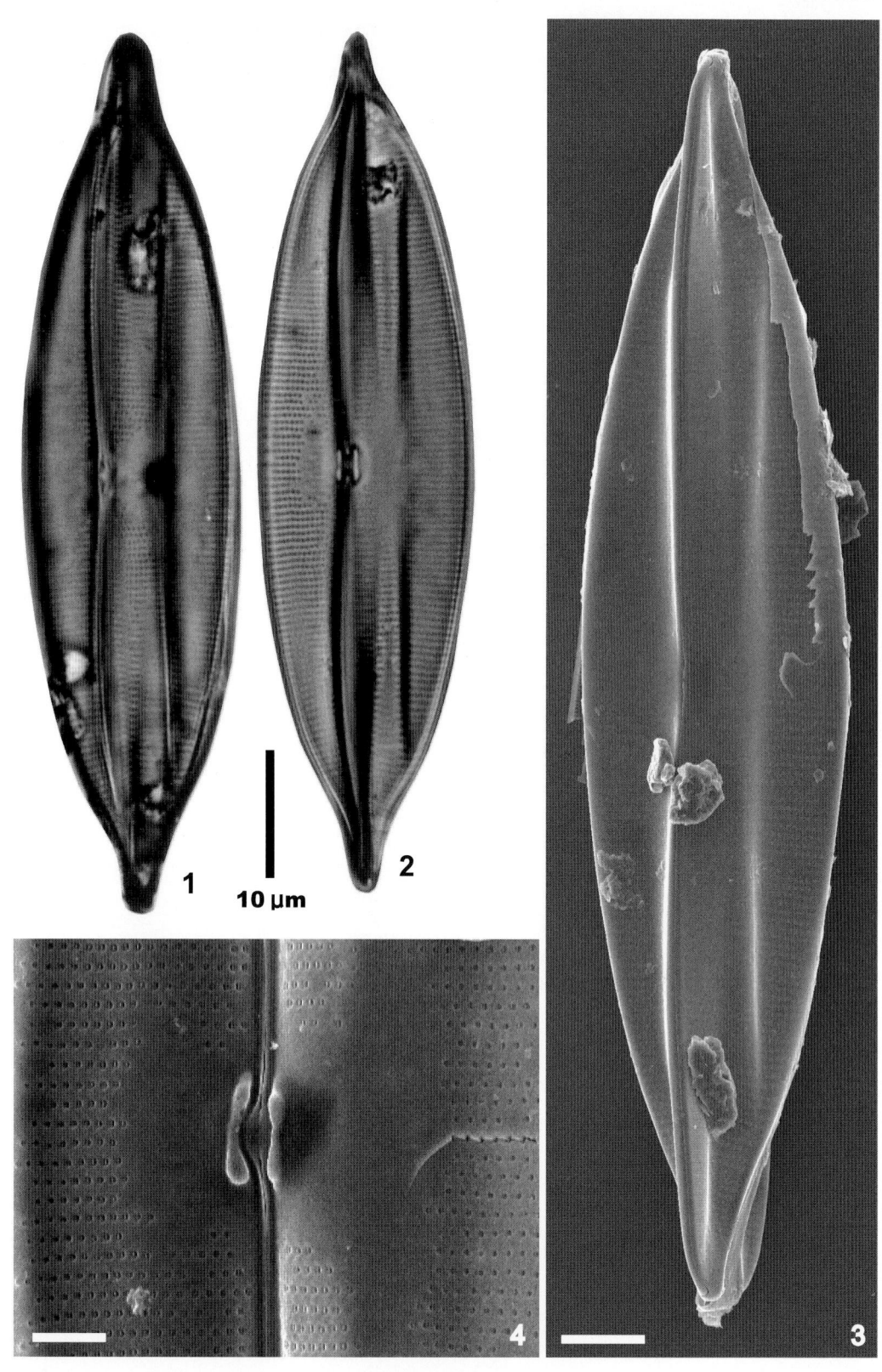

10 µm

Plate 239.   Biraphid, pennate diatoms: Naviculales
Sellaphora bacillum (Ehrenberg) D.G. Mann 1989

文　献：Patrick, R. & Reimer, C.W. 1966.　The diatoms of the United States 1. Monographs of the Academy of Natural Sciences of Philadelphia, no. 13. pp. 688. Philadelphia.
　　Mann, D.G. 1989.　The diatom genus *Sellaphora*: Separation from *Navicula*. British Phycologocal Journal **24**: 1-20.

形　態：殻は幅の広い棒状，殻端は広円形，軸域は狭いが中心域はやや円形で広くなる，殻長 34-61 μm，殻幅約 10 μm．条線は 10 μm に 20-21 本で，軸域近くでは弱い放射状であるがすぐに方向が変化し殻縁付近では平行に近くなる．また中央部では条線密度が小さい．極域では縦溝末端の両側条線に平行な黒線が存在する．軸域から中心域の縦溝の両側には帯状に灰色域が観察できるが，SEM 観察によると外側縦溝両側の軸域・中心域は縦溝に沿って窪んでおり，光顕での灰色域に対応している．外裂溝中央末端は極裂とはわずか反対方向に曲がって中心孔で終了する．両極域には極節の両側に短軸方向へ向かう肥厚部があり，光顕で黒線に見えたものである．また極節の外側には小さな丸い窪みが観察できる．

ノート：本分類群は Mann(1989) により *Sellaphora* 属が設立され，*Navicula* 属から本属へ組み合わされた．
　　Patrick & Reimer(1966) で *Navicula bacillum* は Plate 47, Fig. 4-5 の 2 図ある．このうち Fig. 4 は横子珪藻土の分類群と類似しており，著者は Fig. 4 を念頭に *N. bacillum* と同定したが，Fig. 5 は条線が Fig. 4 よりも強い放射状である点でやや異なっている．最近の本種へ同定している図を見ると Patrick & Reimer(1966) の Fig. 5 タイプのものも多いが，Hustedt(1934) による Schmidt の Atlas (T. 396, Fig. 40) では条線の向きが横子の分類群に近い．

産　出：横子珪藻土：前期更新世・群馬県，琵琶湖底堆積物（Mori 1975）：更新世・滋賀県，野上層（Okuno 1952）：中期更新世・大分県．

図　版：横子珪藻土標本．

Figs 1-7.　*Sellaphora bacillum* (Ehrenberg) D.G. Mann
LM. Figs 1-2. SEM. Figs 3-7. Material from diatomite of Yokogo (Early Pleistocene), Numata City, Gunma, Japan. Scale bars: Figs 3-4＝5 μm, Figs 6-7＝2 μm, Fig. 5＝0.5 μm.

Figs 1-2.　Two different size valves.
Fig. 3.　External view of whole valve.
Fig. 4.　Internal oblique view of whole valve.
Fig. 5.　Detailed internal view of areolae rows of Fig. 4.
Fig. 6.　Enlarged oblique view of a terminal area of Fig. 3.
Fig. 7.　Enlarged oblique view of a terminal area of Fig. 4 showing a cavity (arrow).

Plate 239

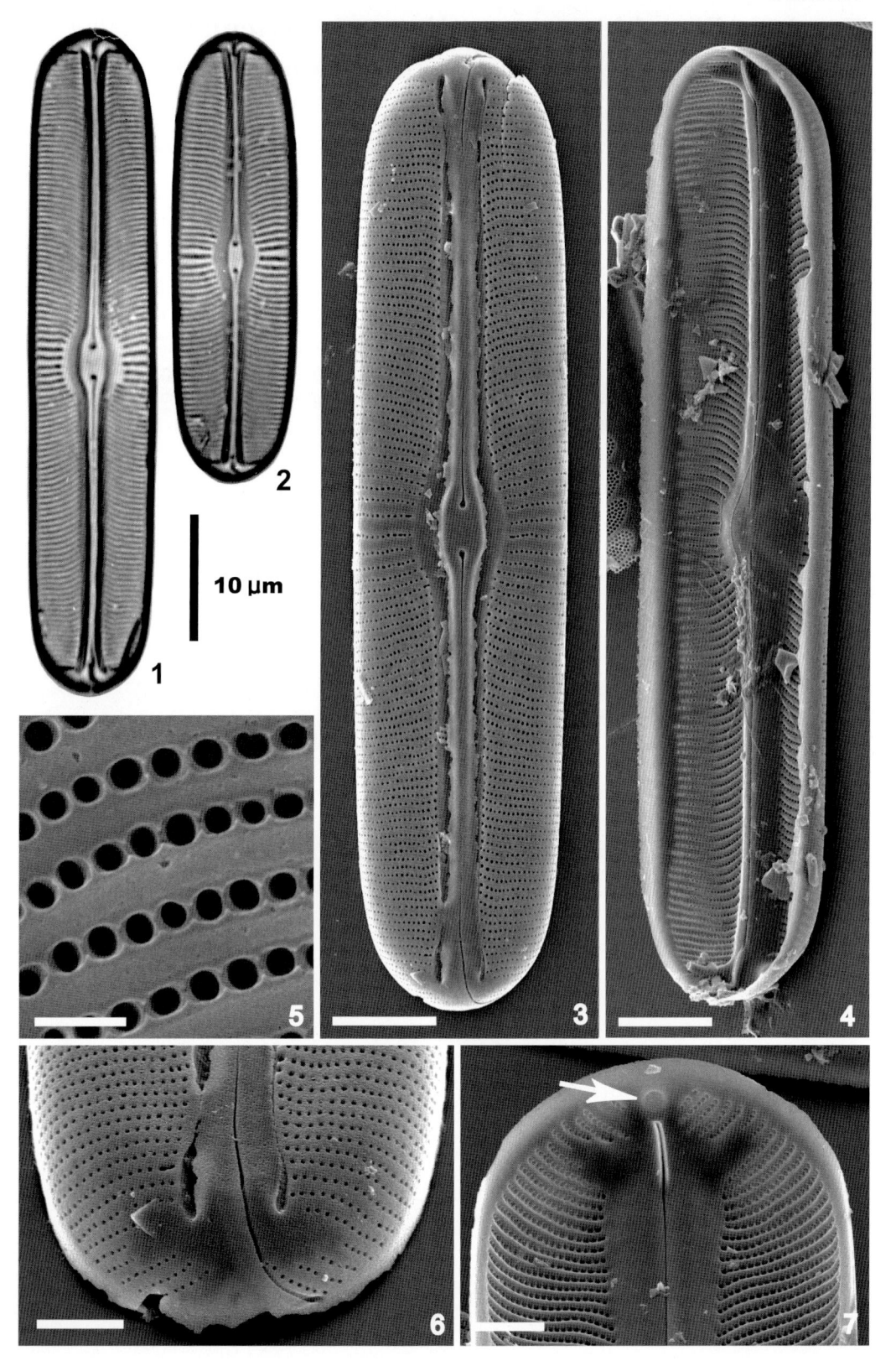

Plate 240.   Biraphid, pennate diatoms: Naviculales
Sellaphora laevissima (Kützing) D.G. Mann 1989

基礎異名：*Navicula laevissima* Kützing 1844

文　　献：Kützing, F.T. 1844.　Die kieselschaligen Bacillarien oder Diatomeen. Nordhsusen, W. Kohne. 152 pp. pls 1-30.

　　　Mann, D.G. 1989.　The diatom genus *Sellaphora*: Separation from *Navicula*. British Phycologocal Journal **24**: 1-20.

形　　態：殻は幅の広い棒状，殻端は広円形，軸域は狭いが中心域はやや広くなる，殻長32-46 μm，殻幅9-11 μm．条線は 10 μm に 16-22 本で，放射状である．また殻中央部では条線数が少なくなる．SEM 観察では外裂溝中央末端はわずか極裂とは反対方向に曲がって中心孔で終了すること，胞紋列の軸域側は窪んで軸域に沿って溝になっていること，胞紋は円形の開口であること，極節の縦溝末端延長上に円形の窪みがあることが観察できる．

産　　出：横子珪藻土：前期更新世・群馬県．

図　　版：横子珪藻土標本．

Figs 1-8. *Sellaphora laevissima* (Kützing) D.G. Mann
LM. Figs 1-3. SEM. Figs 4-8. Material from diatomite of Yokogo (Early Pleistocene), Numata City, Gunma, Japan. Scale bars: Figs 4-5＝5 μm, Figs 6-8＝2 μm.

Figs 1-3.　Three different size valves.

Fig. 4.　External view of whole valve.

Fig. 5.　Internal oblique view of whole valve.

Fig. 6.　Enlarged view of central area of Fig. 4.

Fig. 7.　Enlarged oblique view of an apex of Fig. 4.

Fig. 8.　Enlarged view of an apex of Fig. 5 showing a cavity (arrow).

Plate 240

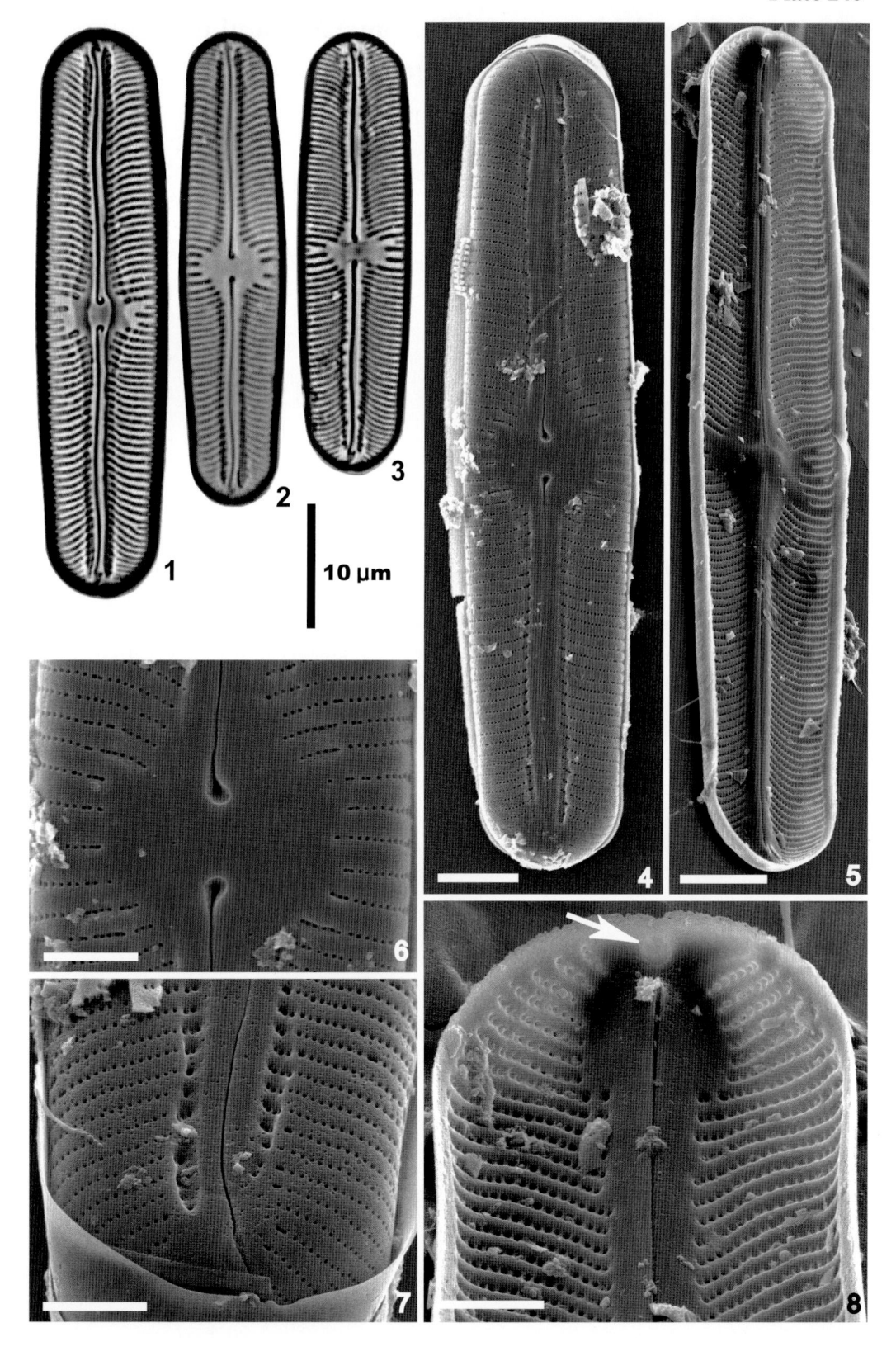

10 μm

Plate 241. Biraphid, pennate diatoms: Naviculales
Stauroneis acuta W. Smith var. *acuta* 1853

文　　献：Smith, W. 1853. Synopsis of British Diatomaceae. Vol. **1**. pp 89. pls 1-31. John van Voorst, London.
　　　　小林　弘・安藤一男 1978. 日本産 *Stauroneis* 属ケイソウ．東京学芸大学紀要第 4 部門 **30**: 273-292.

形　　態：殻は大形で狭皮針形であるが中央部が膨らみ，中央と殻端の間がわずか凹む．殻長 142 μm，殻幅 21 μm．条線は放射状で 10 μm に約 14 本．条線を構成する点紋は 10 μm に約 16 個．外側殻面／殻套境界は肥厚している．中心域は内側では肥厚しており，横に伸びて殻縁に達している．両殻端には偽隔壁がある．

ノ ー ト：小林・安藤(1978)が本種のみでなく，日本の *Stauroneis* 属全般について解説している．

産　　出：由布院珪藻土（Okuno 1952）：鮮新世～更新世・大分県，津森層：中期更新世・熊本県．

図　　版：津森層標本．

Figs 1-4. *Stauroneis acuta* W. Smith var. *acuta*
LM. Fig. 1. SEM. Figs 2-4. Material from Tsumori Formation (Middle Pleistocene), Mashiki Town, Kumamoto, Japan. Scale bars: Figs 2-4＝10 μm.

Figs 1-4.　Same valve, LM and SEM photographs.
Fig. 1.　Whole valve view. Photomontage made by three photographs of same valve.
Fig. 2.　External view of whole valve.
Fig. 3.　Oblique view of Fig. 2 showing ridge (arrow).
Fig. 4.　Internal oblique view of whole valve, stauros (arrow) and pseudosepta (arrowheads).

Plate 241

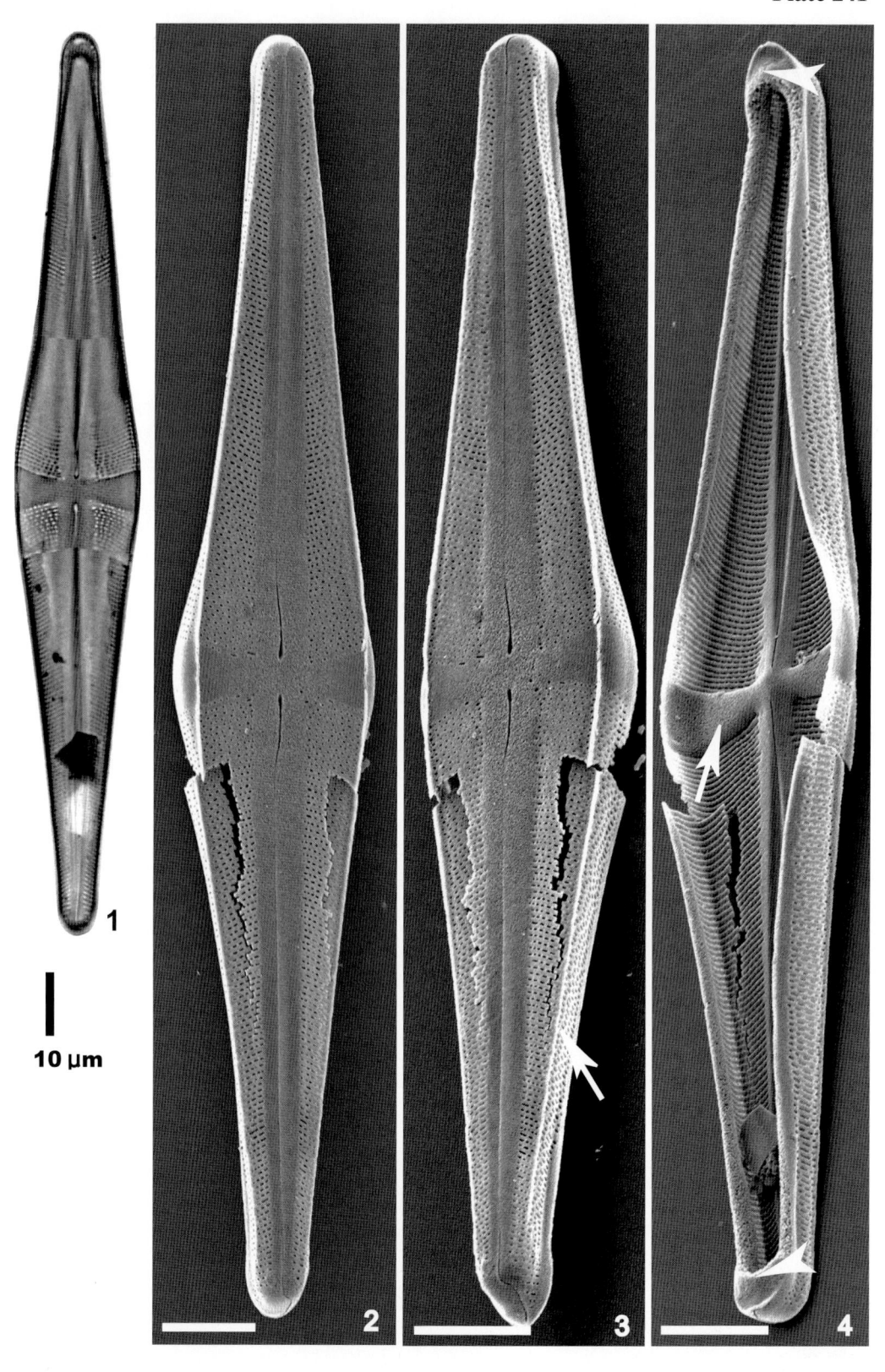

10 µm

Plate 242.　Biraphid, pennate diatoms: Naviculales
Stauroneis acuta var. *terryana* Tempère in Cleve 1894

文　　献：Cleve, P.T. 1894. Synopsis of the naviculoid diatoms, Part 1. Konglica Svenska Vetenskaps-Akademiens. Handlingar **26**: 1-194, pls 1-5.
　　　　　小林　弘・安藤一男 1978. 日本産 *Stauroneis* 属ケイソウ. 東京学芸大学紀要第4部門 **30**: 273-292.
形　　態：殻は大形で皮針形，殻端は細く伸長するが丸みを帯びる．殻長 180-235 μm，殻幅 30-35 μm．条線は 10 μm に約 12 本で，殻端では強い放射状になる．縦溝は両殻端で同方向へ曲がる．外側の殻面／殻端境界は肥厚し，殻面と殻套の条線は連続しない．条線を構成する点紋は 10 μm に 12-14 個．中心節は内側では強く肥厚し，横に伸びて殻縁に達し十字節を形成している．内側軸域も肥厚している．両殻端には偽隔壁がある．
ノ ー ト：小林・安藤(1978)が本変種の命名上の混乱について解説している．
産　　出：沼田湖成層：中期更新世・群馬県．
図　　版：沼田湖成層標本.

Figs 1-5. *Stauroneis acuta* var. *terryana* Tempère
　　LM. Figs 1-2. SEM. Figs 3-5. Material from Numata Lacustrine Deposit (Middle Pleistocene), Numata City, Gunma, Japan. Scale bars: Figs 3-4＝10 μm, Fig. 5＝2 μm.

Figs 1-5.　Same valve, LM and SEM photographs.
Figs 1-2.　Same valve at different focal planes.
Fig. 3.　External view of whole valve.
Fig. 4.　Internal oblique view of whole valve, pseudoseptum (arrow).
Fig. 5.　Part of enlarged view of an apex of Fig. 4.

Plate 242

Plate 243.   Biraphid, pennate diatoms: Naviculales
Stauroneis phoenicenteron (Nitzsch) Ehrenberg var. *phoenicenteron* 1843

基礎異名：*Bacillaria phoenicenteron* Nitzsch 1817

文　　献：小林　弘・安藤一男 1978．日本産 *Stauroneis* 属ケイソウ．東京学芸大学紀要第 4 部門 **30**: 273-292．

　　河島綾子・真山茂樹 1997．阿寒湖の珪藻（5. 羽状類-縦溝類：*Aneumastus, Craticula, Diatomella, Diploneis, Frustulia, Gyrosigma, Luticola, Neidium, Sellaphora, Stauroneis*）．自然環境科学研究 **10**: 35-52.

形　　態：殻は大形で菱形〜皮針形，殻端は細く伸長するが丸みを帯びる．殻長 205 μm，殻幅 21 μm．条線は 10 μm に約 14 本で，中央ではわずかであるが殻端では強い放射状になる．中心節は内側では強く肥厚し，横に伸びて殻縁に達し十字節を形成している．縦溝は両殻端で同方向へ曲がる．外側の殻面／殻端境界は肥厚し，殻面と殻套の条線は途切れる．条線を構成する点紋は 10 μm に約 14 個．

ノ ー ト：化石・現生とも各地から産出報告がある．形態についての記述も多くの論文で見られるが，上記文献欄に記したものが参考になる．

産　　出：宮田層：中新-鮮新世・秋田県，鬼首層（Ichikawa 1955）：後期更新世・宮城県，和村珪藻土（Skvortzov 1937）：新第三紀・長野県，由布院珪藻土（Okuno 1952）：鮮新世〜更新世・大分県，野原層（Tanaka et al. 2004）：前期更新世・大分県，人吉層（Okuno 1964）：後期鮮新世・熊本県．

図　　版：人吉層標本．

Figs 1-4. *Stauroneis phoenicenteron* (Nitzsch) Ehrenberg var. *phoenicenteron*

LM. Fig. 1. SEM. Figs 2-4. Material from Hitoyoshi Formation (Late Pliocene), Hitoyoshi City, Kumamoto, Japan. Scale bars: Figs 2-4＝10 μm.

Fig. 1.　Whole valve view. Photomontage made by two photographs of same valve.
Fig. 2.　External view of whole valve.
Fig. 3.　Oblique view of Fig. 2 showing external stauros and ridge of valve face/mantle boundary.
Fig. 4.　Internal oblique view of whole valve.

Plate 243

Plate 244.　Biraphid, pennate diatoms: Naviculales
Stauroneis phoenicenteron var. *hattorii* Tsumura 1955

文　　献：Tsumura, K. 1955. A contribution to the knowledge of diatoms found in the clod from the pond on Mt. Shichimen-zan in Japan. Journal of Yokohama Municipal (City) University, series C-12, No. 43: 1-32, 11 pls.

形　　態：殻は皮針形，殻端はくちばし状に突出する．殻長 76.5-128 μm，殻幅 16.5-22 μm．条線は放射状であり，10 μm に 15-17 本．条線を構成する点紋は 10 μm に 16-20 本．中心節に接する点紋は線状になる．

ノ　ー　ト：山梨県七面山の"七面山のお池"の試料から記載された種である．本種は十字節 (stauros) に接する条線が線状になることが他と区別する重要な形質である．

産　　出：和村（上村・小林 1983）：新第三系・長野県，横子珪藻土：前期更新世・群馬県，津森層（田中ら 2005）：中期更新世・熊本県．

図　　版：横子珪藻土標本．

Figs 1-4. *Stauroneis phoenicenteron* var. *hattorii* Tsumura
LM. Fig. 1. SEM. Figs 2-4. Material from diatomite of Yokogo (Early Pleistocene), Numata City, Gunma, Japan. Scale bars: Figs 2-4=5 μm.

Fig. 1.　Whole valve view. Photomontage made by two photographs of same valve.
Fig. 2.　External view of whole valve: linear areolae (arrow).
Fig. 3.　Oblique view of part of Fig. 2: linear areolae (arrow).
Fig. 4.　Internal oblique view of whole valve.

Plate 244

Plate 245. Biraphid, pennate diatoms: Rhopalodiales
Epithemia adnata (Kützing) Brébisson 1838

基礎異名：*Frusturia adnata* Kützing 1833

文　献：小林　弘・吉田　稔 1984. コンクリート池のケイソウとその優れた教材性について．東京学芸大学紀要第4部門 **36**: 115-143.

　　河島綾子・真山茂樹 2002. 阿寒湖の珪藻（9. 羽状類-縦溝類：*Amphora, Epithemia, Rhopalodia*）．自然環境科学研究 **15**: 47-58.

形　態：殻は背側に膨れ，腹側では中心付近の2分の1程度がやや凹む．殻端は頭状に突出する．殻長 34-88 μm，殻幅 8-10 μm．肋線は 10 μm に約3本，肋線間の条線は 3-8 本．縦溝は腹側にあるが，殻中心では腹側と背側の中間付近まで背側による．

ノート：小林・吉田(1984)で報告があるように，*Epithemia zebra* (Ehr.) Kütz. は本種の異名である．各地から産出報告がある汎布種であり，形態もかなり変化する種類である．上記2文献が本種について詳しく記してある．

産　出：横子珪藻土：前期更新世・群馬県，大鷲湖沼性堆積物（田中ら 2011）：鮮新世・岐阜県，俣水層（田中・鹿島 2007）：前期更新世・大分県．

図　版：大鷲湖沼性堆積物（1-2, 5-8 図），横子珪藻土（3-4 図）標本．

Figs 1-8. *Epithemia adnata* (Kützing) Brébisson

LM. Figs 1-4. SEM. Figs 5-8. Materials, Figs 1-2, 5-8, from lacustrine deposit of Owashi (Pliocene), Gujo City, Gifu and Figs 3-4, from diatomite of Yokogo (Early Pleistocene), Numata City, Gumna, Japan. Scale bars: Figs 5, 7＝5 μm, Figs 6, 8＝2 μm.

Figs 1-4.　Four different size valves.
Fig. 5.　External view of whole valve.
Fig. 6.　Internal view of whole valve.
Fig. 7.　Oblique view of central area of Fig. 5 showing outer fissure of raphes (arrowheads).
Fig. 8.　Oblique view of central area of Fig. 6 showing inner fissure of raphes (arrowheads).

Plate 245

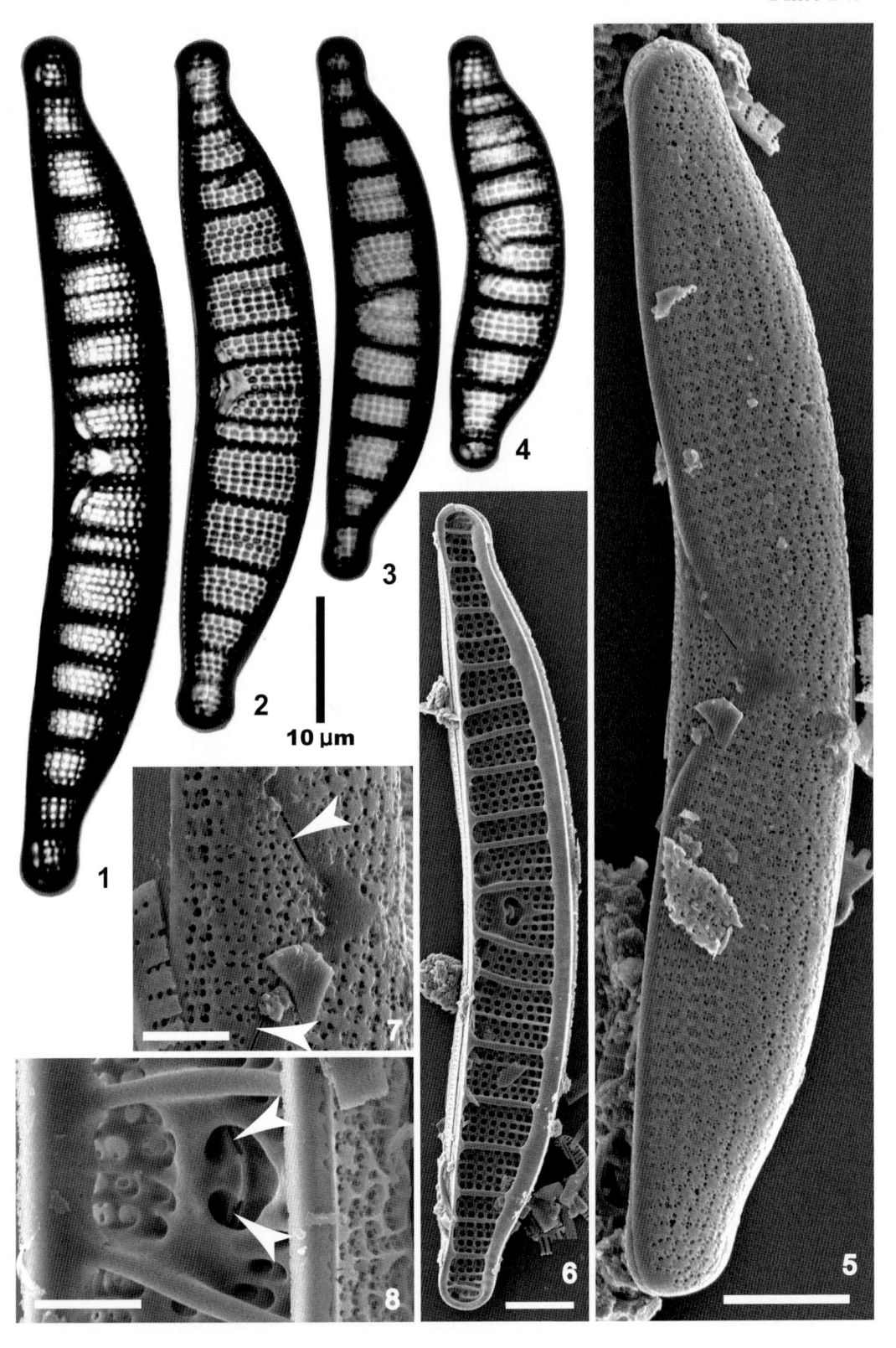

10 μm

Plate 246. Biraphid, pennate diatoms: Rhopalodiales
Epithemia cistula **(Ehrenberg) Ralfs** in Pritchard 1861

基礎異名：*Eunotia cistula*？Ehrenberg 1854
文　　献：Pritchard, A. 1861. A history of Infusoria, including the Desmidiaceae and Diatomaceae, British and foreign. Ed. Ⅳ. 998 pp. 40 pl., London.
形　　態：殻は三日月形で，殻長 20.5-50.5 μm，殻幅 8.5-10.5 μm．肋線は 10 μm に約 3 本．肋線間の条線は 4-8 本である．縦溝は殻端では腹側にあるが，直線的に背側に向かい殻中央では背側殻縁に達する．
ノ ー ト：渡辺ら(2005)により *Epithemia reichelti* Fricke との形態の類似が指摘されている．
産　　出：人形峠層（田中ら 2008）：後期中新世～鮮新世・岡山-鳥取県境，本種の変種 var. *lunaris* Grunow は，八束珪藻土（Okuno 1952）：更新世・岡山県，および嬉野珪藻土（Skvortzov 1937）：新第三紀・佐賀県から報告がある．
図　　版：人形峠層標本．

Figs 1-8.　*Epithemia cistula* (Ehrenberg) Ralfs
LM. Figs 1-4. SEM. Figs 5-8. Material from Ningyo-toge Formation (Late Miocene-Pliocene), boundary of Okayama and Tottori Prefectures, Japan. Scale bars: Figs 5, 7＝5 μm, Figs 6, 8＝0.5 μm.

Figs 1-4.　Different size valves.
Fig. 5.　External oblique view of frustule.
Fig. 6.　Detailed view of central area of Fig. 5 showing raphe branches (arrowheads).
Fig. 7.　Internal view of whole valve.
Fig. 8.　Enlarged view of part of Fig. 7 showing internal raphe branches (arrowheads).

Plate 246

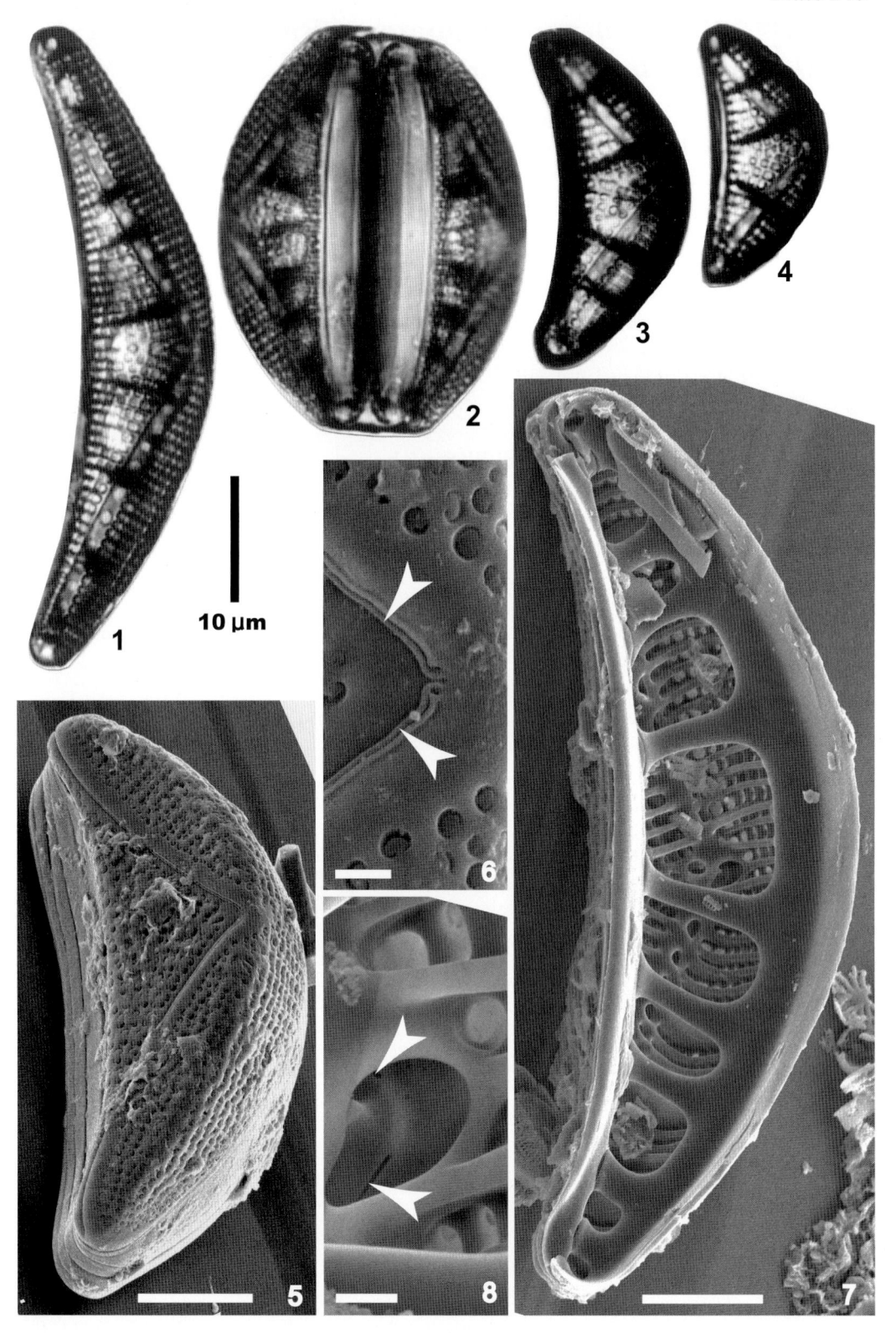

Plate 247. Biraphid, pennate diatoms: Rhopalodiales
Epithemia hyndmanii W. Smith 1850

文　　献：Smith, W. 1853. A synopsis of the British Diatomaceae. Vol. **1**. 89 pp. 31 pls. John van Voorst, London.

形　　態：殻は大形で弓状に曲がるが，殻幅はあまり変化しない．殻長 80-165 μm，殻幅 17-20 μm．肋線は 10 μm に約 4 本．肋線間の条線は 2-3 本である．縦溝は殻中心において外側では途切れるが，内側においては連続している．

ノ　ー　ト：田中ら (2005) が本種および類似種について比較的詳しく記している．

産　　出：鬼首層（Ichikawa 1955）：更新世・宮城県，和村珪藻土（上村・小林 1983）：新第三系・長野県，瓜生坂層（窪田ら 1976）：前期更新世・長野県，大鷲湖沼性堆積物：鮮新世・岐阜県，投入堂凝灰角礫岩層（赤城ら 1984）：中新世・鳥取県，中山香珪藻土（Okuno 1952）：更新世・大分県，津森層（田中ら 2005）：中期更新世・熊本県．

図　　版：大鷲湖沼性堆積物（1, 3-6 図），津森層（2 図）標本．

Figs 1-6. *Epithemia hyndmanii* W. Smith

LM. Figs 1-2. SEM. Figs 3-6. Materials, Figs 1, 3-6, from lacustrine deposit of Owashi (Pliocene), Gujo City, Gifu and Fig. 2, from Tsumori Formation (Middle Pleistocene), Mashiki Town, Kumamoto, Japan. Scale bars: Fig. 4＝10 μm, Figs 3, 5＝2 μm, Fig. 6＝1 μm.

Figs 1-2.　Two different size valves.
Fig. 3.　Enlarged view of part of Fig. 4, showing outer fissure of raphes (arrowheads).
Fig. 4.　External view of whole valve.
Fig. 5.　Oblique view of an apex showing raphe ending.
Fig. 6.　Internal view of valve center showing uninterrrupted raphe (arrowhead).

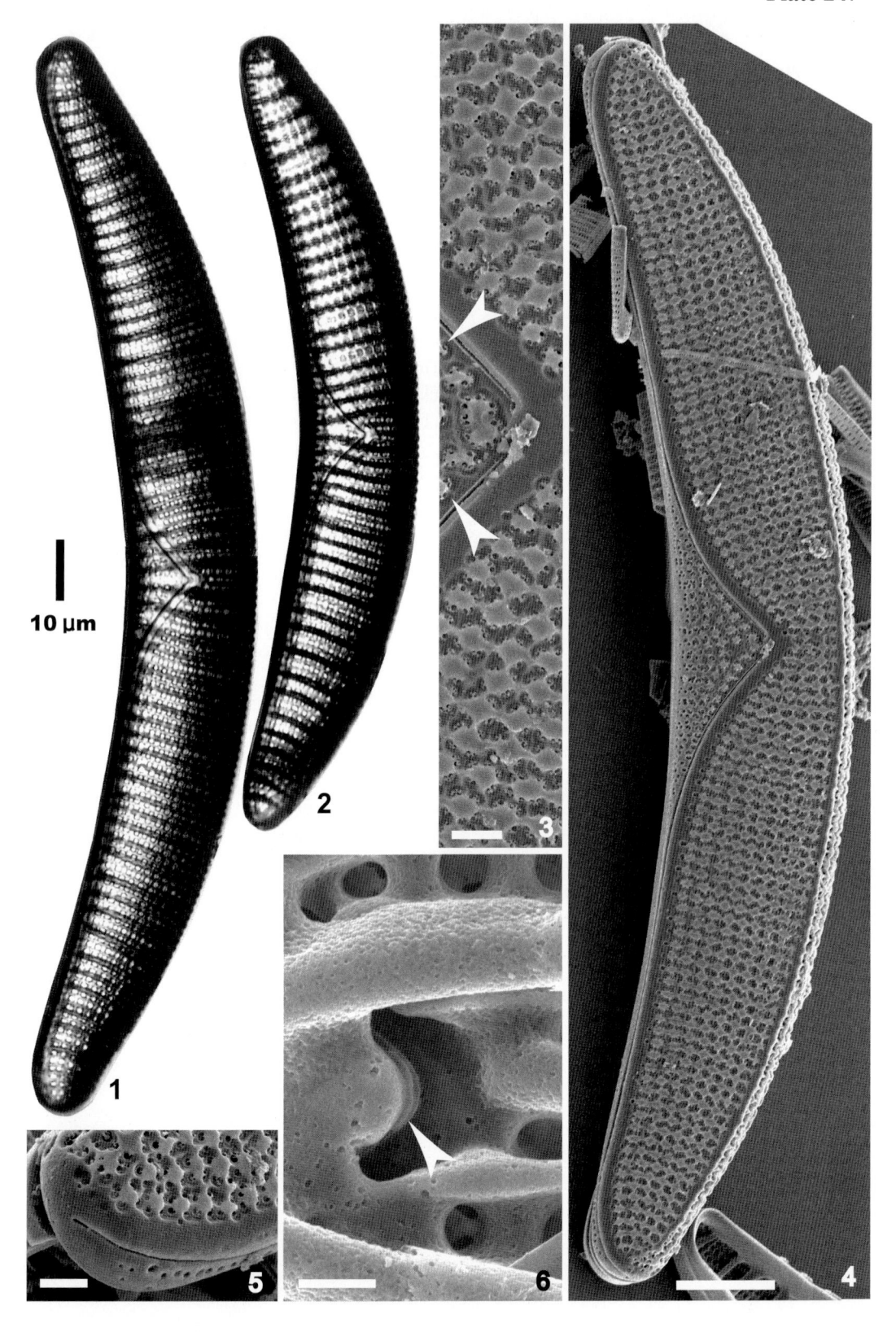

Plate 247

Plates 248-249. Biraphid, pennate diatoms: Rhopalodiales
Epithemia numatensis H. Tanaka sp. nov.

新　種：記載文（英文）は 16 頁参照.
形　態：殻はわずか弓形で，背側に膨れ腹側では凹むが，殻幅はほとんど同じで殻端近くまで達する．殻端は強く頭状に突出し，その基部が非常に狭くなる．殻長 49-88 μm，殻幅 10-14 μm．肋線は 10 μm に 3-4 本，肋線間の条線は(2)3-7(8)本．条線は 10 μm に 13-15 本で，条線を構成する点紋は 10 μm に約 12 個である．縦溝は腹側を走るが，中心および両殻端では腹側・背側の中間付近のやや腹側にある．内側では縦溝は中心節で途切れる．胞紋は外側では肉趾状の師板で閉塞され，内側では円形の胞口である.

ノート：本分類群に類似するものとしては *E. adnata* var. *porcellus*（Kütz.）Patr. と *E. adnata* var. *proboscidea*（Kütz.）Patr. がある．両分類群とも Kützing(1844)で原記載され，Grunow(1862)により *Epithemia zebra* の変種へ組み合わせされている．Patrick & Reimer(1975)はさらに *Epithemia adnata* の変種へ組み合わせを行っている．しかし両者とも殻端の形状は横子産のように極端な頭状ではない．Krammer & Lange-Bertalot(1988)で *E. adnata*（Kütz.）Brébisson として掲載している写真のうちの一つに類似した殻形がある．しかし横子産のように，殻端が強く頭状に突出し首の部分が非常に狭い形の個体のみで一つの分類群を形成しているのではない．Rumrich et al.(2000)は *Epithemia* spec. cf. *adnata* として 4 個体の LM 写真を示している，このうち 2 個体は *E. numatensis* に殻端の形が似ているが，横子産の個体は，特徴のある殻形の個体のみで一つの分類群を形成しているので，新種と考えるのが適当と思う.

産　出：横子珪藻土：前期更新世・群馬県.
図　版：横子珪藻土標本.

Plate 248, Figs 1-6.　*Epithemia numatensis* H. Tanaka sp. nov.

LM. Figs 1-3. SEM. Figs 4-6. Material from diatomite of Yokogo (Early Pleistocene), Numata City, Gunma, Japan. Scale bars: Fig. 4＝5 μm, Figs 5-6＝2 μm.

Figs 1-2.　Holotype specimen at different focal planes.
Figs 1-3.　Two different size valves.
Fig. 4.　External view of whole valve.
Fig. 5.　Enlarged oblique view of an apex of Fig. 4 showing raphe ending (arrowhead).
Fig. 6.　Deatiled view of valve center of Fig. 4 showing central pore (arrowhead) and areolae with occluded volae.

Plate 248

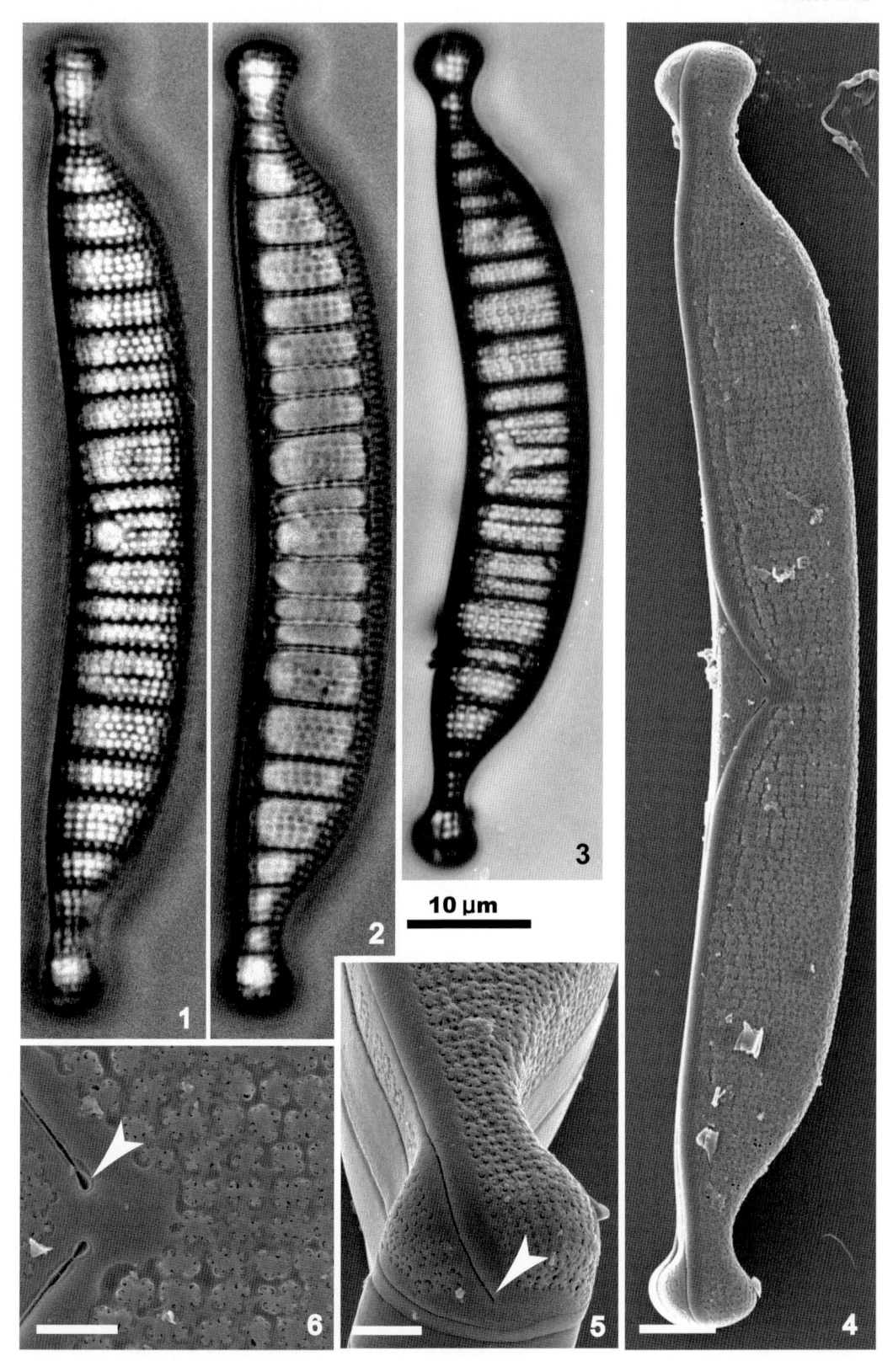

10 µm

Plate 249, Figs 1–5. *Epithemia numatensis* **H. Tanaka** sp. nov.

SEM. Figs 1–5. Material from diatomite of Yokogo (Early Pleistocene), Numata City, Gunma, Japan. Scale bars: Figs 1–2=5 μm, Fig. 4=2 μm, Figs 3, 5=1 μm.

Fig. 1. Internal view of whole valve.
Fig. 2. Oblique view of Fig. 1.
Fig. 3. Detailed view of valve center showing central nodule.
Fig. 4. Enlarged view of part of Fig. 2.
Fig. 5. Enlarged view of part of Fig. 1 showing internal areolae openings between two transapical costae.

Plate 249

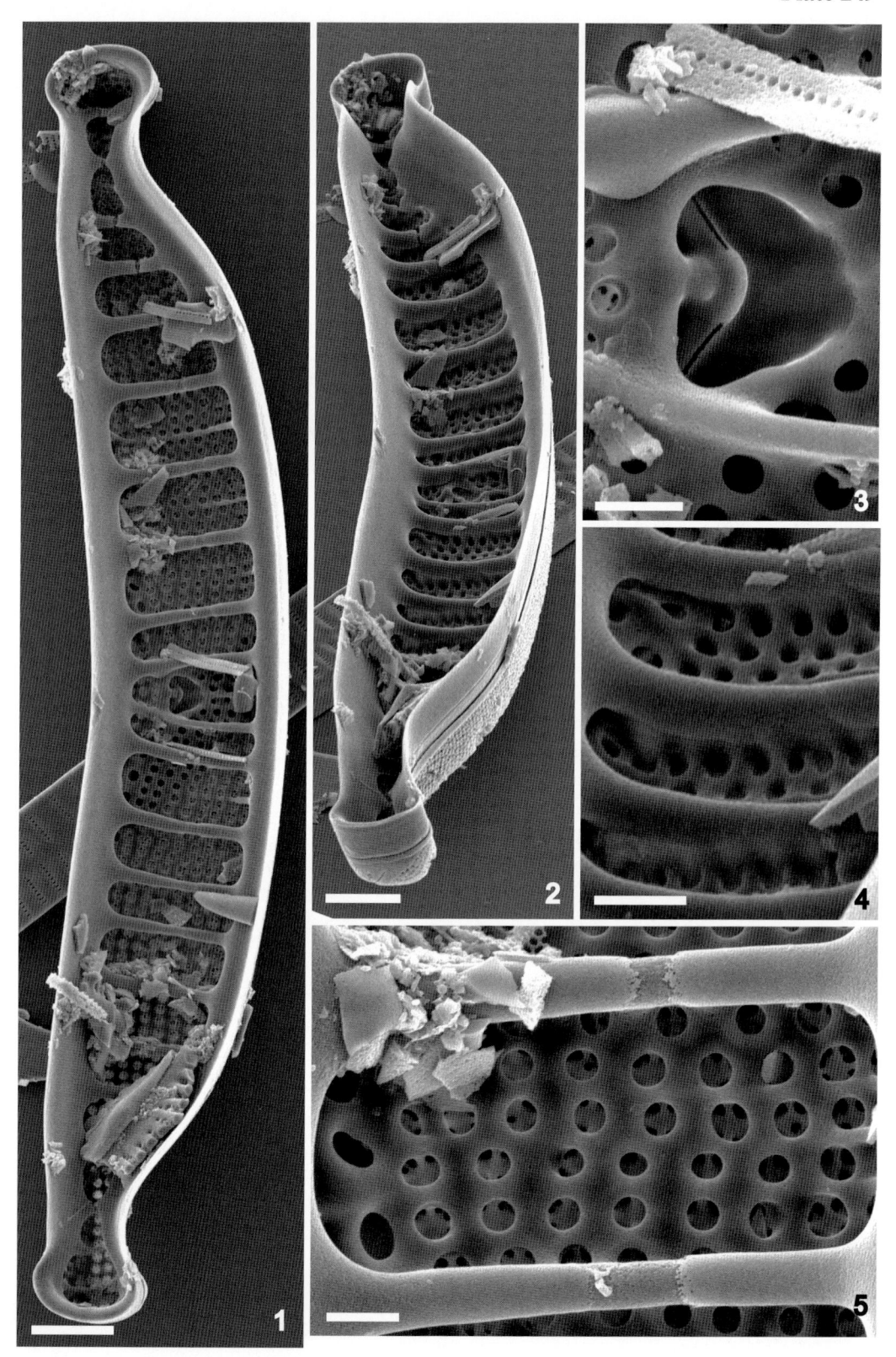

Plate 250. Biraphid, pennate diatoms: Rhopalodiales
Epithemia reticulata Kützing 1849

文　献：Kobayasi, H. & Kobayashi, H. 1988. *Epithemia amphicephale*(Östr.)comb. et stat. nov. and *E. reticulata* Kütz. with special reference to the areolae occlusion. *In*: Round, F.E.(ed) Proceedings of the 9th International Diatom Symposium. pp. 459-466. Biopress Ltd., Bristol and Koeltz Scientific Books, Koenigstein.

形　態：殻はゆるく弓なりに曲がり，両側は平行で殻端は広円状，殻長22-91 μm，殻幅9-15 μm．肋線は10 μmに(3)4(5)本．肋線間の条線は3-5本である．縦溝はほぼ腹側にあるが，中心では腹側と背側の中間付近にまで達する．

ノ ー ト：ここで示した分類群に類似する種としては，*Epithemia muellerii* Fricke と *Epithemia goeppertiana* Hilse がある．このうち *E. muellerii* は Krammer & Lange-Bertlot (1988)および Hartley *et al.*(1996)によると *E. goeppertiana* の異名である．ところが Sims (1983)によると *E. muellerii* は *E. reticulata* の異名である．宮田層の分類群は *E. goeppertiana* か *E. reticulata* のいずれかに同定できると思われるが，発表年の古い *E. reticulata* に同定する．Kobayasi & Kobayashi(1988)はフィンランド産の *E. reticulata* の詳細を報告している．

産　出：宮田層：中新-鮮新世・秋田県．

図　版：宮田層標本．

Figs 1-8. *Epithemia reticulata* Kützing

LM. Figs 1-4. SEM. Figs 5-8. Materials from Miyata Formation (Mio-Pliocene), Senboku City, Akita, Japan. Scale bars: Figs 7-8＝5 μm, Figs 5-6＝2 μm.

Figs 1-2.　Same valve at different focal planes.
Figs 1-4.　Three different size valves.
Fig. 5.　External view of whole valve.
Fig. 6.　Enlarged view of central valve surface of Fig. 5.
Fig. 7.　Internal oblique view of whole valve.
Fig. 8.　Detailed view of an apex of Fig. 5 showing outer fissure of raphe (arrowhead).

Plate 250

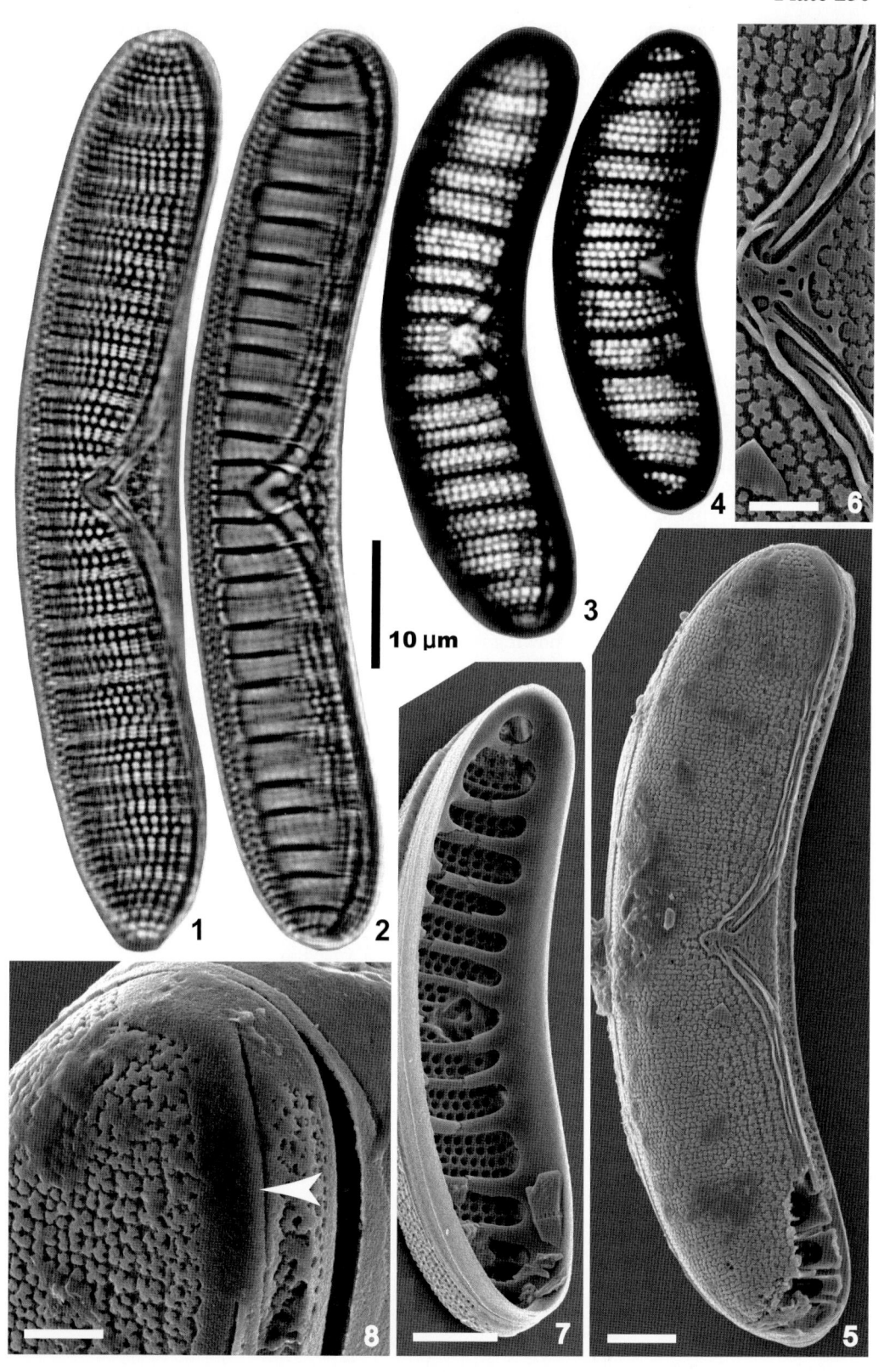

Plate 251.　Biraphid, pennate diatoms: Rhopalodiales
Epithemia sorex Kützing 1844

文　　献：Kützing, F.T. 1844. Die kieselschaligen Bacillarien oder Diatomeen. Nordhausen, W. Kohne. 152 pp. pls 1-30.

形　　態：殻は三日月状で両端は頭状に突出する．殻長27-48 μm，殻幅7.5-10.5 μm．肋線は10 μmに6-8本．肋線間の条線は2(3)本である．縦溝は殻端においては腹側であるが中央では背側の殻縁近くまで達する．SEM観察では胞紋の外側は特徴的な開口があり，縦溝は中央において外側は連続しないが内側は連続している．

産　　出：喜茂別・真狩村珪藻土（奥野　1961）：更新世〜完新世・北海道，須巻層（塩原層群）（Akutsu 1964）：前期更新世・栃木県，横子珪藻土：前期更新世・群馬県，瓜生坂層（小諸層群）（窪田ら　1976）：前期更新世・長野県，大鷲湖沼性堆積物（田中ら　2011）：鮮新世・岐阜県，津森層（田中ら　2005）：中期更新世・熊本県．

図　　版：横子珪藻土標本．

Figs 1-9.　*Epithemia sorex* Kützing

LM. Figs 1-4. SEM. Figs 5-9. Materials from diatomite of Yokogo (Early Pleistocene), Numata City, Gunma, Japan. Scale bars: Figs 5, 7-8＝5 μm, Fig. 6＝2 μm, Fig. 9＝1 μm.

Figs 1-4.　Four different size valves.
Fig. 5.　External view of whole valve.
Fig. 6.　Enlarged view of Fig. 5 showing raphe branches (arrowheads) and caps of areolae.
Fig. 7.　Oblique view of Fig. 5.
Fig. 8.　Internal oblique view of whole valve.
Fig. 9.　Enlarged view of different angle of Fig. 8 showing raphe fissure continuing across a small central nodule (arowhead).

Plate 251

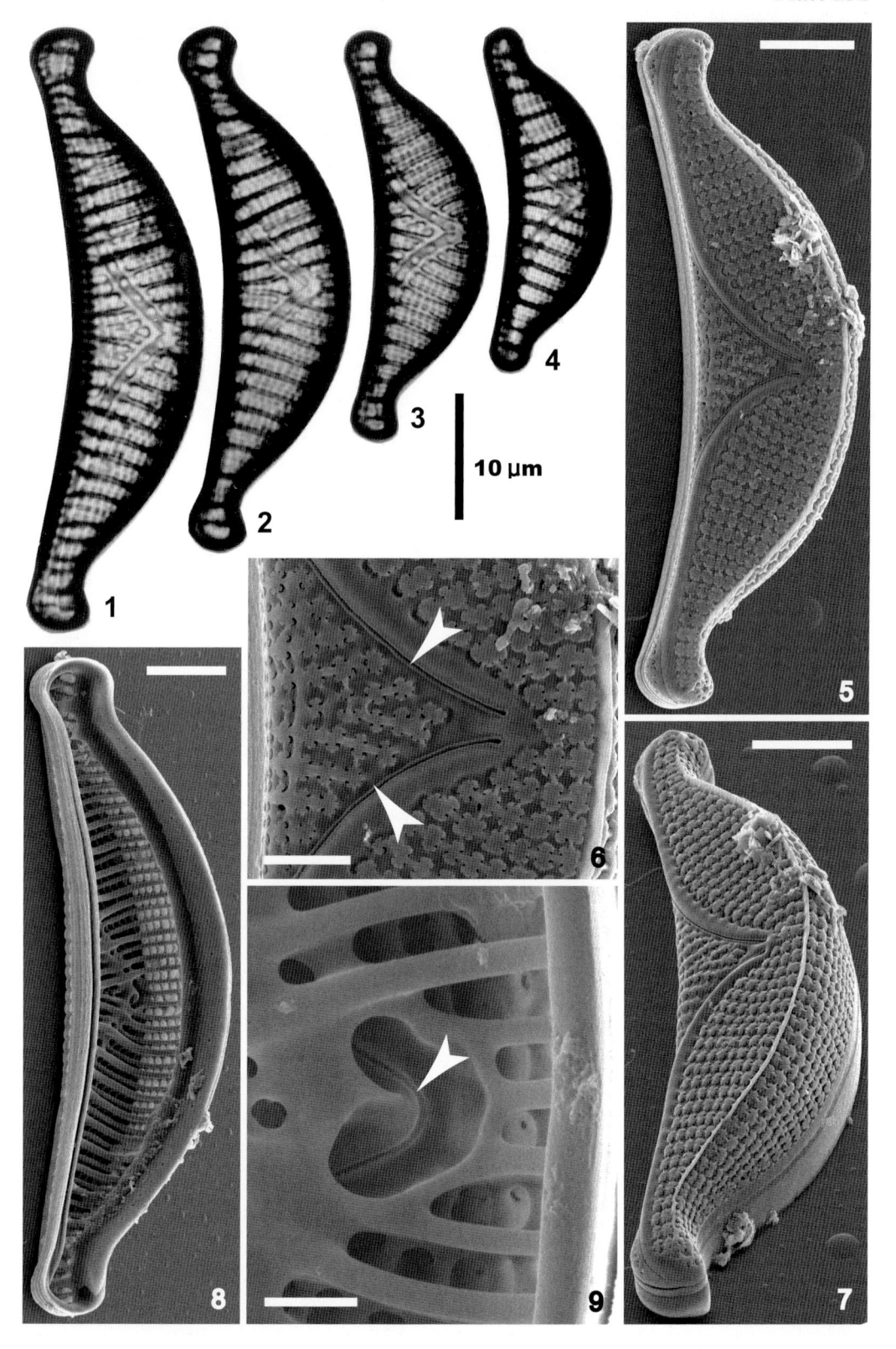

Plate 252.   Biraphid, pennate diatoms: Rhopalodiales
Epithemia turgida var. *porcellus* Héribaud 1903

文　献：Héribaud, F.J. 1903.　Les Diatomées d'Auvergne. Librairie des Sciences Naturelles, Paris. deuxieme mémoire. pp. 1-155. pls 9-12.

形　態：殻は台形を縦にした形で，背側は直線に近く腹側はわずか凹み，殻端は腹側にありやや突出する．殻長32-54 μm，殻幅9-15 μm．肋線は10 μmに3-5本でやや放射状．肋線間の条線は(2)3-5(6)本である．SEM観察では内側縦溝は中央で途切れる．

ノート：宮田層産の本種はHéribaud(1903)の図と比較すると，肋間の条線数が多いこと，縦溝の中央部での位置が腹側に寄っていること，殻形が殻端で急に細くなること等の違いはあるが，既存の図の中では最も本種に近い．

産　出：宮田層：中新−鮮新世・秋田県．

図　版：宮田層標本．

Figs 1-8.   *Epithemia turgida* var. *porcellus* Héribaud
LM. Figs 1-4. SEM. Figs 5-8. Material from Miyata Formation (Mio-Pliocene), Senboku City, Akita, Japan. Scale bars: Figs 7-8＝5 μm, Fig. 6＝2 μm, Fig. 5＝0.5 μm.

Figs 1-4.   Four different size valves.
Fig. 5.   External view of whole valve.
Fig. 6.   Enlarged view of part of Fig. 5 showing external openings of areolae with caps.
Fig. 7.   Internal oblique view of valve.
Fig. 8.   Detailed internal view of valve center showing costae and inner fissures of raphe branches (arrowheads).

Plate 252

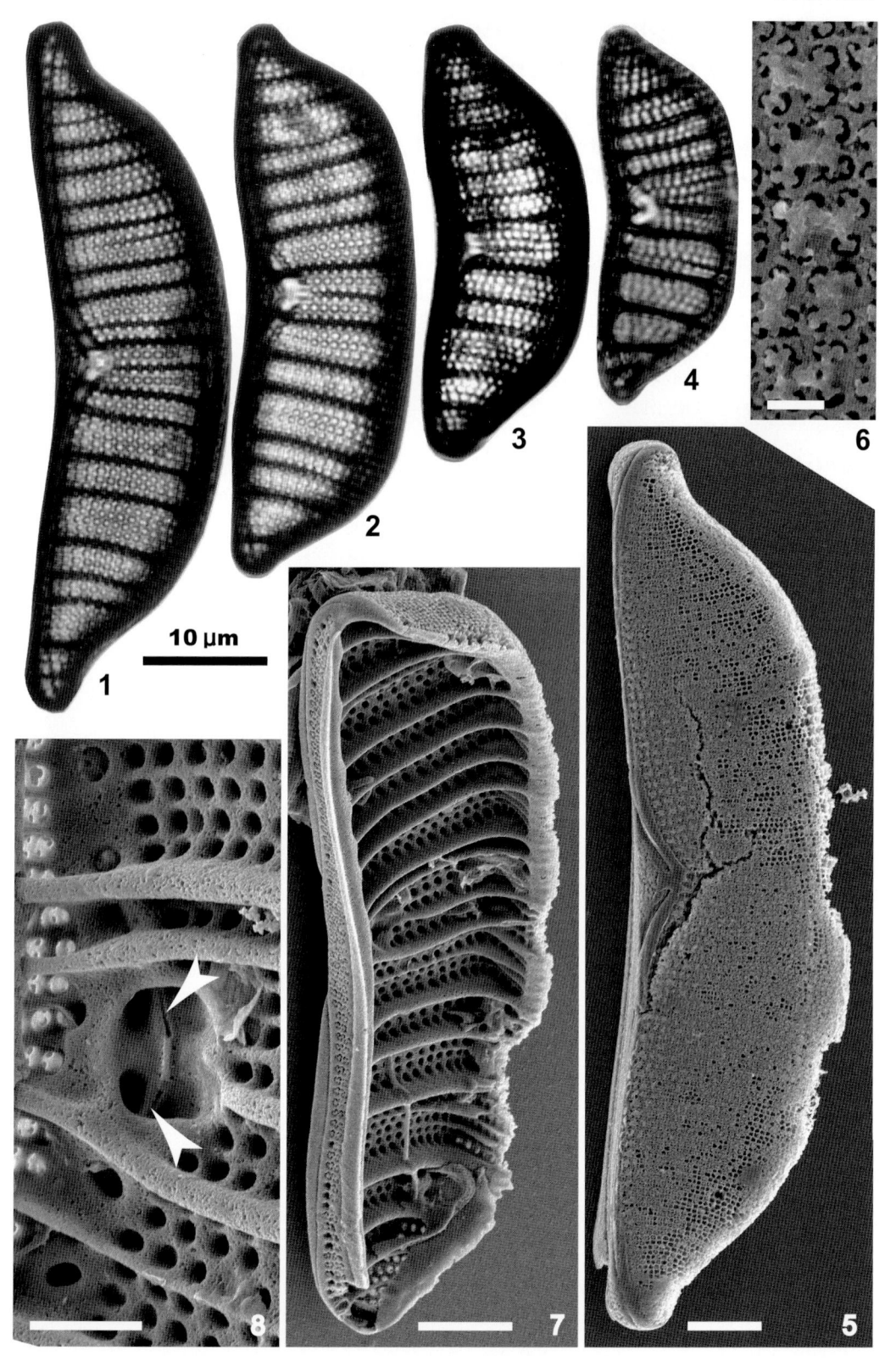

10 µm

Plate 253.　Biraphid, pennate diatoms: Rhopalodiales
Rhopalodia gibba (Ehrenberg) O. Müller 1895

基礎異名：*Navicula gibba* Ehrenberg 1830
文　　献：河島綾子・真山茂樹 2002. 阿寒湖の珪藻（9.羽状類-縦溝類：*Amphora, Epithemia, Rhopalodia*）. 自然環境科学研究 **15**: 47-58.
形　　態：殻は細長く中央では背側が膨れ，腹側はほぼ直線状で殻端では腹側に曲がる．殻長 60-116 μm，肋線は中央では平行であるが，殻端では背側の間隔が広くなる，10 μm に 7-8 本．各肋線間には 2(3)本の条線がある．
ノ ー ト：和文の中では上記の文献が本種の形態を理解する上で役立つと思われる．
産　　出：鬼首層（Ichikawa 1955）：後期更新世・宮城県，琵琶湖底堆積物（Mori 1975）：更新世・滋賀県，奄芸層群（根来・後藤 1981）：鮮新世・三重県，人形峠層：後期中新世〜鮮新世・岡山-鳥取県境，津森層（田中ら 2005）：中期更新世・熊本県．
図　　版：津森層（Fig. 1），人形峠層（Figs 2-6）標本．

Figs 1-6.　*Rhopalodia gibba* (Ehrenberg) O. Müller

LM. Figs 1-3. SEM. Figs 4-6. Materials, Fig. 1, from Tsumori Formation (Middle Pleistocene), Mashiki Town, Kumamoto, Japan and Figs 2-6, from Ningyo-toge Formation (Late Miocene-Pliocene), boundary between Okayama and Tottori Prefectures, Japan. Scale bars: Fig. 4＝5 μm, Fig. 5＝1 μm, Fig. 6＝0.5 μm.

Figs 1-2.　Different size frustules.
Fig. 3.　Whole valve view.
Fig. 4.　External view of whole valve.
Fig. 5.　Enlarged view of valve center of Fig. 4.
Fig. 6.　Enlarged view of part of Fig. 5 showing external areolae covered by veluma.

Plate 253

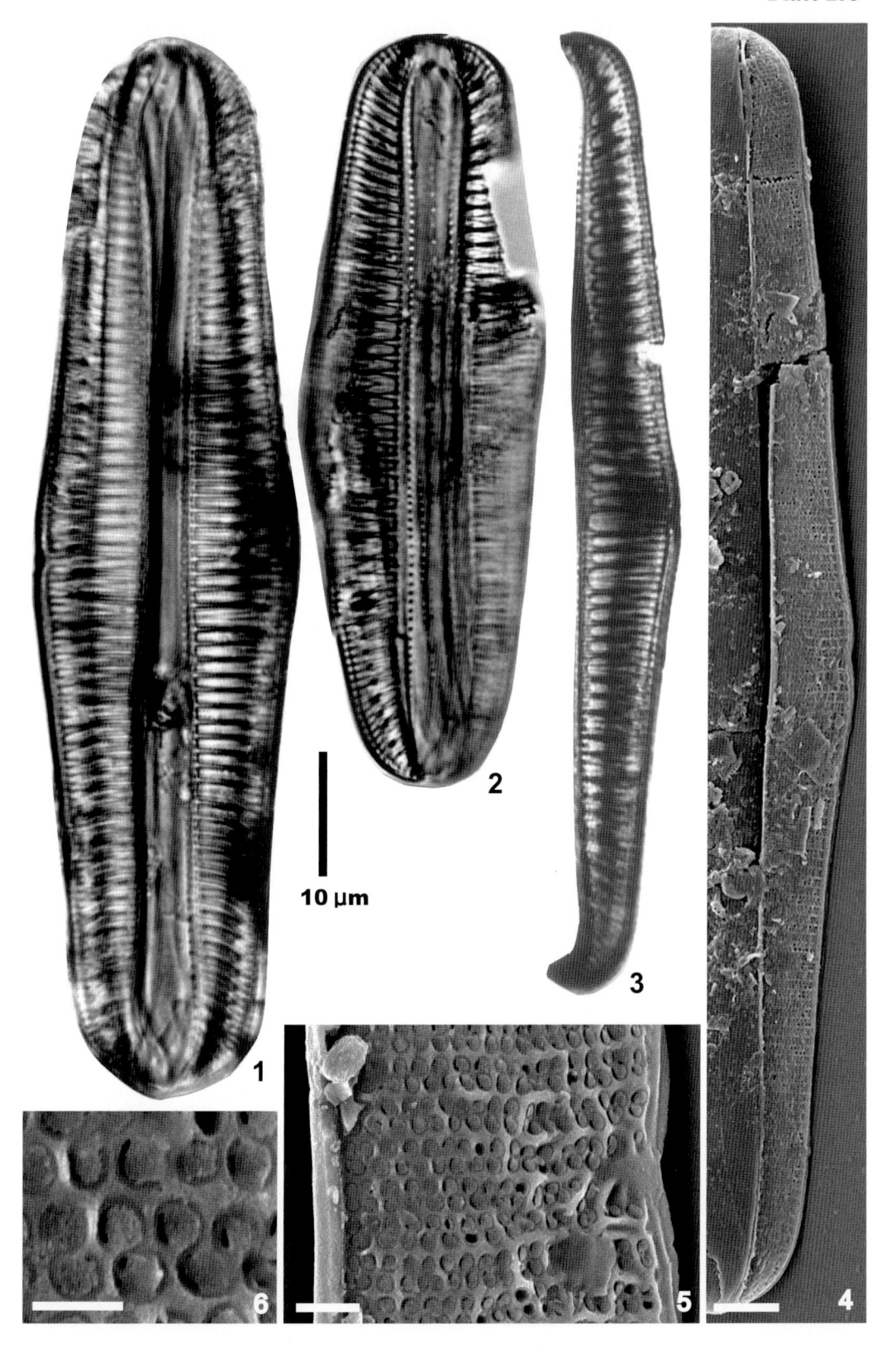

Plate 254. Biraphid, pennate diatoms: Rhopalodiales
Rhopalodia gibberula (Ehrenberg) O. Müller 1895

基礎異名：Eunotia gibberula Ehrenberg 1841
文　　献：Müller, O. 1895. *Rhopalodia*ein neues genus der Bacillariaceen.(Engler's) botanische jahrbücher für systematik, Pflanzengeschichte und Pflanzengeographie, Band **22**: 54-71, 2 Taf.
形　　態：殻は三日月～半月状で，背側は膨れるが腹側はほぼ直線で，殻端では腹側に曲がる．殻長 41-47 μm，肋線は中央では平行であるが，殻端では背側の間隔が広くなる，10 μm に 4-5 本．各肋線間には 3-7 本の条線がある．縦溝は背側にある．
ノ ー ト：著者が本書で使用した，利根川源流（群馬県）のサンプルからはほぼ同じ大きさの個体しか産出しなかったが，原口ら(1998)の埼玉県産の個体は殻長 40-70 μm と記されている．
産　　出：野殿層（中島・南雲 1999）：中期更新世・群馬県，琵琶湖底堆積物（Mori 1975）：更新世・滋賀県．
図　　版：利根川源流標本（現生）．

Figs 1-7. *Rhopalodia gibberula* (Ehrenberg) O. Müller
LM. Figs 1-2. SEM. Figs 3-7. Material from the uppermost part of the Tone River (Recent), Gunma, Japan. Scale bars: Figs 3-5＝5 μm, Fig. 7＝2 μm, Fig. 6＝1 μm.

Figs 1-2.　Two different size valves.
Fig. 3.　External view of whole valve.
Fig. 4.　Oblique view of Fig. 3.
Fig. 5.　Internal oblique view of whole valve.
Fig. 6.　Enlarged view of central nodule of Fig. 5 (arrowhead).
Fig. 7.　Enlarged view of Fig. 3 of raphe fissure (arrowhead) and veluma of areolae.

Plate 254

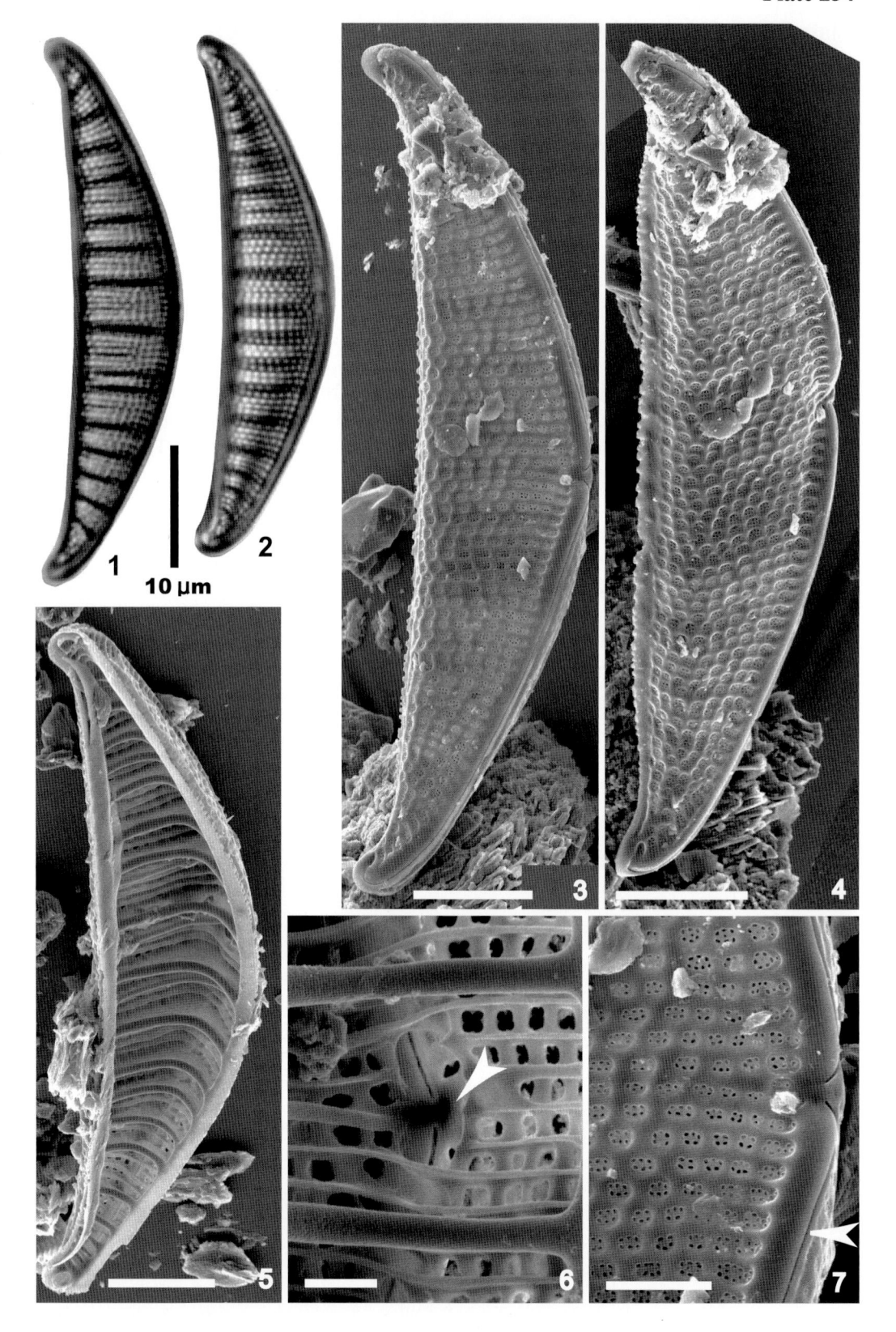

Plate 255. Biraphid, pennate diatoms: Rhopalodiales
Rhopalodia novae-zelandiae Hustedt 1913

文　献：Hustedt, F. 1913. *In*: A. Schmidt's Atlas der Diatomaceen-Kunde(1874-1959), pl. 294. O.R. Reisland, Leipzig.
　　　Simonsen, R. 1987. Atlas and catalogue of the diatom types of Friedrich Hustedt 3 vols. Crammer, Berlin & Stuttgart.
形　態：殻は細長く中央の背側が膨らみ，腹側はほぼ直線状で，殻端は腹側に曲がる．殻長72-113 µm，肋線は10 µmに約14本，各肋線間には2本の胞紋列がある．
産　出：大戸湖成堆積物（田中 1991）：中期更新世・群馬県，津森層（田中ら 2005）：中期更新世・熊本県．
図　版：津森層標本．

Figs 1-7. *Rhopalodia novae-zelandiae* Hustedt
LM. Figs 1-2. SEM. Figs 3-7. Material from Tsumori Formation (Middle Peistocene), Mashiki Town, Kumamoto, Japan. Scale bars: Figs 3-4=5 µm, Fig. 6=2 µm, Fig. 5=1 µm, Fig. 7=0.5 µm.

Figs 1-2.　Two different size valves.
Fig. 3.　External slightly oblique view of whole valve.
Fig. 4.　Internal oblique view of whole valve.
Fig. 5.　Enlarged view of an apex of Fig. 4.
Fig. 6.　Enlarged view of valve center of Fig. 3.
Fig. 7.　Detailed internal view, central inner fissure of raphe branch endings (arrowheads).

Plate 255

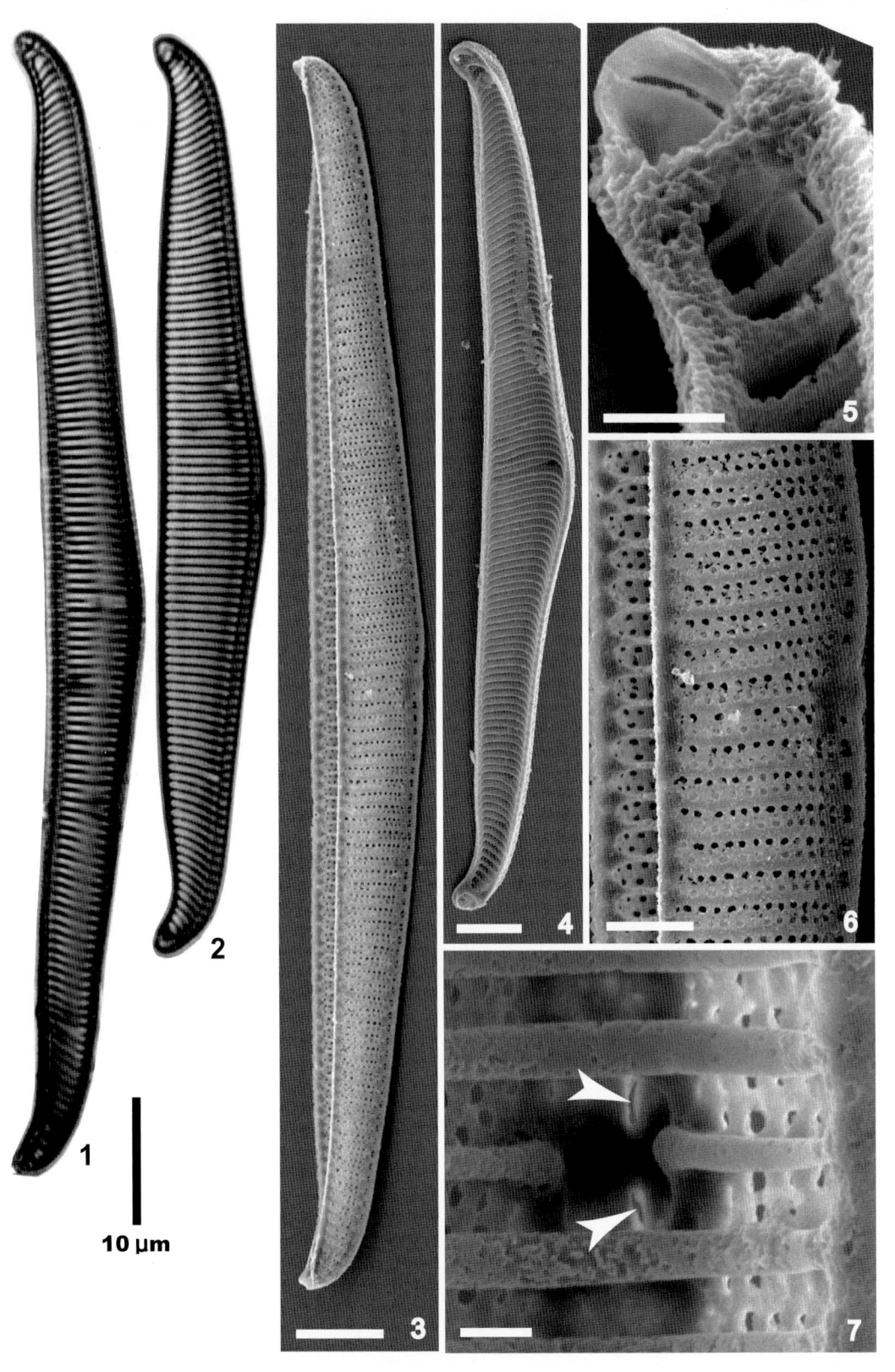

10 μm

Plate 256.　Biraphid, pennate diatoms: Rhopalodiales
Rhopalodia rupestris (W. Smith) Krammer
in Lange-Bertalot & Krammer 1987

基礎異名：*Epithemia rupestris* W. Smith 1853
文　　献：Lange-Bertalot, H. & Krammer, K. 1987. Bacillariophyceae Epithemiaceae Surirellaceae. Bibliotheca Diatomologica. Bd. **15**: 1-289.
形　　態：殻は細長く，中央では背側が膨れ腹側は直線的であるが，殻端では腹側に曲がる．殻長 22-52.5 μm，殻幅 5-7 μm，肋線は中央では平行であるが，殻端では背側の間隔が広くなる，10 μm に 3-5 本．各肋線間には(2)3-6 本の条線がある．
ノート：本種は W. Smith(1853)によって *Epithemia* 属として原記載されたが，O. Müller (1899)は *Rhopalodia gibberula* の変種へ組み合わせをし，Krammer (in Lange-Bertalot & Krammer 1987) は変種から種へランクの変更を行った．筆者はかつて奥平温泉の珪藻群集（現生）の報告を行った際（田中・中島 1985），本分類群を *R. gibberula* に同定した．今回は胞紋が小さく光顕では識別できないことから *R. rupestris* とする．化石で保存の良い標本が得られなかったので，現生標本で図版に示す．
産　　出：含ナウマンゾウ臼歯化石粘土層から（田中 1983）筆者が *R. gibberula* として報告した分類群は本種である：後期更新世・群馬県．
図　　版：奥平鉱泉（新治村）標本（現生・群馬県）．

Figs 1-8.　*Rhopalodia rupestris* (W. Smith) Krammer
LM. Figs 1-3. SEM. Figs 4-8. Material from Okudaira spring (Recent), Niiharu Village, Gunma, Japan. Scale bars: Fig. 6＝5 μm, Figs 4-5, 7＝2 μm, Fig. 8＝1 μm.

Figs 1-2.　Two different size valves.
Fig. 3.　A frustule (two valves).
Fig. 4.　External view of whole valve.
Fig. 5.　Oblique view of Fig. 4.
Fig. 6.　A frustule (two valves).
Fig. 7.　Internal oblique view.
Fig. 8.　Detailed internal view of valve center.

Plate 256

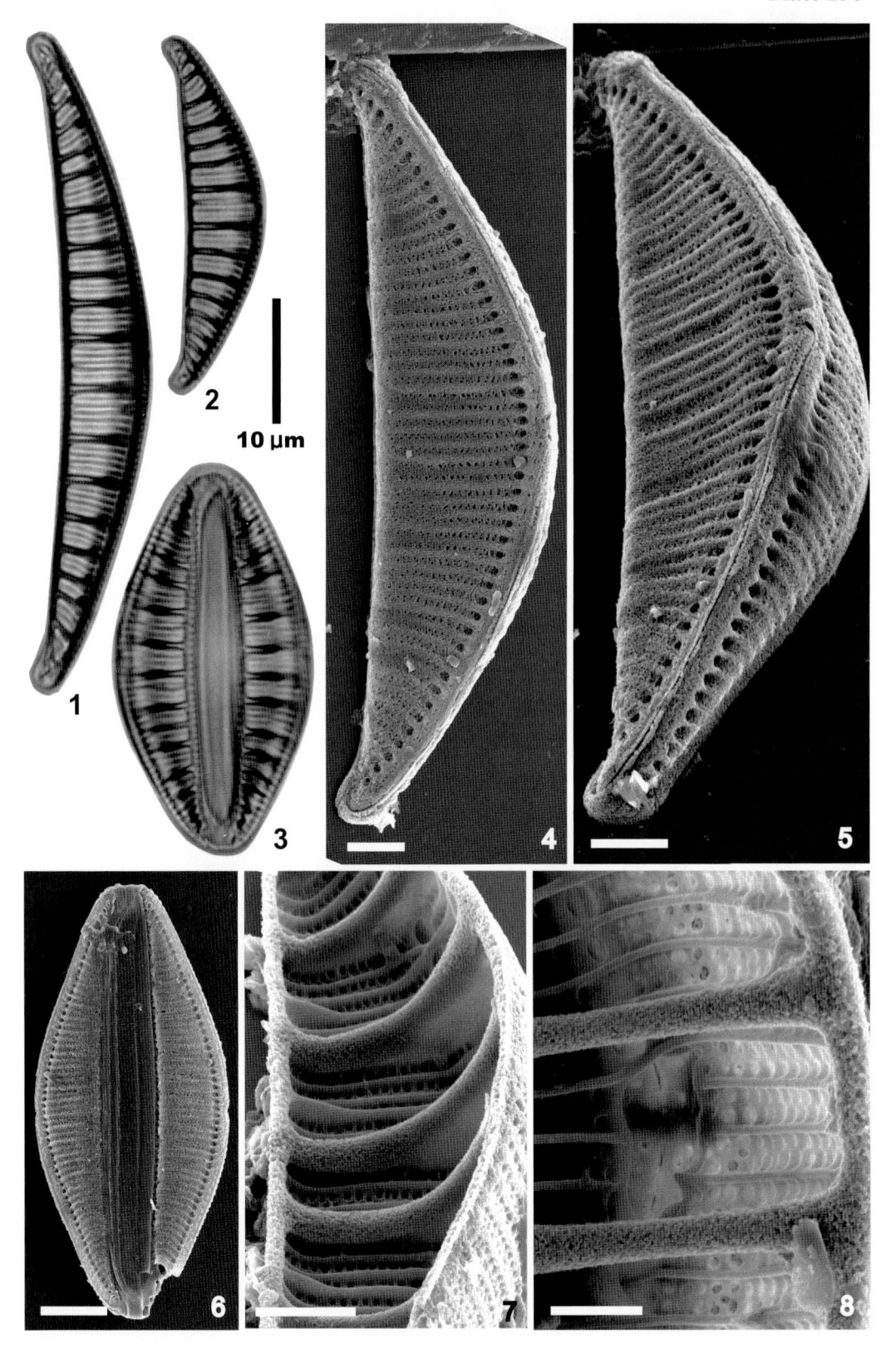

Plate 257. Biraphid, pennate diatoms: Surirellales
Campylodiscus echeneis **Ehrenberg** 1840

文　　献：Schmidt, A. 1899. *In*: A. Schmidt's Atlas der Diatomaceen-Kunde, Series 2, Hft. 14, pl. 54. O.R. Reisland, Leipzig.
　　Okuno, H. 1954. Electron-microscopic fine structure of fossil diatoms Ⅱ. Transactions and Proceedings of the Palaeontological Society of Japan, N.S., 14: 143-148, pls 17-18.

形　　態：殻はほぼ円形であるが強く鞍状に曲がる．殻径約 80 μm．縦溝は殻面／殻套境界に所在する．殻面には丸～横長の胞紋がおよそ放射状に分布し，殻套では縦長の胞紋が 1 列に並ぶ．

ノ　ー　ト：入手しやすい資料としては Schmidt(1899)，Okuno(1954) がある．Okuno(1954) は海水～汽水生，沿岸生と記している．

産　　出：稲城層：前期更新世・東京都，大阪層群（Okuno 1954）：更新世・大阪府．

図　　版：稲城層標本．

Figs 1-6. *Campylodiscus echeneis* **Ehrenberg**
LM. Figs 1-2. SEM. Figs 3-6. Materials from Inagi Formation (Early Pleistocene), Fuchu City, Tokyo, Japan. Scale bars: Figs 3-4＝10 μm, Fig. 6＝5 μm, Fig. 5＝2 μm.

Figs 1-2.　Same valve at different focal planes.
Fig. 3.　Side view of whole valve.
Fig. 4.　Side view, at different angle of Fig. 3, external raphe fissure (arrowhead).
Fig. 5.　Enlarged internal view of areolae.
Fig. 6.　Enlarged view of part of Fig. 4, raphe fissure (arrowhead).

Plate 257

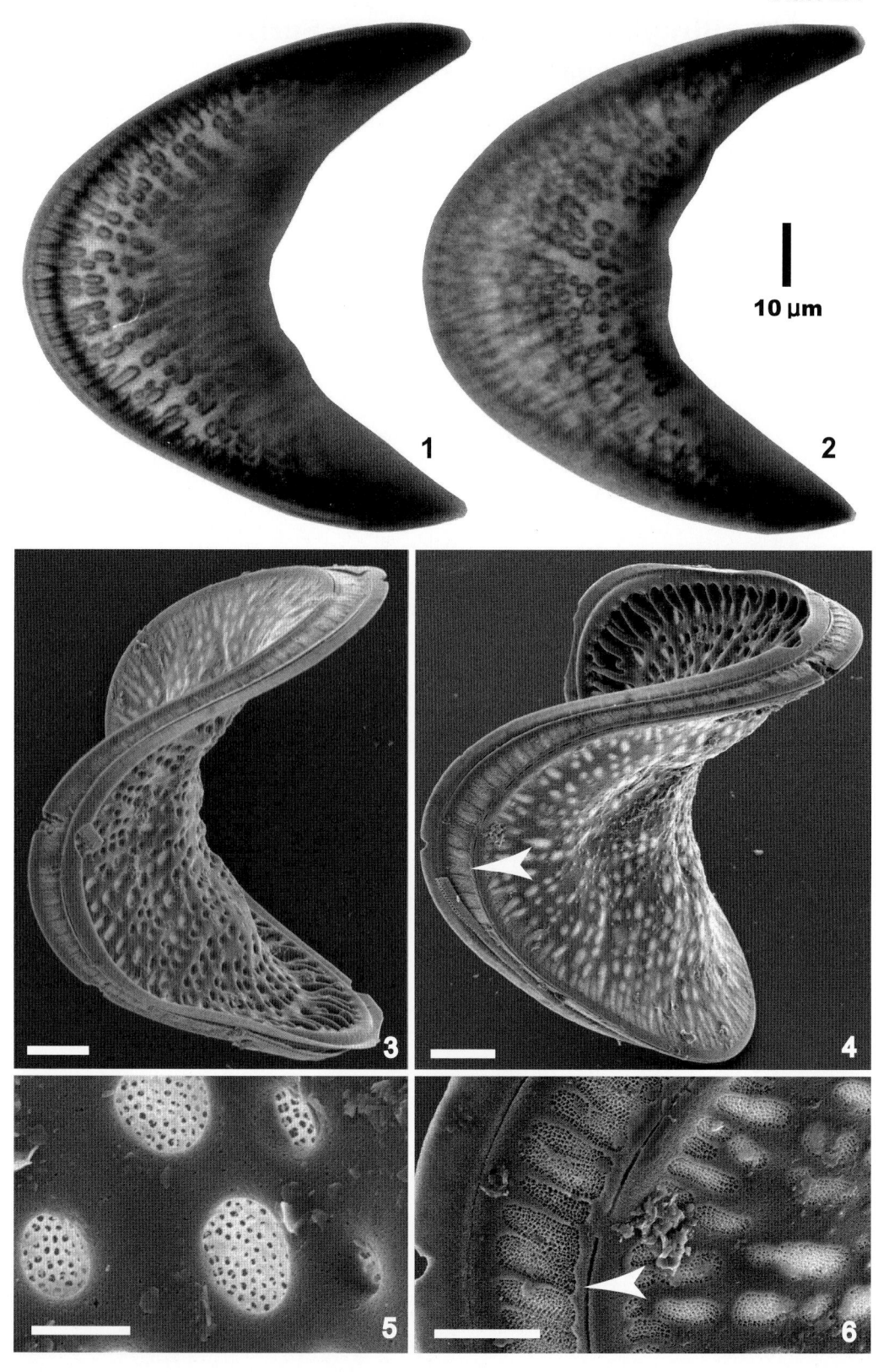

10 µm

Plate 258. Biraphid, pennate diatoms: Surirellales
Campylodiscus levanderi Hustedt 1925

文　献：Simonsen, R. 1987. Atlas and catalogue of the diatom types of Friedrich Hustedt 3 vols. Crammer, Berlin & Stuttgart.

形　態：殻はほぼ円形であるが強く鞍状に曲がる．殻径約 100 μm．光顕では殻面に殻縁方向へ向かう凹凸が観察できる．SEM 観察によるとこの凹凸には胞紋が分布している．また全体にわたって細かい針が分布する．

ノート：本属は淡水成層からも産出している．Negoro(1981)は未同定ではあるが古琵琶湖層群（佐山層・鮮新世）から，Ichikawa(1951)は *Campylodiscus cribrosus* を茅山粘土質頁岩から報告している．

産　出：屈斜路湖：北海道（現生），阿寒湖（河島・真山 2004）：北海道（現生）．

図　版：屈斜路湖標本．

Figs 1-5. *Campylodiscus levanderi* Hustedt
LM. Figs 1-2. SEM. Figs 3-5. Materials from Lake Kussharo (Recent), Hokkaido, Japan. Scale bars: Fig. 3＝10 μm, Fig. 5＝2 μm, Fig. 4＝1 μm.

Figs 1-2.　Same valve at different focal planes.
Figs 3.　Whole valve view.
Fig. 4.　Detailed internal view showing raphe endings, raphe (arrowhead).
Fig. 5.　Valve margin showing fenestrae and areolae rows.

Plate 258

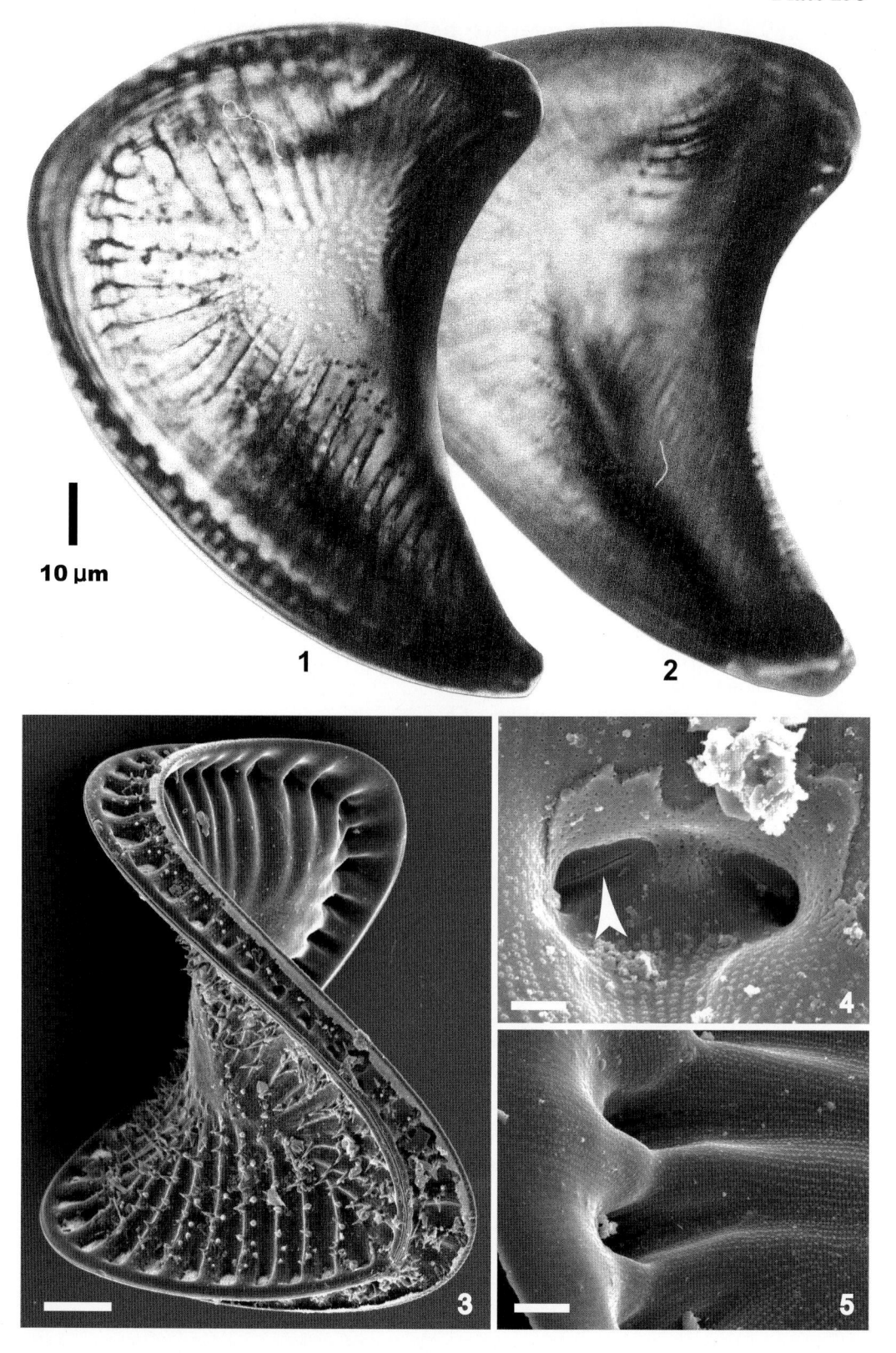

10 μm

Plate 259.   Biraphid, pennate diatoms: Surirellales
Cymatopleura elliptica (Brébisson) W. Smith 1851

基礎異名：*Surirella elliptica* Brébisson 1894
文　　献：Smith, W. 1851.   Notes on the Diatomaceae with descriptions of British species included in the genera *Campylodiscus, Surirella, Cymatopleura*. Annals and Magazine of Natural History **7**, ser. 2: 1-14, pls 1-3.
形　　態：殻は大形で楕円形，両殻端はややくさび形，殻長 110-180 μm，殻幅 58-94 μm．殻面は長軸方向に波打つ．管状縦溝が殻面／殻套境界に分布している．殻縁内側には肋がある．
ノ ー ト：大形で特徴のある形態の殻なので同定は容易である．本書では殻面観において殻中央部が膨れ，殻全体の形が楕円形の個体の写真を使用したが，Krammer & Lange-Bertalot (1988)および Hartley *et al.*(1996)では，殻中央で両側が平行の個体も本分類群に含めている．日本においては俣水層産の個体が中央で両側が平行であった．
産　　出：鬼首層（Ichikawa 1955）：更新世・宮城県，小野上層（田中・小林 1995）：前期更新世・群馬県，俣水層：前期更新世・大分県．
図　　版：小野上層標本．

Figs 1-3.   *Cymatopleura elliptica* (Brébisson) W. Smith
SEM. Figs 1-3. Materials from Onogami Formation (Early Pleistocene), Shibukawa City, Gunma, Japan. Scale bars: Fig. 1＝20 μm, Fig. 2＝10 μm, Fig. 3＝5 μm.

Fig. 1.   Whole valve view.
Fig. 2.   Oblique view of Fig. 1.
Fig. 3.   Enlarged oblique view of an apex of Fig. 1.

Plate 259

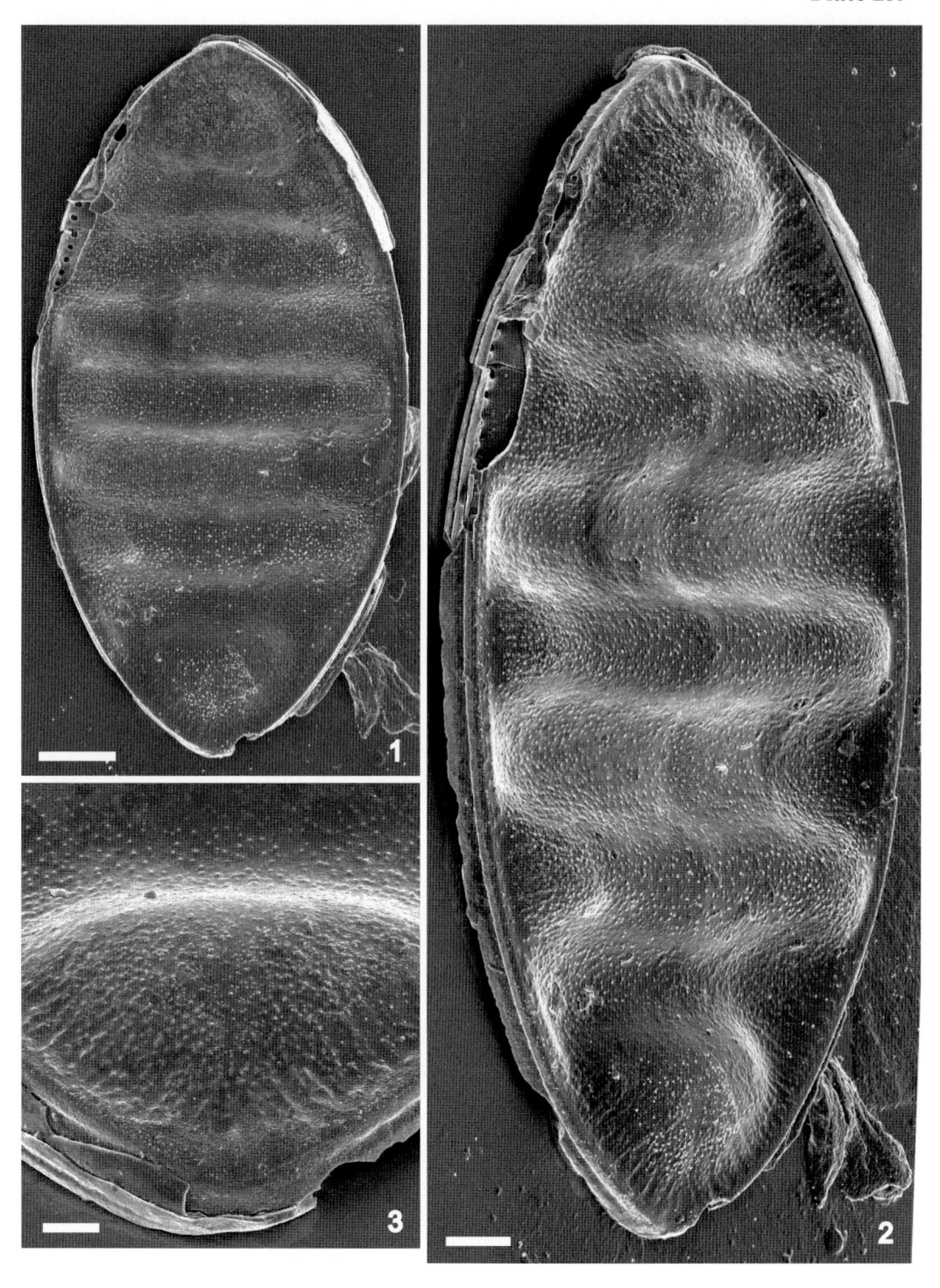

Plate 260. Biraphid, pennate diatoms: Surirellales
Cymatopleura solea (Brébisson) W. Smith 1851

基礎異名：*Cymbella solea* Brébisson 1838
文　　献：Smith, W. 1851. Notes on the Diatomaceae, with descriptions of British species included in the genera *Campylodiscus, Surirella, Cymatopleura.* Annals and Magazine of Natural History **7** ser. 2: 1-14, pls 1-3.
形　　態：殻は大形で線状〜幅の広い棒状，両殻端はくさび形，中央部がくびれる．殻長 109-189 μm，殻幅 19-25 μm．殻面は長軸方向に波打つ．条線は中央では平行で殻端では放射状になる．殻縁には肋線が観察され，その数 10 μm に 7-8 本．
産　　出：和村珪藻土（Skvortzov 1937）：新第三紀・長野県，津森層：中期更新世・熊本県．
図　　版：津森層標本．

Figs 1-6.　*Cymatopleura solea* (Brébisson) W. Smith
LM. Figs 1-2. SEM. Figs 3-6. Material from Tsumori Formation (Middle Pleistocene), Mashiki Town, Kumamoto, Japan. Scale bars: Figs 2, 4=10 μm, Fig. 6=5 μm, Fig. 5=2 μm.

Figs 1, 3-6.　Same valve, LM and SEM photographs.
Figs 1-2.　Two different size valves.
Fig. 3.　External view of whole valve.
Fig. 4.　Internal oblique view of whole valve.
Fig. 5.　Enlarged view of valve margin of Fig. 4.
Fig. 6.　Enlarged view of valve center of Fig. 3.

Plate 260

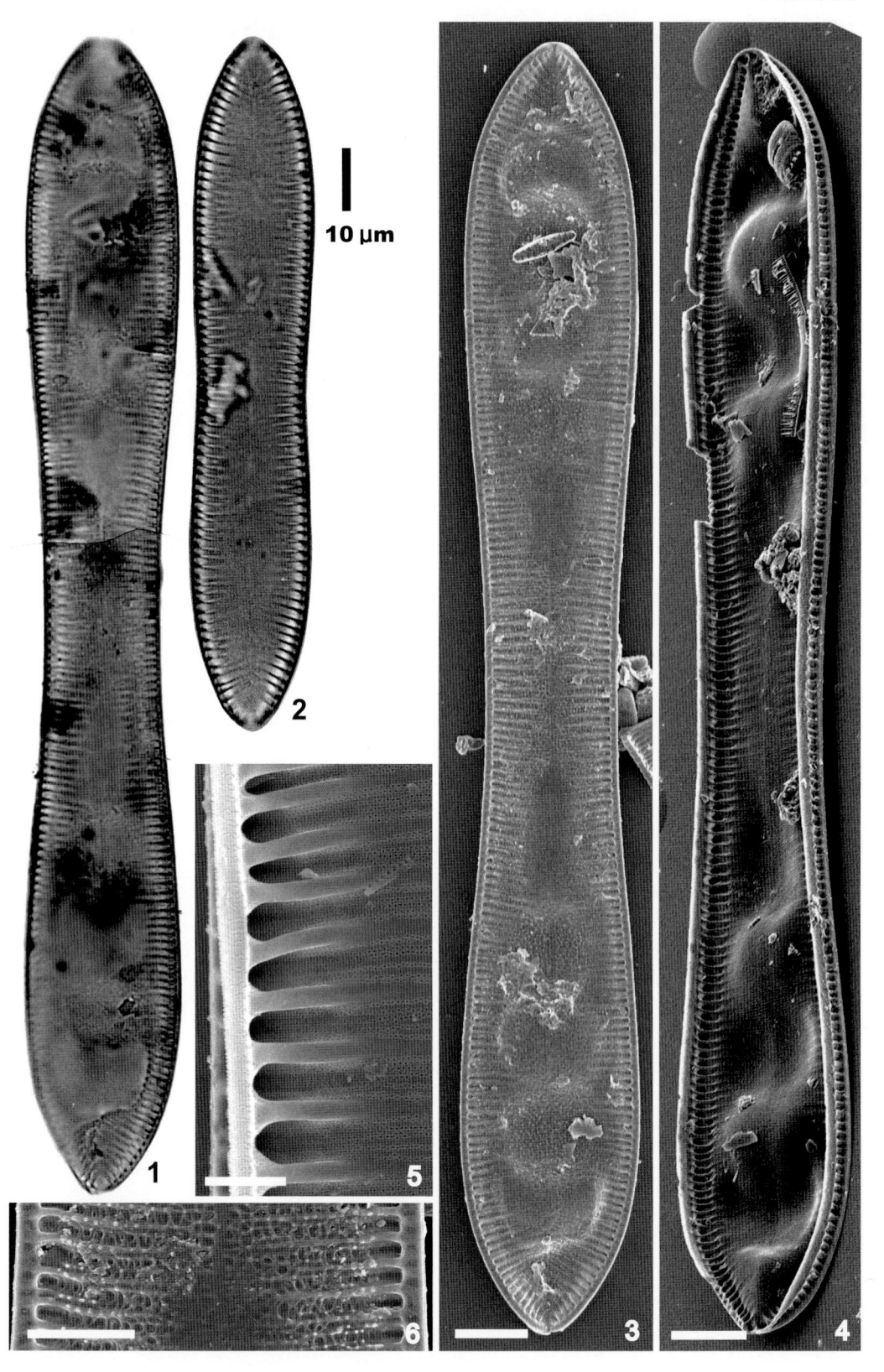

10 µm

Plate 261.　Biraphid, pennate diatoms: Surirellales
Surirella bifrons Ehrenberg 1843

文　　献：Krammer, K. & Lange-Bertalot, H. 1988.　Bacillariophyceae. Teil 2. Bacillariaceae, Epithemiaceae, Surirellaceae. 596 pp. Süßwasserflora von Mitteleuropa, Bd. **2/2**, Begründet von A. Pascher. Gustav Fischer Verlag, Stuttgart, New York.

形　　態：殻は大形で，殻面観は全体に丸みを帯びて膨らみ，上下対称である．両端は共に細いくちばし状で突出する．殻長 123-172 μm，殻幅 56-78 μm．殻面中央には長軸方向の線状隆起があり，両側からの翼管は放射状にこの隆起に向かうが，両殻端では放射の角度が大きくなる，翼管数 100 μm に 12-16 本．

ノ ー ト：Cleve-Euler(1952)による *Surirella biseriata* var. *tumida* Müller の図にも似るが，Krammer & Lange-Bertalot(1988)の *S. bifrons* Ehrenb. の図にも類似するので，最近の出版物による同定に揃えた．

産　　出：尾本層：前期更新世・大分県．

図　　版：尾本層標本．

Figs 1-5.　*Surirella bifrons* Ehrenberg

LM. Fig. 1. SEM. Figs 2-5. Material from Omoto Formation (Early Pleistocene), Kitsuki City, Oita, Japan. Scale bars: Figs 2, 5＝10 μm, Fig. 4＝5 μm, Fig. 3＝2 μm.

Fig. 1.　Whole valve view.
Fig. 2.　Internal view of whole valve.
Fig. 3.　Enlarged view of valve margin of Fig. 2 showing a fenestra.
Fig. 4.　External oblique view of an apex, raphe fissure (arrowhead).
Fig. 5.　External oblique view of whole valve, raphe fissure (arrowhead) and fenestra (arrow).

Plate 261

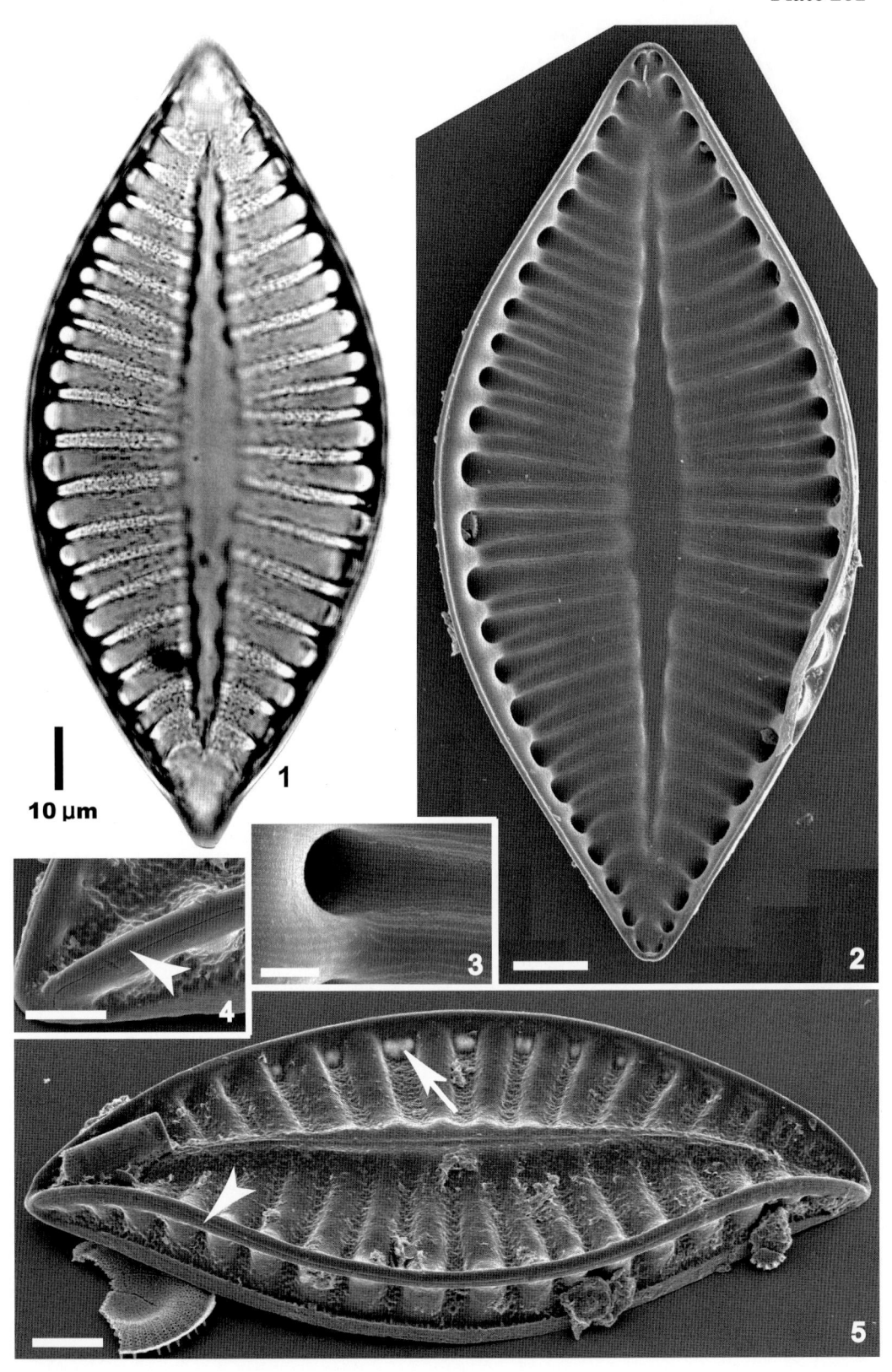

Plate 262.   Biraphid, pennate diatoms: Surirellales
Surirella splendida (Ehrenberg) Kützing 1844

基礎異名：*Navicula*（？）*splendida* Ehrenberg 1832
文　　献：Kützing, F.T. 1844. Die kieselschaligen Bacillarien oder Diatomeen. Nordhausen, W. Kohne. 152 pp. pls 1-30.
形　　態：殻は大形で，頭部は丸みを帯びたくさび形，殻面観は長卵形である．殻長 70-280 μm，殻幅 70-85 μm．殻面中央には長軸方向の線状隆起があり，両側からの翼管はこの隆起に向かい，中央では平行であるが両殻端では放射状になる，その数 100 μm に約 8 本である．
産　　出：鬼首層：後期更新世・宮城県．
図　　版：鬼首層標本．

Figs 1-6. *Surirella splendida* (Ehrenberg) Kützing

LM. Fig. 1. SEM. Figs 2-6. Material from Onikobe Formation (Late Pleistocene), Osaki City, Miyagi, Japan. Scale bars: Figs 2-3, 5-6=20 μm, Fig. 4=5 μm.

Fig. 1.　Whole valve view.
Fig. 2.　External view of whole valve.
Fig. 3.　Oblique view of Fig. 2.
Fig. 4.　Enlarged view of Fig. 3 showing fenestrae with fenestral bars and alar canal, fenestra (arrow).
Fig. 5.　Internal view of whole valve.
Fig. 6.　Oblique view of Fig. 5.

Plate 262

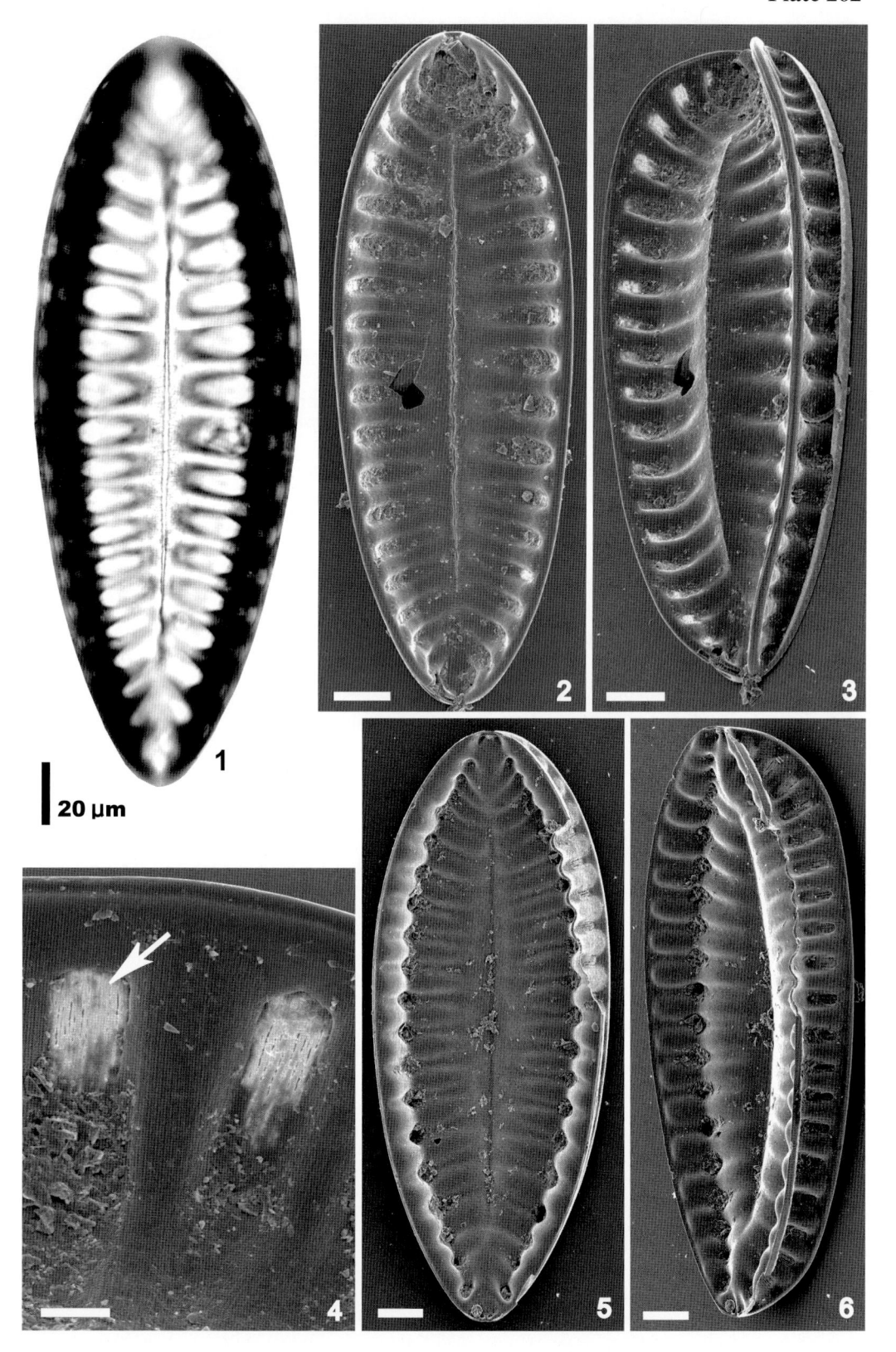

20 μm

Plate 263. Biraphid, pennate diatoms: Surirellales
Surirella robusta var. *splendida* f. *constricta* Hustedt in A. Schmidt *et al.* 1912

文　献：Hustedt, F. 1912. *In*: A. Schmidt's Atlas der Diatomaceen-Kunde (1874-1959), pl. 283. O.R. Reisland, Leipzig.

　　　　Simonsen, R. 1987. Atlas and catalogue of the diatom types of Friedrich Hustedt 3 vols. Crammer, Berlin & Stuttgart.

形　態：殻は大形で，殻面観は幅の広い線状であるが中央の左右が凹む，両殻端はやや丸みを帯びたくさび形である．殻長 250 μm，殻幅 60 μm．殻面中央には長軸方向の線状隆起があり，両側からの翼管はこの隆起に向かうが両殻端ではやや放射状になる，その数 100 μm に約 10 本．翼管は隆起しており，その間の翼窓からの線状部は相対的に凹む．

産　出：鬼首層：後期更新世・宮城県．

図　版：鬼首層標本．

Figs 1-5.　*Surirella robusta* var. *splendida* f. *constricta* Hustedt
LM. Fig. 1. SEM. Figs 2-5. Material from Onikobe Formation (Late Pleistocene), Osaki City, Miyagi, Japan. Scale bars: Figs 3, 5=20 μm, Fig. 2=10 μm, Fig. 4=2 μm.

Figs 1-5.　Same valve, LM and SEM photographs.
Fig. 1.　Whole valve view.
Fig. 2.　External view of whole valve.
Fig. 3.　Oblique view of Fig. 2.
Fig. 4.　Internal oblique view of whole valve.
Fig. 5.　Enlarged view of an apex of Fig. 4, raphe ending (arrowhead).

Plate 263

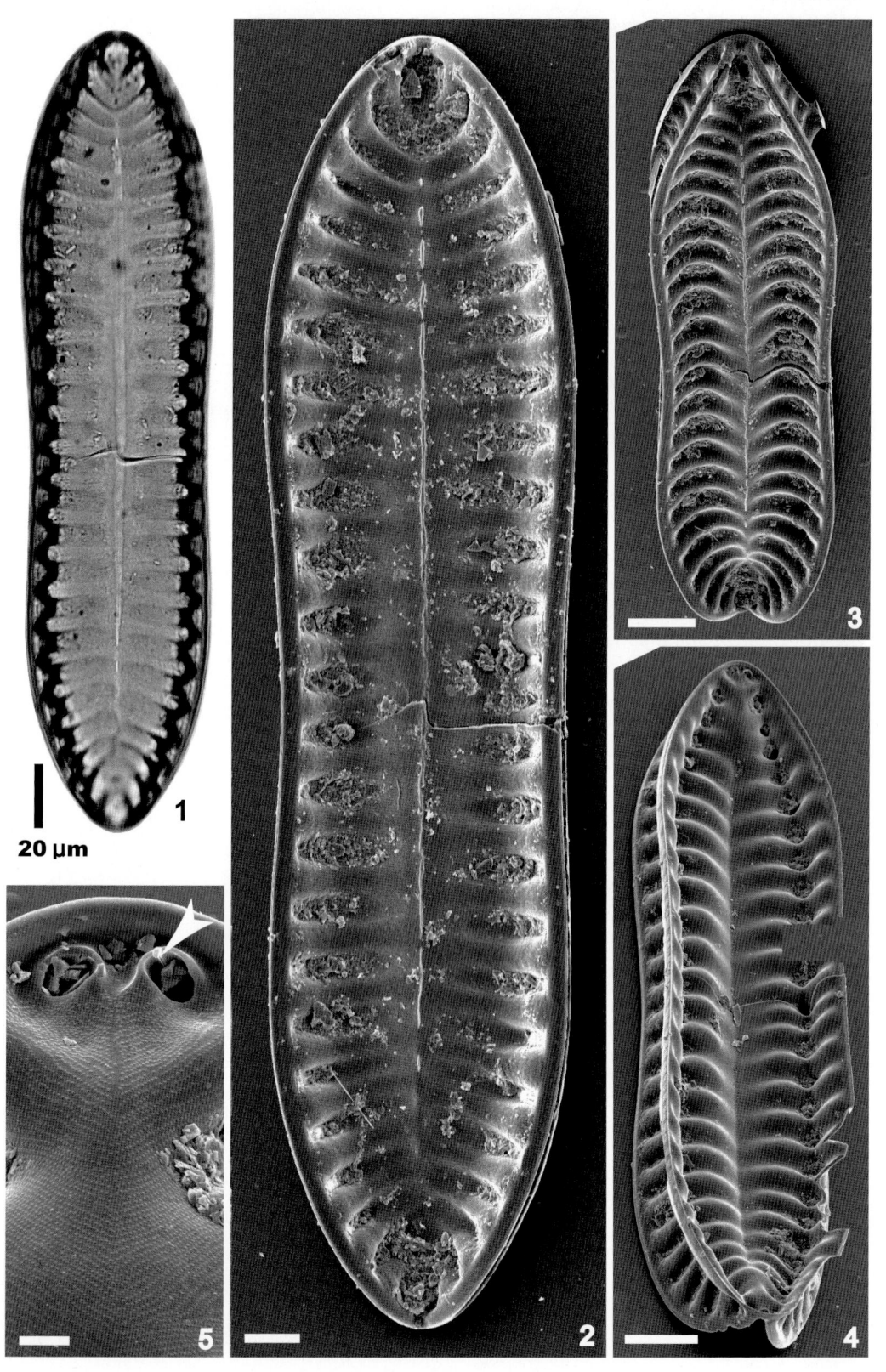

20 μm

Plate 264.   Biraphid, pennate diatoms: Surirellales
Surirella tenera W. Gregory 1856

文　　献：Schmidt, A. 1875. *In*: A. Schmidt's Atlas der Diatomaceen-Kunde(1874-1959), pl. 23. O.R. Reisland, Leipzig.

形　　態：殻は大形で，殻面観は異極性の細長い卵形，頭部は広円形，足部は幅の広いくさび形である．殻長 250 μm，殻幅 60 μm．殻面中央には長軸方向の線状隆起があり，両側からの翼管はこの隆起に向かうが両殻端では放射状になる，その数 100 μm に約 10 本．翼管は隆起しており，その間の翼窓からの線状部は相対的に凹むが両者とも幅はほぼ同じである．

産　　出：鬼首層：後期更新世・宮城県．

図　　版：鬼首層標本．

Figs 1-6.　*Surirella tenera* W. Gregory
LM. Fig. 1. SEM. Figs 2-6. Material from Onikobe Formation (Late Pleistocene), Osaki City, Miyagi, Japan. Scale bars: Figs 2-4＝20 μm, Figs 5-6＝5 μm.

Figs 1-6.　Same valve, LM and SEM photographs.
Fig. 1.　Whole valve view.
Fig. 2.　Internal view of whole valve.
Fig. 3.　External oblique view of whole valve.
Fig. 4.　Oblique view of Fig. 2.
Fig. 5.　Enlarged oblique view of an apex of Fig. 2.
Fig. 6.　Enlarged view of Fig. 3 showing fenestrae and alar canals, fenestra (arrow).

Plate 264

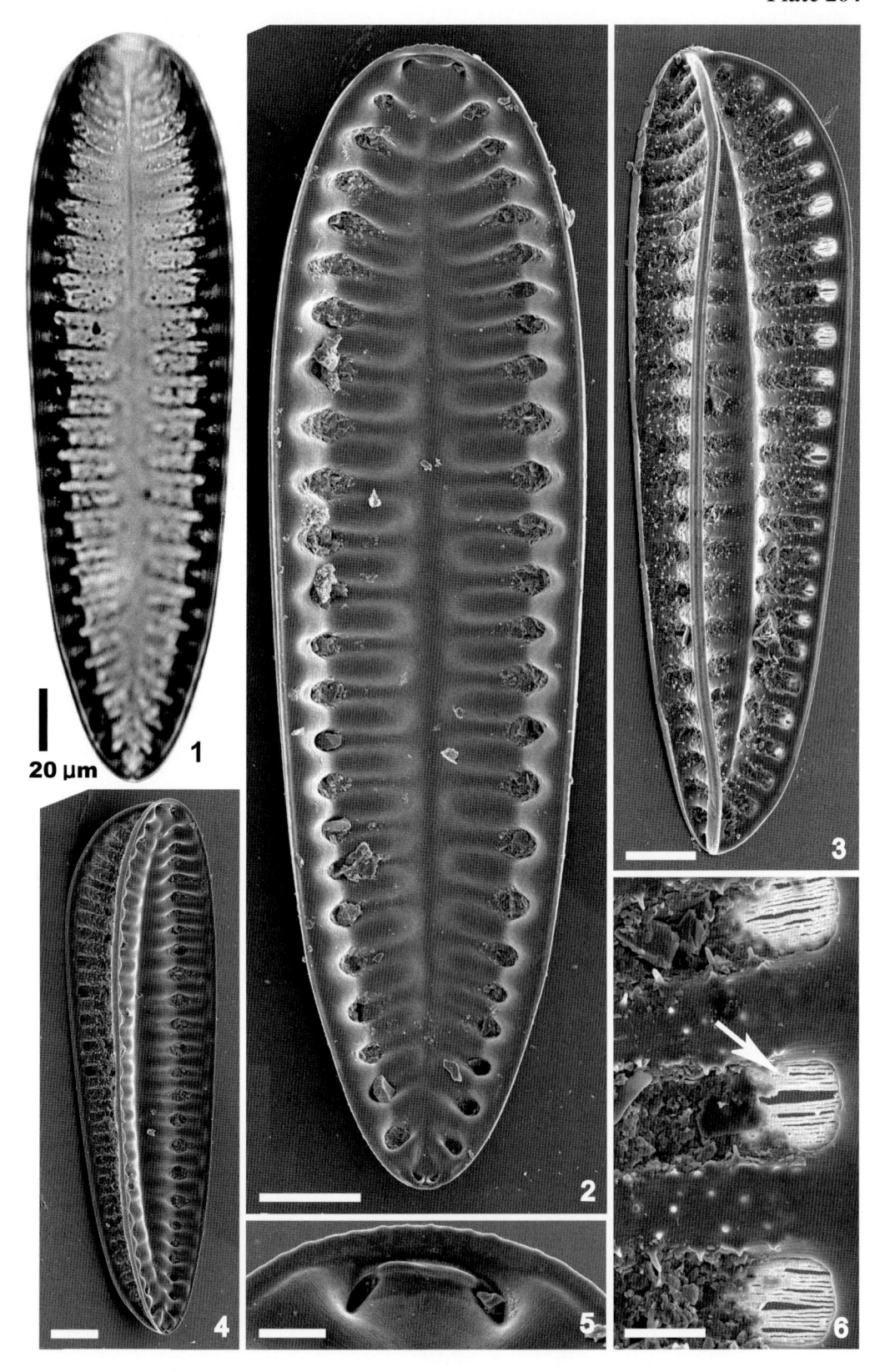

20 μm

Plate 265.   Biraphid, pennate diatoms: Thalassiophysales
Amphora copulata (Kützing) Schoeman & Archibald 1986

基礎異名：*Frustulia copulata* Kützing 1833

文　　献：Schoeman, F.R. & Archibald, R.E.M. 1986.   Observations on *Amphora* species (Bacillariophyceae) in the British Museum (Natural History). **V**. Some species from the subgenus *Amphora*. South African Journal of Botany **52**: 425-437.

Nagumo, T. 2003.   Taxonomic studies of the subgenus *Amphora* Cleve of the genus *Amphora* (Bacillariophyceae) in Japan. Bibliotheca Diatomologica **49**: 1-265.

形　　態：殻はやや細長い半月形で，背側が膨れ，腹側はやや凹むが中央ではわずか凸状になる．殻長 24-48 µm，殻幅 5.5-11 µm，外面での縦溝両末端は背側に曲がる．軸域は狭い．条線は背側で 10 µm に 12-14 本，腹側の中心域は殻面から殻端まで広がる．縦溝の背側では中心付近に条線が途切れる無紋域がある．胞紋は横に長く，長軸方向に数本の肋がある．

ノート：*A. affinis* Kütz, *A. ovalis* var. *affinis* (Kütz) Van Heurck, *A. ovalis* var. *libyca* (Ehrenb.) Cleve は本種の異名である（Nagumo 2003）．現生種についてではあるが，Nagumo (2003) で日本産の本属について詳しい記載がある．

産　　出：三徳層（Tanaka & Nagumo 2006）：後期中新世・鳥取県，鬼首層（Ichikawa 1995）：後期更新世・宮城県，琵琶湖ボーリングコア（Mori 1975）：更新世・滋賀県，和村珪藻土（上山・小林 1893）：新第三紀・長野県，嬬恋湖成層（高橋ら 1981）：中期更新世・群馬県．

図　　版：横子珪藻土標本．

Figs 1-6. *Amphora copulata* (Kützing) Schoeman & Archibald
LM. Figs 1-3. SEM. Figs 4-6. Material from diatomite of Yokogo (Early Pleistocene), Numata City, Gunma, Japan. Scale bars: Fig. 4＝5 µm, Figs 5-6＝2 µm.

Figs 1-3.   Three different size valves.
Fig. 4.   External view of whole valve.
Fig. 5.   Oblique view of part of Fig. 6.
Fig. 6.   Internal view of whole valve.

Plate 265

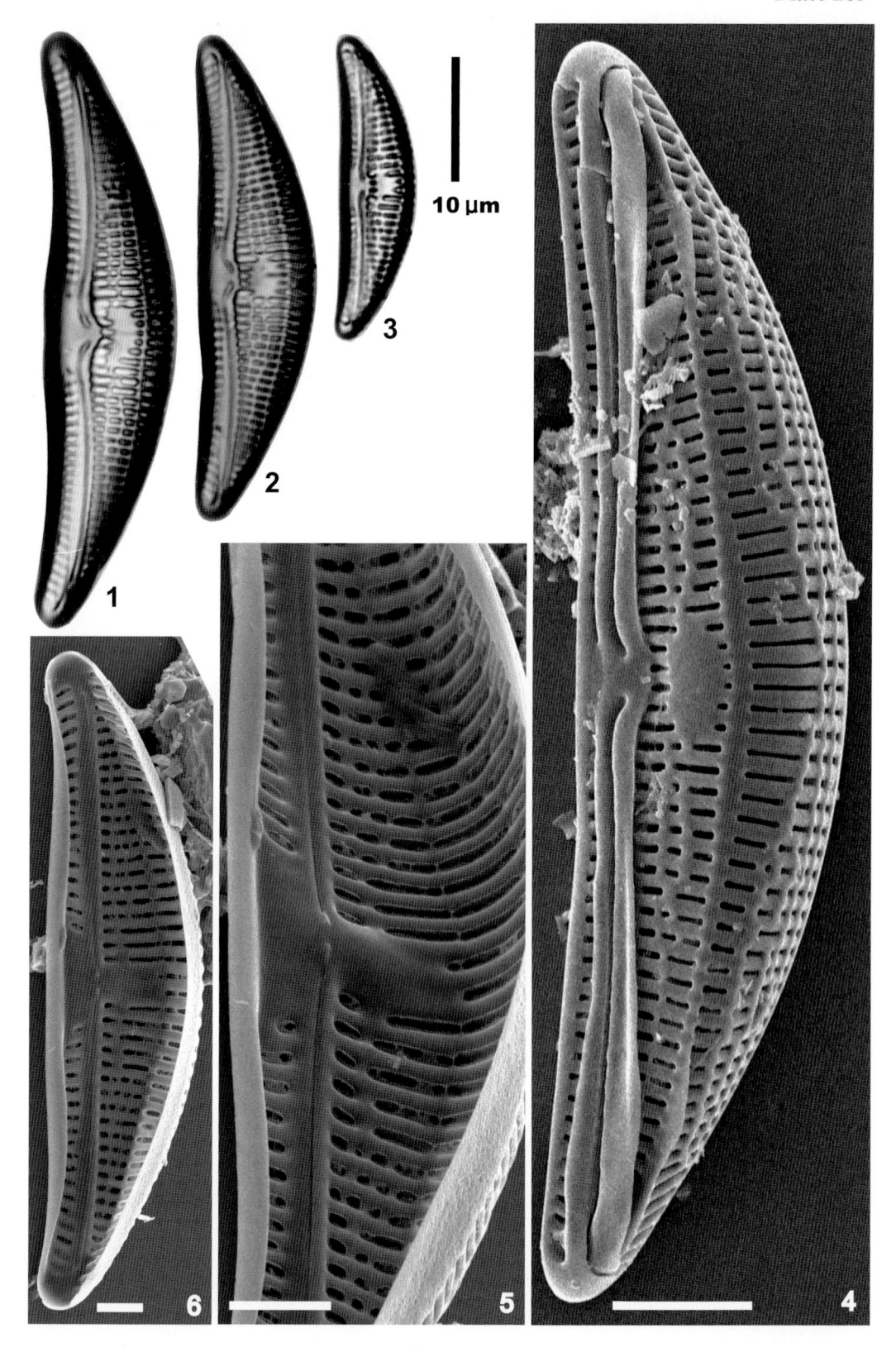

10 μm

Plate 266.   Biraphid, pennate diatoms: Thalassiophysales
Amphora veneta Kützing 1844

文　　献：Kützing, F.T. 1844. Die kieselschaligen Bacillarien oder Diatomeen. 152 pp. 30 pls. W. Köhne, Nordhausen.

形　　態：小形で被殻は卵形であるが，1個の殻の殻面は三日月形である．殻長 13-40 μm，殻幅 4-7 μm，横長の点紋から構成される条線は放射状に配列し，10 μm に 22-28 本であるが，中心付近ではやや粗くなる．

ノ ー ト：小林(1995)は金魚用水槽の本種について報告している．小林・吉田(1984)は本種を淡水にも汽水にも出現する広塩性着生種であると記している．

産　　出：星尾鉱泉（田中 1985）：群馬県（現生）．

図　　版：星尾鉱泉標本．

Figs 1-10.   *Amphora veneta* Kützing
LM. Figs 1-5. SEM. Figs 6-10. Material from Hoshio mineral spring (Recent), Nanmoku Village, Gunma, Japan. Scale bars: Figs 6-7, 9=2 μm, Figs 8, 10=1 μm.

Figs 1-4.   Four different size valves.
Fig. 5.   A frustule.
Fig. 6.   External view of whole valve.
Fig. 7.   Different angle view of Fig. 6.
Fig. 8.   Enlarged view of valve apex of Fig. 9.
Fig. 9.   Internal oblique view of whole valve.
Fig. 10.   Detailed internal view of valve center showing central nodule (arrow) and striae.

Plate 266

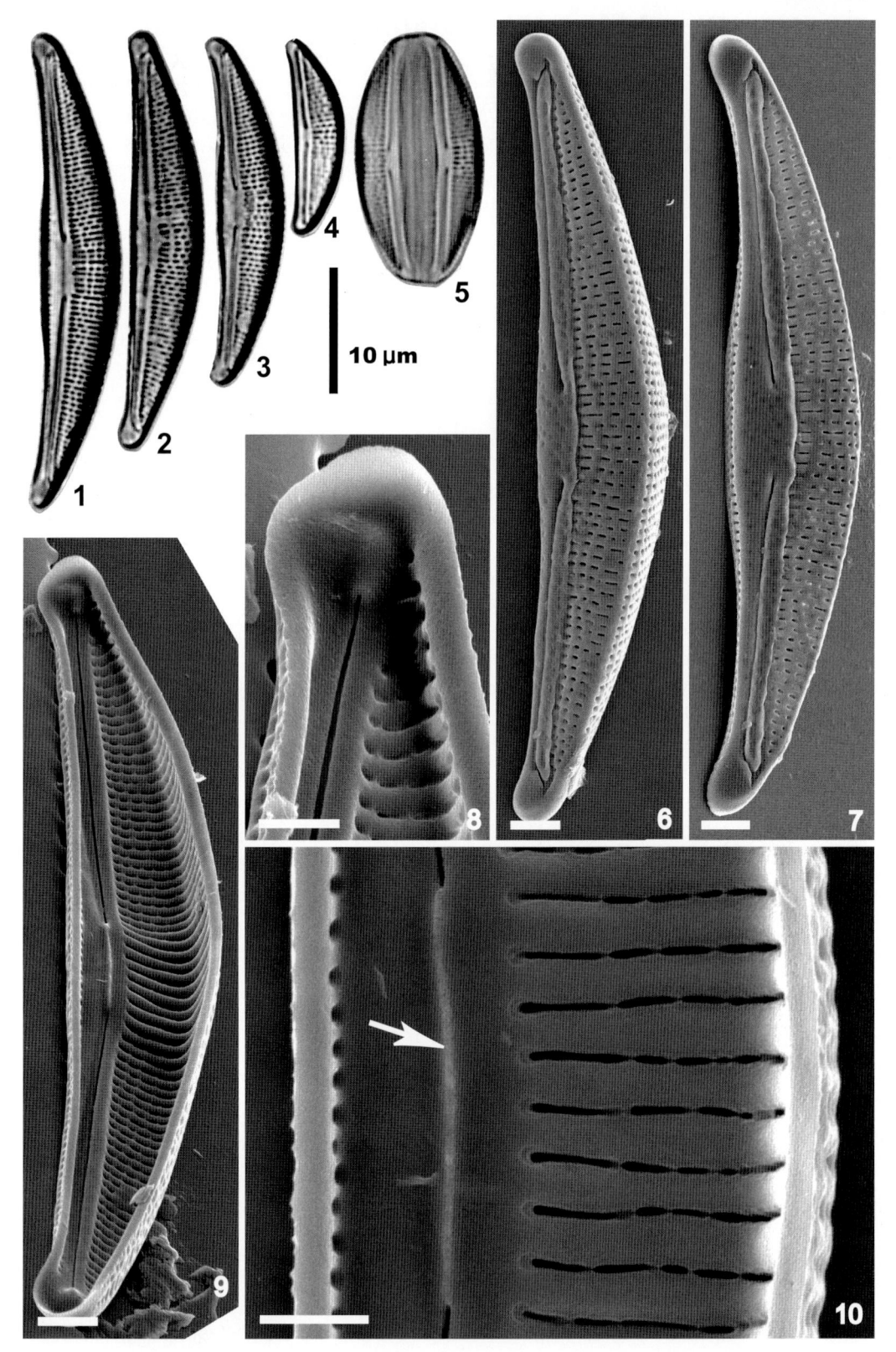

Plate 267.　Biraphid, pennate diatoms: Thalassiophysales
Denticula elegans f. *valida* Pedicino 1867

文　　献：Patrick, R. & Reimer, C.W. 1975. The diatoms of the United States 2(1). Monographs of the Academy of Natural Sciences of Philadelphia, no. 13. pp. 213. Philadelphia.

形　　態：小形で菱形に近い皮針形である，殻長 12.5-26.0 μm，殻幅 4.5-6.0 μm，肥厚線（横断肋）は 10 μm に 5-8 本，条線は 10 μm に約 18 本である．殻面は縦溝の所在する側が高くなっているので，ふつう斜めになっている．

ノ ー ト：海生の *Denticula* 属は化石・現生種ともよく研究されており，特に化石の本属は地層の堆積年代決定に重要である．淡水の化石は報告が少なく，本分類群は現生では各地の温泉（鉱泉）から知られているが，化石の報告は見当たらない．
　　　　本書では肥厚線（横断肋）の英文表記は *Denticula* 属で従来使用されてきた Pseudoseptum を使用している．

産　　出：赤久縄鉱泉（田中・中島 1985）：群馬県（現生）．

図　　版：赤久縄鉱泉標本．

Figs 1-11.　*Denticula elegans* f. *valida* Pedicino

LM. Figs 1-5. SEM. Figs 6-11. Material from Akaguna mineral spring (Recent), Kanna Town, Gunma, Japan. Scale bars: Figs 6-10＝2 μm, Fig. 11＝1 μm.

Figs 1-4.　Valve views of four different size valves.
Fig. 5.　Girdle view.
Fig. 6.　External view of whole valve.
Fig. 7.　Oblique view of Fig. 6, raphe (arrowhead).
Fig. 8.　Internal oblique view of whole valve, pseudoseptum (arrow).
Fig. 9.　Internal view of whole valve with band.
Fig. 10.　Oblique view of Fig. 9.
Fig. 11.　Enlarged view of Fig. 10, pseudoseptum (arrow).

Plate 267

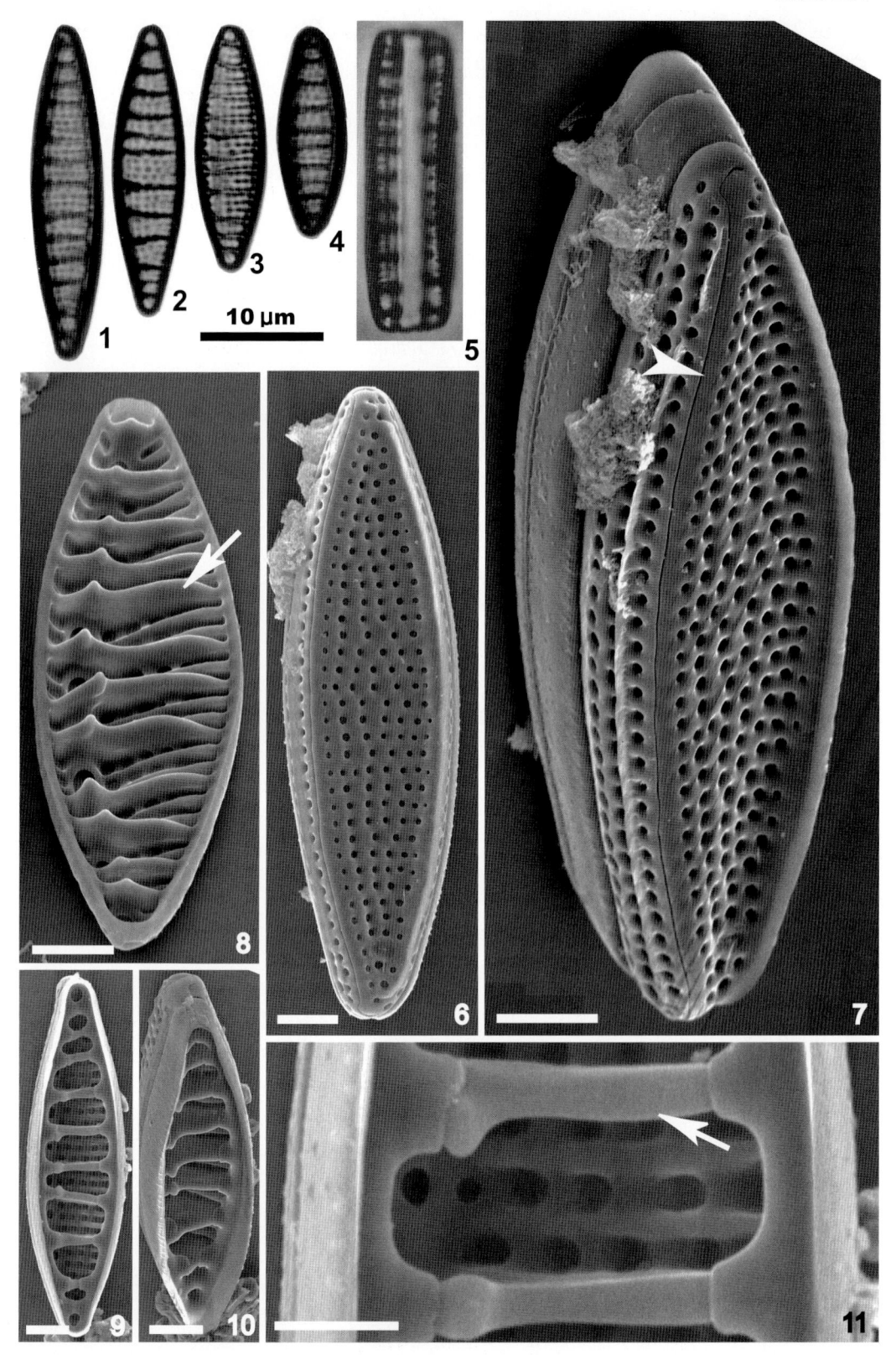

Plate 268. Biraphid, pennate diatoms: Thalassiophysales
Denticula tenuis Kützing 1844

文　献：Kützing, F.T. 1844. Die kieselschaligen Bacillarien oder Diatomeen. 52 pp. 30 pls. W. Köhne, Nordhausen.

形　態：殻形は小形で，殻端が丸みを帯びた細い皮針形である．殻長 22-27 μm，殻幅 4-5.5 μm，短軸方向に走る肥厚線（横断肋）は 10 μm に 5-6 本，条線は 10 μm に約 24 本である．殻面は短軸方向にうねる．縦溝は短軸方向へ偏って走る．

ノート：本種は冷水を好むとされる（Patric & Reimer 1975）．淡水からの *Denticula* 属産出は少ないが，野殿層（中期更新世・群馬県）から *Denticula kuetzingii* Grunow が報告されている（中島・南雲 1999）．

産　出：大戸湖沼性堆積物（田中 1991）：中期更新世・群馬県．

図　版：大戸湖沼性堆積物標本.

Figs 1-11. *Denticula tenuis* Kützing

LM. Figs 1-6. SEM. Figs 7-11. Material from lacustrine deposit of Odo (Middle Pleistocene), Higashiagatsuma Town, Gunma, Japan. Scale bars: Figs 7-10＝2 μm, Fig. 11＝1 μm.

Figs 1-2, 3-4, 5-6.　Three valves, different focal planes.
Fig. 7.　External view of whole valve, raphe (arrowhead).
Fig. 8.　Oblique view of Fig. 7.
Fig. 9.　Internal view of whole valve.
Fig. 10.　Oblique view of Fig. 9.
Fig. 11.　Enlarged view of Fig. 10, pseudoseptum (arrow).

Plate 268

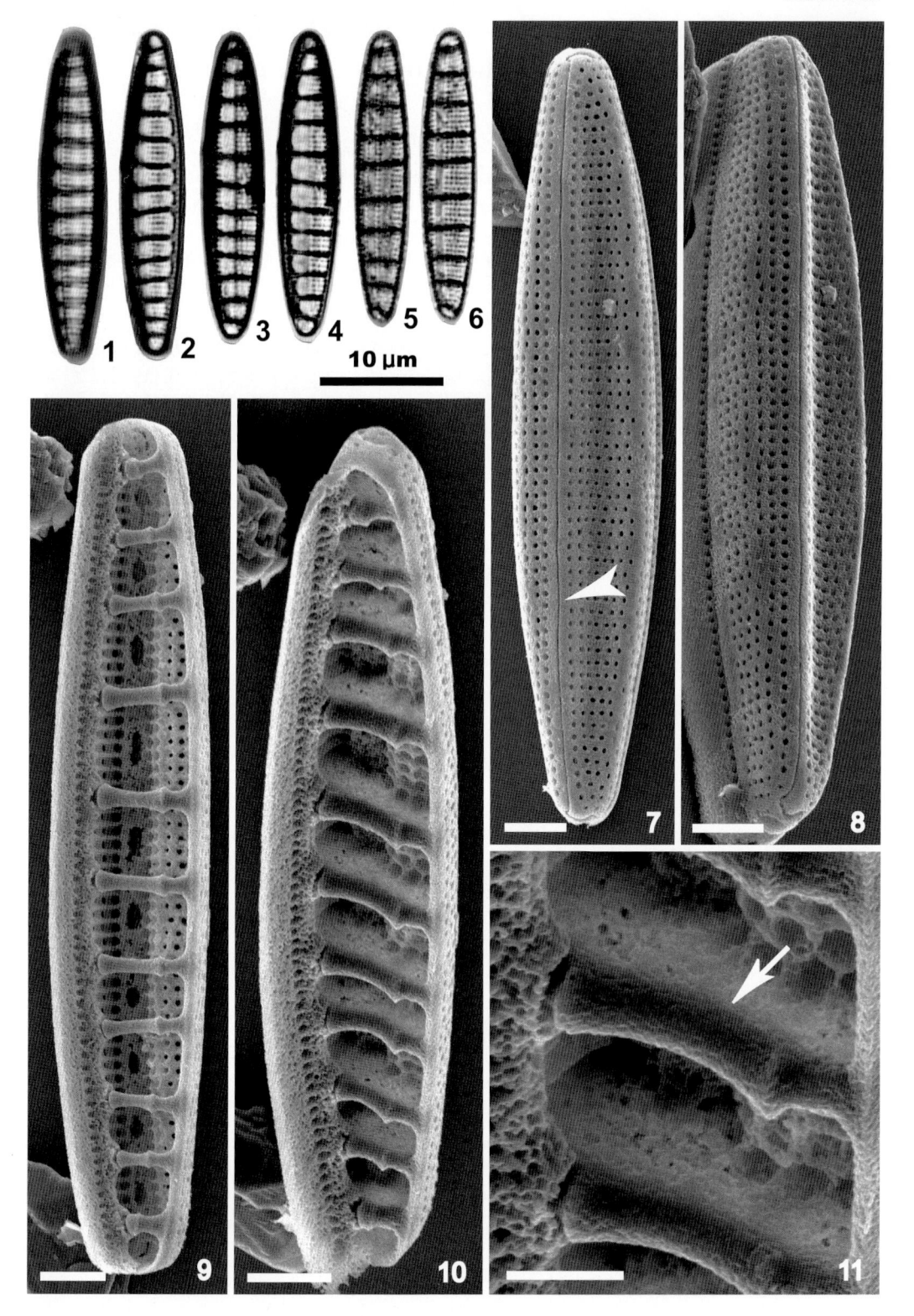

10 μm

Plate 269.　Biraphid, pennate diatoms: Thalassiophysales
Halamphora normanii (Rabenhorst) Levkov 2009

基礎異名：*Amphora normanii* Rabenhorst 1864
文　　献：安藤一男　1981．日本産コケ付着ケイソウ(4)．藻類 **29**: 201-207．
　　　　　Levkov, Z. 2009．*Amphora* sensu lato．*In*: Lange-Bertalot, H.(ed.) Diatoms of Europe. Diatoms of the Europian inland waters and comparable habitats **5**: 916 pp. A.R.G. Gantner, Verlag K.G.
形　　態：殻面はかなり細長い三日月形で先端が頭状に突出する．殻長 26-33 μm，殻幅 4-5 μm，やや横長の点紋から構成される条線はわずか放射状に配列し，背側で 10 μm に 18-20 本である．外中心裂溝は背側に曲がる．
ノ ー ト：本種は長く *Amphora* 属へ所属していたが，Levkov(2009)により *Halamphora* 属が設立され，本属へ組み合わせになった．*Amphora* 属としてであるが，安藤(1981)に，現生の日本における産出地・産出場所について詳しい記載があり，光顕写真も添えられている．
産　　出：野殿層（中島・南雲 1999）：中期更新世・群馬県，星尾鉱泉（田中 1985）：群馬県（現生）．
図　　版：星尾鉱泉標本．

Figs 1-8.　*Halamphora normanii* (Rabenhorst) Levkov
LM. Figs 1-3. SEM. Figs 4-8. Material from Hoshio mineral spring (Recent), Nanmoku Village, Gunma, Japan. Scale bars: Figs 4-5, 7＝5 μm, Fig. 8＝1 μm, Fig. 6＝0.5 μm.

Figs 1-2.　Valve views of two different size valves.
Fig. 3.　　A frustule.
Fig. 4.　　Oblique view of Fig. 5.
Fig. 5.　　External view of whole valve.
Fig. 6.　　Detailed view of areolae of Fig. 7.
Fig. 7.　　Internal slightly oblique view of whole valve.
Fig. 8.　　Enlarged internal oblique view of valve center, central nodule (arrow).

Plate 269

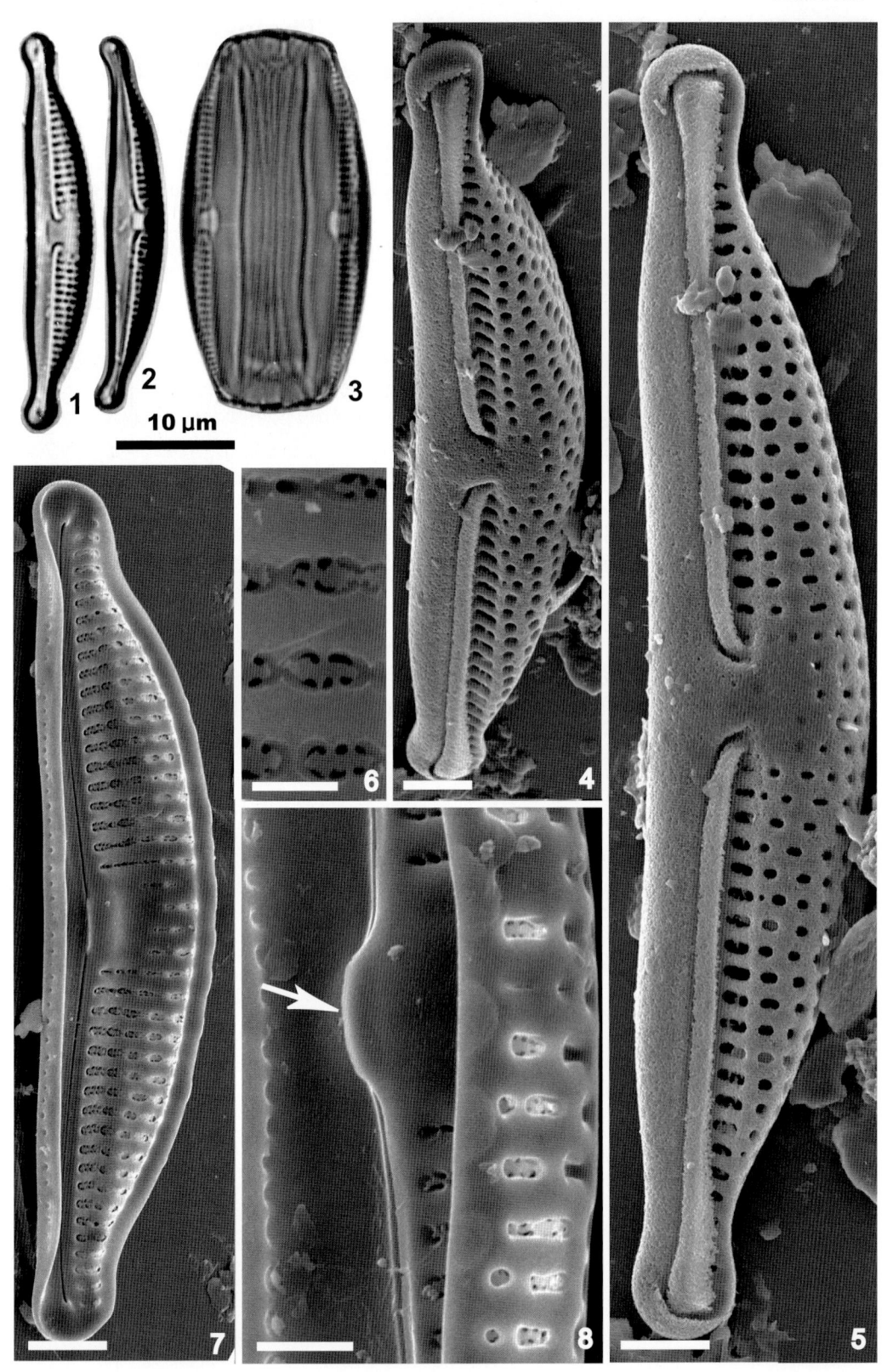

10 μm

Plate 270. Biraphid, pennate diatoms: Thalassiophysales
Nitzschia heidenii (F. Meister) Hustedt in A. Schmidt *et al.* 1924

基礎異名：*Nitzschia moissacensis* var. *heidenii* F. Meister 1914

文　　献：Meister, F. 1914. Beiträge zur Bacillariaceenflora, Japan. Archiv für Hydrobiologie und Planktonkunde Bd. **9**: S. 226-232.

　　　　Hustedt, F. 1924. *In*: A. Schmidt's Atlas der Diatomaceen-Kunde(1874-1959), Tafel 351. O.R. Reisland, Leipzig.

形　　態：殻は狭皮針形で殻端は小頭状．光顕での観察では，殻の片側に短軸方向へ太い黒線が殻幅の 1/3-1/2 程度にあり，条線はやや細長い点紋の列から構成されている．殻長 23-66 μm，殻幅約 8 μm．光顕で太い黒線に見えた間板は 10 μm に(3)4-6(7)個．条線は 10 μm に約 16 本．縦溝は管状縦溝で短軸方向へ偏って走るが，この部分は強く突出している．

ノ ー ト：原記載の Meister(1914)の図では両殻端が強く頭状に突出している．横子珪藻土産の個体はそれに比べると突出が弱いが，最も類似した形態をしている．東京大学植物園が原産地の日本固有種（小林 1960）であり，池田湖（Skvortzow 1937）や尾瀬沼等広く産し，化石としても見出されている．

産　　出：横子珪藻土：前期更新世・群馬県，大鷲湖沼性堆積物（田中ら 2011）：鮮新世・岐阜県，有井の珪藻土（三重県・完新世）および山川の珪藻土（鹿児島県・更新世）（Okuno 1952）．

図　　版：横子珪藻土標本．

Figs 1-8.　*Nitzschia heidenii* (F. Meister) Hustedt

LM. Figs 1-3. SEM. Figs 4-8. Material from diatomite of Yokogo (Early Pleistocene), Numata City, Gunma, Japan. Scale bars: Figs 4-6＝2 μm, Figs 7-8＝1 μm.

Figs 1-3.　Three different size valves.
Fig. 4.　External oblique view of whole valve, raphe (arrowhead).
Fig. 5.　Internal view of whole valve.
Fig. 6.　Oblique view of Fig. 5.
Fig. 7.　Enlarged oblique view of an apex of Fig. 4, raphe (arrowhead).
Fig. 8.　Detailed view of part of Fig. 5, raphe (arrowhead).

Plate 270

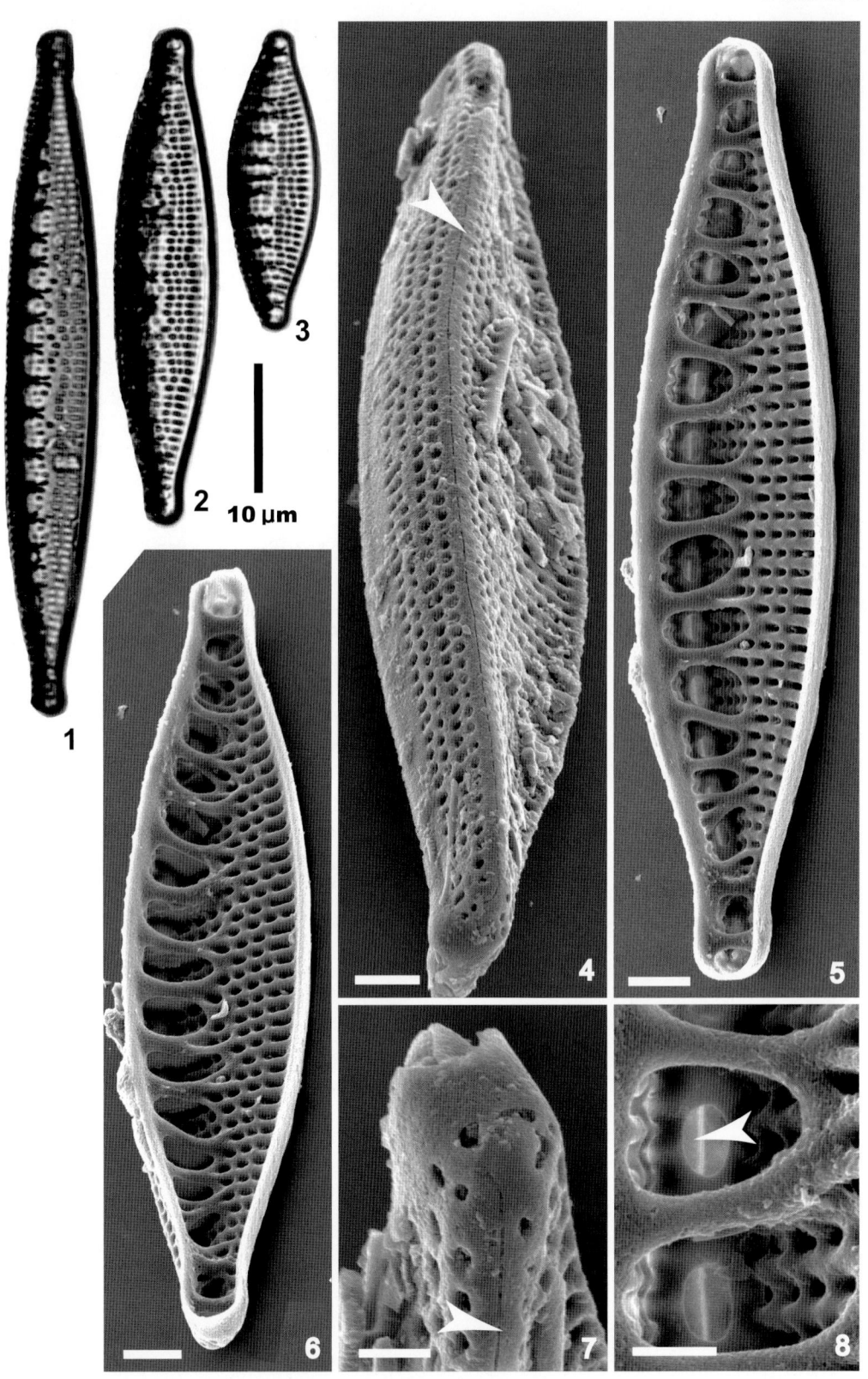

Plate 271. Biraphid, pennate diatoms: Thalassiophysales *Nitzschia tabellaria* (Grunow) Grunow in Cleve & Grunow 1880

基礎異名：*Denticula tabellaria* Grunow 1862

文　　献：Cleve, P.T. & Grunow, A. 1880.　Beiträge zur Kenntnis der arctischen Diatomeen. Kungliga Svenska Vetenskaps-Akademiens Handlingar **17**(2): 121 pp, 7 Taf.
　Kobayasi, H., Kobori, S. & Sunaga, S. 1994.　Taxonomy and morphology of two forms of the Nitzschia sinuata complex. pp. 281-289. *In*: J.P. Kociolek (ed.) Proceedings of the 11th International Diatom Symposium. California Axademy of Sciences, San Francisco.

形　　態：小形で殻中央部が短軸方向へ強く膨れる，また殻端は小頭状である．殻長 22-24 μm，殻幅約 8 μm．短軸方向の黒線（間板）は 10 μm に 6-7 個．条線は点紋列からなり，10 μm に 19-22 本．縦溝は管状縦溝で短軸方向へ偏って走るが，この部分は強く突出している．

ノ ー ト：Kobayasi *et al.*(1994)に日本産（一部外国産を含む）標本を使用して本種の形態が詳述されている．*Nitschia sinuata* var. *tabellaria*（Grunow）Grunow は本種の異名である（Kobayasi *et al.* 1994）．

産　　出：大戸湖沼性堆積物（田中 1991）：中期更新世・群馬県．

図　　版：大戸湖沼性堆積物標本．

Figs 1-9.　*Nitzschia tabellaria* (Grunow) Grunow

LM. Figs 1-4. SEM. Figs 5-9. Material from lacustrine deposit of Odo (Middle Pleistocene), Higashiagatsuma Town, Gunma, Japan. Scale bars: Figs 5-8＝2 μm, Fig. 9＝1 μm.

Figs 1-2.　External view, same valve at different focal planes.
Figs 3-4.　Internal view, same valve at different focal planes.
Fig. 5.　External view of whole valve, raphe (arrowhead).
Fig. 6.　Internal view of whole valve.
Fig. 7.　Oblique view of Fig. 6.
Fig. 8.　Oblique view of Fig. 5.
Fig. 9.　Enlarged view of part of Fig. 6 showing inner fissure of raphe (arrowhead).

Plate 271

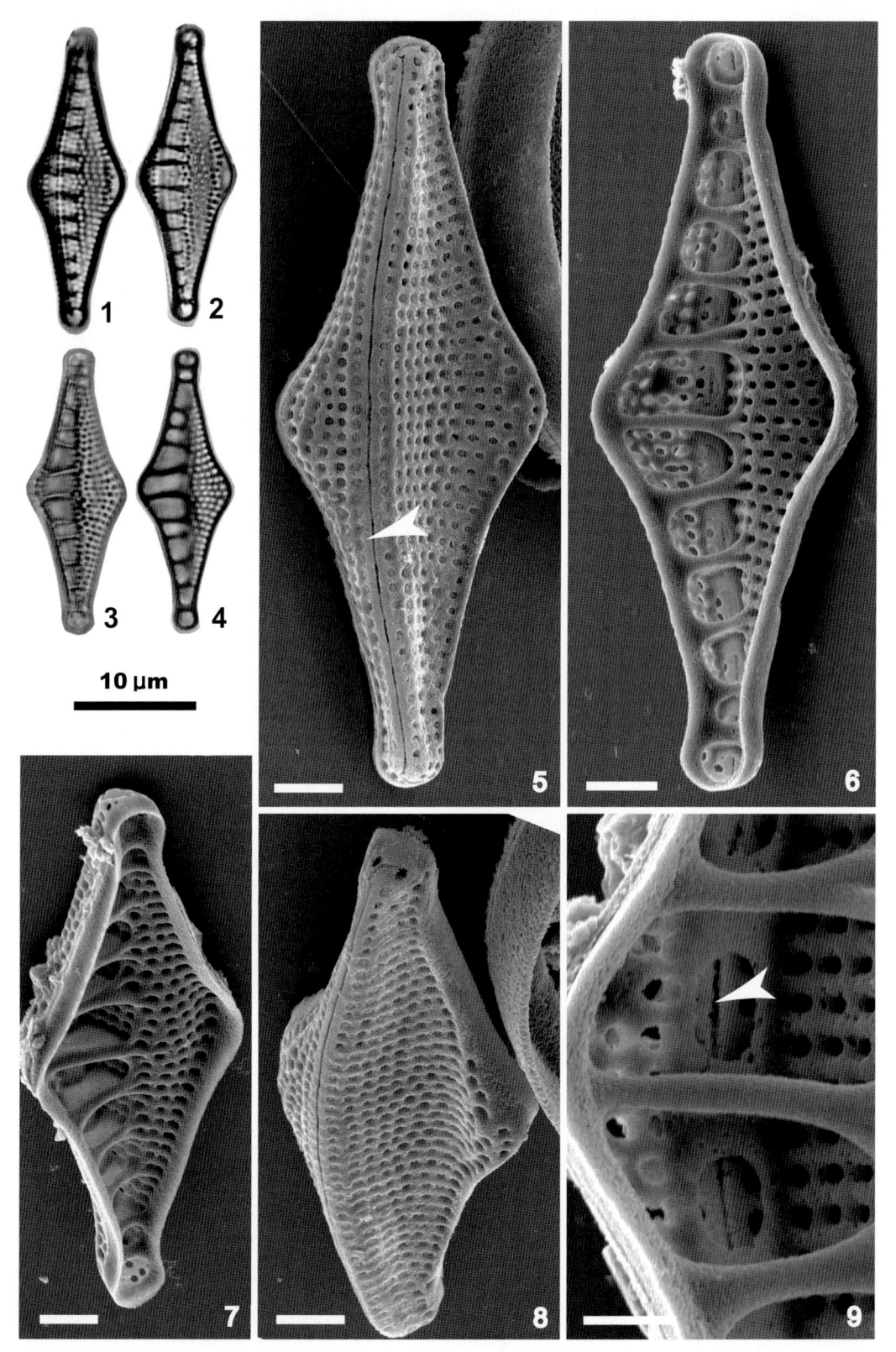

10 μm

VIII. 引用文献：List of References

赤城三郎・山名　巖・平尾澄昌・広田昌昭・衣笠弘直 1984. 鳥取県三朝町成より産する後期中新世の植物化石. 鳥取大学教育学部研究報告 自然科学 **33**: 49-69, 11 pls.

秋葉文雄 2008. 幻の仙台産石灰質団塊―珪藻化石層序の話―. 地質ニュース 648 号: 62-71.

Akutsu, J. 1964. The geology and paleontology of Shiobara and its vicinity, Tochigi Prefecture. Science Reports, Tohoku University 2nd Series (Geology) **35**: 211-293, pls 57-66.

安藤一男 1969. 埼玉県百穴湖・吉見湖・和名湖のケイソウ. 秩父自然科学博物館研究報告 **1969** (15): 55-63.

安藤一男 1981. 日本産コケ付着ケイソウ(4). 藻類 **29**: 201-207.

安藤一男・原口和夫・小林　弘 1971. 埼玉県仙女ガ池のケイソウ. 秩父自然科学博物館研究報告 **16**: 57-79.

Andrews, G.W. 1970. Late Miocene nonmarine diatoms form the Kilgore area, Cherry County, Nebraska: U.S. Geologocal Survey Professional Paper 683-A: 24, 3 pls.

Brant, L.A. & Furey, P.C. 2011. Morphological variation in *Eunotia serra*, with a focus on the rimoportula. Diatom Research **26**: 221-226.

Brummitt, R.K. & Powell, C.E. 1992. Authors of plant names. A list of authors of scientific names of plants, with recommended standard forms of thier names, including abbreviations. 732 pp. Royal Botanic Gardens, Kew.

Brun, J. 1891. Diatomées espèces nouvelles marines, fossiles ou pélagiques. Mémoires de la Société de Physique et D'Histoire Naturelle de Genève **31**: 1-47, pls 11-22.

Brun, J. & Tempére, J. 1889. Diatomées fossiles du Japon, espèces marines & nouvelles des calcaires argileux de Sendaï & de Yedo. Mémoires de la Société de Physique et D'Histoire Naturelle de Genève **30**: 1-75, pls 1-9.

Cleve, P.T. 1895. Synopsis of the naviculoid diatoms. Part II. Konglica Svenska Vetenskaps-Akademiens Handlingar, Bd. **27**: pp. 1-219, pls 1-4.

Cleve, P.T. & Grunow, A. 1880. Beiträge zur Kenntniss der arctischen Diatomeen. Kongliga Svenska Vetenskaps-Akademiens Handlingar **17** (2): 1-121, pls 1-7.

Cleve-Euler, A. 1952. Die diatomeen von Schweden und Finnland. Kungl. Svenska Vetenskapsademiens Handlingar Band **3** (3): pp. 1-153, Figs 1318-1583, Taf. 7.

Compère, P. 1982. Taxonomic revision of the diatom genus *Pleurosira* (Eupodiscaceae) *Bacillaria* **5**: 165-190.

Compère, P. 2001. *Ulnaria* (Kützing) Compère, a new genus name for *Fragilaria* subgen. *Alterasynedra* Lange-Bertalot with comments on the typification of *Synedra* Ehrenberg. *In*: Jahn, R., Kociolek, J.P., Witkowski, A. & Compère, P. (eds) Lange-Bertalot-Festschrift. pp. 97-101. A.R.G. Gantner, Ruggel.

Crawford, R.M. 1988. A reconsideration of *Melosira arenaria* and *M. teres*; resulting in a proposed new genus *Ellerbeckia*. *In*: Round, F.E. (ed.) Algae and the aquatic environment. pp. 413-433. Biopress, Bristol.

Crawford, R.M., Likhoshway, Y.V. & Jahn, R. 2003. Morphology and identity of *Aulacoseira italica* and typification of *Aulacoseira* (Bacillariophyta). Diatom Research **18**: 1-19.

Cremer, H. & Vijver, B.V. 2006. On *Pliocaenicus costatus* (Bacillariophyceae) in Lake El'gygytgyn, East Siberia. European Journal of Phycology **41**: 169-178.

Cremer, H. & Vijver, B.V. 2007. *Pliocaenicus costatus*: an emended species description. *In*: Kusber, W.-H. & Jahn, R. (eds) Proceedings of the 1st Central European Diatom Meeting 2007. pp. 35-38. Botanic Garden and Botanical Museum Berlin-Dahlem, Freie Universität Berlin, Berlin.

Edlund, M.B., Williams, R.M. & Soninkhishig, N. 2003. The planktonic diatom diversity of ancient Lake Hovsgol, Mongolia. Phycologia **42**: 232-260.

Ehrenberg, C.G. 1838. Die Infusionsthierchen als vollkommende Organismen: Ein Blick in das tiefere organische Leben der Natur. 548 pp. Voss, Leipzig.

Ehrenberg, C.G. 1843. Verbreitung und Einfluss des mikroskopischen Lebens in Süd- und Nord-Amerika. Abh. K. Akad. Wiss. Berlin 1841 (1): 291-446.

Fricke, F. 1904. *In*: A. Schmidt's Atlas der Diatomaceen-Kunde, pl. 247. O.R. Reisland, Leipzig.

Fricke, F. 1906. *In*: A. Schmidt's Atlas der Diatomaceen-Kunde, pl. 267. O.R. Reisland, Leipzig.

藤田　崇 1973. 鳥取県中部の新第三系について. 地質学論集 **9**: 159-171.

福島　博・小林艶子・寺尾公子 1984. 羽状ケイ藻 *Navicula confervacea* (Kütz.) Grunow の分類学的検討(1). 日本水処理生物学会誌 **20**: 20-33.

Fukushima, H., Ko-Bayashi, T., Terao, K. & Yoshitake, S. 1988. Morphological variability of *Diatoma vulgare* Bory var. *grande*. *In*: Round, F.E. (ed.) Proceedings of the 9th International Diatom Symposium, Bristol 1986. pp. 377-389. Biopress, Bristol.

福島　博・小林艶子・吉武佐紀子 2002. 温泉産新種珪藻, *Navicula tanakae* Fukush., Ts. Kobay. & Yoshit. nov. sp. について. Diatom **18**: 13-21.

Gaiser, E.E. & Johansen, J. 2000. Freshwater diatoms from Carolina Bays and other isolated wetlands on the Atlantic costal plain of South Carolina, U.S.A., with descriptions of seven taxa new to science. Diatom Research **15**: 75-130.

後藤敏一 2003. 学名の著者名の標準的な略号. Diatom **19**: 71-74.

Grunow, A. 1862. Die österreichischen Diatomaceen, nebst Anschluβ einiger neuen Arten von anderen Lokalitäten und einer kritischen Übersicht der bisher bekannten Gattungen und Arten. Verhandlungen der Kaiserlich-Königlichen Zoologisch-Botanischen Gesellschaft Wien **12**: 315-472, 545-588, 7 pls.

Håkannson, H. 2002. A compilation and evaluation of species in the general *Stephanodiscus*, *Cyclostephanos* and *Cyclotella* with a new genus in the family Stepanodiscaceae. Diatom Research **17**: 1-139.

Håkannson, H. & Stoermer, E.F. 1984. Observations on the type material of *Stephanodiscus hantzschii* Grunow in Cleve & Grunow. Nova Hedwigia **39**: 477-495.

原口和夫 2001. 余呉湖の珪藻. Diatom **17**: 141-148.

原口和夫・三友清史・小林　弘 1998. 埼玉の藻類 珪藻類. *In*: 伊藤　洋（編）1998年版埼玉県植物誌, pp. 527-600. 埼玉県教育委員会.

Hartley, B., Barber, H.G. & Carter, J.R. 1996. An Atlas of British diatoms. Sims, P.A. (ed.) pp. 1-601. Biopress, Bristol.

Hasegawa, Y. 1975. Significance of diatom thanatocoenoses in the Neolithic sea-level change problem. Pacific Geology **10**: 47-78.

秦　光男・長谷川康雄 1970. 北海道奥尻島南部新第三系の地質と化石珪藻群. 地球科学 **24**: 93-107.

Haworth, E.Y. & Hurley, M.A. 1986. Comparison of the stelligeroid taxa of the centric diatom genus *Cyclotella*. *In*: Ricard, M. (ed.) Proceedings 8th International Diatom Symposium, Paris 1984. pp. 43-58. Koeltz, Königstein.

Hayashi, T., Tanimura, Y. & Sakai, H. 2007. *Thalassiosira*, *T. inlandica* sp. nov. (Bacillariophyta), with semicontinuous cribra and elongated marginal fultoportulae. Phycologia **46**: 353-362.

Hendey, N.I. 1964. An introductory account of the smaller algae of British coastal waters. Part V: Bacillariophyceae (Diatoms). 317 pp. 45 pls. Her Majesty's Stationery Office, London.

Héribaud, F.J. 1903. Les Diatomées d'Auvergne. Librairie des Sciences Naturelles, Paris. deuxieme mémoire. 155 pp. pls 9-12.

平中宏典・柳沢幸夫・黒川克己 2004. 新潟県中条地域中新統内須川層のテフラ層序. 地球科学 **58**: 31-46.

Houk, V. & Klee, R. 2004. The stelligeroid taxa of the genus *Cyclotella* (Kützing) Brébisson (Bacillariophyceae) and their transfer into the new genus *Discostella* gen. nov. Diatom Research **19**: 203–228.

Houk, V. & Klee, R. 2007. Atlas of freshwater centric diatoms with a brief key and descriptions. Part II. Melosiraceae and Aulacoseiraceae (Supplement to Part I). Fottea, Olomouc **7**: 85–255.

Houk, V., Klee, R. & Tanaka, H. 2010. Atlas of freshwater centric diatoms with a brief key and descriptions. Part III. Stephanodiscaceae A: *Cyclotella, Tertiarius, Discostella*. In: Poulíčková, A. (ed.): Fottea **10** Supplement, 498 pp.

Houk, V., Klee, R. & Tanaka, H. 2014. Atlas of freshwater centric diatoms with a brief key and descriptions. Part IV. Stephanodiscaceae B: *Stephanodiscus, Cyclostephanos, Pliocaenicus, Hemistephanos, Stephanocostis, Mesodictyon & Spicaticribra*. In: Poulíčková, A. (ed.): Fottea **14** Supplement. (in press)

Hustedt, F. 1911. *In*: A. Schmidt's Atlas der Diatomaceen-Kunde, pl. 271. O.R. Reisland, Leipzig.

Hustedt, F. 1912. *In*: A. Schmidt's Atlas der Diatomaceen-Kunde, pl. 281. O.R. Reisland, Leipzig.

Hustedt, F. 1913. *In*: A. Schmidt's Atlas der Diatomaceen-Kunde, pls 287, 290, 291. O.R. Reisland, Leipzig.

Hustedt, F. 1924. *In*: A. Schmidt's Atlas der Diatomaceen-Kunde, pl. 351. O.R. Reisland, Leipzig.

Hustedt, F. 1927. Bacillariales aus dem Aokiko-see in Japan. Archiv für Hydrobiologie **18**: 155–172.

Hustedt, F. 1928. Die Kieslalgen Deutschlands, ökologische und der Schweiz unter Berucksichtigung der übrigen Länder Europas sowie der angrenzenden Meers gebiete. *In*: Rabenhorst, Kryptogamen-Flora von Deutschland, Österreich und der Schweiz **7**, Teil 1. pp. 273–464.

Hustedt, F. 1930. Bacillariophyta (Diatomeae). *In*: Pascher, A. (ed.), Die Süsswasser-Flora Mitteleuropas. Heft 10. 466 pp. Gustav Fischer Verlag, Jena.

Hustedt, F. 1934. *In*: A. Schmidt's Atlas der Diatomaceen-Kunde, pls 392, 396. O.R. Reisland, Leipzig.

Hustedt, F. 1937. Systematische und ökologische Untersuchungen über die Diatomeen-flora von Java, Bali und Sumatra nach dem Material Deutschen Limnologischen Sunda-Expedition. Archiv für Hydrobiologie, Supplement Band **15**: 131–177.

Hustedt, F. 1938. Systematische und ökologische Untersuchungen über die Diatomeenflora von Java, Bali und Sumatra nach dem Material der Deutschen Limnologischen Sunda-Expedition. Archiv für Hydrobiologie (Suppl.) **15**: 393–506.

Hustedt, F. 1939. Die Diatomeenflora des Küstengebietes der Nordseen von Dollart bis zur Elbemündung. I. Abhanlungen, naturwissenschaftl. Verein zu Bremen, Bd. **31**: Heft 3, S. 571–677 pp. 123 Textfig.

Hustedt, F. 1942. Süßwasserdiatomeen des indomalayischen A rchipels und der Hawaii-Inseln. Internationale Revue der gesamten Hydrobiologie **42**: 1–252.

Hustedt, F. 1944. *In*: A. Schmidt's Atlas der Diatomaceen-Kunde, pl. 377. O.R. Reisland, Leipzig.

Hustedt, F. 1957. Die Diatomeenflora des Fluss-systems der Weser im Gebiet der Hansestadt Bremen. Abhandlungen herausgegeben vom Naturwissenschaftlichen Verein zu Bremen **34**: 181–440.

Hustedt, F. 1961-1966. Die Kieselalgen Deutschlands, Österreichs und der Schweiz unter Berücksichtigung der übrigen Länder Europas sowie der angrenzenden Meeresgebiete. *In*: Rabenhorst, L. (ed.) Kryptogamen-Flora III. 816 pp. Leipzip.

Ichikawa, W. 1951. Fossil diatoms found in the Kayama clay-shale on the outskirts of Kanazawa City. Transactions and Proceedings of the Palaeontological Society of Japan, New Series **4**: 97–111.

Ichikawa, W. 1955. On fossil diatoms from the Onikoube Basin, Miyagi-prefecture, collected by Dr. N. Katayama. The Science Reports of the Kanazawa University IV: 151–175.

Idei, M. & Kobayasi, H. 1989. The fine structure of *Diploneis finnica* with special reference to the marginal openings. Diatom Research **4**: 25-37.

出井雅彦・南雲　保 1995. 無縦溝珪藻 *Fragilaria* 属(狭義の)とその近縁属. 藻類 **43**: 227-239.

出井雅彦・鈴木照男 1999. 能登山戸田珪藻泥岩の珪藻及び能登珪藻土の利用. 文教大学女子短期大学部紀要 **43**: 21-33.

石塚吉浩・水野清秀・松浦浩久・星住英夫 2005. 豊後杵築地域の地質. 地域地質研究報告 (5 万分の 1 地質図幅), 83 pp. 産総研地質調査総合センター.

岩橋八洲民 1935. 日本淡水産中心型硅藻(其四). 植物研究雑誌 **11**: 768-771.

岩橋八洲民 1936. 日本淡水産中心型硅藻(其五). 植物研究雑誌 **12**: 121-127.

Juhlin-Dannfelt, H. 1882. On the diatoms of the Baltic Sea. Bihang till Kongliga Svenska Vetenskaps-Akademiens Handlingar Bd. **6**: 1-52, 6 Taf.

Julius, M.L., Curtin, M. & Tanaka, H. 2006. *Stephanodiscus kusuensis*, sp. nov a new Pleistocene diatom from southern Japan. Phycological Research **54**: 294-301.

Jüttner, I., Krammer, K., Vijver, B.V., Tuji, A., Simkhada, B., Gurung, S., Sharma, S., Sharma, C. & Cox, E.J. 2010. *Oricymba* (Cymbellales, Bacillariophyceae), a new cymbelloid genus and three new species from the Nepalese Himalaya. Phycologia **49**: 407-423.

加藤君雄・小林　弘・南雲　保 1977. 八郎潟調整池のケイソウ類. *In*: 八郎潟調整池生物相調査会(編), 63-137 pp. 秋田県.

Kato, M., Tanimura, Y., Fukusawa, H. & Yasuda, Y. 2003. Intraspecific variation during the life cycle of a modern *Stephanodiscus* species (Bacillariophyceae) inferred from the fossil record of Lake Suigetsu, Japan. Phycologia **42**: 292-300.

河島綾子・小林　弘 1993. 阿寒湖の珪藻(1. 中心類). 自然環境科学研究 **6**: 41-58.

河島綾子・小林　弘 1995. 阿寒湖の珪藻(3. 羽状類-広義の *Fragilaria* を除く無縦溝類). 自然環境科学研究 **8**: 35-49.

河島綾子・真山茂樹 1997. 阿寒湖の珪藻(5. 羽状類-縦溝類: *Aneumastus*, *Craticula*, *Diatomella*, *Diploneis*, *Frustulia*, *Gyrosigma*, *Luticola*, *Neidium*, *Sellaphora*, *Stauroneis*). 自然環境科学研究 **10**: 35-52.

河島綾子・真山茂樹 1998. 阿寒湖の珪藻 (6. 羽状類-縦溝類: *Cavinula*, *Diadesmis*, *Geissleria*, *Hippodonta*, *Navicula*, *Placoneis*). 自然環境科学研究 **11**: 23-41.

河島綾子・真山茂樹 2000. 阿寒湖の珪藻 (7. 羽状類-縦溝類: *Caloneis*, *Pinnularia*). 自然環境科学研究 **13**: 67-83.

河島綾子・真山茂樹 2001. 阿寒湖の珪藻 (8. 羽状類-縦溝類: *Cymbella*, *Encyonema*, *Gomphoneis*, *Gomphonema*, *Gomphosphenia*, *Reimeria*). 自然環境科学研究 **14**: 89-109.

河島綾子・真山茂樹 2004. 阿寒湖の珪藻 (11. 羽状類-縦溝類: *Campylodiscus*, *Cymatopleura*, *Surirella* and additional 10 taxa). 自然環境科学研究 **17**: 1-21.

Khursevich, G.K. 1994. Morphology and taxonomy of some centric diatom species from the Miocene sediments of the Dzhilinda and Tunkin Hollows. *In*: Kociolek, J.P. (ed.) Proceedings of the 11th International Diatom Symposium, San Francisco 1990. pp. 269-280. California Academy of Sciences, San Francisco.

Khursevich, G.K. & Stachura-Suchoples, K. 2008. The genus *Pliocaenicus* Round & Håkansson (Bacillariophyta): Morphology, taxonomy, classification and biogeography. Nova Hedwigia **86**: 419-444.

Klee, R. & Houk, V. 1996. Morphology and ultrastructure of *Cyclotella woltereckii* Hustedt (Bacillariophyceae). Archiv für Protistenkunde **147**: 19-27.

Kociolek, J.P. & Stoermer, E.F. 1988. Taxonomy and systematic position of the *Gomphoneis quadripunctata* species complex. Diatom Research **3**: 95-108.

Kozyrenko, T.F., Strelnikova, N.I., Khursevich, G.K., Tsoy, I.B., Jakovschikova, T.K., Muchina, V.V., Olshtynskaja, A.P. & Semina, G.I. 2008. The diatom of Russia and adjacent countries. Fossil and recent. 171 pp. St. Petersburg University Press, St. Petersburg.

小林　弘 1960. 長瀞自然岩石園の珪藻類．秩父自然科学博物館研究報告 **1960** (10): 67-76, pl. 1-10.

小林　弘・原口和夫 1969. 川越近郊の湧泉池から得たケイソウについて．秩父自然科学博物館研究報告 **1969** (15): 27-54.

小林　弘 1995. 金魚用水槽から得た *Amphora veneta* Kützing の大きさの範囲．Diatom **10**: 89.

Kobayasi, H. & Ando, K. 1977. Diatoms from irrigation ponds in Musashikyuryo-shinrin Park, Saitama Prefecture. Bulletin of Tokyo Gakugei University ser. 4. **29**: 231-263.

小林　弘・安藤一男 1978. 日本産 *Stauroneis* 属ケイソウ．東京学芸大学紀要第 4 部門 **30**: 273-292.

Kobayasi, H., Ando, K. & Nagumo, T. 1981. On some endemic species of the genus *Eunotia* in Japan. *In*: Ross, R. (ed.) Proceedings of the 6th Symposium on recent and fossil diatoms. pp. 93-114. Koenigstein, Otto Koeltz.

小林　弘・吉田　稔 1984. コンクリート池のケイソウとその優れた教材性について．東京学芸大学紀要第 4 部門 **36**: 115-143.

Kobayasi, H. & Kobayashi, H. 1986. Fine structure and taxonomy of the small and tiny *Stephanodiscus* (Bacillariophyceae) species in Japan 4. *Stephanodiscus costatilimbus* sp. nov. The Japanese Journal of Phycology **34**: 8-12.

Kobayasi, H. & Kobayashi, H. 1988. *Epithemia amphicephale* (Östr.) comb. et stat. nov. and *E. reticulata* Kütz. with special reference to the areolae occlusion. *In*: Round, F.E. (ed) Proceedings of the 9th International Diatom Symposium, Bristol 1986. pp. 459-466. Biopress Ltd., Bristol and Koeltz Scientific Books, Koenigstein.

Kobayasi, H., Kobori, S. & Sunaga, S. 1994. Taxonomy and morphology of two forms of the *Nitzschia sinuata* complex. *In*: Kociolek, J.P. (ed.) Proceedings of the 11th International Diatom Symposium, San Francisco 1990. pp. 281-289. California Axademy of Sciences, San Francisco.

小林　弘・出井雅彦・真山茂樹・南雲　保・長田敬五 2006. 小林弘珪藻図鑑「第 1 巻」．531 pp. 内田老鶴圃，東京．

Kociolek, J.P. & Stoermer, E.F. 1988. Taxonomy and systematic position of the *Gomphoneis quadripunctata* species complex. Diatom Research **3**: 95-108.

小藤美樹・長田敬吾・南雲　保 2004. 徳島県那賀川の珪藻類．日本歯科大学紀要（一般教育系）**33**: 73-80.

Krammer, K. 1992. *Pinnularia* eine monographie der europäischen taxa. Bibliotheca Diatomologica **26**: 1-353.

Krammer, K. 1997. Die cymbelloiden Diatomeen. Eine Monographic der weltweit bekannten Taxa. Teil 1. Allgemeines und *Encyonema* part. Bibliotheca Diatomologica **36**: 1-382.

Krammer, K. 1999. Validierung von *Cymbopleura* nov. gen. Iconographia Diatomologica **6**: 292.

Krammer, K. 2000. The genus *Pinnularia*. *In*: Lange-Bertalot, H. (ed.) Diatoms of Europe **1**: 703 pp. A.R.G. Gantner Verlag, Ruggell.

Krammer, K. 2002. *Cymbella*. *In*: Lange-Bertalot, H. (ed.) Diatoms of Europe. Diatoms of the European inland waters and comparable habitats **3**: 1-584. A.R.G. Gantner Verlag, Ruggell.

Krammer, K. 2003. Diatoms of the European inland waters and comparable habitats. Vol. **4**. *Cymbopleura, Delicata, Navicymbula, Gomphocymbellopsis Afrocymbella*. *In*: Lange-Bertalot, H. (ed.) Diatoms of Europe. 530 pp., 164 pls. A.R.G. Gantner Verlag, Ruggell.

Krammer, K. & Lange-Bertalot, H. 1985. Naviculaceae, Neue und wenig bekannte taxa, neue Kombinationen und Synonyme sowie Bemerkungen zu einigen Gattungen. Bibliotheca Diatomologica **9**: 5-230, 43 pls.

Krammer, K. & Lange-Bertalot, H. 1986. Bacillariophyceae. Teil 1. Naviculaceae. 876 pp. Süßwasserflora von Mitteleuropa, Bd. **2/1**: Begründet von A. Pascher. Gustav Fischer Verlag, Stuttgart, New York.

Krammer, K. & Lange-Bertalot, H. 1988. Bacillariophyceae. Teil 2. Bacillariaceae, Epithemiaceae,

Surirellaceae. 596 pp. Süßwasserflora von Mitteleuropa, Bd. **2/2**: Begründet von A. Pascher. Gustav Fischer Verlag, Stuttgart, New York.

Krammer, K. & Lange-Bertalot, H. 1991. Bacillariophyceae. Teil 3. Centrales, Fragilariaceae, Eunotiaceae. 576 pp. Süßwasserflora von Mitteleuropa, Bd. **2/3**: Begründet von A. Pascher. Gustav Fischer Verlag, Stuttgart, Jena.

窪田英夫・小林一恵・山崎　博　1976. 長野県小諸市南方瓜生坂峠における化石珪藻群集について. 地学研究 **27**: 71-88.

Kützing, F.T. 1844. Die kieselschaligen Bacillarien oder Diatomeen. Nordhausen, W. Köhne. 152 pp, pls 1-30.

Lange-Bertalot, H. 1995. *Gomphosphenia paradoxa* nov. spec. et nov. gen. und Vorschlag zur Lösung taxonomischer Probleme infolge eines veränderten Gattungskonzepts von *Gomphonema* (Bacillariophyceae). Nova Hedwigia **60**: 241-252.

Lange-Bertalot, H. & Krammer, K. 1987. Bacillariophyceae Epithemiaceae Surirellaceae. Bibliotheca Diatomologica **15**: 1-289.

Lee, J.H., Gotoh, T. & Chung, J. 1992. A study of diatom species *Gomphonema vibrio* Ehr. var. *subcapitatum* (Mayer) Lee, comb. nov. The Korean Journal of Phycology **7**: 79-87.

Lee, J.H., Gotoh, T. & Chung, J. 1993. *Cymbella orientalis* sp. nov., a freshwater diatom from the Far East. Diatom Research **8**: 99-108.

Levkov, Z. 2009. *Amphora* sensu lato. In: Lange-Bertalot, H. (ed.) Diatoms of Europe: Diatoms of the European inland waters and comparable habitats **5**. 916 pp. A.R.G. Gantner Verlag K.G.

Likhoshway, Y.V. & Crawford, R.M. 2001. The rimoportula—a neglected feature in the systematics of *Aulacoseira*. In: Economou—Amilli, A. (ed.) 16th International Diatom Symposium, Athens 2000. pp. 33-47. Amvrosiou Press, Athens.

Ludwig, T.A.V., Tremarin, P.I., Becker, V. & Torgan, L.C. 2008. *Thalassiosira rudis* sp. nov. (Coscinodiscophyceae): a new freshwater species. Diatom Research **23**: 389-400.

Lund, J.W.G. 1951. Contributions to our knowledge of British algae. Hydrobiologica **3**: 93-100.

Lupikina, E.G. 1965. Diatomeae novae et curiosae e stratis ermanicis partis Kamczatkae occidentalis. Novitates Systematicae Plantarum non Vascularium **2**: 15-22.

Mann, D.G. 1989. The diatom genus *Sellaphora*: Separation from *Navicula*. British Phycologocal Journal **24**: 1-20.

松本幡郎・村田正文・今中啓喜　1984. 大分県北部の上部新生界の火山層序. I. 宇佐・耶馬溪地域. 熊本大学理学部紀要(地学) **13**: 1-24.

Matsuo, H. 1968. A study on the Neogene plants in the inner side of Central Honshu, Japan. II: On the Minoshirotori Flora (Pliocene) of the Paleovolcano-lake deposits. Annals of Science the College of Liberal Arts, Kanazawa Univ. **5**: 29-77, pls 1-8.

Mayama, S. 2001. Valuable taxonomic characters in the valve mantle and girdle of some *Eunotia* species. In: Jahn, R., Kociolek, J.P., Witkowski, A. & Compère, P. (eds) Lange-Bertalot-Festschrift. pp. 119-130. A.R.G. Gantner Verlag, Ruggell.

Mayama, S. & Kobayasi, H. 1990. Studies of *Eunotia* species in the classical Degernas materials housed in the Swedish Museum of Natural History. Diatom Research **5**: 351-366.

Mayama, S. & Kobayasi, H. 1991. Observations of *Eunotia arcus* Ehr., type species of the genus *Eunotia* (Bacillariophyceae). The Japanese Journal of Phycology **39**: 131-141.

Mayama, S., Idei, M., Osada, K. & Nagumo, T. 2002. Nomenclatural changes for 20 diatom taxa occurring in Japan. Diatom **18**: 89-91.

Meister, F. 1913. Beiträge zur Bacillariaceenflora Japan. Archif Für Hydrobiologie und Planktonkunde **8**: 305-312.

Meister, F. 1934. Seltene und neue Kieselalgen. I. Berichte der Schweizerischen Botanischen Gesellschaft (Zürich). Bd. **44**: S. 87-106.

Metzeltin, D. & Lange-Bertalot, H. 1995. Kritische wertung der taxa in *Didymosphrnia* (Bacillariophyceae). Nova Hedwigia **60**: 381-405.

Metzeltin, D., Lange-Bertalot, H. & Nergui, S. 2009. Diatoms in Mongolia. *In*: Lange-Bertalot, H. (ed.) Iconographia Diatomologica **20**: 1-686.

Moisseva, A.I. 1971. Atlas of Neogene diatom algae of the primorye. 152 pp. Leningrad, Nedra.

Mori, S. 1975. Vertical distribution of diatoms in core samples from Lake Biwa. Paleolimnology of Lake Biwa and the Japanese Pleistocene **3**: 368-391.

森　忍 1981. 濃尾平野の沖積層のケイソウ群集．瑞浪市化石博物館研究報告 **8**: 127-138, pls 48-50.

Mori, S. 1986. Diatom assemblages and Late Quaternary environmental changes in the Nobi Plain, central Japan. The Journal of Earth Sciences **34**: 109-138.

Müller, O. 1899. Bacillariaceen aus den Natrontälern von El Kab (Ober-Ägypten). Hedwigia **38**: 274-321.

Naguno, T. 2003. Taxonomic studies of the subgenus *Amphora* Cleve of the genus *Ampholra* (Bacillariophyceae) in Japan. Bibliotheca Diatomologica **49**: 1-263.

南雲　保・小林　弘 1977. 光顕及び電顕的研究に基く *Melosira arentii* (Kolbe) comb. nov. について．藻類 **25**: 182-183.

南雲　保・安藤一男 1984. 埼玉県荒川低地沖積層のケイソウ(2)．日本歯科大学紀要 **13**: 123-134.

南雲　保・小林　弘 1985. 淡・汽水産珪藻 *Cyclotella* 属の3種，*C. atomus*, *C. caspia*, *C. meduanae* の微細構造．日本プランクトン学会報 **32**: 101-109.

南雲　保・三橋扶佐子・田中宏之 1998. 群馬県倉渕地域の更新世湖沼性堆積物の珪藻．日本歯科大学紀要 **27**: 167-178.

南雲　保・真山茂樹 2000. 珪藻類の分類と系統．月刊海洋/号外 No. 21: 35-45.

南雲　保・田中宏之 2001. 長野県真田町東部に分布する更新世湖沼性堆積物中の珪藻．日本歯科大学紀要(一般教育系) **30**: 191-198.

南雲　保・田中宏之 2006. 南西諸島奄美大島名瀬市泥染公園の珪藻．日本歯科大学紀要(一般教育系) **35**: 51-59.

中島啓治・南雲　保 1999. 群馬県安中市野殿に分布する中期更新世野殿層の珪藻．群馬県立自然史博物館研究報告 **3**: 25-36.

納谷友規 2008. 関東平野中央部のボーリングコアにおける *Puncticulata rhomboideo-elliptica* の出現層準．Diatom **24**: 91(講演要旨).

Naya, T., Tanimura, Y., Nakazato, R. & Amano, K. 2007. Modern distribution of diatoms in the surface sediments of Lake Kiraura, central Japan. Diatom **23**: 55-70.

Negoro, K. 1981. Fossil diatoms of the Kobiwako Group viz. ancient deposit of Lake Biwa. Acta Phytotaxonomica et Geobotanica **32**: 90-104.

根来健一郎 1981. 古琵琶湖層群の堅田累層の化石硅藻．日本植物分類学会誌 **32**: 183-191.

根来健一郎・後藤敏一 1981. 奄芸層群の化石硅藻(第1報)．瑞浪化石博物館研究報告 **8**: 77-103, pls 17-36.

野尻湖珪藻グループ 1980. 野尻湖層の珪藻遺骸群集．*In*: 歌代　勤 (編) 野尻湖周辺の人類遺跡と古環境．pp. 75-100. 地質学論集19号，日本地質学会，上越市．

Nygaard, G. 1956. Ancient and recent flora of diatoms and Chrysophyceae in Lake Gribso. Folia Limnologica Scandinavica **8**: 32-262, pls 1-12.

小川カホル 1990. 手賀沼に出現するタラシオシーラ科の浮遊珪藻．Diatom **5**: 59-68.

Ohtsuka, T. 2002. Checklist and illustration of diatoms in the Hii River. Diatom **18**: 23-56.

奥野春雄 1943. 日本珪藻土鉱床の植物分類学的研究(第1報)．植物学雑誌 **57**: 364-370.

奥野春雄 1944. 日本珪藻土鉱床の植物分類学的研究(第2報)．植物学雑誌 **58**: 8-14.

Okuno, H. 1952. Atlas of fossil diatoms from Japanese diatomite deposits. 51 pp. 29 pls. Botanical Institute, Faculty of Textile Fibers, Kyoto University of Industrial Arts and Textile Fibers, Kyoto.

Okuno, H. 1954. Electron-microscopic fine structure of fossil diatoms II. Trans. Proc. Palaeont. Soc. Japan, N.S., No. 14, pp. 143-148, pls 17-18.

Okuno, H. 1955. Electron-microscopic fine structure of fossil diatoms III. Trans. Proc. Palaeont. Soc. Japan, N.S., No. 19, pp. 53-58, pls 8, 9.

奥野春雄 1958. 北海道瀬棚町の珪藻土について(2). 植物研究雑誌 **33**: 193-198, pls 1, 2.

奥野春雄 1959. 北海道瀬棚町の珪藻土について(3). 植物研究雑誌 **34**: 25-29.

奥野春雄 1959. 北海道瀬棚町の珪藻土について(4). 植物研究雑誌 **34**: 272-277.

奥野春雄 1959. 北海道瀬棚町の珪藻土について(5). 植物研究雑誌 **34**: 353-360.

奥野春雄 1961. 北海道瀬棚町の珪藻土について(2). 植物研究雑誌 **36**: 394-400, pls 7, 8.

Okuno, H. 1964. Fossil diatoms. *In*: Helmcke, J.G. & Krieger, W. (eds) Diatomeenschalen im elektronenmikroskopischen Bild. Teil. V. 1-48. pls 414-513. J. Cramer, Weinheim.

長田敬五・南雲　保 1983. 新潟県，郡殿ノ池および男池のケイソウ. 日本歯科大学紀要 **12**: 203-238.

Pantocsek, J. 1905. Beiträge zur Kenntnis der Fossilen Bacillarien Ungarns. 3 Teil. 118 pp, pls 1-42. W. Junk, Berlin.

Patrick, R. & Reimer, C.W. 1966. The diatoms of the United States 1. Monographs of the Academy of Natural Sciences of Philadelphia, no. 13. 688 pp. Philadelphia.

Patrick, R. & Reimer, C.W. 1975. The diatoms of the United States 2 (1). Monographs of the Academy of Natural Sciences of Philadelphia, no. 13. 213 pp. Philadelphia.

Round, F.E. 1982. *Cyclostephanos*—a new genus within the Sceletonemaceae. Archiv für Protistenkunde **125**: 323-329.

Round, F.E., Crawford, R.M. & Mann, D.G. 1990. The diatoms. Biology & morphology of the genera. 747 pp. Cambridge University Press, Cambridge.

Round, F.E. & Håkansson 1992. Cyclotelloid species from a diatomite in the Harz Mountains, Germany, including *Pliocaenicus* gen. nov. Diatom Research **7**: 109-125.

Rumrich, U., Lange-Bertalot, H. & Rumrich, M. 2000. Diatoms of the Andes. Ichonographia Diatomologica **9**: 1-650.

佐竹俊子・小林　弘 1991. 淡水産中心類珪藻 *Aulacoseira valida* (Grunow in Van Heurck) Krammer の微細構造. 自然環境科学研究 **4**: 45-57.

Schmidt, A. 1885. *In*: A. Schmidt's Atlas der Diatomaceen-Kunde, pl. 9. O.R. Reisland, Leipzig.

Schmidt, A. 1889. *In*: A. Schmidt's Atlas der Diatomaceen-Kunde, pl. 54. O.R. Reisland, Leipzig.

Simonsen, R. 1979. The diatom system: ideas on phylogeny. Bacillaria **2**: 9-71.

Simonsen, R. 1987. Atlas and catalogue of the diatom types of Friedrich Hustedt. Vol. **1**: Catalogue. pp. 1-525. Vol. **2**: Atlas, Taf. 1-395. Vol. **3**: Atlas, Taf. 396-772. J. Cramer Berlin/Stuttgart.

Sims, P.A. 1983. A taxonomic study of the genus *Epithemia* with special reference to the type species *E. turgida* (Ehrenb.) Kütz. Bacillaria **6**: 211-235.

Siver, P.A. & Kling, H. 1997. Morphological observations of *Aulacoseira* using scanning electron microscopy. Canadian Journal of Botany **75**: 1807-1835.

Siver, P.A., Hamilton, P.B., Stachura-Suchoples, K. & Kociolek, J.P. 2003. Morphological observations of *Neidium* species with sagittate apices, including the description of *N. cape-codii* sp. nov. Diatom Research **18**: 131-148.

Skabichevskij, A.P. 1983. A new species of *Didymosphenia lineata* (Bacillariophyta) and its variability. Bot Zhurn. (Moscow & Leningrad) **68**: 1254-1260.

Skvortzow, B.W. 1936. Diatoms from Biwa Lake, Honshu Island, Nippon. Philipppine Journal of Science **61**: 253-291, 8 pls.

Skvortzow, B.W. 1936. Diatoms from Kizaki Lake, Honshu Island, Nippon. Philippine Journal of Science **61**: 9-73, 16 pls.

Skvortzov, B.V. 1937. Neogene diatoms from Wamura, Nagano Prefecture, Central Nippon. Memoirs of the College of Science, Kyoto Imperial University, Ser. B. **7**: 137-156, 4-8 pls.

Skvortzow, B.W. 1937. Diatoms from Ikeda Lake, Satsuma Province Kiusiu Island, Nippon. Philippine Journal of Science **62**: 191-218, 4 pls.

Skvortzov, B.V. 1937. Neogene diatoms from Saga Prefecture, Kiushiu Island, Nippon. Memoirs of the College of Science, Kyoto Imperial University, Ser. B. **7**: 157-174, 9-11 pls.

Skvortsov, B.V. 1938. Diatoms collected by Mr. Yoshikazu Okada in Nippon. The Journal of Japanese Botany **14**: 52-65.

Smith, W. 1853. A synopsis of British Diatomaceae. Vol. **1**: 1-89, pls 1-31. John van Voorst, London.

Stachura-Suchoples, K. & Khursevich, G. 2007. On the genus *Pliocaenicus* Round & Håkansson (Bacillariophyceae) from the Northern Hemisphere. *In*: Kusber, W.-H. & Jahn, R. (eds) Proceedings of the 1st Central European Diatom Meeting. pp. 155-158. Botanic Garden and Botanical Museum Berlin-Dahlem, Freie Universität Berlin, Berlin.

高橋　清・下野　洋 1980. 岐阜県美濃白鳥湖成層産植物性プランクトンについて．長崎大学教養部紀要 **20**: 7-18.

高橋啓一・神谷英利・黒岩俊明・小林将喜・山岸勝治・礒田喜義・中島啓治・田中宏之 1981. 群馬県嬬恋村産のゾウ化石，および産出地の地質について．群馬県立歴史博物館紀要2号: 1-23.

田中宏之 1983. 群馬県太田市，含ナウマンゾウ臼歯化石粘土層の珪藻化石．地学研究 **34**: 123-141.

田中宏之 1985. 群馬県南牧村星尾鉱泉の珪藻．地学研究 **36**: 223-239.

田中宏之 1989. 南西諸島，喜界島湧泉池の珪藻．群馬県立歴史博物館紀要10号: 31-42.

田中宏之 1991. 群馬県吾妻町大戸に分布する中期更新世湖沼性堆積物の珪藻．群馬県立歴史博物館紀要12号: 1-10.

Tanaka, H. 2000. *Stephanodiscus komoroensis* sp. nov., a new Pleistocene diatom from central Japan. Diatom Research **15**: 149-157.

Tanaka, H. 2003. *Cyclostephanos kyushuensis* sp. nov., from Pliocene sediments in southwestern Japan. Diatom Research **18**: 357-364.

田中宏之 2007. 秋田県，宮田層から産出した *Pliocaenicus nipponicus* H. Tanaka & Nagumo について．Diatom **23**: 119-120.

Tanaka, H. 2007. Taxonomic studies of the genera *Cyclotella* (Kützing) Brébisson, *Discostella* Houk et Klee and *Puncticulata* Håkansson in the Family Stephanodiscaceae Glezer et Makarova (Bacillariophyta) in Japan. Bibliotheca Diatomologica **53**: 1-205.

田中宏之 2009. 新潟県，高田城外堀から見出された *Cyclotubicoalitus undatus* Stoermer, Kociolek & Cody (Centrales, Bacillariophyceae). Diatom **25**: 164-165.

田中宏之 2010. 沖縄県，福上湖から見出された *Spicaticribra kingstonii* Johansen, Kociolek et Lowe. Diatom **26**: 44-45.

田中宏之 2012. 恩原および辰巳峠地域（岡山・鳥取県境）の人形峠層から見出された淡水生珪藻化石．Diatom **28**: 14-18.

田中宏之・中島啓治 1983. 尾瀬沼の珪藻—二ツ岳降下軽石層以降の珪藻群集—．群馬県立歴史博物館紀要4号: 1-28.

田中宏之・中島啓治 1985. 群馬県老神・奥平・梨木・嶺・赤久縄温泉及び福島県元温泉小屋温泉のケイソウ．群馬県立歴史博物館紀要6号: 1-22.

田中宏之・小林　弘 1992. 群馬県，中之条上部湖成層（中期更新世）の珪藻．群馬県立歴史博物館紀要13号: 17-38.

田中宏之・小林　弘 1995. 前期更新世湖沼性堆積物小野上累層の珪藻．Diatom **10**: 75-86.

田中宏之・小林　弘 1996. 香坂礫岩層（前期鮮新統）の珪藻．地学研究 **45**: 21-25.

Tanaka, H. & Kobayasi, H. 1996. A new species of *Cyclotella*, *C. kohsakaensis* sp. nov., from a Pliocene deposit, Central Japan. Diatom **12**: 1-6.

Tanaka, H. & Kobayasi, H. 1996. The fine structure of *Cyclotella rhomboideo-elliptica* Skuja, from a Middle Pleistocene deposit, Gunma Prefecture, central Japan. Diatom **12**: 35-41.

Tanaka, H. & Kobayasi, H. 1999. *Pliocaenicus costatus* and *P. omarensis* found in Japan. *In*: Mayama, S., Idei, M. & Koizumi, I. (eds) Proceedings of the 14th Diatom Symposium, Tokyo 1996. pp. 135-143. Koeltz, Koenigstein.

田中宏之・南雲　保　2000. 化石珪藻*Stephanodiscus komoroensis* Tanakaのタイプ試料(前期更新世)の珪藻群集と古環境. 地学研究 **49**: 67-75.

田中宏之・南雲　保　2000. 本邦新産属珪藻 *Cyclotubicoalitus undatus* Stoermer, Kociolek & Cody (Centrales, Bacillariophyceae). 藻類 **48**: 105-108.

Tanaka, H. & Nagumo, T. 2000. *Cyclostephanos numataensis* sp. nov., a new Pleistocene diatom from central Japan. Diatom **16**: 19-25.

Tanaka, H. & Nagumo, T. 2002. *Cyclotella pliostelligera* sp. nov., a new fossil freshwater diatom from Japan. *In*: John, J. (ed.) Proceedings of the 15th Diatom Symposium, Perth 1998. pp. 351-358. Gantner Verlag, Ruggell/Liechtenstein.

Tanaka, H. & Nagumo, T. 2004. *Pliocaenicus nipponicus* sp. nov., a new freshwater fossil diatom from central Japan. Diatom **20**: 105-111.

Tanaka, H., Nagumo, T., Kashima, K. & Mitsuhashi, F. 2004. Pliocene diatoms from freshwater diatomite in Yamaga Town, Kyushu, Japan. Diatom **20**: 113-122.

Tanaka, H. & Nagumo, T. 2005. *Puncticulata ozensis* sp. nov., a new freshwater diatom in Lake Oze, Japan. Diatom **21**: 47-55.

田中宏之・南雲　保・鹿島　薫　2005. 熊本県益城町に分布する津森層(中期更新世)の淡水生化石珪藻群集. Diatom **21**: 119-130.

Tanaka, H. & Nagumo, T. 2006. *Stephanodiscus miyagiensis* sp. nov. from Pleistocene sediment in northeastern Japan. Diatom Research **21**: 371-378.

Tanaka, H. & Nagumo, T. 2006. Late Miocene freshwater diatoms from Mitoku area in Misasa Town, Tottori Prefecture, Japan. Diatom **22**: 17-25.

田中宏之・南雲　保　2007. 波志江沼(群馬県伊勢崎市)の中心類珪藻. 自然環境科学研究 **20**: 25-39.

田中宏之・南雲　保　2007. 福島県鎌沼から見出した *Aulacoseira distans* (Ehrenb.) Simonsen var. *nivalis* (W. Sm.) E.Y. Haw. の形態について. Diatom **23**: 113-116.

田中宏之・鹿島　薫　2007. 大分県杵築市西俣水に分布する珪藻土から見出された前期更新世淡水生珪藻群集. 地学研究 **56**: 137-145.

Tanaka, H. & Nagumo, T. 2008. *Thalassiocyclus pankensis* sp. nov., a new diatom from the Panke Swamp, northern Japan (Bacillariophyta). Phycological Research **56**: 83-88.

Tanaka, H., Nagumo, T. & Akiba, F. 2008. *Aulacoseira hachiyaensis* sp. nov., a new Early Miocene freshwater fossil diatom from the Hachiya Formation, Japan. *In*: Likhosway, Y. (ed.) Proceedings of the 19th International Diatom Symposium, Listvyanka 2006. pp. 115-123. Biopress, Bristol.

田中宏之・鈴木秀和・南雲　保　2008. 岡山・鳥取県境に分布する人形峠層(上部中新統～鮮新統)から見出された淡水生珪藻化石群集. Diatom **24**: 51-62.

田中宏之・鹿島　薫・南雲　保　2008. 大分県九重町右田に分布する珪藻土から見出された *Stephanodiscus kusuensis* Julius, Tanaka & Curtin. 地学研究 **57**: 137-141.

田中宏之・南雲　保　2009. 池田湖から見出された本邦新産中心類珪藻 *Spicaticribra kingstonii* Johansen, Kociolek & Lowe 及び共産した同類珪藻(Bacillariophyta). 藻類 **57**: 86-92.

Tanaka, H. & Nagumo, T. 2009. Two new *Mesodictyopsis* species, *M. akitaensis* sp. nov. and *M. miyatanus* sp. nov., from a Late Miocene to Pliocene freshwater sediment, Japan. Acta Botanica Croatica **68**: 221-230.

Tanaka, H. & Nagumo, T. 2009. *Stephanodiscus uemurae*, a new Mio-Pliocene diatom species from the Miyata Formationm Akita Prefecture, northern Honshu, Japan. Diatom **25**: 45-51.

Tanaka, H. & Nagumo, T. 2009. *Pliocaenicus omarensis* (Kupts.) Stachura-S. & Khur. found from Pliocene sediments of the Koriyama Formation, Kagoshima Prefecture, Japan. Diatom **25**: 86-90.

Tanaka, H. & Nagumo, T. 2010. *Aulacoseira satsumaensis*, a new Pliocene diatom species with

Tanaka, H., Nagumo, T. & Kashima, K. 2010. *Tertiariopsis nipponicus* sp. nov., from a Pliocene freshwater deposit in southwestern Japan. Diatom Research **25**: 175-183.

Tanaka, H. & Kashima, K. 2010. *Cyclotella kitabayashii* sp. nov., a new fossil diatom species from Pliocene sediment in southwestern Japan. Diatom **26**: 10-16.

Tanaka, H. & Nagumo, T. 2010. Fine structure of the *Aulacoseira crassipunctata* Krammer in Japan. Diatom **26**: 40-43.

田中宏之・渡辺 剛・南雲 保 2011. 岐阜県郡上市北部大鷲に分布する湖成堆積物から見出された珪藻化石群集. 瑞浪化石博物館研究報告 **37**: 123-134.

田中宏之・南雲 保 2011. 赤城山大沼(群馬県)の珪藻. 日本歯科大学紀要 **40**: 47-56.

田中宏之・北林栄一 2011. 芳野層(中部更新統, 熊本県)から見出された淡水生中心類珪藻. Diatom **27**: 52-57.

Tanaka, H. & Nagumo, T. 2011. *Aulacoseira iwakiensis* sp. nov., a new elliptical *Aulacoseira* species, from an Early Miocene sediment, Japan. Diatom **27**: 1-8.

Tanaka, H. & Nagumo, T. 2011. *Aulacoseira houki*, a new Early Miocene freshwater diatom from Hiramaki Formation, Gifu Prefecture, Japan. Diatom Research **26**: 161-165.

Tanaka, H. & Saito-Kato, M. 2011. *Pliocaenicus tanimurae*, a new Pliocene diatom species from Aguni Island, Okinawa Prefecture, southwestern Japan. Diatom Research **26**: 155-160.

Tanaka, H. & Nagumo, T. 2012. *Cyclotella iwatensis* sp. nov. from Mio-Pliocene freshwater sediment, Iwate Prefecture, Japan. Diatom Research **27**: 121-126.

Tanaka, H. & Nagumo, T. 2012. *Tertiariopsis costatus*, a new diatom species from the Mio-Pliocene freshwater sediment of Masuzawa Formation, Iwate Prefecture, Japan. Diatom **28**: 1-6.

Tanaka, H. & Nagumo, T. 2013. *Dimidialimbus bungoensis*, gen. nov. and sp. nov. (Stephanodiscaceae) a new diatom genus from Early Pleistocene sediment, Kyushu, Japan. Diatom **29**: 13-19.

Tanaka, M., Matsuoka, K. & Takagi, Y. 1984. The genus *Melosira* (Bacillariophyceae) from the Pliocene Iga Formation of the Kobiwako Group in Mie Prefecture, central Japan. Bulletin of the Mizunami Fossil Museum **11**: 55-68.

田中正明・松岡敬二 1985. 古琵琶湖層群伊賀累層の珪藻化石群集. 地団研専報 **29**: 89-100.

田中正明・松岡敬二 1985. 滋賀県甲賀・阿山地域の鮮新世淡水生珪藻化石. 瑞浪市化石博物館研究報告 **12**: 57-69, pls 26-31.

Tanimura, Y., Kato, M., Fukusawa, H., Mayama, S. & Yokoyama, K. 2006. Cytoplasmic mass preserved in Early Holocene diatoms: a possible taphonomic process and its Paleo-ecological implications. Journal Phycology **42**: 270-279.

Theriot, E., Håkansson, H., Kociolek, J.P., Round, F.E. & Stoermer, E.F. 1987. Validation of the centric diatom genus name *Cyclostephanos*. British Phycological Journal **22**: 345-347.

津村孝平 1967. 珪藻類の珍奇な数種について. 関東学院工学部研究報告 **12**: 231-245.

津村孝平 1973. 日本の珪藻類研究史, 関東学院大学経済学部一般教育論集「自然・人間社会」**1973** (1): 25-37.

Tuji, A. 2002. Observations on *Aulacoseira nipponica* from Lake Biwa, Japan, and *Aulacoseira solida* from North America (Bacillariophyceae). Phycological Research **50**: 313-316.

Tuji, A. 2004. The diatom type materials of Haruo Okuno 1. Five diatom species described by Okuno (1943, 1944) from the Yatuka deposit. Bulletin of the National Museum of Nature and Science, Ser. B **30**: 79-88.

Tuji, A. & Nergui, S. 2008. *Didymosphenia geminata* (Lyngb.) Mart. Schmidt の北海道での産出, およびアイスランド産, モンゴル産の個体との比較. Diatom **24**: 80-85.

Tuji, A. & Kociolek, J.P. 2000. Morphology and taxonomy of *Stephanodiscus suzukii* sp. nov. and *Stephanodiscus pseudosuzukii* sp. nov. (Bacillariophyceae) from Lake Biwa, Japan and *S. carconensis* from North America. Phycological Research **48**: 231-239.

辻　彰洋・伯耆晶子 2001. 琵琶湖の中心目珪藻. *In*: 中村ら (編) 琵琶湖研究モノグラフ **7**. 90 pp. 滋賀県琵琶湖研究所.

Tuji, A., Kawashima, A., Julius, M.L. & Stoermer, E.F. 2003. *Stephanodiscus akanensis* sp. nov., a new species of extant diatom from Lake Akan, Hokkaido, Japan. Bulletin of the National Museum, Ser. B **29**: 1-8.

Tuji, A. & Houki, A. 2004. Taxonomy, ultrastructure, and biogeography of the *Aulacoseira subarctica* species complex. Bulletin of the National Museum of Nature and Science, Ser. B **30**: 35-54.

Tuji, A. & Williams, D.M. 2007. Type examination of Japanese diatoms described by Friedrich Meister (1913) from Lake Suwa. Bulletin of the National Museum of Nature and Science, Ser. B **33**: 69-79.

上山　敏・小林　弘 1983. 同一古典試料による和村産化石ケイソウのスクフォルツォフ論文との比較. 東京学芸大学紀要 第4部門 **35**: 71-94.

渡辺仁治・浅井一視・大塚泰介・辻　彰洋・伯耆晶子 2005. 淡水珪藻生態図鑑. 666 pp. 内田老鶴圃, 東京.

柳沢幸夫・平中宏典・黒川克己 2003a. 新潟県津川地域の中部〜上部中新統の珪藻化石層序およびテフラ層序に基づく年代層序. 地球科学 **57**: 205-220.

柳沢幸夫・平中宏典・黒川克己 2003b. 新潟県新発田市北東部地域の中新統の珪藻化石層序とテフラ層序の対応関係. 地球科学 **57**: 299-313.

柳沢幸夫・田中裕一郎・高橋雅紀・岡田利典・須藤　斎 2004. 常磐地域日立市に分布する中新統多賀層群の複合年代層序. 地球科学 **58**: 17-30.

柳沢幸夫・平中宏典・黒川克己 2010. 新潟県津川地域の中部〜上部中新統野村層と常浪層の珪藻化石層序. 地質調査研究報告 **61**: 417-443.

柳沢幸夫・渡辺真人 2011. 5万分の1地質図幅「戸賀及び船川」地域 (男鹿半島) の新第三紀及び第四期の珪藻化石層序資料. 地質調査総合センター研究資料集 no. 533: 1-17.

柳沢幸夫・工藤　崇 2011. 5万分の1地質図幅「加茂」地域 (新潟県) の新第三紀及び第四期の珪藻化石層序資料. 地質調査総合センター研究資料集 no. 537: 1-59.

Yoshikawa, S. 2007. Sedimentary diatoms in Sawano-ike Pond, Kyoto City. Diatom **23**: 91-104.

Van Heurck, H. 1880-1885. Synopsis des diatomées de Belgique. Atlas: 132 pls. Text: 235 pp. 3 pls. Ducaju et Cie. Anvers.

Williams, D.M. 1985. Morphology, taxonomy and inter-relationships of the ribbed araphid diatoms from the genera *Diatoma* and *Meridion* (Diatomaceae: Bacillariophyta). Bibliotheca Diatomologica **8**: 1-228, 27 pls.

Williams, D.M. 1987. Observations on the genus *Tetracyclus* Ralfs (Bacillariophyta) Ⅰ. Valve and girdle structure of the extant species. British Phycological Journal **22**: 383-399.

Williams, D.M. 1989. Observations on the genus *Tetracyclus* Ralfs (Bacillariophyta) Ⅱ. Morphology and Taxonomy of some fossil species previously classified in *Stylobilium* Ehrenberg. British Phycological Journal **24**: 317-327.

Williams, D.M. 1990. Examination of auxospore valves in *Tetracyclus* from fossil specimens and the establishment of their identity. Diatom Research **5**: 189-194.

Williams, D.M. & Round, F.E. 1987. Revision of the genus *Fragilaria*. Diatom Research **2**: 267-288.

Williams, D.M. & Round, F.E. 1988. *Fragilariforma*, nom. nov., a new generic name for *Neofragilaria* Williams & Round. Diatom Research **3**: 265-267.

Xie, S., Lin, B. & Cai, S. 1985. Studies by means of LM and EM on a new species, *Cyclotella asterocostata* Lin, Xie et Cai. Acta Phytotaxonomica Sinica **23**: 473-475.

Zimmermann, C., Poulin, M. & Pienitz, R. 2010. Diatoms of North America: The Pliocene-Pleistocene freshwater flora of Bylot Island, Nunavut, Canadian High Arctic. *In*: Lange-Bertalot, H. (ed.) Iconographia Diatomologica **21**: 1-407.

IX. 学名索引：Index of Scientific Names
（太字は新種，または新組み合わせを行った分類群）

A

Achnanthes exigua var. *angustirostrata* (Krasske) Lange-Bertalot ·····································336
Achnanthes obliqua (Gregory) Hustedt ··338
Achnanthes okunoi (Hustedt) H. Tanaka comb. nov. ·································16, 340, 342
Actinella brasiliensis Grunow ···424
Actinocyclus normanii f. *subsalsa* (Juhlin-Dannfelt) Hustedt ··100
Actinocyclus octonarius Ehrenberg s.l. ···102, 104
Amphora copulata (Kützing) Schoeman & Archibald ··566
Amphora veneta Kützing ···568
Aneumastus tusculus (Ehrenberg) D.G. Mann & Stickle ··450
Aulacoseira ambigua (Grunow) Simonsen ···38
Aulacoseira cataractarum (Hustedt) Simonsen ···40
Aulacoseira crassipunctata Krammer ···42
Aulacoseira fukushimae H. Tanaka sp. nov. ·····································1, 44, 46
Aulacoseira granulata (Ehrenberg) Simonsen var. *granulata* ···48
Aulacoseira granulata var. *angustissima* (O. Müller) Simonsen ·······································50
Aulacoseira hachiyaensis H. Tanaka ···52
Aulacoseira houki H. Tanaka ···54
Aulacoseira italica (Ehrenberg) Simonsen ···56
Aulacoseira iwakiensis H. Tanaka ···58
Aulacoseira longispina (Hustedt) Simonsen ···60
Aulacoseira miosiris H. Tanaka sp. nov. ···2, 62, 64
Aulacoseira nipponica (Skvortsov) Tuji ···66
Aulacoseira nivalis (W. Smith) English & Potapova ···68
Aulacoseira polispina H. Tanaka sp. nov. ·································3, 70, 72, 74
Aulacoseira pusilla (F. Meister) Tuji & Houki ···76
Aulacoseira satsumaensis H. Tanaka ···78, 80
Aulacoseira subarctica (O. Müller) Haworth ···82
Aulacoseira tenella (Nygaard) Simonsen ···84
Aulacoseira tsugaruensis H. Tanaka sp. nov. ·································4, 86, 88
Aulacoseira valida (Grunow) Krammer ···90

B

Brevisira arentii (Kolbe) Krammer ···92

C

Caloneis schumanniana (Grunow) Cleve ··452
Campylodiscus echeneis Ehrenberg ···550
Campylodiscus levanderi Hustedt ···552
Cavinula pseudoscutiformis (Hustedt) D.G. Mann & Stickle ···454
Cocconeis jimboites VanLandingham ··344
Cocconeis placentula Ehrenberg var. *placentula* ···346
Cocconeis placentula var. *lineata* (Ehrenberg) Van Heurck ······································348
Craticula ambigua (Ehrenberg) D.G. Mann ···456
Craticula cuspidata (Kützing) D.G. Mann ···458
Cyclostephanos costatilimbus (H. Kobayasi & Kobayashi) Stoermer, Håkansson & Theriot ·······120

593

Cyclostephanos dubius (Fricke) Round ··122
Cyclostephanos kyushuensis H. Tanaka ··124, 126
Cyclostephanos numataensis H. Tanaka & Nagumo···128
Cyclotella atomus Hustedt ···130
Cyclotella cyclopuncta Håkansson & Carter ··132
Cyclotella iris Brun & Héribaud s.l. ···134
Cyclotella iwatensis H. Tanaka···136
Cyclotella kitabayashii H. Tanaka···138, 140
Cyclotella kohsakaensis H. Tanaka & H. Kobayasi··142
Cyclotella meneghiniana Kützing ···144
Cyclotella mesoleia (Grunow) Houk, Klee & Tanaka ··146, 148
Cyclotella nogamiensis H. Tanaka sp. nov.··5, 150, 152
Cyclotella notata Loseva···154
Cyclotella ocellata Pantocsek ···156
Cyclotella oitaensis H. Tanaka sp. nov.··6, 158, 160
Cyclotella ozensis (H. Tanaka & Nagumo) H. Tanaka ···162, 164
Cyclotella praetermissa Lund···166
Cyclotella radiosa (Grunow) Lemmermann ··168
Cyclotella rhomboideo-elliptica Skuja var. *rhomboideo-elliptica*··170, 172
Cyclotella rhomboideo-elliptica var. *rounda* Qi & Yang···174
Cyclotella satsumaensis H. Tanaka & Houk··176
Cyclotella schumannii (Grunow) Håkansson ···178
Cyclotubicoalitus undatus Stoermer, Kociolek & Cody···180
Cymatopleura elliptica (Brébisson) W. Smith···554
Cymatopleura solea (Brébisson) W. Smith ··556
Cymbella cymbiformis C. Agardh ··350
Cymbella neoleptoceros Krammer ···352
Cymbella ocellata H. Tanaka sp. nov.··10, 354, 356
Cymbella okunoi H. Tanaka sp. nov.··12, 358, 360
Cymbella orientalis Lee··362
Cymbella peraspera Krammer··364
Cymbella proxima Reimer ···366
Cymbella stuxbergii var. *robusta* Okuno···368
Cymbella tsumurae H. Tanaka sp. nov.···13, 370, 372, 374
Cymbopleura apiculata Krammer··376
Cymbopleura inaequalis (Ehrenberg) Krammer··378, 380
Cymbopleura naviculiformis (Auerswald) Krammer···382
Cymbopleura subaequalis (Grunow) Krammer ··384

D

Denticula elegans f. *valida* Pedicino ··570
Denticula tenuis Kützing···572
Diadesmis confervacea Kützing ··460
Diatoma anceps (Ehrenberg) Kirchner ···282
Diatoma ehrenbergii Kützing ···284
Diatoma hyemalis (Roth) Heiberg ···286
Diatoma mesodon (Ehrenberg) Kützing···286
Diatoma vulgaris Bory···284
Didymosphenia curvata (Skvortsov & Meyer) Metzeltin & Lange-Bertalot··386
Didymosphenia fossils Horikawa & Okuno··388
Didymosphenia geminata (Lyngbye) M. Schmidt··390

Didymosphenia nipponica **H. Tanaka sp. nov.** ·· 14, 392, 394
Dimidialimbus bungoensis H. Tanaka ·· 182
Diploneis elliptica (Kützing) Cleve ·· 462
Diploneis finnica (Ehrenberg) Cleve ··· 464
Diploneis ovalis (Hilse) Cleve ·· 466
Diploneis smithii var. *rhombica* Mereschkowsky ··· 468
Diploneis subovalis Cleve ·· 470
Discostella asterocostata (Xie, Lin & Cai) Houk & Klee ··································· 184
Discostella kitsukiensis **H. Tanaka sp. nov.** ·· 7, 186, 188
Discostella pliostelligera (H. Tanaka & Nagumo) Houk & Klee ························· 190, 192
Discostella pseudostelligera (Hustedt) Houk & Klee ······································· 194
Discostella stelligera (Cleve & Grunow) Houk & Klee ···································· 196
Discostella woltereckii (Hustedt) Houk & Klee ·· 198

E

Ellerbeckia arenaria (Moore) R.M. Crawford f. *arenaria* ································· 114
Ellerbeckia arenaria f. *teres* (Brun) R.M. Crawford ·· 116, 118
Encyonema geisslerae Krammer ·· 396
Encyonema vulgare Krammer ··· 398
Epithemia adnata (Kützing) Brébisson ··· 526
Epithemia cistula (Ehrenberg) Ralfs ··· 528
Epithemia hyndmanii W. Smith ··· 530
Epithemia numatensis **H. Tanaka sp. nov.** ·· 16, 532, 534
Epithemia reticulata Kützing ·· 536
Epithemia sorex Kützing ·· 538
Epithemia turgida var. *porcellus* Héribaud ··· 540
Eunotia arcus Ehrenberg ·· 426
Eunotia biareofera f. *linearis* H. Kobayasi ·· 428
Eunotia bidens Ehrenberg ·· 430
Eunotia clevei Grunow ··· 432
Eunotia diadema Ehrenberg ··· 434
Eunotia duplicoraphis H. Kobayasi, Ando & Nagumo ····································· 436
Eunotia epithemioides Hustedt ··· 438
Eunotia formica Ehrenberg ··· 440
Eunotia incisa W. Gregory ·· 442
Eunotia monodon var. *tropica* (Hustedt) Hustedt ·· 444
Eunotia nipponica Skvortsov ·· 446
Eunotia serra Ehrenberg ·· 448

F

Fragilaria neoproducta Lange-Bertalot ··· 288
Fragilaria vaucheriae (Kützing) Petersen ·· 290
Fragilariforma fossilis **(Pantocsek) H. Tanaka comb. nov. et stat. nov.** ············· 17, 292
Fragilariforma kamczatica **(Lupikina) H. Tanaka comb. nov. et stat. nov.** ········· 17, 294
Frustulia rhomboides var. *amphipleuroides* (Grunow) De Toni ························ 472
Frustulia rhomboides var. *saxonica* (Rabenhorst) De Toni ······························ 474

G

Gomphoneis okunoi Tuji ·· 400
Gomphoneis tumida (Skvortsov) Kociolek & Stoermer ···································· 402
Gomphonema augur var. *gautieri* Van Heurck ·· 404

Gomphonema biceps F. Meister ················406
Gomphonema coronatum Ehrenberg ················408
Gomphonema nipponicum Skvortsov ················410
Gomphonema truncatum Ehrenberg ················412
Gomphonema vastum Hustedt ················414
Gomphopleura frickei Reichelt ················416
Gomphosphenia grovei var. *lingulata* (Hustedt) Lange-Bertalot ················418
Gyrosigma spencerii (Quekett) Griffith & Henfrey ················476

H
Halamphora normannii (Rabenhorst) Levkov ················574
Hannaea arcus (Ehrenberg) R.M. Patrick var. *arcus* ················296
Hannaea arcus var. *hattoriana* (F. Meister) Ohtsuka ················296
Hannaea arcus var. *recta* (Cleve) M. Idei ················296
Hydrosera whampoensis (A.F. Schwarz) Deby ················96

M
Melosira undulata (Ehrenberg) Kützing var. *undulata* ················108
Melosira undulata var. *producta* A. Schmidt ················110
Melosira varians C. Agardh ················112
Meridion circulare var. *constrictum* (Ralfs) Van Heurck ················298
Mesodictyon yanagisawae H. Tanaka (in press) ················200
Mesodictyopsis akitaensis H. Tanaka & Nagumo ················202
Mesodictyopsis miyatanus H. Tanaka ················204
Miosira tscheremissinovae (Khursevich) Khursevich ················94

N
Navicula americana Ehrenberg ················478
Navicula anthracis Cleve & Brun ················480
Navicula cari Ehrenberg ················482
Navicula hasta Pantocsek ················484
Navicula radiosa Kützing ················486
Navicula reinhardtii (Grunow) Grunow ················488
Navicula tanakae Fukushima, Ts. Kobayashi & Yoshitake ················490
Neidium ampliatum (Ehrenberg) Krammer ················492
Neidium gracile Hustedt ················494
Nitzschia heidenii (F. Meister) Hustedt ················576
Nitzschia tabellaria (Grunow) Grunow ················578

O
Oricymba cunealjaponica H. Tanaka sp. nov. ················15, 420, 422
Orthoseira asiatica (Skvortsov) H. Kobayasi ················106

P
Pinnularia episcopalis Cleve ················496
Pinnularia esoxiformis Fusey ················498
Pinnularia higoensis Okuno ················500
Pinnularia lignitica Cleve ················502
Pinnularia macilenta (Ehrenberg) Ehrenberg ················504
Pinnularia rivularis Hustedt ················506
Pinnularia senjoensis H. Kobayasi ················508

Pinnularia subgibba var. *lanceolata* Gaiser & Johansen ················510
Plagiotropis lepidoptera var. *proboscidea* (Cleve) Reimer ················512
Pleurosira laevis (Ehrenberg) Compère ················280
Pliocaenicus costatus (Loginova, Lupikina & Khursevich) Flower, Ozornina & A.I. Kuzmina
················206, 208
Pliocaenicus nipponicus H. Tanaka & Nagumo ················210
Pliocaenicus omarensis (Kuptsova) Stachura-Suchoples & Khursevich ················212
Pliocaenicus radiatus H. Tanaka (in press) ················214
Pliocaenicus tanimurae H. Tanaka & Saito-Kato ················216
Pseudostaurosira brevistriata var. *nipponica* (Skvortsov) H. Kobayasi ················300

R
Rhopalodia gibba (Ehrenberg) O. Müller ················542
Rhopalodia gibberula (Ehrenberg) O. Müller ················544
Rhopalodia novae-zelandiae Hustedt ················546
Rhopalodia rupestris (W. Smith) Krammer ················548

S
Sellaphora bacillum (Ehrenberg) D.G. Mann ················514
Sellaphora laevissima (Kützing) D.G. Mann ················516
Spicaticribra kingstonii Johansen, Kociolek & Lowe ················218
Stauroneis acuta W. Smith var. *acuta* ················518
Stauroneis acuta var. *terryana* Tempère ················520
Stauroneis phoenicenteron (Nitzsch) Ehrenberg var. *phoenicenteron* ················522
Stauroneis phoenicenteron var. *hattorii* Tsumura ················524
Staurosira construens Ehrenberg var. *construens* ················302
Staurosira construens var. *binodis* (Ehrenberg) Hamilton ················304
Staurosira construens var. *triundulata* (H. Reichelt) H. Kobayasi ················306
Staurosirella lapponica (Grunow) D.M. Williams & Round ················308
Stephanodiscus akanensis Tuji, Kawashima, Julius & Stoermer ················220, 222
Stephanodiscus hashiensis H. Tanaka (in press) ················224
Stephanodiscus iwatensis H. Tanaka sp. nov. ················8, 226, 228
Stephanodiscus kobayasii H. Tanaka (in press) ················230, 232
Stephanodiscus komoroensis H. Tanaka ················234, 236
Stephanodiscus kusuensis Julius, Tanaka & Curtin ················238, 240
Stephanodiscus kyushuensis H. Tanaka (in press) ················242, 244
Stephanodiscus minutulus (Kützing) Cleve & Möller ················246
Stephanodiscus miyagiensis H. Tanaka & Nagumo ················248, 250
Stephanodiscus nagumoi H. Tanaka (in press) ················252, 254
Stephanodiscus rotula (Kützing) Hendey ················256, 258
Stephanodiscus suzukii Tuji & Kociolek ················260, 262
Stephanodiscus tenuis Hustedt ················264
Stephanodiscus uemurae H. Tanaka ················266
Stoermeria trifoliata (Cleve) Kociolek, Escobar & Richardson ················98
Surirella bifrons Ehrenberg ················558
Surirella robusta var. *splendida* f. *constricta* Hustedt ················562
Surirella splendida (Ehrenberg) Kützing ················560
Surirella tenera W. Gregory ················564

T
Tabellaria fenestrata (Lyngbye) Kützing ················312

***Tabellaria japonica* H. Tanaka sp. nov.** ··· 10, 314, 316
Tertiariopsis costatus H. Tanaka ··· 268
Tertiariopsis nipponicus H. Tanaka ··· 270
***Tertiarius agunensis* H. Tanaka sp. nov.** ··· 9, 272, 274
Tetracyclus castellum (Ehrenberg) Grunow ··· 318
Tetracyclus cruciformis Andrews ··· 320
Tetracyclus ellipticus (Ehrenberg) Grunow var. *ellipticus* ··· 322
Tetracyclus ellipticus var. *constricta* Hustedt ··· 324
Tetracyclus ellipticus var. *lancea* f. *subrostrata* Hustedt ··· 326
Tetracyclus ellipticus var. *latissima* f. *minor* Hustedt ··· 328
Tetracyclus emarginatus (Ehrenberg) W. Smith ··· 330
Tetracyclus glans (Ehrenberg) Mills ··· 332
Tetracyclus lacustris Ralfs ··· 334
Thalassiocyclus pankensis H. Tanaka & Nagumo ··· 276
Thalassiosira lacustris (Grunow) Hasle ··· 278

U
Ulnaria capitata (Ehrenberg) Compère ··· 310

X．地層・産地別一覧：List of Formations or Localities and Taxa

　図版で使用した標本（写真）は次の地層または産地から採取したものである（地層名等の番号は 19-20 頁の表と一致する．太字は新種，または新組み合わせを行った分類群）．

1．中新世・鮮新世・更新世試料採取地

1　小松沢層：*Staurosirella lapponica*
2　網走湖底ボーリングコア：*Cyclostephanos dubius*
3　太櫓層（瀬棚の珪藻土）：*Diatoma anceps*, ***Fragilariforma fossilis***, ***F. kamczatica***, *Meridion circulare* var. *constrictum*, *Tetracyclus cruciformis*, *T. ellipticus* var. *constricta*, *T. ellipticus* var. *latissima* f. *minor*, *T. lacustris*, ***Achnanthes okunoi***, *Cocconeis jimboites*, *Cymbella stuxbergii* var. *robusta*, *Gomphopleura frickei*, *Actinella brasiliensis*, *Eunotia bidens*, *E. clevei*
4　六角沢層：***Aulacoseira tsugaruensis***
5　宮田層：***Aulacoseira miosiris***, *Miosira tscheremissinovae*, *Stoermeria trifoliata*, *Ellerbeckia arenaria* f. *teres*, *Discostella pliostelligera*, *D. stelligera*, *Mesodictyopsis akitaensis*, *M. miyatanus*, *Stephanodiscus uemurae*, *Tetracyclus emarginatus*, *Eunotia formica*, *Diploneis elliptica*, *Frustulia rhomboides* var. *amphipleuroides*, *Epithemia reticulata*, *E. turgida* var. *porcellus*
6　舛沢層：*Cyclotella iwatensis*, ***Stephanodiscus iwatensis***, *Tertiariopsis costatus*
7　鬼首層：*Cyclotella ocellata*, *Stephanodiscus miyagiensis*, *Cymbella peraspera*, *Surirella splendida*, *S. robusta* var. *splendida* f. *constricta*, *S. tenera*
8　白沢層：*Mesodictyon yanagisawae*, *Navicula cari*
9　向山層：*Navicula anthracis*
10　円田層：*Aulacoseira granulata* var. *granulata*
11　紫竹層：***Aulacoseira fukushimae***, *A. italica*, *A. iwakiensis*, *Melosira undulata* var. *producta*, ***Tabellaria japonica***, *Tetracyclus ellipticus* var. *ellipticus*, *T. ellipticus* var. *lancea* f. *subrostrata*, *Eunotia biareofera* f. *linearis*
12　塩原湖成層：*Stephanodiscus minutulus*
13　小野上層：*Ellerbeckia arenaria* f. *teres*, *Cyclotella radiosa*, *Cymatopleura elliptica*
14　沼田湖成層：*Cyclostephanos numataensis*, *Diatoma ehrenbergii*, *D. hyemalis*, *D. mesodon*, *Hannaea arcus* var. *arcus*, *H. arcus* var. *hattoriana*, *H. arcus* var. *recta*, *Didymosphenia geminata*, *Gomphoneis okunoi*, *G. tumida*, *Pinnularia episcopalis*, *Stauroneis acuta* var. *terryana*
15　横子珪藻土：*Staurosira construens* var. *binodis*, *Achnanthes exigua* var. *angustirostrata*, *Cocconeis placentula* var. *placentula*, ***Cymbella tsumurae***, *Cymbopleura apiculata*, *C. naviculiformis*, *C. subaequalis*, *Gomphonema augur* var. *gautieri*, *G. coronatum*, *Eunotia diadema*, *E. monodon* var. *tropica*, *Craticula cuspidata*, *Sellaphora bacillum*, *S. laevissima*, *Stauroneis phoenicenteron* var. *hattorii*, *Epithemia adnata*, ***E. numatensis***, *E. sorex*, *Amphora copulata*, *Nitzschia heidenii*
16　中之条湖成層：*Aulacoseira subarctica*, *Cyclotella rhomboideo-elliptica* var. *rhomboideo-*

elliptica, Stephanodiscus rotula, Cymbella cymbiformis, Gomphosphenia grovei var. *lingulata, Eunotia arcus, E. duplicoraphis, Diploneis finnica*

17 嬬恋湖成層：*Eunotia nipponica, Pinnularia subgibba* var. *lanceolata*
18 大戸湖沼性堆積物：*Cyclotella rhomboideo-elliptica* var. *rounda, Pseudostaurosira brevistriata* var. *nipponica, Denticula tenuis, Nitzschia tabellaria*
19 倉渕湖沼性堆積物：*Cyclotella rhomboideo-elliptica* var. *rhomboideo-elliptica*
20 兜岩層：*Pliocaenicus nipponicus*
21 稲城層：*Thalassiosira lacustris, Pleurosira laevis, Didymosphenia curvata, Encyonema geisslerae, Diploneis smithii* var. *rhombica, Campylodiscus echeneis*
22 渋沢湖沼性堆積物：*Cyclotella cyclopuncta*
23 香坂礫岩層：*Ellerbeckia arenaria* f. *teres, Cyclotella kohsakaensis*
24 瓜生坂層：*Stephanodiscus komoroensis*
25 美ヶ原層：*Pliocaenicus costatus*
26 大鷲湖沼性堆積物：*Aulacoseira ambigua, Melosira undulata* var. *undulata, Cyclotella meneghiniana, Fragilaria neoproducta, F. vaucheriae, Staurosira construens* var. *construens, Tabellaria fenestrata, Achnanthes exigua* var. *angustirostrata, Cocconeis placentula* var. *lineata, Cymbella neoleptoceros, Didymosphenia fossils, Encyonema vulgare, Gomphonema truncatum, G. vastum, Eunotia incisa, Cavinula pseudoscutiformis, Epithemia adnata, E. hyndmanii*
27 蜂屋層：*Aulacoseira hachiyaensis*
28 平牧層：*Aulacoseira houki*
29 伊賀層：*Staurosira construens* var. *triundulata*
30 大阪層群：*Cyclotella mesoleia*
31 春来層：*Pliocaenicus omarensis*
32 三徳層：*Navicula cari*
33 人形峠層：*Melosira undulata* var. *undulata, Ellerbeckia arenaria* f. *teres, Cyclotella iris* s.l., *Ulnaria capitata, Tetracyclus castellum, Epithemia cistula, Rhopalodia gibba*
34 長者原層：*Aulacoseira ambigua*
35 津房川層：**Aulacoseira polispina**, *Cyclotella kitabayashii, Tertiariopsis nipponicus, Diadesmis confervacea*
36 俣水層：**Didymosphenia nipponica**
37 野原層：*Ellerbeckia arenaria* f. *teres, Cyclostephanos kyushuensis*
38 野上層：*Aulacoseira granulata* var. *angustissima*, **Cyclotella nogamiensis**, *Stephanodiscus kusuensis*, **Cymbella okunoi**
39 阿蘇野層：*Cyclotella schumannii*
40 尾本層：*Cyclotella notata*, **C. oitaensis**, *Dimidialimbus bungoensis, Stephanodiscus kobayasii, S. nagumoi, Aneumastus tusculus, Navicula reinhardtii, Surirella bifrons*
41 加貫層：**Discostella kitsukiensis**
42 津森層：*Staurosira construens* var. *construens, Cymbella proxima, Cymbopleura inaequalis*, **Oricymba cunealjaponica**, *Gomphonema nipponicum, Caloneis schumanniana, Gyrosigma spencerii, Navicula hasta, N. radiosa, Neidium ampliatum, Pinnularia lignitica, Stauroneis acuta* var. *acuta, Epithemia hyndmanii, Rhopalodia gibba, R. novae-zelandiae, Cymatopleura solea*
43 人吉層：*Ellerbeckia arenaria* f. *teres, Achnanthes obliqua, Eunotia epithemioides,*

Diploneis finnica, Navicula americana, N. hasta, Pinnularia higoensis, P. macilenta, Stauroneis phoenicenteron var. *phoenicenteron*

44 永野層：*Stephanodiscus kyushuensis*
45 郡山層：*Aulacoseira satsumaensis, Cyclotella satsumaensis,* **Cymbella ocellata**
46 筆ん崎層：*Pliocaenicus radiatus, P. tanimurae,* **Tertiarius agunensis**

2. 完新世・現生試料採取地

1 パンケ沼：*Thalassiocyclus pankensis*
2 屈斜路湖：*Aulacoseira crassipunctata, Campylodiscus levanderi*
3 阿寒湖：*Stephanodiscus akanensis*
4 八郎潟調整池：*Cyclostephanos costatilimbus*
5 鎌沼：*Aulacoseira nivalis*
6 毘沙門沼：*Aulacoseira crassipunctata*
7 元温泉小屋温泉：*Aulacoseira cataractarum, Pinnularia esoxiformis*
8 尾瀬沼・尾瀬地域：*Aulacoseira valida, Cyclotella ozensis, Tetracyclus glans, Eunotia serra*
9 中禅寺湖：*Aulacoseira longispina*
10 大沼（赤城山）：*Cyclotella praetermissa*
11 池の岳の池塘：*Frustulia rhomboides* var. *saxonica*
12 利根川源流：*Diatoma mesodon, Rhopalodia gibberula*
13 奥平鉱泉：*Rhopalodia rupestris*
14 桐生川：*Gomphonema biceps*
15 波志江沼：*Aulacoseira granulata* var. *angustissima, Cyclotubicoalitus undatus, Discostella asterocostata, D. pseudostelligera, Stephanodiscus hashiensis, S. tenuis*
16 多々良沼：*Discostella asterocostata*
17 城沼：*Cyclotubicoalitus undatus, Discostella asterocostata*
18 磯部鉱泉：*Navicula tanakae*
19 赤久縄鉱泉：*Denticula elegans* f. *valida*
20 星尾鉱泉：*Amphora veneta, Halamphora normannii*
21 神流川：*Orthoseira asiatica, Diatoma vulgaris*
22 北浦：*Actinocyclus normanii* f. *subsalsa, Discostella woltereckii*
23 霞ヶ浦：*Actinocyclus normanii* f. *subsalsa*
24 手賀沼：*Cyclotella atomus*
25 諏訪湖：*Aulacoseira pusilla, Stephanodiscus tenuis*
26 山中湖：*Ellerbeckia arenaria* f. *arenaria*
27 琵琶湖：*Aulacoseira nipponica, Stephanodiscus suzukii*
28 沢の池：*Aulacoseira crassipunctata*
29 藺牟田池：*Brevisira arentii*
30 吹上浜（泥炭層）：*Hydrosera whampoensis, Actinocyclus octonarius* s.l., *Melosira varians, Gomphosphenia grovei* var. *lingulata, Diploneis ovalis, D. subovalis*
31 池田湖：*Spicaticribra kingstonii*
32 泥染公園の泥田：*Craticula ambigua, Neidium gracile, Pinnularia rivularis, P. senjoensis, Plagiotropis lepidoptera* var. *proboscidea*

33　上嘉鉄湧泉池：*Cymbella orientalis*
34　大正池：*Diadesmis confervacea*
35　福上湖：*Aulacoseira tenella*

著者紹介

田中 宏之　Hiroyuki TANAKA
青森県青森市生まれ
群馬県大泉町立北中学校卒
群馬県立太田高等学校卒
群馬大学教育学部（自然科学科）卒
群馬県公立中学校（10年）
群馬県立博物館（含準備室）（19年）
群馬県立高等学校（9年）
博物館学芸員，学術博士，水産学博士

Atlas of Freshwater Fossil Diatoms in Japan ― Including related recent taxa ―

2014年3月10日　第1版発行

著者の了解により検印を省略いたします

日本淡水化石珪藻図説
―関連現生種を含む―

著　者 © 田　中　宏　之
発行者　内　田　　　学
印刷者　山　岡　景　仁

発行所　株式会社　内田老鶴圃（ろうかくほ）　〒112-0012 東京都文京区大塚3丁目34-3
電話（03）3945-6781（代）・FAX（03）3945-6782
http://www.rokakuho.co.jp/
印刷／三美印刷 K.K.・製本／榎本製本 K.K.

Published by UCHIDA ROKAKUHO PUBLISHING CO., LTD.
3-34-3 Otsuka, Bunkyo-ku, Tokyo 112-0012, Japan

U.R. No. 603-1

ISBN 978-4-7536-4084-3 C3045

小林 弘 珪藻図鑑　第1巻　H.Kobayasi's Atlas of Japanese Diatoms based on electron microscopy vol.1

小林　弘・出井雅彦・真山茂樹・南雲　保・長田敬五　著　B5判・596頁・定価（本体34,000円＋税）

本書は珪藻の分類学の成書として長く刊行が待たれていた待望の書であり，斯界の第一人者，故小林弘博士の名を冠するものである．プレートとその解説をはじめ，特殊な用語が多く使われる珪藻の殻構造の解説を電顕写真や線画を添えて分かりやすく示す．用語の英語，日本語，ラテン語の一覧表や，学名と和名の対照表などを付し読者の便宜を図った．

淡水珪藻生態図鑑　群集解析に基づく汚濁指数DAIpo，pH耐性能

渡辺仁治　編著　浅井一視・大塚泰介・辻　彰洋・伯耆晶子　著　B5判・784頁・定価（本体33,000円＋税）

日本のみならず世界各地から約1500のサンプルを採集，膨大なサンプルの生態情報を処理検討し，約1000種の珪藻についてその結果を分かり易くまとめる．生態情報の妥当性を期するため，すべてのサンプルを統一条件下で採集し，好清水か好汚濁か＝きれいな水を好むのか，汚れた水を好むのか等を判断する環境指標としての珪藻群集の適性を多くの図版で具体的に示す．

淡　水　藻　類　淡水産藻類属総覧

山岸高旺著　B5判・1444頁・定価（本体50,000円＋税）

本書は淡水における藻類，約1500属を収録した淡水藻類の属の総覧である．配列は淡水藻類を12分類群に分けるBourrellyの分類系を採った．これに加え異名とされるもの，関連するものをさらに約800属所収する．60年に及ぶ著者の淡水藻研究の集大成として，淡水藻類の全体像に迫る大著である．本文は，それぞれの分類群の「細胞・藻体」「生殖・生活史」「分類・分類表」を示した後，それぞれの属の記載が中心となり，線画による基本的な図版を示しながら，属の分類基準とされる形態形質，生殖形質，生育状況を述べる．また類似属との関係や産状など特記事項も詳細に記す．学名総索引をはじめ，和文，欧文の事項索引，また属名のカナ読み索引を付した．

藻類の生活史集成　全3巻

堀　輝三　編　B5判

本書全3巻は，藻類学の分野の中で個別に扱われることの多かった各藻類の生活史を，分かりやすい形で成書としたものである．収録全種について，それぞれ明らかになっている生活史を図示し対面頁に簡潔な解説を添え，見開きで読み取れるように構成．全巻の巻末に，1〜3巻共通の学名総索引，和名索引を付す．

第1巻　緑　色　藻　類　（185種）448頁・定価（本体8,000円＋税）
第2巻　褐藻・紅藻類　（171種）424頁・定価（本体8,000円＋税）
第3巻　単細胞性・鞭毛藻類　（146種）400頁・定価（本体7,000円＋税）

新日本海藻誌　―日本産海藻類総覧―

吉田　忠生　著　B5判・1252頁・定価（本体46,000円＋税）

岡村金太郎著「日本海藻誌」以来，実に60余年ぶりに刊行された海藻学の決定版．斯界の権威が日本の海藻を網羅して書き下ろした歴史的大著．綱，目，科，属，種などの分類階級ごとに，形質の特徴，および他との比較などを詳細に記述．また「綱から目へ，目から科へ・・・」わかりやすい検索表が付く．各種ごとに極めて詳細，細緻な文献リストが付される．さらに種ごとにタイプ産地，タイプ標本，分布地域名が示される．学名，和名の由来，生育地の特徴など，関連する話題も豊富．

有用海藻誌　海藻の資源開発と利用に向けて

大野　正夫　編著　B5判・596頁・定価（本体20,000円＋税）

本書は「生物学編」，「利用編」，「機能性成分編」の3編から構成されどの項目からも必要なところから読むことができる．生物学編は，利用分野ごとに分けて，種名の査定に必要な形態，生活史，分布生態を記述．これらの水産，食用などへの利用や産業的背景，利用の歴史についても詳述する．利用編は，海藻産業の歴史的背景，加工技術から化学構造，品質などにふれ，将来への展望を示す．機能性成分編では，あまり知られていない海藻の成分とその利用範囲を幅広く記述．

http://www.rokakuho.co.jp/